TECHNIQUES FOR NOISE ROBUSTNESS IN AUTOMATIC SPEECH RECOGNITION

TECHNIQUES FOR NOISE ROBUSTNESS IN AUTOMATIC SPEECH RECOGNITION

Editors

Tuomas Virtanen
Tampere University of Technology, Finland

Rita Singh
Carnegie Mellon University, USA

Bhiksha Raj
Carnegie Mellon University, USA

WILEY
A John Wiley & Sons, Ltd., Publication

This edition first published 2013

Registered office
John Wiley & Sons Ltd, The Atrium, Southern Gate, Chichester, West Sussex, PO19 8SQ, United Kingdom

For details of our global editorial offices, for customer services and for information about how to apply for permission to reuse the copyright material in this book please see our website at www.wiley.com.

Library of Congress Cataloging-in-Publication Data

Virtanen, Tuomas.
Techniques for noise robustness in automatic speech recognition / Tuomas Virtanen, Rita Singh, Bhiksha Raj.
p. cm.
Includes bibliographical references and index.
ISBN 978-1-119-97088-0 (cloth)
1. Automatic speech recognition. I. Singh, Rita. II. Raj, Bhiksha. III. Title.
TK7882.S65V57 2012
006.4′54–dc23

2012015742

A catalogue record for this book is available from the British Library.

ISBN: 978-0-470-97409-4

Typeset in 10/12pt Times by Aptara Inc., New Delhi, India
Printed and bound in Singapore by Markono Print Media Pte Ltd

Contents

List of Contributors xv

Acknowledgments xvii

1 Introduction 1
Tuomas Virtanen, Rita Singh, Bhiksha Raj

1.1 Scope of the Book 1
1.2 Outline 2
1.3 Notation 4

Part One FOUNDATIONS

2 The Basics of Automatic Speech Recognition 9
Rita Singh, Bhiksha Raj, Tuomas Virtanen

2.1 Introduction 9
2.2 Speech Recognition Viewed as Bayes Classification 10
2.3 Hidden Markov Models 11
2.3.1 Computing Probabilities with HMMs 12
2.3.2 Determining the State Sequence 17
2.3.3 Learning HMM Parameters 19
2.3.4 Additional Issues Relating to Speech Recognition Systems 20
2.4 HMM-Based Speech Recognition 24
2.4.1 Representing the Signal 24
2.4.2 The HMM for a Word Sequence 25
2.4.3 Searching through all Word Sequences 26
References 29

3 The Problem of Robustness in Automatic Speech Recognition 31
Bhiksha Raj, Tuomas Virtanen, Rita Singh

3.1 Errors in Bayes Classification 31
3.1.1 Type 1 Condition: Mismatch Error 33
3.1.2 Type 2 Condition: Increased Bayes Error 34
3.2 Bayes Classification and ASR 35
3.2.1 All We Have is a Model: A Type 1 Condition 35

	3.2.2	*Intrinsic Interferences—Signal Components that are Unrelated to the Message: A Type 2 Condition*	36
	3.2.3	*External Interferences—The Data are Noisy: Type 1 and Type 2 Conditions*	36
3.3	External Influences on Speech Recordings		36
	3.3.1	*Signal Capture*	37
	3.3.2	*Additive Corruptions*	41
	3.3.3	*Reverberation*	42
	3.3.4	*A Simplified Model of Signal Capture*	43
3.4	The Effect of External Influences on Recognition		44
3.5	Improving Recognition under Adverse Conditions		46
	3.5.1	*Handling the Model Mismatch Error*	46
	3.5.2	*Dealing with Intrinsic Variations in the Data*	47
	3.5.3	*Dealing with Extrinsic Variations*	47
	References		50

Part Two SIGNAL ENHANCEMENT

4	**Voice Activity Detection, Noise Estimation, and Adaptive Filters for Acoustic Signal Enhancement** *Rainer Martin, Dorothea Kolossa*		**53**
4.1	Introduction		53
4.2	Signal Analysis and Synthesis		55
	4.2.1	*DFT-Based Analysis Synthesis with Perfect Reconstruction*	55
	4.2.2	*Probability Distributions for Speech and Noise DFT Coefficients*	57
4.3	Voice Activity Detection		58
	4.3.1	*VAD Design Principles*	58
	4.3.2	*Evaluation of VAD Performance*	62
	4.3.3	*Evaluation in the Context of ASR*	62
4.4	Noise Power Spectrum Estimation		65
	4.4.1	*Smoothing Techniques*	65
	4.4.2	*Histogram and GMM Noise Estimation Methods*	67
	4.4.3	*Minimum Statistics Noise Power Estimation*	67
	4.4.4	*MMSE Noise Power Estimation*	68
	4.4.5	*Estimation of the* A Priori *Signal-to-Noise Ratio*	69
4.5	Adaptive Filters for Signal Enhancement		71
	4.5.1	*Spectral Subtraction*	71
	4.5.2	*Nonlinear Spectral Subtraction*	73
	4.5.3	*Wiener Filtering*	74
	4.5.4	*The ETSI Advanced Front End*	75
	4.5.5	*Nonlinear MMSE Estimators*	75
4.6	ASR Performance		80
4.7	Conclusions		81
	References		82

5 Extraction of Speech from Mixture Signals 87
Paris Smaragdis

5.1 The Problem with Mixtures 87
5.2 Multichannel Mixtures 88
5.2.1 Basic Problem Formulation 88
5.2.2 Convolutive Mixtures 92
5.3 Single-Channel Mixtures 98
5.3.1 Problem Formulation 98
5.3.2 Learning Sound Models 100
5.3.3 Separation by Spectrogram Factorization 101
5.3.4 Dealing with Unknown Sounds 105
5.4 Variations and Extensions 107
5.5 Conclusions 107
References 107

6 Microphone Arrays 109
John McDonough, Kenichi Kumatani

6.1 Speaker Tracking 110
6.2 Conventional Microphone Arrays 113
6.3 Conventional Adaptive Beamforming Algorithms 120
6.3.1 Minimum Variance Distortionless Response Beamformer 120
6.3.2 Noise Field Models 122
6.3.3 Subband Analysis and Synthesis 123
6.3.4 Beamforming Performance Criteria 126
6.3.5 Generalized Sidelobe Canceller Implementation 129
6.3.6 Recursive Implementation of the GSC 130
6.3.7 Other Conventional GSC Beamformers 131
6.3.8 Beamforming based on Higher Order Statistics 132
6.3.9 Online Implementation 136
6.3.10 Speech-Recognition Experiments 140
6.4 Spherical Microphone Arrays 142
6.5 Spherical Adaptive Algorithms 148
6.6 Comparative Studies 149
6.7 Comparison of Linear and Spherical Arrays for DSR 152
6.8 Conclusions and Further Reading 154
References 155

Part Three FEATURE ENHANCEMENT

7 From Signals to Speech Features by Digital Signal Processing 161
Matthias Wölfel

7.1 Introduction 161
7.1.1 About this Chapter 162
7.2 The Speech Signal 162

7.3	Spectral Processing	163
	7.3.1 Windowing	163
	7.3.2 Power Spectrum	165
	7.3.3 Spectral Envelopes	166
	7.3.4 LP Envelope	166
	7.3.5 MVDR Envelope	169
	7.3.6 Warping the Frequency Axis	171
	7.3.7 Warped LP Envelope	175
	7.3.8 Warped MVDR Envelope	176
	7.3.9 Comparison of Spectral Estimates	177
	7.3.10 The Spectrogram	179
7.4	Cepstral Processing	179
	7.4.1 Definition and Calculation of Cepstral Coefficients	180
	7.4.2 Characteristics of Cepstral Sequences	181
7.5	Influence of Distortions on Different Speech Features	182
	7.5.1 Objective Functions	182
	7.5.2 Robustness against Noise	185
	7.5.3 Robustness against Echo and Reverberation	187
	7.5.4 Robustness against Changes in Fundamental Frequency	189
7.6	Summary and Further Reading	191
	References	191
8	**Features Based on Auditory Physiology and Perception** *Richard M. Stern, Nelson Morgan*	**193**
8.1	Introduction	193
8.2	Some Attributes of Auditory Physiology and Perception	194
	8.2.1 Peripheral Processing	194
	8.2.2 Processing at more Central Levels	200
	8.2.3 Psychoacoustical Correlates of Physiological Observations	202
	8.2.4 The Impact of Auditory Processing on Conventional Feature Extraction	206
	8.2.5 Summary	208
8.3	"Classic" Auditory Representations	208
8.4	Current Trends in Auditory Feature Analysis	213
8.5	Summary	221
	Acknowledgments	222
	References	222
9	**Feature Compensation** *Jasha Droppo*	**229**
9.1	Life in an Ideal World	229
	9.1.1 Noise Robustness Tasks	229
	9.1.2 Probabilistic Feature Enhancement	230
	9.1.3 Gaussian Mixture Models	231

9.2 MMSE-SPLICE 232
9.2.1 Parameter Estimation 233
9.2.2 Results 236
9.3 Discriminative SPLICE 237
9.3.1 The MMI Objective Function 238
9.3.2 Training the Front-End Parameters 239
9.3.3 The Rprop Algorithm 240
9.3.4 Results 241
9.4 Model-Based Feature Enhancement 242
9.4.1 The Additive Noise-Mixing Equation 243
9.4.2 The Joint Probability Model 244
9.4.3 Vector Taylor Series Approximation 246
9.4.4 Estimating Clean Speech 247
9.4.5 Results 247
9.5 Switching Linear Dynamic System 248
9.6 Conclusion 249
References 249

10 Reverberant Speech Recognition **251**
Reinhold Haeb-Umbach, Alexander Krueger

10.1 Introduction 251
10.2 The Effect of Reverberation 252
10.2.1 What is Reverberation? 252
10.2.2 The Relationship between Clean and Reverberant Speech Features 254
10.2.3 The Effect of Reverberation on ASR Performance 258
10.3 Approaches to Reverberant Speech Recognition 258
10.3.1 Signal-Based Techniques 259
10.3.2 Front-End Techniques 260
10.3.3 Back-End Techniques 262
10.3.4 Concluding Remarks 265
10.4 Feature Domain Model of the Acoustic Impulse Response 265
10.5 Bayesian Feature Enhancement 267
10.5.1 Basic Approach 268
10.5.2 Measurement Update 269
10.5.3 Time Update 270
10.5.4 Inference 271
10.6 Experimental Results 272
10.6.1 Databases 272
10.6.2 Overview of the Tested Methods 273
10.6.3 Recognition Results on Reverberant Speech 274
10.6.4 Recognition Results on Noisy Reverberant Speech 276
10.7 Conclusions 277
Acknowledgment 278
References 278

Part Four MODEL ENHANCEMENT

11 Adaptation and Discriminative Training of Acoustic Models 285
Yannick Estève, Paul Deléglise

11.1 Introduction 285
11.1.1 Acoustic Models 286
11.1.2 Maximum Likelihood Estimation 287
11.2 Acoustic Model Adaptation and Noise Robustness 288
11.2.1 Static (or Offline) Adaptation 289
11.2.2 Dynamic (or Online) Adaptation 289
11.3 Maximum *A Posteriori* Reestimation 290
11.4 Maximum Likelihood Linear Regression 293
11.4.1 Class Regression Tree 294
11.4.2 Constrained Maximum Likelihood Linear Regression 297
11.4.3 CMLLR Implementation 297
11.4.4 Speaker Adaptive Training 298
11.5 Discriminative Training 299
11.5.1 MMI Discriminative Training Criterion 301
11.5.2 MPE Discriminative Training Criterion 302
11.5.3 I-smoothing 303
11.5.4 MPE Implementation 304
11.6 Conclusion 307
References 308

12 Factorial Models for Noise Robust Speech Recognition 311
John R. Hershey, Steven J. Rennie, Jonathan Le Roux

12.1 Introduction 311
12.2 The Model-Based Approach 313
12.3 Signal Feature Domains 314
12.4 Interaction Models 317
12.4.1 Exact Interaction Model 318
12.4.2 Max Model 320
12.4.3 Log-Sum Model 321
12.4.4 Mel Interaction Model 321
12.5 Inference Methods 322
12.5.1 Max Model Inference 322
12.5.2 Parallel Model Combination 324
12.5.3 Vector Taylor Series Approaches 326
12.5.4 SNR-Dependent Approaches 331
12.6 Efficient Likelihood Evaluation in Factorial Models 332
12.6.1 Efficient Inference using the Max Model 332
12.6.2 Efficient Vector-Taylor Series Approaches 334
12.6.3 Band Quantization 335
12.7 Current Directions 337
12.7.1 Dynamic Noise Models for Robust ASR 338

12.7.2 Multi-Talker Speech Recognition using Graphical Models 339
12.7.3 Noise Robust ASR using Non-Negative Basis Representations 340
References 341

13 Acoustic Model Training for Robust Speech Recognition **347**
Michael L. Seltzer

13.1 Introduction 347
13.2 Traditional Training Methods for Robust Speech Recognition 348
13.3 A Brief Overview of Speaker Adaptive Training 349
13.4 Feature-Space Noise Adaptive Training 351
13.4.1 Experiments using fNAT 352
13.5 Model-Space Noise Adaptive Training 353
13.6 Noise Adaptive Training using VTS Adaptation 355
13.6.1 Vector Taylor Series HMM Adaptation 355
13.6.2 Updating the Acoustic Model Parameters 357
13.6.3 Updating the Environmental Parameters 360
13.6.4 Implementation Details 360
13.6.5 Experiments using NAT 361
13.7 Discussion 364
13.7.1 Comparison of Training Algorithms 364
13.7.2 Comparison to Speaker Adaptive Training 364
13.7.3 Related Adaptive Training Methods 365
13.8 Conclusion 366
References 366

Part Five COMPENSATION FOR INFORMATION LOSS

14 Missing-Data Techniques: Recognition with Incomplete Spectrograms **371**
Jon Barker

14.1 Introduction 371
14.2 Classification with Incomplete Data 373
14.2.1 A Simple Missing Data Scenario 374
14.2.2 Missing Data Theory 376
14.2.3 Validity of the MAR Assumption 378
14.2.4 Marginalising Acoustic Models 379
14.3 Energetic Masking 381
14.3.1 The Max Approximation 381
14.3.2 Bounded Marginalisation 382
14.3.3 Missing Data ASR in the Cepstral Domain 384
14.3.4 Missing Data ASR with Dynamic Features 386
14.4 Meta-Missing Data: Dealing with Mask Uncertainty 388
14.4.1 Missing Data with Soft Masks 388

	14.4.2	*Sub-band Combination Approaches*	391
	14.4.3	*Speech Fragment Decoding*	393
14.5	Some Perspectives on Performance		395
	References		396
15	**Missing-Data Techniques: Feature Reconstruction**		**399**
	Jort Florent Gemmeke, Ulpu Remes		
15.1	Introduction		399
15.2	Missing-Data Techniques		401
15.3	Correlation-Based Imputation		402
	15.3.1	*Fundamentals*	402
	15.3.2	*Implementation*	404
15.4	Cluster-Based Imputation		406
	15.4.1	*Fundamentals*	406
	15.4.2	*Implementation*	408
	15.4.3	*Advances*	409
15.5	Class-Conditioned Imputation		411
	15.5.1	*Fundamentals*	411
	15.5.2	*Implementation*	412
	15.5.3	*Advances*	413
15.6	Sparse Imputation		414
	15.6.1	*Fundamentals*	414
	15.6.2	*Implementation*	416
	15.6.3	*Advances*	418
15.7	Other Feature-Reconstruction Methods		420
	15.7.1	*Parametric Approaches*	420
	15.7.2	*Nonparametric Approaches*	421
15.8	Experimental Results		421
	15.8.1	*Feature-Reconstruction Methods*	422
	15.8.2	*Comparison with Other Methods*	424
	15.8.3	*Advances*	426
	15.8.4	*Combination with Other Methods*	427
15.9	Discussion and Conclusion		428
	Acknowledgments		429
	References		430
16	**Computational Auditory Scene Analysis and Automatic Speech Recognition**		**433**
	Arun Narayanan, DeLiang Wang		
16.1	Introduction		433
16.2	Auditory Scene Analysis		434
16.3	Computational Auditory Scene Analysis		435
	16.3.1	*Ideal Binary Mask*	435
	16.3.2	*Typical CASA Architecture*	438

16.4 CASA Strategies 440
16.4.1 IBM Estimation Based on Local SNR Estimates 440
16.4.2 IBM Estimation using ASA Cues 442
16.4.3 IBM Estimation as Binary Classification 448
16.4.4 Binaural Mask Estimation Strategies 451
16.5 Integrating CASA with ASR 452
16.5.1 Uncertainty Transform Model 454
16.6 Concluding Remarks 458
Acknowledgment 458
References 458

17 Uncertainty Decoding **463**
Hank Liao

17.1 Introduction 463
17.2 Observation Uncertainty 465
17.3 Uncertainty Decoding 466
17.4 Feature-Based Uncertainty Decoding 468
17.4.1 SPLICE with Uncertainty 470
17.4.2 Front-End Joint Uncertainty Decoding 471
17.4.3 Issues with Feature-Based Uncertainty Decoding 472
17.5 Model-Based Joint Uncertainty Decoding 473
17.5.1 Parameter Estimation 475
17.5.2 Comparisons with Other Methods 476
17.6 Noisy CMLLR 477
17.7 Uncertainty and Adaptive Training 480
17.7.1 Gradient-Based Methods 481
17.7.2 Factor Analysis Approaches 482
17.8 In Combination with Other Techniques 483
17.9 Conclusions 484
References 485

Index **487**

List of Contributors

Jon Barker
University of Sheffield, UK

Paul Deléglise
University of Le Mans, France

Jasha Droppo
Microsoft Research, USA

Yannick Estève
University of Le Mans, France

Jort Florent Gemmeke
KU Leuven, Belgium

Reinhold Haeb-Umbach
University of Paderborn, Germany

John R. Hershey
Mitsubishi Electric Research Laboratories, USA

Dorothea Kolossa
Ruhr-Universität Bochum, Germany

Alexander Krueger
University of Paderborn, Germany

Kenichi Kumatani
Disney Research, USA

Jonathan Le Roux
Mitsubishi Electric Research Laboratories, USA

Hank Liao
Google Inc., USA

Rainer Martin
Ruhr-Universität Bochum, Germany

John McDonough
Carnegie Mellon University, USA

Nelson Morgan
International Computer Science Institute and the University of California, Berkeley, USA

Arun Narayanan
The Ohio State University, USA

Bhiksha Raj
Carnegie Mellon University, USA

Ulpu Remes
Aalto University School of Science, Finland

Steven J. Rennie
IBM Thomas J. Watson Research Center, USA

Michael L. Seltzer
Microsoft Research, USA

Rita Singh
Carnegie Mellon University, USA

Paris Smaragdis
University of Illinois at Urbana-Champaign, USA

Richard Stern
Carnegie Mellon University, USA

Tuomas Virtanen
Tampere University of Technology, Finland

DeLiang Wang
The Ohio State University, USA

Matthias Wölfel
Pforzheim University, Germany

Acknowledgments

The editors would like to thank Jort Gemmeke, Joonas Nikunen, Pasi Pertilä, Janne Pylkkönen, Ulpu Remes, Rahim Saeidi, Michael Wohlmayr, Elina Helander, Kalle Palomäki, and Katariina Mahkonen, who have have assisted by providing constructive comments about individual chapters of the book.

Tuomas Virtanen would like to thank the Academy of Finland for financial support; and Professors Moncef Gabbouj, Sourish Chaudhuri, Mark Harvilla, and Ari Visa for supporting his position in the Department of Signal Processing, which has allowed for his editing this book.

1

Introduction

Tuomas Virtanen[1], Rita Singh[2], Bhiksha Raj[2]
[1]*Tampere University of Technology, Finland*
[2]*Carnegie Mellon University, USA*

1.1 Scope of the Book

The term "computer speech recognition" conjures up visions of the science-fiction capabilities of HAL2000 in *2001, A Space Odessey*, or "Data," the anthropoid robot in *Star Trek*, who can communicate through speech with as much ease as a human being. However, our real-life encounters with automatic speech recognition are usually rather less impressive, comprising often-annoying exchanges with interactive voice response, dictation, and transcription systems that make many mistakes, frequently misrecognizing what is spoken in a way that humans rarely would. The reasons for these mistakes are many. Some of the reasons have to do with fundamental limitations of the mathematical framework employed, and inadequate awareness or representation of context, world knowledge, and language. But other equally important sources of error are distortions introduced into the recorded audio during recording, transmission, and storage.

As automatic speech-recognition—or ASR—systems find increasing use in everyday life, the speech they must recognize is being recorded over a wider variety of conditions than ever before. It may be recorded over a variety of *channels*, including landline and cellular phones, the internet, etc. using different kinds of microphones, which may be placed close to the mouth such as in head-mounted microphones or telephone handsets, or at a distance from the speaker, such as desktop microphones. It may be corrupted by a wide variety of *noises*, such as sounds from various devices in the vicinity of the speaker, general background sounds such as those in a moving car or background babble in crowded places, or even competing speakers. It may also be affected by *reverberation*, caused by sound reflections in the recording environment. And, of course, all of the above may occur concurrently in myriad combinations and, just to make matters more interesting, may change unpredictably over time.

Techniques for Noise Robustness in Automatic Speech Recognition, First Edition.
Edited by Tuomas Virtanen, Rita Singh, and Bhiksha Raj.

For speech-recognition systems to perform acceptably, they must be *robust* to the distorting influences. This book deals with techniques that impart such robustness to ASR systems. We present a collection of articles from experts in the field, which describe an array of strategies that operate at various stages of processing in an ASR system. They range from techniques for minimizing the effect of external noises at the point of signal capture, to methods of deriving features from the signal that are fundamentally robust to signal degradation, techniques for attenuating the effect of external noises on the signal, and methods for modifying the recognition system itself to recognize degraded speech better.

The selection of techniques described in this book is intended to cover the range of approaches that are currently considered state of the art. Many of these approaches continue to evolve, nevertheless we believe that for a practitioner of the field to follow these developments, he must be familiar with the fundamental principles involved. The articles in this book are designed and edited to adequately present these fundamental principles. They are intended to be easy to understand, and sufficiently tutorial for the reader to be able to implement the described techniques.

1.2 Outline

Robustnesss techniques for ASR fall into a number of different categories. This book is divided into five parts, each focusing on a specific category of approaches. A clear understanding of robustness techniques for ASR requires a clear understanding of the principles behind automatic speech recognition and the robustness issues that affect them. These foundations are briefly discussed in Part One of the book. Chapter 2 gives a short introduction to the fundamentals of automatic speech recognition. Chapter 3 describes various distortions that affect speech signals, and analyzes their effect on ASR.

Part Two discusses techniques that are aimed at minimizing the distortions in the speech signal itself.

Chapter 4 presents methods for *voice-activity detection* (VAD), *noise estimation*, and *noise-suppression* techniques based on filtering. A VAD analyzes which signal segments correspond to speech and which to noise, so that an ASR system does not mistakenly interpret noise as speech. VAD can also provide an estimate of the noise during periods of speech inactivity. The chapter also reviews methods that are able to track noise characteristics even during speech activity. Noise estimates are required by many other techniques presented in the book.

Chapter 5 presents two approaches for separating speech from noises. The first one uses multiple microphones and an assumption that speech and noise signals are statistically independent of each other. The method does not use *a priori* information about the source signals, and is therefore termed *blind source separation*. Statistically independent signals are separated using an algorithm called *independent component analysis*. The second approach requires only a single microphone, but it is based on *a priori* information about speech or noise signals. The presented method is based on factoring the spectrogram of noisy speech into speech and noise using *nonnegative matrix factorization*.

Chapter 6 discusses methods that apply multiple microphones to selectively enhance speech while suppressing noise. They assume that the speech and noise sources are located in spatially different positions. By suitably combining the signals recorded by each microphone they are able to perform *beamforming*, which can selectively enhance signals from the location of the speech source. The chapter first presents the fundamentals of conventional linear microphone arrays, then reviews different criteria that can be used to design them, and then presents methods that can be used in the case of *spherical microphone arrays*.

Part Three of the book discusses methods that attempt to minimize the effect of distortions on *acoustic features* that are used to represent the speech signal.

Chapter 7 reviews conventional feature extraction methods that typically parameterize the envelope of the spectrum. Both methods based on *linear prediction* and *cepstral* processing are covered. The chapter then discusses *minimum variance distortionless response* or *warping* techniques that can be applied to make the envelope estimates more reliable for purposes of speech recognition. The chapter also studies the effect of distortions on the features.

Chapter 8 approaches the noise robustness problem from the point of view of human speech perception. It first presents a series of auditory measurements that illustrate selected properties of the human auditory system, and then discusses principles that make the human auditory system less sensitive to external influences. Finally, it presents several computational *auditory models* that mimic human auditory processes to extract noise robust features from the speech signal.

Chapter 9 presents methods that reduce the effect of distortions on features derived from speech. These *feature-enhancement* techniques can be trained to map noisy features to clean ones using training examples of clean and noisy speech. The mapping can include a criterion which makes the enhanced features more *discriminative*, i.e., makes them more effective for speech recognition. The chapter also presents methods that use an explicit model for additive noises.

Chapter 10 focuses on the recognition of reverberant speech. It first analyzes the effect of reverberation on speech and the features derived from it. It gives a review of different approaches that can be used to perform recognition of reverberant speech and presents methods for enhancing features derived from reverberant speech based on a model of reverberation.

Part Four discusses methods which modify the statistical parameters employed by the recognizer to improve recognition of corrupted speech.

Chapter 11 presents adaptation methods which change the parameters of the recognizer without assuming a specific kind of distortion. These *model-adaptation* techniques are frequently used to adapt a recognizer to a specific speaker, but can equally effectively be used to adapt it to distorted signals. The chapter also presents training criteria that makes the statistical models in the recognizer more *discriminative*, to improve the recognition performance that can be obtained with them.

Chapter 12 focuses on compensating for the effect of interfering sound sources on the recognizer. Based on a model of interfering noises and a model of the interaction process between speech and noise, these *model-compensation* techniques can be used to derive a statistical model for noisy speech. In order to find a mapping between the models for clean and noisy speech, the techniques use various approximations of the interaction process.

Chapter 13 discusses a methodology that can be used to find the parameters of an ASR system to make it more robust, given any signal or feature enhancement method. These *noise-adaptive-training* techniques are applied in the training stage, where the parameters the ASR system are tuned to optimize the recognition accuracy.

Part Five presents techniques which address the issue that some information in the speech signal may be lost because of noise. We now have a problem of *missing data* that must be dealt with.

Chapter 14 first discusses the general taxonomy of different missing-data problems. It then discusses the conditions under which speech features can be considered reliable, and when they may be assumed to be missing. Finally, it presents methods that can be used to perform robust ASR when there is uncertainty about which parts of the signal are missing.

Chapter 15 presents methods that produce an estimate of missing features (i.e., *feature reconstruction*) using reliable features. Reconstruction methods based on a Gaussian mixture model utilize local correlations between missing and reliable features. The reconstruction can also be done separately for each state of the ASR system. *Sparse representation* methods model the noisy observation as a linear combination of a small number of atomic units taken from a larger dictionary, and the weights of the atomic units are determined using reliable features only.

Chapter 16 discusses methods that estimate which parts of a speech signal are missing and which ones are reliable. The estimation can be based either on the signal-to-noise ratio in each time-frequency component, or on more perceptually motivated cues derived from the signal, or using a binary classification approach.

Chapter 17 presents approaches which enable the modeling of the *uncertainty* caused by noise in the recognition system. It first discusses feature-based uncertainty, which enables modeling of the uncertainty in enhanced signals or features obtained through algorithms discussed in the previous chapters of the book. Model-based *uncertainty decoding*, on the other hand, enables us to account for uncertainties in model compensation or adaptation techniques. The chapter also discusses the use of uncertainties with noise-adaptive training techniques.

We also revisit the contents of the book in the end of Chapter 3, once we have analyzed the types of errors encountered in automatic speech recognition.

1.3 Notation

The table below lists the most commonly used symbols in the book. Some of the chapters deviate from the definitions below, but in such cases the used symbols are explicitly defined.

Symbol	Definition
$a, b, c, \ldots$	Scalar variables
$A, B, C, \ldots$	Constants
$\mathbf{a}, \mathbf{b}, \mathbf{c}, \ldots$	Vectors
$\mathbf{A}, \mathbf{B}, \mathbf{C}, \ldots$	Matrices
$\otimes$	Convolution
$\mathcal{N}$	Normal distribution
$\mathcal{E}\{x\}$	Expected value of x
$\mathbf{A}^\mathsf{T}$	Transpose of matrix $\mathbf{A}$
$x_{i:j}$	Set $x_i, x_{i+1}, \ldots, x_j$
s	Speech signal
n	Additive noise signal
x	Noisy speech signal
h	Response from speaker to microphone
t	Time index

Symbol	Definition
f	Frequency index
$\mathbf{x}_t$	Observation vector of noisy speech in frame t
q	State variable
q_t	State at time t
$\boldsymbol{\mu}$	Mean vector
$\boldsymbol{\Theta}, \boldsymbol{\Sigma}$	Covariance matrix
P, p	Probability

Part One

Foundations

2

The Basics of Automatic Speech Recognition

Rita Singh[1], Bhiksha Raj[1], Tuomas Virtanen[2]
[1] *Carnegie Mellon University, USA*
[2] *Tampere University of Technology, Finland*

2.1 Introduction

In order to understand the techniques described later in this book, it is important to understand how automatic speech-recognition (ASR) systems function. This chapter briefly outlines the framework employed by ASR systems based on hidden Markov models (HMMs).

Most mainstream ASR systems are designed as probabilistic Bayes classifiers that identify the most likely word sequence that explains a given recorded acoustic signal. To do so, they use an estimate of the probabilities of possible word sequences in the language, and the probability distributions of the acoustic signals for each word sequence. Both the probability distributions of word sequences, and those of the acoustic signals for any word sequence, are represented through parametric *models*. Probabilities of word sequences are modeled by various forms of grammars or N-gram models. The probabilities of the acoustic signals are modeled by HMMs.

In the rest of this chapter, we will briefly describe the components and process of ASR as outlined above, as a prelude to explaining the circumstances under which it may perform poorly, and how that relates to the remaining chapters of this book. Since this book primarily addresses factors that affect the *acoustic* signal, we will only pay cursory attention to the manner in which word-sequence probabilities are modeled, and elaborate mainly on the modeling of the acoustic signal.

In Section 2.2, we outline Bayes classification, as applied to speech recognition. The fundamentals of HMMs—how to calculate probabilities with them, how to find the most likely explanation for an observation, and how to estimate their parameters—are given in Section 2.3. Section 2.4 describes how HMMs are used in practical ASR systems. Several issues related to practical implementation are addressed. Recognition is not performed with

Techniques for Noise Robustness in Automatic Speech Recognition, First Edition.
Edited by Tuomas Virtanen, Rita Singh, and Bhiksha Raj.

the speech signal itself, but on *features* derived from it. We give a brief review of the most commonly used features in Section 2.4.1. Feature computation is covered in greater detail in Chapters 7 and 8 of the book. The number of possible word sequences that must be investigated in order to determine the most likely one is potentially extremely large. It is infeasible to explicitly characterize the probability distributions of the acoustics for each and every word sequence. In Sections 2.4.2 and 2.4.3, we explain how we can nevertheless explore all of them by *composing* the HMMs for word sequences from smaller units, and how the set of all possible word sequences can be represented as compact graphs that can be searched.

Before proceeding, we note that although this book largely presents speech recognition and robustness issues related to it from the perspective of HMM-based systems, the fundamental ideas presented here, and many of the algorithms and techniques described both in this chapter and elsewhere in the book, carry over to other formalisms that may be employed for speech recognition as well.

2.2 Speech Recognition Viewed as Bayes Classification

At their core, state-of-art ASR systems are fundamentally *Bayesian classifiers*. The Bayesian classification paradigm follows a rather simple intuition: the best guess for the explanation of any observation (such as a recording of speech) is the most *likely* one, given any other information we have about the problem at hand. Mathematically, it can be stated as follows: let C_1, C_2, $C_3, \ldots$ represent all possible explanations for an observation $\mathbf{X}$. The Bayesian classification paradigm chooses the explanation C_i such that

$$P(C_i|\mathbf{X}, \theta) \geq P(C_j|\mathbf{X}, \theta) \quad \forall j \neq i, \tag{2.1}$$

where $P(C_i|\mathbf{X}, \theta)$ is the conditional probability of class C_i given the observation $\mathbf{X}$, and θ represents all other evidence, or information known *a priori*. In other words, it chooses the *a posteriori* most probable explanation C_i, given the observation and all prior evidence.

For the *ASR* problem, the problem is now stated as follows. Given a speech recording $\mathbf{X}$, the sequence of words $\hat{w}_1, \hat{w}_2, \cdots$ that were spoken is estimated as

$$\hat{w}_1, \hat{w}_2, \cdots = \underset{w_1, w_2, \cdots}{\operatorname{argmax}} P(w_1, w_2, \cdots |\mathbf{X}, \Lambda). \tag{2.2}$$

Here, Λ represents other evidence that we may have about what was spoken. Equation (2.2) states that the "best guess" word sequence $\hat{w}_1, \hat{w}_2 \cdots$ is the word sequence that is *a posteriori* most probable, after consideration of both the recording $\mathbf{X}$ and all other evidence represented by Λ.

In order to implement Equation (2.2) computationally, the problem is refactored using Bayes' rule as follows:

$$\hat{w}_1, \hat{w}_2, \cdots = \underset{w_1, w_2, \cdots}{\operatorname{argmax}} P(\mathbf{X}|w_1, w_2, \cdots)P(w_1, w_2, \cdots |\Lambda). \tag{2.3}$$

In the term $P(\mathbf{X}|w_1, w_2, \cdots)$, we assume that the speech signal $\mathbf{X}$ becomes independent of all other factors, once the sequence of words is given. The *true* distribution of $\mathbf{X}$ for any word sequence is not known. Instead it is typically modeled by a *hidden Markov model* (HMM) [2]. Since the term $P(\mathbf{X}|w_1, w_2, \cdots)$ models the properties of the acoustic speech signal, is it termed an *acoustic model*.

The second term on the right-hand side of Equation (2.3), $P(w_1, w_2, \cdots |\Lambda)$, provides the *a priori* probability of a word sequence, given all other evidence Λ. In theory, Λ may include evidence from our knowledge of the linguistic structure of the language (i.e., how people usually string words together when they speak), about the context of the current conversation, world knowledge, and anything else that one might bring to bear on the problem. However, in practice, the probability of a word sequence is usually assumed to be completely specified by a *language model*. The language model is often represented as a *finite-state* or a *context-free* grammar, or alternatively, as a statistical *N-gram* model.

2.3 Hidden Markov Models

Speech signals are *time-series* data, i.e., they are characterized by a sequence of measurements $\mathbf{x}_0, \mathbf{x}_1, \cdots$, where the sequence represents a progression through time and $\mathbf{x}_t$ represents the tth measurement in the series (the exact nature of the measurement $\mathbf{x}_t$ is discussed in Section 2.4.1). In the case of speech, this time series is *nonstationary*, i.e., its characteristics vary with time, as illustrated by the example in Figure 2.1.

HMMs are statistical models of time-series data. An HMM models a time series as having been generated by a process that goes through a series of *states* following a Markov chain. When in any state, the next state that the process will visit is determined stochastically and is only dependent on the current state. At each time, the process draws an observation from a probability distribution associated with the state it is currently in. Figure 2.2 illustrates the generation of observations by the process.

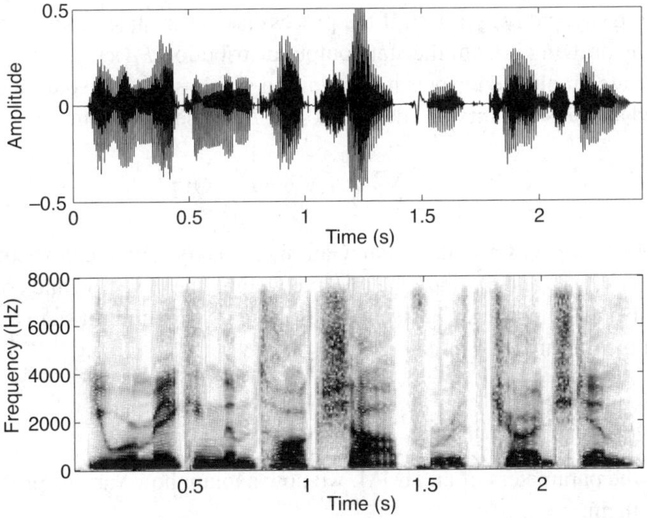

Figure 2.1 *Upper panel*: a speech signal. *Lower panel*: a time-frequency representation, or *spectrogram*, of the signal. In this figure the horizontal axis represents time and the vertical axis represents frequency. The intensity of the picture at any location represents the energy in the time-frequency component represented by the location. The observed time-varying patterns in energy distribution across frequencies are characteristic of the spoken sounds.

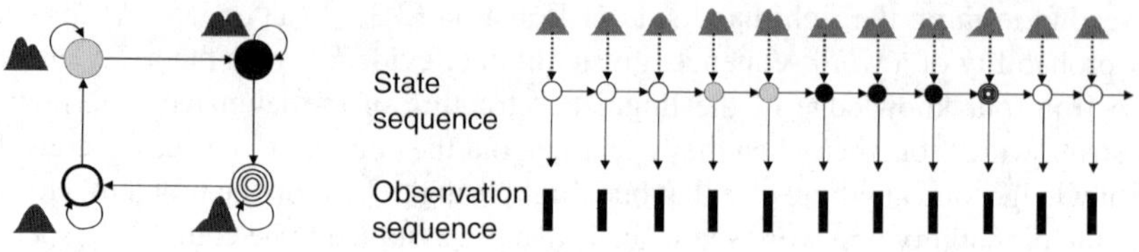

Figure 2.2 *Left panel*: schematic illustration of an HMM. The four circles represent the states of the HMM and the arrows represent allowed transitions. Each HMM state is associated with a state output distribution as shown. *Right panel*: generation of an observation sequence. The process progresses thorough a sequence of states. At each visited state, it generates an observation by drawing from the corresponding state output distribution.

Mathematically, an HMM is described as a probabilistic function of a Markov chain [11], and is a *doubly stochastic* model. The first level of this model is a Markov chain that is specified by an *initial* state probability distribution, usually denoted as π, and a transition matrix, which we will denote as $\mathbf{A}$. π specifies the probability of finding the process in any state at the very first instant. Representing the sequence of states visited by the process as $q_0, q_1, \cdots$, $\pi(i) = P(q_0 = i)$ is the probability that at the very first instant the process will be in state *i*. $\mathbf{A}$ is a matrix whose (i, j)th entry $a_{i,j} = P(q_{t+1} = j|q_t = i)$ represents the probability that the process will *transition* to state *j*, given that the process is currently in state *i*. The Markov chain thus is a probabilistic specification of the manner in which the process progresses through states.

The second level of the model is a set of *state output* probability distributions, one associated with each state. We denote the state output probability distribution associated with any state *i* as $P(\mathbf{x}|i)$, or more succinctly as $P_i(\mathbf{x})$. If the process arrives at state *i* at time *t*, it generates an observation $\mathbf{x}_t$ by drawing it from the state output distribution $P_i(\mathbf{x})$.

When HMMs are employed in speech-recognition systems the state output distributions are usually modeled as Gaussian mixture densities, and $P_i(\mathbf{x})$ has the form

$$P_i(\mathbf{x}) = \sum_{k=1}^{K} w_{i,k} \mathcal{N}(\mathbf{x}; \boldsymbol{\mu}_{i,k}, \boldsymbol{\Theta}_{i,k}), \tag{2.4}$$

where $\mathcal{N}(\mathbf{x}; \boldsymbol{\mu}, \boldsymbol{\Theta})$ represents a multivariate Gaussian density with mean vector $\boldsymbol{\mu}$ and covariance matrix $\boldsymbol{\Theta}$. $w_{i,k}$, $\boldsymbol{\mu}_{i,k}$ and $\boldsymbol{\Theta}_{i,k}$ are the mixture weight, mean vector, and covariance matrix of the *k*th Gaussian in the mixture Gaussian state output distribution for state *i*. *K* is the number of Gaussians in the mixture.

2.3.1 Computing Probabilities with HMMs

Having defined the parameters of an HMM, we now explain how various probabilities can be computed from them.

The Probability of Following a Specific State Sequence

The state sequence that the process follows is governed by the underlying Markov chain (i.e., the *first* level of the doubly stochastic process). The probability that the process follows a state sequence $q_{0:T-1} = q_0, q_1, \cdots, q_{T-1}$ can be written using Bayes' rule as

$$\begin{aligned} P(q_{0:T-1}) &= P(q_0)P(q_1|q_0)P(q_2|q_0,q_1)\cdots P(q_{T-1}|q_0\cdots q_{T-2}) \\ &= P(q_0)P(q_1|q_0)P(q_2|q_1)\cdots P(q_{T-1}|q_{T-2}) \\ &= P(q_0)\prod_{t=1}^{T-1} P(q_t|q_{t-1}) \\ &= \pi_{q_0}\prod_{t=1}^{T-1} a_{q_{t-1},q_t}. \end{aligned} \tag{2.5}$$

Here, we have used the Markovian property of the process: at any time, the future behavior of the process depends only on the current state and not on how it arrived there. Thus, $P(q_t|q_0\cdots q_{t-1}) = P(q_t|q_{t-1})$.

The Probability of Generating a Specific Observation Sequence from a Given State Sequence

We can also compute the probability that the process will produce a specific observation sequence $\mathbf{x}_{0:T-1} = \mathbf{x}_0, \mathbf{x}_1, \cdots, \mathbf{x}_{T-1}$, when it follows a specific state sequence $q_{0:T-1} = q_0, q_1, \cdots, q_{T-1}$. According to the model, the observation generated by the process at any time depends only on the state that the process is currently in, that is $P(\mathbf{x}_t|q_0, q_1, \cdots, q_{T-1}) = P(\mathbf{x}_t|q_t) = P_{q_t}(\mathbf{x}_t)$. Thus,

$$P(\mathbf{x}_{0:T-1}|q_{0:T-1}) = \prod_{t=0}^{T-1} P_{q_t}(\mathbf{x}_t). \tag{2.6}$$

The Probability of Following a Particular State Sequence and Generating a Specific Observation Sequence

The joint probability of following a particular state sequence and generating a specific observation sequence can be factored into two terms: the product of a probability of a state sequence (2.5) and a state-sequence conditional probability of an observation sequence (2.6), as illustrated in Figure 2.3. The probability that the process will proceed through a particular state sequence $q_{0:T-1}$ and generate an observation sequence $\mathbf{x}_{0:T-1}$ can thus be stated as

$$\begin{aligned} P(\mathbf{x}_{0:T-1}, q_{0:T-1}) &= P(\mathbf{x}_{0:T-1}|q_{0:T-1})P(q_{0:T-1}) \\ &= P(q_0)P(\mathbf{x}_0|q_0)\prod_{t=1}^{T-1} P(q_t|q_{t-1})P(\mathbf{x}_t|q_t) \qquad (2.7) \\ &= \pi_{q_0}P_{q_0}(\mathbf{x}_0)\prod_{t=1}^{T-1} a_{q_t,q_{t-1}}P_{q_t}(\mathbf{x}_t). \qquad (2.8) \end{aligned}$$

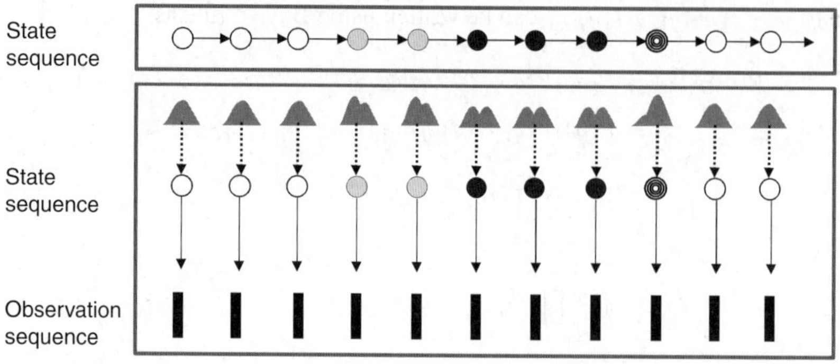

Figure 2.3 An HMM process can be factored into two parts: following a state sequence (top panel) and generating the observation sequence from the state sequence (bottom panel).

The Forward Probability

The probability $P(\mathbf{x}_{0:t}, q_t = i)$ that the process arrives at state i at time t while generating the first t observations $\mathbf{x}_{0:t}$, is often called the *forward* probability and denoted by $\alpha(i,t)$. At $t = 0$, when there have been no transitions, we only need to consider the initial state of the process and the first observation, and therefore

$$\begin{aligned}\alpha(i,0) &= P(\mathbf{x}_0, q_0 = i)\\ &= P(\mathbf{x}_0|q_0 = i)P(q_0 = i)\\ &= P_{q_0}(\mathbf{x}_0)\pi_{q_0}.\end{aligned}$$

Thereafter, $\alpha(i,t)$ can be recursively defined. In order to arrive at state j at time t, the process must be at some state i at $t-1$ and transition to j. Thus, the probability that the process will follow a state sequence that takes it through i at $t-1$ and arrive at j at t and generate the observation sequence $\mathbf{x}_{0:t} = \mathbf{x}_{0:t-1}, \mathbf{x}_t$ is merely the probability that the process will arrive at i at $t-1$ while generating $\mathbf{x}_{0:t-1}$, transition from i to j and finally generate $\mathbf{x}_t$ from j, that is

$$\begin{aligned}P(\mathbf{x}_{0:t}, q_{t-1} = i, q_t = j) &= P(\mathbf{x}_{0:t-1}, q_{t-1} = i)P(q_t = j|q_{t-1} = i)P(\mathbf{x}_t|q_t = j)\\ &= \alpha(i,t-1)a_{i,j}P_j(\mathbf{x}_t).\end{aligned}$$

Since $\alpha(j,t)$ is not a function of the state at $t-1$, i must be integrated out from the above equation:

$$\alpha(j,t) = \sum_{i=1}^{Q} P(\mathbf{x}_{0:t}, q_{t-1} = i, q_t = j) \tag{2.9}$$

$$= P_j(\mathbf{x}_t)\sum_{i=1}^{Q} \alpha(i,t-1)a_{i,j}, \tag{2.10}$$

where Q denotes the total number of states in the model.

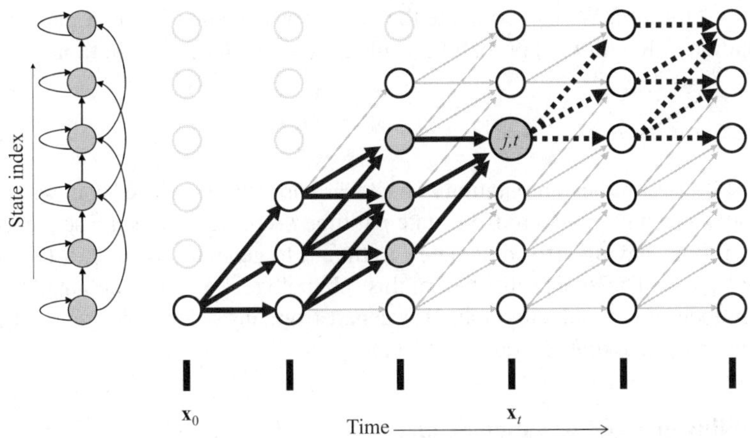

Figure 2.4 A "Trellis" showing all possible state sequences that the HMM on the left may follow to generate the observation sequence shown at the bottom. The thick solid lines show the "forward" subtrellis representing all state sequences that terminate at state *j* at time *t*. The "backward" subtrellis, shown by the thick dotted lines, shows all state sequences that depart from state *j* at time *t*. The union of the two shows all state sequences that visit state *j* at time *t*.

Figure 2.4 gives a graphical illustration of the recursion that can be used to obtain $\alpha(j, t)$. The figure shows a directed acyclic graph, called a trellis, that represents all possible state sequences that an HMM might follow to generate an observation sequence. The HMM is shown on the left, along the vertical axis. The observation sequence $\mathbf{x}_0, \mathbf{x}_1, \cdots$ is represented by the sequence of bars at the bottom. In the trellis, node (j, t) aligned with a state *j* of the HMM and the *t*th observation $\mathbf{x}_t$ of the observation sequence represents the event that the process visits *j* in the *t*th time step and draws the observation $\mathbf{x}_t$ from its state output distribution. All nodes and edges have probabilities associated with them. The node probability associated with node (j, t) is $P_j(\mathbf{x}_t)$. An edge that starts at a state *i* and terminates at *j* is assigned the transition probability $a_{i,j}$. The probabilities of subpaths through the trellis combine multiplicatively in any path. The probabilities of multiple incoming paths to any node combine additively.

The "forward" subgraph, shown by the thick arrows terminating at state *j* at time *t*, represents the set of all state sequences that the process may follow to arrive at state *j* at time *t* when generating $\mathbf{x}_{0:t}$. The total probability for this set of paths including the node at (j, t), is $\alpha(j, t)$. This subgraph is obtained by extending all subgraphs that end at any state *i* at $t - 1$ (shown by the shaded states at $t - 1$) by an edge that terminates at *j*, leading to Equation (2.9) as the rule for computing the total forward probability for node (j, t).

The Backward Probability

The probability $P(\mathbf{x}_{t+1:T-1}|q_t = j)$ that the process generates all future observations $\mathbf{x}_{t+1:T-1}$, given that it departs from state *j* at time *t*, is called the *backward* probability and is denoted by $\beta(j, t)$. Note that $\beta(j, t)$ does *not* include the current observation at time *t*; it only gives the probability of all *future* observations given the state at the current time. We also note that at the final time instant $T - 1$ there are no future observations. Hence, we define

$\beta(j, T-1) = 1$ for all j. We can compute the $\beta(j,t)$ terms recursively in a manner similar to the computation of the forward probability, only now we go *backward* in time:

$$\beta(j,t) = \sum_{i=1}^{Q} a_{j,i}\beta(i,t+1)P_i(\mathbf{x}_{t+1}). \tag{2.11}$$

Figure 2.4 illustrates the computation of backward probabilities. The "backward" subgraph with the dotted arrows emanating from state j at time t represents all state sequences that the HMM may follow, having arrived at state j at time t, to generate the rest of the observation sequence $\mathbf{x}_{t+1:T-1}$. The total probability of this subgraph is $\beta(j,t)$. It is obtained by extending a path from node (j,t) to all subgraphs that depart from any achievable state i at time $t+1$, leading to the recursive rule of Equation (2.11).

The Probability of an Observation Sequence

We can now compute the probability that the process will generate an observation sequence $\mathbf{x}_{0:T-1}$. Since this does not consider the specific state sequence followed by the process to generate the sequence, we must consider *all* possible state sequences. Thus, the probability of producing the observation sequence is given by

$$P(\mathbf{x}_{0:T-1}) = \sum_{q_{0:T-1}} P(\mathbf{x}_{0:T-1}, q_{0:T-1}).$$

Direct computation of this equation is clearly expensive or infeasible. If the process has Q possible states that it can be in at any time, the total number of possible state sequences is Q^T, which is exponential in T. Direct computation of $P(\mathbf{x}_{0:T-1})$ as given above will require summing over an exponential number of state sequences. However, using the forward and backward probabilities computed above, the computation of the probability of an observation sequence becomes trivial.

The probability that the process will generate $\mathbf{x}_{0:T-1}$ while following a state sequence that visits a specific state j at time t is given by

$$\begin{aligned} P(\mathbf{x}_{0:T-1}, q_t = j) &= P(\mathbf{x}_{0:t}, q_t = j)P(\mathbf{x}_{t+1:T-1}|q_t = j) \\ &= \alpha(j,t)\beta(j,t). \end{aligned} \tag{2.12}$$

Figure 2.4 illustrates this computation. The complete subgraph including both the solid and dotted edges represents all state sequences that the process may follow when generating $\mathbf{x}_{0:T-1}$ that visit state j at time t. The total probability of this subgraph is obtained by extending the forward subgraph by the backward subgraph and is given by $\alpha(j,t)\beta(j,t)$.

Since $P(\mathbf{x}_{0:T-1})$ must take into account *all* possible states at any time, it can be obtained from Equation (2.12) by summing over all states:

$$\begin{aligned} P(\mathbf{x}_{0:T-1}) &= \sum_{j=1}^{Q} P(\mathbf{x}_{0:T-1}, q_t = j) \\ &= \sum_{j=1}^{Q} \alpha(j,t)\beta(j,t). \end{aligned} \tag{2.13}$$

Note that the above equation holds for all t.

The Probability that the Process Was in a Specific State at a Specific Time, Given the Generated Observations

We are given that the process has generated an observation sequence $\mathbf{x}_{0:T-1}$. We wish to compute the *a posteriori* probability $P(q_t = i|\mathbf{x}_{0:T-1})$ that it was in a specific state *i* at a given time *t*. The probability is often referred to as $\gamma(i, t)$, and is directly obtained using Equations (2.12) and (2.13):

$$\gamma(i,t) = \frac{P(\mathbf{x}_{0:T-1}, q_t = i)}{P(\mathbf{x}_{0:T-1})} = \frac{\alpha(i,t)\beta(i,t)}{\sum_{j=1}^{Q}\alpha(j,t)\beta(j,t)}. \tag{2.14}$$

Given an observation $\mathbf{x}_{0:T-1}$, we can also compute the *a posteriori* probability $P(q_t = i, q_{t+1} = j|\mathbf{x}_{0:T-1})$ that the process was in state *i* at time *t* and in state *j* at $t + 1$ as

$$\gamma(i,j,t) = P(q_t = i, q_{t+1} = j|\mathbf{x}_{0:T-1}) \tag{2.15}$$

$$= \frac{\alpha(i,t)a_{i,j}P_j(\mathbf{x}_{t+1})\beta(j,t+1)}{P(\mathbf{x}_{0:T-1})}. \tag{2.16}$$

2.3.2 *Determining the State Sequence*

Given an observation sequence $\mathbf{x}_{0:T-1}$, one can estimate the state sequence followed by the process to generate the observations. We do so by finding the *a posteriori* most probable state sequence, i.e., the sequence $\hat{q}_{0:T-1}$ such that $P(\hat{q}_{0:T-1}|\mathbf{x}_{0:T-1})$ is maximum:

$$\hat{q}_{0:T-1} = \underset{q_{0:T-1}}{\operatorname{argmax}}\, P(q_{0:T-1}|\mathbf{x}_{0:T-1}) = \underset{q_{0:T-1}}{\operatorname{argmax}}\, P(q_{0:T-1}, \mathbf{x}_{0:T-1}).$$

In the right-hand side of the above equation, we have used the fact that the state sequence with the maximum *a posteriori* probability given the data also has the largest *joint* probability with the observation sequence. Once again, direct estimation is infeasible since one must evaluate an exponential number of state sequences to find the best one, but a dynamic programming alternative makes it feasible.

The Markov nature of the model ensures that the most likely state sequence $\hat{q}_{0:T-1}$ ending in state *j* at $t + 1$ is simply an extension of a most likely state sequence ending in one of the states $i = 1, \cdots, Q$ at *t*. Let $\delta_t(i)$ denote the probability of the most likely state sequence ending in state *i* at time *t*, that is

$$\delta_t(i) = \max_{q_{0:t-1}} P(\mathbf{x}_{0:t}, q_{0:t-1}, q_t = i). \tag{2.17}$$

Also let $\psi_t(i)$ denote the state at time $t - 1$ in the most likely state sequence ending in *i* at time *t*. For $t = 0$, since there is no previous time instant $t - 1$, we simply have $\psi_0(i) = 0$ and

$$\delta_0(i) = \pi_i P_i(\mathbf{x}_0). \tag{2.18}$$

At subsequent time indices $t = 1, \ldots, T - 1$, we recursively calculate $\delta_t(i)$ by selecting from the extensions of most probable state sequences at $t - 1$ to state *i* at *t*:

$$\psi_t(i) = \underset{j}{\operatorname{argmax}}\, \delta_{t-1}(j)a_{j,i}, \tag{2.19}$$

$$\delta_t(i) = \delta_{t-1}(\psi_t(i))a_{\psi_t(i),i}P_i(\mathbf{x}_t). \tag{2.20}$$

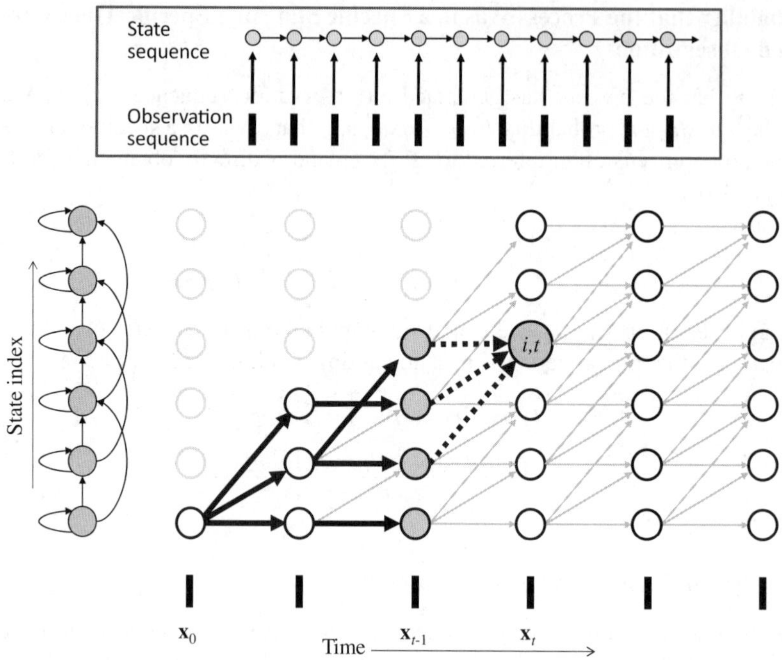

Figure 2.5 *Upper panel*: the inference problem addressed in Viterbi decoding. The state sequence that the process followed while producing an observation sequence is to be inferred from the observations. *Lower panel*: each path shown by the thick solid lines ending at any of the shaded nodes at $t-1$ represents the most probable state sequence ending at the state for that node at $t-1$. The most probable state sequence to node (i, t) is an extension of one of these by the edges shown by the dotted lines.

This recursion is illustrated by Figure 2.5.

The probability $\delta*$ of the most likely overall state sequence is simply the largest among the probabilities of the most likely state sequences ending at any of the states $i = 1, \ldots, Q$ at $T - 1$:

$$\delta* = \max_i \delta_{T-1}(i).$$

The state index $\hat{q}_{T-1}$ of the most likely sequence at time $T - 1$ is obtained as

$$\hat{q}_{T-1} = \underset{i}{\operatorname{argmax}}\, \delta_{T-1}(i).$$

The entire state sequence for times $t = T-2, T-3, \ldots, 0$ is obtained by backtracking as

$$\hat{q}_t = \psi_{t+1}(\hat{q}_{t+1}). \tag{2.21}$$

The above procedure, known as the *Viterbi* Algorithm [5,13], forms the basis for the search employed in ASR systems as we shall see later.

2.3.3 Learning HMM Parameters

The above sections described how to determine various probabilities, and how to identify the most likely state sequence to explain an observation sequence when all HMM parameters are known. Let us now consider a more fundamental problem: how to estimate the parameters of an HMM from a collection of data instances.

In an HMM, the parameters to be estimated are the initial state probabilities $\pi_i = P(q_0 = i)$, the transition probabilities $a_{i,j} = P(q_{t+1} = j|q_t = i)$ and the parameters of the state output distributions $P_i(\mathbf{x})$, which, in speech-recognition systems that assume state output distributions to be Gaussian mixtures, would be the mixture weights $w_{k,i}$, mean vectors $\boldsymbol{\mu}_{k,i}$ and covariance matrices $\boldsymbol{\Theta}_{k,i}$ for each Gaussian k of each state i. We assume that the number of states Q in the HMM and the number of Gaussians K in the Gaussian mixtures are known. In practice K and Q are often set by hand, although techniques do exist to estimate them from data as well.

To learn the parameters of the HMM, we typically use multiple observation sequences, which we refer to as "training" instances. In the equations below, we denote individual training instances by $\mathbf{X}$, and the sum over all the training instances by $\sum_{\mathbf{X}}$. Since individual training instances are of potentially different lengths, we have also used the subscripted value $T_{\mathbf{X}}$ to refer to the length (number of observations) in any $\mathbf{X}$, i.e., $\mathbf{X}$ comprises the observation sequence $\mathbf{x}_{0:T_{\mathbf{X}}-1}$. Let us denote the total number of data instances used to train the HMM by N.

The most common estimation procedure is based on the expectation maximization (EM) algorithm, and is known as the Baum–Welch algorithm. The derivation of the algorithm can be found in various references (e.g., [9]); here, we simply state the actual formulae with an attempt at providing an intuitive explanation. The EM estimation algorithm consists of iterations of the following formulae. In these formulae, $\gamma_{\mathbf{X}}(i,t)$ and $\gamma_{\mathbf{X}}(i,j,t)$ refer to the terms in Equations (2.14) and (2.16) obtained using the current estimates of the HMM parameters. The subscript $\mathbf{X}$ is used to indicate that the term has been computed from a specific training data instance $\mathbf{X} = \mathbf{x}_{0:T_{\mathbf{X}}-1}$:

$$\pi_i = \frac{\sum_{\mathbf{X}} \gamma_{\mathbf{X}}(0,i)}{N}, \tag{2.22}$$

$$a_{i,j} = \frac{\sum_{\mathbf{X}} \sum_{t=0}^{T_{\mathbf{X}}-2} \gamma_{\mathbf{X}}(i,j,t)}{\sum_{\mathbf{X}} \sum_{t=0}^{T_{\mathbf{X}}-2} \sum_{j=1}^{Q} \gamma_{\mathbf{X}}(i,j,t)}, \tag{2.23}$$

$$\gamma_{\mathbf{X}}^{k}(i,t) = \gamma_{\mathbf{X}}(i,t) \frac{w_k \mathcal{N}(\mathbf{x}_t; \boldsymbol{\mu}_{i,k}, \boldsymbol{\Theta}_{i,k})}{\sum_{k'=1}^{K} w_{k'} \mathcal{N}(\mathbf{x}_t; \boldsymbol{\mu}_{i,k'}, \boldsymbol{\Theta}_{i,k'})}, \tag{2.24}$$

$$w_{i,k} = \frac{\sum_{\mathbf{X}} \sum_{t=0}^{T_{\mathbf{X}}-1} \gamma_{\mathbf{X}}^{k}(i,t)}{\sum_{\mathbf{X}} \sum_{t=0}^{T_{\mathbf{X}}-1} \gamma_{\mathbf{X}}(i,t)}, \tag{2.25}$$

$$\boldsymbol{\mu}_{i,k} = \frac{\sum_{\mathbf{X}} \sum_{t=0}^{T_{\mathbf{X}}-1} \gamma_{\mathbf{X}}^{k}(i,t)\mathbf{x}_t}{\sum_{\mathbf{X}} \sum_{t=0}^{T_{\mathbf{X}}-1} \gamma_{\mathbf{X}}^{k}(i,t)}, \tag{2.26}$$

$$\boldsymbol{\Theta}_{i,k} = \frac{\sum_{\mathbf{X}} \sum_{t=0}^{T_{\mathbf{X}}-1} \gamma_{\mathbf{X}}^{k}(i,t)(\mathbf{x}_t - \boldsymbol{\mu}_{i,k})(\mathbf{x}_t - \boldsymbol{\mu}_{i,k})^{\top}}{\sum_{\mathbf{X}} \sum_{t=0}^{T_{\mathbf{X}}-1} \gamma_{\mathbf{X}}^{k}(i,t)}. \tag{2.27}$$

The above equations are easily understood if one thinks of $\gamma_{\mathbf{X}}(i,t)$ as an *expected* count of the number of times state i was visited by the process at time t when generating $\mathbf{X}$.

Thus, $\sum_{\mathbf{X}} \gamma_{\mathbf{X}}(0, i)$ is the expected count of the number of times the process was in state i at time 0 when generating the N observation sequences. Equation (2.22) is simply a ratio of the count of the expected number of sequences where the state of the process at the initial time instant was i and the total number of sequences, an intuitive ratio of counts. Similarly, $\gamma_{\mathbf{X}}(i, j, t)$ is the expected number of times the process transitioned from state i to j at time t. Thus, Equation (2.23) is the ratio of the *total* expected number of transitions from state i to j, and the total expected number of times the process was in state i.

Equation (2.24) represents the expected number of times the observation at time t in $\mathbf{X}$ was drawn from the kth Gaussian of the state output distribution of i. Equation (2.25) is similarly the ratio of the expected number of observations generated from the kth Gaussian of i to the expected total number of observations from i. We note that all of these equations are intuitive extensions of familiar count-based estimation of probabilities in multinomial data.

Equations (2.26) and (2.27) are likely to be somewhat less intuitive, in that they are not ratios of counts. Rather, they are *weighted* averages of first and second-order terms derived from the observations. The numerator in Equation (2.26) is the expected sum of the observations generated by the kth Gaussian of i. The denominator is the expected count of the number of observations generated from the Gaussian. The ratio is strictly analogous to the familiar formula for the mean of a set of vectors. Equation (2.27) computes a similar quantity for the second moment of the data. In order to reduce computational complexity and to avoid singular covariance matrices, the nondiagonal entries of $\mathbf{\Theta}_{i,k}$ are often restricted to be zero.

2.3.4 Additional Issues Relating to Speech Recognition Systems

The basic HMM formalism described above is typically extended in several ways in the implementation of speech-recognition systems. We describe some key extensions below. These extensions are not specific to speech recognition, but are also commonplace in other applications.

The Nonemitting State

A *nonemitting* state in an HMM is a state that has no emission probabilities associated with it. When the process visits this state, it generates no observations, but proceeds on to the next transition. The inclusion of nonemitting states in an HMM separates the progression of the process through the Markov chain underlying the HMM from the progression of time. Visits to nonemitting states do not represent a progression of time; only emitting states where observations are generated represent time progression. To prevent the process from remaining indefinitely within nonemitting states (thereby not producing additional observations and thus effectively "freezing" time), self-transitions are not allowed on nonemitting states. More generally, loops between nonemitting states in the Markov chain are disallowed; any loop must include at least one emitting state.

Nonemitting states serve a number of theoretical and practical purposes in Markov models for processes:

- A Markov process that generates observations of finite duration must *terminate* once the final observation has been generated. In order for a Markov process to terminate, it must have an *absorbing* state that, once arrived at, ceases all further activity. An absorbing state

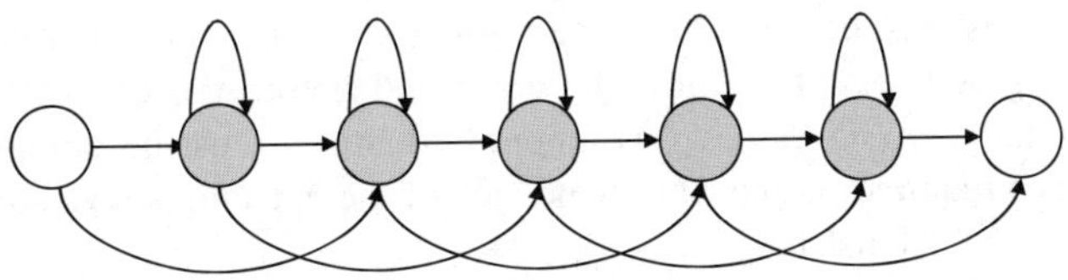

Figure 2.6 An example of an HMM with two nonemitting states. Only the shaded states have state output distributions associated with them. The two extreme states which are not shaded are nonemitting states. No observations are generated from nonemitting states, and they have no self-transitions. In this example, the process is assumed to start at the first nonemitting state and transition out of it immediately. It stops generating observations when it arrives at the terminal nonemitting state.

can be viewed as a nonemitting state with no outgoing transitions. In the absence of such an absorbing state, all outputs generated by the process are infinitely long since there is no mechanism for the process to terminate. As a result, any finite-length observation must per-force be considered to be a partial observation of an infinitely long output generated by the process.

- The conventional specification of HMM parameters includes a set of *initial*-state probabilities $\{\pi_i\}$ that specify the probability that the process will be in any state i at the instant when the first observation is generated. This can instead be reframed through the use of an *initial* queiscent nonemitting state "-1" that has only outgoing transitions, but no incoming transitions. The model now assumes that the process resides in this state until it begins to generate observations. It then transitions to one of the remaining states with a probability $a_{-1,i}$, where $a_{-1,i}$ is identical to the initial state probabilities in the conventional notation, i.e., $a_{-1,i} = \pi_i$.
- Nonemitting states with both incoming and outgoing transitions provide a convenient mechanism for concatenating HMMs of individual symbols (phonemes or words) into longer HMMs (for words, sentences, or grammars) as we explain later in this chapter.

Figure 2.6 shows an example of an HMM that has left-to-right "Bakis topology" HMM [1] which employs both a nonemitting initial (quiescent) state and a nonemitting final (absorbing) state. This topology, which does not permit the process to return to a state once it has transitioned out of it, is most commonly used to represent speech sounds. This constraint can be imposed by defining $a_{i,j} = 0$ for $j < i$.

The inclusion of nonemitting states modifies the various estimation and update formulae in a relatively minor way. We must now consider that the process may visit one or more nonemitting states between any two time instants. Moreover, the set of nonemitting states a process can visit may vary from time instant to time instant. For instance, in the HMM of Figure 2.6, a nonemitting state can only be visited after generating a minimum of two observations.

Let $\mathcal{Q}(t)$ be the set of emitting states that the process may visit at time instant t, and let $\mathcal{U}(t)$ be the set of nonemitting states that it may visit after t, before it advances to time instant $t+1$. The calculation of the forward variable in Equation (2.9) is modified to

$$\alpha(j,t) = \begin{cases} P_j(\mathbf{x}_t)\sum_{i:a_{i,j}>0}\alpha(i,t-1)a_{i,j}, & j \in \mathcal{Q}(t) \\ \sum_{i:a_{i,j}>0}\alpha(i,t)a_{i,j}, & j \in \mathcal{U}(t). \end{cases}$$

The forward variables must be calculated recursively for $t = 0, \ldots, T-1$ as earlier. The $\alpha(i,t)$ values for emitting states $i \in \mathcal{Q}(t)$ must be computed before those for nonemitting states $i \in \mathcal{U}(t)$. Additionally, $\alpha(i,t)$ values for nonemitting states must be computed in such an order that variables required on the right-hand side of the equation above are available when assigning a value for the left-hand side.

The calculation of the backward variable in Equation (2.11) changes to

$$\beta(i,t) = \sum_{j:j \in \mathcal{Q}(t+1)} \beta(j,t+1) a_{i,j} P_j(\mathbf{x}_{t+1}) + \sum_{j:j \in \mathcal{U}(t) \wedge a_{i,j} > 0} \beta(j,t) a_{i,j}.$$

As in the case of forward probability computation, the order of computation of $\beta(i,t)$ terms must be such that the variables on the right-hand side are available when assigning a value for the left-hand side.

The state occupancy probabilities $\gamma(i,t)$ continue to be computed as in Equation (2.14), with the addendum that they can now be computed for both, emitting states $i \in \mathcal{Q}(t)$ and nonemitting states $i \in \mathcal{U}(t)$.

The transition-occupancy probabilities $\gamma(i,j,t)$ can occur between both types of states:

$$\gamma(i,j,t) = \begin{cases} \dfrac{\alpha(i,t) a_{i,j} P_j(\mathbf{x}_{t+1}) \beta(j,t+1)}{P(\mathbf{x}_{0:T-1})} & j \in \mathcal{Q}(t+1) \\ \dfrac{\alpha(i,t) a_{i,j} \beta(j,t)}{P(\mathbf{x}_{0:T-1})} & j \in \mathcal{U}(t). \end{cases}$$

All reestimation formulae for HMM parameters remain unchanged, with the modification that we do not need to estimate state output probability distribution parameters for nonemitting states.

A corresponding modification is also required for the Viterbi algorithm, which is used to find the optimal state sequence. The most likely predecessor to state *i* at *t* is computed according to

$$\psi_t(i) = \begin{cases} \operatorname*{argmax}_{j:a_{i,j}>0} \delta_{t-1}(j) a_{j,i}, & i \in \mathcal{Q}(t) \\ \operatorname*{argmax}_{j:a_{i,j}>0} \delta_t(j) a_{j,i} & i \in \mathcal{U}(t). \end{cases}$$

The probability $\delta_t(i)$ of the most likely state sequence arriving at state *i* while generating the observation sequence $\mathbf{x}_{0:t}$ is now given by

$$\delta_t(i) = \begin{cases} P_i(\mathbf{x}_t) \delta_{t-1}(\psi_t(i)) a_{\psi_t(i),i}, & i \in \mathcal{Q}(t) \\ \delta_t(\psi_t(i)) a_{\psi_t(i),i} & i \in \mathcal{U}(t). \end{cases}$$

If the HMM has absorbing states, the final state of the most likely sequence at time $T-1$ is the absorbing state *i* with the highest probability $\delta_{T-1}(i)$. Otherwise, the final state of the most likely state sequence is the emitting state with the highest $\delta_{T-1}(i)$. The complete, most-probable state sequence can then be obtained by backtracking. The details of the backtracking procedure are similar to Equation (2.21) and are omitted here.

Composing HMMs

The HMM for a compound process comprising sequences of subprocesses can be composed from the HMMs for the subprocesses. This mechanism is often utilized in speech-recognition

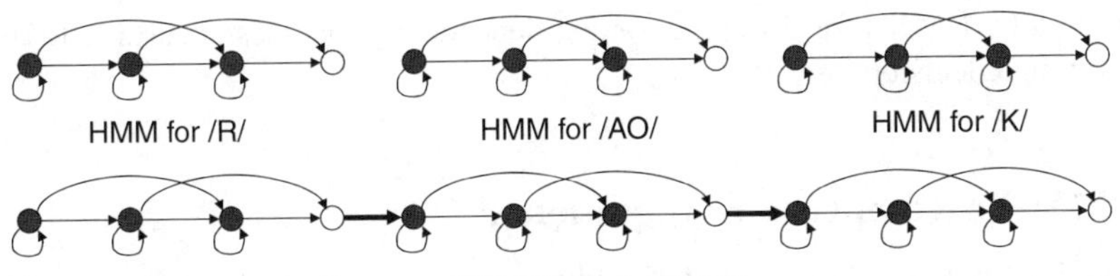

Figure 2.7 The HMM for the word "ROCK" is composed from the HMMs for the phonemes that constitute it—"/R/," "/AO/," and "/K/" in this example. Here, the HMMs for the individual phonemes have Bakis topology with a final nonemitting state (shown by the blank circles). The composed HMM for ROCK has nonemitting states between the phonemes, as well as a terminal nonemitting state. Other ways of composing the HMM for the word, which eliminate the internal nonemitting states, are also possible.

systems. For instance, HMMs for words in a language are often composed by concatenating HMMs for smaller units of sound such as phonemes (or phonemes in context, such as diphones or triphones) that are present in the word. Figure 2.7 illustrates this with an example.

Parameter Sharing

When simultaneously modeling multiple classes, some of which are highly similar, it is often useful to assume that the HMMs for some of the classes obtain their parameters from a common pool of parameters. Consequently, subsets of parameters for several HMMs may be identical. For instance, in speech-recognition systems, it is common to assume that the transition probabilities of the HMMs for all context-dependent versions of a phoneme are identical. Similarly, it is also common to assume that the parameters of the state output distributions of the HMMs for various context-dependent phonemes are shared.

Sharing of parameters does not affect either the computation of the forward and backward probabilities, or the estimation of the optimal state sequence for an observation. The primary effect of sharing is on the *learning* of HMM parameters. We note that each of the parameter estimation rules in Equations (2.22)–(2.27) specifies a single parameter for a specific state i of the HMM, and is of the form

$$\text{parameter}(i) = \frac{\text{numerator}(i)}{\text{denominator}(i)}$$

This is now modified to

$$\text{parameter}(\mathcal{I}) = \frac{\sum_{i\in\mathcal{I}} \text{numerator}(i)}{\sum_{i\in\mathcal{I}} \text{denominator}(i)}$$

$$\text{parameter}(i) = \text{parameter}(\mathcal{I}) \quad \forall i \in \mathcal{I},$$

where $\mathcal{I}$ represents the *set* of states that share the same parameter.

The manner in which parameters are shared, i.e., the set $\mathcal{I}$ for the various parameters, is often determined based on the expected similarity of the parameters of the HMMs that share them. In speech-recognition systems, the parameters of Gaussian-mixture state output distributions are usually shared by HMMs for different context-dependent phonemes according to groupings obtained through decision trees [8].

2.4 HMM-Based Speech Recognition

Having outlined hidden Markov models and the various terms that can be computed from them, we now return to the subject of HMM-based speech recognition. To recap, we restate the Bayesian formulation for the speech-recognition problem: given a speech recording $\mathbf{X}$, the sequence of words $\hat{w}_1, \hat{w}_2, \cdots$ that was spoken is estimated as

$$\hat{w}_1, \hat{w}_2, \cdots = \underset{w_1, w_2, \cdots}{\operatorname{argmax}} P(\mathbf{X}|w_1, w_2, \cdots)P(w_1, w_2, \cdots). \tag{2.28}$$

We are now ready to look at the details of exactly how this classification is implemented.

2.4.1 *Representing the Signal*

The first factor to consider is the representation of the speech signal itself. Speech recognition is not performed directly with the speech signal. The information in speech is primarily in its *spectral* content and its modulation over time, which may often not be apparent from the time-domain signal, as illustrated by Figure 2.1. Accordingly, ASR systems first compute a sequence of *feature vectors* $\mathbf{X} = \mathbf{x}_0, \mathbf{x}_2, \cdots \mathbf{x}_{T-1}$ to capture the salient spectral characteristics of the signal. The probability $P(\mathbf{X}|w_1, w_2, \cdots)$ in Equation (2.28) is computed using these feature vectors.

The most commonly used features are mel-frequency cepstra [4] and perceptual linear prediction cepstral features [7]. A variety of other features have also been proposed, and some of these and the motivations behind them are described in Chapters 7 and 8.

The principal mechanism for deriving feature vectors is as follows: the signal is segmented into analysis frames, typically 25-ms wide. Adjacent analysis frames are typically shifted by 10 ms with respect to one another, resulting in an analysis rate of 100 frames/s. A feature vector is derived from each frame. Specifically, the widely used mel-frequency cepstral features are obtained as follows:

1. The signal is preemphasized using a first-order finite impulse response high-pass filter to boost its high-frequency content.
2. The preemphasized signal is windowed, typically with a Hamming window [6].
3. A power spectrum is derived from it using a discrete Fourier transform (DFT) and squaring the magnitudes of the individual frequency components of the DFT.
4. The frequency components of the power spectrum are then integrated into a small number of bands using a filter bank that mimics the frequency sensitivity of the human auditory system as specified by the *mel* scale [12], to obtain a *mel* spectrum.
5. The mel spectral components are then compressed by a logarithmic function to mimic the loudness perception of the human auditory system.
6. A discrete cosine transform (DCT) is then performed on the log-compressed mel spectrum to obtain a cepstral vector. The first few components of the cepstral vector, typically 13 in number, are finally retained to obtain the mel-frequency cepstral vector for the frame.

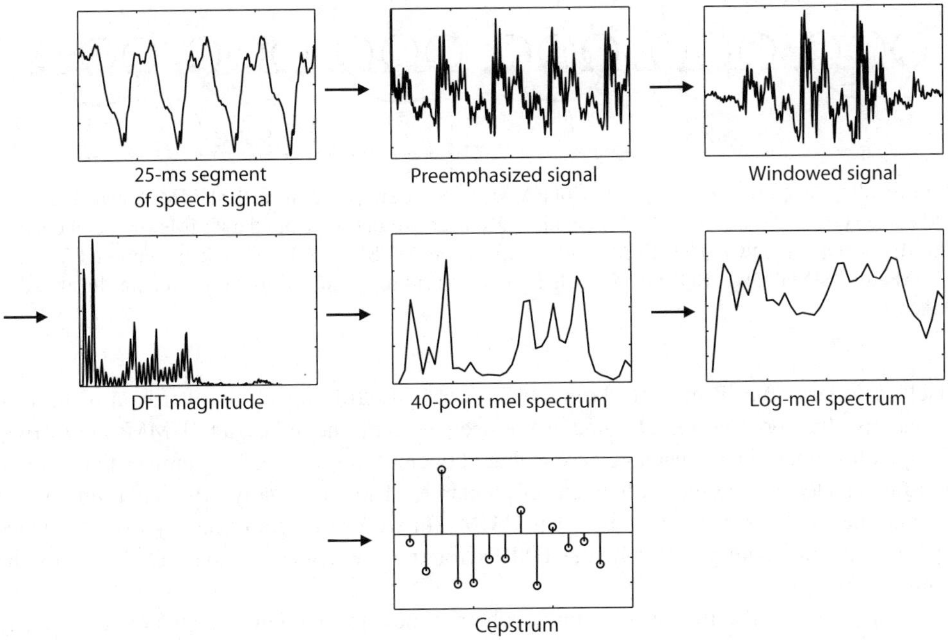

Figure 2.8 A typical example of the sequence of operations for computing mel-frequency cepstra.

An example of the above processing is shown in Figure 2.8.

Often each cepstral vector is augmented with a *velocity* (or *delta*) term, typically computed as the difference between adjacent cepstral vectors, and an *acceleration* (or *double-delta*, or *delta-delta*) term, typically computed as the difference between the velocity features for adjacent frames. The cepstral, velocity and acceleration vectors are concatenated to obtain an extended feature vector. The terms *static* and *dynamic* features are also used to describe the cepstral features and their derivatives, respectively.

2.4.2 The HMM for a Word Sequence

The main term in Equation (2.28) is $P(\mathbf{X}|w_1, w_2, \cdots)$. We will henceforth assume that the speech recording $\mathbf{X}$ is a sequence of feature vectors. The probability distribution $P(\mathbf{X}|w_1, w_2, \cdots)$ is modeled using an HMM.

The HMM for each of the word sequences considered in Equation (2.28) must ideally be learned from example (or training) instances of the word sequence. The number of word sequences to consider in Equation (2.28) is typically very large, and it is usually not possible to obtain a sufficient number of training instances of each word sequence to learn its HMM properly. Therefore, we must factor the problem.

The *vocabulary* of a speech recognizer, i.e., the set of words it can recognize, is finite, although the words can compose infinitely many word sequences. Therefore, we only learn HMMs for the words that the system can recognize. The HMM for any word sequence is *composed* from the HMMs for the words as explained in Section 2.3.4. Often, the vocabulary

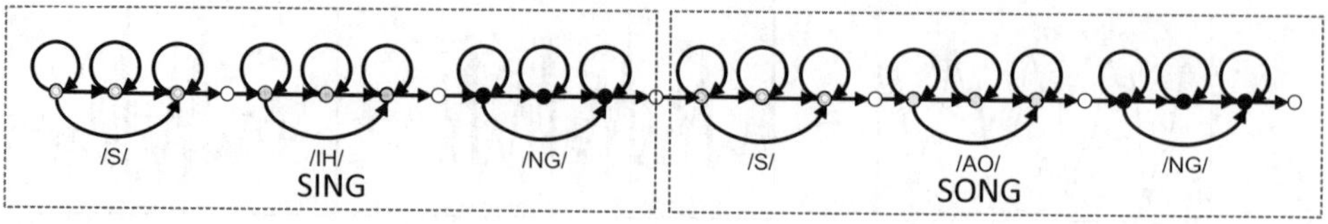

Figure 2.9 Illustrating the composition of HMMs for word sequences from the HMMs for smaller units. Here, the HMMs for the words "SING" and "SONG" are composed from the HMMs for the phonemes /S/, /IH/, /AO/, and /NG/. The HMM for the word sequence "SING SONG" is then composed from the HMMs for "SING" and "SONG." Note that this is essentially identical to the procedure illustrated in Figure 2.7.

itself is so large that it may not be possible to train HMMs for all the words in it. In such situations, the words in turn are modeled as sequences of phonemes, and HMMs are learned for the phonemes. The advantage here is that phonemes are far fewer in number than words: most languages have at most a few tens of phonemes. There is usually sufficient training data to train the HMMs for all phonemes. The HMMs for words that compose any word sequence are now in turn composed from the HMMs for the phonemes. Figure 2.9 illustrates the composition.

The HMMs for the lowest level units, whether they are phonemes or words, are usually assigned a left-to-right Bakis topology, as illustrated in Figures 2.7 and 2.9. Commonly, phonemes are modeled *in context*, for example triphones. For example, a phoneme /AX/, when in the context of a preceding /B/ and a succeeding /D/ (as in the word BUD) may be assigned a separate HMM than when the same phoneme is preceded by /D/ and followed by /B/ (as in the word DUB). However, since a complete set of in-context phonemes may be very large, their parameters are frequently shared, as explained in Section 2.3.4, in order to reduce the total number of parameters required to represent all units.

2.4.3 *Searching through all Word Sequences*

Direct evaluation of Equation (2.28) to perform recognition is clearly an infeasible task under most circumstances: $P(\mathbf{X}|w_1, w_2, \cdots)$ must be computed for every possible word sequence in order to identify the most probable one. However, the classification problem becomes more tractable if we modify it to the following:

$$\hat{w}_1, \hat{w}_2, \cdots = \underset{w_1, w_2, \cdots}{\text{argmax}} \left[\max_{\mathbf{q}} P(\mathbf{X}, \mathbf{q}|w_1, w_2, \cdots) P(w_1, w_2, \cdots) \right], \tag{2.29}$$

where q represents a state sequence through the HMM for $w_1, w_2, \cdots$. When calculating the probability of a word sequence, Equation (2.29) considers only the most probable state sequence corresponding to the word sequence instead of all possible state sequences. The word sequence that has the most probable state sequence of all is chosen.

This rather simple modification converts recognition to a tractable problem. To explain, we begin by considering a simple problem where the spoken utterance may only be one of a restricted set of sentences, such as all sentences specified by a simple finite-state grammar. Although the grammar actually specifies a distinct set of "acceptable" word sequences, the set

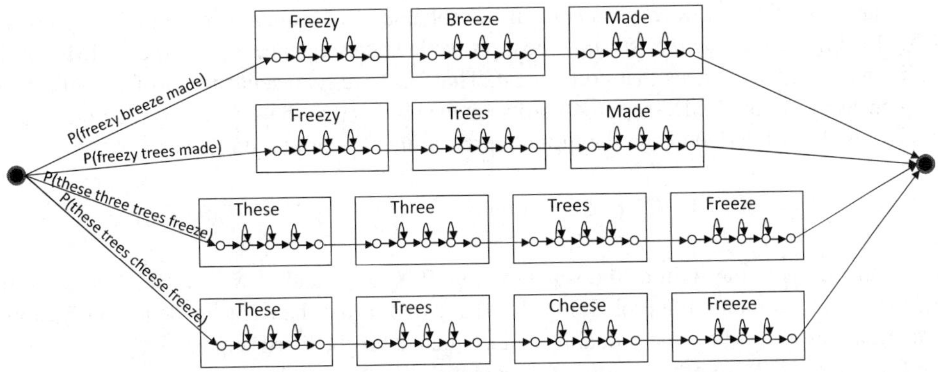

(a) A composite HMM from a word graph representing four sentences

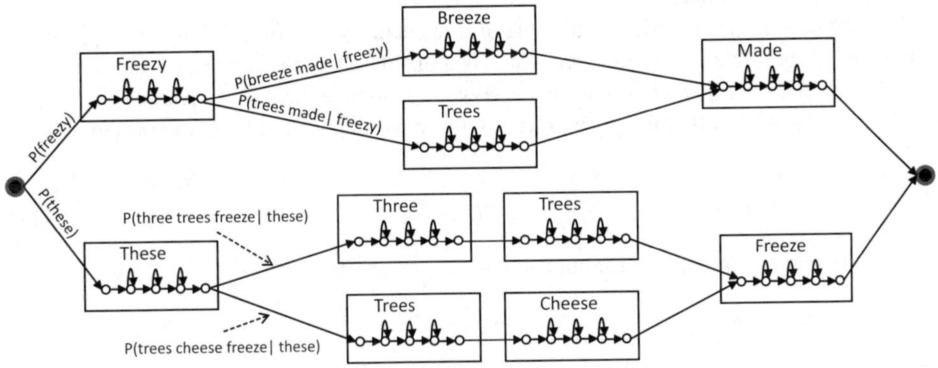

(b) A *compressed* version of the same graph

Figure 2.10 A composite HMM for recognizing four sentences from a poem. Each path from the source state to the final absorbing state represents one of the four sentences. In the upper panel, the *a priori* probability of each sentence is assigned to the transition from the source state to the first word in the word sequence. In the lower panel, the *a priori* probabilities are *spread*, enabling portions of the HMM to be shared between different sentences. The product of word probabilities on any path from source to sink is the *a priori* probability for the corresponding word sequence.

can be represented as a single directed word graph as illustrated by Figure 2.10. The graph has a "source" node and one or more "sinks." Any path from a source to a sink is a valid word sequence.

We can now replace the words on the edges by HMMs for the words, as also illustrated by Figure 2.10. This results in a single "composite" HMM that represents all valid word sequences. The source node now becomes the initial source state for the composite HMM and the sink node (assuming for simplicity that we only have a single sink node) becomes an absorbing terminal state. Any state sequence that begins at the source state and ends at the terminal state also represents a state sequence through the HMM for a single word sequence. The *a priori* probabilities of word sequences can be incorporated into the composite HMM in multiple ways, two of which are shown in Figure 2.10.

It can be shown that the *a posteriori* most probable state sequence through this composite HMM is identical to the *a posteriori* most probable state sequence among the HMMs for the individual word sequences that compose it. That is, if we represent the set of all valid word sequences as $\mathcal{W}$, the HMM for any word sequence $\mathbf{w} \in \mathcal{W}$ in the set as $\mathcal{H}(\mathbf{w})$, and the single composite HMM derived from the graph representing all word sequences as $\mathcal{L}$:

$$\underset{\mathbf{q}}{\operatorname{argmax}}\, P(\mathbf{X}, \mathbf{q}; \mathcal{L}) \cong \underset{\mathbf{q}}{\operatorname{argmax}} \left[\max_{\mathbf{w} \in \mathcal{W}} P(\mathbf{X}, \mathbf{q}; \mathcal{H}(\mathbf{w})) P(\mathbf{w}) \right]. \tag{2.30}$$

Here, the terms to the right of the semicolon in $P(\mathbf{X}, \mathbf{q}; \mathcal{L})$ and $P(\mathbf{X}, \mathbf{q}; \mathcal{H}(\mathbf{w}))$ represent the HMM used to compute the probability. It is easy to see that the right-hand side of the above equation represents the most probable state sequence for the word sequence $\hat{w}_1, \hat{w}_2, \cdots$ given by Equation (2.29). In other words, if we identify the most probable state sequence through the composite HMM, and determine the word sequence that it represents, we would also have found the solution to Equation (2.29).

When the recognizer must recognize natural speech, the set of valid word sequences is infinite, and all possible word sequences must be considered as candidates. In this scenario, it is usual to model the *a priori* probability of a word sequence through an N-gram model. An N-gram model specifies that the probability of any word depends only on the previous $N - 1$ words:

$$P(w_m | w_1 w_2 \cdots w_{m-1}) = P(w_m | w_{m-N+1} \cdots w_{m-1}). \tag{2.31}$$

Thus, the probability of the word sequence $w_1, w_2, \cdots, w_K$ is given by:

$$P(w_1, w_2, \cdots, w_K) = P(w_1|b)P(w_2|b\, w_1) \cdots P(w_{N-1}|b\, w_1 \cdots w_{N-2}) \cdot \left[\prod_{k=N}^{K} P(w_k|w_{k-N+1} \cdots w_{k-1}) \right] P(e|w_{K-N+1} \cdots w_{K-1}),$$

where *b* and *e* are special symbols that indicate the beginning and termination of a word sequence. The various N-gram probabilities can be learned from analysis of text using a variety of methods. We refer the reader to [3] or [10, pp. 191–234] for details on estimating N-gram probabilities.

The N-gram model for a language as given above carries an interesting implication. It states that all instances of word *w* that occur after a specified sequence of $N - 1$ words are statistically equivalent. This permits us to represent the set of all possible word sequences in the language as a graph where each N-gram is represented exactly once. For a vocabulary of V words, the graph will hence have no more than $O(V^N)$ edges, each representing a unique N-gram, and at most $O(V^{N-1})$ nodes representing words. We can represent a given N-gram model by assigning to each edge in the graph the probability of the N-gram that it represents. As before, any path from the source node to a sink node on this graph represents a word sequence. The product of the probabilities of the edges on this path will be exactly the probability for the word sequence as specified by the N-gram model. Figure 2.11 illustrates this for a vocabulary of two words using a trigram model.

We can now compose a composite HMM for the entire language by connecting the HMMs for words according to the graph. Recognition is performed by finding the most probable state sequence through this HMM for a given speech recording. It can be shown that the

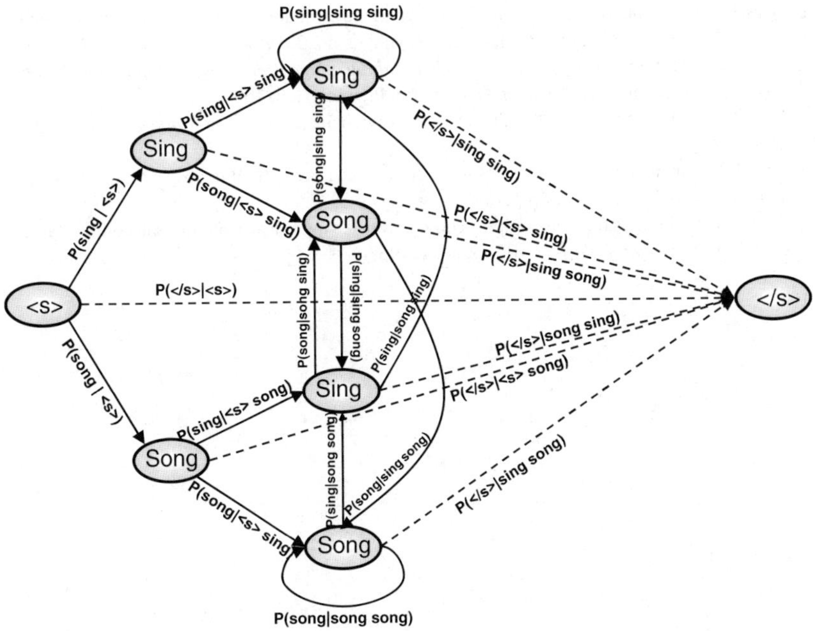

Figure 2.11 A graph representing a trigram LM over a vocabulary of two words in a hypothetical language consisting only of the words "sing" and "song." The symbols "<s>" and "</s>" represent the source node indicating the start of a sentence, and a sink node representing the termination of a sentence respectively. Dotted edges represent transitions into the sink node. In the composite HMM formed from this graph, each oval representing a word would be replaced by the HMM for the word.

word sequence corresponding to the most probable state sequence is identical to that given by Equation (2.29).

References

[1] R. Bakis, "Continuous speech recognition via centisecond acoustic states," *Journal of the Acoustical Society of America*, vol. 59, no. S1, 1976.

[2] L. E. Baum and T. Petrie, "Statistical inference for probabilistic functions of finite state Markov chains," *Annals of Mathematical Statistics*, vol. 37, no. 6, 1966.

[3] S. Chen and J. Goodman, "An empirical study of smoothing techniques for language modeling," *Computer Speech & Language*, vol. 13, no. 4, 1999.

[4] S. Davis and P. Mermelstein, "Comparison of parametric representations for monosyllabic word recognition in continuously spoken sentences," *IEEE Transactions on Acoustics, Speech and Signal Processing*, vol. 28, no. 4, 1980.

[5] G. D. Forney, "The Viterbi algorithm," *Proceedings of the IEEE*, vol. 6, no. 3, 1973.

[6] F. J. Harris, "On the use of windows for harmonic analysis with the discrete Fourier transform," *Proceedings of the IEEE*, vol. 66, no. 1, 1978.

[7] H. Hermansky, "Perceptual linear predictive (PLP) analysis of speech," *Journal of the Acoustical Society of America*, vol. 87, no. 4, 1990.

[8] M.-Y. Hwang, X. Huang, and F. Alleva, "Predicting unseen triphones with senones," *IEEE Transactions on Speech and Audio Processing*, vol. 4, no. 6, 1996.

[9] B. Juang, S. Levinson, and M. Sondhi, "Maximum likelihood estimation for multivariate observations of Markov chains," *IEEE Transactions on Information Theory*, vol. 32, no. 2, 1986.

[10] D. Jurafsky and J. H. Martin, *An Introduction to Natural Language Processing, Computational Linguistics, and Speech Recognition*. Upper Saddle River, New Jersey: Prentice-Hall, 2000.

[11] T. Petrie, "Probabilistic functions of finite state Markov chains," *Annals of Mathematical Statistics*, vol. 40, no. 1, 1969.

[12] S. Stevens, J. Volkman, and E. Newman, "A scale for the measurement of the psychological magnitude pitch," *Journal of the Acoustical Society of America*, vol. 8, no. 3, 1937.

[13] A. J. Viterbi, "Error bounds for convolutional codes and an asymptotically optimum decoding algorithm," *IEEE Transactions on Information Theory*, vol. 13, no. 2, 1967.

3

The Problem of Robustness in Automatic Speech Recognition

Bhiksha Raj[1], Tuomas Virtanen[2], Rita Singh[1]
[1]*Carnegie Mellon University, USA*
[2]*Tampere University of Technology, Finland*

This chapter deals primarily not with what makes automatic speech-recognition systems (ASRs) work, but with some of the factors that make them go *wrong*. As mentioned earlier in Section 1.1, ASR systems often make errors in conditions in which a human listener could continue to hold a conversation effortlessly. Most real-life situations where people converse with one another or with an automated system are fraught with acoustic adversity. The speech that is finally heard may be distorted by a variety of external influences, not related to what was spoken, which affect its characteristics. While humans are not affected by them, ASR systems can be highly sensitive to these distortions. In other words, ASR systems are not *robust* to distortions in the speech signal in the manner that humans are. In this chapter, we discuss some of the reasons for this lack of robustness.

We recall that the problem of automatic speech recognition is fundamentally one of Bayesian classification. Recognition errors in ASR systems are a consequence of misclassification. Therefore, we begin by briefly discussing the rationale behind Bayesian classification and the conditions under which it can perform poorly. Later in the chapter, we relate these to the causes for errors in ASR, describe the various types of distortions that affect speech to evoke these causes, and discuss approaches to mitigate them.

3.1 Errors in Bayes Classification

The problem of classification can be summarized as follows: we have a data instance characterized by a feature $\mathbf{x}$. We know *a priori* that it belongs to one of a set of classes $\mathcal{C} = \{C\}$. Based on the value taken by the feature $\mathbf{x}$, we must determine the class $C \in \mathcal{C}$ to which the instance belongs.

Techniques for Noise Robustness in Automatic Speech Recognition, First Edition.
Edited by Tuomas Virtanen, Rita Singh, and Bhiksha Raj.

The Bayes classification rule [3] can be derived by minimizing the probability of error in a classifier. Defining $R(c, C)$ as the error incurred by assigning a data instance that actually belongs to class C, to the class c

$$R(c, C) = \begin{cases} 0, & \text{if } c = C \\ 1, & \text{if } c \neq C. \end{cases}$$

Now, consider an instance where x takes the value X. If the data instance is assigned to class c by a classifier, the *expected* error is

$$\begin{aligned} L(c, X) &= R(c, c)P(c|X) + R(c, \overline{c})P(\overline{c}|X) \\ &= P(\overline{c}|X) = 1 - P(c|X), \end{aligned} \tag{3.1}$$

where $P(c|X)$ is a short-hand notation for $P(c|\mathbf{x} = X)$, and is the *a posteriori* probability of the class c given that feature x took the value X. In other words, it is the fraction of all data instances for which $\mathbf{x} = X$, that also belonged to c. We denote the set of classes other than c by $\overline{c}$, that is $\overline{c} = C \backslash c$. Equation (3.1) simply states that the expected error of classifying X as c is equal to the fraction of all data instances for which the feature x took the value X which do *not* belong to c. Minimizing the expected error $L(c, X)$ gives us the familiar Bayes classification rule:

$$\hat{c}_P(X) = \text{argmax}_c P(c|X) \tag{3.2}$$

$$= \text{argmax}_c P(c) P_{\mathbf{x}}(X|c), \tag{3.3}$$

where we have used the explicit notation $P_{\mathbf{x}}(X|c)$ to represent the probability that for any instance of c, the feature x will take the value X. We denote the estimated class for the instance X by $\hat{c}_P(X)$. Here, we have used the subscript P to indicate that the selected class maximizes the *true a posteriori* probability $P(c|X)$. In other words, the optimal classifier that minimizes the expected error must choose c such that the largest fraction of all instances for which $\mathbf{x} = X$ are from c.

The *error rate* of a classifier is the expected fraction of all data instances that will be misclassified by the classifier. The error rate of the *optimal* classifier of Equation (3.2) is called the *Bayes error*. It is the statistical mean of the expected error given by Equation (3.1), over all possible values of x, when classification is performed by the optimal Bayes classifier:

$$\mathcal{E}\{L(\hat{c}_P(X), X)\} = 1 - \int_X dX P_{\mathbf{x}}(X) P(\hat{c}_P(X)|X). \tag{3.4}$$

The Bayes error for a feature x is the *lowest* possible error rate for classification with it. Let us represent this as $\mathcal{L}_{\mathbf{x}}$.

We will also alternately represent the Bayes error for x as $\mathcal{L}_{\mathbf{x}}(P, P)$. The first argument P in our notation indicates that the *true* conditional distribution of the classes is $P(c|X)$ (P being used as an abbreviation of $P(c|X)$). The second argument, which is also P here, explicitly indicates that classification too has been performed using $P(c|X)$ in Equation (3.2). The reason for using this notation will become apparent shortly.

Let us now consider the conditions under which the error rate of a classifier can increase. For later reference, we will assign a *type* to each of these conditions.

3.1.1 Type 1 Condition: Mismatch Error

The formulation of Equation (3.2) considers the *true* conditional class-probability distribution $P(c|X)$. In other words, in order to achieve the Bayes error of Equation (3.1), the decisions of the Bayes classifier must be based on the true *a posteriori* probability distributions $P(c|X)$. If the alternate factored form of Equation (3.3) is used, the classifier must use the true class probability $P(c)$ and the true class-conditioned data probability, $P_{\mathbf{x}}(X|c)$.

In practice, the distributions used by a Bayes classifier are frequently not the true distributions. There are two common reasons for this discrepancy:

- *Model mismatch error*: It is usually difficult, if not impossible to characterize the true distribution $P_{\mathbf{x}}(X|c)$ of any data. Instead, a proxy, typically a model $\widehat{P}_{\mathbf{x}}(X|c)$, is used. The parameters of the model are usually learned from data to minimize the divergence between $\widehat{P}_{\mathbf{x}}(X|c)$ and $P_{\mathbf{x}}(X|c)$. Differences between the two nevertheless remain, and can sometimes be large. Similar modeling errors also occur for $P(c)$.
- *Data-mismatch error*: Data-mismatch errors happen when the value X of the feature $\mathbf{x}$ used for classification is modified for some reason. This might happen, for instance, when it has been affected by noise. As a result, instead of obtaining X as the feature value, we obtain $Z = g(X)$ for some function $g()$. In effect, the actual feature obtained is $\mathbf{z} = g(\mathbf{x})$ whose true distribution is $P_{\mathbf{z}}(Z|c)$, whereas classification is performed assuming that the feature is $\mathbf{x}$, that is with $P_{\mathbf{x}}(Z|c)$.

Discrepancies between the true and assumed distributions, regardless of the cause, have a common consequence: the effective *a posteriori* class probability used for classification differs from the true posterior.

Now consider that classification is performed assuming a distribution $\widehat{P}(c|X)$, which may not be the true distribution $P(c|X)$:

$$\hat{c}_{\widehat{P}}(X) = \text{argmax}_c \widehat{P}(c|X). \tag{3.5}$$

Its easy to see that $L(\hat{c}_{\widehat{P}}(X), X) \geq L(\hat{c}_P(X), X)$. Let us represent the error rate $\mathcal{E}\{L(\hat{c}_{\widehat{P}}(X), X)\}$ for this classifier by $\mathcal{L}(P, \widehat{P})$. The notation in $\mathcal{L}(P, \widehat{P})$ represents the fact that data with distribution $P(c|X)$ are classified by a Bayes classifier that assumes a distribution $\widehat{P}(c|X)$. This error rate is given by

$$\begin{aligned}\mathcal{L}(P, \widehat{P}) &= \mathcal{E}\{L(\hat{c}_{\widehat{P}}(X), X)\} \\ &= 1 - \int_X \mathrm{d}X P_{\mathbf{x}}(X) P(\hat{c}_{\widehat{P}}(X)|X).\end{aligned}$$

Since optimal classification with minimum error, that is the Bayes error, is only guaranteed if classification is performed using the true distribution $P(c|X)$:

$$\mathcal{L}(P, \widehat{P}) \geq \mathcal{L}(P, P). \tag{3.6}$$

More generally, if classification is performed using the factored formulation of Equation (3.3), performing classification based on a distribution $\widehat{P}_{\mathbf{x}}(X|c)$ instead of the true distribution $P_{\mathbf{x}}(X|c)$ will result in increased classification error rate. Thus, any mismatch between

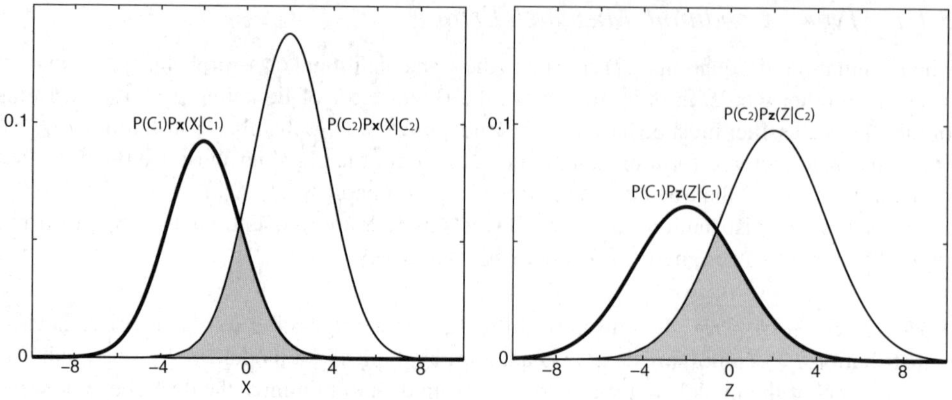

(a) Classification with X.

(b) Classification with Z.

Figure 3.1 Illustration of Bayes error through a two-class problem. *Left panel*: the two curves represent the scaled class-conditional distributions $P(c)P_{\mathbf{x}}(X|c)$ for both classes, which are assumed to be Gaussian. The shaded region is the area under the lower of the two curves. It represents the probability of drawing a feature from either class that is more likely to be obtained from the other class, and will be misclassified. This is the Bayes error. *Right panel*: the curves represent $P(c)P_{\mathbf{z}}(Z|c)$, where $\mathbf{z} = \mathbf{x} + \mathbf{y}$, and $\mathbf{y}$ is a zero-mean Gaussian random variable that is independent of the class c. The variance of $\mathbf{z}$ is the sum of the variances of $\mathbf{x}$ and $\mathbf{y}$, and is greater than the variance of $\mathbf{x}$ alone. Consequently, the area of the shaded region increases. Generally, even if the distributions are not Gaussian, the Bayes error for $\mathbf{z}$ will be greater than that for $\mathbf{x}$.

the true distribution of data and the distribution assumed by a Bayes classifier will result in suboptimal classification.

3.1.2 *Type 2 Condition: Increased Bayes Error*

We first note that for any one-to-one (injective) function $f()$ with the property that for any X the transformed value $Y = f(X)$ is unique, $P(c|\mathbf{x} = X) = P(c|f(\mathbf{x}) = f(X))$[1]. Thus, monotonic transformations of X such as $\log()$, $\exp()$ etc. will not affect the Bayes error, that is

$$\mathcal{L}_{f(\mathbf{x})} = \mathcal{L}_{\mathbf{x}}. \tag{3.7}$$

Consider a random variable $\mathbf{x}$ that is dependent on a class variable c (e.g., measurements of observations from the class). Let $\mathbf{y}$ be a random variable that is independent of the class (e.g., noise). Let $\mathbf{z} = \mathbf{x} + \mathbf{y}$. The Bayes error of a classifier based on $\mathbf{z}$ will be no lower than the Bayes error of $\mathbf{x}$ alone, that is $\mathcal{L}_{\mathbf{z}} \geq \mathcal{L}_{\mathbf{x}}$. In practice, $\mathcal{L}_{\mathbf{z}}$ will usually be greater than $\mathcal{L}_{\mathbf{x}}$. Figure 3.1 illustrates this pictorially.

[1] Here, we have explicitly used the notation $P(c|\mathbf{x} = X)$ to represent the *a posteriori* probability of c given that the feature $\mathbf{x}$ takes the value X, and $P(c|f(\mathbf{x}) = f(X))$ to represent the *a posteriori* probability of c given that the $f(\mathbf{x})$ takes the value $f(X)$. This is to avoid the potential confusion of the notation $P(c|f(X))$ which may be interpreted either as $P(c|\mathbf{x} = f(X))$ or $P(c|f(\mathbf{x}) = f(X))$.

If $\mathbf{z}$ is obtained by combining $\mathbf{x}$ and $\mathbf{y}$ through some function $g()$, that is, $\mathbf{z} = g(\mathbf{x}, \mathbf{y})$, such that we can write $U(\mathbf{z}) = V(\mathbf{x}) + W(\mathbf{y})$ for some injective $U()$, $V()$ and $W()$, $\mathcal{L}_{\mathbf{z}} \geq \mathcal{L}_{\mathbf{x}}$ still holds as a consequence of Equation (3.7). In practice, this will usually hold for most functions $g()$ that we encounter, even if we cannot express the relation between $\mathbf{z}$, $\mathbf{x}$ and $\mathbf{y}$ as an additive decomposition of injective functions.

Each of the two condition types described above has multiple consequences vis-à-vis automatic speech recognition.

3.2 Bayes Classification and ASR

ASR is generally performed through Bayes classification. We recount the classification rule here for reference. Given a speech signal $\mathbf{X}$, the sequence of words $\hat{w}_1, \hat{w}_2, \cdots$ is recognized according to

$$\hat{w}_1, \hat{w}_2, \cdots = \underset{w_1, w_2, \cdots}{\text{argmax}} \, P(w_1, w_2, \cdots | \mathbf{X}) \tag{3.8}$$

$$= \underset{w_1, w_2, \cdots}{\text{argmax}} \, P(\mathbf{X} | w_1, w_2, \cdots) P(w_1, w_2, \cdots). \tag{3.9}$$

The set of classes here is the set of all possible word sequences. We have dropped the subscript used to distinguish between the feature and the value it takes (such as in $P_{\mathbf{X}}(X|c)$) in $P(\mathbf{X}|w_1, w_2, \cdots)$ for brevity.

We now consider the issues that affect speech-recognition performance. We identify three key factors.

3.2.1 *All We Have is a Model: A Type 1 Condition*

As mentioned in Section 3.1, a Bayes classifier is only optimal if the distributions employed are the *true* distributions of the data. In the formulation of Equation (3.9), we would require $P(w_1, w_2, \cdots)$ to represent the true probability of the word sequence $w_1, w_2, \cdots$, and $P(\mathbf{X}|w_1, w_2, \cdots)$ to represent the true probability distribution of all feature vector sequences derived from recordings of the word sequence $w_1, w_2, \cdots$.

However, in practice, we do not have the true probability distributions for the data—we only have *models* that attempt to represent these true distributions. The probability of a word sequence $P(w_1, w_2, \cdots)$ is typically represented by a finite-state or context-free grammar, or an N-gram language model. The probability of a speech signal given a specific word sequence $P(\mathbf{X}|w_1, w_2, \cdots)$ is modeled by an hidden Markov model (HMM) (or some other model if the recognizer is not HMM based). The actual process that produces speech signals is significantly more complex than HMMs, and unlikely to be representable exactly by an HMM.

Thus, we have a Type 1 condition represented by Equations (3.5) and (3.6)—classification is performed using a distribution that is different from the true conditional probability distribution of the classes. The classification error rate is greater than the Bayes error. This is a fundamental restriction that affects any statistical classification paradigm: the true nature of the distribution of the data can never be known and can only be guessed.

3.2.2 Intrinsic Interferences—Signal Components that are Unrelated to the Message: A Type 2 Condition

A person's speech carries a significant amount of information in addition to the actual words being spoken. For instance, information about the person's identity, their emotional state, the emphasis used, etc. are all present in the signal. These features are largely unrelated to the lexical content of the speech. Even if they are related to the underlying words in a specific situation, they are usually highly variable in their manifestations, making them poor general predictors of what was spoken. For instance, the pitch patterns or spectral harmonics in a particular speaker's speech may well be characteristic of the specific sounds he or she is producing, but it is highly unlikely that any other speaker will employ exactly the same pitch patterns or harmonic structures when producing those same sounds. Thus, while these patterns are effective cues for identifying the spoken sounds for that specific speaker, they are useless as cues for identifying the same sounds uttered by a different speaker.

These *lexical-content-independent* attributes of the speech signal can affect the feature vectors derived from it, increasing their intrinsic variability, thereby increasing the minimum (Bayes) recognition error that may be obtained from them. Alternately viewed, each of these nonlexical attributes can be considered to be a separate random variable, and the observed signal itself can be viewed as a composition of these variables and the lexical-content-related features. This results in the Type 2 condition described in Section 3.1.2, and we may expect an increased recognition error with respect to what we would obtain if the speech signal comprised only lexical-content-related features.

3.2.3 External Interferences—The Data are Noisy: Type 1 and Type 2 Conditions

In addition to the above, speech signals are frequently influenced by *external* factors, not related to what was uttered by the speaker. These influences can result in two kinds of effects. First, they can introduce a mismatch between the distributions of the data being recognized and those employed for classification (a Type 1 condition). Second, even if the classification is done with the appropriate distributions, they can cause an increase in the minimum (Bayes) recognition error (a Type 2 condition).

The effect of external influences on the speech signal, and techniques to mitigate them are the primary focus of this book. Below, we briefly consider the nature of these influences.

3.3 External Influences on Speech Recordings

In order to understand what external influences may affect a recording and how they might do so, we begin by considering the nature of the speech signal itself and how it is captured. The speech signal is a pressure wave—the speaker's mouth causes minute variations in the pressure of the surrounding air. These vibrations must be sensed and converted to a digital signal that the recognizer can operate on.

Figure 3.2 illustrates the overall signal capture process that delivers the pressure waves produced by the speaker as a digital signal to the recognizer. The first component in this process is a microphone. The primary sensing element in the microphone is a membrane that

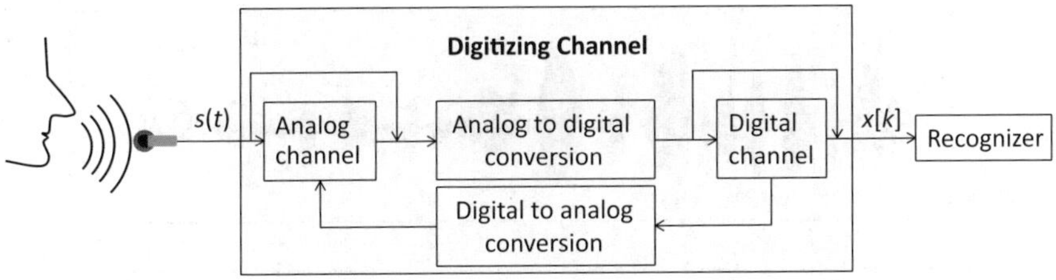

Figure 3.2 Signal-capture procedure. The speech uttered by a speaker is converted by the microphone to an analog signal $s(t)$. The signal is transmitted through a variety of digital and analog channels and eventually delivered to the recognizer as the digital signal $x[k]$.

is caused to vibrate by the pressure wave [4]. The vibrations of the membrane are converted to an analog electrical signal $s(t)$. The analog signal $s(t)$ is then conveyed through a *digitizing channel* which delivers it as the digital signal $x[k]$ to the recognizer. In representing the analog and digital signals here, we have employed the convention of representing continuous time t within parentheses for continuous time signals, as in $s(t)$, and using square brackets to represent the discrete sample index k for digital signals, as in $x[k]$.

The digitizing channel includes zero or more analog transmission channels, analog-to-digital conversion, and zero or more digital transmission channels. Our definition of digital transmission channels include storage and retrieval schemes that may incorporate compression. The actual combination of analog and digital transmission channels depends on the particular application in which ASR is deployed. Often, the exact configuration of the digitizing channel, i.e., the number and type of analog and digital transmission channels between the speaker and the recognizer, is known. This happens, for instance, when speech is recognized on a local computer, or transmitted over a dedicated link to a recognizer. At other times, for example, for recognition of speech recorded over telephone channels, even this may not be known, since land-line, cell-phone, and internet telephony all employ different transmission schemes.

Ideally, the signal that is eventually delivered to the recognizer would be a faithful facsimile of the signal produced by the speaker. In practice, though, both analog and digital transmission channels will frequently modify or distort the signal that is transmitted. Separately, extraneous noises may also interfere with the speech signal. As a result, the actual signal that is delivered to the recognizer is a distorted version of what would ideally be delivered.

Below, we discuss how these various influences affect the speech signal. Initially, we will discuss the distortions introduced by various components of the signal-capture process, following which we will consider the effect of extraneous noises. For illustrative purposes, we use as an example an undistorted speech signal and its spectrogram shown in Figure 3.3.

3.3.1 Signal Capture

By the term "signal capture," we refer to the entire process of capturing and digitizing a signal to prepare it for recognition, as shown in Figure 3.2. Let us now examine the various components of this process.

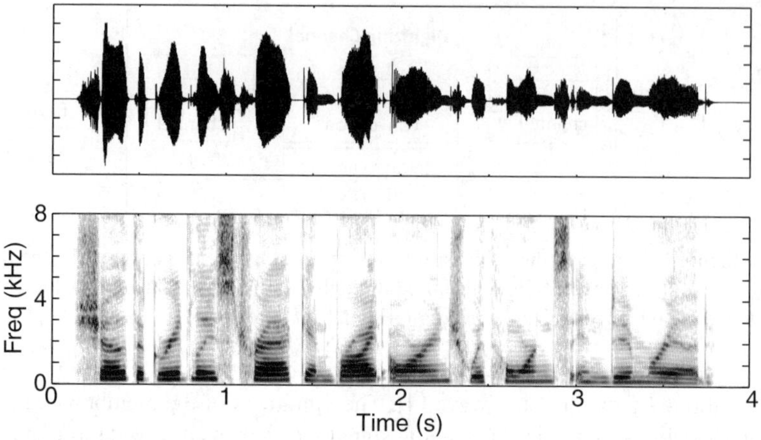

Figure 3.3 *Top.* A "clean" speech signal that has been recorded over a close-talking microphone in a noise-free environment. The signal is sampled at 16 000 samples per second and has frequency components up to 8 kHz. *Bottom.* The spectrogram of the signal.

The Microphone

At the leading end of the sound-capture process is a microphone that converts pressure waves to analog electrical signals. For the purpose of this discussion, we consider the microphone to comprise all components of the system that are involved in converting the acoustic pressure wave into the analog electrical signal $s(t)$. This includes the actual microphone element that transduces the pressure wave to an electrical signal, and any preamplifier that is required to boost the captured signals to a level that is acceptable for transmission or digitization.

Ideally, the microphone would respond equally to *all* the frequencies in the speech signal, i.e., it would have a flat frequency response. In reality, the response of the microphone is usually nonuniform across all frequencies. Although high-quality microphones tend to have a relatively flat response across most of the frequency spectrum, more typically the response tends to be variable. Moreover, no two microphones will have exactly the same frequency response, even if they are manufactured identically. The frequency characteristics of the signal captured by the microphone will also vary due to other external factors, such as the direction and distance of the speaker with respect to the microphone, and even the atmospheric conditions in the space that the speech is being recorded in.

In addition, the response of the microphone itself is not always perfectly linear—an increase in the energy with which the speaker speaks may not result in a proportional increase in the energy of the captured signal. When the captured signals are boosted by a preamplifier, the linearity of the response of the preamplifier also becomes a factor. A preamplifier's response is generally linear at low to medium gain levels, but at high gain levels it becomes nonlinear. This can cause saturation of the signal values, with concomitant spectral distortions including aliasing in the final digitized signal as illustrated in Figure 3.4.

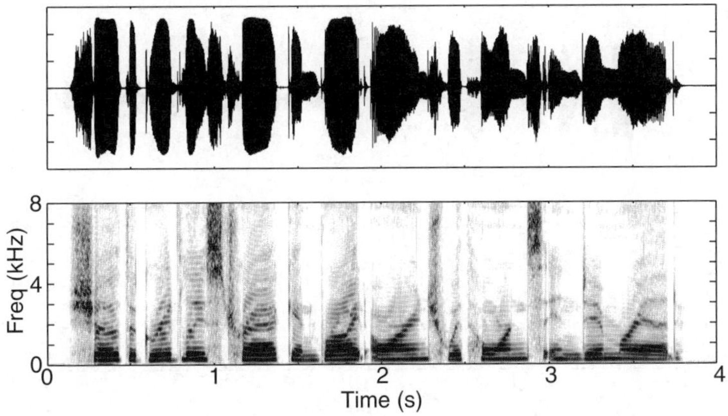

Figure 3.4 The effect of saturation. *Top*: The signal from Figure 3.3 when it has been distorted by saturation resulting from excessive gain in the amplifier. Saturation is often visually apparent in the envelope of the signal. *Bottom*: The spectrogram of the saturated signal, after it has been digitized at 16 kHz. Saturation results in high-frequency components that get aliased into spurious spectrographic patterns in the 0–8 kHz frequency band. Note that the spurious patterns are chiefly visible in regions where the signal amplitude was high enough to be affected by the saturation. Other forms of nonlinearities also result in similar distortion.

The Channel

The channel can often introduce a variety of distortions as discussed below:

- *Analog transmission*. If the analog signal is transmitted in analog form, for example over an analog radio or television channel, some form of encoding scheme, typically amplitude or frequency modulation, is used. These schemes, while nominally nondestructive to the transmitted signal, will nevertheless introduce distortions into it.
- *Bandwidth restrictions*. The bandwidth permitted to the signal is often limited. For instance, older analog telephone lines only transmit the 300–3400-Hz frequency band of the signal—the remaining frequencies are filtered out prior to transmission. Modern cellphone and VoIP channels can transmit higher bandwidths (e.g., 8 kHz), but still the most commonly used cut-off frequency is 4 kHz. This can result in significant loss of useful spectral information as shown by the example in Figure 3.5.
- *Digitization*. The analog signal must eventually be digitized for processing with a computer. Digitizing requires sampling the speech signal in time. According to the Nyquist sampling theorem [8] no information is lost in sampling, provided that the sampling frequency is at least twice as high as the highest frequency component in the signal. Spectral components at frequencies above half the sampling frequency will get *aliased* into spurious spectral patterns in lower frequencies. To prevent this, the analog signal must be bandlimited by an *antialiasing* filter that attenuates all frequencies above half the sampling rate, prior to digitization. Poor attenuation of these frequencies will show up as aliasing artifacts in the digital signal. In addition, the digital signal must be *quantized* to the bit-resolution employed

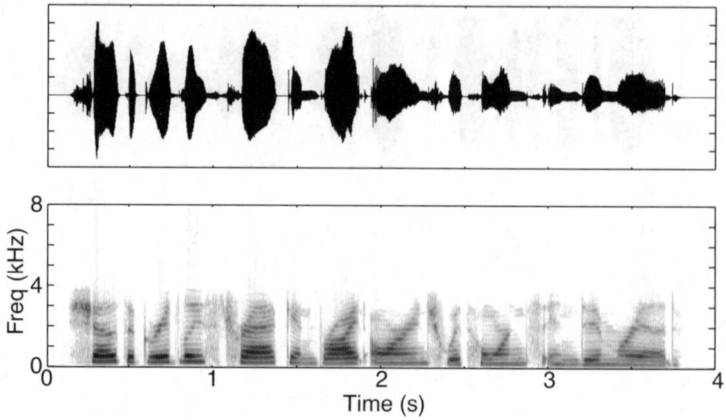

Figure 3.5 The signal from Figure 3.3 when it has been bandlimited to the frequencies present in a typical telephone signal (top), and its spectrogram (bottom). Information-bearing spectrographic patterns in the higher frequencies have been erased.

(i.e., the number of bits used to represent each sample). This inevitably results in quantization errors, which show up as low-energy noise in the digitized signal.

- *Clipping*. Analog-to-digital converters operate on a limited range of incoming signal values. Signal amplitudes that exceed these limits are truncated to the largest value that can be represented by the analog-to-digital converter, resulting in distortion of the signal and its spectral content as seen in Figure 3.6.

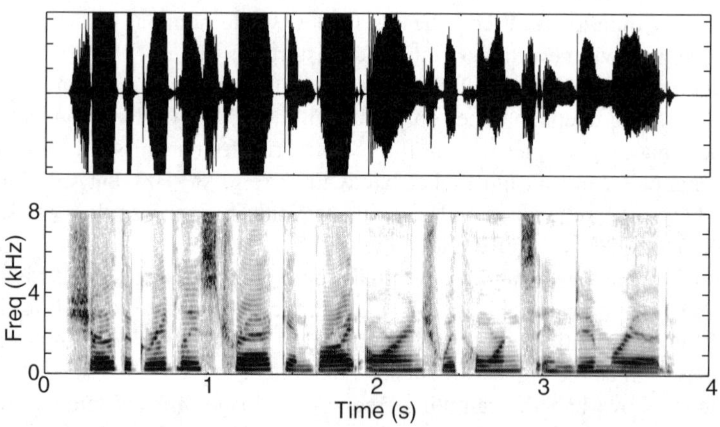

Figure 3.6 The effect of clipping that results when the dynamic range of the analog signal exceeds the operating range of the analog-to-digital converter. The signal in the upper panel has several segments where its amplitude is the maximum allowed by the 16-bit digitization used in this example. The spectrogram of the signal, shown in the lower panel, exhibits clear indications of aliasing and other distortions in the regions where the signal has been clipped.

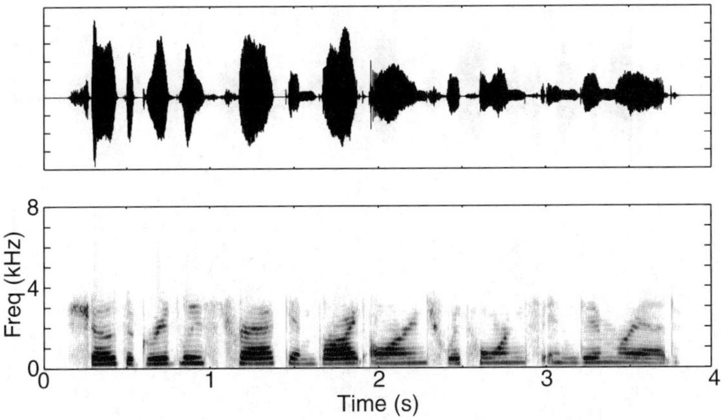

Figure 3.7 An example of a signal that has been corrupted by a coded transmission channel. The speech signal in Figure 3.3 has been coded by the ITU-T G.723.1 codec [6] at 6.3 kbits/s. The upper panel shows the signal and the bottom panel shows its spectrogram. The codec operates on reduced-bandwidth (4 kHz) speech. Additional distortions introduced by the codec can be seen by comparing this spectrogram with Figure 3.5.

- *Coding distortions.* Digitized speech data are commonly also transmitted over wired or wireless channels. In these cases, it is usual to compress the digital signal through one of a large variety of lossy coding schemes. The received coded data is then decoded to recreate a digital signal. The coding schemes are designed to retain the *intelligibility* of clean speech signals, but the decoded signals are not an exact replica of the originally encoded data. As a consequence, the coding schemes modify the spectral content of the speech signal as seen in Figure 3.7. The coding schemes are frequently designed to function optimally on speech signals, but will badly distort other types of signals. In particular, when the speech signal is corrupted by noises that the coding scheme is not designed for, this can magnify the effect of these noises on the intelligibility of the decoded signal.
- *Data loss.* Digital channels typically transmit the signal as packets of encoded data. Frequently, some of these packets are "lost" during transmission and are not received at the destination. Consequently, the recomposed signal will have gaps in it. Coding schemes for digital channels institute mechanisms to smooth over these gaps to reduce discontinuities in the signal; nevertheless the information in these gaps is lost.

3.3.2 Additive Corruptions

In addition to the various distortions and changes inflicted by the signal capture process, speech signals can also be corrupted by undesired signals that get recorded along with the speech (Figure 3.8). These signals are usually from external sound sources which produce signals that are also incident on the microphone. These might include sounds from localized or "point" sources such as a radio, air conditioner or even a competing speaker, or diffuse sound sources such as the hum in an automobile that cannot be localized. These noises are primarily *additive*—the recorded signal is a direct sum of the signals from the different sound

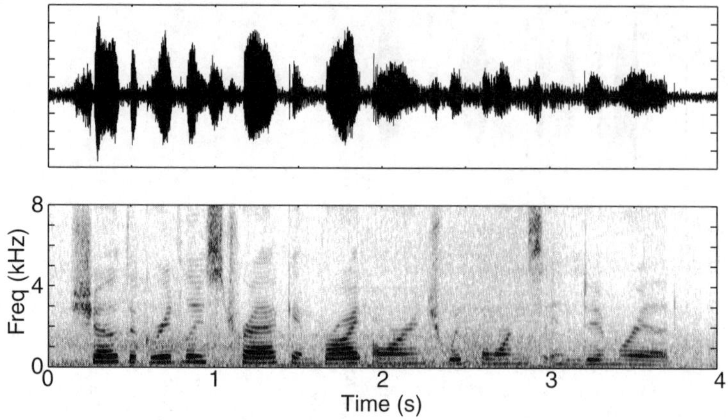

Figure 3.8 The signal of Figure 3.3 when it has been corrupted by street traffic noise to 10 dB. In addition to introducing various spectral features of its own, the noise also obscures many of the spectral patterns in the speech.

sources. Other noises are introduced by the recording equipment itself even when there is no external noise source. All of these noises, in addition to being unrelated to the actual speech signal, are also often time varying, introducing an extra degree of variability that complicates matters further.

The level of undesired noise in a noisy signal is usually quantified through the *signal-to-noise ratio*, or SNR of the signal, and expressed in *decibels*. If we represent the *clean* speech signal that we would capture in the absence of noise as $s[k]$ and the noise that corrupts it as $n[k]$, the SNR in decibels is given by

$$\mathrm{SNR(dB)} = 10\log_{10}\frac{\sum_k s[k]^2}{\sum_k n[k]^2}. \tag{3.10}$$

The lower the SNR, the higher the level of the corrupting noise in comparison to the speech. An SNR of 0 dB implies that speech and noise have equal power in the noisy signal.

3.3.3 Reverberation

Most recordings are performed in closed spaces with walls and other objects that can reflect the sound signal. In these situations, the signal generated by the speaker not only travels directly from the speaker's mouth to the microphone but also arrives at the microphone through reflections from walls, reflections of reflections, etc., as illustrated in Figure 3.9. Since the reflections must travel a longer distance than the direct signal itself, their arrival at the microphone is delayed with respect to the direct signal. These delayed signals are consequently combined with the direct signal to result in the phenomenon known as reverberation.

The effect of reverberation is to "smear" sounds in the recording—even a sharp click produced by the speaker gets recorded as an extended signal, since the reflections continue to bring delayed, but attenuated copies of the signal to the microphone long after the speaker has produced the click. The smearing of the signal causes a loss of signal quality, since the

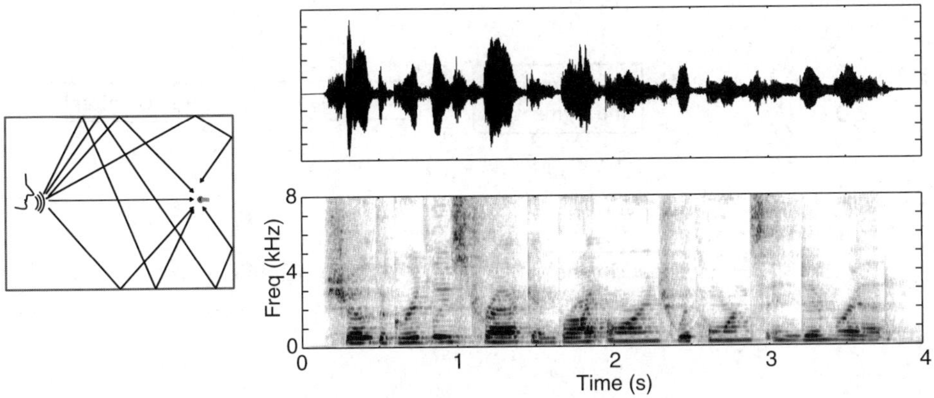

Figure 3.9 *Left*: in reverberant recording environments the signals arriving at the microphone after one or more reflections from walls and other reflecting surfaces interfere with the direct signal from the speaker to the microphone. *Right*: the signal in Figure 3.3 when it has been reverberated in a room with a reverberation time T_{60} of 373 ms (top) and its spectrogram (bottom). The signal is smeared in time, which makes temporal details less clearly perceivable.

microphone now captures the currently spoken sound along with other sounds spoken in the past.

The amount of reverberation in a recording environment is typically characterized by *reverberation time* T_{60} [4], which is defined as the time taken for the energy in the reflections of a signal to decrease to 60 dB below that in the initial direct signal. The longer the T_{60} time, the greater the reverberation.

3.3.4 A Simplified Model of Signal Capture

All of the above effects introduce artifacts in the digitized signal processed by the recognizer, which are largely unrelated to the actual spoken message, i.e., the words uttered by the speaker. To explain how they affect recognition, it is convenient to consider the "idealized" model of the signal capture process shown in Figure 3.10a. The speaker generates a pristine signal in an ideal environment with no external interferences. A perfect microphone transduces this signal to an ideal signal $s(t)$ which is digitized by a perfect analog-to-digital converter to the "clean" digital signal $s[k]$. $s[k]$ captures the signal generated by the speaker faithfully. Thereafter, a noisy channel corrupts the clean signal, distorting it and adding various noises to it to generate the distorted signal $x[k]$ which is delivered to the recognizer. We attribute all corrupting phenomena to this channel. Ideally, the channel would not distort the signal at all and $x[k] = s[k]$. However, in reality the relationship is more complex and has the form $x[k] = f(s[0]..s[k], \mathbf{H}, \mathbf{N})$, where $f()$ is a possibly nonlinear, usually unknown function, $\mathbf{N}$ represents the aggregate of the various noises that affect the signal and $\mathbf{H}$ represents the characteristics of the channel that delivers the digitized signal to the recognizer. The characteristics of the channel can be time varying.

Often, for purposes of analysis and mathematical tractability, the noisy channel is assumed to follow the simplified model shown in Figure 3.10b. It comprises a linear time-invariant filter

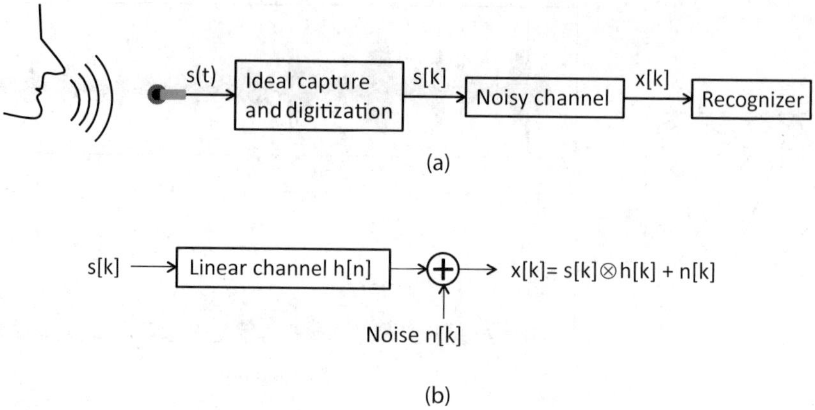

Figure 3.10 (a) Idealized model of the signal capture process. The speech uttered by the speaker is converted to the pristine analog signal $s(t)$ by the microphone. This signal is digitized to obtain the pristine digital signal $s[k]$ which is transmitted through a noisy channel to arrive as the noisy signal $x[k]$ at the recognizer. (b) A simplified model that is often assumed for the noisy channel that transforms the ideal digital signal $s[k]$ to the noisy digital signal $x[k]$. It comprises a linear filter $h[k]$ followed by addition of noise $n[k]$.

$h[k]$ that modifies the spectral characteristics of the signal, followed by an addition of noise $n[k]$. Thus,

$$x[k] = s[k] \otimes h[k] + n[k], \tag{3.11}$$

where $\otimes$ represents the convolution operator. In this model, both reverberation and any spectral shaping effect from the recording setup are attributed to $h[k]$.

3.4 The Effect of External Influences on Recognition

How do the various factors described above affect automatic speech recognition? To understand, we recall the old adage that a picture is worth a thousand words. Figure 3.11 shows recognition accuracy on speech corrupted by different types of external influences. For purposes of this illustration, we have assumed the corruption model of Figure 3.10b. We separately evaluate the effect of the two components of this model, the linear channel $h[k]$ and the additive noise $n[k]$.

Figure 3.11a shows the effect of additive noise on recognition error. In this experiment, noisy speech was created by adding corrupting music digitally to a clean speech signal, with no linear filtering involved. By the term "clean" speech here, we refer to speech that has been recorded over a close-talking microphone and can be assumed to be unaffected by any linear filter or additive noise, and is a close approximation to the ideal signal $s[k]$. Since there is no linear filtering, the noisy speech is simply $x[k] = s[k] + n[k]$. The curves show recognition accuracy on speech corrupted by noise as a function of the SNR of the signal. The lower curve shows the recognition accuracy obtained on noisy speech when the recognizer has been trained on clean speech. The upper curve shows recognition performance on noisy speech, when the recognizer too has been trained on the same kind of noisy speech.

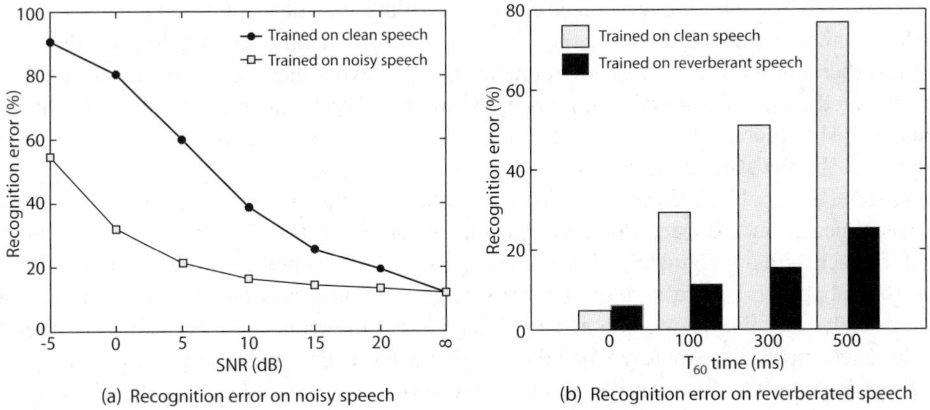

Figure 3.11 (a) The effect of different levels of background music on speech-recognition accuracy. The noise level is measured in terms of the signal-to-noise ratio of the noisy signal expressed in decibels. (b) The effect of reverberation on speech recognition. The different bars show recording environments with different reverberation times. The speech data were drawn from the *Wall Street Journal* (WSJ) corpus [9] in both cases.

We pause here to explain the distinction between the two curves. A recognizer comprises a collection of HMMs for various sound units (words or phonemes). The parameters of the HMMs are trained from a collection of speech recordings. For simplicity, let us assume that the HMMs represent the *true* class-conditional distributions of the data they are trained from.

In the upper curve of Figure 3.11a the HMMs are trained from clean speech. They represent $P_{\mathbf{S}}(\mathbf{S}|c)$, the true class conditional distribution of feature vectors **S** derived from clean speech $s[k]$. In the notation here the subscript **S** indicates that $P_{\mathbf{S}}(\mathbf{S}|c)$ is the probability distribution of feature vectors of clean speech, while the argument, also **S** here, indicates that probability values are also computed for feature vectors of clean speech. The feature vectors **X** of the *noisy* test speech (i.e., the speech to be recognized) $x[k]$ on the other hand, have a different distribution $P_{\mathbf{X}}(\mathbf{X}|c)$. This gives us a *data-mismatch* variety of Type 1 condition, since the recognizer, having been trained on clean speech, performs classification based on $P_{\mathbf{S}}(\mathbf{X}|c)$. As a result, the recognition accuracy is compromised. As the level of noise increases (i.e., as the SNR decreases), the difference between $P_{\mathbf{S}}(\mathbf{X}|c)$ and $P_{\mathbf{X}}(\mathbf{X}|c)$ increases, resulting in increased recognition error as shown by the curve.

In the lower curve of Figure 3.11a, the HMMs have been trained on noisy speech that has been corrupted by the same kind and level of noise as the test speech. As a result they represent the distribution $P_{\mathbf{X}}(\mathbf{X}|c)$. Such a recognizer is called a "matched" recognizer for the test data—the distributions used by the recognizer for Bayes classification are the same as the distribution of the test data, and we have an optimal Bayes classifier. This may be considered to be the "best" case-recognition accuracy that can be obtained with the noisy test data. Nevertheless, since the noise is not related to the underlying sound classes, we get a Type 2 condition—the Bayes error is greater than what can be obtained with the clean signal $s[k]$. Furthermore, the error, while optimal for the data, increases as the noise level increases, although it increases much less than when the recognizer is trained on clean speech. Thus, increasing noise levels degrade recognition accuracy in all cases.

Figure 3.11b shows recognition accuracies on data that have only been affected by a linear channel, but no noise. Here, $x[k] = s[k] \otimes h[k]$. Specifically, here the linear filter $h[k]$ is the reverberation in a recording environment. For this experiment, clean data were digitally reverberated by passing them through a digital filter $h[k]$ that was obtained using the "image method" [1] to represent the reverberant room response of a simulated $5m \times 4m \times 3m$ room. Figure 3.11b shows recognition accuracy on reverberated speech as a function of reverberation time. Here, the gray bars show recognition performance when the recognizer has been trained on clean speech, but the speech to be recognized is reverberant. The black bars show recognition performance with a recognizer, when the recognizer itself has been trained on a collection of reverberant speech from a variety of reverberant recording conditions. Once again, we note that in both cases the recognition accuracy degrades with increasing reverberation, although it plummets much more rapidly when the recognizer has been trained on clean speech.

Although the above results only consider simulations of the simplified model of Figure 3.10, similar patterns are also observed for other more realistic signal degradations. Recognition error generally increases as the level of degradation increases (as measured by any appropriate quantitative metric), even when the recognizer is matched to the test data. In practice, the data used to train the recognizer are rarely matched to the test data and this results in further increased error.

In the examples above, we have not considered the other two effects mentioned earlier, i.e., (1) that the recognizer is only a *model* for the true distribution of speech and (2) the speech signal itself contains various additional components that are unrelated to the spoken words and interfere with our ability to recognize it. These effects also affect recognition accuracy.

3.5 Improving Recognition under Adverse Conditions

It is clear that the various problems described in Sections 3.2 and 3.3 affect speech-recognition accuracy and must be addressed. The chapters in this book present a number of solutions. Let us briefly review possible mechanisms for addressing the problems and how the chapters relate to them.

3.5.1 Handling the Model Mismatch Error

Let us begin by considering the most basic problem of all, described in Section 3.2.1—the HMMs in a speech recognizer are only models of the true distributions of speech from the various sound classes, and hence do not guarantee optimal recognition. As we saw in Section 3.1, the reason for this is that $\mathcal{L}(\widehat{P}, P)$, the error rate for a classifier that performs classification based on an assumed distribution $\widehat{P}_{\mathbf{X}}(X|c)$, is greater than the Bayes error $\mathcal{L}(P, P)$.

One solution to this problem is to simply train the recognizer on a large amount of speech, and have a sufficient number of parameters (i.e., Gaussians in state output distributions) in the HMMs to approximate the true distributions of speech as well as possible within the limitations of the model.

Another solution is to estimate the distribution $\widehat{P}_{\mathbf{X}}(X|c)$ to directly minimize $\mathcal{L}(\widehat{P}, P)$, rather than to match the distribution of the data from the class. In the context of speech-recognition systems, this would imply learning HMM parameters to minimize recognition error, rather than to model the distributions of the individual sound units as faithfully as possible. Such techniques are usually called *discriminative* training techniques.

In Chapter 11, Estève and Deléglise describe MMI and MPE training, two discriminative learning formalisms that explicitly attempt to minimize recognition error based on this principle.

3.5.2 Dealing with Intrinsic Variations in the Data

The second, fundamental problem that affects speech recognition is that the signal carries information beyond what is spoken, and this can interfere with recognition as explained in Section 3.2.2. The solution to this is to utilize feature-extraction techniques that can somehow capture only the characteristics of the signal that relate to the underlying lexical message, without representing remaining characteristics that are irrelevant to recognition. While we do not explicitly address this problem in this book, the feature-extraction techniques described by Stern and Morgan in Chapter 8 are based on modeling aspects of human perception that key in on phonetic, and consequently lexical contents of the signal, implicitly reducing, although not eliminating the representation of other lexical-content-independent aspects of the signal.

3.5.3 Dealing with Extrinsic Variations

However, the bulk of this book is aimed at addressing the problem of external influences. Specifically, we consider additive noise, which might indeed be the most vexing problem of all, although many of the techniques are applicable to generically degraded speech.

Normalizing the Data

The primary issue that arises from external influences is that they induce a data-mismatch error in the recognizer—the distributions of the speech to be recognized differ from those used to train the classifier. Furthermore, even the individual utterances in the training data may vary in their distributions, both because the recording conditions and the distortions affecting the signals may have varied from recording to recording, and because of *instrinsic* variations in the speech.

The solution, of course, is to somehow modify the distribution of the features derived from each test recording, so that they conform to the distributions used by the recognizer. Indeed, this is what some of the techniques presented in later chapters do.

However, a rather large benefit can be obtained by simply matching the *moments* of the distributions of the features obtained from test and training recordings. *Feature-normalization* techniques take this approach.

They assume that individual speech recordings, both from the training and test data, are perturbed samples drawn from an unknown common underlying distribution. Therefore, they attempt to modify the distributions of each recording to return them to this underlying distribution. This distribution itself cannot be known, but for the purposes of normalization it can be assumed to be a zero-mean Gaussian with a variance of unity for each component. A variety of normalization schemes are commonly used in ASR systems, based on this assumption.

- *Mean normalization* [5] subtracts the mean of the sequence of feature vectors for an utterance from the individual feature vectors, such that the resulting sequence of vectors has zero mean:

$$\mu_{\mathbf{x}} = \frac{1}{T}\sum_k \mathbf{x}_k,$$
$$\mathbf{x}_k = \mathbf{x}_k - \mu_{\mathbf{x}} \quad \forall\, k. \tag{3.12}$$

- *Variance normalization* [10] also normalizes the variance of the individual components of the feature vectors to unity:

$$\sigma_{\mathbf{x}}(i) = \sqrt{\frac{1}{T}\sum_k (\mathbf{x}_k(i) - \mu_k(i))^2},$$
$$\mathbf{x}_k(i) = \frac{\mathbf{x}_k(i) - \mu_{\mathbf{x}}(i)}{\sigma_{\mathbf{x}}(i)} \quad \forall\, k, i, \tag{3.13}$$

  where $\mathbf{x}_k(i)$ represents the *i*th component in the *k*th feature vector $\mathbf{x}_k$.
- *Histogram equalization* [2] goes beyond matching moments. It further maps the cumulative histogram for individual components of the feature vectors to the cumulative distribution function of a zero-mean, unit variance Gaussian.

Normalization is surprisingly effective and has become a staple part of ASR systems. Indeed, any "baseline" evaluation of a recognizer naturally includes, at the minimum, mean normalization of features. Nevertheless, it is not a panacea. Noise and other distortions modify the distributions of the data in a manner that cannot be completely accounted for by simple normalization techniques, and considerable degradation of recognition performance still occurs. More sophisticated techniques are needed to deal with them.

Advanced Techniques

Clearly, the best way to deal with degraded speech is not to have degradation in the speech signal at all. In the context of additive noise, recognizers are best served if the recorded speech simply does not contain high levels of noise.

In Chapter 6, McDonough and Kumatani discuss how this may be achieved even in noisy recording environments using a microphone array. As they explain, by suitably combining signals captured by multiple microphones, it becomes possible to selectively enhance signals from the location of the speaker, effectively suppressing signals from other locations.

Noise may also be eliminated by not considering regions of the speech signal that do not contain speech at all. In Chapter 4, Martin and Kolossa present methods for *voice activity detection* to accurately identify the locations of speech carrying regions of the signal, thereby eliminating unnecessary noise-carrying regions from consideration for recognition.

In spite of our best efforts, the captured signal is likely to contain noise anyway. In Chapter 4, Martin and Kolossa also describe techniques for estimating the noise and suppressing it in the noisy signal. Smaragdis describes techniques for *separating* speech out from noise using single- or multiple-microphone recordings in Chapter 5.

Feature computation is another stage in the capture and characterization of speech signals for recognition, where the problem of noise may be directly addressed. Noise robustness may be achieved by deriving features from the signal that, while retaining all the characteristics of speech, are relatively unaffected by noise. Stern and Morgan in Chapter 8 and Wölfel in Chapter 7 describe techniques for extracting features from speech signals that approach the problem of robust feature extraction from human-auditory and signal processing perspectives, respectively.

This far, we have considered what happens to a signal or features derived from it *before* it arrives at the speech recognizer. Nevertheless, we note that the above procedures do have an effect on the recognizer. In cases where the recognizer has been trained on clean speech, they have the effect of reducing the mismatch between the probability distributions employed by the recognizer and those of the the test data to be recognized. If the recognizer is trained on noisy speech there is no mismatch. Nevertheless, by reducing the variations among the speech signals in the training data due to noise, they can improve the minimum recognition error (Bayes error) achievable with the data. Nevertheless the recognizer itself was not explicitly considered in developing these solutions.

However, once the features arrive at the recognizer, the classifier must explicitly be considered. As we know by now, the primary reason for the increased recognition error is mismatch between the test data and those used to train the recognizer. Here, we have two options—we may either modify the features derived from the signal such that their distributions better match those used by the recognizer, or we may modify the distributions employed by the recognizer.

Droppo describes feature-enhancement techniques in Chapter 9 that attempt to modify features derived from noisy signals, such that their distributions are closer to that of clean speech. In Chapter 10, Häb-Umbach describes a technique for enhancing features derived from reverberant speech.

In Chapter 11, Estève and Deléglise describe *adaptation* methods, that modify the distributions in a recognizer to better match those of incoming speech features using generic affine transforms. In Chapter 12, Hershey *et al.* describe more detailed methods of modifying recognizer parameters using models of the interaction process between speech and noise.

Here, a generic rule of thumb must be kept in mind. Techniques that modify the features only change the *average* characteristics of the features, without eliminating the variations introduced by noise. Thus, the features will typically retain variations that are not represented in the recognizer. On the other hand, techniques that modify the distributions of the recognizer to match the incoming data can do so in such a manner that they also represent the variation in the data correctly. Thus, methods that modify the distributions in the recognizer to match the data may generally be expected to result in better recognition accuracy than those that modify the features to match the distributions in the recognizer. We refer the reader to Appendix A of [7] for a more detailed explanation of this issue.

This factor must hence explicitly be taken in to account. In Chapter 13, Seltzer describes a methodology for learning speech recognizer parameters in a manner that makes it more amenable to features that have been enhanced in any given manner. Liao describes techniques for explicitly incorporating the uncertainty in noise-compensated or otherwise enhanced features into a recognizer in Chapter 17.

A completely different perspective is derived from the fact that speech has time-varying spectral characteristics. As a result, when a speech signal is corrupted by noise, some spectro-temporal components nevertheless retain their fidelity since they have much greater energy than the noise, while others may get obliterated. *Missing-feature* techniques key in on this characteristic. They attempt to identify the *reliable* spectro-temporal components that have been relatively unaffected by noise, and use the partial information in these regions to perform recognition.

The key to these methods, of course, is identifying which spectro-temporal components of a signal may be considered to reliably belong to the speech signal. Narayanan and Wang describe methods to identify these components using various techniques in Chapter 16.

In Chapter 14, Barker describes conditions under which speech features may be considered reliable, and describes methods to perform recognition with partial data. He also describes methods to perform recognition when there is uncertainty about which parts of the signal are missing.

Gemmeke and Remes, on the other hand, take a different approach in Chapter 15. They describe methods of *reconstructing* the unreliable spectro-temporal components of speech. The resulting complete spectro-temporal characterization can now be used to recognize speech.

Together the techniques presented in this book address the problem of robust speech recognition from a variety of perspectives. We invite the reader to read on about them in greater detail in the following chapters.

References

[1] J. Allen and D. Berkley, "Image method for efficiently simulating small-room acoustics," *Journal of the Acoustical Society of America*, vol. 65, no. 4, 1979.

[2] Ángel de la Torre, A. M. Peinado, J. C. Segura, J. L. Pérez-Córdoba, M. C. Benítez, and A. J. Rubio, "Histogram equalization of speech representation for robust speech recognition," *IEEE Transactions on Speech and Audio Processing*, vol. 13, no. 3, 2005.

[3] R. O. Duda, P. E. Hart, and D. G. Stork, *Pattern Classification*. New York: Wiley-Interscience Publication, 2000.

[4] J. Eargle, *Handbook of Recording Engineering, 4th Ed.* New York: Springer, 2005.

[5] S. Furui, "Cepstral analysis technique for automatic speaker verification," *IEEE Transactions on Acoustics, Speech and Signal Processing*, vol. 29, no. 2, 1981.

[6] *Dual rate speech coder for multimedia communications transmitting at 5.3 and 6.3 kbit/s*, http://www.itu.int/rec/T-REC-G.723.1, International Telecommunication Union Std. T-REC-G.723.1, 1996.

[7] P. J. Moreno, "Speech recognition in noisy environments," PhD dissertation, Carnegie Mellon University, 1996.

[8] H. Nyquist, "Certain topics in telegraph transmission theory," *Transactions of the AIEE*, vol. 47, 1928.

[9] D. Paul, "The design for the wall street journal-based csr corpus," *HLT '91 Proceedings of the Workshop on Speech and Natural Language*, 1991.

[10] O. Viikki and K. Laurila, "Cepstral domain segmental feature vector normalization for noise robust speech recognition," *Speech Communication*, vol. 25, no. 1, 1998.

Part Two

Signal Enhancement

4

Voice Activity Detection, Noise Estimation, and Adaptive Filters for Acoustic Signal Enhancement

Rainer Martin, Dorothea Kolossa
Ruhr-Universität Bochum, Germany

4.1 Introduction

The presence of acoustic noise degrades automatic speech-recognition (ASR) performance as it adds irrelevant information to the target signal. In the best case, this irrelevant information does not disturb the speech recognizer; in the worst case, it leads to a complete mismatch of the acoustic signal and the signal model of the recognizer. One widely used approach to improve the performance of ASR is to filter the acoustic signal such that the amount of irrelevant information is reduced and the match of the signal with its model is improved.

In the past 20-some years, many different filtering methods for noise reduction have been proposed, either using a single signal or multiple microphone signals. Although beamforming methods based on multiple microphone signals yield larger improvements than single-microphone processing methods, the latter are very widely used.

On the one hand, single-channel approaches are relatively easy to apply as their microphone arrangement requires less space and they need in general less hardware and computational resources. On the other hand, single-channel methods do not provide a spatial selectivity and are restricted in their ability to remove time-varying noise components. Therefore, the complete restoration of the undisturbed speech signal, as desirable as it would be, is hard to achieve with a single-microphone approach: The removal of broadband noise goes along with a degradation of the target signal. Therefore, most single-microphone noise-reduction systems are adjusted to achieve a suppression of noise power in the order of 10–20 dB.

Single-microphone noise-reduction methods use a variety of different processing approaches. The simplest approach is to employ a voice activity detector (VAD) and to discard

Techniques for Noise Robustness in Automatic Speech Recognition, First Edition.
Edited by Tuomas Virtanen, Rita Singh, and Bhiksha Raj.

those signal segments which contain noise only. In the context of ASR, this procedure is also known as *frame dropping*. Others use adaptive filters in the time domain such as the Wiener filter, the Kalman filter and/or linear prediction techniques. However, the largest class of methods employs some form of spectral decomposition in conjunction with an adaptive spectral gain function. The spectral decomposition may be based on a block transformation, for example, discrete Fourier transform or Karhunen–Loève transform, or a bank of filters. These transforms or filters may provide either a uniform or nonuniform frequency resolution; the latter is often based on a model of the filters in the human auditory system. In all of these cases, the processing model must enable the reconstruction of the enhanced time-domain signal with small reconstruction errors. The computation of the adaptive gain is in turn closely linked to the VAD and to the estimation of the noise power spectral density.

The objective of this chapter is to review the most prominent techniques for voice activity detection, noise power estimation and single-channel noise reduction and to provide insights into the tradeoff between noise reduction and target signal distortion. Throughout this chapter, we will focus exclusively on additive noise, that is, consider disturbed signals $x[\ell]$ which are a sum of the target speech signal $s[\ell]$ and the noise signal $n[\ell]$, $x[\ell] = s[\ell] + n[\ell]$. Here, all signals are discrete-time signals sampled at a rate f_s with sampling index ℓ. Furthermore, $s[\ell]$ and $n[\ell]$ are assumed to be zero-mean signals and to be statistically independent, which also implies $\mathrm{E}\{s[\ell_1]n(\ell_2)\} = 0 \forall \ell_1, \ell_2$, where $\mathrm{E}\{\cdot\}$ denotes the statistical expectation. Waveforms of a clean speech signal $s[\ell]$ and two noisy versions $x[\ell]$ thereof are shown in Figure 4.1. The

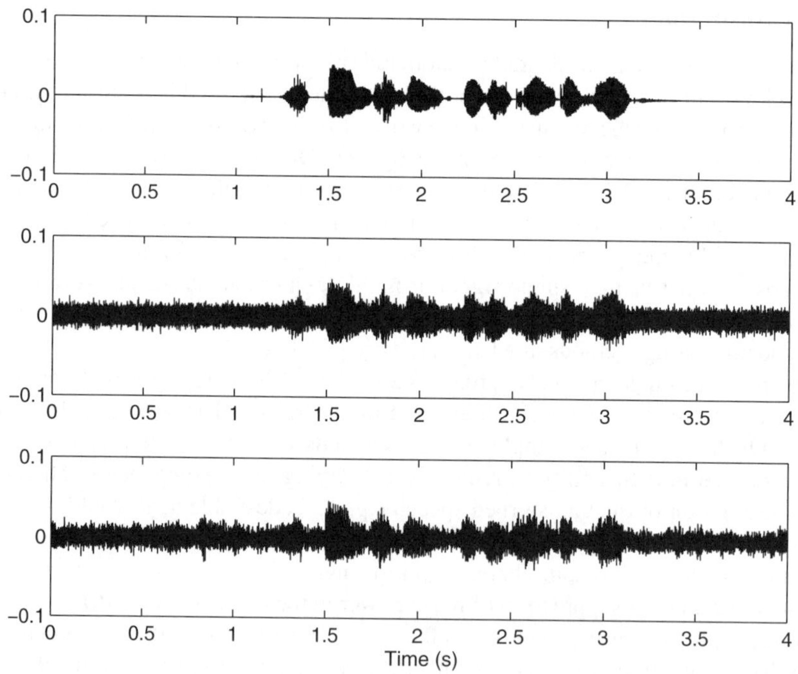

Figure 4.1 Waveforms of a clean speech signal (top), of the signal disturbed by white Gaussian noise (center), and of the signal disturbed by babble noise (bottom). The sentence "Spring street is straight ahead" is spoken by a female speaker. The SNR is 2 dB in both noisy cases.

signal in the center plot is contaminated by stationary white Gaussian noise and the signal in the bottom plot by nonstationary speaker babble noise. The signal-to-noise ratio (SNR) is 2 dB in both cases. These three signals will be used throughout this chapter for illustration purposes.

4.2 Signal Analysis and Synthesis

Many noise-reduction approaches process the signal in a spectral domain. Spectral analysis-synthesis schemes based on invertible block transformations provide very attractive processing models for signal enhancement as they can easily be adjusted to provide perfect or near-perfect reconstruction of the input signal. For signal-independent block transformations such as the discrete Fourier transform (DFT), an additional advantage lies in their high computational efficiency. Furthermore, spectral components of natural signals show less overlap in the spectral domain than in the time domain and are less correlated than the time domain signal samples. Therefore, each spectral component may be modeled and processed independently, which greatly facilitates the algorithm design and, in a first-order approximation, allows for a relatively straightforward scaling of these algorithms in terms of sampling rate or bandwidth. For these reasons we will base the discussion of VAD and noise-reduction techniques on the widely used DFT and a perfect reconstruction overlap-add signal synthesis approach. Most of these techniques are equally useful when other types of decompositions or filter banks, for instance with a nonuniform frequency resolution, are employed.

4.2.1 *DFT-Based Analysis Synthesis with Perfect Reconstruction*

The DFT provides a uniform spectral resolution which may be controlled by choosing an appropriate analysis window $w_A[\ell]$. Likewise, reconstruction artifacts can be controlled by the synthesis window $w_S[\ell]$. In fact, using the DFT, it is straightforward to design a perfect reconstruction analysis-synthesis system, where in the analysis stage we obtain the Fourier transform of the tth signal frame:

$$\begin{aligned} X(t,f) &= \sum_{\ell=0}^{M-1} w_A[\ell] x[tR+\ell] \exp\left(-j\frac{2\pi f\ell}{M}\right) \\ &= \sum_{\ell=0}^{M-1} w_A[\ell] s[tR+\ell] \exp\left(-j\frac{2\pi f\ell}{M}\right) + \sum_{\ell=0}^{M-1} w_A[\ell] n(tR+\ell) \exp\left(-j\frac{2\pi f\ell}{M}\right) \\ &= S(t,f) + N(t,f). \end{aligned} \tag{4.1}$$

M is the length of the DFT, which is assumed to be an even number, R is the frame shift, and t and f are the frame and frequency bin indices. Given the Fourier coefficients $\widehat{S}(t,f)$ of the enhanced signal $\widehat{s}[\ell]$, the overlap-add synthesis can be written as

$$\widehat{s}[\ell] = \sum_{t=-\infty}^{\infty} w_S[\ell - tR] \frac{1}{M} \sum_{f=0}^{M-1} \widehat{S}(t,f) \exp\left(j\frac{2\pi f(\ell - tR)}{M}\right). \tag{4.2}$$

Table 4.1 Window choices for DFT-based perfect-reconstruction spectral analysis-synthesis systems.

$w_A[\ell]$	$w_S[\ell]$	R
Hamming	Boxcar	$M/4$
Hann	Boxcar	$M/2$
Square-root Hann	Square-root Hann	$M/2$

The window functions are defined in the text. Boxcar denotes the rectangular window.

The window functions $w_A[\ell]$ and $w_S[\ell]$ are defined to have nonzero samples in the interval $\ell = 0 \ldots M-1$ only. Typical analysis windows are the Hamming, the Hann, and the square-root Hann window, which, for $\ell = 0 \ldots M-1$, are written as follows:

$$\text{Hamming window:} \quad w_A[\ell] = 0.54 - 0.46\cos\left(\frac{2\pi\ell}{M-1}\right), \tag{4.3}$$

$$\text{Hann window:} \quad w_A[\ell] = 0.5\left(1 - \cos\left(\frac{2\pi\ell}{M-1}\right)\right), \tag{4.4}$$

$$\text{square-root Hann window:} \quad w_A[\ell] = \sqrt{0.5\left(1 - \cos\left(\frac{2\pi\ell}{M-1}\right)\right)}. \tag{4.5}$$

These window functions and the corresponding synthesis windows and frame shifts R that result in a perfect-reconstruction analysis-synthesis system are summarized in Table 4.1.

Frequently, we will also use the magnitude $|X(t,f)|$ or the squared magnitude $|X(t,f)|^2$ of the complex DFT coefficients $X(t,f)$. The latter, with an appropriate normalization of the window function $w_A[\ell]$, is often referred to as the periodogram (see [15]). However, the periodogram is a short-time estimate of the power spectral density with a relatively large variance. Depending on its application, it requires further smoothing as outlined in Section 4.4. Nevertheless, it is highly instructive to plot the temporal succession of $20\log_{10}(|X(t,f)|)$ in a color or gray-level plot. This is known as the *spectrogram*. Figure 4.2 depicts the spectrograms of the signals shown in Figure 4.1. The succession of different phones as well as the harmonics

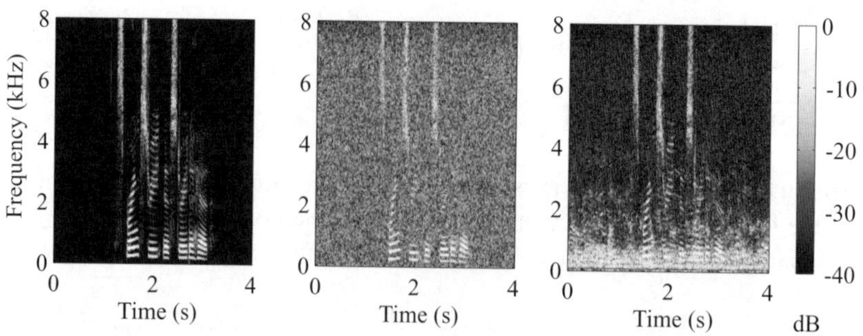

Figure 4.2 Spectrograms of the signals in Figure 4.1: the clean speech signal (left), the signal disturbed by white Gaussian noise (center), and the signal disturbed by babble noise (right).

of the fundamental frequency during voiced sounds can clearly be observed. Furthermore, although the two noise signals can be hardly distinguished in the time domain plots in Figure 4.1, they exhibit their significant differences in the corresponding spectrograms.

Speech and noise signals are typically modeled as stochastic processes. Therefore, in the next section, we will briefly discuss the statistics of discrete Fourier coefficients. Whenever possible, we will drop the frame index t for improved readability.

4.2.2 Probability Distributions for Speech and Noise DFT Coefficients

Discrete Fourier transform coefficients and the quantities derived thereof have a number of statistical properties that facilitate the design of optimal detection and estimation algorithms. For large transform lengths, the complex Fourier coefficients are known to be asymptotically complex-Gaussian distributed [15]. Thus, for the real and the imaginary parts of $X(f)$, $\Re\{X(f)\}$ and $\Im\{X(f)\}$, the probability density functions (PDFs) are given by

$$p_{\Re\{X(f)\}}(a) = \frac{1}{\sqrt{\pi\sigma_X^2(f)}} \exp\left(-\frac{a^2}{\sigma_X^2(f)}\right)$$

$$p_{\Im\{X(f)\}}(b) = \frac{1}{\sqrt{\pi\sigma_X^2(f)}} \exp\left(-\frac{b^2}{\sigma_X^2(f)}\right), \tag{4.6}$$

where $\sigma_X^2(f) = \mathrm{E}\{|X(f)|^2\}$ denotes the signal power in frequency bin f. The power $\sigma_X^2(f)$ is equal to the noise power $\sigma_N^2(f)$ during speech pauses and to $\sigma_S^2(f) + \sigma_N^2(f)$ during speech activity. For the joint probability density of real and imaginary parts we obtain for $f \notin \{0, M/2\}$

$$p_{\Re\{X(f)\},\Im\{X(f)\}}(a, b) = \frac{1}{\pi\sigma_X^2(f)} \exp\left(-\frac{a^2+b^2}{\sigma_X^2(f)}\right). \tag{4.7}$$

As a consequence of the Gaussian model the magnitude $B(f) = |X(f)|$ and the magnitude-squared DFT coefficients $B^2(f) = |X(f)|^2$ follow a Rayleigh distribution

$$p_{B(f)}(a) = \begin{cases} \frac{2a}{\sigma_X^2(f)} \exp\left(-\frac{a^2}{\sigma_X^2(f)}\right) & a \geq 0 \\ 0 & a < 0 \end{cases} \tag{4.8}$$

and an exponential distribution

$$p_{B^2(f)}(a) = \begin{cases} \frac{1}{\sigma_X^2(f)} \exp\left(-\frac{a}{\sigma_X^2(f)}\right) & a \geq 0 \\ 0 & a < 0 \end{cases} \tag{4.9}$$

respectively.

The asymptotic distributions are a good approximation to the observed data when the span of correlation of the signal is much smaller than the length M of the DFT. This is a valid assumption for many noise signals. However, for speech signals, the duration of highly correlated vowels is frequently larger than the transform length. In these cases, the observed distribution is more closely approximated by *super-Gaussian* distributions. These distributions are more peaky at small amplitudes than the Gaussian distribution and are heavy tailed. While sometimes specialized PDFs like the gamma and Laplace densities have been employed (see, e.g., [32,48,49]), parametric distributions are in general more practical. For example, the χ-distribution of speech amplitudes $A(f) = |S(f)|$ is given by

$$p_{A(f)}(a) = \frac{2}{\Gamma(\mu)} \left(\frac{\mu}{\sigma_S^2(f)} \right)^{\mu} a^{2\mu-1} \exp\left(-\frac{\mu}{\sigma_S^2(f)} a^2 \right), \tag{4.10}$$

where $\Gamma(\cdot)$ denotes the complete gamma function and μ is a "shape" parameter which may be adjusted to fit the empirical data or to optimize estimation results.

4.3 Voice Activity Detection

To improve the robustness of the ASR, the detection of talk spurts is of critical importance. This can be accomplished by the recognizer itself, for instance, by including a silence or a background noise model. In this way, the recognizer can deal with roughly end-pointed utterances that may still contain long noise-only segments. Carrying this idea to the extreme means to do no end-pointing at all, as explored for instance in [68].

However, in many cases, it is more practical to evaluate the voice activity at the acoustic front-end and to exclude noise-only segments from further processing. For this purpose, a VAD is used, the design of which is not trivial if the noise is nonstationary or speech-like. As for any detection device, its design has to balance missed hits with false alarms.

Voice activity detection is employed not only in the context of ASR but also in mobile communications for the control of discontinuous transmission schemes (a.k.a. DTX, where a mobile device sends a radio signal only when its user talks) and in many noise tracking algorithms for speech enhancement. As a result, a large variety of different approaches are available. Most of them are tailored to their specific types of application and might not be ideally suited for robust ASR. For example, unlike the application in mobile communication, the latency of the detector is less critical in ASR applications. Voice activity detection methods for ASR and DTX have in common that the speech signal should be detected with high probability. For ASR applications, some noise-only segments can be admitted, especially when the recognizer includes a well-adapted background model.

4.3.1 VAD Design Principles

Most VAD approaches are composed of a feature extraction stage, a detector, and some form of state tracking.

Feature Extraction

A variety of short-term features have been proposed for application in VAD devices. Early approaches were mostly based on the short-time energy of the signal (e.g., [64]). Obviously, these approaches are not robust to high levels of noise. Therefore, most state-of-the-art detectors consider SNR in spectral subbands (see, e.g., [65,71]). In addition, some approaches use features which are tailored to the properties of speech signals, such as linear prediction coefficients or prediction residuals, the periodicity of the signal, or properties of the human auditory systems, see [55] for a range of such features. In any case, the features come along with their specific statistical properties. Thus, the selection of features has a significant impact on optimal detector designs.

In the Fourier domain, the distribution and dependencies of spectral coefficients have been thoroughly investigated, some results are summarized in Section 4.2.2. Therefore, Fourier coefficients serve as a good basis for designing VAD algorithms. For example, the statistical properties of the instantaneous SNR which is also known as the *a posteriori* SNR,

$$\gamma(t,f) = \frac{|X(t,f)|^2}{\sigma_N^2(t,f)}, \tag{4.11}$$

where $\sigma_N^2(t,f)$ denotes the noise power in signal frame t and frequency bin f, make the *a posteriori* SNR a very suitable detection statistics. Under the assumption of a complex Gaussian model for the Fourier coefficients $X(t,f)$ the *a posteriori* SNR $\gamma(t,f)$ follows an exponential distribution (4.9). Then, during speech pause the mean and the variance of $\gamma(t,f)$ are equal to one [76, page 127]. Since the mean and the variance are significantly larger than one for speech activity, the *a posteriori* SNR or an average thereof over frequency is well suited to construct a threshold test for VAD (see, e.g., [70,76, page 426].

A significant challenge in the design of a VAD arises when the noise is nonstationary and non-Gaussian. In this case, the statistics of the noise is more similar to the statistics of the speech signal than for Gaussian noise. As a consequence, a more precise statistical model is required. To differentiate between the two signals, it is advantageous to adapt a statistical model to the observed noise. Breithaupt and Martin [10] present an approach that evaluates the outlier statistics of the *a posteriori* SNR. An outlier is detected whenever $\gamma(t,f) > \gamma_{\text{th}}$, where a typical value is $\gamma_{\text{th}} = 4$. The PDF of the *a posteriori* SNR during speech pause is modeled by a Rayleigh inverse Gaussian distribution

$$p_{\gamma(t,f)}(a) = \sqrt{\frac{2}{\pi}}\,\alpha^{3/2}\,\delta\,\exp(\delta|\alpha|)\frac{a}{(\delta^2+a^2)^{3/4}}\mathrm{K}_{3/2}(\alpha\sqrt{\delta^2+a^2}), \tag{4.12}$$

where $\mathrm{K}_{3/2}(\cdot)$ is a modified Bessel function of the second kind. The shape parameter α determines the heavy tailedness of the distribution of the *a posteriori* SNR during speech pause and is used here to model the different noise types. The scale parameter δ is determined by the variance. Both α and δ are estimated by the expectation-maximization algorithm presented in [21] using the first few signal frames, which implicitly assumes that there is no speech activity at the beginning of the signal. Using this model, the expected number of outliers during speech pause can be computed. Since speech in the short-time Fourier domain is also heavy tailed and typically has more outliers than noise, speech presence can be assumed when the observed number of outliers is greater than the expected outlier count resulting from the above noise model.

As an alternative to DFT-based measures, the wavelet coefficients of a signal may also serve to distinguish speech from noise segments. The rationale lies in the fact that detail coefficients of a multiresolution wavelet decomposition are mainly determined by speech, while exhibiting smaller values for noise (see [46]). Therefore, comparing the energy ratios between appropriate sets of wavelet coefficients can lead to an effective VAD when a suitable energy measure is used (see, e.g., [59,69]).

Detection Principles

In the previous section, a number of features, such as the *a posteriori* SNR $\gamma(f)$, have been discussed. Having decided on the set of features $\mathbf{c}$ to be used, a statistically optimal decision can now be reached, for instance, by the maximum likelihood (ML) or the maximum *a posteriori* (MAP) principles.

Given a feature set $\mathbf{c}$, the ML detector entails comparing the likelihood $p(\mathbf{c}(t)|M_1)$ of speech presence and the likelihood $p(\mathbf{c}(t)|M_0)$ of speech absence, where M_1 and M_0 denote the statistical model parameters of the features under speech presence and speech absence, respectively.

Using these likelihood values, a decision for the hypothesis that speech is present, $\mathrm{H}^{(1)}(t)$, can be taken if $p(\mathbf{c}(t)|M_1) > p(\mathbf{c}(t)|M_0)$, or equivalently, if $\frac{p(\mathbf{c}(t)|M_1)}{p(\mathbf{c}(t)|M_0)} > 1$, and for the nonspeech hypothesis $\mathrm{H}^{(0)}(t)$ otherwise. The MAP detector uses the posterior densities $p(M_1|\mathbf{c}(t))$ and $p(M_0|\mathbf{c}(t))$, which leads to the MAP criterion for speech presence $p(\mathbf{c}(t)|M_1)P(M_1) > p(\mathbf{c}(t)|M_0)P(M_0)$.

Obviously, for equal *a priori* probabilities of speech presence and absence, $P(M_1)$ and $P(M_0)$, respectively, the ML and MAP estimators lead to the same result. In general, the MAP estimator will test whether $\frac{p(\mathbf{c}(t)|M_1)}{p(\mathbf{c}(t)|M_0)} > \frac{P(M_0)}{P(M_1)}$.

For K independent feature vector components $c_k(t)$, and after replacing the threshold of $\frac{P(M_0)}{P(M_1)}$ by the parameter $\exp(\eta')$, the MAP detector corresponds to the decision rule

$$\Lambda(t) = \prod_{k=0}^{K-1} \frac{p(c_k(t)|M_1)}{p(c_k(t)|M_0)} \overset{\text{def}}{=} \prod_{k=0}^{K-1} \Lambda_k(t) \underset{\mathrm{H}^{(1)}(t)}{\overset{\mathrm{H}^{(0)}(t)}{\lessgtr}} e^{\eta'}. \tag{4.13}$$

Rather than thresholding the product of all likelihood ratios $\Lambda_k(t)$ directly, we may also work with $\log \prod_k (\Lambda_k) = \sum_k \log(\Lambda_k)$, which is often preferable as many useful densities belong to the exponential family. Obviously, this is equivalent to a test on the mean of log-likelihood ratios

$$\log\left(\Lambda(t)\right)^{\frac{1}{K}} = \frac{1}{K}\sum_{k=0}^{K-1} \log(\Lambda_k(t)) \underset{\mathrm{H}^{(1)}(t)}{\overset{\mathrm{H}^{(0)}(t)}{\lessgtr}} \frac{\eta'}{K} = \eta. \tag{4.14}$$

In [70], the likelihoods of speech presence and speech absence are based on the distribution (4.7) of the complex Fourier coefficients. In this case, $\Lambda_k(t)$ can be approximated by the *a posteriori* SNRs of all $K = M/2$ frequency bands. Other features are used in conjunction with the same likelihood ratio test in [31,73]. In [73], it is suggested to use only harmonic frequency components as features for this test if the current frame is voiced, which reduces speech clipping under low-SNR conditions. More elaborate decision functions using temporal characteristics of the power envelope are introduced in [55,66], and a likelihood test using multiple signal frames is described in [67].

Smoothing Detector Decisions Using a Finite State Machine

Short-time decisions may exhibit a lot of fluctuations. These may be due to short speech pauses between syllables and words but also to wrong decisions. Especially in low-SNR conditions, such decision errors may lead to significant clipping of talk spurts. A simple method to cope

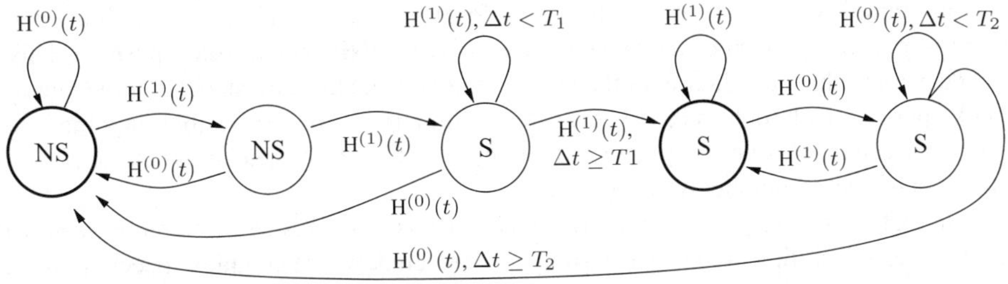

Figure 4.3 A five-state Markov model for smoothing instantaneous voice activity decisions $\mathrm{H}(t)$. The labels of the states show, whether the current frame is classified as speech (S) or nonspeech (NS). T_1 and T_2 denote the minimum length of speech segments and the hangover time, respectively. Δt is the time that has passed since a state has been entered [10].

with clipping at the end of words is to include a hangover procedure which holds the speech active decision for a period of 100 ms or more. Offline methods can also add a number of hangover frames to the beginnings of detected utterances (see [64]), and more elaborate schemes use a full-fledged Markov model to take the temporal dynamics of the speech signal into account. An example for the latter strategy of smoothing the frame decisions for the hypothesis noise, $\mathrm{H}^{(0)}(t)$, or speech-plus-noise, $\mathrm{H}^{(1)}(t)$, is shown in Figure 4.3. Figure 4.4

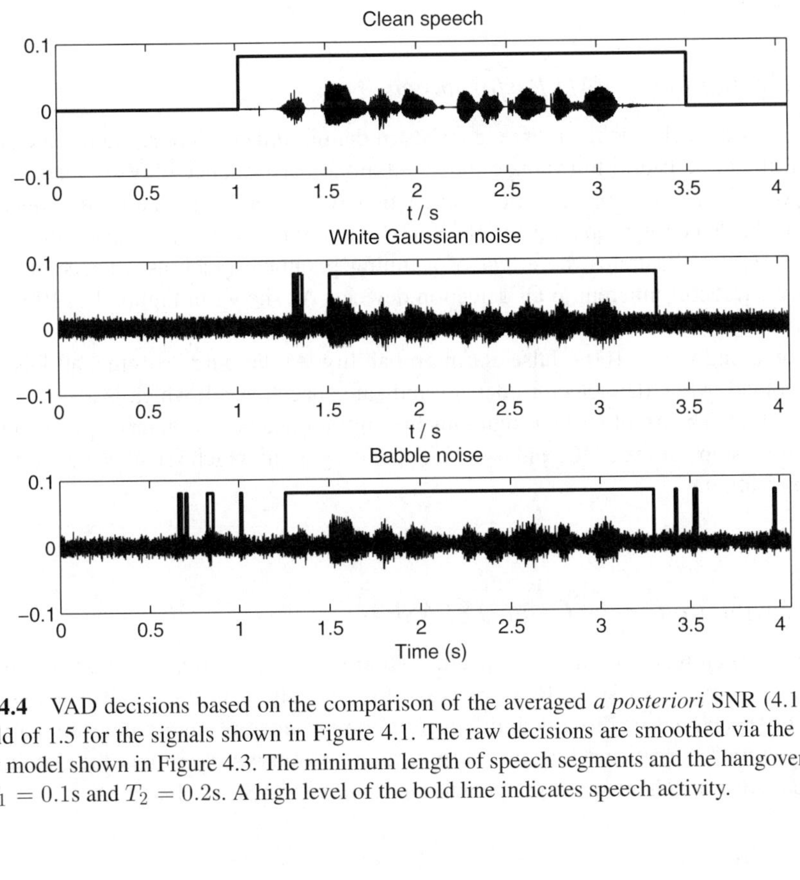

Figure 4.4 VAD decisions based on the comparison of the averaged *a posteriori* SNR (4.11) with a threshold of 1.5 for the signals shown in Figure 4.1. The raw decisions are smoothed via the five-state Markov model shown in Figure 4.3. The minimum length of speech segments and the hangover time are set to $T_1 = 0.1\mathrm{s}$ and $T_2 = 0.2\mathrm{s}$. A high level of the bold line indicates speech activity.

shows an example of typical VAD performance for the speech signals shown in Figure 4.1. In the first step, a comparison of the averaged *a posteriori* SNR with a fixed threshold delivers initial decisions which are then smoothed using the five-state Markov model. For the most part the talk spurt is nicely detected. However, some limitations are also visible: in white noise, the initial fricative /s/ is not fully detected while in babble noise, speech-like noise segments result in an increased number of false detections.

An alternative to such explicit hangover mechanisms is suggested by Ramírez *et al.* [67]. Here, the idea of multiple hypothesis testing is considered, where complex speech/nonspeech hypotheses are defined regarding a number of observation windows simultaneously. A MAP decision over these multiframe hypotheses is shown to be helpful for speech/nonspeech discrimination under severe noise conditions.

Most recently, another statistical model has also been applied to improve VAD decisions in nonstationary noise. This model, a *partially observable Markov decision process* or POMDP, is also quickly adaptable for short-time sporadic noises (see [58]). For this purpose, the system includes prior knowledge on temporal and feature-space noise characteristics in an elaborate state-space model, including such states as "breath" or "click" explicitly. On the basis of this model, the VAD decisions are reached by a so-termed *agent*, which not only realizes the frame-by-frame decision function, but has additional actions it can take to improve discriminance for critical frames, such as that of computing additional features when necessary. Decisions are reached by optimizing a reward function, which penalizes wrong decisions but also time delays, thus attempting to attain the best trade-off between decision quality and efficiency.

4.3.2 Evaluation of VAD Performance

In the context of ASR, the final target consists in the minimization of recognition error rates. However, it can be useful to evaluate the stand-alone performance of VAD algorithms. For this purpose, the receiver operating characteristic (ROC) (see [20]), is a useful concept. The ROC plots the detection probability $P(\mathrm{H}^{(1)}(t)|M_1)$ as a function of the false alarm probability $P(\mathrm{H}^{(1)}(t)|M_0)$. It illustrates the effect of a tuning parameter such as a decision threshold and thus the tradeoff inherent to all detection devices. As shown in Figure 4.5, 100% speech detection can be achieved if the detector indicates speech at all times. However, this goes necessarily along with a 100% false alarm probability. At the other extreme, all false alarms can be avoided, when the detector does not indicate speech at all, which is also not a useful operating point. The art of VAD design thus lies in finding a detection principle which yields a steep initial slope of the ROC, and to set an operating point which satisfies the requirements of the application.

4.3.3 Evaluation in the Context of ASR

Automatic speech recognition poses specific requirements for VAD, which go beyond those measurable by means of static ROC curves. Most notably, it is important that low-energy portions of speech not be discarded, not even in the presence of highly nonstationary noise. This is even more significant for low-energy frames within words, so that the state machine and the hangover mechanism are of special importance for ASR applications.

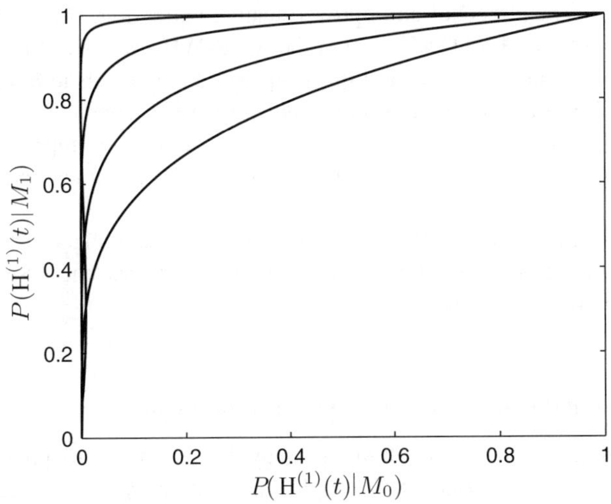

Figure 4.5 Receiver operating characteristic (ROC) for several detection algorithms. The ROC represents the $P(\mathrm{H}^{(1)}(t)|\text{speech is present})$ versus $P(\mathrm{H}^{(1)}(t)|\text{speech is absent})$ tradeoff. The closer the ROC approaches the upper left corner the better the detector.

But while these requirements may make VAD more challenging, at the same time, ASR systems can also be used to inform the VAD and to thus arrive at better results than may be possible in a stand-alone application. This is discussed in more detail below:

- The VAD essentially aims at distinguishing the feature distribution of noise from that of noisy speech. Any ASR system contains a clean speech model with a high level of detail. This speech model can well form the basis of a noisy speech model, either by construction, or by retraining or adapting the model to the noisy speech characteristics in the operating environment of the system.
- ASR already contains a decision logic designed to classify temporal sequences of phonemes. When the ASR is equipped with a model for noise alone, this model can be included in the recognition process, thus performing VAD as part of the operation of ASR. However, the success of such a strategy depends strongly on the quality of the noise model and while it may be promising for ASR trained under matched conditions, an ASR trained on clean data will profit greatly from an added VAD under noisy conditions.

As can be seen from these considerations, it is important to measure the performance of a VAD not only in terms of its classification accuracy or ROC but also by means of the finally obtainable recognition rate. Since there is a wide variety of tasks and experimental setups, in the following, we will present some exemplary results and attempt to draw conclusions for more general setups as well.

Effects of Improving Static Classification Performance

For improving static performance, all approaches that are described in Section 4.3.1 for feature extraction and detection apply.

As an example for the effects of an improved feature extraction, [41] compares the results of wavelet-coefficient-based VAD with that of the AMR-WB (Adaptive Multirate-Wide band) VAD specified in [1], which uses standard spectrum-based features in addition to a correlation-based periodicity measure. Depending on the noise type and characteristics, it can be seen in [41] that the improved, wavelet-based VAD features can lead to improved VAD decisions under highly noisy conditions, both in terms of classification accuracy and ASR recognition rate.

However, while great gains in ASR performance are due to a suitable VAD under mismatched conditions, the effect of an external VAD is not pronounced once the ASR is trained on the specific noise from the test scenario.

Effects of Improving the State Machine and Decision Logic

For stationary noise, a simple smoothing strategy for a slowly adaptive statistical noise model may be sufficient. In contrast, for highly nonstationary noise conditions, it is necessary to find ways of rapidly adapting the internal noise or noisy-speech models to new situations.

This leads to the idea of using switching dynamic models, as described, for example, by Fujimoto and Ishizuka [30]. In their work, a model of clean and a model of noisy speech, realized in the form of Gaussian mixture models (GMM), are decoded with the help of a switching Kalman filter. For this purpose, a switching state space model is defined, allowing transitions from a "noise" to a "speech plus noise" state, and vice versa, at any time frame. The computations are then carried out in two iterated steps:

1. A state estimation, which is in effect the VAD decision.
2. The noise update, which is carried out using the Kalman filtering update formulas.

In this approach, a clean speech model is given in the form of a GMM, pretrained on 5050 utterances which were parameterized as 24th-order log-Mel spectra. For this method, the evaluation was carried out on a database of Japanese digit sequences. The results show a significant reduction of error rates relative to the baseline of using no VAD at all. The greatest factor here for improved speech-recognition accuracies was found to be the quality of the speech model [30].

However, all reported results here are obtained under mismatched conditions. In contrast, [67], also using connected digits recognition as an example, evaluates both scenarios—matched and mismatched conditions. As already seen above, having an adaptive decision strategy with an appropriate hangover mechanism to allow for time varying noise is again a significant advantage for ASR performance.

In contrast, under matched conditions the ASR performance reported in [67] is quite similar for many of the presented strategies, again illustrating the effect of a precise internal noise model in the ASR system. This should not mislead one into disregarding the design of VAD for matched conditions, though—in this paper as in many other publications, the VAD recommended in G.729 for DTX-purposes [40], gives very low word accuracies in conjunction with frame dropping. This effect may well be hypothesized to correspond to its high rate of false rejections, that is to its erroneous classification of a significant percentage of speech frames as noise.

Conditions for Using Implicit VAD

When the ASR system contains a silence model, as it is almost always true for ASR beyond isolated command word recognition, the VAD can be carried out as a part of the ASR search.

However, the quality of such systems is vitally dependent on the quality of the nonspeech model that the ASR system uses. This can be seen in [41], where the performance of ASR after multicondition training is hardly influenced by the quality of the VAD, whereas the performance of an ASR after clean training is shown to suffer significantly from bad VAD decisions in the front end. Thus, an internal silence model adaptation of the ASR can be seen to be of great importance when implicit VAD is desired, and adaptation strategies such as those described in Chapter 11 may be considered for the purpose.

If adaptation is possible, or if training and deployment take place in statistically very similar noise backgrounds, a simple VAD is generally sufficient, or VAD may be carried out completely as part of the ASR search with no noticeable loss in accuracy (see [41,68]). The only requirement that is truly significant in such cases is that the employed VAD not have a high miss rate, as this will degrade ASR performance in all cases.

4.4 Noise Power Spectrum Estimation

Most noise-reduction algorithms and model compensation methods (see Chapter 12) require an estimate of the background noise power and in most cases, the noise power needs to be estimated in frequency subbands corresponding to the spectral analysis scheme of the noise-reduction or feature-extraction algorithm. In contrast to applications in mobile telephony, the latency of the noise-reduction processing in ASR is not of critical concern. In most systems, short utterances are processed as a whole and the noise power spectral density of a particular utterance is obtained from the first few signal frames which are assumed to be noise only. This simple off-line processing approach is certainly suitable for short utterances and fairly stationary noise. However, in general the noise power needs to be tracked over time.

To obtain estimates of the noise power or average noise magnitudes, voice activity decisions are not explicitly required. Nevertheless, many approaches make use of a VAD and acquire the noise power during periods of speech absence. However, the design of a reliable VAD for low SNR and nonstationary noise conditions is a difficult task. Therefore, many widely used methods for noise power estimation and power tracking do not rely on binary decisions of a VAD but use a soft-decision noise power update scheme, the minimum power tracking principle, or a combination of these. These methods have the advantage that the noise power estimate can be updated to some extent also during speech activity and are therefore more appropriate when the noise is nonstationary.

A critical issue of any noise power spectral density estimator is the balance of estimation errors and tracking speed. In general, a fast-tracking algorithm is more susceptible to random fluctuations in the observed noisy signal. To cope with these fluctuations, some smoothing is necessary. Therefore, we will begin with a brief review of smoothing methods before we discuss the tracking methods in more detail.

4.4.1 Smoothing Techniques

As the noise waveform during voice activity is random and unknown, the spectral magnitude and the phase of the noise components also exhibit random fluctuations. Therefore, to reduce the estimation error on average, the noise power estimator has to deliver a smoothed version of the noise components in the observed signal. In what follows, we will briefly review several options for smoothing periodograms $|X(t-i,f)|^2$ of the noisy signal. Obviously, during speech pause these will approximate the noise power spectral density.

The Moving Average

Given a succession of periodograms $|X(t-i,f)|^2$, $i = 0 \ldots I-1$, and no speech activity, the noise power spectral density may be estimated via a moving average

$$\overline{P}_X(t,f) = \frac{1}{I}\sum_{i=0}^{I-1} |X(t-i,f)|^2 \tag{4.15}$$

over I frames as proposed in [77]. When successive DFT spectra $X(t-i,f)$ are statistically independent and complex Gaussian distributed, the resulting power estimate $\overline{P}_X(t,f)$ follows a χ^2-distribution with $2I$ degrees of freedom. A disadvantage of the moving average method is the necessity to store the past $I-1$ signal frames. Moving average smoothing can also be interpreted as a convolution of the temporal succession of periodograms with a rectangular kernel function.

The above smoothing method cannot be applied when only a single frame of spectral data is given, for example at the beginning of a speech utterance. Then, smoothing over frequency is the method of choice. In this case, the single periodogram may be convolved with a nonnegative spectral kernel function $\kappa(f)$, or, equivalently, after an inverse Fourier transform the corresponding autocorrelation function may be multiplied with a window function. Therefore, the Fourier transform of this window function is the spectral kernel. When a triangular or a Parzen window is used in the autocorrelation domain, the corresponding spectral kernel is nonnegative. Then, its convolution with the sequence of periodgrams will always be nonnegative, a highly desirable property of power estimates.

In a variation of this scheme spectral kernel functions are used which vary in their spectral width and / or vary as a function of time. In the simplest case of a rectangular spectral kernel

$$\kappa(f) = \begin{cases} \frac{1}{2L(t,f)+1} & f = -L(t,f), \ldots, L(t,f) \\ 0 & \text{else} \end{cases} \tag{4.16}$$

the smoothing procedure is described as

$$\widetilde{P}_X(t,f) = \sum_{q=-L(t,f)}^{L(t,f)} \kappa(q)|X(t,q)|^2 = \frac{1}{2L(t,f)+1} \sum_{q=f-L(t,f)}^{f+L(t,f)} |X(t,q)|^2, \tag{4.17}$$

where $2L(t,f)+1$ indicates the number of bins which are averaged at frame index t and frequency bin index f. Thus, the smoothing process may be adapted to the speech or noise signal, as proposed, for example in [27] for the suppression of undesirable fluctuations ("musical noise", see Section 4.5). For complex Gaussian data the distribution of $\widetilde{P}_X(t,f)$ may again be approximated by a χ^2-PDF. However, the correlation of adjacent bins, which depends on the spectral analysis window $w_A[\ell]$, needs to be taken into account. The resulting degrees of freedom are discussed in [54].

First-Order Recursive Smoothing

First-order recursive smoothing

$$P_X(t, f) = \alpha P_X(t-1, f) + (1-\alpha)|X(t, f)|^2 \tag{4.18}$$

is more memory efficient than the moving average as it needs to store only the previous average $P_X(t-1, f)$. The smoothed data can be also written in terms of an infinite sum

$$P_X(t, f) = (1-\alpha)\sum_{i=0}^{\infty} \alpha^i |X(t-i, f)|^2. \tag{4.19}$$

It follows that recent data contributes most to the average, as the implicit exponential weighting reduces the influence of past data. The probability distribution of the recursive average may be approximated by a χ^2-distribution where for statistically independent complex Gaussian DFT spectra the number of degrees of freedom is given by $2(1+\alpha)/(1-\alpha)$. However, when the χ^2-distribution is used as an approximation, the tails of the true distribution may not be well represented (see [51]).

4.4.2 *Histogram and GMM Noise Estimation Methods*

The standard smoothing methods such as first-order recursive smoothing require explicit information about the presence of speech, and therefore a VAD and / or an intelligent control of the smoothing parameter α. Other methods, such as the minimum tracking methods (see Section 4.4.3), are more robust in this sense but will require some form of estimation error compensation. In any case, it is beneficial to acquire knowledge about the statistics of the observed signal and to derive VAD parameters (see Section 4.3.1) or noise power estimates based on these statistics. The latter approach has been followed in [38] where the histogram of the noisy magnitudes is tracked, or in [74] which fits a GMM to the observed noisy data. For SNR above 0 dB the histogram of the noisy data clearly exhibits a bimodal structure where the peak at lower amplitudes corresponds to noise and the peak at higher amplitudes corresponds to speech-plus-noise components. In [74], a two-component GMM is fitted to the observed log-magnitude data. Then, a speech-versus-noise decision threshold is derived from the point of equal probabilities of the two component densities. Based on the resulting decisions, the noise power can be estimated using one of the above smoothing techniques. In the approach proposed in [38] a noise estimate is derived from a histogram of magnitude DFT coefficients which belong to noise-only segments. The histogram is acquired over a period of 400 ms and the noise magnitude corresponding to the peak of the histogram is extracted. Then, to attenuate outliers, successive peaks are smoothed over time using a first-order recursive system. The acquisition of the noise histogram is stopped whenever the observed noisy DFT magnitudes are larger than the product of the last noise estimate by a factor in the range of 1.5–2.5.

4.4.3 *Minimum Statistics Noise Power Estimation*

The *minimum statistics* noise power estimator has been originally proposed for full-band noise power estimation in [52] and subsequently for estimating the noise power in frequency subbands in [53]. It does not rely on binary voice activity decisions. The basic principle

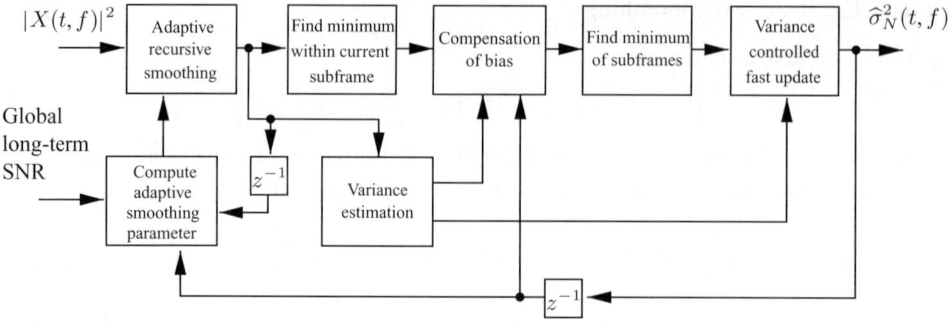

Figure 4.6 The minimum statistics noise power estimation approach [47]. Its main ingredients are a minimum search on short segments of typically 200 ms, a bias compensation, a minimum search over longer segments of 1–2 s, a fast noise power update mechanism, and a variance estimator. z^{-1} denotes a delay element.

is detailed in the block diagram in Figure 4.6. The main idea is to search for the minima of recursively smoothed periodograms within a time window of 1.5–2 s. Toward this end, the "adaptive recursive smoothing" block in Figure 4.6 implements a first-order recursive, temporal smoothing (see Section 4.4.1) where the smoothing parameter α is controlled by a smoothed *a posteriori* SNR and the global long-term SNR (summarized in block "compute adaptive smoothing parameter"). Little smoothing is applied during speech activity and much smoothing during speech absence. As a result, the variance of the smoothed signal power is high during speech activity and low during speech absence. Consecutive power estimates are compiled into a subframe of about 200 ms and the minimum within these subframes is determined. Since the power of noisy speech drops to the level of the noise during speech pauses and in between words and syllables, the observed minimum may serve as an initial noise power estimate. However, the fluctuations in the power due to nonstationary noise and random variations around the mean may render the minimum substantially smaller than the mean value. This bias will depend on the amount of smoothing applied to the periodograms. Therefore, a compensation of this systematic error (in block "compensation of bias") is necessary. In the next block, the minimum over an extended duration of several subframes, corresponding to 1.5–2 s, is searched. Further details, such as the online variance estimation and a fast update mechanism which acts on the subframe level can be found in [47]. The minimum tracking method has also become an ingredient to many other soft-decision tracking approaches such as the improved minimum-controlled recursive averaging (IMCRA) method (see [16]).

4.4.4 MMSE Noise Power Estimation

The minimum statistics method can cope with stationary and mildly nonstationary noise. It cannot follow noise power variations on a time scale much below 2 s. Therefore, noise-tracking methods have been developed which respond faster to nonstationary noise. However, improving the tracking speed may also lead to more speech leakage, that is the noise estimate will also contain some speech components.

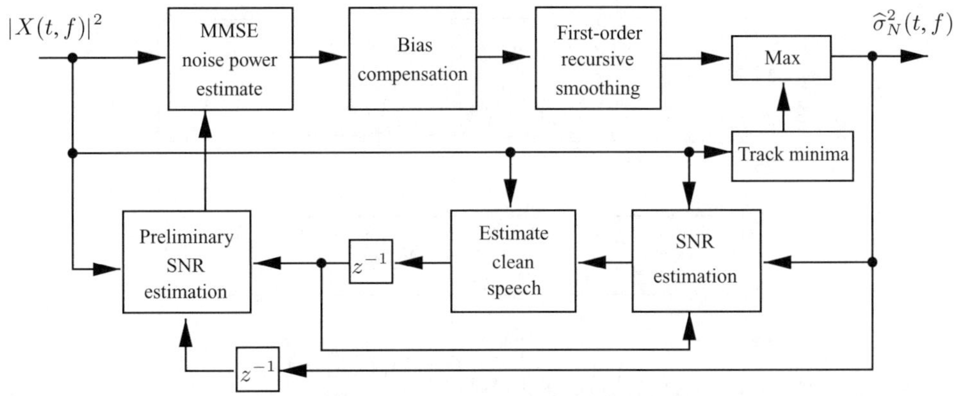

Figure 4.7 The minimum mean-square error (MMSE) noise estimation approach [37]. It comprises an MMSE noise power estimator, a bias compensation, and a signal-to-noise ratio (SNR) and clean speech estimator. z^{-1} denotes a delay element.

Some of the recently developed methods are based on short-time estimation principles in the DFT domain as explained in Section 4.5.5. For example, the methods proposed by Yu [81] and by Hendriks *et al.* [37] employ an estimator of the noise periodogram $|N(f)|^2$ under a Gaussian model and minimize the mean-square error (MMSE) $\mathrm{E}\{(|\widehat{N(f)}|^2 - |N(f)|^2)^2\}$, the solution of which is the conditional expectation

$$|\widehat{N(f)}|^2 = \mathrm{E}\left\{|N(f)|^2|X(f)\right\}. \tag{4.20}$$

This estimator depends on the *a priori* SNR, the estimates of which are typically biased. Thus, to derive an exact noise power estimate the bias needs to be compensated in an accurate fashion. The main components of this algorithm are shown in Figure 4.7. In addition to the optimal noise power estimator the method employs a simple minimum tracking mechanism and takes the maximum of the two estimates in order to avoid estimates that are too small. In a recent evaluation, the approach of [37] has been shown to outperform many other approaches on highly nonstationary noise (see [72]). Figure 4.8 depicts the result of the MMSE noise estimation process applied to the two noisy signals in Figure 4.1. The graphs show the signal power as well as the estimated noise floor versus time. For the stationary white noise (top graph) the noise power estimate matches the power of the noise very well. For the speech signal disturbed by babble noise, we find that the noise power is slightly underestimated. This is, in fact, desirable as the noise power is fluctuating and a noise-power overestimation would lead to speech signal distortions.

4.4.5 *Estimation of the* A Priori *Signal-to-Noise Ratio*

Because of its inherent normalization, it is often more practical to use the SNR instead of the noise power. In fact, most noise-reduction approaches are formulated in terms of an

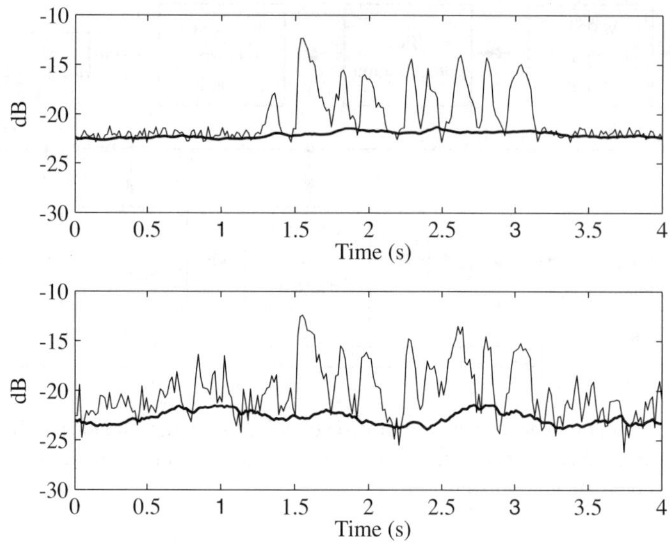

Figure 4.8 Estimated signal power and average noise floor for the signal disturbed by white Gaussian noise (top) and for the signal disturbed by babble noise (bottom) as shown in Figure 4.1. The noise power is averaged over all frequencies and estimated using the MMSE approach of Hendriks *et al.* [37].

a posteriori SNR as introduced in Equation (4.11) and an *a priori* SNR

$$\xi(t,f) = \frac{\sigma_S^2(t,f)}{\sigma_N^2(t,f)}, \tag{4.21}$$

where $\sigma_N^2(t,f)$ and $\sigma_S^2(t,f)$ denote the noise power and speech power in frame *t* and frequency bin *f*. While the *a posteriori* SNR is straightforward to compute, the *a priori* SNR is not directly accessible and must be estimated. Two successful methods are outlined below.

The Decision-Directed Approach

The *decision-directed* approach for *a priori* SNR estimation as introduced in [23] uses an estimate of the speech spectral components of the previous frame $t-1$ in conjunction with the current observation of frame *t*. Both contributions to the estimated *a priori* SNR $\widehat{\xi}_{\mathrm{DD}}(t,f)$ are formulated in terms of signal-to-noise ratios and are now functions of frequency *f* and frame index *t*. They are weighted with α_d and $(1-\alpha_d)$, respectively,

$$\widehat{\xi}_{\mathrm{DD}}(t,f) = \alpha_d \frac{G(t-1,f)^2 |X(t-1,f)|^2}{\sigma_N^2(t,f)} + (1-\alpha_d)\max\left(0, \gamma(t,f)-1\right), \tag{4.22}$$

where $G(t-1,f)$ denotes a spectral gain function, see Section 4.5, and $\gamma(t,f)$ the *a posteriori* SNR as before. Note, that $\gamma(t,f)-1$ is also the maximum-likelihood (ML) estimate of $\xi(t,f)$ for a χ^2-distributed *a posteriori* SNR as shown in [33]. α_d controls the tradeoff between a fast reaction to speech transients and a low error variance during steady state. As a fast reaction to transient speech sounds also emphasizes fluctuations in the residual noise, the best setting for the smoothing parameter is not easy to determine and may depend on the application, see [11] for a thorough analysis. Several improvements to the basic decision-directed approach have been reported (e.g., in [22,60]).

Temporal Cepstrum Smoothing

In [13], an approach for the *a priori* SNR estimation has been proposed which is based on a temporal smoothing of the cepstrum of the ML estimate of $\xi(t, f)$. The cepstrum is well suited to represent speech sounds and has been shown to be a highly effective domain for smoothing noisy speech spectra and related quantities. The method has been shown to outperform the decision-directed approach [13].

The basic idea of temporal cepstrum smoothing is to remove random fluctuations and outliers in the spectrum while preserving the salient speech features. This requires the identification of the cepstral bins which are primarily related to speech. In general, these are the low-order cepstral coefficients describing the spectral envelope and the cepstral peak corresponding to the fundamental frequency during voiced speech frames. The latter needs to be tracked in an adaptive fashion. Then, using a first-order recursive smoothing, the cepstral components not corresponding to these speech features can be smoothed over time. In this way random fluctuations in the spectrum are significantly reduced. Note that this method can also be applied to arbitrary spectral gain functions (see [12]).

4.5 Adaptive Filters for Signal Enhancement

While the VAD enables the end pointing and dropping of noise-only frames, the enhancement of speech segments requires an adaptive filter which, in most cases, is based on noise power and SNR estimates. The most prominent techniques are briefly discussed below. All of these methods can be cast in a spectral modification framework which achieves noise reduction through the application of a spectral gain function. Extended discussions of these methods are found, for example in [19,50,76].

4.5.1 Spectral Subtraction

The spectral subtraction technique represents one of the first successful noise-reduction approaches in speech signal processing. It was pioneered by Boll [7], Berouti *et al.* [6], Preuss [63], and others and is based on the idea of subtracting the noise from the noisy signal in the autocorrelation or power spectral density domains. In its simplest form it uses a VAD and first-order recursive noise power and signal power estimates

$$P_X(t, f) = \alpha_X P_X(t-1, f) + (1-\alpha_X)|X(t, f)|^2 \tag{4.23}$$

$$P_N(t, f) = \alpha_N P_N(t-1, f) + (1-\alpha_N)|N(t, f)|^2, \tag{4.24}$$

where the smoothing parameters α_X and α_N are typically in the range of $0 \leq \alpha_X \leq 0.5$ and $0.5 \leq \alpha_N < 1$. $P_X(t, f)$ and $P_N(t, f)$ denote the estimates of the power of noisy speech and of noise, respectively. As these short-time estimates are subject to random fluctuations, a simple subtraction of estimated powers may yield negative results. Thus, a limitation is necessary, and an estimate of the clean speech power may be obtained via

$$\begin{aligned}|\widehat{S(t,f)}|^2 &= \max\left(P_X(t,f) - P_N(t,f), 0\right)\\ &= P_X(t,f)\max\left(1 - \frac{P_N(t,f)}{P_X(t,f)}, 0\right) = P_X(t,f)|G_{\mathrm{SS}}(t,f)|^2,\end{aligned} \tag{4.25}$$

where the $\max(\cdot,\cdot)$ function guarantees nonnegative results. Equation (4.25) links the estimated power spectra at the input and the output of a noise-reduction filter. Therefore, the spectral subtraction method may be interpreted in terms of a time-variant linear filter with magnitude response

$$G_{\mathrm{SS}}(t,f) = \sqrt{\max\left(1 - \frac{P_N(t,f)}{P_X(t,f)}, 0\right)}. \tag{4.26}$$

Since we subtract in the power spectral density domain, this approach is called *power subtraction*. Many variations of this basic principle have been proposed, such as the *magnitude subtraction*

$$\begin{aligned}|\widehat{S(t,f)}| &= \max\left(\sqrt{P_X(t,f)} - \sqrt{P_N(t,f)}, 0\right)\\ &= \sqrt{P_X(t,f)}\max\left(1 - \frac{\sqrt{P_N(t,f)}}{\sqrt{P_X(t,f)}}, 0\right) = \sqrt{P_X(t,f)}\,|G_{\mathrm{MS}}(t,f)|\end{aligned} \tag{4.27}$$

or a generalized form

$$|\widehat{S(t,f)}| = \sqrt{P_X(t,f)}\left[\max\left(1 - \left(\frac{P_N(t,f)}{P_X(t,f)}\right)^{\beta}, 0\right)\right]^{\alpha}, \tag{4.28}$$

where the parameters α and β control the shape of the spectral gain function. They can be kept either fixed or adapted to the characteristics of the speech and the noise signals (see, i.e. [35]). Obviously, the general subtraction rule includes the power subtraction ($\alpha = 0.5$ and $\beta = 1$), the magnitude subtraction ($\alpha = 1$ and $\beta = 0.5$), as well as an approximation to the Wiener gain ($\alpha = 1$ and $\beta = 1$). In general, the subtraction parameters control the tradeoff between a maximum amount of noise reduction, random fluctuations in the residual noise and target signal distortions. For a more thorough discussion, we refer the reader to [18,39,75].

Spectral subtraction techniques as discussed above typically achieve a fairly good speech quality. However, the residual noise after processing is characterized by many spectral outliers (see [75]). These outliers appear randomly in all spectral bins and excite short sinusoidal tones during the synthesis process. In listening experiments, these random fluctuations are perceived as rapid fluctuations also known as *musical noise* or *musical tones*. As a result, the overall quality of the processed signal may not be acceptable. The impact of random outliers on ASR performance is less clear. If the ASR is trained on the processed signal, it can be assumed that they will not significantly degrade the performance. In other cases, they might lead to a mismatch with the silence or background model as they increase the variance of the processed signal and of the features derived thereof. In general, it is advisable to avoid musical tones.

In the context of spectral subtraction, two simple techniques have been proposed to reduce the variance of the processed spectral components during speech absence. The first is the introduction of an *oversubtraction* or *noise overestimation* factor. This factor emphasizes the estimated noise power by a factor in the range of 1–3 and thus leads to a significant suppression of spectral components in low-SNR bins. To a lesser extent, it also leads to a degradation of high-SNR bins and thus needs to be applied with care. Therefore, if the voice quality is of great concern, noise overestimation should be avoided. The second technique which is frequently applied in conjunction with noise overestimation is the introduction of a *spectral floor*. The spectral floor is defined in proportion to the input signal power, it can be set, for example to a value of 10 dB below the spectral power of the noisy input signal. Thus, Equation (4.25) may be modified into

$$|\widehat{S(t,f)}|^2 = \max\left(P_X(t,f) - P_N(t,f), 0.1P_X(t,f)\right). \tag{4.29}$$

This measure reduces the variance of the processed residual noise but also limits the maximum noise reduction. As a result the musical tones are less audible and the residual noise appears to be more natural.

4.5.2 *Nonlinear Spectral Subtraction*

The limitations of the noise overestimation with spectral floor approach as described above has triggered the development of a plethora of heuristic schemes which exercise some adaptive control on the noise overestimation factor and/or the spectral floor. A highly successful implementation of such schemes, known as *nonlinear spectral subtraction* (NSS), was proposed for robust speech recognition in [43,44]. The NSS employs a frequency-dependent nonlinear mapping $\phi(\overline{X(t,f)}, \alpha(t,f), \overline{N(t,f)})$ in the subtraction gain

$$G_{\mathrm{NSS}}(t,f) = 1 - \frac{\phi(\overline{X(t,f)}, \alpha(t,f), \overline{N(t,f)})}{\overline{X(t,f)}}, \tag{4.30}$$

where the temporal averages of speech and noise magnitude $\overline{X(t,f)}$ and $\overline{N(t,f)}$ are computed, similarly to the temporal power averages in Equations (4.23) and (4.24), as

$$\overline{X(t,f)} = \alpha_X \overline{X(t-1,f)} + (1-\alpha_X)|X(t,f)| \tag{4.31}$$

$$\overline{N(t,f)} = \alpha_N \overline{N(t-1,f)} + (1-\alpha_N)|N(t,f)| \tag{4.32}$$

and $\alpha(t,f)$ is the overestimated noise magnitude. The overestimated noise magnitude $\alpha(t,f)$ is computed during speech pauses as the maximum of magnitudes within temporal intervals of about 0.65 s, corresponding to about 40 signal frames at a rate of 62.5 frames per second. Several nonlinear mapping functions ϕ have been proposed (see [44]). The following was found to be useful for robust ASR

$$\phi(\overline{X(t,f)}, \alpha(t,f), \overline{N(t,f)}) = \frac{\alpha(t,f)}{1 + \rho\frac{\overline{X(t,f)}}{\overline{N(t,f)}}}, \tag{4.33}$$

where ρ is a scaling factor. The nonlinear functions follow the general idea of applying a small overestimation factor in high SNR regions and a large factor in low SNR regions. Recognition experiments and substantial improvements using the NSS technique are reported for both 8 and 16 kHz data (see [45]).

4.5.3 Wiener Filtering

The Wiener filter is the MMSE estimator of the desired signal, here, the clean speech signal, subject to a linear constraint. Note that the optimal linear estimator is identical to the unconstrained optimal filter for Gaussian distributed input signals.

The Wiener filter was originally formulated in the time domain and assumes wide-sense stationary input signals [78]. When the output signal, the clean speech estimate, is computed via an infinite-impulse response (IIR) filter

$$\widehat{s}[\ell] = \sum_{\kappa=-\infty}^{\infty} h[\kappa]\, x[\ell-\kappa] \tag{4.34}$$

the coefficients $h_{\mathrm{W}}[\kappa]$ of the Wiener filter are the solution to

$$h_{\mathrm{W}}[\kappa] = \underset{h[\kappa]}{\operatorname{argmin}}\ \mathrm{E}\left\{\left(s[\ell] - \sum_{\kappa=-\infty}^{\infty} h[\kappa]\, x[\ell-\kappa]\right)^2\right\}. \tag{4.35}$$

When the speech and the noise signals are statistically independent and additive, the frequency response of the optimal filter $h_W[\kappa]$ is given (see, e.g., [76, Chapter 11]), by

$$H_{\mathrm{W}}(e^{j\Omega}) = \frac{\Phi_{ss}(e^{j\Omega})}{\Phi_{ss}(e^{j\Omega}) + \Phi_{nn}(e^{j\Omega})}, \tag{4.36}$$

where $\Phi_{ss}(e^{j\Omega})$ and $\Phi_{nn}(e^{j\Omega})$ are the power spectral densities of speech and noise, respectively. Then, the filter coefficients are obtained by an inverse Fourier transform. The frequency response of the Wiener filter allows a nice intuitive interpretation: When the power of the speech signal at a given frequency is much larger than the power of the noise signal, the frequency response is close to one. When the noise power is much larger than the speech power, the frequency response is close to zero. In this way, the Wiener filter suppresses signal components which are dominated by the noise.

Inspired by the above frequency domain solution of the IIR Wiener filter, a corresponding "Wiener" gain function may be also computed for DFT coefficients. In this case, only a single signal frame is considered and the mean-square error is minimized according to

$$G_{\mathrm{W}}(t,f) = \underset{G(t,f)}{\operatorname{argmin}}\ \mathrm{E}\left\{\left|S(t,f) - G(t,f)X(t,f)\right|^2\right\} \tag{4.37}$$

to yield

$$G_{\mathrm{W}}(t,f) = \frac{\sigma_S^2(t,f)}{\sigma_S^2(t,f) + \sigma_N^2(t,f)} = \frac{\xi(t,f)}{1+\xi(t,f)}. \tag{4.38}$$

Thus, the Wiener gain function depends solely on the *a priori* SNR, for the estimation of which the considerations in Section 4.4.5 apply.

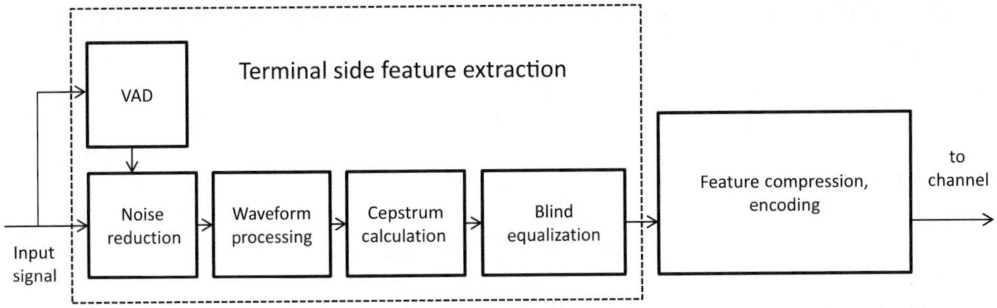

Figure 4.9 Block diagram of the ETSI AFE as described in [28].

4.5.4 *The ETSI Advanced Front End*

The ETSI advanced front end (AFE) [28], was specified for distributed ASR in 2002, and it has since proven its suitability for ASR under very noisy conditions (e.g., in [57] and [80]). The specification is separated into two sections: terminal-side and server-side processing.

On the terminal side, it is based on a standard approach for extracting mel-frequency cepstrum coefficients (MFCCs), but includes two additional components, namely, a two-stage mel-warped Wiener-like filter for signal preprocessing and a blind equalization of cepstrum features (which is similar in principle to a cepstral mean normalization). For noise power estimation the AFE uses a VAD based on an SNR threshold and a hangover mechanism. Furthermore, a *decision-directed* approach is employed for the estimation of the SNR, the square-root of which is then used in the gain function.

In addition, a low-bit-rate coder for speech transmission over wireless networks and a server-side bitstream decoder and feature processor are also specified, with the server-side feature processor being responsible for the calculation of the energy feature and of first and second derivatives. An overview of the AFE feature extraction stage is shown in Figure 4.9.

4.5.5 *Nonlinear MMSE Estimators*

Flexibility beyond the above solutions is introduced if the noise-reduction problem is considered in an estimation-theoretic framework. The estimation problem may be formulated for either the complex DFT coefficients $S(t, f)$ or their magnitude $A(t, f)$. Additional degrees of freedom are available since specific non-Gaussian probability distributions for speech and noise coefficients may be assumed and also functions $c(A(t, f))$ of the DFT magnitude $A(t, f)$ may be defined as the estimation target. Thus, a better fit to the observed probability distributions is achieved and perceptually more meaningful error measures can be introduced. In what follows, we will discuss the most prominent instantiations of this concept and conclude with a flexible parametric amplitude estimation approach. Again, we will drop the frame index t for improved readability.

Amplitude Estimation: The Gaussian Case

The MMSE short-time spectral amplitude (MMSE-STSA) estimator minimizes the mean quadratic error in the spectral amplitudes $\mathrm{E}\{(A(f) - \widehat{A}(f))^2\}$. The optimal estimator [23] can be expressed in terms of the complete Γ-function and the confluent hypergeometric function ${}_1F_1(a, c; x)$ [34, Theorem 9.210.1]

$$\widehat{A}(f) = \mathrm{E}\{A(f)|X(f)\} = \frac{\sqrt{v(f)}}{\gamma(f)}\Gamma(1.5)\,{}_1F_1(-0.5, 1; -v(f))|X(f)|, \tag{4.39}$$

where $v(f)$ is defined as

$$v(f) = \frac{\xi(f)}{1+\xi(f)}\gamma(f). \tag{4.40}$$

Therefore, using the phase of the noisy DFT coefficient, we have

$$\widehat{S}(f) = \frac{\sqrt{v(f)}}{\gamma(f)}\Gamma(1.5)_1F_1(-0.5, 1; -v(f))X(f) = G_{\mathrm{STSA}}X(f), \tag{4.41}$$

where G_{STSA} denotes the corresponding spectral gain function.

MMSE Magnitude-Squared Estimation

As a second example, we consider the estimation of $c(A(f)) = A(f)^2$. The MMSE estimator then minimizes $\mathrm{E}\{(A(f)^2 - \widehat{A}(f)^2)^2\}$. When the probability distribution of $A(f)^2$ is specified in terms of the PDFs for the real and imaginary parts, it is convenient to decompose the optimal estimator into the real and the imaginary parts

$$\widehat{A}(f)^2 = \mathrm{E}\left\{A(f)^2|X(f)\right\} = \int_{-\infty}^{\infty}\int_{-\infty}^{\infty}(\Re\{S(f)\}^2 + \Im\{S(f)\}^2)$$
$$\cdot\; p\left(\Re\{S(f)\}, \Im\{S(f)\} \mid \Re\{X(f)\}, \Im\{X(f)\}\right) \mathrm{d}\Re\{S(f)\}\, \mathrm{d}\Im\{S(f)\}.$$

For Gaussian real and imaginary parts of speech and noise, the solution (see [2]) is related to the Wiener filter and is given by

$$\widehat{A}(f)^2 = \left(\frac{\sigma_S^2(f)}{\sigma_S^2(f)+\sigma_N^2(f)}\right)^2 |X(f)|^2 + \frac{\sigma_S^2(f)\sigma_N^2(f)}{\sigma_S^2(f)+\sigma_N^2(f)}. \tag{4.42}$$

Solutions for non-Gaussian statistical models are described in [8].

MMSE Log-Spectral Amplitude Estimation

Small speech signal amplitudes are very important for speech intelligibility. Therefore, it is sensible to use an error measure which places more emphasis on small signal amplitudes, for example by using a compressive function such as $c(A(f)) = \log(A(f))$.

The MMSE log-spectral amplitude (MMSE-LSA) estimator minimizes the mean-square error of the logarithmically weighted amplitudes $\mathrm{E}\{(\log(A(f)) - \log(\widehat{A}(f)))^2\}$ and thus improves the estimation of small amplitudes. The MMSE-LSA estimator was derived in [24] and is given by

$$\widehat{A}(f) = \exp\left(\mathrm{E}\{\log(A(f))|X(f)\}\right) = \frac{\xi(f)}{1+\xi(f)}\exp\left(\frac{1}{2}\int_{v(f)}^{\infty}\frac{\exp(-\tau)}{\tau}d\tau\right)|X(f)| \tag{4.43}$$

with $v(f)$ as defined in Eq. (4.40). For a practical implementation, the exponential integral function in Equation (4.43) can be tabulated as a function of $v(f)$. Then, the enhanced complex coefficient and the corresponding gain function G_{LSA} are given by

$$\widehat{S}(f) = \frac{\xi(f)}{1+\xi(f)} \exp\left(\frac{1}{2}\int_{v(f)}^{\infty} \frac{\exp(-\tau)}{\tau} d\tau\right) X(f) = G_{\mathrm{LSA}} X(f). \tag{4.44}$$

A General MMSE Amplitude Estimator

While for most noise signals the Gaussian distribution is appropriate, it turns out that the speech components are more closely modelled by a super-Gaussian distribution. Therefore, the estimators based on the Gaussian speech and noise model have been extended to more general distribution models (see [4,26,36,48,49,61]). Furthermore, a compressive cost function, as used, for example, in the derivation of the MMSE-LSA estimator, is beneficial in many applications. In [14], an MMSE estimator based on the χ-density of Equation (4.10) and a parametric compressive function

$$c(A(f)) = A(f)^{\beta} \tag{4.45}$$

has been developed. Starting with the general MMSE solution

$$\widehat{c(A(f))} = \mathrm{E}\left\{c(A(f))\middle| X(f)\right\} \tag{4.46}$$

and substituting Equations (4.10) and (4.45) into Equation (4.46), the estimate $A(f)$ of the clean speech magnitude becomes

$$\begin{aligned}\widehat{A}(f) &= c^{-1}(\widehat{c(A(f))}) \\ &= \sqrt{\frac{\xi(f)}{\mu+\xi(f)}}\left[\frac{\Gamma(\mu+\frac{\beta}{2})}{\Gamma(\mu)}\frac{{}_1F_1(1-\mu-\frac{\beta}{2},1;-\nu(f))}{{}_1F_1(1-\mu,1;-\nu(f))}\right]^{\frac{1}{\beta}}\sqrt{\sigma_N^2(f)}\end{aligned} \tag{4.47}$$

with $c^{-1}(\cdot)$ the inverse of $c(\cdot)$. We have $\nu(f) = \gamma(f)\,\xi(f)/(\mu+\xi(f))$ and $\gamma(f)$ the *a posteriori* SNR as before. Equation (4.47) is valid for $\mu > 0$ and $\mu + \beta/2 > 0$ with $\beta \neq 0$. Note that this implies that $\beta < 0$ can be a valid choice. The estimator (4.47) can be tuned by its two parameters μ and β, and yields several known estimators depending on the choice of μ and β (see Table 4.2). The compression in Equation (4.45) contains the power, the magnitude,

Table 4.2 List of magnitude estimators that are contained in Equation (4.47) as special cases.

β	μ	Estimator
1	1	STSA [23, Equation (7)]
$\beta \to 0$	1	LSA [24, Equation (20)]
$\beta > 0$	1	[79, Equation (14)], [24, Equation (13)]
1	$\mu > 0$	[3, Equation (6)], [26, Equation (12)]
2	1	[2, Equation (20)]

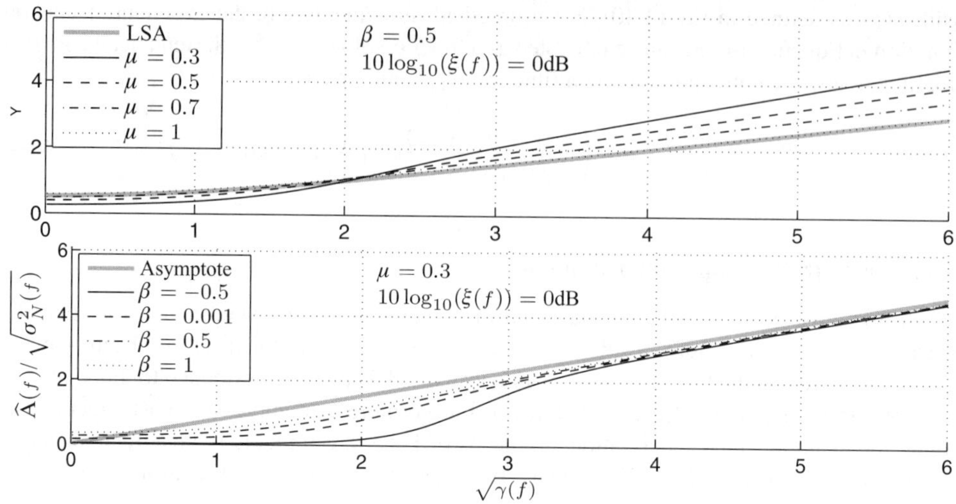

Figure 4.10 Input-output mapping characteristics for an *a priori* SNR of 0 dB. The graphs show the estimated speech amplitudes versus the noisy input amplitudes, both quantities being normalized on the square root of the noise power. Therefore, values $\sqrt{\gamma(f)} \approx 1$ are typical for speech pause and values $\sqrt{\gamma(f)} \gg 1$ are typical for speech activity. Upper graph: variation of the shape parameter μ for $\beta = 0.5$. Lower graph: variation of the compression parameter β for $\mu = 0.3$. The asymptote of Equation (4.48) is also shown.

and the root estimator of Porter and Boll [62]. Furthermore, in [79], it was shown for $\mu = 1$ and $\beta \to 0$ that the solution (4.47) approaches that of the MMSE-LSA estimator in [24].

A high input SNR results in $\nu(f) \gg 1$. Then, the mapping characteristics can be shown with [56, Equation (2.17)] to asymptotically approach

$$\left.\frac{\hat{A}(f)}{\sqrt{\sigma_N^2(f)}}\right|_{\nu(f)\gg 1} = \frac{\xi(f)}{\mu + \xi(f)}\sqrt{\gamma(f)}. \tag{4.48}$$

For $\mu = 1$, this is the Wiener solution. Figure 4.10 plots the normalized input–output characteristics of this estimator where the input is the square root of the *a posteriori* SNR $\gamma(f)$. When the shape parameter μ is tuned toward low values, more noise reduction is achieved for small normalized amplitudes while less attenuation is applied to larger normalized amplitudes. When using a shape parameter $\mu = 0.3$, corresponding to a super-Gaussian speech PDF, a reduction of the compression parameter β leads to a significant attenuation of small normalized amplitudes, especially for negative values of β.

The time domain waveforms and the corresponding spectrograms of the clean signal and two enhanced signals are shown in Figures 4.11 and 4.12. The signals have been processed using the log-spectral amplitude estimator, the MMSE noise estimator, and the decision-directed SNR estimator. Clearly, the noise level is significantly reduced but also the speech signal is somewhat attenuated. However, the most salient features of the speech signal, such as harmonic structures in vowels and high-frequency components in consonants are preserved.

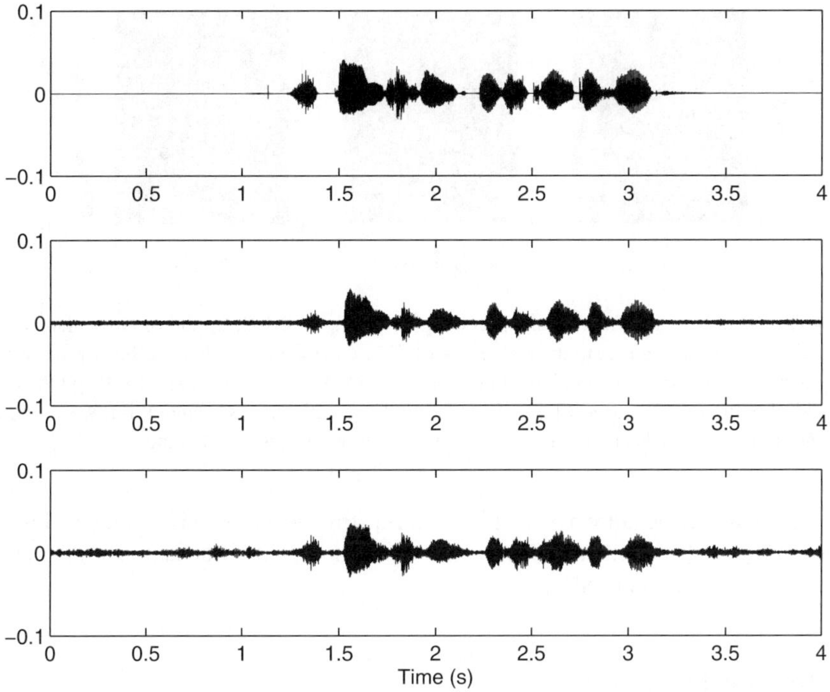

Figure 4.11 Waveforms of a clean speech signal (top) and of two enhanced signals (center: white Gaussian noise, bottom: babble noise). The corresponding noisy signals are shown in Figure 4.1. The enhanced signals have been processed using the log-spectral amplitude estimator Equation (4.44), the MMSE noise estimator [37], and the decision-directed SNR estimator Equation (4.22).

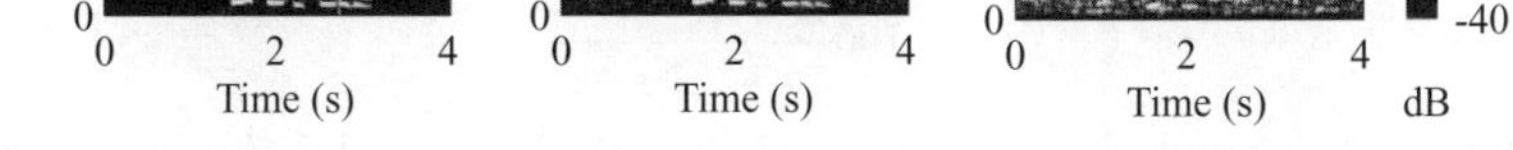

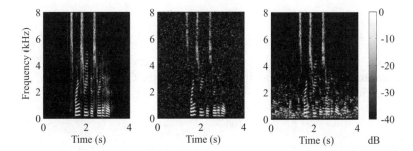

Figure 4.12 Spectrograms of the clean speech signal in Figure 4.1 (left) and of two enhanced signals (center: white Gaussian noise, right: babble noise). The corresponding noisy signals and their spectrograms are shown in Figures 4.1 and 4.2. The enhanced signals have been processed using the log-spectral amplitude estimator Equation (4.44), the MMSE noise estimator [37], and the decision-directed SNR estimator Equation (4.22).

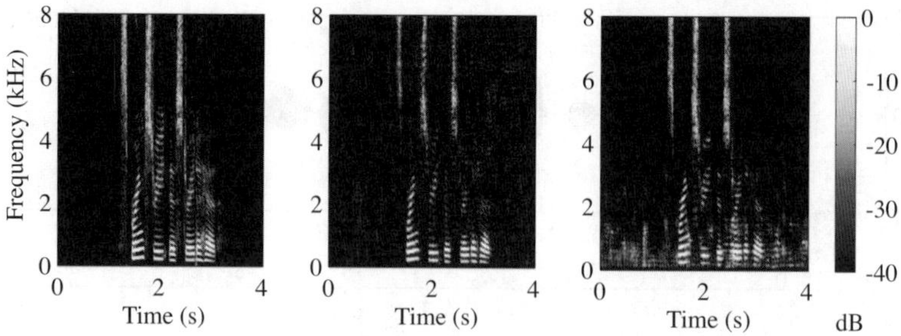

Figure 4.13 Spectrogram of the clean speech signal in Figure 4.1 (left) and of two enhanced signals (center: white Gaussian noise, right: babble noise). The enhanced signals have been processed using the log-spectral amplitude estimator Equation (4.44), the MMSE noise estimator [37], and the decision-directed SNR estimator Equation (4.22). In addition, temporal cepstrum smoothing was applied. The corresponding noisy signals and their spectrograms are shown in Figures 4.1 and 4.2.

Figure 4.13 shows the result when the temporal cepstrum smoothing is employed. From these spectrograms, we observe that for both noise types the random fluctuations in the enhanced signal are significantly reduced.

4.6 ASR Performance

To give an overview of effects, in the following, we show recognition results for a number of noise-reduction techniques. We have tested the LSA, STSA, and Wiener magnitude estimators on speech with artificially added white and babble noise. The speech data was taken from the GRID database, a small-vocabulary audiovisual command-and-control corpus (see [17]). For each utterance, the noise signal $n(\ell)$ was scaled in such a way as to yield the SNRs 0, 10, 20, and 30 dB according to

$$\mathrm{SNR} = 10 \log_{10} \frac{\sum_{\ell=\ell_0}^{\ell_1} x^2(\ell)}{\sum_{\ell=\ell_0}^{\ell_1} n^2(\ell)}, \tag{4.49}$$

where ℓ_0 and ℓ_1 are the first and the last sample of the speech signal within the file. These were obtained from the transcription files provided with the GRID corpus.

The word accuracy in percent, abbreviated by PA, was measured by

$$\mathrm{PA} = 100 \frac{N - D - S - I}{N}. \tag{4.50}$$

Here, N denotes the number of reference labels, and D, S, and I signify the number of deletions, substitutions and insertions, respectively.

ASR tests were carried out using simple left-right HMMs trained with the JASPER system [42], using feature vectors with 13 MFCCs and first and second derivatives. Furthermore, cepstral mean subtraction was applied per utterance. The output distributions were diagonal covariance GMMs with five mixture components. Training was carried out on 6000 sentences from the GRID database, using speakers $s1$ through $s10$, with Baum–Welch reestimations carried out as many times as necessary to maximize recognition rates on a 2000-sentence

Table 4.3 Recognition accuracy (PA) for unprocessed data (baseline) and after signal processing. Highly significant improvements ($P = 0.01$) are indicated in bold print.

Estimator	Noise	0 dB	10 dB	20 dB	30 dB	Clean
Baseline						
	Babble	40.9	76.9	92.9	97.4	99.0
	White	19.9	47.8	85.3	96.2	99.0
WIENER						
	Babble	**44.3**	**81.2**	**94.4**	97.8	98.9
	White	**46.0**	**80.3**	**94.4**	**98.0**	98.8
STSA						
	Babble	**49.2**	**83.2**	**94.9**	97.9	99.0
	White	**36.6**	**76.8**	**94.3**	**98.1**	99.0
LSA						
	Babble	**44.5**	**80.4**	**94.4**	97.9	98.9
	White	**43.1**	**80.0**	**95.2**	**98.3**	98.9

development set. The training was carried out on clean data without any preprocessing. Final recognition accuracies were obtained on 2000 held-out sentences from the same 10 speakers.

Table 4.3 shows recognition results for unprocessed speech, and for speech after noise reduction using the Wiener, STSA and LSA estimators, respectively. For all noise-reduction methods, the minimum *a priori* SNR was set to –20dB.

For all scenarios up to 30 dB in white and up to 20 dB in babble noise, the improvements of all three estimators are significant at the 1% level, when considering the null-hypothesis that the recognition performance is unchanged by preprocessing.

Additionally, one can observe one trend at least for the two noise types considered—whereas the STSA estimator shows the best overall performance for the nonstationary babble noise, the Wiener and LSA estimators generally reach better recognition rates for the white noise case.

Most importantly, there is no need for retraining the model on processed speech, as none of the considered estimators lead to significant losses of accuracy even when testing the system on clean data.

4.7 Conclusions

In this chapter, we have introduced some of the most prominent techniques for VAD, noise estimation, and signal enhancement. In the preprocessing stages of the ASR system, these algorithms go hand in hand and are used either for frame dropping or adaptive noise-reduction filtering. An advantage of these lies in the fact that they, at least to some extent, alleviate the need to train the recognizer on the noisy speech and thus increase the range of applications for ASR. Typical improvements that can be achieved with single-channel noise-reduction preprocessing are in the order of $20 \pm 10\%$ in an SNR range of 0–20 dB.

It is interesting to note that noise-reduction preprocessing for human consumption and for ASR does follow similar goals. For both applications, highly intelligible speech is important.

Therefore, distortions to the target signal must be avoided, especially when there is no opportunity to train either the auditory system or the ASR on processed speech data. Secondly, undesirable random fluctuations in the processed signals must be avoided. In the acoustic signal, these manifest in the form of annoying *musical noise* while they may lead to a mismatch of variances in the ASR. In fact, it has been shown that by a proper control of the variance of the processed signal both a high acoustic quality and a high ASR performance can be achieved [9].

Despite all this progress, a lot of work remains to be done, especially when scenarios with nonstationary noise or competing speakers are considered. Some recent developments such as data-driven approaches, [25,29], or the use of uncertainty-of-observation techniques, [5], have shown promising results in such contexts.

References

[1] 3GPP, *TS 26.194 Adaptive Multi-Rate - Wideband speech codec, Voice Activity Detector, V6.0.0, 3GPP*, 2004.

[2] A. Accardi and R. Cox, "A Modular Approach to Speech Enhancement with an Application to Speech Coding," in *Proceedings of the IEEE International Conference on Acoustics, Speech Signal Processing (ICASSP)*, vol. 1, pp. 201–204, March 1999.

[3] I. Andrianakis and P. White, "MMSE Speech Spectral Amplitude Estimators with Chi and Gamma Speech Priors," in *Proceedings of the IEEE International Conference on Acoustics, Speech Signal Processing (ICASSP)*, vol. 3, pp. 1068–1071, 2006.

[4] I. Andrianakis and P. White, "Speech Spectral Amplitude Estimators using Optimally Shaped Gamma and Chi Priors," *Speech Communication*, vol. 51, pp. 1–14, 2009.

[5] R. F. Astudillo, D. Kolossa, P. Mandelartz, and R. Orglmeister, An uncertainty propagation approach to robust ASR using the ETSI advanced front-end, *IEEE JSTSP Special Issue on Speech Processing for Natural Interaction with Intelligent Environments*, vol. 4, pp. 824–833, 2010.

[6] M. Berouti, R. Schwartz, and J. Makhoul, "Enhancement of Speech Corrupted by Acoustic Noise," in *Proceedings of the IEEE International Conference on Acoustics, Speech Signal Processing (ICASSP)*, pp. 208–211, 1979.

[7] S. Boll, "Suppression of Acoustic Noise in Speech Using Spectral Subtraction," *IEEE Transactions on Acoustics, Speech and Signal Processing*, vol. 27, pp. 113–120, 1979.

[8] C. Breithaupt and R. Martin, "MMSE Estimation of Magnitude-Squared DFT Coefficients with Supergaussian Priors," in *Proceedings of the IEEE International Conference on Acoustics, Speech Signal Processing (ICASSP)*, vol. I, pp. 896–899, 2003.

[9] C. Breithaupt and R. Martin, "Statistical Analysis and Performance of DFT Domain Noise Reduction Filters for Robust Speech Recognition," in *Proceedings of the 9th International Conference on Spoken Language Processing (ICSLP)*, pp. 365–368, 2006.

[10] C. Breithaupt and R. Martin, "Voice Activity Detection in the DFT Domain Based on a Parametric Noise Model," in *Proceedings of the International Workshop on Acoustic Echo and Noise Control (IWAENC)*, 2006.

[11] C. Breithaupt and R. Martin, "Analysis of the Decision-Directed SNR Estimator for Speech Enhancement With Respect to Low-SNR and Transient Conditions," *IEEE Transactions on Audio, Speech and Language Processing*, vol. 19, no. 2, pp. 277–289, 2011.

[12] C. Breithaupt, T. Gerkmann, and R. Martin, "Cepstral Smoothing of Spectral Filter Gains for Speech Enhancement without Musical Noise," *IEEE Signal Processing Letters*, vol. 14, no. 12, pp. 1036–1039, 2007.

[13] C. Breithaupt, T. Gerkmann, and R. Martin, "A Novel A Priori SNR Estimation Approach Based on Selective Cepstro-Temporal Smoothing," in *Proceedings of the IEEE International Conference on Acoustics, Speech Signal Processing (ICASSP)*, pp. 4897–4900, 2008.

[14] C. Breithaupt, M. Krawczyk, and R. Martin, "Parameterized MMSE Spectral Magnitude Estimation for the Enhancement of Noisy Speech," in *Proceedings of the IEEE International Conference on Acoustics, Speech Signal Processing (ICASSP)*, pp. 4037–4040, 2008.

[15] D. Brillinger, *Time Series: Data Analysis and Theory*. San Francisco, CA: Holden-Day, 1981.

[16] I. Cohen, "Noise Estimation in Adverse Environments: Improved Minima Controlled Recursive Averaging," *IEEE Transactions on Speech and Audio Processing*, pp. 466–475, 2003.

[17] M. Cooke, J. Barker, S. Cunningham, and X. Shao, "An audio-visual corpus for speech perception and automatic speech recognition," *Journal of the Acoustical Society of America*, vol. 120, no. 5, pp. 2421–2424, November 2006.

[18] J.R. Deller, J.H.L. Hansen, and J.G. Proakis, "Discrete-time processing of speech signals." New York, NY: IEEE Press, 2000.

[19] E. Diethorn, "Subband Noise Reduction Methods for Speech Enhancement," in *Audio Processing for Next-Generation Multimedia Communication Systems*, Boston, MA: Kluwer, 2004.

[20] J. Egan, *Signal Detection Theory and ROC Analysis*. New York, NY: Academic, 1975.

[21] T. Eltoft, "The Rician Inverse Gaussian Distribution: A New Model for Non-Rayleigh Signal Amplitude Statistics," *IEEE Transactions on Image Processing*, vol. 14, no. 11, pp. 1722–1735, 2005.

[22] Y. Ephraim and I. Cohen, "Recent Advancements in Speech Enhancement," in *The Electrical Engineering Handbook*, R. Dorf, Ed. Boca Raton, FL: CRC Press, Ch. 15, 2006.

[23] Y. Ephraim and D. Malah, "Speech Enhancement Using a Minimum Mean-Square Error Short-Time Spectral Amplitude Estimator," *IEEE Transactions on Acoustics, Speech and Signal Processing*, vol. 32, no. 6, pp. 1109–1121, Dec. 1984.

[24] Y. Ephraim and D. Malah, "Speech Enhancement Using a Minimum Mean-Square Error Log-Spectral Amplitude Estimator," *IEEE Transactions on Acoustics, Speech and Signal Processing*, vol. 33, no. 2, pp. 443–445, April 1985.

[25] J. Erkelens, J. Jensen, and R. Heusdens, "A data-driven approach to optimizing spectral speech enhancement methods for various error criteria," *Speech Communication*, vol. 49, no. 7–8, pp. 530–541, 2007.

[26] J. Erkelens, R. Hendriks, R. Heusdens, and J. Jensen, "Minimum Mean-Square Error Estimation of Discrete Fourier Coefficients with Generalized Gamma Priors," *IEEE Transactions on Audio, Speech and Language Processing*, vol. 15, no. 6, pp. 1741–1752, 2007.

[27] T. Esch and P. Vary, "Efficient Musical Noise Suppression for Speech Enhancement Systems," in *Proceedings of the IEEE International Conference on Acoustics, Speech Signal Processing (ICASSP)*, pp. 4409–4412, 2009.

[28] *ETSI Standard document, Speech Processing, Transmission and Quality Aspects (STQ); Distributed Speech recognition; Front-end feature extraction algorithm; Compression algorithms, ETSI ES 202 050 v1.1.5 (2007-01)*, ETSI, January 2007.

[29] T. Fingscheidt, S. Suhadi, and K. Steinert, "Environment-optimized Speech Enhancement," *IEEE Transactions on Audio, Speech and Language Processing*, vol. 16, no. 4, pp. 825–834, 2008.

[30] M. Fujimoto and K. Ishizuka, "Noise robust voice activity detection based on switching Kalman filter," O*IEICE Transactions on Information and System*, vol. E91-D, pp. 467–477, 2008.

[31] T. Fukuda, O. Ichikawa, and M. Nishimura, "Improved voice activity detection using static harmonic features," in *Proceedings of the IEEE International Conference on Acoustics, Speech Signal Processing (ICASSP)*, pp. 4482–4485, Mar. 2010.

[32] S. Gazor and W. Zhang, "Speech Probability Distribution," *IEEE Signal Processing Letters*, vol. 10, no. 7, pp. 204–207, 2003.

[33] T. Gerkmann, "Statistical Analysis of Cepstral Coefficients and Applications in Speech Enhancement," Ph.D. dissertation, Ruhr-Universität Bochum, 2010.

[34] I. Gradshteyn and I. Ryzhik, *Table of Integrals, Series, and Products* (6th ed). Waltham, MA: Academic Press, 2000.

[35] J.H.L. Hansen, "Speech Enhancement Employing Adaptive Boundary Detection and Morphological Based Spectral Constraints," in *Proceedings of the IEEE International Conference on Acoustics, Speech Signal Processing (ICASSP)*, pp. 901–905, May 1991.

[36] R. Hendriks, J. Erkelens, J. Jensen, and R. Heusdens, "Minimum Mean-Square Error Amplitude Estimators for Speech Enhancement under the Generalized Gamma Distribution," in *Proceedings of the International Workshop on Acoustic Echo and Noise Control (IWAENC)*, 2006.

[37] R. Hendriks, R. Heusdens, and J. Jensen, "MMSE based Noise PSD Estimation with Low Complexity," in *Proceedings of the IEEE International Conference on Acoustics, Speech Signal Processing (ICASSP)*, pp. 4266–4269, 2010.

[38] H. Hirsch and C. Ehrlicher, "Noise Estimation Techniques for Robust Speech Recognition," in *Proceedings of the IEEE International Conference on Acoustics, Speech Signal Processing (ICASSP)*, vol. 1, pp. 153–6, 1995.

[39] T. Inoue, Y. Takahashi, H. Saruwatari, K. Shikano, and K. Kondo, "Theoretical Analysis of Musical Noise in Generalized Spectral Subtraction: Why Should Not Use Power/Amplitude Subtraction?" in *Proceedings of the European Signal Processing Conference (EUSIPCO)*, pp. 994–998, 2010.

[40] ITU-T, *Recommendation G.729 Annex B. A silence compression scheme for G.729 optimized for terminals conforming to recommendation V.70*, 1996.

[41] M. Jeub, D. Kolossa, R. Astudillo, and R. Orglmeister, "Performance analysis of wavelet-based voice activity detection," in *Proceedings of the DAGA*, pp. 407–410, 2009.

[42] D. Kolossa, J. Chong, S. Zeiler, and K. Keutzer, "Efficient manycore CHMM speech recognition for audiovisual and multistream data," in *Proceedings of Interspeech*, Makuhari, Japan, pp. 2698–2701, September 2010.

[43] P. Lockwood and J. Boudy, "Experiments with a Nonlinear Spectral Subtractor (NSS), Hidden Markov Models and the Projection, for Robust Speech Recognition in Cars," in *Proceedings of the European Conference on Speech Communication and Technology (EUROSPEECH)*, pp. 79–82, 1991.

[44] P. Lockwood and J. Boudy, "Experiments with a Nonlinear Spectral Subtractor (NSS), Hidden Markov Models and the Projection, for Robust Speech Recognition in Cars," *Speech Communication*, vol. 11, no. 2–3, pp. 215–228, 1992.

[45] P. Lockwood, J. Boudy, and M. Blanchet, "Non-Linear Spectral Subtraction (NSS) and Hidden Markov Models for Robust Speech Recognition," in *Proceedings of the IEEE International Conference on Acoustics, Speech Signal Processing (ICASSP)*, vol. I, 1992, pp. 265–268.

[46] S. Mallat, *A Wavelet Tour of Signal Processing*. San Diego: Academic Press, 1998.

[47] R. Martin, "Noise Power Spectral Density Estimation Based on Optimal Smoothing and Minimum Statistics," *IEEE Transactions on Speech and Audio Processing*, vol. 9, no. 5, pp. 504–512, Jul. 2001.

[48] R. Martin, "Speech Enhancement Using MMSE Short Time Spectral Estimation with Gamma Distributed Speech Priors," in *Proceedings of the IEEE International Conference on Acoustics, Speech Signal Processing (ICASSP)*, vol. I, pp. 253–256, 2002.

[49] R. Martin, "Speech Enhancement based on Minimum Mean Square Error Estimation and Supergaussian Priors," *IEEE Transactions on Speech and Audio Processing*, vol. 13, no. 5, pp. 845–856, 2005.

[50] R. Martin, "Statistical Methods for the Enhancement of Noisy Speech," in *Speech Enhancement*, J. Benesty, S. Makino, and J. Chen, Eds. Orlando, FL: Springer-Verlag, 2005.

[51] R. Martin, "Bias Compensation Methods for Minimum Statistics Noise Power Spectral Density Estimation," *Signal Processing, Elsevier*, vol. 86, no. 6, pp. 1215–1229, 2006.

[52] R. Martin, "An Efficient Algorithm to Estimate the Instantaneous SNR of Speech Signals," in *Proceedings of the European Conference on Speech Communication and Technology (EUROSPEECH)*, 1993, pp. 1093–1096.

[53] R. Martin, "Spectral Subtraction Based on Minimum Statistics," in *Proceedings of the European Signal Processing Conference (EUSIPCO)*, pp. 1182–1185, 1994.

[54] R. Martin and T. Lotter, "Optimal recursive smoothing of non-stationary periodograms," in *Proceedings of the International Workshop on Acoustic Echo and Noise Control (IWAENC)*, pp. 167–170, 2001.

[55] M. Marzinzik and B. Kollmeier, "Speech Pause Detection for Noise Spectrum Estimation by Tracking Power Envelope Dynamics," *IEEE Transactions on Speech and Audio Processing*, vol. 10, no. 2, pp. 109–118, 2002.

[56] K. Muller, "Computing the Confluent Hypergeometric Function, M(a,b,x)," *Numerical Mathematics*, 2001.

[57] C. Neves, A. Veiga, L. Sa, and F. Perdigao, "Efficient noise-robust speech recognition front-end based on the etsi standard," in *Ninth International Conference on Signal Processing*, pp. 609–612, 2008.

[58] C. Park, N. Kim, J. Cho, and J. Kim, "Integration of sporadic noise model in pomdp-based voice activity detection," in *Proceedings of the IEEE International Conference on Acoustics, Speech Signal Processing (ICASSP)*, pp. 4486–4489, March 2010.

[59] T. Pham, "Wavelet analysis for robust speech processing and applications," PhD dissertation, Graz University of Technology, Austria, 2007.

[60] C. Plapous, C. Marro, L. Mauuary, and P. Scalart, "A Two-step Noise Reduction Technique," in *Proceedings of the IEEE International Conference on Acoustics, Speech Signal Processing (ICASSP)*, pp. 289–292, 2004.

[61] E. Plourde and B. Champagne, "Integrating the Cochlea's Compressive Nonlinearity in the Bayesian Approach for Speech Enhancement," in *Proceedings of the European Signal Processing Conference (EUSIPCO)*, pp. 70–74, 2007.

[62] J. Porter and S. Boll, "Optimal Estimators for Spectral Restoration of Noisy Speech," in *Proceedings of the IEEE International Conference on Acoustics, Speech Signal Processing (ICASSP)*, pp. 18A.2.1–18A.2.4, 1984.

[63] R. Preuss, "A Frequency Domain Noise Cancelling Preprocessor for Narrowband Speech Communication Systems," in *Proceedings of the IEEE International Conference on Acoustics, Speech Signal Processing (ICASSP)*, pp. 212–215, 1979.

[64] L. R. Rabiner and M. R. Sambur, "An algorithm for determining the endpoints of isolated utterances," *Bell System Technical Journal*, vol. 54, no. 2, pp. 297–315, Feb. 1975.

[65] J. Ramirez, J. Segura, C. Benitez, A. de la Torre, and A. Rubio, "An Effective Subband OSF-Based VAD with Noise Reduction for Robust Speech Recognition," *IEEE Transactions on Speech and Audio Processing*, vol. 13, no. 6, pp. 1119–1129, 2005.

[66] J. Ramírez, J. C. Segura, C. Benítez, Á. de la Torre, and A. Rubio, "Efficient Voice Activity Detection Algorithms using Long-term Speech Information," *Speech Communication*, vol. 42, pp. 271–287, 2004.

[67] J. Ramírez, J. C. Segura, J. M. Górriz, and L. García, "Improved voice activity detection using contextual multiple hypothesis testing for robust speech recognition," *IEEE Transactions on Audio, Speech and Language Processing*, vol. 15, no. 8, pp. 2177–2189, 2007.

[68] O. Segawa, K. Takeda, and F. Itakura, "Continuous speech recognition without end-point detection," *IEEE International Conference on Acoustics, Speech and Signal Processing*, vol. 1, pp. 245–248, 2001.

[69] J. Shaojun, G. Haitao, and Y. Fuliang, "A new algorithm for voice activity detection based on wavelet transform," in *Proceedings of the International Symposium on Intelligent Multimedia, Video and Speech Processing*, pp. 222–225, October 2004.

[70] J. Sohn and W. Sung, "A Voice Activity Detector Employing Soft Decision Based Noise Spectrum Adaptation," in *Proceedings of the IEEE International Conference on Acoustics, Speech Signal Processing (ICASSP)*, vol. 1, pp. 365–368, 1998.

[71] J. Sohn, N. Kim, and W. Sung, "A Statistical Model-Based Voice Activity Detector," *IEEE Signal Processing Letters*, vol. 6, no. 1, pp. 1–3, 1999.

[72] J. Taghia, J. Taghia, N. Mohammadiha, J. Sang, V. Bouse, and R. Martin, "An Evaluation of Noise Power Spectral Density Estimation Algorithms in Adverse Acoustic Environments," in *Proceedings of the IEEE International Conference on Acoustics, Speech Signal Processing (ICASSP)*, pp. 4640–4643, 2011.

[73] L. N. Tan, B. Borgstrom, and A. Alwan, "Voice activity detection using harmonic frequency components in likelihood ratio test," in *Proceedings of the IEEE International Conference on Acoustics, Speech Signal Processing (ICASSP)*, pp. 4466–4469, 2010.

[74] D. Van Compernolle, "Noise adaptation in a hidden Markov model speech recognition system," *Computer Speech & Language*, vol. 3, pp. 151–167, 1989.

[75] P. Vary, "Noise Suppression by Spectral Magnitude Estimation - Mechanism and Theoretical Limits," *Signal Processing, Elsevier*, vol. 8, pp. 387–400, 1985.

[76] P. Vary and R. Martin, *Digital Speech Transmission: Enhancement, Coding and Error Concealment*. Chichester, England: John Wiley & Sons, 2006.

[77] P. Welch, "The use of fast fourier transform for the estimation of power spectra: A method based on time averaging over short, modified periodograms," *IEEE Transactions on Audio and Electroacoustics*, vol. 15, no. 2, pp. 70–73, 1967.

[78] N. Wiener, *Extrapolation, Interpolation, and Smoothing of Stationary Time Series with Engineering Applications*. Cambridge, MA: The MIT Press, 1949.

[79] C. You, S. Koh, and S. Rahardja, "β-order MMSE Spectral Amplitude Estimation for Speech Enhancement," *IEEE Transactions on Audio, Speech and Language Processing*, vol. 13, no. 4, pp. 475–486, 2005.

[80] D. Yu, L. Deng, J. Droppo, J. Wu, Y. Gong, and A. Acero, "Robust speech recognition using a cepstral minimum-mean-square-error-motivated noise suppressor," *IEEE Transactions on Audio, Speech and Language Processing*, vol. 16, no. 5, pp. 1061–1070, 2008.

[81] R. Yu, "A Low-complexity Noise Estimation Algorithm Based on Smoothing of Noise Power Estimation and Estimation Bias Correction," in *Proceedings of the IEEE International Conference Acoustics, Speech Signal Processing (ICASSP)*, pp. 4421–4424, 2009.

5

Extraction of Speech from Mixture Signals

Paris Smaragdis
University of Illinois at Urbana-Champaign, USA

5.1 The Problem with Mixtures

Traditionally, signal-processing and pattern-recognition algorithms tend to look at signals under the assumption of little, if any, interference. The vast majority of algorithms for speech recognition, pitch detection, phonetic classification, etc., assume that the input is a relatively clean speech signal, potentially contaminated by a simple noise term such as additive Gaussian noise. The reason for that tendency is partially pedagogical; one should not only know how to treat a clean signal before moving to more complex cases but also a result of being limited in our abilities to mathematically analyze signals. Once we are confronted with mixture signals a lot of our signal processing intuition and mathematical foundations no longer apply directly, and there is little algorithmic basis to map well-defined operations on clean speech to cases of speech plus interference.

A practical way out of this problem is that of considering preprocessing steps that attempt to separate mixed signals and provide a reasonably clean version of the speech component. Once that is obtained, and under the assumption that the interference is reasonably well removed, one can perform operations which assume a clean speech input which we can now provide.

Historically the field of source separation has seen many approaches based on a varying range of schools of thought ranging from logic-based systems, to neuroscience-inspired models, to pure mathematical operations and many more. In this chapter we will not try to make any extensive survey of these approaches, there is simply no space to do them all justice. Instead we will examine two successful approaches to extracting speech from mixtures, which have been well received because of their relative simplicity and strong performance. In the remaining of this chapter we will focus on multichannel source separation based on Independent Component Analysis (ICA), and on monaural source separation based on spectral factorization methods.

Techniques for Noise Robustness in Automatic Speech Recognition, First Edition.
Edited by Tuomas Virtanen, Rita Singh, and Bhiksha Raj.

5.2 Multichannel Mixtures

Multichannel sound mixtures are the types of recordings we obtain when we employ microphone arrays. Using a set of microphones which are sampled synchronously we can capture an audio scene from multiple locations and then take advantage of the differences between all the individual microphone recordings to try to isolate any speech in the recorded mixture. There is a rich literature on approaches that employ microphone arrays as described in Chapter 6. Here, we will examine a specific subset of that work, one that is called *blind source separation*. The name of this area arises from the fact that, unlike standard microphone array methods, we do not assume to know anything about the recording situation (such as microphone positions, source statistics, etc.) and in effect we operate blindly. This is, of course, a very constraining situation which we will later relax for our purposes; however, the basic principle of providing minimal information and constructing a system that "sorts things out" on its own will still remain.

5.2.1 Basic Problem Formulation

The initial formulation of the multichannel mixture case will be one that is known as the *instantaneous mixture case*. The instantaneous mixing model is an extreme simplification of the mixing process, but one that helps us form a strong foundation to attack the more complex problem at hand.

An instantaneous mixture of N sounds $s_i[t]$ as recorded by N microphones is assumed to be

$$x_i[t] = \sum_j a_{i,j} s_j[t]. \tag{5.1}$$

The signals $x_i[t]$ are what each microphone records and the values $a_{i,j}$ represent gain factors that model the attenuation of each sound as it gets recorded by each microphone. Pictorially this is explained in Figure 5.1 where we sketch the case of two sound sources are recorded from two microphones. All microphones will get to record all the sounds, but there will be a difference on how loud each sound is captured from every microphone depending on their relative distance. Sound sources that are close to a microphone will be get recorded louder, whereas more distance sources will be softer.

In order to make the notation more convenient we will rewrite the above model in matrix form as

$$\mathbf{x}[t] = \mathbf{A} \cdot \mathbf{s}[t], \tag{5.2}$$

where the vector $\mathbf{x}[t] = [x_1[t], x_2[t], \ldots, x_N[t]]^\top$ represents the recorded multichannel signal, the matrix $\mathbf{A}(i, j) = a_{i,j}$ contains the source gains and is known as the mixing matrix, and the vector $\mathbf{s}[t] = [s_1[t], s_2[t], \ldots, s_N[t]]^\top$ are the signals of the original sources. Given this form we can clearly see the problem. We will have access to what the microphones record ($\mathbf{x}[t]$), but that would be a mixed set of sources. The true sources $\mathbf{s}[t]$ are obviously not observed, nor do we know the gain values in the mixing matrix $\mathbf{A}$ since the location of the sources relative to the microphones is also unknown. Ideally if we are able to somehow estimate the values of the mixing matrix $\mathbf{A}$, then we can use an estimate of its inverse $\mathbf{W}$ to obtain an estimate of the sources by

$$\hat{\mathbf{s}}[t] = \mathbf{W} \cdot \mathbf{x}[t], \tag{5.3}$$

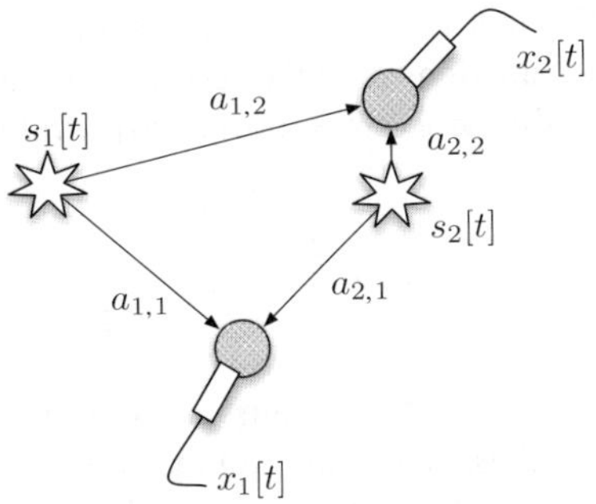

Figure 5.1 An illustrative example of a 2 × 2 instantaneous mixture. The sound sources are denoted by $s_j[t]$ and the microphone recordings by $x_i[t]$. Depending on the distance between the sources and the microphones a gain change is imposed and denoted by $a_{i,j}$. Due to proximity, $s_2[t]$ will be recorded louder than $s_1[t]$ in microphone 2; therefore, the value of $a_{2,2}$ will be relatively larger than the rest.

and obtain a clean version of the speech signal(s) that are in the recording. The matrix **W** is often referred to as the unmixing matrix. Alas the above process will involve solving an equation with three variables in which two are unknown, so there is little hope of using conventional algebra to help here. A solution to this problem came by clever use of statistics and is now known as ICA.

Independent Component Analysis

The above problem in its pure form is one that is found in many applications in signal processing and its solution was sought after for a long time. The term itself was first coined in [1], although similar approaches had been presented prior to that. In the mid-1990s after a series of publications on efficient and robust approaches to achieve ICA [2,3,4] the above problem was effectively solved to a satisfying degree and some of its first successful applications were on the problem of separating mixed sounds as presented above.

The key observation we need to make in order to solve the above equation is that the time series $s_i[t]$, being sounds that emanate from physically decoupled systems such as different speakers, are going to be *statistically independent*. As such if try to solve the problem by finding the maximally statistically independent set of $\hat{\mathbf{s}}[t]$ then we stand a chance at recovering the original sources. However, this optimization is not straightforward. In this chapter we will go through a simple sketch of the basic idea behind ICA, but we will note that there are dozens of different ways to solve the ICA problem and there are far more rigorous approaches to describing the process than what we will use here. The interested reader can refer to [5] for an excellent collection of information on the history and wide variety of this field.

We start by a simple way to state statistical independence for two random variables. We will consider two random variables x_1 and x_2 to be independent if

$$E\{g_1(x_1)g_2(x_2)\} - E\{g_1(x_1)\}E\{g_2(x_2)\} = 0, \tag{5.4}$$

for all element-wise continuous functions $g_i(\cdot)$ that are zero outside a finite interval. Therefore as a starting point we would want to find an algorithm that finds values for **W** such that the above constraint is satisfied (or at least best approximated) between all our source estimates

$\hat{s}_i[t]$. Taking advantage of the fact that most normal sound recordings are zero mean and constraining ourselves to odd functions $g_i(\cdot)$ the second term in the above equation becomes zero and we are left with a simple expression that can evaluate the independence across multiple variables:

$$D(\mathbf{s}) = E\left\{\prod_i g_i(\hat{s}_i[t])\right\}. \tag{5.5}$$

In the above expression minimizing $D(\mathbf{s})$ results in increasing the statistical independence between the signals in $\mathbf{s}[t]$. Needless to say the fact that we require this to hold for an infinite set of functions $g_i(\cdot)$ is not making the problem easier. However, in practice, if we specify a well-chosen set of $g_i(\cdot)$ it is sufficient to achieve our goal. This type of optimization is known as *nonlinear decorrelation* and there are various ways to solve it. Curiously, whether we start with this objective in mind or many other ones that also strive for independence, we find that a specific form of solution always comes up. This solution is in the form of an iterative algorithm performing gradient descent shown in Algorithm 1.

Algorithm 1: ICA via nonlinear decorrelation

ε {Value defining when parameter updates are not significant anymore}
$\mathbf{W} \leftarrow \mathbf{I}$ {Initial conditions assume no transform is needed}
$0 < \mu < 1$ {Learning rate parameter}
repeat
  for all t do
    $\mathbf{y}[t] = \mathbf{W} \cdot \mathbf{x}[t]$
    $\Delta\mathbf{W} = \left[\mathbf{I} - g(\mathbf{y}[t]) \cdot \mathbf{y}[t]^{\mathsf{T}}\right] \cdot \mathbf{W}$
    $\mathbf{W} \leftarrow \mathbf{W} + \mu\Delta\mathbf{W}$
  end for
until $||\Delta\mathbf{W}|| < \epsilon$

In practice choosing $g(x) = \tanh(x)$ yields good results for sound sources, although other function combinations can also be used, often depending on our knowledge of the statistics of the sources, or one could estimate the optimal nonlinearities as well [6]. The parameter μ is a learning rate parameter that tempers the gradient updates to avoid jerky updates. The parameter ε is a threshold that specifies when the update to $\mathbf{W}$ is not significant and further iterations are unnecessary.

ICA for Separating Sound Sources

Let us now demonstrate an example with speech signals. Consider the two signals $s_1[t]$ and $s_2[t]$ in Figure 5.2. These are spoken sentences from a male and female speaker. We mix these two recordings by performing

$$\mathbf{x}[t] = \mathbf{A} \cdot \mathbf{s}[t] = \begin{pmatrix} x_1[t] \\ x_2[t] \end{pmatrix} = \begin{pmatrix} 2 & 1 \\ 1 & 1 \end{pmatrix} \cdot \begin{pmatrix} s_1[t] \\ s_2[t] \end{pmatrix}. \tag{5.6}$$

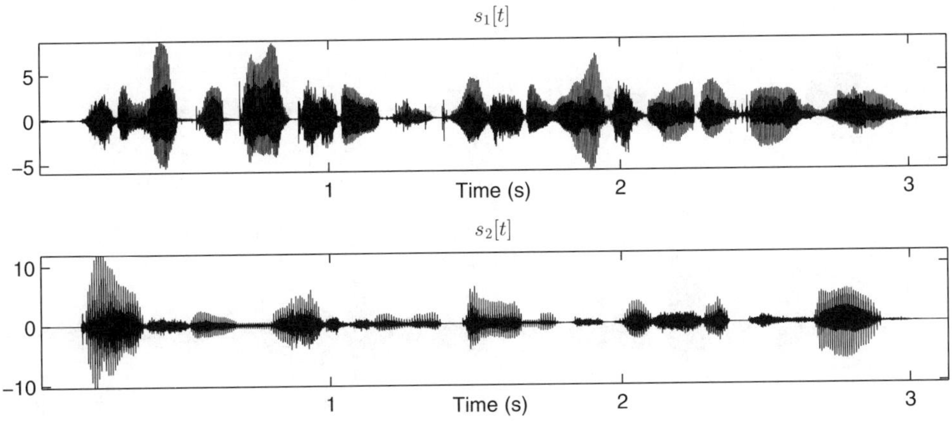

Figure 5.2 The two original speech recordings used in the instantaneous separation example.

The mixed time series $x_i[t]$ are shown in Figure 5.3. Based only on these two signals we use the ICA algorithm presented above and after 200 iterations we estimate the unmixing matrix

$$\mathbf{W} = \begin{pmatrix} 1.0378 & -1.0363 \\ -1.2425 & 2.4847 \end{pmatrix}, \tag{5.7}$$

which when multiplied by the mixing matrix $\mathbf{A}$ results in

$$\mathbf{W} \cdot \mathbf{A} = \begin{pmatrix} -1.2425 & 2.4847 \\ 1.0378 & -1.0363 \end{pmatrix} \cdot \begin{pmatrix} 2 & 1 \\ 1 & 1 \end{pmatrix} = \begin{pmatrix} -0.0003 & 1.2422 \\ 1.0393 & 0.0015 \end{pmatrix}, \tag{5.8}$$

a matrix with approximately only one significant value per row (or column). This means that multiplication with the unmixing matrix "undoes" the mixing effects of the mixing matrix and results in an estimate of the original sources $s_i[t]$. That estimate is shown in Figure 5.4,

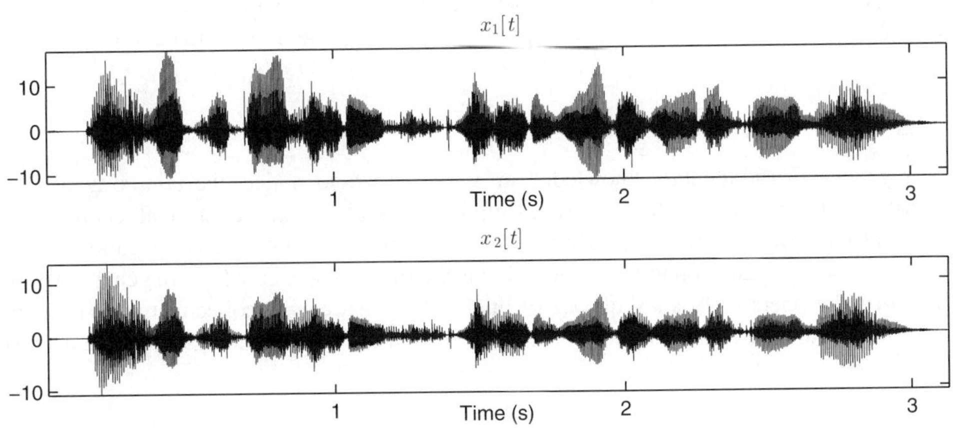

Figure 5.3 The multichannel mixed speech recording used in the instantaneous separation example.

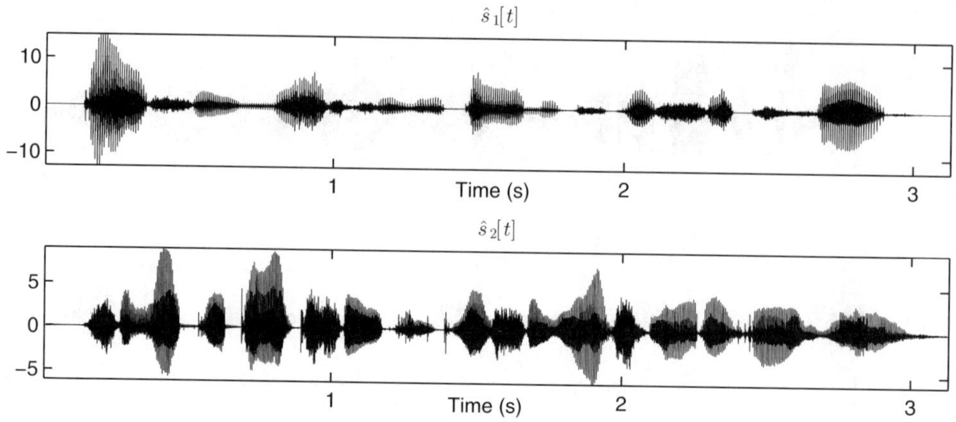

Figure 5.4 The separated speech signals used in the instantaneous separation example.

where by visual inspection one can (perhaps) see that the two original speech signals have been separated. However there are two disturbing observations. We note that the order of the two signals has been swapped, that is $\hat{s}_1[t] \approx s_2[t]$ and $\hat{s}_2[t] \approx s_1[t]$, and furthermore that the gain of the separated version of $s_2[t]$ has been increased (something we can also see in the above equation where the elements of significant magnitude are not both close to 1). These two observations illustrate problem that plagues ICA algorithms, the output estimates will be arbitrarily permuted and scaled. The reason why this happens is that as defined the independence criterion that we optimize is invariant to these two operations. If a set of signals are mutually statistically independent, changing their order or gain will not change that relationship and all possible scalings and permutations will be just as independent as any other. In this particular example the scaling issue is not a problem since we can always renormalize the estimated outputs to a gain that we prefer. However, the source permutation problem is more significant because in cases where we would try to separate, say, speech from noise we would not know which of the two output signals is the speech signal and which is the noise. This is, of course, a problem that can be easily resolved by using simple classifiers, but as we see later on it can introduce more severe problems which are not as easy to deal with.

5.2.2 *Convolutive Mixtures*

As anyone with rudimentary knowledge in acoustics would notice, the unmixing process described in the previous section is far too simplistic to work in real life. In a real recording the original sound sources will go through a series of transformations such as propagation delays, room reverberation and nonuniform microphone responses (see Figure 5.5). In general we can think of these operations as a sequence of linear filters that can be collectively described by one convolution operation. Taking this in mind we can update the mixing model we used so far as

$$x_i[t] = \sum_{j}^{N} \sum_{k}^{T} h_{i,j}[k] s_j[t-k], \tag{5.9}$$

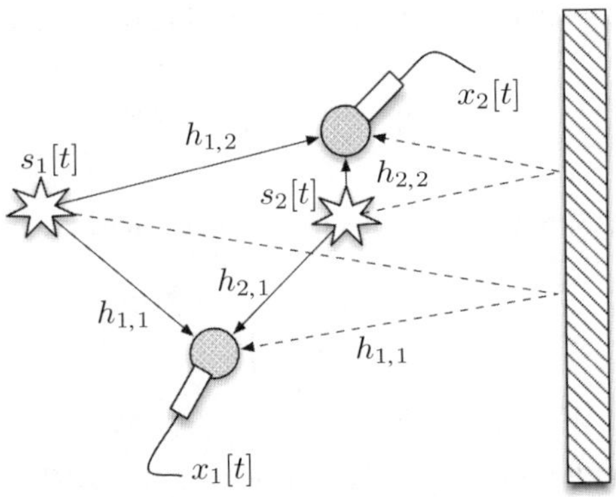

Figure 5.5 An illustrative example of a 2×2 convolutive mixing. The sound sources are denoted by $s_j[t]$ and the microphone recordings by $x_i[t]$. Each microphone records not only the direct source signal but also its reflections from other objects in that setting. In addition to that there is filtering that takes place in the microphone, as well a propagation delay due to the fact that sound transfer is not instantaneous. All these parameters are encapsulated in the mixing filters $h_{i,j}$. The dashed lines are example reflections from the two sources as they bounce of a wall (striped structure on the right). In a real situation there will be a high number of reflections from all objects in the setting, most of them being higher order reflections (i.e., reflections of already reflected signals).

where the T-point length filters $h_{i,j}$ encapsulate the transformations that source signal j undergoes as it travels from its origin to microphone i. This model is considerably more complex than the instantaneous case, but we note that it is still a linear problem, and that we can use the same principles as before. This time we will strive to find a set of *unmixing filters* that will "undo" the effects of the mixing filter $h_{i,j}$, but we still use independence as a objective to strive for.

Although there are ways to solve this problem in the time domain by directly inverting the equation above [7,8], we will instead transform this problem into a collection of instantaneous mixing problems. Consider the application of a length L short-time DFT [9] on one of the sources:

$$S_j[f,t] = \text{DFT}\{S_j[t,\ldots,t+L-1]^{\mathsf{T}}\}, \tag{5.10}$$

where $S_j[f,t]$ is the Fourier coefficient for frequency f at time t. In that domain we can express the convolution with a corresponding filter $h_{i,j}$ as an element-wise multiplication in the frequency domain:

$$\sum_k^T h_{i,j}[k]s_j[t-k] \equiv H_{ij}[f]S_j[f,t], \tag{5.11}$$

where $H_{ij}[f]$ is the fth frequency of the DFT of h_{ij}. We have to make sure that the DFT size is long enough to accommodate the convolution operation. This is done by the use of zero padding when taking the Fourier transform in order to avoid circular convolution [10]. With this new viewpoint we can rewrite the convolutive mixing problem as:

$$X_i[f,t] = \sum_j^N H_{ij}[f] S_j[f,t] \tag{5.12}$$

or we can use matrix notation to rewrite this mixing once more as:

$$\begin{pmatrix} X_1[f,t] \\ \vdots \\ X_N[f,t] \end{pmatrix} = \begin{pmatrix} H_{1,1}[f] & \cdots & H_{1,N}[f] \\ \vdots & \ddots & \vdots \\ H_{N,1}[f] & \cdots & H_{N,N}[f] \end{pmatrix} \cdot \begin{pmatrix} S_1[f,t] \\ \vdots \\ S_N[f,t] \end{pmatrix} \Rightarrow \mathbf{x}_f[t] = \mathbf{H}_f \cdot \mathbf{s}_f[t], \tag{5.13}$$

where the f subscript indicates the fth frequency of the DFT of the respective signal. Therefore, $\mathbf{x}_f[t]$ is a vector containing the fth frequency bin of the DFT of each recorded signal at time t, $\mathbf{H}_f$ is a matrix containing the fth bin of all the filters, and $\mathbf{s}_f[t]$ contains that frequency bin of the original sources.

We can now make the observation that the equation above is the same as that of the instantaneous mixing model, except that now we have to consider this mixing on each possible frequency independently (Figure 5.6). Since we use an L-point DFT this means that we will have $L/2+1$ unique instantaneous mixing problems to solve. Solving these problems will result in estimating a set of frequency unmixing matrices $\mathbf{G}_f$ which will be the inverse of their corresponding $\mathbf{H}_f$. Multiplying the mixed frequency components with these matrices will yield a set of separated frequency components which we can then transform back to the time domain with an inverse short-time Fourier transform (STFT).

The benefits of performing this operation in the frequency domain are twofold. First we decompose a complex problem to a set of simpler problems which we can easily solve, and second we can take advantage of the fast Fourier transform to efficiently perform all the necessary convolutions that this operation implies. In realistic scenarios we are likely to require long convolutions, in the thousands of filter taps, where time-domain operations can

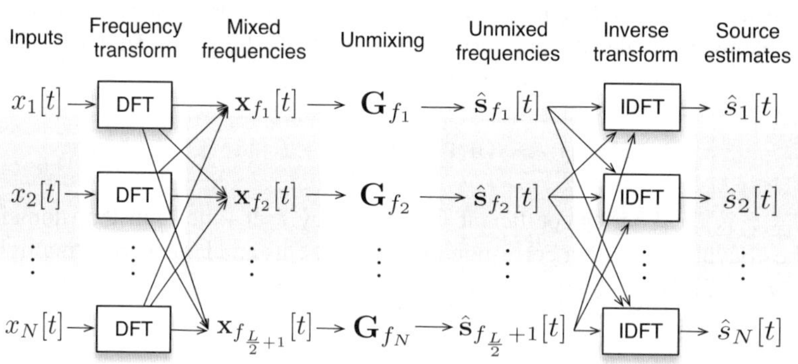

Figure 5.6 Illustration of ideal convolutive unmixing in the frequency domain. The input signals are being decomposed to multiple frequency bands which are then individually separated using the instantaneous unmixing model we introduced in Section 5.2.1. Once this unmixing has been performed in the frequency domain we can transform the unmixed frequencies back to the time domain to obtain the estimated source signals.

be comparatively very inefficient. By making use of the frequency domain we can speed up these computations considerably.

Complications in the Frequency Domain

After we reformulated the problem of convolutive mixing as a set of instantaneous mixing problems, it appears that we are now able to solve this problem since it has been reduced to something we already know how to solve. However there are a few complications that we need to address, both relating the ambiguities in the ICA algorithm.

As we mentioned previously there is no guarantee regarding the scaling and the order of the outputs of ICA. In our frequency domain separation context this means that each frequency band will be arbitrarily scaled and that even though the frequency components will be separated, we can not be sure that their order will be the same across all of the frequency bands. In the case of the scaling issue this will result in an arbitrary equalization of the outputs where the spectral character of the original sources can be severely distorted. In the case of the permutation we will notice that some frequency bands of an output will contain one source, but others will contain another source. Resolving these problems is easy for scaling, but hard for the permutations.

In order to resolve the scaling problem we can take a number of different steps. A simple approach is to ensure that all the the unmixing matrices $\mathbf{G}_f$ preserve the energy of their input signals. This implies that the determinant of all $\mathbf{G}_f$ is set to be 1; a scaling we can easily perform once we have estimated these matrices [11]. Performing this operation ensures that the energy coming out of each ICA transform is comparable to what came in; therefore we alleviate cases in which some unmixing matrices would boost their input by a large factor, whereas others would suppress it. Although this does not guarantee that we will have no scaling issues, it will alleviate most obvious audible problems by ensuring that no frequency bands get a disproportionate boost of attenuation imposed on them. More sophisticated approaches have also been explored, such as scalings of $\mathbf{G}_f$ that try to maximize the similarity of the output source spectra to the microphone input spectra, or postprocessing that shapes the spectra of the output sources according to predetermined priors (e.g., if we know that the source is a speech signal we can then equalize the source estimate to have the average spectrum of speech).

The permutation problem is a more complex issue which is not as easy to address. Finding the correct source permutation for all output frequencies is a combinatorial optimization problem which quickly becomes practically intractable in a realistic situation. There have been multiple solutions proposed to resolve this problem, ranging from simple approximations to more sophisticated systems. In terms of simple solutions one thing to take advantage of is the continuity of real-world objects in the frequency domain. This statement implies that the sequence of unmixing matrices has some smoothness, that is $\mathbf{G}_f$ is somewhat similar to $\mathbf{G}_{f+1}$. An obvious way to take advantage of this property is to permute each unmixing matrix so that we maximize its similarity to its neighboring unmixing matrices across frequencies. Although practical for two sources this quickly becomes an inefficient approach when we have more sources and we require more extensive searching. Another approach along these lines is to bias the adaptation of the unmixing matrices so that neighboring structure is taken into account [12]. An example of such an operation is to average the current estimate of $\mathbf{G}_f$ with $\mathbf{G}_{f-1}$ and $\mathbf{G}_{f+1}$

so at to bias their convergence towards similar solutions thus achieving a permutation with a smooth spectrum. Although this can work in practice it results in significant contamination of the proper estimates and a substantial loss of separation quality. More sophisticated approaches take into account additional information such as the location of the sensors. Armed with that information one can treat the situation at hand as a microphone array problem and examine the spatial locations that the filters $\mathbf{G}_f$ steer the array towards. Upon doing so we can cluster the separating matrices in terms of the directions that they are pointing the microphone array to. By appropriately rearranging them so that all the directions are consistent for each source estimate we can obtain a solution that is geometrically correct and does not focus on multiple sources for each output. More details of this approach and frequency domain ICA-based separation are shown in [13], and the effects of these approaches on speech recognition are examined in [14].

An example of this process is shown in Figures 5.7 and 5.8. The input is a simple convolutive mixture of two speech recordings. In Figure 5.7 we display the estimated and the corrected directional responses that highlight the problem with the inconsistent permutations across frequencies. These plots show how the response of each source estimator across frequency and angle of incidence. Ideally a source estimator would focus consistently on the same angle

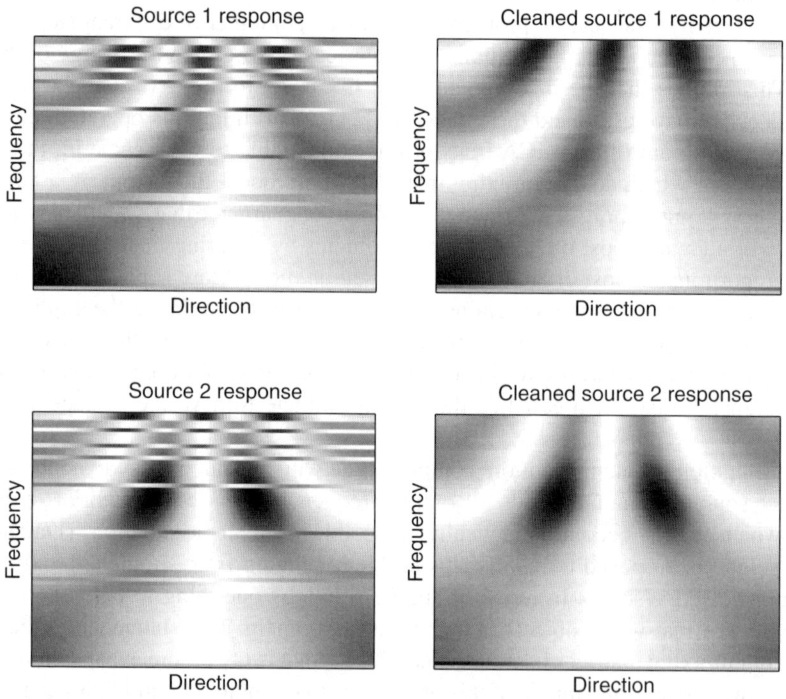

Figure 5.7 Learned directional responses from an example convolutive mixture. The two patterns on the left denote the array's response to a sound as it pertains to the two discovered sources across all frequencies and sound incidence angles. Note how the array's response changes abruptly across certain frequencies indicating permutation problems. In the right we display the beampatterns after we correct for permutation problems in our unmixing matrices.

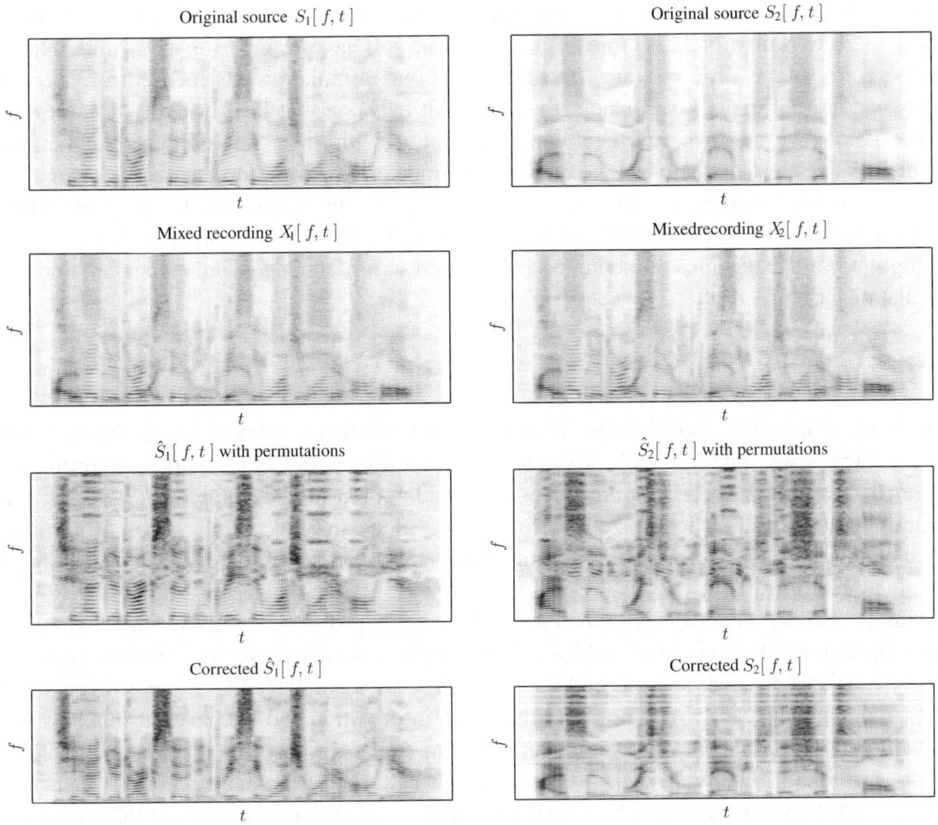

Figure 5.8 A convolutive separation example. The top two spectrograms show the original input sounds, the plot pair under that shows the recorded mixtures from two microphones. Note how elements from both sounds exist in both of the recordings now. The third row displays the result of performing ICA in each frequency without correcting for permutation errors. Note how in the left plot elements of the interfering sound are present in some frequency bands. Likewise these bands are erroneous in the second source estimate. The bottom row shows the corrected source estimates when attempting to correct the beam patterns to point to consistent directions across all frequencies. Also note how the source estimates have more high frequency energy in the case of the first source, and a set of frequency notches in the second source. These are effects that are due to the scaling problem when estimating each frequency component separately.

throughout all the frequencies. Due to physical constraints in the design of arrays we will see sidelobes that form in the response for higher frequencies, but we should always expect to see a consistent vertical pattern that implies a frequency-wide response towards a fixed angle which is where the targeted source is expected to be. As shown in Figure 5.7 the response patterns when we have permutation problems create an inconsistent pattern which disappears once we choose the right permutation. In Figure 5.8 we display the spectrograms of all the involved signals where again we can see the effects of permutation errors and some of the effects of incorrect scaling.

Although these approaches can work well for reasonable mixing situations they do suffer from a serious limitation. It is quite unlikely that one will have access to as many microphones as there are sources in a mixture. Because so far our formulation has focused on invertible cases we did not explicitly consider that case, but it is one that is very likely to confront a practitioner. The case where one attempts separation with fewer microphones than sources (which is the most realistic case), is called the undetermined mixtures case and is one that has been studied widely, and attacked with a variety of approaches. Due to space limitations we will not be covering the details of these approaches, but we will note that they are rather straightforward extensions of the material presented so far. The interested reader can find a lot of that material in [17].

5.3 Single-Channel Mixtures

In the previous sections we considered the case where we had multiple microphones sampling an auditory scene and we took advantage of the differences between their recordings in order to focus on a specific source. However a limitation of these approaches is that they require multiple time-synchronized microphones, which can be a costly proposition for many applications. An alternative problem formulation is that of attempting separation from a single-channel recording.

Single-channel separation is a much more challenging problem because its objective cannot be easily defined. In the case of multiple-channel signals we could use the differences between the channels to discover sources and the entire process was reduced to a linear inversion. In the single-channel case we do not have that extra information and we have to somehow pick out only elements that belong to one source from a single waveform, a mathematically ill-defined problem. Historically this problem has been attacked from the perception perspective (see Chapter 16) where researchers attempted to extract parts of the mixture that seemed to be correlated enough to imply that they belong to the same source. How and when these parts matched was based on principles of psychoacoustics, the study of human hearing. Although intuitive, this approach makes it hard to construct compact mathematical models and complicates any subsequent implementations. A more recent trend in single-channel separation is one that uses training data to assist an algorithm in selecting the proper source in a mixture. In this chapter we will examine a specific area of this approach, that based on nonnegative spectral factorizations.

5.3.1 Problem Formulation

Let us consider the example mixture shown as a spectrogram in Figure 5.9. It contains two sounds that of a speaker and that of a siren. Using a time-frequency representation of that mixture, we can visualize the two sounds in, the wavy pattern being the siren and the finer texture in the background being the speech signal. The mixture $x[t]$ itself is defined as

$$x[t] = \sum_{i}^{N} s_i[t], \tag{5.14}$$

where the time series $s_i[t]$ are the actual sources that comprise it (in the presented case the speech and the siren). Note that unlike the multichannel case we only observe one recording.

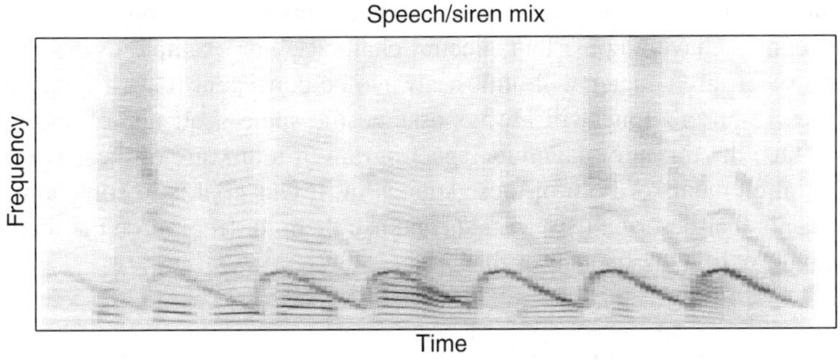

Figure 5.9 A mixture of a speech and a siren sound. The wavy pattern in this spectrogram is the siren sound as it oscillates in frequency. The finer texture superimposed on that is the recording of a female speaker. Single-channel separation is the process of automatically picking components of such an input that belong to one source only. Although this seems like a tractable task in in this case, in many other situations it becomes a very difficult problem to resolve manually.

We also do not make use of scalar gains for each source since for our purposes an arbitrarily scaled version of any source is the same. The estimation problem at hand is now to directly estimate all the $s_i[t]$ from only $x[t]$. Obviously this is a very ill-defined problem since we are asked to estimate N times more data than we observe.

Theoretically one can manually select the parts of that spectrogram that belong to the undesirable sound, set their energy to zero and then invert that representation in order to obtain a cleaned version of the target sound (see use of binary masks in Chapter 16). However this kind of matching can be extremely tedious and in more realistic mixture cases very hard to perform. In order to get some help in performing this task we will take advantage of examples of how the sources in the mixture sound like, and then use that information to help us perform the separation. Unlike the approach presented in the previous sections this is not "blind" anymore since we will use a lot of information outside of the input data.

We will start by examining this problem directly in the time-frequency domain. We will use a STFT representation of the mixture sound $x[t]$:

$$X[f,t] = \mathrm{DFT}\{x[t,\ldots,t+L-1]^{\mathsf{T}}\}. \tag{5.15}$$

Since the STFT representation is linear it will hold that

$$x[t] = \sum_i^N s_i[t] \Rightarrow X[f,t] = \sum_i^N S_i[f,t], \tag{5.16}$$

where $S_i[f,t]$ are the STFT's of the original sources $s_i[t]$. We will now make an assumption that it also holds that

$$|X[f,t]| = \sum_i^N |S_i[f,t]|. \tag{5.17}$$

Even though this is not a mathematically correct statement it is in practice approximately correct and a widely used approximation in this field.

We can now make a few key observations about this new mixing model. In general a sound class tends to have a consistent spectral character. For example in the above mixture the speaker's spectral character will almost always be consistent (e.g., a nasal high-pitched voice), whereas a siren's tone will always assume the same sequence of spectral states in succession. Thus, the mixture magnitude spectrogram of a mixture of these two sounds will contain spectral elements of both of these sounds in it. Our goal is to somehow capture the spectral elements that describe these sounds and use them to decompose the mixture in parts that spectrally look like the speaker or the siren.

5.3.2 Learning Sound Models

In order to learn sound models we need some training data. We thus obtain sound examples of the classes of sounds that we expect to find in the mixture we try to resolve. For example, in the above case where we know that the mixture consists of a speech and a siren signal we can obtain some clean recordings of speech and sirens. These recordings are only supposed to be indicative of the timbre of the intended outputs and are of course different instances from the ones in the mixture. We denote these training sounds as $r_i[t]$ and we obtain their magnitude spectrograms:

$$R_i[f,t] = |\mathrm{DFT}\{r_i[t,\ldots,t+L-1]^{\mathsf{T}}\}|. \tag{5.18}$$

We will additionally denote the magnitude spectrograms of the training data as a set of matrices $\mathbf{R}_i = R_i[f,t]$.

We now wish to somehow learn the salient spectral elements of these two sounds. In order to do so we will use a matrix-factorization technique called Nonnegative Matrix Factorization (NMF) [16]. This matrix decomposition takes a nonnegative $M \times N$ matrix $\mathbf{X}$ and approximates it as a product of a nonnegative $M \times K$ matrix $\mathbf{W}$ and a nonnegative $K \times N$ matrix $\mathbf{H}$. The user needs to specify the value of K which is the rank of the decomposition. The matrix $\mathbf{W}$ contains a set of column vectors that describe the vertical structure of $\mathbf{X}$ and the matrix $\mathbf{H}$ contains a set or row vectors that describe the horizontal structure. Alternatively we can think of the $\mathbf{W}$ matrix as a set of bases and $\mathbf{H}$ as their corresponding activations. In the audio case the matrix $\mathbf{W}$ contains a set of spectral shapes in its columns and matrix $\mathbf{H}$ contains their corresponding activation across time in its rows. This problem can be solved in a variety of ways; however for our purposes the most broad interpretation comes from [15]. A central problem in the estimation the two matrix factors is that we need to define a proper cost function for NMF, that is specify what we mean by approximating $\mathbf{X}$. In order to get maximal breadth we can define the difference between $\mathbf{X}$ and $\mathbf{W} \cdot \mathbf{H}$ as

$$D(\mathbf{X}|\mathbf{W}\mathbf{H}) = \left\| \frac{\mathbf{X}^{\beta} + (\beta-1)(\mathbf{W}\cdot\mathbf{H})^{\beta} - \beta\mathbf{X} \odot (\mathbf{W}\cdot\mathbf{H})^{\beta-1}}{\beta(\beta-1)} \right\|, \tag{5.19}$$

where all the exponentiations are element-wise and $\odot$ denotes element-wise multiplication. The parameter β is key here since it help shape the cost function. Some notable cases are when $\beta = 2$, in which case the above expression becomes proportional to the Euclidean distance between $\mathbf{X}$ and $\mathbf{W} \cdot \mathbf{H}$, when $\beta = 1$ where the resulting cost function becomes something akin to the Kullback–Leibler divergence, and when $\beta = 0$ where the distance becomes the Itakura–Saito divergence. Consequently, the choice of β informs the noise model that we assume in this model which for the above cases would be Gaussian, Dirichlet and Gamma distributed, respectively. The choice of β is up to the user, and in theory it dictates the kind of input that $\mathbf{X}$ represents, magnitude spectra for $\beta = 1$, power spectra for $\beta = 0$ as shown in [15]. In this chapter we generate all subsequent examples using magnitude spectra and $\beta = 1$. Somewhat varying results can be achieved with other values of β, although qualitatively there is no significant change. In order to estimate $\mathbf{W}$ and $\mathbf{H}$ the iterative algorithm shown in Algorithm 2 is used.

Algorithm 2: NMF algorithm

β {Parameter to specify the type of cost function used}
ε {Desired maximum value for $D(\mathbf{X}|\mathbf{WH})$}
k {Rank of approximation}
$\mathbf{X} \in \mathbb{R}_{\geq 0}^{m,n}$ {Input to approximate}
$\mathbf{W} \in \mathbb{R}_{\geq 0}^{m,k}, \mathbf{W}_{i,j} \sim U[0,1]$ {Random initial conditions}
$\mathbf{H} \in \mathbb{R}_{\geq 0}^{k,n}, \mathbf{H}_{i,j} \sim U[0,1]$ {Random initial conditions}
repeat

$$\mathbf{H} \leftarrow \mathbf{H} \odot \frac{\mathbf{W}^\top((\mathbf{W}\cdot\mathbf{H})^{\beta-2} \odot \mathbf{X})}{\mathbf{W}^\top(\mathbf{W}\cdot\mathbf{H})^{\beta-1}}$$

$$\mathbf{W} \leftarrow \mathbf{W} \odot \frac{((\mathbf{W}\cdot\mathbf{H})^{\beta-2} \odot \mathbf{X})\mathbf{H}^\top}{(\mathbf{W}\cdot\mathbf{H})^{\beta-1}\mathbf{H}^\top}$$

Nornalize columns of $\mathbf{W}$ to sum to 1
until $D(\mathbf{X}|\mathbf{WH}) < \epsilon$

To best illustrate how this decomposition works consider the nonnegative factorization of the siren spectrogram in Figure 5.10 using $K = 5$. This is a different recording of the same sound that is in the mixture. Upon factorizing it we see that the matrix $\mathbf{W}_{\mathrm{siren}}$ contains a set of spectral bases, and the matrix $\mathbf{H}_{\mathrm{siren}}$ tells us how these get activated in time. Similarly we can decompose the speech recording from the speaker in the original mixture. The results of that are shown in Figure 5.11. Note how the two $\mathbf{W}$ matrices encapsulate the spectral structure of the two sound classes. The components in $\mathbf{W}_{\mathrm{speech}}$ describe different aspects of the speaker such as pitched harmonic parts (the three middle components), or wide-band elements (first and last component). Likewise $\mathbf{W}_{\mathrm{siren}}$ contains a simple series of tones which when put in order can describe the siren sound. We note here that given ample training data, the $\mathbf{W}$ spectral basis matrices are presumed to be indicative of all spectral states that a learned sound class assumes and can constitute a model of that sound class. In general that information can be obtained by using a few seconds of training data (usually 30–60 seconds for most cases), although this is something that can vary depending on the nature of the sounds to model and the diversity of the training data.

5.3.3 Separation by Spectrogram Factorization

If we are confronted with a mixture of two sounds and we already have learned a set of dictionaries for the constituent sound classes, we can speculate that the elements contained

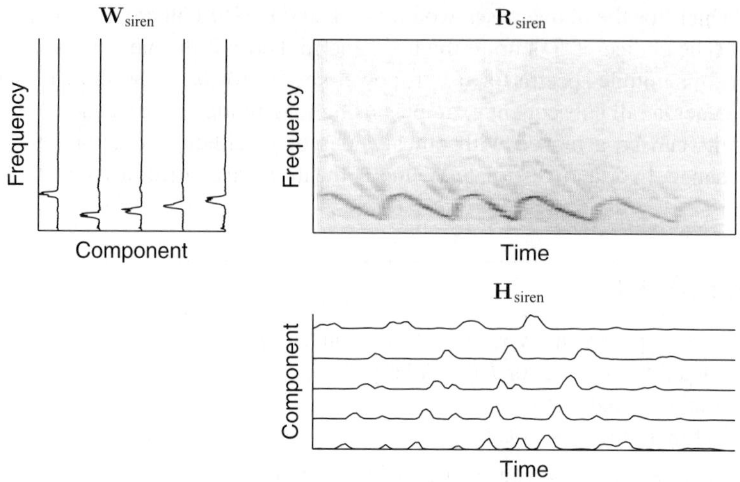

Figure 5.10 Nonnegative spectral factorization of a siren sound. The input sound is the top right plot, the columns of the spectral bases matrix $\mathbf{W}_{\mathrm{siren}}$ are on the top left and the rows of the matrix $\mathbf{H}_{\mathrm{siren}}$ are shown at the bottom plot. Note how the columns of $\mathbf{W}_{\mathrm{siren}}$ capture some of the spectral states of the siren sound. Likewise the $\mathbf{H}_{\mathrm{siren}}$ matrix tell us how we can mix the elements of $\mathbf{W}$ in order to approximate the input $\mathbf{R}_{\mathrm{siren}}$.

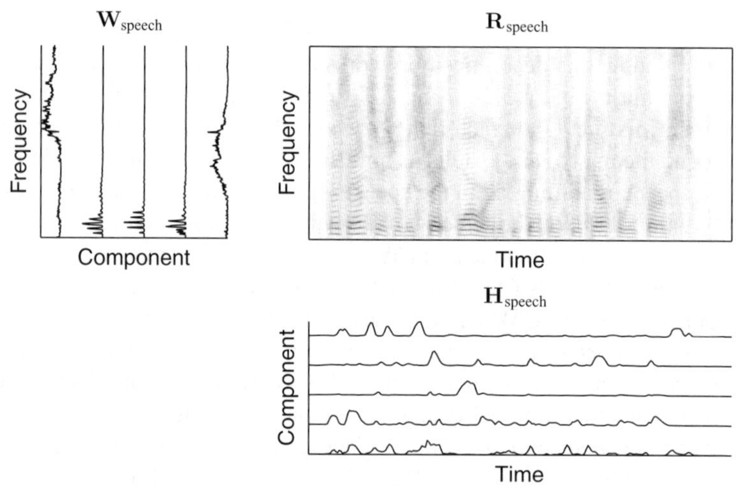

Figure 5.11 Nonnegative spectral factorization of a speech sound. The input sound is the top right plot, the columns of the spectral bases matrix $\mathbf{W}_{\mathrm{speech}}$ are on the top left and the rows of the matrix $\mathbf{H}_{\mathrm{speech}}$ are shown at the bottom plot. Note how the columns of $\mathbf{W}_{\mathrm{speech}}$ capture some of the spectral states of the speech sound.

in the learned $\mathbf{W}_i$ matrices can be used to reconstruct the mixture. In order to perform this approximation we need to estimate $\mathbf{H}$ in the following equation:

$$\mathbf{X} = (\mathbf{W}_1, \ldots, \mathbf{W}_N) \cdot \mathbf{H}, \tag{5.20}$$

where $\mathbf{X}$ is the magnitude spectrogram of the mixture sound. This is of course a simpler instance of NMF in which we only have to estimate the $\mathbf{H}$ matrix. We can do this with the update rule shown previously by keeping the values in $\mathbf{W}$ fixed and updating only $\mathbf{H}$. Just as the matrix $\mathbf{W}$ is segmented by blocks that correspond to spectral bases for each sound source, likewise the matrix $\mathbf{H}$ contains a similar segmentation. In order to keep the correspondence with the bases in $\mathbf{W}$ the matrix $\mathbf{H}$ can be interpreted as

$$\mathbf{H} = \begin{pmatrix} \mathbf{H}_1 \\ \vdots \\ \mathbf{H}_N \end{pmatrix}, \tag{5.21}$$

where $\mathbf{H}_i$ contains the activations for source i.

Given this segmentation we can rewrite the mixing model above as

$$\mathbf{X} = \left(\mathbf{W}_1, \ldots, \mathbf{W}_N \right) \cdot \begin{pmatrix} \mathbf{H}_1 \\ \vdots \\ \mathbf{H}_N, \end{pmatrix} = \sum_i \mathbf{W}_i \cdot \mathbf{H}_i. \tag{5.22}$$

If we find the optimal values for the matrix $\mathbf{H}$ we will essentially know where in time and how much of each classes' components we would have to use to approximate the input mixture. Given such a source-specific explanation of the mixture we could then use only one source's bases and activations in order to reconstruct only the part of the input that this particular source contributed. Thus, in order to estimate the contribution of source i we would perform

$$\mathbf{X}_i \approx \mathbf{W}_i \cdot \mathbf{H}_i, \tag{5.23}$$

where $\mathbf{X}_i$ would correspond to the estimated magnitude spectrogram of the isolated source i. However, this estimate is potentially incomplete and is not guaranteed to contain all of the input's energy. Since this is a low-rank approximation of the input it is very likely that low-energy sections have not been adequately represented and might be missing when we reconstruct. This problem can be addressed by using the following source approximation instead which allocates all of the input's energy to the resulting source spectrograms:

$$\mathbf{X}_i \approx (\mathbf{W}_i \cdot \mathbf{H}_i) \odot \left(\frac{\mathbf{X}}{\sum_j \mathbf{W}_j \cdot \mathbf{H}_j} \right). \tag{5.24}$$

Once the individual source magnitude spectrograms $\mathbf{X}_i$ are obtained we can combine them with the original phase spectrogram of the mixture in order to invert this representation back to a time domain signal. This can be done naively by simply modulating the original phase spectrogram $\mathbf{P}$ of the mixture with the source magnitude spectrogram approximations:

$$\mathbf{F}_i = \mathbf{X}_i \odot \left(e^{\mathbf{P}\sqrt{-1}} \right), \tag{5.25}$$

Algorithm 3: NMF-based separation

$\mathbf{X} \leftarrow$ magnitude spectrogram of mixture
$\mathbf{P} \leftarrow$ phase spectrogram of mixture
$\mathbf{X}_i \leftarrow$ magnitude spectrogram of training data for source i
$\mathrm{NMF}() \leftarrow$ Algorithm 2
for $i = 1$ to N **do**
  $\mathbf{W}_i \leftarrow \mathrm{NMF}[\mathbf{X}_i]$
end for
$\mathbf{W} = [\mathbf{W}_1, ..., \mathbf{W}_N]$
$\mathbf{H} \leftarrow \mathrm{NMF}[\mathbf{X}]$ given $\mathbf{W}$
for $i = 1$ to N **do**
  $\mathbf{H}_i \leftarrow$ rows of $\mathbf{H}$ corresponding to $\mathbf{W}_i$
  $\mathbf{F}_i = (\mathbf{W}_i \cdot \mathbf{H}_i) \odot \left(e^{\mathbf{P}\sqrt{-1}}\right) \odot \left(\frac{\mathbf{X}}{\mathbf{W} \cdot \mathbf{H}}\right)$
  $source_i \leftarrow \mathrm{ISTFT}[\mathbf{F}_i]$
end for

---

where the exponentiation is element-wise. The complex-valued spectrograms $\mathbf{F}_i$ can then be inverted back to the time domain using the inverse short-time Fourier transform in order to produce the waveforms of the separated sounds. The overall process is shown in Algorithm 3.

Let us now examine how this operation performs in the case of the siren and the speech mixture in Figure 5.9. In this example we used 50 bases for each of the sounds. In general if we do not use any regularization, more than a few hundred bases will result in an increased risk that the bases will assume a very basic shape that can also model other sources and thus result in source confusion during the separation process. Using too few bases will not give the models the required flexibility to approximate the sources and the result will be suboptimal. Choosing 50 bases per source as we did is a good conservative estimate which works for most cases. The training data consisted of 20 seconds of speech from the same speaker as the one in the mix, and 2 seconds of the siren sound. The sounds that made up the mixture were not used in the training data. The purpose of the training data is to represent the spectral characteristics of the two sources in the mixture. Because of the repetitive and almost deterministic spectral character of the siren we only needed 2 seconds, whereas the more varying speech signal necessitated a larger training set. The estimated outputs are shown in Figure 5.12. Using our knowledge of the true inputs to the mixture we find that the signal to interference ratio was about 14 dB while maintaining the proper spectral character of the target source.

This is, of course, a contrived example which is designed to illustrate how this approach works. As realistic as it might have been it contained two sources which are spectrally very different thus making their separation an easy task. Had we been confronted with a mixture of two speech recordings then the task would be harder since the two basis sets that describe each source would be very similar and discerning between the two sources would be a more ambiguous process. However, even in that case it is not uncommon to obtain a signal to interference ratio of 10 dB or more, assuming that the two speakers do not have largely identical voices [18].

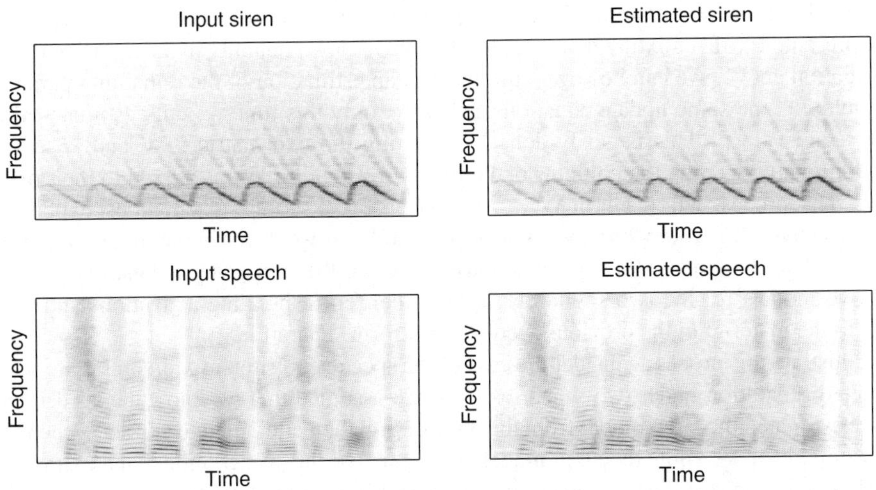

Figure 5.12 The results of performing NMF-based source separation on the mixture in Figure 5.9. The left plots show the magnitude spectrograms of the two sources before they were mixed, and the right plots show their estimates by observing only the mixture and some training data. Although one can see minor traces of the siren sound in the estimated speech spectrogram, for all practical purposes the sources were accurately recovered.

5.3.4 Dealing with Unknown Sounds

A more serious problem that arises in the case of this single-channel model is one that comes with practical deployment. In is often unrealistic to expect to have training data for all the sources that comprise a mixture. One way to address this problem is to make a couple of observations. In most applications, we will know either the kind of target source that we are interested in (e.g., to extract speech from some unknown background noise), or we will know the kind of interference to expect (e.g., a plane cockpit system will expect a specific type of ambient noises). Additionally, unlike the multichannel case it is not required that we separate all sources in a mixture. In the case of a speaker in a street we only need to treat the problem as a two sound class case where one class is the speaker and the other class includes all the other sounds (cars, background babble, etc). Based on these observation we can simplify the overall problem as that of source separation between two sources, one of which is known but the other is not.

Despite its apparent difficulty this problem can be solved in the single-channel mixture framework. The basic setup here is that we have in our disposal a training set for one of the two implied sound classes in the mixture (it can be either the target or the interference), and a mixture containing that sound class plus one more. In terms of the decomposition we used previously, we now have

$$\mathbf{X} = \mathbf{W} \cdot \mathbf{H} = \left(\mathbf{W}_{\text{target}}, \mathbf{W}_{\text{other}} \right) \cdot \begin{pmatrix} \mathbf{H}_{\text{target}} \\ \mathbf{H}_{\text{other}} \end{pmatrix}, \tag{5.26}$$

where $\mathbf{X}$ is the magnitude spectrogram of the mixture and the $\mathbf{W}$ and $\mathbf{H}$ matrices are the bases and activations of the target (subscript $target$) and the interference (subscript $other$)

respectively. If we have training data for, say, the target sound we can learn $\mathbf{W}_{\text{target}}$ offline. Once presented with a mixture that would be the only known quantity in the right-hand side of the equation above. We can however still perform an estimation of the remaining parameters by employing the same update equations as before, only this time we only estimate $\mathbf{W}_{\text{other}}$, $\mathbf{H}_{\text{target}}$ and $\mathbf{H}_{\text{other}}$. This is easily done by updating the two matrices $\mathbf{W}$ and $\mathbf{H}$ as usual, but additionally keeping the columns of $\mathbf{W}$ that correspond to $\mathbf{W}_{\text{target}}$ fixed to their original values. This process will result in $\mathbf{W}_{\text{other}}$ to converge to a set of bases that best explain what $\mathbf{W}_{\text{target}}$ can not. In other words, we will learn a model of whatever in the input mixture does not sound like the already known sound class. After all these matrices are estimated we will have two models for the two sound classes and their corresponding activations and we can revert to the material in the previous section to perform separation.

To illustrate this process consider the case of the speech and siren mixture in Figure 5.9. We attempted separation under two scenarios. First we assumed that we know the speech model and we estimated the siren model from the mixture and then attempted to extract speech. Then we assumed that we know the siren model and estimated the speech model and extracted the speech again. The resulting suppression of the siren was around 10 and 14 db, respectively. The magnitude spectrogram of the extracted speech is shown in Figure 5.13.

As these examples show this approach can be quite flexible and used in a variety of situations where we only know what the target source will be, or only what the interference source will be like. Especially in speech situations, this is a realistic expectation where a system can be optimized for only speech (or a specific speaker), or be optimized for the particular setting where the recording takes place, such as a plane cockpit, a car, etc.

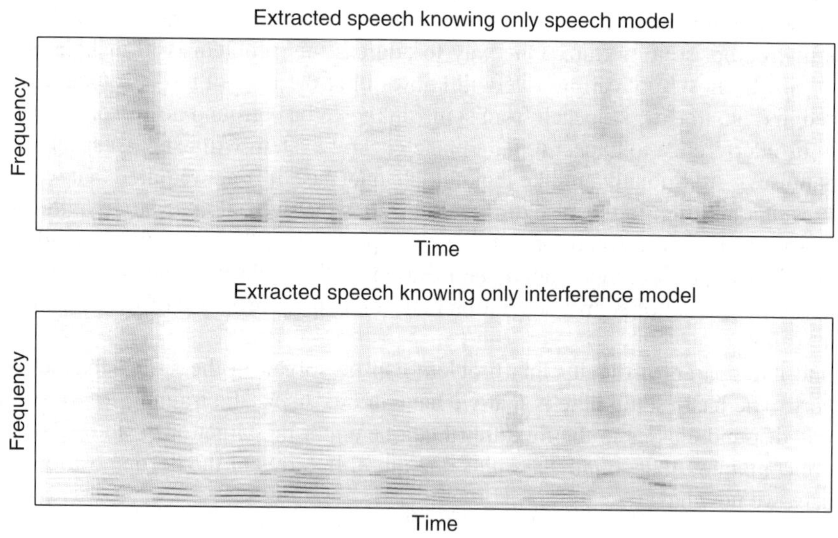

Figure 5.13 Extracted speech from the mixture in Figure 5.9. The top plot shows the extracted speech when the speech model is known and the interference is not, and the bottom plot shows the extracted speech when the interference model was known and the speech model was not. Note how the two approaches yield slightly different results, but do manage to separate the speech signal well enough.

5.4 Variations and Extensions

So far we described a simple approach to separation of speech from noise using nonnegative models. More elaborate approaches have been developed, and they can often provide additional advantages depending on the problem at hand. These approaches include convolutive models [19,20] that are able to model time-frequency elements of sources, Markovian models [21,22] which can statistically model time evolution of the sources, sparse and overcomplete models [18] that use overcomplete dictionaries to model sources and tensor methods that generalize the matrix factorization to tensors and allow multichannel formulations [23]. Other interesting directions include the statistical underpinnings of these approaches are exposed in [15,24,25,26]. Techniques such as these have been used in the field of speech recognition in order to remove unwanted sources, or to directly perform recognition on noisy signals. Some of this work is described in [27,28,29,30].

5.5 Conclusions

In this chapter we described two popular approaches in the audio source separation literature, the ICA-based multichannel approaches and the NMF-based single-channel approaches. As many fields of speech and audio research, these areas are constantly evolving and improving at an impressive rate. In this chapter, we only covered the foundations of these techniques up to a point where a reasonably simple implementation can be devised. As in all speech research the devil is in the details and it often takes considerable engineering to obtain that useful handful of dB of improvement from the otherwise easy to get to baseline. It is out hope that this chapter we set the proper foundations so that the interested reader can endeavor in this field and experiment with this very interesting area of research.

References

[1] Pierre Comon, "Independent component analysis, A new concept?," *Signal Processing*, vol. 36, no. 3, pp. 287–314, 1994.

[2] Anthony J. Bell and Terrence J. Sejnowski, "An Information-Maximization Approach to Blind Separation and Blind Deconvolution," *Neural Computation*, vol. 7, no. 6, pp. 1129–1159, 1995.

[3] Shun-Ichi Amari, Andrzej Cichocki, and Howard H. Yang, "A New Learning Algorithm for Blind Signal Separation," *Advances in Neural Information Processing Systems*, vol. 8, pp. 757–763, 1996.

[4] Jean-Francois Cardoso, "High-Order Contrasts for Independent Component Analysis," *Neural Computation*, vol. 11, no. 1, pp. 157–192, 1996.

[5] Aapo Hyvarinen. "Survey on independent component analysis," *Neural Computing Surveys*, vol. 2, pp. 94–128, 1999.

[6] Richard Everson and Stephen Roberts, "Independent Component Analysis: A Flexible Nonlinearity and Decorrelating Manifold Approach," *Neural Computation*, vol. 11, no. 8, pp. 1957–1983, 1999.

[7] Te-Won Lee, Anthoy J. Bell, and Reinhold Orglmeister, "Blind Source Separation of Real World Signals," *IEEE International Conference Neural Networks*, Houston, 1997.

[8] Douglas Scott C., Andrzej Cichocki, and Shun-Ichi Amari, "Multichannel blind separation and deconvolution of sources with arbitrary distributions," *Proceedings of the IEEE Workshop on Neural Networks for Signal Processing*, pp. 436–445, 1997.

[9] Lawrence Rabiner, Jont Allen, "Short-time Fourier analysis tecniques for FIR system identification and power spectrum estimation," *IEEE Transactions on Acoustics, Speech and Signal Processing Volume*, vol. 27, no. 2, pp. 182–192, 1979.

[10] Alan V. Oppenheim and Ronald W. Schafer, "Discrete-time signal processing," Prentice Hall Signal Processing Series. Upper Saddle River, NJ, USA: Prentice-Hall, Inc., 1989.
[11] Paris Smaragdis. "Information Theoretic Approaches to Source Separation," Masters Thesis, MAS Department, Massachusetts Institute of Technology, 1997.
[12] Paris Smaragdis, "Blind Separation of Convolved Mixtures in the Frequency Domain," *International Workshop on Independence and Artificial Neural Networks,* University of La Laguna, Tenerife, Spain, February 9–10, 1998.
[13] H. Sawada, R. Mukai, S. Araki, S. Makino, "A Robust and Precise Method for Solving the Permutation Problem of Frequency-Domain Blind Source Separation," *IEEE Transactions on Speech and Audio Processing*, vol. 12, no. 5, pp. 530–538, September 2004.
[14] Hiroshi Saruwatari, Satoshi Kurita, Kazuya Takeda, Fumitada Itakura, Tsuyoki Nishikawa, and Kiyohiro Shikano, "Blind source separation combining independent component analysis and beamforming," *EURASIP Journal on Applied Signal Processing*, vol. 2003, pp. 1135–1146, January 2003.
[15] Cedric Fevotte, Nancy Bertin, and Jean-Louis Durrieu, "Nonnegative matrix factorization with the Itakura-Saito divergence: With application to music analysis," *Neural Computation*, vol. 21, no. 3, pp. 793–830, March 2009.
[16] Daniel D. Lee and H. Sebastian Seung, "Algorithms for Non-negative Matrix Factorization," in *Neural Information Processing Systems*, pp. 556–562, 2000.
[17] Shoji Makino, Te-Won Lee, and Hiroshi Sawada, "Blind Speech Separation," The Netherlands: Springer, September 2007.
[18] Paris Smaragdis, Madhusudana Shashanka, and Bhiksha Raj, "A sparse non-parametric approach for single channel separation of known sounds," in *Neural Information Processing Systems*, pp. 1705–1713, 2009.
[19] Tuomas Virtanen, "Separation of Sound Sources by Convolutive Sparse Coding," *ISCA Tutorial and Research Workshop on Statistical and Perceptual Audio Processing, SAPA*, 2004.
[20] Paris Smaragdis, "Convolutive Speech Bases and Their Application to Supervised Speech Separation," *IEEE Transactions on Audio, Speech and Language Processing*, vol. 15, no. 1, pp. 1–12, 2007.
[21] Gautham. J. Mysore, Paris Smaragdis, and Bhiksha Raj. "Non-negative Hidden Markov Modeling of Audio with Application to Source Separation," in *Proceedings of the International Conference on Latent Variable Analysis and Signal Separation (LVA / ICA)*, St. Malo, France, September 2010.
[22] Alexey Ozerov, Cedric Fevotte, and Maurice Charbit, "Factorial scaled hidden Markov model for polyphonic audio representation and source separation," *IEEE Workshop on Applications of Signal Processing to Audio and Acoustics (WASPAA'09)*, Mohonk, NY, October 18–21, 2009.
[23] Alexey Ozerov, Cedric Fevotte, Raphael Blouet, and Jean-Louis Durrieu, "Multichannel nonnegative tensor factorization with structured constraints for user-guided audio source separation," *IEEE International Conference on Acoustics, Speech and Signal Processing (ICASSP'11)*, Prague, May 2011, pp. 257–260.
[24] Madhusudana V. Shashanka, Bhiksha Raj, and Paris Smaragdis, "Probabilistic Latent Variable Models as Non-Negative Factorizations," *Computational Intelligence and Neuroscience*, May 2008.
[25] Ali Taylan Cemgil, "Bayesian Inference for Nonnegative Matrix Factorisation Models," *Computational Intelligence and Neuroscience*, vol. 1, no. 2, 2009.
[26] Ali Taylan Cemgil, Umut Simsekli, and Yusuf Cem Subakan, "Probabilistic Latent Tensor Factorization Framework for Audio Modeling," in *IEEE WASPAA*, 2011.
[27] Bhiksha Raj, Tuomas Virtanen, Sourish Chaudhuri, and Rita Singh, "Non-negative matrix factorization based compensation of music for automatic speech recognition," in *Interspeech*, 2010.
[28] Jort F. Gemmeke, Tuomas Virtanen, and Antti Hurmalainen, "Exemplar-based sparse representations for noise robust automatic speech recognition," in *IEEE Transactions on Audio, Speech and Language Processing*, vol. 19, no. 7, pp. 2067–2080, 2011.
[29] Felix Weninger, Jurgen Geiger, Martin Wollmer, Bjorn Schuller, and Gerhard Rigoll, "The Munich 2011 CHiME Challenge Contribution: NMF-BLSTM Speech Enhancement and Recognition for Reverberated Multisource Environments," in *International Workshop on Machine Listening in Multisource Environments*, Florence, Italy, September 1, 2011.
[30] Paris Smaragdis and Bhiksha Raj, "The Markov selection model for concurrent speech recognition," *IEEE International Workshop on Machine Learning for Signal Processing (MLSP)*, Kittila, Finland, August 2010.

6

Microphone Arrays

John McDonough[1], Kenichi Kumatani[2]
[1] *Carnegie Mellon University, USA*
[2] *Disney Research, USA*

This contribution takes as its objective the class of techniques suitable for performing speech recognition, not on the signal capture by a single microphone, but on that obtained by combining the signals from several microphones. The techniques discussed here differ from those presented in Chapter 5 in that they are based on the pair of assumptions that:

1. The geometry of the array of microphones is fixed and known.
2. The position of the active speakers relative to the array are known or can be accurately estimated.

Such techniques—known collectively as *beamforming*—have been the subject of intense interest in recent years within the acoustic array processing research community. Unfortunately, such techniques have been largely ignored in the mainstream automatic speech-recognition field, although this may rapidly change given the recent release and widespread popularity of the Microsoft Kinect® platform. The simplest of beamforming algorithms, the *delay-and-sum beamformer*, uses only this geometric knowledge—that is the arrangement of the microphones and the speaker's position—to compensate for the time delays of the signals arriving at each sensor and then additively combine them. More sophisticated *adaptive beamformers* minimize the total output power of the array under the constraint that the desired source must be unattenuated. The conventional adaptive beamforming algorithms attempt to minimize a quadratic optimization criterion related to signal-to-noise ratio (SNR). However, recent research has revealed that such quadratic criteria are not optimal for acoustic beamforming of human speech. Hence, we also present beamformers based on nonconventional optimization criteria that have appeared more recently in the literature. In particular, recent research has revealed that useful optimization criteria can be devised by attempting to restore the non-Gaussian statistical characteristics present in uncorrupted or "clean" speech. As these characteristics are diminished through the introduction of noise or reverberation, the use of adaptive beamforming

Techniques for Noise Robustness in Automatic Speech Recognition, First Edition.
Edited by Tuomas Virtanen, Rita Singh, and Bhiksha Raj.

techniques to restore the original statistical characteristics reduces the effect of these distortions, and hence improves speech-recognition performance.

A second research trend upon which we will report is the growing use of *spherical* microphone arrays. The literature on array processing with spherical arrays differs from the "conventional" array processing literature in that it attempts to explicitly account for diverse acoustic phenomena, namely, the diffraction of sound around a solid sphere, as well its scattering from such an object. While diffraction and scattering are present in all acoustic array processing applications, the conventional literature takes them into account only through the calculation of second-order statistics (SOS) between pairs of sensors. In the spherical array literature, on the other hand, these effects are incorporated into the theoretical analysis.

A third trend upon which we report is the combination of adaptive beamforming techniques developed for conventional arrays with the acoustic theory developed for spherical arrays. This all important research direction, which has only very recently appeared, will, in the opinion of the current authors, dominate the field in the coming years and decades.

The balance of this contribution is organized as follows. We begin in Section 6.1 by discussing speaker tracking based on the use of Bayesian filters. In Section 6.2, we review the basics of the conventional array-processing literature. Beginning with the theory of linear apertures, we investigate the effects of processing with discrete arrays as well as array steering. Two important concepts introduced in this section are those of poor low-frequency directivity, which arises from the finite extent of an array, and high-frequency spatial aliasing that arises from the necessity of sampling an aperture at discrete points. Our discussion of adaptive array processing begins in Section 6.3. This includes both array processing based on SOS, as introduced in Section 6.3.1, as well as that based on higher order statistics (HOS) or *non-Gaussian* criteria, as discussed in Section 6.3.8. Other topics covered in Section 6.3 include theoretical models for noise fields in Section 6.3.2, subband analysis and synthesis for adaptive filtering and beamforming in Section 6.3.3, beamforming performance criteria in Section 6.3.4, the generalized sidelobe canceller (GSC) in Section 6.3.5 as well as its recursive implementation in Section 6.3.6, and other conventional beamformers in Section 6.3.7. The first set of distant speech-recognition (DSR) results is presented in Section 6.3.10; these compare the performance of several different beamforming optimization criteria. Section 6.4 takes up our presentation of spherical array processing; we discuss acoustic diffraction and scattering, as well as their effects on the sensitivity of a spherical array to plane waves. We also introduce the concept of decomposing a sound field into spherical harmonics, which can be used for beamforming much like the output of a single microphone is used in conventional beamforming techniques. Section 6.5 describes how beamforming techniques based on the SOS discussed in Sections 6.3.1 can be profitably applied to spherical array processing. A comparison of conventional, linear, and spherical arrays is presented in Section 6.6 based on the beamforming performance criteria defined in Section 6.3.4. Thereafter, our second set of DSR results comparing the two arrays is presented. Finally, in Section 6.8, we present our conclusions.

6.1 Speaker Tracking

Before beamforming can be effectively used to enhance the speech of a desired speaker in a DSR application, the speaker's position, denoted as $\mathbf{x}$, relative to the microphone array must be known or estimated. Hence, let us begin our discussion by briefly examining how such an estimation can be performed.

The *time delay of arrival* (TDOA) between the microphones at positions $\mathbf{m}_1$ and $\mathbf{m}_2$ can be expressed as

$$T(\mathbf{m}_1, \mathbf{m}_2, \mathbf{x}) \triangleq \frac{\|\mathbf{x} - \mathbf{m}_1\| - \|\mathbf{x} - \mathbf{m}_2\|}{c}, \tag{6.1}$$

where c is the speed of sound, which is approximately 344 m/s at sea level. The definition (6.1) can be rewritten as

$$T(\mathbf{m}_m, \mathbf{m}_n, \mathbf{x}) \triangleq \frac{D_m - D_n}{c}, \tag{6.2}$$

where

$$D_n \triangleq \|\mathbf{x} - \mathbf{m}_n\| \; \forall \; n = 0, \ldots, S-1 \tag{6.3}$$

is the distance from the speaker to the microphone located at $\mathbf{m}_n$ and S is the total number of microphones, as shown in Figure 6.1.

Let $\hat{\tau}_{mn}$ denote the *observed* TDOA for the mth and nth microphones. The TDOA can be observed or estimated with a variety of well-known techniques. Perhaps the most popular method involves the *phase transform* (PHAT) [7], a variant of the *generalized cross-correlation* (GCC), which can be expressed as:

$$\rho_{mn}(\tau) \triangleq \frac{1}{2\pi} \int_{-\pi}^{\pi} \frac{Y_m(e^{j\omega\tau}) Y_n^*(e^{j\omega\tau})}{|Y_m(e^{j\omega\tau}) Y_n^*(e^{j\omega\tau})|} \, e^{j\omega\tau} \, d\omega, \tag{6.4}$$

where $Y_n(e^{j\omega\tau})$ denotes the short-time Fourier transform (STFT) of the signal arriving at the nth sensor in the array [54]. The definition of the GCC in (6.4) follows directly from the frequency domain calculation of the cross-correlation of two sequences. The normalization term $\left|Y_m(e^{j\omega\tau}) Y_n^*(e^{j\omega\tau})\right|$ in the denominator of the integrand is intended to weight all frequencies equally. It has been shown that such a weighting leads to more robust TDOA estimates in noisy and reverberant environments [12]. Once $\rho_{mn}(\tau)$ has been calculated, the TDOA estimate is

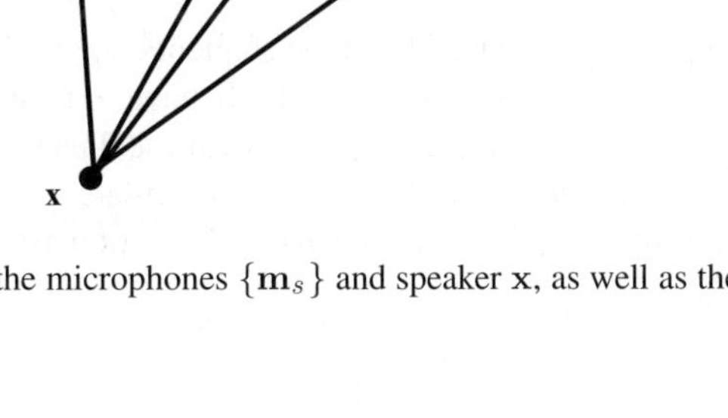

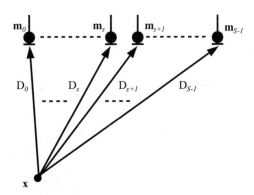

Figure 6.1 Positions of the microphones $\{\mathbf{m}_s\}$ and speaker $\mathbf{x}$, as well as the distances between them $\{D_s\}$.

obtained from

$$\hat{\tau}_{mn} = \max_{\tau} \rho_{mn}(\tau). \tag{6.5}$$

In other words, the "true" TDOA is taken as that which maximizes the value of the PHAT. As (6.5) is typically calculated with an inverse *discrete Fourier transform* (DFT), a parabolic interpolation is often performed to overcome the granularity in the estimate due to the digital sampling interval [54]. Usually $Y_n(e^{j\omega_k})$ appearing in (6.4) is calculated with a Hamming analysis window of 15–25 ms in duration [12].

Let us assume that the microphones are divided into a number S_2 of distinct microphone pairs. Consider two microphones located at $\mathbf{m}_{s1}$ and $\mathbf{m}_{s2}$ comprising the sth microphone pair, and once more define the TDOA as in (6.2), where $\mathbf{x}$ represents the position of an active speaker, and define $T_s(\mathbf{x}) \triangleq T(\mathbf{m}_{s1}, \mathbf{m}_{s2}, \mathbf{x})$. Source localization based on the maximum likelihood (ML) criterion [31] proceeds by minimizing the error function

$$\epsilon(\mathbf{x}) = \sum_{s=0}^{S_2-1} \frac{[\hat{\tau}_s - T_s(\mathbf{x})]^2}{\sigma_s^2}, \tag{6.6}$$

where σ_s^2 denotes the error covariance associated with this observation, and $\hat{\tau}_s$ is the observed TDOA as in (6.4) and (6.5).

Although (6.6) implies we should find that $\mathbf{x}$ minimizing the instantaneous error criterion, we would be better advised to attempt to minimize such an error criterion over a series of time instants. In so doing, we exploit the fact that the speaker's position cannot change instantaneously; thus, both the present and past TDOA estimates are potentially useful in estimating a speaker's current position. Klee *et al.* [32] proposed to recursively minimize the least square error position estimation criterion (6.6) with a variant of the *extended Kalman filter* (EKF). This was achieved by first associating the *state* $\mathbf{x}_k$ of the EKF with the speaker's position at time k, and the kth observation with a vector of TDOAs. In keeping with the formalism of the EKF, Klee *et al.* then postulated a *state* and *observation equation*:

$$\mathbf{x}_k = \mathbf{F}_{k|k-1}\mathbf{x}_{k-1} + \mathbf{u}_{k-1} \text{ and} \tag{6.7}$$

$$\mathbf{y}_k = \mathbf{H}_{k|k-1}(\mathbf{x}_k) + \mathbf{v}_k, \tag{6.8}$$

respectively, where

- $\mathbf{F}_{k|k-1}$ denotes the *transition matrix*,
- $\mathbf{u}_{k-1}$ denotes the *process noise*,
- $\mathbf{H}_{k|k-1}(\mathbf{x})$ denotes the vector-valued *observation function*, and
- $\mathbf{v}_k$ denotes the *observation noise*.

The unobservable state $\mathbf{x}_k$ is to be inferred from the sequence $\mathbf{y}_k$ of observations. The process $\mathbf{u}_k$ and observation $\mathbf{v}_k$ noises are unknown, but both have zero-mean Gaussian pdfs and known covariance matrices. Associating $\mathbf{H}_{k|k-1}(\mathbf{x})$ with the TDOA function (6.1) with one component per microphone pair, it is straightforward to calculate the appropriate linearization about the current state estimate required by the EKF [67, Section 10.2]:

$$\mathbf{H}_k(\mathbf{x}) \triangleq \nabla_{\mathbf{x}} \mathbf{H}_{k|k-1}(\mathbf{x}). \tag{6.9}$$

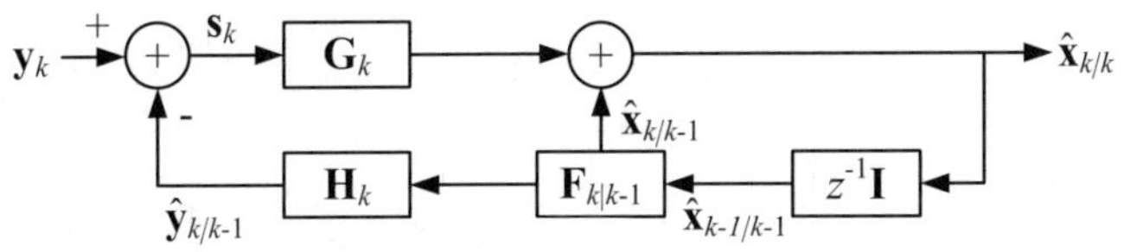

Figure 6.2 Predictor-corrector structure of the Kalman filter.

By assumption $\mathbf{F}_{k|k-1}$ is known so that the *predicted state estimate* is obtained from

$$\hat{\mathbf{x}}_{k|k-1} = \mathbf{F}_{k|k-1}\hat{\mathbf{x}}_{k-1|k-1}, \tag{6.10}$$

where $\hat{\mathbf{x}}_{k-1|k-1}$ is the *filtered state estimate* from the prior time step; the calculation of $\hat{\mathbf{x}}_{k|k-1}$ as in (6.10) is known as *prediction*. Let us define the *innovation* as

$$\mathbf{s}_k \triangleq \mathbf{y}_k - \mathbf{H}_{k|k-1}\left(\hat{\mathbf{x}}_{k|k-1}\right).$$

The innovation is called as such because it represents the component of the response of the system that could not be predicted from the state equation (6.7). The new filtered state estimate is calculated from

$$\hat{\mathbf{x}}_{k|k} = \hat{\mathbf{x}}_{k|k-1} + \mathbf{G}_k\,\mathbf{s}_k, \tag{6.11}$$

where $\mathbf{G}_k$ denotes the *Kalman gain*, which can be calculated through a well-known recursion [67, Section 4.3]. A block diagram illustrating the prediction and correction steps in the state estimate update of a conventional Kalman filter is shown in Figure 6.2.

6.2 Conventional Microphone Arrays

We will now analyze the characteristics of conventional apertures and arrays. As we will learn in Section 6.4, several of these characteristics are shared by the less conventional spherical apertures and arrays.

The relationship between the spherical coordinates (r, θ, ϕ) and Cartesian coordinates (x, y, z) is shown in Figure 6.3; the *polar angle* θ and *azimuth* ϕ are measured from the z- and x-axes, respectively, and have ranges $0 \leq \theta \leq \pi$ and $-\pi \leq \phi \leq \pi$ where $\phi = \pi/2$ corresponds to the y-axis. In the figure, a *plane wave* is impinging on an array of microphones located along the x-axis; the vector $\mathbf{a}$ indicates the *direction of arrival* of the plane wave. A plane wave is named as such because any locus of constant phase—or *wavefront*—is a plane; the plane-wave assumption is most accurate when the sources are relatively distant from the array as compared to the *aperture length*, which by definition is the maximum physical extent of the aperture.

Before taking up the case of conventional microphone arrays, let us consider the *linear aperture* of length L shown in Figure 6.4. The unit normal vector perpendicular to the wavefront can be expressed in Cartesian coordinates as

$$\mathbf{a} = -\begin{bmatrix} \sin\theta\cos\phi \\ \sin\theta\sin\phi \\ \cos\theta \end{bmatrix}. \tag{6.12}$$

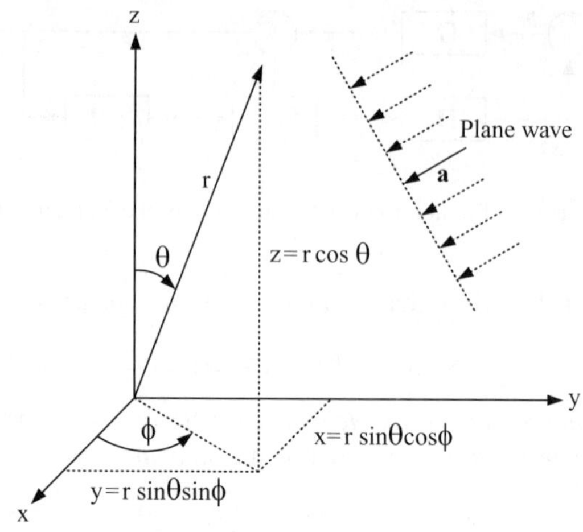

Figure 6.3 Relation between the spherical coordinates (r, θ, ϕ) and Cartesian coordinates (x, y, z).

The *vector wavenumber*

$$\mathbf{k} \triangleq \frac{2\pi}{\lambda}\mathbf{a}, \tag{6.13}$$

where λ is the length of the propagating wave, indicates both the direction of arrival and frequency of the propagating wave; the direction of arrival is given by $\mathbf{a} \triangleq \mathbf{k}/|\mathbf{k}|$, while the *scalar wavenumber*—defined as

$$k \triangleq \|\mathbf{k}\| = \frac{2\pi}{\lambda} = \frac{\omega}{c}, \tag{6.14}$$

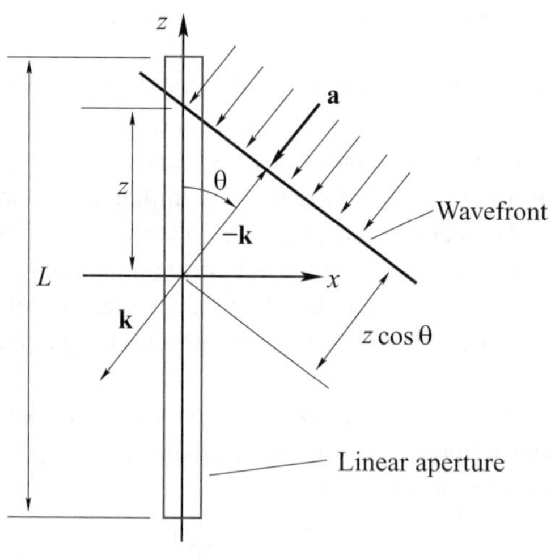

Figure 6.4 A plane wave with normal vector $\mathbf{a}$ and wavenumber $\mathbf{k}$ impinging on a linear aperture of length L lying along the z-axis.

where c is the speed of sound—is the angular frequency of the plane wave. Both k and $\mathbf{k}$ are often referred to as simply the *wavenumber*; we will also adopt this practice where the difference between the two is clear from context. For an arbitrary point $\mathbf{x}$, the TDOA with respect to the origin of the coordinate system is

$$\tau(\mathbf{x}) = \frac{\mathbf{a}^T \mathbf{x}}{c}. \tag{6.15}$$

Assuming now that all points on the linear aperture lie on the z-axis, then (6.12) and (6.15) imply

$$\tau(z) = -\frac{z \cos\theta}{c} = -\frac{uz}{c}, \tag{6.16}$$

where $u \triangleq \cos\theta$ is the *direction cosine* for the z-axis. The component of $\mathbf{k}$ along the z-axis is given by:

$$k_z \triangleq -\|\mathbf{k}\| \cos\theta = -\frac{\omega}{c} u = -\frac{2\pi}{\lambda} u. \tag{6.17}$$

From (6.16) and (6.17) it then follows

$$\omega\tau(z) = k_z z. \tag{6.18}$$

Consider now a narrow-band source signal $f(t)$ with spectrum $F(\omega)$. Given that a delay $\tau(z)$ in the time domain corresponds to a linear phase shift $e^{-i\tau(z)\omega}$ in the frequency domain, the Fourier transform of the signal component arriving at point z can be expressed as:

$$F(\omega, k_z, z) = F(\omega)\mathrm{e}^{-i\tau(z)\omega} = F(\omega)\mathrm{e}^{-ik_z z}, \tag{6.19}$$

where[1] $i \triangleq \sqrt{-1}$. If the signal components arriving along the entire aperture are *weighted* with a function $w_{\mathrm{a}}^*(z)$ and then *combined*, then the result is the *frequency wavenumber response function*

$$\Upsilon(\omega, k_z) \triangleq \int_{-\infty}^{\infty} w_{\mathrm{a}}^*(z)\mathrm{e}^{-ik_z z}\, \mathrm{d}z. \tag{6.20}$$

Let us initially assume that

$$w_{\mathrm{a}}(z) = \frac{1}{L} \begin{cases} 1, & \forall\ -L/2 \leq z \leq L/2, \\ 0, & \text{otherwise.} \end{cases} \tag{6.21}$$

Substituting (6.21) into (6.20), we find

$$\Upsilon(\omega, k_z) = \int_{-L/2}^{L/2} \mathrm{e}^{-ik_z z}\, \mathrm{d}z = \mathrm{sinc}\left(\frac{L}{2} k_z\right),$$

where

$$\mathrm{sinc}(x) \triangleq \frac{\sin x}{x}. \tag{6.22}$$

[1] For present purposes, we break with the signal processing convention of defining $j \triangleq \sqrt{-1}$, as j must be reserved to denote the spherical Bessel function in the sequel.

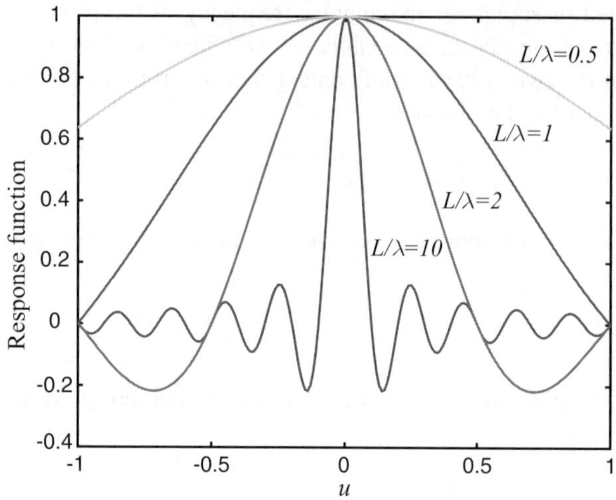

Figure 6.5 Beam patterns for the linear aperture with $L/\lambda = 0.5, 1, 2,$ and 10.

Equivalently, given that $\text{sinc}(x)$ is an even function

$$\Upsilon(\omega, k_z) = \text{sinc}\left(-\frac{L}{2} \cdot \frac{2\pi}{\lambda} u\right) = \text{sinc}\left(\frac{\pi L}{\lambda} \cdot u\right). \tag{6.23}$$

A plot of $\Upsilon(\omega, k_z)$ for several values of L/λ is shown in Figure 6.5. In order to properly analyze the curves in the figure, we must introduce several new terms. With our initial analysis, we strive primarily for intuition rather than mathematical precision; the latter will follow in subsequent sections. First of all, as the x-axis of the plot in Figure 6.6 corresponds to $u = \cos\theta$, we recognize

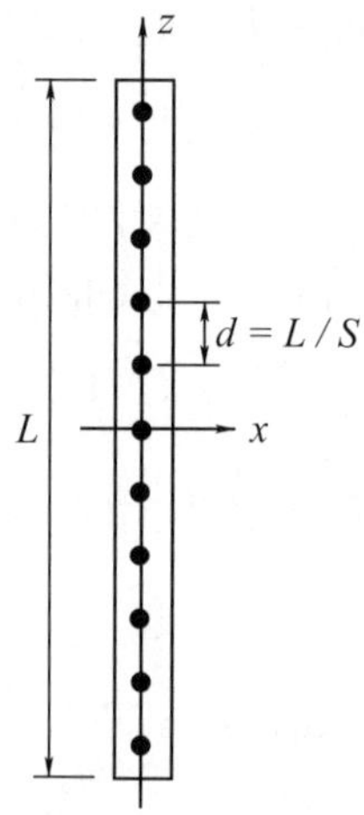

Figure 6.6 A linear aperture of length L and its approximation with an array of $S = 11$ elements with a uniform spacing of $d = L/S$.

that these curves represent the sensitivity of the array to plane waves impinging from various directions. In general, we will refer to such curves as *beam patterns*. The maximum sensitivity for all beam patterns is achieved at $u = 0$, which corresponds to $\theta = \pi/2$. We will refer to this angle of maximum sensitivity as the *look direction*, because it will presumably align with the direction of the desired source. All beam patterns attain a value of unity in the look direction, which implies that a plane wave impinging from this direction is neither amplified nor attenuated; we will refer to this condition by saying that any beam pattern fulfilling it satisfies the *distortionless constraint* in the look direction. We will refer to the broad lobe around the look direction as the *main lobe*, and the smaller lobes on either side of the main lobe as *side lobes*. Finally, we will refer to the capacity of a given beamformer to maximize the ratio of its sensitivity in the look direction to its average sensitivity over all directions as its *directivity*; high directivity is associated with focussing on a desired signal impinging from the look direction while suppressing noise and interference from other directions.

Now that we have equipped ourselves with the proper vocabulary, we can proceed with the analysis of the beam patterns in Figure 6.5. The figure indicates that for very low frequencies in which $L \leq \lambda$, the directivity of the linear aperture is poor. However, the directivity improves with increasing frequency. Clearly, all beam patterns satisfy the distortionless constraint for a look direction of $u = 0$. The size of the main lobe grows broader with decreasing frequency and increasing wavelength. For higher frequencies with shorter wavelengths, the beam pattern exhibits a marked sibe-lobe structure. As we will learn in Section 6.3, more advanced adaptive beamforming algorithms attempt to reduce the effects of ambient noise and interfering signals by controlling the structure of these side lobes, a process known as *null steering*.

As a uniformly sensitive aperture is difficult or impossible to construct, let us consider sampling the aperture at *S* points:

$$z_s = \left(s - \frac{S-1}{2}\right) d \ \forall \ s = 0, 1, \ldots, S-1, \tag{6.24}$$

where $d \triangleq L/S$ is the *intersensor spacing* of the array elements as shown in Figure 6.6. This sampling is accomplished by defining the *sampled* sensitivity function:

$$w_S(z) \triangleq \frac{1}{S} \sum_{s=0}^{S-1} \delta(z - z_s) \tag{6.25}$$

Substituting (6.25) into (6.20), provides

$$\Upsilon_s(\omega, k_z) = \frac{1}{S} \exp\left\{ik_z d\left(\frac{S-1}{2}\right)\right\} \sum_{s=0}^{S-1} e^{-ik_z sd},$$

which can be readily simplified to [67, Section 13.1.3]

$$\Upsilon_S(\omega, k_z) = \frac{1}{S} \cdot \frac{\sin\left(S\frac{d}{2}k_z\right)}{\sin\left(\frac{d}{2}k_z\right)} = \text{sinc}_S\left(\frac{d}{2} \cdot \frac{2\pi}{\lambda} \cdot u\right) = \text{sinc}_S\left(\frac{\pi d}{\lambda} \cdot u\right), \tag{6.26}$$

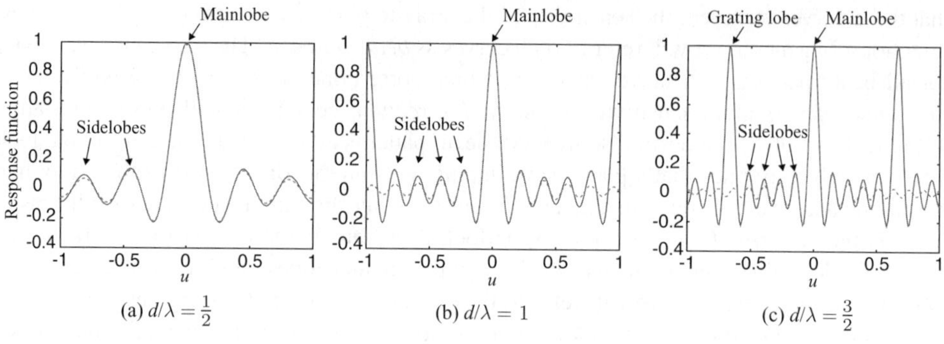

Figure 6.7 Beam patterns for the linear aperture (dotted line) and linear array (solid line) with $S = 11$ and (a) $d/\lambda = \frac{1}{2}$, (b) $d/\lambda = 1$, and (c) $d/\lambda = \frac{3}{2}$.

where

$$\text{sinc}_S(x) \triangleq \frac{1}{S} \cdot \frac{\sin Sx}{\sin x}.$$

Note that unlike (6.23), the beam pattern (6.26) is *periodic* with period λ/d. Beam patterns $\Upsilon_S(\omega, k_z)$ for several values of d/λ are shown in Figure 6.7. From the figure, it is apparent that for $d/\lambda \leq 1/2$, the behavior of the array is a very good approximation of that of the continuous aperture throughout the entire working range $-1 \leq u \leq 1$. On the other hand, while the behavior of the main lobe around $u =$ is good for $d/\lambda = 1, 3/2$, large spurious lobes with the same magnitude as the main lobe arise at points well removed from the look direction; these are known as *grating lobes*.

Clearly, the look direction for the beam patterns in Figures 6.5 and 6.7 is given by $(\theta_L, \phi_L) = (\pi/2, 0)$, which is typically referred to as *broadside*. Setting the look direction to broadside is achieved with a uniform weighting of the linear aperture as in (6.21), or the uniform weighting of the sensor outputs in (6.25). The process of setting the look direction is known as *beam steering* or simply *steering*. The look direction can readily be set to any desired direction $k = k_L$ by setting the sensor weights to

$$w_s(z; k_L) \triangleq \frac{1}{S} \sum_{s=0}^{S-1} e^{-ik_L d} \delta(z - z_s). \tag{6.27}$$

Doing so yields the beam pattern:

$$B(k_z, \omega; k_L) \triangleq \mathbf{v}_\mathbf{k}^H(k_L)\mathbf{v}_\mathbf{k}(k_z), \tag{6.28}$$

where the *array manifold vector* is defined as

$$\mathbf{v}_\mathbf{k}(k_z) \triangleq \left[e^{i\left(\frac{S-1}{2}\right)k_z d} \quad e^{i\left(\frac{S-1}{2}-1\right)k_z d} \quad \cdots \quad e^{-i\left(\frac{S-1}{2}\right)k_z d} \right]^T. \tag{6.29}$$

The array manifold vector is nothing more than a vector of phase shifts induced by the propagation delay for each sensor.

While the *visible region* is by definition $-1 \leq u \leq 1$, it is customary to conceptualize *u* as extending over the entire real line. This is done to facilitate the visualization of grating lobes

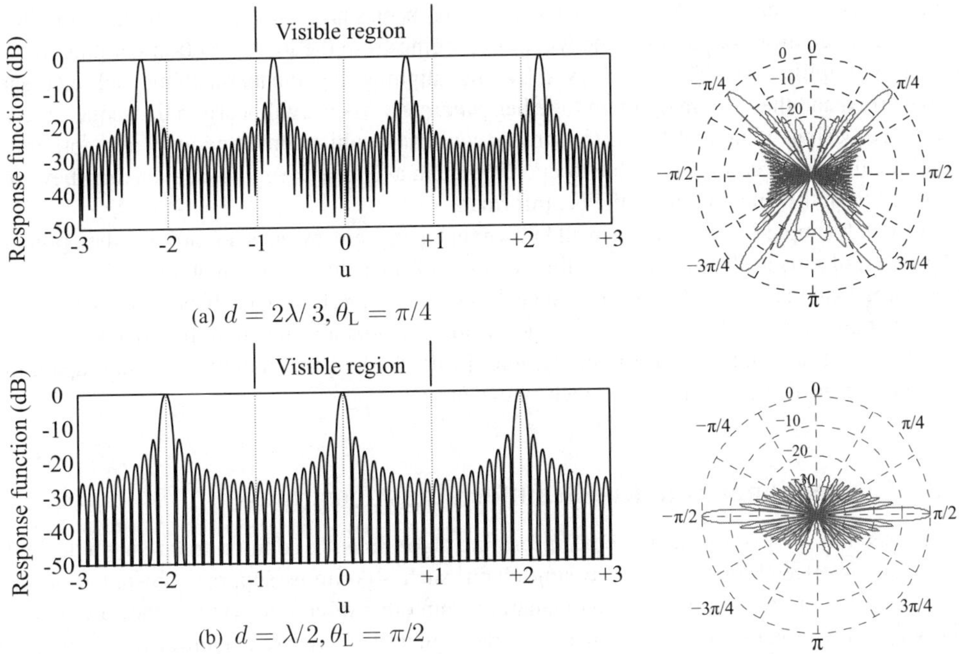

Figure 6.8 Effect of steering on the grating lobes for $S = 20$ plotted in Cartesian and polar coordinates.

moving into the visible region as the result of beam steering. The effect of steering is shown in Figure 6.8, from which it is apparent that the grating lobes recur at regular intervals whether or not they are within the visible region. From the figure it is apparent that steering can cause grating lobes to enter the beam pattern for $d > \lambda/2$. This phenomenon is known as *spatial aliasing*, and occurs when the propagating wave is not sampled sufficiently often *in space*. The *half wavelength rule* [67, Section 13.1.4] states that avoiding spatial aliasing—even when steering over the entire front half plane—requires that

$$\frac{d}{\lambda} \leq \frac{1}{2},$$

which is to say, the wave must be sampled as least twice along its length. This rule is analogous to the *Nyquist sampling theorem* [55, Section 4] from conventional signal processing.

The beam pattern (6.28) leads to the definition of the *delay-and-sum* beamformer weights as:

$$\mathbf{w}_{\mathrm{DS}}^H(\omega, k_z) = \frac{\mathbf{v}_{\mathbf{k}}^H(\omega, k_z)}{\mathbf{v}_{\mathbf{k}}^H(\omega, k_z)\mathbf{v}_{\mathbf{k}}(\omega, k_z)}. \tag{6.30}$$

The delay-and-sum beamformer is the simplest fixed design that satisfies the *distortionless constraint*, which can be expressed as

$$\mathbf{w}^H \mathbf{v}_{\mathbf{k}} = 1, \tag{6.31}$$

where for convenience the dependence on (ω, k_z) has been suppressed. Equation (6.31) implies that such a beamformer passes plane waves impinging from the look direction without attenuation or amplification. This is achieved by time aligning the signals reaching each element in the array, and then summing them together *coherently*. As we will learn in the sequel, more advanced designs maintain this distortionless constraint, while simultaneously attempting to combine the signal components due to interference in a destructive manner, such that they are suppressed in the final output of the beamformer.

Although simple to analyze, it is well known that a linear array does not provide the optimal placement of sensors. The analysis of this section indicates that designing a microphone array for a broadband signal such as human speech involves a careful trade-off between achieving sufficient directivity at low frequencies, and avoiding spatial aliasing at high frequencies (see, e.g., [63, Section 3.9.2] and Gazor and Grenier [20]). We will encounter these issues again in considering the design of spherical microphone arrays.

6.3 Conventional Adaptive Beamforming Algorithms

In this section, we discuss adaptive beamforming algorithms. In addition to passing a desired signal undistorted through the processing chain, such algorithms suppress unwanted noise, reverberation, or overlapping speech emanating from other directions. Hence, they are potentially far more effective at enhancing the desired signal than any fixed beamformer design.

6.3.1 Minimum Variance Distortionless Response Beamformer

The delay-and-sum beamformer can emphasize a wave emanating from a desired or look direction, and to some degree suppress waves impinging from other directions. However, as it is a *fixed* design, it does not provide optimal suppression for strong, coherent sources of interference. In contrast, the *adaptive beamformers* can effectively place a null on any interference by controlling the sidelobe structure of the beam pattern, which is achieved by minimizing the variance of beamformer's outputs while maintaining a distortionless constraint in the look direction. This section describes one of the most basic adaptive beamforming methods, the *minimum variance distortionless response* (MVDR) beamformer. The MVDR beamformer is based on the use of second order statistics (SOS), that is it requires only the knowledge of the covariance or *spatial spectral matrix* of the inputs to the microphone array.

For reasons of computational efficiency, modern adaptive filtering or beamforming algorithms are usually implemented in the frequency or—better yet—subband domain [24, Section 7]. Section 6.3.3 briefly presents subband analysis and synthesis. Here, we consider beamforming in the subband domain. Let us define the *subband domain snapshot* for an array of S discrete sensors as

$$\mathbf{X}(\omega) \triangleq \begin{bmatrix} X_0(\omega) & X_1(\omega) & \cdots & X_{S-1}(\omega) \end{bmatrix}^T, \tag{6.32}$$

where $X_s(\omega)$ is the subband component for sensor s. For present purposes, let us assume that the complete snapshot consists of the sum

$$\mathbf{X}(\omega) = \mathbf{F}(\omega) + \mathbf{N}(\omega), \tag{6.33}$$

where $\mathbf{F}(\omega)$ is the component contributed by the desired signal and $\mathbf{N}(\omega)$ is due to the ambient noise and interference. Let us define the desired signal $F(\omega)$, which we assume to be transmitted on a plane wave with wavenumber $\mathbf{k}_s$ impinging on the sensor array. Letting $\mathbf{m}_s$ denote the position of the sensor s leads to the representation

$$\mathbf{F}(\omega) \triangleq F(\omega)\mathbf{v_k}(\mathbf{k}_s), \tag{6.34}$$

where the array manifold vector in this case is defined as

$$\mathbf{v_k}(\mathbf{k}_s) \triangleq \left[\mathrm{e}^{-\mathbf{k}_s^T \mathbf{m}_0} \quad \mathrm{e}^{-\mathbf{k}_s^T \mathbf{m}_1} \quad \cdots \quad \mathrm{e}^{-\mathbf{k}_s^T \mathbf{m}_{S-1}} \right]^T. \tag{6.35}$$

Alternatively, we can express the array manifold vector as

$$\mathbf{v}(\omega) \triangleq \left[\mathrm{e}^{-i\omega\tau_0} \quad \mathrm{e}^{-i\omega\tau_1} \quad \cdots \quad \mathrm{e}^{-i\omega\tau_{S-1}} \right], \tag{6.36}$$

where

$$\omega\tau_s = \mathbf{k}_s^T \mathbf{m}_s \ \forall\, s = 0, 1, \ldots, S-1.$$

Equation (6.36) is actually a more general definition of the array manifold vector than (6.35), in as much as it encompasses spherical as well as plane waves; it is only necessary to modify the way in which τ_s is calculated. The output of the beamformer can then be expressed as

$$Y(\omega) = \mathbf{w}^H(\omega)\mathbf{X}(\omega), \tag{6.37}$$

where $\mathbf{w}^H(\omega)$ are the frequency-dependent sensor weights.

In order to calculate the optimal MVDR sensor weights, the covariance matrix of the outputs of the array sensors must be known or estimated. Here, we assume they are known such that

$$\mathbf{\Sigma_X}(\omega) \triangleq \mathcal{E}\left\{ \mathbf{X}(\omega)\mathbf{X}^H(\omega) \right\}, \tag{6.38}$$

where $\mathcal{E}\{-\}$ is the probabilistic expectation operator [57, Sections 5–3]. We then determine the optimum weight vector that minimizes the variance of the beamformer's outputs:

$$\Sigma_Y \triangleq \mathcal{E}\left\{ |Y(\omega)|^2 \right\} = \mathbf{w}^H(\omega)\mathbf{\Sigma_X}(\omega)\mathbf{w}(\omega), \tag{6.39}$$

subject to the distortionless constraint (6.31). The well-known solution is the MVDR beamformer [67, Section 13.3.1]. The weight vector of the MVDR beamformer can be expressed as

$$\mathbf{w}_{\mathrm{MVDR}}^H(\omega) = \frac{\mathbf{v}^H(\omega)\mathbf{\Sigma_X}^{-1}(\omega)}{\mathbf{v}^H(\omega)\mathbf{\Sigma_X}^{-1}(\omega)\mathbf{v}(\omega)}. \tag{6.40}$$

In practice, acoustic beamforming applications update the covariance matrix $\mathbf{\Sigma_X}$ only during periods of inactivity of the desired source in order to avoid cancellation of the desired signal, which is known as *signal cancellation* [65]. Van Trees [63, Section 6.2.4] refers to the beamformer that uses the entire input for computation of the covariance matrix as the *minimum power distortionless response* (MPDR) beamformer, although both beamformers are commonly referred to as MVDR beamformers in the literature.

In order to avoid excessively large side lobes in the beam pattern, small weights are typically added to the main diagonal of $\mathbf{\Sigma_X}$, which is known as *diagonal loading* [67, Section 13.3.7].

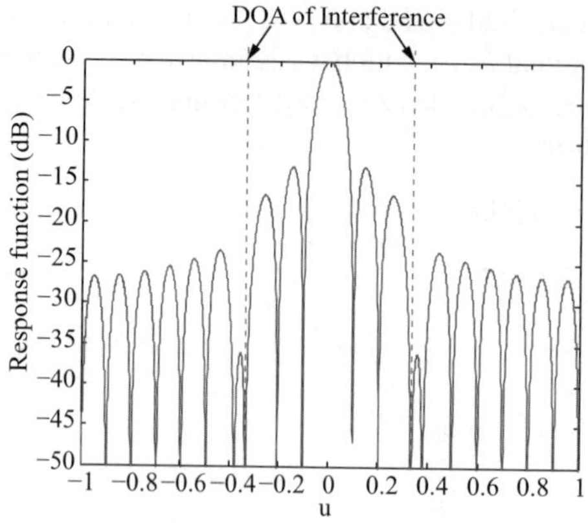

(a) Beam pattern in Cartesian coordinates

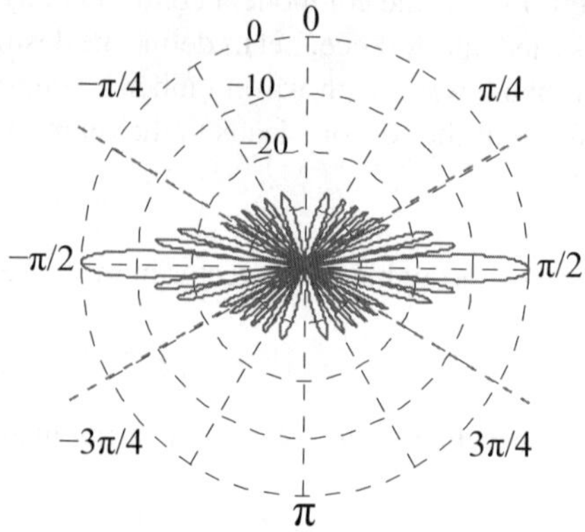

(b) Polar plot of the beam pattern

Figure 6.9 Beam patterns of the MVDR beamformer with $N = 20$ sensors and $d/\lambda = 1/2$ in the case of the look direction of $u = 0$ for two interference signals.

Letting σ_{d}^2 denote the amount of diagonal loading, the weight vector of the MVDR beamformer can be written as

$$\mathbf{w}_{\mathrm{MVDR}}^H = \frac{\mathbf{v}^H \left(\boldsymbol{\Sigma}_{\mathbf{X}} + \sigma_{\mathrm{d}}^2 \mathbf{I}\right)^{-1}}{\mathbf{v}^H \left(\boldsymbol{\Sigma}_{\mathbf{X}} + \sigma_{\mathrm{d}}^2 \mathbf{I}\right)^{-1} \mathbf{v}}, \tag{6.41}$$

where the frequency ω is omitted here for the sake of clarity.

Figure 6.9 shows the beam patterns of the MVDR beamformer constructed from the linear array with twenty equally spaced sensors and an intersensor spacing of $d = \lambda/2$. In Figure 6.9, the diagonal loading is $\sigma_{\mathrm{d}}^2 = 0.01$, and the look direction is set as $u = 0$; two interference signals are assumed to come from $u = -0.3$ and $u = 0.3$ as indicated by the dotted lines. It is clear from the figure that the MVDR beamformer can maintain unity gain in the look direction at $u = 0$, while placing deep nulls on the directions of arrival of the interference at $u = \pm 0.3$.

6.3.2 *Noise Field Models*

Although the MVDR beamformer can effectively suppress the interference signals by computing the noise covariance matrix from actual observations, it is often better to use a theoretical noise field model in practice. Two models that appear frequently in the literature are the incoherent and diffuse noise models.

In the case that a noise field is spatially uncorrelated (incoherent), the correlation of noise signals received at microphones at any given spatial location is zero. It was shown in [6, Section 4] that, under that condition, the noise covariance matrix becomes an identity matrix, that is, $\boldsymbol{\Sigma}_{\mathbf{X}}(\omega) = \sigma_N^2 \mathbf{I}$. In that case, the MVDR solution for the sensor weights becomes equivalent to those of the delay-and-sum beamformer. The incoherent noise model is often appropriate when the distance between microphones is large and there are no coherent noise sources.

If the sensors of an array receive plane wave noise signals uniformly distributed on the surface of a sphere with random phase, a *spherically isotropic noise field* results. In this case, the components of the noise covariance matrix can be expressed as

$$\Sigma_{\mathbf{N}_{s,s'}}(\omega) = \sigma_N^2 \,\mathrm{sinc}\left(\frac{\omega d_{s,s'}}{c}\right), \tag{6.42}$$

where $d_{s,s'}$ is the distance between microphones s and s'. When the weights (6.41) of the MVDR beamformer are estimated based on (6.42), the *superdirective beamformer* is obtained [67, Section 13.3.4]. Another theoretical noise model frequently used in acoustic beamforming is the *cylindrically isotropic noise field*, which yields a sensor covariance matrix of

$$\Sigma_{\mathbf{N}_{s,s'}}(\omega) = \sigma_N^2 \, J_0\left(\frac{\omega\, d_{s,s'}}{c}\right), \tag{6.43}$$

where J_0 is the *cylindrical Bessel function* of order zero [53, Section 10.2]. The cylindrically isotropic noise field is a good approximation for *babble noise* [5, Section 2.3.3].

6.3.3 Subband Analysis and Synthesis

Because of its computational efficiency, adaptive filtering and beamforming operations are often performed in the *frequency* or *subband* domain [67, Section 11], which provides additional advantages in terms of speed of convergence. Frequency domain analysis is typically performed by applying a windowing sequence $w[n]$, such as the *Hamming window*, to isolate a segment of the input, then performing a *discrete Fourier transform* (DFT) to this windowed sequence. This is equivalent to calculating STFT of the segment [56, Section 10.3]. In principal, the same steps are also used for subband analysis. However, in the latter, if *M* subbands are to be used for analysis, the length of the window is typically *mM* for some integer $m > 1$, which implies that the windowed signal must be time aliased. The advantage afforded by the longer window is that the stop band suppression can be much greater than that achieved by frequency domain analysis [67, Section 11.8]. This is a desirable characteristic for both adaptive filtering and beamforming as the outputs of all subbands can be treated as statistically independent; this independence is violated if there is significant spectral overlap between adjacent subbands. Moreover, the design of a subband analysis bank can be paired with that of a subband synthesis bank such that the combination is able to reconstruct the original input signal to arbitrary accuracy; that is it is able to achieve *perfect reconstruction*. Subband analysis and synthesis filter banks that are optimally suited to adaptive filtering and beamforming applications achieve perfect reconstruction through *oversampling* rather than *aliasing cancellation* [67, Section 11].

A system for performing subband analysis and synthesis is shown in Figure 6.10. The set of transfer functions $\{H_m(z)\}$ comprises the *analysis filter bank*, which splits the input $x[n]$ into *M* subband signals $\{X_m[n]\}_{m=0}^{M-1}$. The set $\{G_m(z)\}$ of transfer functions comprises the *synthesis filter bank*, which recombines the *M* subband signals $\{Y_m[n]\}_{m=0}^{M-1}$ into a single output $\hat{x}[n]$. Each $Y_m[n]$ is obtained by multiplying $X_m[n]$ with a complex constant, which is determined with an adaptive filtering or beamforming algorithm. In the former case, there are a single analysis bank and a single synthesis bank; in the latter, there is one analysis bank for

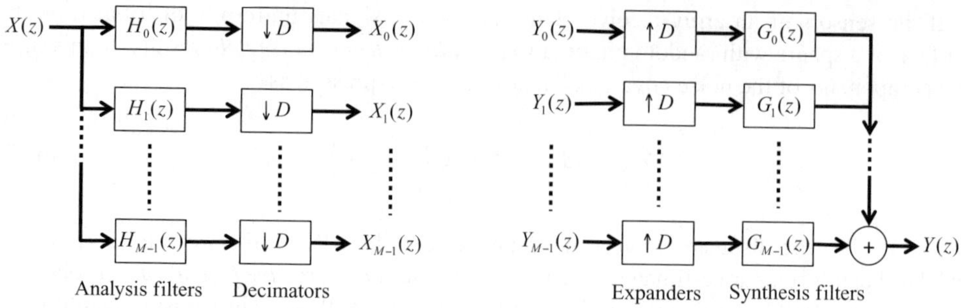

(a) Analysis processing for the input of the sth channel.

(b) Synthesis processing for the beamformer's output.

Figure 6.10 Schematic of subband processing.

each element in a sensor array, but only a single synthesis bank as all the samples for a given subband are combined with a beamforming operation prior to synthesis.

A common class of filter banks is that wherein the impulse response of each filter is obtained by modulating a prototype impulse response $h_0[n]$ according to

$$h_m[n] = h_0[n]\, e^{j2\pi nm/M}\, \forall m = 0, \ldots, M-1, \tag{6.44}$$

which implies that the impulse responses for all the filters in the bank are obtained from a single prototype. The processes of windowing and filtering are then equivalent provided that $h_0[n] = w[-n]$. Applying the z-transform to both sides of (6.44), we obtain

$$H_m(z) \triangleq H_0(zW_M^m), \tag{6.45}$$

where $W_M = e^{-j2\pi/M}$ is the Mth root of unity. Equation (6.45) implies that $H_m(e^{j\omega})$ is a *shifted version* of the frequency response of $H_0(e^{j\omega})$ according to

$$H_m(e^{j\omega}) = H_0(e^{j(\omega - 2\pi m)/M)}). \tag{6.46}$$

Similarly, for the synthesis bank, the impulse responses of the individual filters are related by

$$g_m[n] = g_0[n]\, e^{j2\pi nm/M}\ \forall\ m = 0, \ldots, M-1,$$

so that we can write

$$G_m(z) \triangleq G_0(zW_M^m). \tag{6.47}$$

We will now introduce two important operations in the filter bank system, *decimation* and *expansion*. Figure 6.10 also illustrates two corresponding blocks referred to as the D-fold *decimator* and the D-fold *expander*. The D-fold decimator with input $x[n]$ produces the output

$$x_D[n] = x[nD] \tag{6.48}$$

for integer D. In the frequency domain, the output of the decimator can be written as

$$X_D(e^{j\omega}) = \frac{1}{D}\sum_{k=0}^{D-1} X(e^{j(\omega-2\pi k)/D}); \tag{6.49}$$

see Vaidyanathan [62, Section 4.1]. From (6.49), the operation of the decimator can be interpreted as follows: (1) stretch the input spectrum $X(e^{j\omega})$ by a factor of D in order to form $X(e^{j\omega/D})$, (2) create $D-1$ copies of the stretched spectrum by shifting it with an amount of 2π, (3) sum the original spectrum and all the stretched versions together, and (4) divide it by D. Normally, each stretched version is overlapped with the other shifted copies. The effect of the overlap is known as *frequency aliasing*. In order to control such aliasing, the decimation factor D is set according to $D = M/2^r$ for some integer $r > 1$, which implies that the subbands are *oversampled*.

The D-fold expander takes input $y[n]$ and interpolates as

$$y_E[n] = \begin{cases} y[n/D] & \text{if } n \text{ is an integer multiple of } D, \\ 0 & \text{otherwise.} \end{cases} \tag{6.50}$$

Based on (6.50), we can write

$$Y_E(z) = \sum_{n=-\infty}^{\infty} y_E[n]\, z^{-n} = \sum_{k=-\infty}^{\infty} y_E[kD]\, z^{-kD} = \sum_{k=-\infty}^{\infty} y[k]\, z^{-kD}. \tag{6.51}$$

Upon setting $z = e^{j\omega}$ for the last equality, we have

$$Y_E(e^{j\omega}) = Y(e^{j\omega D}). \tag{6.52}$$

It is clear from (6.52) that the expander scales the frequency axis, which creates images of the compressed spectrum of $Y(e^{j\omega})$; this is known as *imaging*.

The filter bank obtained in the fashion described above is known as a *uniform DFT filter bank*, where DFT refers to the discrete Fourier transform used in its implementation. The uniform DFT analysis and synthesis filter banks are typically implemented in *polyphase* form in order to achieve maximal computational efficiency [67, Section 11]. The task of designing a uniform DFT filter bank devolves to that of designing the analysis and synthesis prototypes $h_0[n]$ and $g_0[n]$, respectively.

In the class of cosine modulated filter banks, perfect reconstruction is achieved through aliasing cancellation [62, Section 5.6]. However, during adaptive filtering or beamforming, the perfect reconstruction property can be destroyed as arbitrary magnitude scalings and phase shifts are applied to the subband samples. De Haan *et al.* [11] abandoned aliasing cancellation and designed analysis and synthesis prototypes based on minimization of the individual aliasing components for each subband. De Haan *et al.* also demonstrated that adaptive beamforming with their filter banks provides superior speech enhancement due to better suppression of aliasing effects; frequency distortion effect in such filter banks can be eliminated by imposing a *Nyquist(M)* constraint on the filter bank prototypes [33].

Use of a digital filter bank requires that the array manifold vector (6.36) be redefined as

$$\mathbf{v}(\omega_m) \triangleq \left[e^{-i\omega_m \tau_0 f_s} \quad e^{-i\omega_m \tau_1 f_s} \quad \dots \quad e^{-i\omega_m \tau_{S-1} f_s} \right], \tag{6.53}$$

where f_s is the digital sampling frequency.

6.3.4 *Beamforming Performance Criteria*

Before continuing our discussion of adaptive array processing algorithms, we introduce three measures of beamforming performance, namely, *array gain* (AG), *white noise gain* (WNG), and *directivity index* (DI). These criteria will prove useful in our performance comparisons of conventional, linear and spherical arrays in Section 6.5.

Array Gain

The array gain is defined as the ratio of the SNR at the output of the beamformer to the SNR at the input of a single channel of the array. Hence, array gain is a useful measure of how much a particular acoustic array processing algorithm enhances the desired signal. In this section, we formalize the concept of the array gain, and calculate it for both the delay-and-sum and MVDR beamformers given in (6.30) and (6.40), respectively.

As in Section 6.3.1, let us assume that the component of the desired signal reaching each component of a sensor array is $F(\omega)$ and the component of the noise and interference reaching each sensor is $N(\omega)$. This implies that the SNR at the input of the array can be expressed as

$$\mathrm{SNR}_{\mathrm{in}}(\omega) \triangleq \frac{\Sigma_F(\omega)}{\Sigma_N(\omega)}, \tag{6.54}$$

where $\Sigma_F(\omega) \triangleq \mathcal{E}\{|F(\omega)|^2\}$ and $\Sigma_N(\omega) \triangleq \mathcal{E}\{|N(\omega)|^2\}$. Then, for the vector of beamforming weights $\mathbf{w}^H(\omega)$, the output of the array is given by

$$Y(\omega) = \mathbf{w}^H(\omega)\,\mathbf{X}(\omega) = Y_F(\omega) + Y_N(\omega), \tag{6.55}$$

where $Y_F(\omega) \triangleq \mathbf{w}^H(\omega)\,\mathbf{F}(\omega)$ and $Y_N(\omega) \triangleq \mathbf{w}^H(\omega)\,\mathbf{N}(\omega)$ are, respectively, the signal and noise components in the output of the beamformer. Let us define the *spatial spectral covariance matrices*:

$$\mathbf{\Sigma_F}(\omega) \triangleq \mathcal{E}\{\mathbf{F}(\omega)\mathbf{F}^H(\omega)\},$$
$$\mathbf{\Sigma_N}(\omega) \triangleq \mathcal{E}\{\mathbf{N}(\omega)\mathbf{N}^H(\omega)\}.$$

Then, upon assuming the $F(\omega)$ and $N(\omega)$ are statistically independent, the variance of the output of the beamformer can be calculated according to

$$\Sigma_Y(\omega) = \mathcal{E}\{|Y(\omega)|^2\} = \Sigma_{Y_F}(\omega) + \Sigma_{Y_N}(\omega), \tag{6.56}$$

where

$$\Sigma_{Y_F}(\omega) \triangleq \mathbf{w}^H(\omega)\,\mathbf{\Sigma_F}(\omega)\,\mathbf{w}(\omega) \tag{6.57}$$

is the variance of the signal component of the beamformer's output, and

$$\Sigma_{Y_N}(\omega) \triangleq \mathbf{w}^H(\omega)\,\mathbf{\Sigma_N}(\omega)\,\mathbf{w}(\omega) \tag{6.58}$$

is the variance of the noise component. Expressing the snapshot of the desired signal once more as in (6.32), we find that the spatial spectral matrix $\mathbf{F}(\omega)$ of the desired signal can be written as

$$\mathbf{\Sigma_F}(\omega) = \Sigma_F(\omega)\,\mathbf{v_k}(\mathbf{k}_s)\,\mathbf{v_k}^H(\mathbf{k}_s). \tag{6.59}$$

Substituting (6.59) into (6.57), we can calculate the variance of the output signal spectrum as

$$\Sigma_{Y_F}(\omega) = \mathbf{w}^H(\omega)\,\mathbf{v}_\mathbf{k}(\mathbf{k}_s)\,\Sigma_F(\omega)\,\mathbf{v}_\mathbf{k}^H(\mathbf{k}_s)\,\mathbf{w}(\omega). \tag{6.60}$$

If we now assume that $\mathbf{w}(\omega)$ satisfies the distortionless constraint (6.31), then (6.60) reduces to

$$\Sigma_{Y_F}(\omega) = \Sigma_F(\omega),$$

which holds for both the delay-and-sum and MVDR beamformers.

Substituting (6.30) into (6.58) it follows that the noise component present at the output of the delay-and-sum beamformer (DSB) is given by

$$\Sigma_{Y_N}(\omega) = \frac{1}{N^2}\mathbf{v}_\mathbf{k}^H(\mathbf{k}_s)\,\boldsymbol{\Sigma}_\mathbf{N}(\omega)\,\mathbf{v}_\mathbf{k}(\mathbf{k}_s) \tag{6.61}$$

$$= \frac{1}{N^2}\mathbf{v}_\mathbf{k}^H(\mathbf{k}_s)\boldsymbol{\rho}_\mathbf{N}(\omega)\mathbf{v}_\mathbf{k}(\mathbf{k}_s)\Sigma_N(\omega), \tag{6.62}$$

where the *normalized spatial spectral matrix* $\boldsymbol{\rho}_\mathbf{N}(\omega)$ is defined through the relation

$$\boldsymbol{\Sigma}_\mathbf{N}(\omega) \triangleq \Sigma_N(\omega)\,\boldsymbol{\rho}_\mathbf{N}(\omega). \tag{6.63}$$

Hence, the SNR at the output of the beamformer is given by

$$\mathrm{SNR}_{\mathrm{out}}(\omega) \triangleq \frac{\Sigma_{Y_F}(\omega)}{\Sigma_{Y_N}(\omega)} = \frac{\Sigma_F(\omega)}{\mathbf{w}^H(\omega)\,\boldsymbol{\Sigma}_\mathbf{N}(\omega)\mathbf{w}(\omega)}. \tag{6.64}$$

Then based on (6.54) and (6.64), we can calculate the array gain of the DSB as

$$A_{\mathrm{dsb}}(\omega, \mathbf{k}_s) \triangleq \frac{\Sigma_{Y_F}(\omega)}{\Sigma_{Y_N}(\omega)} \Big/ \frac{\Sigma_F(\omega)}{\Sigma_N(\omega)} = \frac{N^2}{\mathbf{v}_\mathbf{k}^H(\mathbf{k}_s)\,\boldsymbol{\rho}_\mathbf{N}(\omega)\,\mathbf{v}_\mathbf{k}(\mathbf{k}_s)}. \tag{6.65}$$

Repeating the foregoing analysis for the MVDR beamformer (6.40), we arrive at

$$A_{\mathrm{mvdr}}(\omega, \mathbf{k}_s) = \mathbf{v}_\mathbf{k}^H(\mathbf{k}_s)\,\boldsymbol{\rho}_\mathbf{N}^{-1}(\omega)\,\mathbf{v}_\mathbf{k}(\mathbf{k}_s). \tag{6.66}$$

If noise at all sensors are spatially uncorrelated, then $\boldsymbol{\rho}_\mathbf{N}(\omega)$ is the identity matrix and the MVDR beamformer reduces to the DSB. From (6.65) and (6.66), it can be seen that in this case, the array again is

$$A_{\mathrm{mvdr}}(\omega, \mathbf{k}_s) = A_{\mathrm{dsb}}(\omega, \mathbf{k}_s) = N. \tag{6.67}$$

In all other cases

$$A_{\mathrm{mvdr}}(\omega, \mathbf{k}_s) > A_{\mathrm{dsb}}(\omega, \mathbf{k}_s). \tag{6.68}$$

The MVDR beamformer is of particular interest because it comprises the preprocessing component of two other important beamforming structures. Firstly, the MVDR beamformer followed by a suitable postfilter yields the *maximum SNR* beamformer [63, Section 6.2.3]. Secondly, and more importantly, by placing a Wiener filter [24, Section 2.2] on the output of the MVDR beamformer, the *minimum mean-square error* (MMSE) beamformer is obtained [63, Section 6.2.2]. Such *postfilters* are important because it has been shown that they can yield significant reductions in error rate [46,47]. If only a single subband is considered, the MVDR beamformer without modification will uniformly provide the highest SNR, as indicated by (6.68), and hence the highest array gain; we will return to this point in Section 6.5.

White Noise Gain

The WNG is by definition [10]:

$$G_{\mathrm{w}}(\omega) \triangleq \frac{\left|\mathbf{w}^H(\omega)\,\mathbf{v}(\mathbf{k}_{\mathrm{s}})\right|^2}{\mathbf{w}^H(\omega)\,\mathbf{w}(\omega)}. \tag{6.69}$$

The numerator of (6.69), which will be unity for any beamformer satisfying the distortionless constraint (6.31), represents the power of the desired signal at the output of the beamformer, while the denominator is equivalent to the array's sensitivity to self-sensor noise. Gilbert and Morgan [22] explain that WNG also reflects the sensitivity of the array to random variations in its components, including the positions and response characteristics of its sensors. Hence, WNG is a useful measure of system robustness.

It can be shown that uniform weighting of the sensor outputs provides the highest WNG [63, Section 2.6.3]. Hence, we should expect the delay-and-sum beamformer to provide the highest WNG in all conditions; we will reexamine this assumption in Section 6.5.

Directivity Index

We now describe our third beamforming performance metric. Let us begin by defining the *power pattern* as

$$P(\theta,\phi) \triangleq |B(\theta,\phi)|^2, \tag{6.70}$$

where $B(\theta,\phi)$ is the beam pattern described in Section 6.2 as a function of the spherical coordinates $\Omega \triangleq (\theta,\phi)$; see Figure 6.3. Let $\Omega_0 \triangleq (\theta_0,\phi_0)$ denote the look direction. The *directivity* is typically defined in the traditional (i.e., nonacoustic) array processing literature as [63, Section 2.6.1]

$$D(\omega) \triangleq \frac{4\pi P(\theta_0,\phi_0)}{\int_{\Omega_{\mathrm{sph}}} P(\theta,\phi)\,d\Omega}, \tag{6.71}$$

where Ω_{sph} represents the surface of a sphere with differential area $d\Omega$; we will consider such spherical integrals in detail in Sections 6.4 and 6.5.

Assuming that the beamforming coefficients satisfy the distortionless constraint (6.31) implies $P(\Omega_0) = 1$ such that (6.71) can be simplified and expressed in decibels as the *directivity index*

$$\begin{aligned} \mathrm{DI} &\triangleq -10\log_{10}\left[\frac{1}{4\pi}\int_{\Omega} P(\theta,\phi)\,d\Omega\right] \\ &= -10\log_{10}\left[\frac{1}{4\pi}\int_0^{2\pi}\int_0^{\pi} P(\theta,\omega)\sin\theta d\theta d\phi\right]. \end{aligned} \tag{6.72}$$

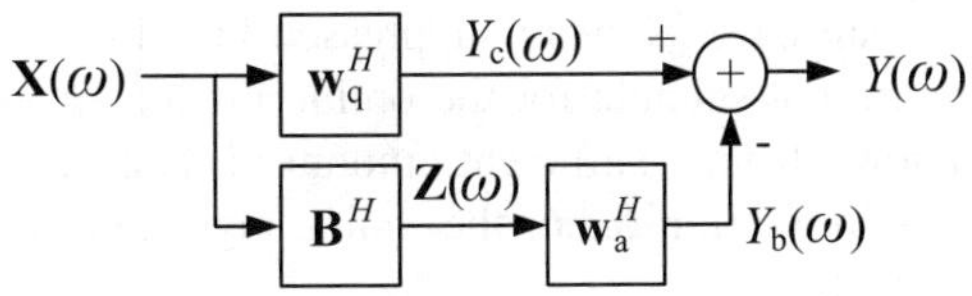

Figure 6.11 Generalized sidelobe canceller.

Note the critical difference between array gain and directivity index. While the former requires specific knowledge of the acoustic environment in which a given beamformer operates, the latter is the ratio of the sensitivity of the array in the look direction to that averaged over the surface of the sphere. Hence, the directivity index is independent of the acoustic environment once the beamforming weights have been specified.

In the acoustic array-processing literature, directivity is more often defined as SNR in the presence of a spherically isotropic diffuse noise field with sensor covariance matrix defined in (6.42); see Bitzer and Simmer [5]. Under this definition, the directivity index can be expressed as

$$\text{DI} \triangleq -10\log_{10} \frac{\left|\mathbf{w}^H\mathbf{v}(\mathbf{k}_S)\right|^2}{\mathbf{w}^H\boldsymbol{\Gamma}_{SI}\mathbf{w}}. \tag{6.73}$$

The superdirective beamformer mentioned in Section 6.3.2 will uniformly provide the highest directivity index, although this may not be the case when the covariance matrix (6.42) is diagonally loaded to achieve greater robustness. We will return to this point in Section 6.5.

6.3.5 Generalized Sidelobe Canceller Implementation

The MVDR beamformer can be also realized with a GSC. Figure 6.11 illustrates the beamformer in GSC configuration.

Henceforth, we will suppress the frequency index ω for the sake of convenience. The weights of the GSC beamformer consists of three components, the quiescent weight vector $\mathbf{w}_q$, the blocking matrix $\mathbf{B}$, and the active weight vector $\mathbf{w}_a$. The output of the beamformer at frame k for a given subband can be expressed as

$$Y(k) = (\mathbf{w}_q - \mathbf{B}\mathbf{w}_a)^H\ \mathbf{X}(k). \tag{6.74}$$

In keeping with the GSC formalism, $\mathbf{w}_q$ is chosen to give unity gain in the desired look direction [67, Section 13.3.7]. The blocking matrix is chosen to be orthogonal to the quiescent vector such that $\mathbf{B}^H\mathbf{w}_q = \mathbf{0}$. The blocking matrix can be, for example, calculated with an orthogonalization technique such as the modified Gram—Schmidt method, QR decomposition or singular value decomposition (SVD) [23]. The orthogonality implies that the distortionless constraint will be satisfied for any choice of $\mathbf{w}_a$. Note that the blocking matrix is not unique.

In the case that the position of a sound source is static, the active weight vector is typically adjusted so that the variance of the GSC beamformer's outputs is minimized. Without diagonal loading, the solution of the active weight vector can be expressed as

$$\mathbf{w}_a^H = \mathbf{w}_q^H\boldsymbol{\Sigma}_\mathbf{X}\mathbf{B}\left(\mathbf{B}^H\boldsymbol{\Sigma}_\mathbf{X}\mathbf{B}\right)^{-1}, \tag{6.75}$$

where $\boldsymbol{\Sigma}_\mathbf{X}$ is the covariance matrix of the input vectors.

The interference signal can be effectively suppressed based on Equations (6.40), (6.41), or (6.75). However, they are not suitable for the online implementation because it assumes that the SOS, $\Sigma_{\mathbf{X}}$, are known from a sufficient amount of data. It is preferably updated on a sample-by-sample basis. In Section 6.3.6, the online algorithm of the conventional GSC beamformer will be described.

6.3.6 Recursive Implementation of the GSC

In many applications, the active weight vector of the GSC beamformer is updated at each frame by using a *recursive least squares* (RLS) method or a *least mean square* (LMS) algorithm. Here, we review, without formal proof, the RLS algorithm and briefly comment on the differences between the RLS and LMS algorithms. Further details can be found in [63, Section 7.4] and [63, Section 7.6], respectively.

In order to recursively update the active weight vector at frame T while retaining the information from frames $t = 0, 1, \ldots, T-1$, we first introduce *forgetting factor* $0 < \mu < 1$ and define the *exponentially weighted spatial spectral matrix*

$$\mathbf{\Phi}_X(T) \triangleq \sum_{t=1}^{T} \mu^{T-t} \mathbf{X}(t)\mathbf{X}^H(t). \tag{6.76}$$

Similarly, the exponentially weighted spatial spectral matrix of the output $\mathbf{Z}(t)$ of the blocking matrix is

$$\mathbf{\Phi}_Z(T) \triangleq \sum_{t=1}^{T} \mu^{T-t} \mathbf{Z}(t)\mathbf{Z}^H(t) = \mathbf{B}^H \mathbf{\Phi}_X(T)\mathbf{B}. \tag{6.77}$$

Finally, the cross-correlation between the blocking matrix's output $\mathbf{Z}(t)$ and the quiescent vector's output $Y_c(t)$ is given by

$$\mathbf{\Phi}_{ZY_c^*}(T) \triangleq \sum_{t=1}^{T} \mu^{T-t} \mathbf{Z}(t)Y_c^*(t) = \mathbf{B}^H \mathbf{\Phi}_X(T)\mathbf{w}_q. \tag{6.78}$$

Now, let us define the precision matrix and Kalman gain respectively as

$$\mathbf{P}_Z(T) = \mathbf{\Phi}_Z^{-1}(T) \text{ and} \tag{6.79}$$

$$\mathbf{g}_Z(T) = \frac{\mu^{-1}\mathbf{P}_Z(T-1)\mathbf{Z}(T)}{1+\mu^{-1}\mathbf{Z}^H(T)\mathbf{P}_Z(T-1)\mathbf{Z}(T)}. \tag{6.80}$$

The notations are deliberately chosen by considering the relationship between the RLS algorithm and the Kalman filter [24, Section 10.8]. In this case, the well-known Riccati equation [8, Section 7.4] can be expressed as

$$\mathbf{P}_Z(T) = \mu^{-1}\mathbf{P}_Z(T-1) - \mu^{-1}\mathbf{g}_Z(T)\mathbf{Z}^H(T)\mathbf{P}_Z(T-1). \tag{6.81}$$

Postmultiplying both sides of (6.81) by $\mathbf{Z}(T)$ and substituting the resulting equality into (6.80), we find the gain vector

$$\mathbf{g}_Z(T) = \mathbf{P}_Z(T)\mathbf{Z}(T). \tag{6.82}$$

The goal of the RLS method is to minimize a weighted sum of array outputs $Y(t)$ defined as

$$\Phi_Y(T) \triangleq \sum_{t=1}^{T} \mu^{T-t}|Y(t)|^2. \tag{6.83}$$

Note that the importance of the past outputs decreases exponentially with time T. Upon taking the derivative of (6.83) with respect to $\mathbf{w}_\mathrm{a}(T)$ and setting the result to zero, we find

$$\mathbf{w}_\mathrm{a}(T) = \mathbf{\Phi}_Z^{-1}(T)\mathbf{\Phi}_{ZY_c^*}(T) = \mathbf{P}_Z(T)\mathbf{\Phi}_{ZY_c^*}(T). \tag{6.84}$$

Substituting (6.78) and (6.81) into (6.84), we obtain

$$\mathbf{w}_\mathrm{a}(T) = \mathbf{w}_\mathrm{a}(T-1) + \mathbf{g}_Z(T)\mathrm{e}_p^*(T). \tag{6.85}$$

where

$$\mathrm{e}_p(T) = Y_c(T) - \mathbf{w}_\mathrm{a}^H(T-1)\mathbf{Z}(T). \tag{6.86}$$

At each frame, we can add the diagonal component σ_z^2 to $\mathbf{Z}(t)\mathbf{Z}^H(t)$. The RLS update formula with diagonal loading can be then written as

$$\mathbf{w}_\mathrm{a}(T) = \left[\mathbf{I} - \sigma_z^2\mathbf{P}_Z(T)\right]\mathbf{w}_\mathrm{a}(T-1) + \mathbf{g}_Z(T)\mathrm{e}_p^*(T), \tag{6.87}$$

Notice that (6.87) does not directly load the diagonal component of the sample spectral matrix in contrast to (6.41).

The update algorithms of (6.86) and (6.87) are suitable for on-line operation because the active weight vector can be adapted with the instantaneous input vector.

The LMS algorithm is a stochastic gradient procedure where a small step in the direction of the instantaneous gradient is taken at each time step. The difference between the RLS and LMS implementations is the step size parameter for the update formula. The LMS algorithm has a choice to adjust the step size parameter whereas the GSC-RLS algorithm uses $\mathbf{\Phi}_Z^{-1}$ as the step size. Multiplying the gradient with $\mathbf{\Phi}_Z^{-1}$ enables each component of the active weight vector to converge at the same rate. The disadvantage would be additional computation which can be non-trivial.

6.3.7 *Other Conventional GSC Beamformers*

In theory, the conventional MVDR beamfomers described above can eliminate interfering signals. However, in practice, they are prone to the signal cancellation problem whenever there is an interfering signal that is correlated with the desired signal. In real acoustic environments, interfering signals are highly correlated with the desired signal, as the latter is reflected from hard surfaces such as walls and tables and thereafter impinges on the sensor array from directions that are distinct from the look direction. Beam steering errors as well as magnitude and phase errors in the frequency responses of the individual sensors in an array can also cause signal cancellation to occur.

To avoid the signal cancellation, many algorithms have been proposed in the literature. Such approaches can be classified into the following categories:

1. Updating the active weight vector only when noise signals are dominant [9,26,52].
2. Constraining the update formula for the active weight vector [8,28,51].

3. Blocking the leakage of desired signal components into the sidelobe canceller by designing the blocking matrix [25,27,28,64].
4. Taking speech distortion due to the the leakage of a desired signal into account using a multichannel Wiener filter which aims at minimizing a weighted sum of residual noise and speech distortion terms [13].
5. Using acoustic transfer functions from the desired source to each sensor in an array instead of merely compensating for time delays [9,19,59,64].

As mentioned before, the algorithms discussed in this section minimize nearly the same criterion based on the SOS such as the variance of the beamformer's outputs. In the following section, we describe beamforming algorithms which adjust the active weight vector based on HOS; these algorithms have appeared more recently in the literature.

6.3.8 Beamforming Based on Higher Order Statistics

A multidimensional Gaussian pdf is completely characterized once its mean vector and covariance matrix are known. Hence, speech-enhancement techniques that assume—either implicitly or explicitly—that speech is a Gaussian random process are said to be second-order methods. For any non-Gaussian pdf on the other hand, the higher order moments have a great deal of influence on the fine structure of the pdf. Thus, enhancement techniques that take into account the deviation from Gaussianity inherent in human speech are said to based on HOS. Such techniques are the subject of this section.

Statistical Characteristics of Human Speech

In order to avoid the signal cancellation problem, HOS recently have been introduced to the field of acoustic beamforming. HOS have long been used in the field of *independent component analysis* (ICA) [29].

The entire field of ICA is founded on the assumption that all signals of interest are not Gaussian distributed [30]. Briefly, the reasoning is grounded on two points:

1. The *central limit theorem* states that the pdf of the sum of independent random variables (RVs) will approach Gaussian in the limit as more and more components are added, *regardless* of the pdfs of the individual components. This implies that the sum of several RVs will be closer to Gaussian than any of the individual components.
2. The *entropy* for a complex-valued RV Y is defined as

$$H(Y) \triangleq -\mathcal{E}\{\log p_Y(v)\} = -\int p_Y(v)\log p_Y(v)\mathrm{d}v, \tag{6.88}$$

where $p_Y(.)$ is the pdf of Y. The integral form of the entropy for the continuous RVs with the pdfs is referred to the *differential entropy*, or simply *entropy*, and distinguished from an ensemble average for samples. The entropy is the basic measure of information in *information theory* [18]. It is well known that a Gaussian RV has the highest entropy of all RVs with a given variance [18, Theorem 7.4.1], which also holds for complex Gaussian RVs [50, Theorem 2]. Hence, a Gaussian RV is, in some sense, the least *predictable* of all RVs. Information-bearing signals, on the other hand, are redundant and thus contain structure that makes them more predictable than Gaussian RVs.

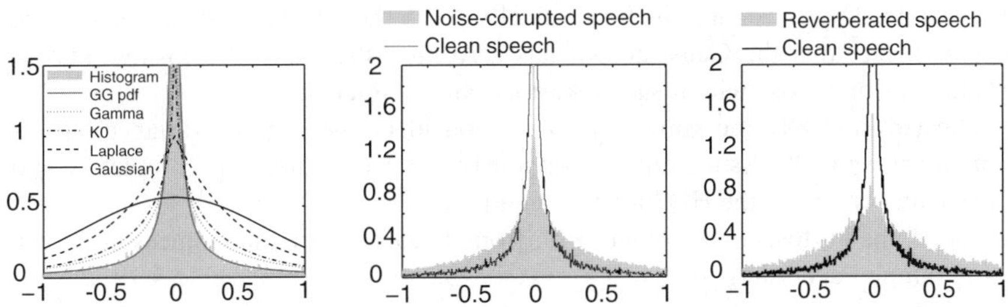

(a) Super-Gaussian pdf (b) Noise-corrupted speech (c) Reverberated speech

Figure 6.12 Histograms of real parts of subband frequency components of clean speech and (a) pdfs, (b) noise-corrupted speech, and (c) reverberated speech.

These points suggest that one must look for the *least* Gaussian RV in order to obtain the information-bearing signals. The fact that the pdf of speech is super-Gaussian has often been reported in the literature [34,45,16]. Noise, on the other hand, is more nearly Gaussian distributed. In fact, the pdf of the sum of several super-Gaussian RVs becomes closer to Gaussian. Thus, a mixture consisting of a desired signal and several interfering signals can be expected to be nearly Gaussian distributed.

Figure 6.12a shows a histogram of the real parts of subband samples of speech at frequency 800 Hz. To generate these histograms, the authors used 43.9 min of clean speech recorded with a close-talking microphone (CTM) from the development set of the Speech Separation Challenge, Part 2 (SSC2) [43]. The Gaussian, Laplace, K_0, Γ [34], and generalized Gaussian (GG) pdfs [35] are also shown in Figure 6.12a. In Figure 6.12a, the parameters of each pdf were estimated from training data based on the ML criterion. It is clear from Figure 6.12 that the distribution of clean speech is not Gaussian but super-Gaussian. Figure 6.12a suggests that the GG pdfis well suited for modeling subband samples of speech. From Figure 6.12 a, it is also clear that the Laplace, K_0, Γ, and GG pdfs exhibit the spikey and heavy-tailed characteristics. Super-Gaussian pdfs have a sharp concentration of probability mass at the mean, relatively little probability mass as compared with the Gaussian at intermediate values of the argument, and a relatively large amount of probability mass in the tail; that is far from the mean.

Figure 6.12b shows histograms of real parts of subband components calculated from clean speech and noise-corrupted speech. The figure indicates that the pdf of the noise-corrupted signal, which is in fact the sum of the speech and noise signals, is closer to Gaussian than that of clean speech.

Figure 6.12c shows histograms of clean speech and reverberant speech in the subband domain. In order to produce the reverberant speech, a clean speech signal was convolved with an impulse response measured in a room; see Lincoln *et al.* [43] for the configuration of the room. We can observe from Figure 6.12c that the pdf of reverberated speech is also closer to Gaussian than the original clean speech.

Maximum Kurtosis Beamforming

The *excess kurtosis* or simply *kurtosis* of a RV Y with zero mean is defined as

$$\text{kurt}(Y) \triangleq \mathcal{E}\{|Y|^4\} - \beta(\mathcal{E}\{|Y|^2\})^2, \quad (6.89)$$

where β is a positive constant, which is typically set to three for kurtosis of real-valued RVs in order to ensure that the Gaussian pdf has zero kurtosis; pdfs with positive kurtosis are super-Gaussian, and those with negative kurtosis are sub-Gaussian.

As indicated in (6.89), the kurtosis measure considers not only the variance but also the fourth moment, an HOS. As mentioned previously, any Gaussian pdf is completely specified by its mean and variance; the HOS are not required.

In practice, the kurtosis of T outputs $Y(t)$ from a beamformer can be measured by simply averaging samples according to

$$J_{\text{kurt}}(Y) \triangleq \frac{1}{T}\sum_{t=0}^{T-1} |Y(t)|^4 - \beta \left(\frac{1}{T}\sum_{t=0}^{T-1} |Y(t)|^2 \right)^2, \tag{6.90}$$

where the frequency index ω has been omitted for clarity. The kurtosis criterion does not require any explicit assumption as to the exact form of the pdf. Due to its simplicity, it is widely used as a measure of non-Gaussianity. As demonstrated in Kumatani *et al.* [34,35], maximizing the degree of super-Gaussianity yields an active weight vector $\mathbf{w}_\text{a}$ capable of canceling interference—including incoherent noise that leaks through the sidelobes—without the signal cancellation problem encountered in conventional beamforming.

As discussed in Section 6.3.1, diagonal loading is typically used in beamforming with SOS in order to reduce the norm of the active weight vector and thereby improve robustness by inhibiting the formation of excessively large sidelobes [67, Section 13.3.8]. Such a regularization term can be also applied to maximum kurtosis beamforming by defining the modified optimization criterion of (6.90) with a weight parameter α as

$$\mathcal{J}_{\text{kurt}}(Y;\alpha) \triangleq J_{\text{kurt}}(Y) - \alpha \|\mathbf{w}_\text{a}\|^2 \quad \alpha > 0. \tag{6.91}$$

In Kumatani *et al.* [35], the sensitivity of the weight parameter α was investigated in terms of speech recognition. The best recognition performance was obtained with $\alpha = 0.01$ although the effect was not significant.

Unfortunately, there is no closed-form solution for the active weight vector which provides the maximum kurtosis. Thus, we have to resort to numerical optimization algorithms such as gradient descent. Upon substituting (6.74) into (6.91) and taking the partial derivative with respect to $\mathbf{w}_\text{a}$, we obtain

$$\frac{\partial \mathcal{J}_{\text{kurt}}(Y;\alpha)}{\partial \mathbf{w}_\text{a}^*} = \frac{2}{T}\sum_{t=0}^{T-1} \left\{ -|Y(t)|^2 + \beta\sigma_Y^2 \right\} \mathbf{B}^H(t)\mathbf{X}(t)Y^*(t) - \alpha \mathbf{w}_\text{a}, \tag{6.92}$$

where σ_Y^2 is the variance of beamformer's outputs.

Equation (6.92) is sufficient to implement a numerical optimization algorithm based, for example, on the method of steepest descent [4, Section 1.6], whereby kurtosis of beamformer's outputs can be maximized. The norm of the active weight vector is usually normalized in addition to the regularization term because it tends to become large.

Maximum Negentropy Beamforming

Another criterion for measuring the degree of super-Gaussianity is negentropy. The negentropy of a complex-valued RV Y is defined as

$$\text{neg}(Y) \triangleq H(Y_{\text{gauss}}) - \beta H(Y), \tag{6.93}$$

where Y_{gauss} is a Gaussian variable with the same variance σ_Y^2 as Y, β is a positive constant for adjusting the equilibrium condition and normally set to unity for negentropy. The entropy of Y_{gauss} can be expressed as

$$H(Y_{\text{gauss}}) = \log\left|\sigma_Y^2\right| + (1 + \log \pi). \tag{6.94}$$

Note that negentropy is nonnegative, and zero if and only if Y has a Gaussian distribution. Clearly, it can measure how far the desired distribution is from the Gaussian pdf. Computing the entropy of the super-Gaussian variables $H(Y)$ requires knowledge of their specific pdf. Thus, it is important to find a family of pdfs capable of closely modeling the distributions of actual speech signals.

However, the value calculated for kurtosis can be strongly influenced by a few samples with a low observation probability. Hyvärinen and Oja [30] demonstrates that negentropy is generally more robust in the presence of outliers than kurtosis.

As shown in Figure 6.12a, the distribution of the subbands of clean speech can be represented with the generalized Gaussian pdf. In the case that the complex-valued RV Y possesses circular symmetry, the complex-valued GG pdf can be expressed as

$$p_{\text{GG}}(Y) = \frac{f}{2\pi B^2(f)\Gamma(2/f)\hat{\sigma}^2} \exp\left[-\left|\frac{Y}{\hat{\sigma}B(f)}\right|^f\right], \tag{6.95}$$

where

$$B(f) = \left[\frac{\Gamma(2/f)}{\Gamma(4/f)}\right]^{1/2} \quad \text{and} \tag{6.96}$$

$\Gamma(.)$ is the gamma function [3, Section 5.2]. Note that the GG with $f = 2$ corresponds to the Gaussian pdf, whereas the GG pdf converges to a uniform distribution in the case of $f \rightarrow +\infty$.

The parameters of the GG pdf can be, for example, estimated using the ML criterion, as in Wölfel and McDonough [67, Section 13.5.2] and Kumatani *et al.* [37]. The shape parameters are estimated independently for each subband, as the optimal pdf is frequency dependent.

In order to develop maximum negentropy beamforming, we compute an ensemble average of negative log-likelihoods instead of differential entropy. In this case, negentropy of T frames of output from the array can be calculated according to

$$J_{\text{neg}}(Y) = -\frac{1}{T}\sum_{t=0}^{T-1} \log p_{\text{gauss}}\left(Y(t)\right) + \beta\frac{1}{T}\sum_{t=0}^{T-1} \log p_{\text{GG}}\left(Y(t)\right), \tag{6.97}$$

where $p_{\text{gauss}}(.)$ is the complex Gaussian pdf. Similar to (6.91), a regularization term can be added to the empirical negentropy to provide the modified optimization criterion

$$\mathcal{J}_{\text{neg}}(Y;\alpha) = J_{\text{neg}}(Y) - \alpha\|\mathbf{w}_{\text{a}}\|^2 \quad \alpha > 0. \tag{6.98}$$

Upon substituting (6.74) into (6.98) and taking the partial derivative, we obtain

$$\frac{\partial \mathcal{J}_{\text{neg}}(Y;\alpha)}{\partial \mathbf{w}_{\text{a}}^*} = \frac{1}{T}\sum_{t=0}^{T-1}\left\{-\frac{1}{\sigma_Y^2} + \beta\frac{f|Y(t)|^{f-2}}{2(B(f)\hat{\sigma})^f}\right\}\mathbf{B}^H(t)\mathbf{X}(t)Y^*(t) - \alpha\mathbf{w}_{\text{a}}. \tag{6.99}$$

The HOS-based beamformers, maximum kurtosis and maximum negentropy beamformers, do not suffer from signal cancellation encountered in the SOS-based adaptive beamformers. Therefore, the active weight vector can be adapted even when the desired source is active. Indeed, Kumatani *et al.* [34,35] demonstrated through acoustic simulations that beamformers based on HOS can emphasize the desired signal by coherently adding its reflections after a suitable phase shift in the subband domain. Hence, adaptation of the active weight vector is best performed *while* the desired speaker is active.

6.3.9 Online Implementation

Adaptive beamforming algorithms require a certain amount of data for stable estimation of the active weight vector. In the case of HOS-based beamforming, this problem becomes significant because it entirely relies on numerical optimization algorithms.

In order to achieve efficient estimation, a subspace (eigenspace) filter [63, Section 6.8] can be used as a preprocessing step for estimation of the active weight vector. Motivations behind this idea are to (1) reduce the dimensionality of the active weight vector and (2) improve speech enhancement performance based on decomposition of the outputs of the blocking matrix into spatially correlated and ambient signal components. Such decomposition can be achieved by performing an eigendecomposition on the covariance matrix of blocking matrix's outputs. Then, we select the eigenvectors corresponding to the largest eigenvalues as the dominant modes [63, Section 6.8.3]. The dominant modes are associated with the spatially correlated signals and the other modes are averaged as a signal model of ambient noise. By doing so, we can readily subtract the averaged ambient noise component from the beamformer's output. Moreover, the reduction of the dimension of the active weight leads to computationally efficient and reliable estimation. Notice that we adjust the active weight vector based on the maximum kurtosis criterion in contrast to the normal dominant-mode rejection (DMR) beamformers [63, Section 6.8.3]. It is also worth noting that subspace filtering here is analogous to whitening used as a preprocessing measure in the field of ICA [30].

In the following sections, we first discuss the subspace method for maximum kurtosis beamforming and then describe its online implementation. We will continue to omit the frequency index ω for the sake of convenience.

Subspace Method

In the case that there are neither steering errors nor mismatches between microphones, the blocking matrix's output $\mathbf{Z}(t)$ only contains the spatially correlated (coherent) interference and ambient (incoherent) noise signals. However, in the real environments, it also includes the desired signal components due to those errors as well as reverberation effects.

Let us first denote the D spatially correlated signal components contained in the output of the $N \times (N-1)$ blocking matrix as

$$\mathbf{V}(t) = [V_0(t), \cdots, V_d(t), \cdots, V_{D-1}(t)]^T . \tag{6.100}$$

Then, the output of the blocking matrix can be expressed as

$$\mathbf{Z}(t) = \mathbf{A}\mathbf{V}(t) + \mathbf{N}(t), \tag{6.101}$$

where $\mathbf{A}$ and $\mathbf{N}(t)$ represent a mixing matrix and the ambient noise signals, respectively. The direct path from the desired source signal to each microphone is assumed to be excluded from $\mathbf{A}$ because of the distortionless constraint imposed with the blocking matrix. Thus, in the case that there is neither reverberation nor error such as a microphone array mismatch and steering error, $\mathbf{Z}$ consists of the interference signals only.

Assuming that $\mathbf{V}(t)$ and $\mathbf{N}(t)$ are uncorrelated, we can write the covariance matrix of $\mathbf{Z}$ as:

$$\Sigma_{\mathbf{Z}} = \mathcal{E}\left[\mathbf{Z}(t)\mathbf{Z}^H(t)\right] = \mathbf{A}\Sigma_{\mathbf{V}}\mathbf{A}^H + \Sigma_{\mathbf{N}}, \tag{6.102}$$

where

$$\Sigma_{\mathbf{V}} = \mathcal{E}\left[\mathbf{V}(t)\mathbf{V}^H(t)\right] \text{ and } \Sigma_{\mathbf{N}} = \mathcal{E}\left[\mathbf{N}(t)\mathbf{N}^H(t)\right].$$

The subspace method seeks a set of D linearly independent vectors contained in the subspace, $\Re\{\mathbf{A}\}$, spanned by the column vectors of $\mathbf{A}$. The first step for obtaining such set of the vectors is to solve the generalized eigenvalue (GE) decomposition problem as in Van Trees [63], Roy and Kailath [58, Section 6.8],

$$\Sigma_{\mathbf{Z}}\mathbf{E} = \Sigma_{\mathbf{N}}\mathbf{E}\mathbf{\Lambda},$$

where $\mathbf{\Lambda}$ is a diagonal matrix of the eigenvalues sorted in the descending order:

$$\mathbf{\Lambda} = \text{diag}\left[\lambda_0, \cdots, \lambda_{D-1}, \cdots, \lambda_{N-2}\right] \tag{6.103}$$

and $\mathbf{E}$ is a matrix of the corresponding eigenvectors:

$$\mathbf{E} = \left[\mathbf{e}_0, \cdots, \mathbf{e}_{D-1}, \cdots, \mathbf{e}_{N-2}\right]. \tag{6.104}$$

Here, we assume that $\Sigma_{\mathbf{N}}$ is an identity matrix. Then, we select the eigenvectors with the D largest eigenvalues, $\mathbf{E}_V = [\mathbf{e}_0, \cdots, \mathbf{e}_{D-1}]$. In the similar manner, we define the subspace for the ambient noise as $\mathbf{E}_N = [\mathbf{e}_D, \cdots, \mathbf{e}_{N-2}]$.

The theoretical properties of the eigenvectors and eigenvalues can be summarized as follows:

- The subspace spanned by the eigenvectors is equal to that of $\mathbf{A}$, that is $\mathbb{R}\{\mathbf{E}_V\} = \mathbb{R}\{\mathbf{A}\}$.
- The power of the D spatially correlated signals is associated with the D largest eigenvalues.
- The power of $\mathbf{N}(k)$ is equally distributed over all the eigenvalues and $N - D - 1$ smallest eigenvalues are all equal to σ_N^2, that is the noise floor.
- $\mathbb{R}\{\mathbf{E}_N\}$ is the orthogonal complement of $\mathbb{R}\{\mathbf{E}_V\}$, that is $\mathbb{R}\{\mathbf{E}_N\} = \mathbb{R}\{\mathbf{E}_V\}^{\perp}$.

In order to cluster the eigenvectors for the ambient noise, we have to determine the number of the dominant eigenvalues D. Figure 6.13 illustrates actual eigenvalues sorted in descending order over frequencies. In order to generate the plots of the figures in this section, we computed the eigenvalues from the outputs of the blocking matrix on the real data described in Kumatani *et al.* [38].

As shown in Figure 6.13, it is relatively easy to determine the number of the dominant modes, D, especially in the case that the number of microphones is much larger than the number of the spatially correlated signals. We determined D by setting a threshold on the contribution ratio, $\lambda_i / \sum_{j=0}^{N-2} \lambda_j$. Figure 6.14 shows the number of the contribution ratios exceeding thresholds, 10^{-2}, 10^{-3}, and 10^{-4}, at each frequency. Figure 6.14 indicates how many dominant modes are used for subspace filtering when we ignore the eigenvectors associated with the lower

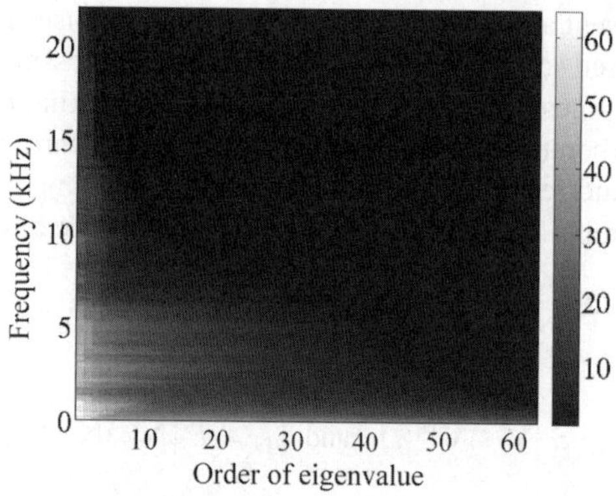

Figure 6.13 Three-dimensional representation of the eigenvalue distribution as a function of the order and frequencies.

contribution ratio than the threshold. It is clear from Figure 6.14 that the lower the threshold for the contribution ratio is set, the more eigenvectors are used. Generally, the lower threshold leads to accurate representation of the spatially correlated signals at the expense of computational efficiency.

In the optimization of the active weight vector, we estimate each component corresponding to the ambient noise signal separately. Accordingly, we use the sum of the eigenvectors for the ambient noise space as $\tilde{\mathbf{e}}_n = \sum_{d=D}^{N-2} \mathbf{e}_d$. Our subspace filter can be now written as

$$\mathbf{U} = [\mathbf{e}_0, \cdots, \mathbf{e}_{D-1}, \tilde{\mathbf{e}}_n,] . \tag{6.105}$$

Note that we assume the covariance matrix in (6.102) can be approximated as

$$\Sigma_{\mathbf{Z}} \approx \sum_{\mathrm{d}=0}^{D-1} \lambda_{\mathrm{d}} \mathbf{e}_{\mathrm{d}} \mathbf{e}_{\mathrm{d}}^{H} + \bar{\sigma}_N^{2} \tilde{\mathbf{e}}_n \tilde{\mathbf{e}}_n^{H} , \tag{6.106}$$

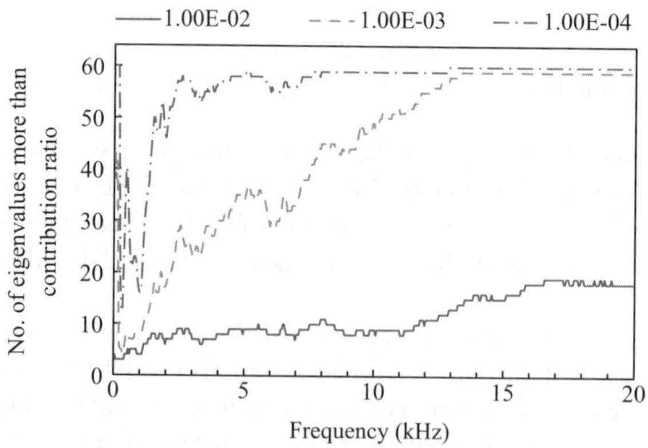

Figure 6.14 Number of contribution ratios exceeding threshold of 10^{-2}, 10^{-3} and 10^{-4} as a function of frequency.

where

$$\bar{\sigma}_N^2 = \frac{1}{N-D-1}\sum_{\mathrm{d}=D}^{N-2}\lambda_{\mathrm{d}}.$$

With the outputs of the subspace filter, we estimate the active weight vector providing the maximum kurtosis value. If the output of the subspace filter is a noise signal, the corresponding component of the active weight vector should be adjusted so as to subtract the noise component from the output of the quiescent vector. If it is an echo of the desired signal, the active weight vector could shift the phase and add the component to the desired signal in order to strengthen it. These operations would be easier by separating the echo from the ambient noise component with the subspace filter.

Block-Wise Adaptation of Maximum Kurtosis Beamforming with Subspace Filtering

Figure 6.15 shows configuration of the MK beamformer with the subspace filter. The beamformer's output can be expressed as

$$Y(t) = \left[\mathbf{w}_{\mathrm{q}}(t) - \mathbf{B}(t)\mathbf{U}(t)\mathbf{w}_{\mathrm{a}}(t)\right]^H \mathbf{X}(t). \tag{6.107}$$

The active weight vector is adjusted so as to achieve the maximum kurtosis of the beamformer's outputs. The difference between (6.74) and (6.107) is the subspace filter between the blocking matrix and active weight vector. The subspace filter can decompose the output vector into the spatially correlated signal and ambient noise components. Therefore, we only need to estimate the phase shifts of the active weight vector on the constrained subspace [30, Section 7.4]. Moreover, the solution of the general eigenvector decomposition is less dependent of the initial values than that of the gradient algorithm for multidimensional maximization or minimization.

Based on equation (6.107), the kurtosis of the outputs is computed from a block of input subband samples at each block instead of using the entire utterance data. We incrementally update the dominant modes and active weight vector at each block b consisting of L_b samples here. Accordingly, the kurtosis at block b can be expressed as

$$J(Y(b)) = \left(\frac{1}{L_b}\sum_{t=(b-1)L_b}^{bL_b-1}|Y(t)|^4\right) - \beta\left(\frac{1}{L_b}\sum_{k=(b-1)L_b}^{bL_b-1}|Y(t)|^2\right)^2. \tag{6.108}$$

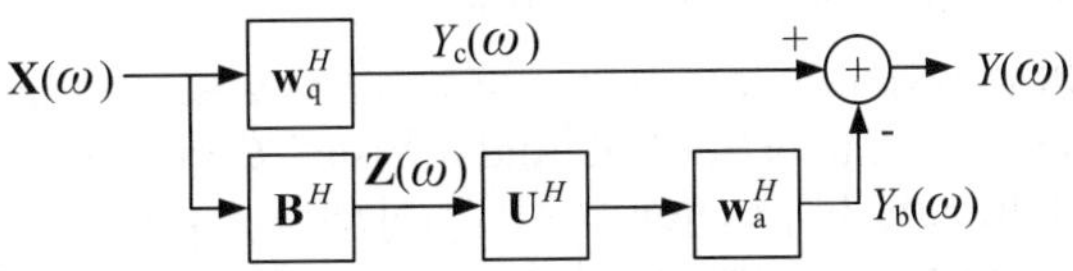

Figure 6.15 Maximum kurtosis beamformer with the subspace filter.

In order to inhibit the formation of excessively large sidelobes, a regularization term is applied to the cost function as follows:

$$\mathcal{J}(Y(b);\alpha) = J(Y(b)) - \alpha\|\mathbf{w}_{\mathrm{a}}(b)\|^2. \tag{6.109}$$

In addition to the regularization term, a unity constraint is imposed on a norm of the active weight vector so as to prevent it from exceeding that of the quiescent vector.

Then, the active weight vector is adjusted so as to maximize the sum of the kurtosis and regularization terms (6.109) under the norm constraint at each block. Again, we have to resort to the gradient-based numerical optimization algorithm. Upon substituting (6.107) and (6.108) into (6.109) and taking the partial derivative with respect to the active weight vector, we obtain

$$\frac{\partial \mathcal{J}(Y(b);\alpha)}{\partial \mathbf{w}_{\mathrm{a}}(b)^*} = -\frac{2}{L_b}\left(\sum_{t=(b-1)L_b}^{bL_b-1} |Y(t)|^2 \mathbf{U}^H(b)\mathbf{B}^H(t)\mathbf{X}(t)Y^*(t)\right)$$
$$+\frac{2\beta}{L_b^2}\left(\sum_{t=(b-1)L_b}^{bL_b-1} |Y(t)|^2\right)\left(\sum_{t=(b-1)L_b}^{bL_b-1} \mathbf{U}^H(b)\mathbf{B}^H(t)\mathbf{X}(t)Y^*(t)\right) - \alpha\mathbf{w}_{\mathrm{a}}(b). \tag{6.110}$$

The gradient (6.110) is iteratively calculated with a block of subband samples until the kurtosis value of the beamformer's outputs converges. For the gradient algorithm, the active weight vectors are initialized with the estimates at the previous block. The active weight vector of the first block is initialized with $\mathbf{w}_{\mathrm{a}} = [0, \cdots, 0, 1]^T$ because the last component corresponds to the ambient noise which should be subtracted from the output of the quiescent vector.

The beamforming algorithm can be summarized as follows:

1. Initialize the active weight vector with $\mathbf{w}_{\mathrm{a}}(0) = [0, \cdots, 1]^T$.
2. Given estimates of time delays, calculate the quiescent vector and blocking matrix.
3. For each block of input subband samples, recursively update the covariance matrix as $\Sigma_{\mathbf{Z}}(b) = \mu\Sigma_{\mathbf{Z}}(b-1) + (1-\mu)\Sigma_{\mathbf{Z}}(b)$ where μ is the forgetting factor, calculate the dominant modes $\mathbf{U}^H(b)$ and estimate the active weight vector $\mathbf{w}_{\mathrm{a}}(b)$ based on the gradient information computed with (6.110) subject to the norm constraint until the kurtosis value of the beamformer's outputs converges.
4. Initialize the active weight vector obtained in step 3 for the next block and go to the step 2.

This block-wise method is able to track a nonstationary sound source, and provides a more accurate gradient estimate than *sample-by-sample* gradient estimation algorithms.

6.3.10 *Speech-Recognition Experiments*

The results of the DSR experiments reported in this section were obtained on speech material from the Multichannel Wall Street Journal Audio Visual Corpus (MC-WSJ-AV) recorded by the Augmented Multiparty Interaction (AMI) project; see Lincoln *et al.* [43] for details of the data-collection apparatus. The size of the recording room was 650 × 490 × 325 cm and the reverberation time T_{60} was approximately 380 ms. In addition to reverberation, some recordings include significant amounts of background noise produced by computer fans and air conditioning. The far-field speech data were recorded with two circular, equispaced eight-channel microphone arrays with the diameter of 20 cm. Additionally, each speaker was equipped with a CTM to provide a reference signal for speech recognition. The sampling rate of the recordings was 16 kHz. For the experiments, we used a portion of data from the *single speaker stationary* scenario where a speaker was asked to read sentences from six positions, four seated around the table, one standing at the white board and one standing at a presentation screen. The test data set for the experiments contains recordings of 10 speakers where each speaker reads approximately 40 sentences taken from the 5000 word vocabulary WSJ task. This provided a total of 352 utterances for a total 39.2 minutes of speech.

Prior to beamforming, we first estimated the speaker's position with a source tracking system [21]. Based on the average speaker position estimated for each utterance, active weight vectors $\mathbf{w}_a$ were estimated for the source. After beamforming, we perform *Zelinski postfiltering* [44] which uses the auto- and cross-power spectrums of the input signals to estimate the target signal and noise power spectrums under the assumption of zero cross-correlation between noise on different sensors. The parameters of the GG pdf were trained with 43.9 min of speech data recorded with the CTM in the SSC2 development set. The training data set for the GG pdf contains recordings of 5 speakers.

We performed four decoding passes on the waveforms obtained with various beamforming algorithms including ones described in prior sections. The details of our ASR system used in the experiments are given in Kumatani *et al.* [35]. Each pass of decoding used a different acoustic model or speaker adaptation scheme. The speaker adaptation parameters were estimated using the word lattices generated during the prior pass, as in Uebel and Woodland [61]. A description of the four decoding passes follows:

1. Decode with the unadapted, conventional ML acoustic model and bigram language model (LM).
2. Estimate vocal tract length normalization (VTLN) [67, Section 9] parameters and *constrained maximum likelihood linear regression* (CMLLR) parameters for each speaker as discussed in Section 11.4.2, then redecode with the conventional ML acoustic model and bigram LM.
3. Estimate VTLN, CMLLR, and MLLR parameters for each speaker as discussed in Section 11.4, then redecode with the conventional model and bigram LM.
4. Estimate VTLN, CMLLR, and MLLR parameters for each speaker, then redecode with the ML-SAT model [67, Section 8.1] and bigram LM.

Table 6.1 shows the WERs for every beamforming algorithm. As references, WERs in recognition experiments on speech data recorded with a *single array channel* (SAC) and CTM are also presented in the table. It is clear from these results that the maximum kurtosis beamforming (MK BF) and maximum negentropy beamforming (MN BF) methods can provide better recognition performance than the SOS-based beamformers, such as the superdirective beamformer (SD BF) [67, Section 13.3.4], the MVDR beamformer (MVDR BF) described in Section 6.3.1, and the generalized eigenvector beamformer (GEV BF) [64]. This is because the HOS-based beamformers can use echoes to enhance the desired signal, as mentioned previously. Adaptation of the HOS-based beamformers was performed while the desired source was active. This fact implies that the maximum kurtosis and negentropy beamformers do not

Table 6.1 Word error rates for each beamforming algorithm after every decoding pass.

	Pass (%WER)			
Beamforming (BF) algorithm	1	2	3	4
Single array channel (SAC)	87.0	57.1	32.8	28.0
Delay-and-sum (D&S) BF	79.0	38.1	20.2	16.5
Superdirective (SD) BF	71.4	31.9	16.6	14.1
Minimum variance distortionless response (MVDR) BF	78.6	35.4	18.8	14.8
Generalized eigenvector (GEV) BF	78.7	35.5	18.6	14.5
Maximum kurtosis (MK) BF	75.7	32.8	17.3	13.7
Maximum negentropy (MN) BF	75.1	32.7	16.5	13.2
SD MN BF	75.3	30.9	15.5	12.2
Close talking microphone (CTM)	52.9	21.5	9.8	6.7

suffer the signal cancellation. It is also clear from Table 6.1 that every adaptive beamformer achieved better recognition performance than the delay-and-sum beamformer (D&S BF).

The SOS-based and HOS-based beamformers can be profitably combined because maximum kurtosis and negentropy beamformers employ different criteria for estimation of the active weight vector. For example, the superdirective beamformer's weight can be used as the quiescent weight vector in GSC configuration [36]. We observe from Table 6.1 that the maximum negentropy beamformer with superdirective beamformer (SD MN BF) provided the best recognition performance in this task.

6.4 Spherical Microphone Arrays

In this section, we discuss the fundamentals of spherical arrays. This includes the acoustic phenomena that occur when a plan wave scatters from the surface of a rigid sphere. We also develop the concept of expanding a function defined on the surface of a sphere in spherical harmonics; such series expansions will play a key role in our development of beamforming algorithms for spherical arrays. This material is intended to provide the theoretical underpinning to the emprical studies presented in the following sections.

Meyer and Elko [48] and Abhayapala and Ward [1] were among the first to propose the use of spherical microphone arrays for beamforming. The state-of-the-art theory of beamforming with spherical microphone arrays explicitly takes into account two phenomena of sound propagation, namely, *diffraction* and *scattering*; see Kutruff [39, Section 2] and Williams [66, Section 6.10]. While these phenomena are certainly present in all acoustic array-processing applications, no particular attempt is typically made to incorporate them into conventional beamforming algorithms; rather, they are simply assumed to contribute to the room impulse response. The development in this section is based loosely on Meyer and Elko [49, Section 2] with interspersed elements from Teutsch [60].

To begin our discussion, let us express a plane wave impinging with a polar angle of θ on an array of microphones as

$$G_{\text{pw}}(kr, \theta, t) \triangleq \mathrm{e}^{i(\omega t + kr\cos\theta)}, \tag{6.111}$$

where $k \triangleq 2\pi/\lambda$ is the wavenumber as before, and *r* is the range at which the wave is observed. The definition (6.111) can be rewritten as

$$G_{\text{pw}}(kr, \theta, t) = \sum_{n=0}^{\infty} i^n \,(2n+1)\, j_n(kr)\, P_n(\cos\theta)\, \mathrm{e}^{i\omega t}, \tag{6.112}$$

where j_n and P_n are, respectively, the *spherical Bessel function* of the first kind [53, Section 10.47] and the *Legendre polynomial*, both of order *n* [66, Section 6.10.1]. A similar expansion of spherical waves—such as would be required for near-field analysis—is provided by Williams [66, Section 6.7.1]. If the plane wave encounters a rigid sphere with a radius of *a* it is scattered [66, Section 6.10.3] to produce a wave with the pressure profile

$$G_{\text{s}}(kr, ka, \theta, t) = -\sum_{n=0}^{\infty} i^n \,(2n+1)\, \frac{j_n'(ka)}{h_n'(ka)}\, h_n(kr)\, P_n(\cos\theta)\, \mathrm{e}^{i\omega t}, \tag{6.113}$$

where $h_n = h_n^{(1)}$ denotes the *Hankel function* [53, Section 10.47] of the first kind[2] while the prime indicates the derivative of a function with respect to its argument. Combining (6.112) and (6.113) yields the total sound pressure field [66, Section 6.10.3]:

$$G(kr, ka, \theta) = \sum_{n=0}^{\infty} i^n \,(2n+1)\, b_n(ka, kr)\, P_n(\cos\theta), \tag{6.114}$$

where the *n*th *modal coefficient* is defined as

$$b_n(ka, kr) \triangleq j_n(kr) - \frac{j_n'(ka)}{h_n'(ka)} h_n(kr). \tag{6.115}$$

In principle, *ka* and *kr* need not be equivalent, but in practice they are; that is the sensors of the array are located on the surface of the scattering sphere. Hence, in the sequel, we will uniformly replace *kr* with *ka*. Note that the time dependence of (6.114) through the term $e^{i\omega t}$ has been suppressed for convenience. Plots of $|b_n(ka, ka)|$ for $n = 0, \ldots, 8$ are shown in Figure 6.16.

Let us now define the *spherical harmonic* of order *n* and degree *m* as [14]

$$Y_n^m(\Omega) \triangleq \sqrt{\frac{(2n+1)}{4\pi} \frac{(n-m)!}{(n+m)!}}\, P_n^m(\cos\theta)\, \mathrm{e}^{im\phi}, \tag{6.116}$$

where $\Omega \triangleq (\theta, \phi)$ and P_n^m is the *associated Legendre function* of order *n* and degree *m* [15, Section 14.3]. The spherical harmonics fulfill the same role in the decomposition of square-integrable functions defined on the surface of a sphere as that played by the complex exponential $\mathrm{e}^{i\omega n t}$ for Fourier analysis of periodic functions defined on the real line. Let γ represent the angle between the points $\Omega \triangleq (\theta, \phi)$ and $\Omega_s \triangleq (\theta_s, \phi_s)$ lying on a sphere, such that

$$\cos\gamma = \cos\theta_s \cos\theta + \sin\theta_s \sin\theta \cos(\phi_s - \phi). \tag{6.117}$$

[2] Note that Meyer and Elko [49] incorrectly used the Hankel function of the second kind in (6.113) and (6.115).

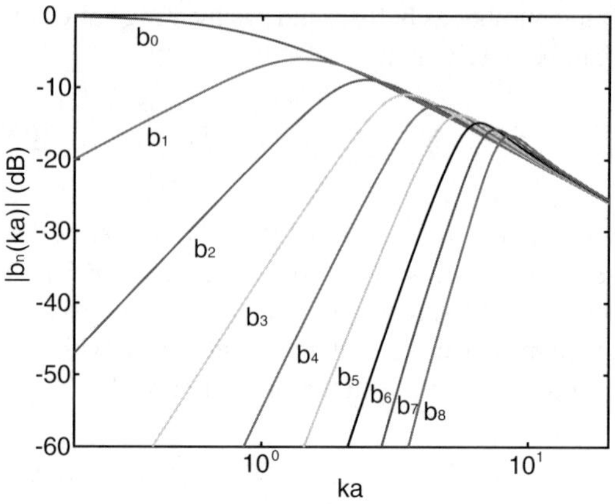

Figure 6.16 Magnitudes of the modal coefficients $b_n(ka, ka)$ for $n = 0, 1, \ldots, 8$, where a is the radius of the scattering sphere and k is the wavenumber.

Then, we can express the *addition theorem for spherical harmonics* [2, Theorem 12.8] as

$$P_n(\cos\gamma) = \frac{4\pi}{2n+1} \sum_{m=-n}^{n} Y_n^m(\Omega_s)\bar{Y}_n^m(\Omega), \tag{6.118}$$

where $\bar{Y}$ denotes the complex conjugate of Y. Upon substituting (6.118) into (6.114), we find

$$G(\Omega_s, ka, \Omega) = 4\pi \sum_{n=0}^{\infty} i^n\, b_n(ka) \sum_{m=-n}^{n} Y_n^m(\Omega_s)\bar{Y}_n^m(\Omega). \tag{6.119}$$

The spherical harmonics$Y_0 \triangleq Y_0^0$, $Y_1 \triangleq Y_1^0$, $Y_2 \triangleq Y_2^0$, and $Y_3 \triangleq Y_3^0$ are shown in Figure 6.17 The spherical harmonics possess the all important property of *orthonormality*, which implies

$$\begin{aligned} \delta_{n,n'}\,\delta_{m,m'} &= \int_{\Omega} Y_n^m(\Omega)\,\bar{Y}_{n'}^{m'}(\Omega)\,d\Omega \\ &= \int_0^{2\pi}\int_0^{\pi} Y_n^m(\Omega)\,\bar{Y}_{n'}^{m'}(\Omega)\sin\theta d\theta d\phi, \end{aligned} \tag{6.120}$$

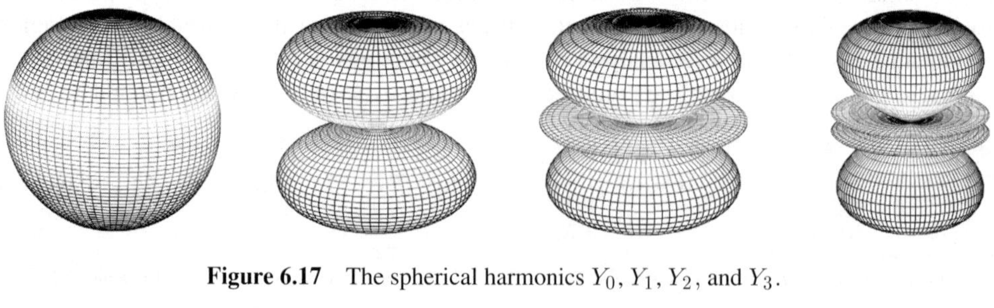

Figure 6.17 The spherical harmonics Y_0, Y_1, Y_2, and Y_3.

where Ω indicates the surface of the sphere of integration and the *Kronecker delta* is defined as

$$\delta_{n,m} \triangleq \begin{cases} 1, & n, m = 0, \\ 0, & \text{otherwise.} \end{cases} \tag{6.121}$$

The plot of Y_0 in Figure 6.17 and magnitudes of the modal coefficients in Figure 6.16 make it clear why the spherical array—like its linear, conventional counterpart—suffers from poor directivity at low frequencies. For $ka = 0.2$, which corresponds to 220 Hz for $a = 5$ cm, only Y_0 is truly useful for beamforming, as the levels of all other modes are 20 dB or more below that of Y_0; amplifying the other modes sufficiently to use them in beamforming would introduce a great deal of self-noise from the array components into the final signal. But Y_0 is completely isotropic; that is, it has no directional characteristics whatsoever, and hence provides no improvement in directivity over a single microphone.

The implication of (6.120) is that the individual terms of the series expansion are orthonormal. Hence, any sound field $V(ka, \Omega_s)$, which is square integrable over a sphere with radius a, admits the modal decomposition [60, Section A.3]:

$$V(ka, \Omega_s) = \sum_{n=0}^{\infty} \sum_{m=-n}^{n} V_n^m(ka)\, Y_n^m(\Omega_s), \tag{6.122}$$

when observed at (a, Ω_s), where

$$V_n^m(ka) \triangleq \int_{\Omega_s} V(ka, \Omega_s)\, \bar{Y}_n^m(\Omega_s)\, d\Omega_s \tag{6.123}$$

is the (n, m)th coefficient of the decomposition. Equation (6.122) is readily verified by substituting (6.122) into (6.123) and applying the orthonormality property (6.120). The coefficients $V_n^m(ka)$ represent a transform domain much like the Fourier coefficients of a periodic function.

Specializing the above by substituting $V(\Omega_s) = G(\Omega_s, ka, \Omega)$ from (6.119) into (6.123) yields

$$G_n^m(\Omega, ka) = \int_{\Omega_s} G(\Omega_s, ka, \Omega)\, \bar{Y}_n^m(\Omega_s)\, d\Omega_s \tag{6.124}$$

$$= 4\pi \sum_{n'=0}^{\infty} i^{n'} b_{n'}(ka) \sum_{m'=-n'}^{n'} \bar{Y}_{n'}^{m'}(\Omega) \int_{\Omega_s} Y_{n'}^{m'}(\Omega_s) \bar{Y}_n^m(\Omega_s)\, d\Omega_s$$

$$= 4\pi\, i^n\, b_n(ka)\, \bar{Y}_n^m(\Omega), \tag{6.125}$$

where the latter equality follows directly from (6.120).[3] These results will shortly prove useful in deriving the modal analog of the array manifold vector defined in Section 6.2.

Equation (6.122) can be interpreted as the decomposition of an arbitrary square-integrable sound field into an infinite series of spherical harmonics or *modes*. Equation (6.124) is then a specialization for a plane wave. We will now consider how those spherical harmonics can be used for beamforming.

[3] Teutsch [60, Section 5.1.2] incorrectly reports the leading coefficient of (6.125) as $\sqrt{4\pi}$.

For the case of a discrete array of microphones as opposed to a continuous, pressure-sensitive surface, it is necessary to reformulate (6.120) as [42]

$$\frac{4\pi}{S}\sum_{s=0}^{S-1} Y_n^m(\Omega_s)\,\bar{Y}_{n'}^{m'}(\Omega_s) = \delta_{n,n'}\delta_{m,m'}, \tag{6.126}$$

where S is the number of sensors, each of which is located at some (Ω_s) for $s = 0, 1, \ldots, S-1$. Similarly, we can define a discrete version of (6.123) as

$$V_n^m(ka) \triangleq \frac{4\pi}{S}\sum_{s=0}^{S-1} V(ka, \Omega_s)\,\bar{Y}_n^m(\Omega_s) \tag{6.127}$$

and of the modal decomposition of the plane wave (6.125) as

$$G_n^m(\Omega, ka) = \frac{4\pi}{S}\sum_{s=0}^{S-1} G(\Omega_s, ka, \Omega)\,\bar{Y}_n^m(\Omega_s). \tag{6.128}$$

Note that the presence of the leading coefficient of $4\pi/S$ in (6.126)–(6.128) ensures that (6.123) and (6.127) are equivalent for $V_n^m(ka) \equiv 1$.

Equations (6.127) and (6.128) are of fundamental importance in that they define the *modal decomposition* that is typically performed prior to beamforming with a spherical array. One of the primary challenges in designing realizable and effective spherical arrays is choosing the set of sensor locations $\{(\Omega_s)\}$ such that (6.126) is satisfied as nearly as possible; see Li *et al.* [42,40]. Li and Duraiswami [41] discuss the use of variable *quadrature weights* [17] to minimize the orthonormality error between (6.120) and (6.126). The early work by Meyer and Elko [49] reported two sensor placement schemes for arrays of $S = 24$ and 32 elements; the deviation of the modes of these discrete arrays from orthonormality are illustrated in Figure 6.18, wherein lighter colored squares indicate a higher sensitivity reported in a decibel scale. From Figure 6.18a, it is clear that the 24-element array maintains orthogonality only through mode $n = 2$, while Figure 6.18b indicates that the 32-element array maintains orthogonality through

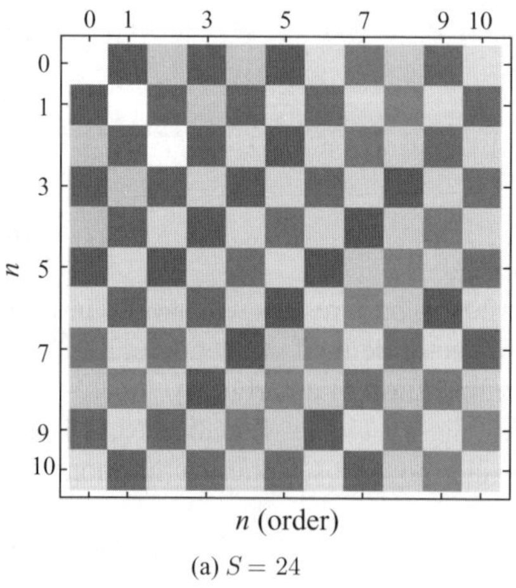

(a) $S = 24$

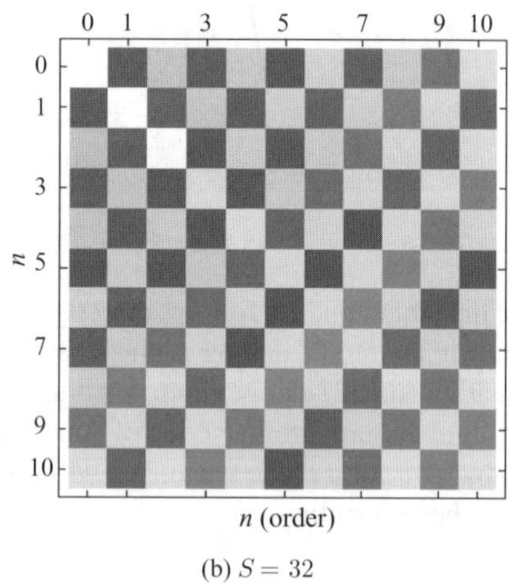

(b) $S = 32$

Figure 6.18 Deviation of the spherical harmonics from orthonormality for (a) $S = 24$ and (b) $S = 32$ elements.

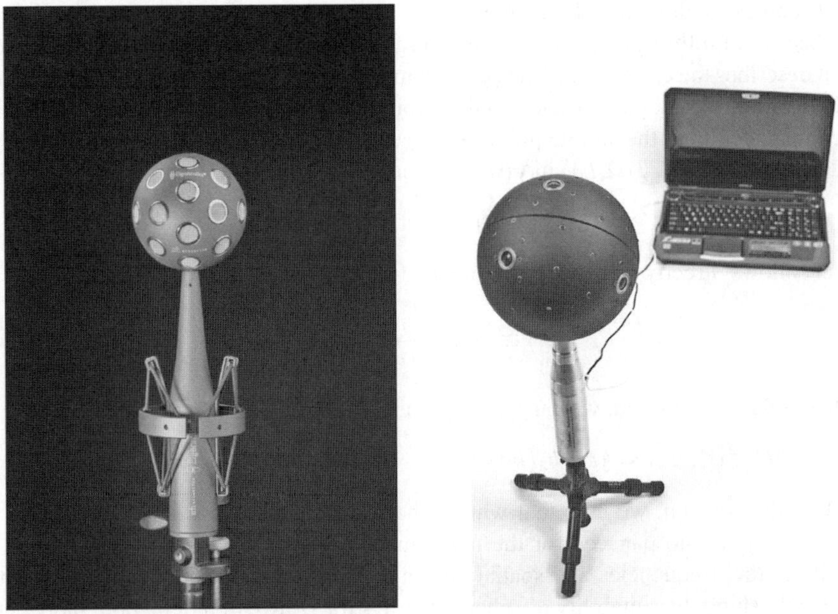

(a) (b)

Figure 6.19 (a) A 32-channel Eigenmike® spherical array; photo reproduced courtesy of *mh acoustics, Inc*. (b) A 64-channel spherical array with five integrated video cameras; photo reproduced courtesy of *VisiSonics Corporation*.

$n = 8$. However, typically, the order is truncated such that $(N + 1)^2 \leq S$ to avoidspatial aliasing, implying that a 32-element array can support a maximum order of $N = 4$.

Shown in Figure 6.19 are a spherical, 32-element *Eigenmike*® manufactured by *mh acoustics* and a 64-element, spherical array with five integrated video cameras manufactured by *VisiSonics Corporation*. Both instruments are available commercially and have attracted a great deal of interest within the acoustic beamforming research community.

Stacking the modal components (6.125) together provides the *modal array manifold vector*

$$\mathbf{v}(\Omega, ka) \triangleq \begin{bmatrix} G_0^0(\Omega, ka) \\ G_1^{-1}(\Omega, ka) \\ G_1^0(\Omega, ka) \\ G_1^1(\Omega, ka) \\ G_2^{-2}(\Omega, ka) \\ G_2^{-1}(\Omega, ka) \\ G_2^0(\Omega, ka) \\ \vdots \\ G_N^{-N}(\Omega, ka) \\ \vdots \\ G_N^N(\Omega, ka) \end{bmatrix}, \tag{6.129}$$

where Ω denotes the direction of arrival of a plane wave. The modal array manifold vector is so dubbed because it fulfills precisely the same role as the array manifold vector defined in (6.29); that is, it describes the excitation of the (n, m)th mode—as opposed to the *n*th *microphone*—of the spherical array by a plane wave arriving from the direction Ω; clearly this interaction is more complicated than the simple phase shift seen in the conventional array. Evaluating any individual component $G_n^m(\Omega, ka)$ in $\mathbf{v}(\Omega, ka)$ requires (6.125) as well as the identity

$$Y_n^{-m}(\Omega) \equiv (-1)^m \bar{Y}_n^m(\Omega);$$

the latter follows directly from the definition (6.116) and the identity [15, Section 14.9]

$$P_n^{-m}(x) \equiv (-1)^m \frac{(n-m)!}{(n+m)!} P_n^m(x).$$

With these relations in mind, we can trivially evaluate $G_3^{-2}(\Omega, ka)$, for example, as

$$G_3^{-2}(\Omega, ka) = 4\pi\, i^3 b_3(ka) \cdot (-1)^2\, Y_3^2(\Omega) = -4\pi\, i\, b_3(ka)\, Y_3^2(\Omega).$$

To close this section, we note that while spherical arrays possess several attractive characteristics, they are no panacea for the many maladies of DSR; they too suffer from poor directivity at low frequencies and spatial aliasing at high frequencies just like conventional arrays. Establishing the suitability of spherical arrays for DSR will require both more detailed analysis and empirical studies; we turn our attention to both of these tasks in the coming sections.

6.5 Spherical Adaptive Algorithms

To begin this section, we investigate the well-known MVDR beamformer for spherical arrays. The solution for the MVDR beamforming weights with diagonal loading is given by (6.41). As discussed in Section 6.4, in the case of a spherical array, we treat each modal component as a microphone, and apply the beamforming weights directly to the output of each mode. In so doing, we are adhering to the decomposition of the entire beamformer into *eigenbeamformer* followed by a *modal beamformer* as initially proposed by Meyer and Elko [48,49].

For use in the GSC discussed in Section 6.3.5, we can set

$$\mathbf{w}_\mathrm{q}(\Omega, ka) = \frac{1}{C}\mathbf{v}(\Omega, ka), \tag{6.130}$$

where C is a normalization constant that ensures satisfaction of the distortionless constraint,

$$\mathbf{w}_\mathrm{q}^H(\Omega_0, ka)\, \mathbf{v}(\Omega_0, ka) = 1, \tag{6.131}$$

for the look direction Ω_0. The blocking matrix can then be derived in the normal way from $\mathbf{w}_\mathrm{q}^H(\theta, \phi, ka)$.

With formulations of the relevant array manifold vector (6.129), we can immediately write the solution (6.30) for the *delay-and-sum* beamformer. Another popular fixed design for spherical array processing is the hypercardioid [49]. The beam patterns obtained with the delay-and-sum and hypercardioid designs are shown in Figure 6.20a and 6.20b, respectively. The MVDR design both with and without a radial symmetry constraint are shown in Figure 6.20 and 6.20d, respectively.

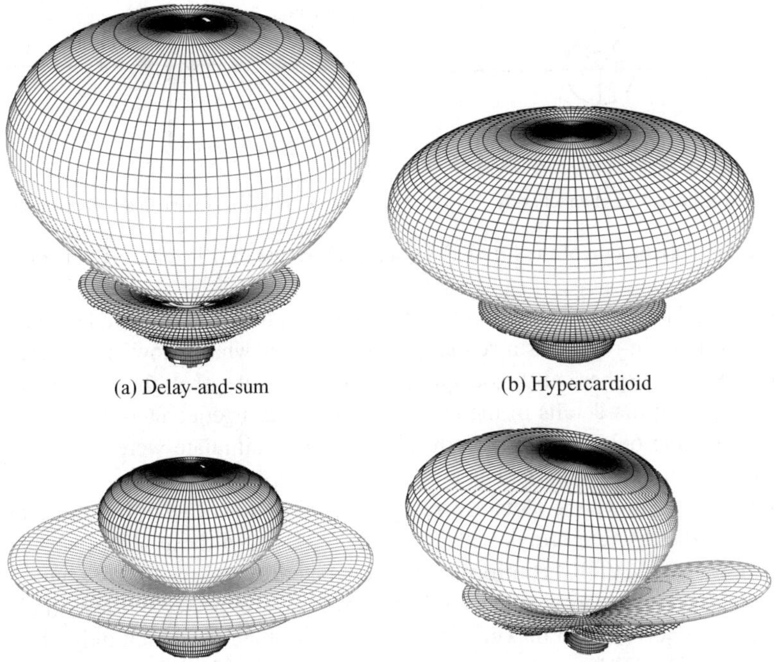

(a) Delay-and-sum (b) Hypercardioid

(c) Radially symmetric MVDR (d) Asymmetric MVDR

Figure 6.20 Spherical beam patterns for $ka = 10.0$: (a) delay-and-sum beam pattern; (b) hypercardioid beam pattern $H = Y_0 + \sqrt{3}\,Y_1 + \sqrt{5}\,Y_2$; (c) symmetric MVDR beam pattern obtained with spherical harmonics Y_n for $n = 0, 1, \ldots, 5$, diagonal loading $\sigma_{\mathrm{D}}^2 = 10^{-2}$ for a plane wave interferer $\pi/6$ rad from the look direction; and (d) asymmetric MVDR beam pattern obtained with spherical harmonics Y_n^m for $n = 0, 1, \ldots, 5, m = -n, \ldots, n$, diagonal loading $\sigma_{\mathrm{D}}^2 = 10^{-2}$ for a plane wave interferer $\pi/6$ rad from the look direction.

Let us now consider the case where the look direction is $(\theta, \phi) = (0, 0)$ and there is a single strong interference signal impinging on the array from $(\theta_{\mathrm{I}}, \phi_{\mathrm{I}}) = (\pi/6, 0)$ with a magnitude of $\sigma_{\mathrm{I}}^2 = 10^{-1}$. In this case, the covariance matrix of the array input is

$$\Sigma_{\mathbf{X}}(ka) = \mathbf{v}(\Omega, ka)\,\mathbf{v}^H(\Omega, ka) + \sigma_{\mathrm{I}}^2\,\mathbf{v}(\theta_{\mathrm{I}}, \phi_{\mathrm{I}}, ka)\,\mathbf{v}^H(\theta_{\mathrm{I}}, \phi_{\mathrm{I}}, ka). \tag{6.132}$$

6.6 Comparative Studies

In this section, we present a set of comparative studies for a conventional, linear array and a spherical array. We first compare the arrays on the basis of the theoretical performance metrics introduced in Section 6.3.4, namely array gain, WNG, and directivity index. Thereafter, we compare the arrays on a metric of more direct interest to those researchers on the forefront of DSR technology, namely, WER.

The orientation of the conventional, linear and spherical arrays shown in Figure 6.21 were used as the basis for evaluating array gain, WNG, and directivity index; these configurations were intended to simulate the condition wherein the arrays are mounted at head height for a

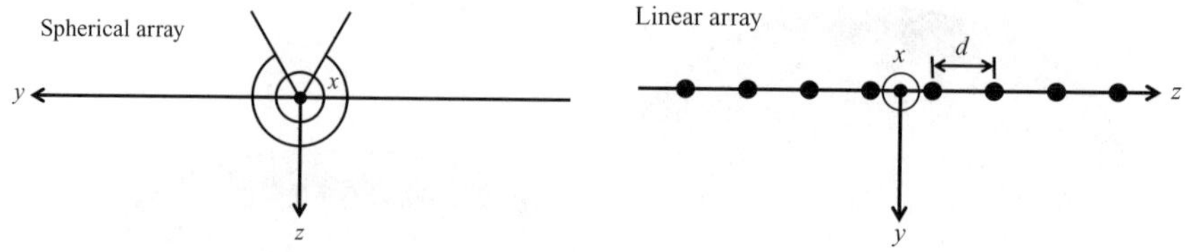

(a) Spherical 32-channel array (b) Conventional, 64-channel, linear array

Figure 6.21 Orientation of the (a) Mark IV linear array and (b) Eigenmike® spherical array.

standing speaker. The acoustic environment we simulated involved a desired speaker, a source of discrete interference—such as a screen projector—somewhat below and to the left of the desired source, and a spherically isotropic noise field—such as might be created by an air conditioning system; the details of the environment, which is equivalent for both arrays, are summarized in Table 6.2 The specific arrays we chose to simulate were the Mark IV linear array and the Eigenmike® spherical array.

In Figure 6.22 are shown the plots of array gain as a function of *ka* of for the ideal spherical array as well as the discrete arrays with $S = 24$ and 32. From these plots two facts become apparent. Firstly, the MVDR beamformer, as anticipated by the theory presented in Section 6.3.1, provides the highest array gain overall. This was to be expected because minimizing the noise variance is equivalent to maximizing SNR if performed over each individual subband; in order to maximize SNR over the entire subband, the subband signals must be weighted by a Wiener filter prior to their combination. Secondly, the figures for $S = 24$ and 32 indicate that the array gain of the ideal array is reduced when the array must be implemented in hardware with discrete microphones.

Figure 6.23 shows the WNG for the ideal spherical array, as well as its discrete counterparts for $S = 24$ and 32. Once more, as predicted by the theory, the uniform (i.e., D&S) beamformer provides the best performance according to this metric. The SD and MVDR beamformers provide substantially lower WNG at low frequencies, but essentially equivalent performance for $ka \geq 30$.

As described in Section 6.2, the beam pattern is the sensitivity of the array to a plane wave arriving from some direction Ω. By weighting each spherical mode (6.125) by $\bar{w}_n^m$, the beam pattern for the ideal array can be expressed as

$$B(\Omega, ka) = 4\pi \sum_{n=0}^{N} i^n b_n(ka) \sum_{m=-n}^{n} \bar{w}_n^m \bar{Y}_n^m(\Omega).$$

Table 6.2 Acoustic environment for comparing the Mark IV linear array with the Eigenmike® spherical microphone array.

Source	Position $\Omega \triangleq (\theta, \phi)$ Mark IV	Eigenmike	Level (dB)
Desired	$(3\pi/8, 0)$	$(\pi/2, -\pi/8)$	0
Discrete interference	$(3\pi/4, \pi/8)$	$(3\pi/8, \pi/4)$	−10
Diffuse noise	—	—	−10

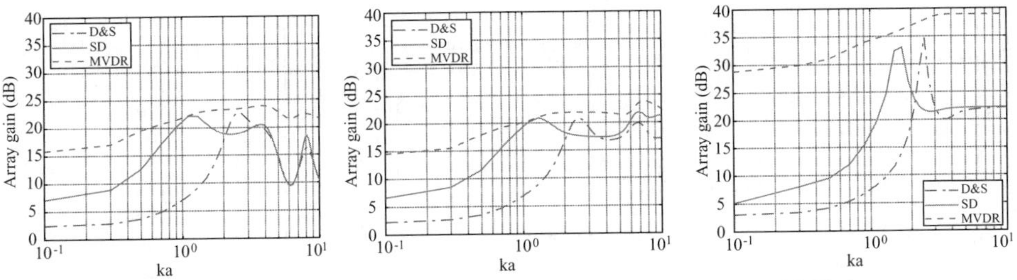

(a) $S = 24$ (b) $S = 32$ (c) Ideal array

Figure 6.22 Array gain as a function of *ka* for (a) $S = 24$, (b) $S = 32$, and (c) the Ideal array.

This implies that the power pattern (6.70) is given by

$$P(\Omega) \triangleq |B(\Omega, ka)|^2 \tag{6.133}$$
$$= 16\pi^2 \sum_{n,n'=0}^{\infty} i^n \,\bar{i}^{n'} \, b_n(ka)\, \bar{b}_{n'}(ka) \sum_{m,m'=-n,-n'}^{n,n'} \bar{w}_n^m \, w_{n'}^{m'} \, \bar{Y}_n^m(\Omega)\, Y_{n'}^{m'}(\Omega).$$

Substituting (6.133) into (6.72) and applying (6.120) then provides [68]

$$\mathrm{DI}_{\mathrm{ideal}}(ka, \mathbf{w}) = -10\log_{10}\left\{4\pi \sum_{n=0}^{N} |b_n(ka)|^2 \sum_{m=-n}^{m} |w_n^m|^2\right\}. \tag{6.134}$$

The directivity index as a function of *ka* for both ideal and discrete arrays is plotted in Figure 6.24. These figures reveal that—as anticipated by the theory of Section 6.3.1—the superdirective (SD) beamformer provides the highest directivity save in the very low frequency region where the sensor covariance matrix (6.42) is dominated by the diagonal loading.

Now, we come to an equivalent set of plots for the Mark IV linear array; these are shown in Figure 6.25, where each metric is shown as a function of d/λ, the ratio of intersensor spacing to wavelength. Once more, the MVDR beamformer provides the highest array gain, the D&S beamformer the highest WNG, and the superdirective beamformer the highest directivity index. What is unsurprising is that the Mark IV provides a higher array gain than the Eigenmike overall, given its greater number of sensors. What is somewhat surprising is the drastic drop

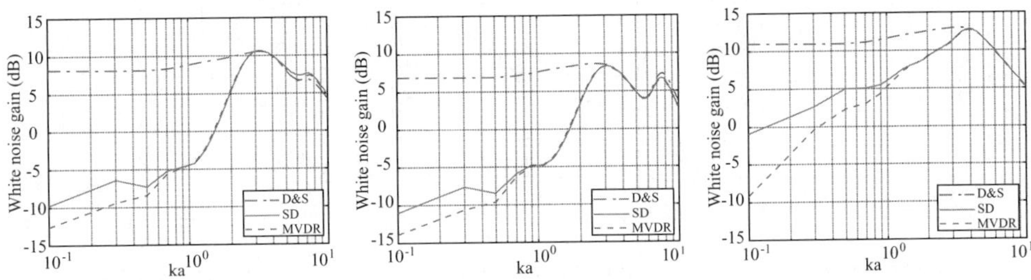

(a) $S = 24$ (b) $S = 32$ (c) Ideal array

Figure 6.23 White noise gain as a function of ka for (a) $S = 24$, (b) $S = 32$, and (c) the Ideal array.

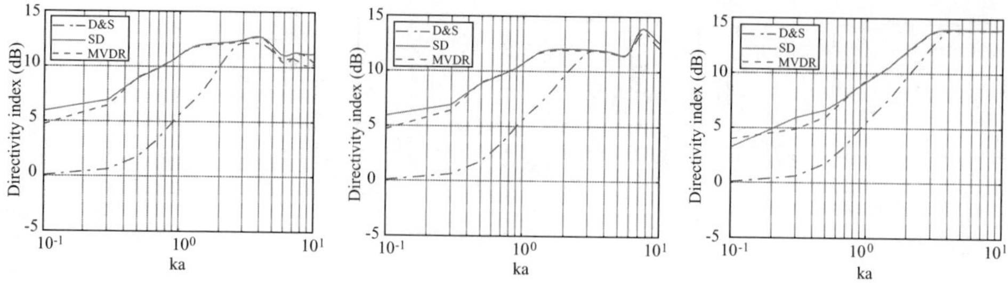

(a) $S = 24$ (b) $S = 32$ (c) Ideal array

Figure 6.24 Directivity index as a function of *ka* for (a) $S = 24$, (b) $S = 32$, and (c) the Ideal array.

in all metrics just below the point $d/\lambda = 1$; this stems from the fact that this is the point where the first grating lobe crosses the source of discrete interference. A grating lobe cannot be suppressed given that—due to spatial aliasing—it is indistinguishable from the main lobe and hence subject to the distortionless constraint as discussed in Section 6.2.

6.7 Comparison of Linear and Spherical Arrays for DSR

As a spherical microphone array has—to the best knowledge of the current authors—never before been applied to DSR, our first step in investigating its suitability for such a task was to capture some prerecorded speech played into a real room through a loudspeaker, then perform beamforming and subsequently speech recognition. Figure 6.26 shows the configuration of room used for these recordings. As shown in the figure, the loudspeaker was placed in two different positions; the locations of the sensors and loudspeaker were measured with *OptiTrack*, a motion-capture system manufactured by *NaturalPoint*. For data capture, we used an Eigenmike® which consists of 32 microphones embedded in a rigid sphere with a radius of 4.2 cm; for further details see the website of mh acoustics, http://www.mhacoustics.com. Each sensor of the Eigenmike® is centered on the face of a truncated icosahedron. We simultaneously captured the speech data with a 64-channel, uniform linear Mark IV microphone array with an intersensor spacing of 2 cm for a total aperture length of 126 cm. Speech data from the

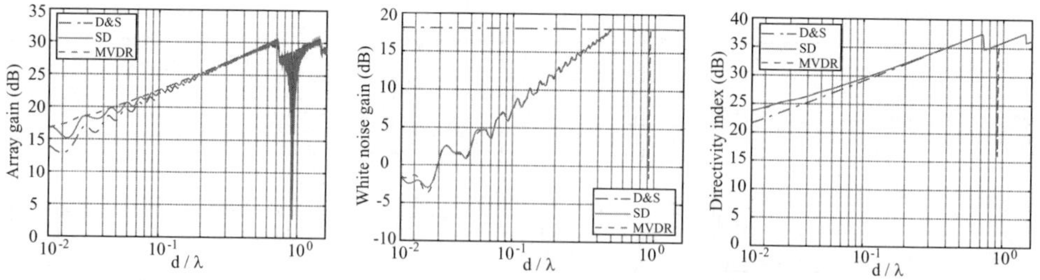

(a) Array gain (b) White noise gain (c) Directivity index

Figure 6.25 (a) Array gain, (b) white noise gain, and (c) directivity index as a function of d/λ for the 64-element, linear Mark IV micophone array with an intersensor spacing of $d = 2$ cm.

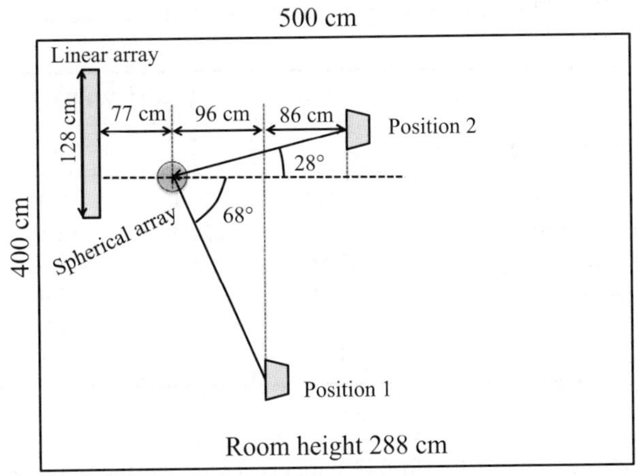

Figure 6.26 The layout of the recording room.

TIMIT corpus were used as test material. The test set consisted of 3,241 words uttered by 37 speakers for each recording position. The far-field data were sampled at a rate of 44.1 kHz. The reverberation time T_{60} in the recording room was approximately 525 ms.

We used the speech recognizer described in Section 6.3.10. Tables 6.3 and 6.4 show WERs for each beamforming algorithm for the cases wherein the incident angles of the target signal to the array were 28° and 68°, respectively. As a reference, the WERs obtained with a SAC and the clean data played through the loudspeaker (Clean data) are also reported. It is clear from the tables that every beamforming algorithm provides superior recognition performance to the SAC after the last adapted pass of recognition. It is also clear from the tables that superdirective beamforming with the small spherical array of radius 4.2 cm (Spherical SD BF) can achieve recognition performance very comparable to that obtained with the same beamforming method with the linear array (SD BF with linear array). In the case that the speaker position is nearly in front of the array, superdirective beamforming with the linear array (SD BF with linear array) can still achieve the best result among all the algorithms. This is because of the highest

Table 6.3 WERs for each beamforming algorithm in the case that the incident angle to the array is 28°.

	Pass (%WER)			
Beamforming (BF) algorithm	1	2	3	4
Single array channel (SAC)	47.3	18.9	14.3	13.6
D&S BF with linear array	44.7	17.2	11.1	9.8
SD BF with linear array	45.5	16.4	10.7	9.3
SD BF with spherical array	43.9	14.2	12.1	10.5
Spherical D&S BF	47.3	16.8	13.0	12.0
Spherical SD BF	42.8	14.5	11.5	10.2
CTM	16.7	7.5	6.4	5.4

Table 6.4 WERs for each beamforming algorithm in the case that the incident angle to the array is 68°.

	Pass (%WER)			
Beamforming (BF) algorithm	1	2	3	4
Single array channel (SAC)	57.8	25.1	19.4	16.6
D&S BF with linear array	53.6	24.3	16.1	13.3
SD BF with linear array	52.6	23.8	16.6	12.8
Spherical D&S BF	57.6	22.7	14.9	13.5
Spherical SD BF	44.8	15.5	11.3	9.7
CTM	16.7	7.5	6.4	5.4

directivity index can be achieved with 64-channels, twice as many as the sensors as in the spherical array. However, in the other configuration, wherein the desired source is at an oblique angle to the array, the spherical superdirective beamformer (Spherical SD BF) provides better results than the linear array because they it is able to maintain the same beam pattern regardless of the incident angle. In these experiments, spherical D&S beamforming (Spherical D&S BF) could not improve the recognition performance significantly because of its poor directivity.

6.8 Conclusions and Further Reading

In this contribution, we have examined the application of spherical microphone arrays to beamforming and compared this with the use of conventional arrays. As we have explained, the primary difference between the conventional and spherical array-processing literature is that the latter makes an explicit attempt to account for and model two phenomena that are present in all acoustic array processing applications, namely, diffraction, which is the tendency of sound to bend around fixed obstacles, and scattering, which is the tendency of sound to be dispersed through reflection from nonplanar surfaces. By modeling both phenomena, spherical array processing typically begins by decomposing the sound field into a number of orthogonal components, which are subsequently weighted and combined, much like the outputs of single microphones in conventional array processing. The advantage of the spherical configuration is that it is spatially isotropic, implying that the look direction can be set to nearly any spherical coordinate with equal ease.

We have found that both conventional and spherical arrays suffer from two primary problems:

1. Poor low-frequency directivity, which is a result of the finite physical aperture that a realizable array must necessarily have.
2. High-frequency spatial aliasing, which arises from the necessity of sampling a continuous aperture at discrete locations through the placement of microphones.

The detrimental effects of both these problems can be minimized, and ongoing attempts to do so are the topics of a great deal of current research. However, neither effect can be eliminated entirely, and hence must be taken into account in developing effective beamforming algorithms for real microphone arrays.

Finally, as we have shown here, the adaptive beamforming algorithms developed in the conventional literature can be effectively applied to spherical arrays. This includes the second-order methods such as MVDR, but would also certainly include the algorithms based on the optimization of non-Gaussian and HOS such as maximum kurtosis and maximum negentropy beamforming. In the view of the present authors, further investigation of these techniques in the context of spherical arrays is likely to prove one of the most promising topics of acoustic beamforming research in the coming years and decades. Moreover, the effect of this research on applications involving DSR is likely to be dramatic.

References

[1] T. D. Abhayapala and D. B. Ward, "Theory and design of high order sound field microphones using spherical microphone array," in *Proceedings of the ICASSP*, Orlando, FL, May 2002.

[2] G. B. Arfken and H. J. Weber, *Mathematical Methods for Physicists*. Boston: Elsevier, 2005.

[3] R. A. Askey and R. Roy, "Gamma function," in *NIST Handbook of Mathematical Functions*, F. W. J. Olver, D. W. Lozier, R. F. Boisvert, and C. W. Clark, Eds. New York: Cambridge University Press, 2010.

[4] D. P. Bertsekas, *Nonlinear Programming*. Belmont, MA, USA: Athena Scientific, 1995.

[5] J. Bitzer and K. U. Simmer, "Superdirective microphone arrays," in *Microphone Arrays*, M. Branstein and D. Ward, Eds. Heidelberg: Springer, 2001, pp. 19–38.

[6] M. Brandstein and D. Ward, "Microphone arrays," *Springer*, Berlin, 2000.

[7] G. C. Carter, "Time delay estimation for passive sonar signal processing," *IEEE Transactions on Acoust., Speech Signal Processing*, vol. ASSP-29, pp. 463–469, 1981.

[8] I. Claesson and S. Nordholm, "A spatial filtering approach to robust adaptive beaming," *IEEE Transactions on Antennas and Propagation*, vol. 19, pp. 1093–1096, 1992.

[9] I. Cohen, S. Gannot, and B. Berdugo, "An integrated real-time beamforming and postfiltering system for nonstationary noise environments," *EURASIP Journal on Applied Signal Processing*, vol. 2003, pp. 1064–1073, 2003.

[10] H. Cox, R. M. Zeskind, and M. M. Owen, "Robust adaptive beamforming," *IEEE Transactions Acoustics, Speech and Signal Processing*, vol. ASSP-35, no. 10, pp. 1365–1376, October 1987.

[11] J. M. De Haan, N. Grbic, I. Claesson, and S. E. Nordholm, "Filter bank design for subband adaptive microphone arrays," *IEEE Transactions on SAP*, vol. 11, no. 1, pp. 14–23, January 2003.

[12] J. H. DiBiase, H. F. Silverman, and M. S. Brandstein, "Robust localization in reverberant rooms," in *Microphone Arrays*, M. Brandstein and D. Ward, Eds. Heidelberg, Germany: Springer-Verlag, 2001, Ch. 4.

[13] S. Doclo, A. Spriet, J. Wouters, and M. Moonen, "Frequency-domain criterion for the speech distortion weighted multichannel Wiener filter for robust noise reduction," *Speech Communication, special issue on Speech Enhancement*, vol. 49, pp. 636–656, 2007.

[14] J. R. Driscoll and J. Dennis M. Healy, "Computing Fourier transforms and convolutions on the 2-sphere," *Advances in Applied Mathematics*, vol. 15, pp. 202–250, 1994.

[15] T. M. Dunster, "Legendre and related functions," in *NIST Handbook of Mathematical Functions*, F. W. J. Olver, D. W. Lozier, R. F. Boisvert, and C. W. Clark, Eds. New York: Cambridge University Press, 2010.

[16] J. S. Erkelens, R. C. Hendriks, R. Heusdens, and J. Jensen, "Minimum mean-square error estimation of discrete Fourier coefficients with generalized Gamma priors," *IEEE Transactions on Audio, Speech and Language Processing*, vol. 15, pp. 1741–1752, 2007.

[17] J. Fliege and U. Maier, "The distribution of points on the sphere and corresponding cubature formulae," *IMA Journal of Numerical Analysis*, vol. 19, pp. 317–334, 1999.

[18] R. G. Gallager, *Information Theory and Reliable Communication*. John Wiley & Sons, New York, 1968.

[19] S. Gannot and I. Cohen, "Speech enhancement based on the general transfer function GSC and postfiltering," *IEEE Transactions on SAP*, vol. 12, pp. 561–571, 2004.

[20] S. Gazor and Y. Grenier, "Criteria for positioning of sensors for a microphone array," *IEEE Transactions on Speech Audio Processing*, vol. 3, pp. 294–303, 1995.

[21] T. Gehrig, U. Klee, J. McDonough, S. Ikbal, M. Wölfel, and C. Fügen, "Tracking and beamforming for multiple simultaneous speakers with probabilistic data association filters," in *Interspeech*, September 2006.

[22] E. N. Gilbert and S. P. Morgan, "Optimum design of antenna arrays subject to random variations," *Bell System Technical Journal*, vol. 34, pp. 637–663, May 1955.

[23] G. H. Golub and C. F. Van Loan, *Matrix Computations*, 3rd ed. Baltimore: The Johns Hopkins University Press, 1996.

[24] S. Haykin, *Adaptive Filter Theory*, 4th ed. New York: Prentice-Hall, 2002.

[25] W. Herbordt and W. Kellermann, "Frequency-domain integration of acoustic echo cancellation and a generalized sidelobe canceller with improved robustness," *European Transactions on Telecommunications (ETT)*, vol. 13, pp. 123–132, 2002.

[26] W. Herbordt and W. Kellermann, "Adaptive beamforming for audio signal acquisition," in *Adaptive Signal Processing—Applications to Real-World Problems*, J. Benesty and Y. Huang, Eds. Berlin, Germany: Springer, 2003, pp. 155–194.

[27] W. Herbordt, H. Buchner, S. Nakamura, and W. Kellermann, "Multichannel bin-wise robust frequency-domain adaptive filtering and its application to adaptive beamforming," *IEEE Transactions on Audio, Speech and Language Processing*, vol. 15, pp. 1340–1351, 2007.

[28] O. Hoshuyama, A. Sugiyama, and A. Hirano, "A robust adaptive beamformer for microphone arrays with a blocking matrix using constrained adaptive filters," *IEEE Transactions on Signal Processing*, vol. 47, pp. 2677–2684, 1999.

[29] A. Hyvärinen, "Survey on independent component analysis," *Neural Computing Surveys*, vol. 2, pp. 94–128, 1999.

[30] A. Hyvärinen and E. Oja, "Independent component analysis: Algorithms and applications," *Neural Networks*, vol. 13, pp. 411–430, 2000.

[31] S. Kay, *Fundamentals of Statistical Signal Processing: Estimation Theory*. Englewood Cliffs, NJ: Prentice-Hall, 1993.

[32] U. Klee, T. Gehrig, and J. McDonough, "Kalman filters for time delay of arrival–based source localization," *Journal of Advanced Signal Processing, Special Issue on Multi–Channel Speech Processing*, August 2005.

[33] K. Kumatani, J. McDonough, S. Schact, D. Klakow, P. N. Garner, and W. Li, "Filter bank design based on minimization of individual aliasing terms for minimum mutual information subband adaptive beamforming," in *Proceedings of the ICASSP*, Las Vegas, NV, USA, April 2008.

[34] K. Kumatani, T. Gehrig, U. Mayer, E. Stoimenov, J. McDonough, and M. Wölfel, "Adaptive beamforming with a minimum mutual information criterion," *IEEE Transactions on ASLP*, vol. 15, pp. 2527–2541, 2007.

[35] K. Kumatani, J. McDonough, D. Klakow, P. N. Garner, and W. Li, "Adaptive beamforming with a maximum negentropy criterion," *IEEE Transactions on ASLP*, August 2008.

[36] K. Kumatani, L. Lu, J. McDonough, A. Ghoshal, and D. Klakow, "Maximum negentropy beamforming with superdirectivity," in *European Signal Processing Conference (EUSIPCO)*, Aalborg, Denmark, 2010.

[37] K. Kumatani, J. McDonough, B. Rauch, and D. Klakow, "Maximum negentropy beamforming using complex generalized gaussian distribution model," in *Asilomar Conference on Signals, Systems and Computers (ASILOMAR)*, Pacific Grove, CA, USA, 2010.

[38] K. Kumatani, J. McDonough, J. F. Lehman, and B. Raj, "Channel selection based on multichannel cross-correlation coefficients for distant speech recognition," in *Joint Workshop on Hands-free Speech Communication and Microphone Arrays (HSCMA)*, Edinburgh, UK, 2011.

[39] H. Kutruff, *Room Acoustics*, 5th ed. New York: Spoon Press, 2009.

[40] Z. Li and R. Duraiswami, "Hemispherical microphone arrays for sound capture and beamforming," in *Proceedings of the WASPAA*, New Paltz, NY, October 2005.

[41] Z. Li and R. Duraiswami, "Flexible and optimal design of spherical microphone arrays for beamforming," *IEEE Transactions on Audio Speech Language Processing*, vol. 15, no. 2, pp. 702–714, February 2007.

[42] Z. Li, R. Duraiswami, E. Grassi, and L. S. Davis, "Flexible layout and optimal cancellation of the orthonormality error for spherical microphone arrays," in *Proceedings of the ICASSP*, Montreal, CA, May 2004.

[43] M. Lincoln, I. McCowan, I. Vepa, and H. K. Maganti, "The multi–channel Wall Street Journal audio visual corpus (mc–wsj–av): Specification and initial experiments," in *Proceedings of the ASRU*, 2005, pp. 357–362.

[44] C. Marro, Y. Mahieux, and K. U. Simmer, "Analysis of noise reduction and dereverberation techniques based on microphone arrays with postfiltering," *Transactions of the SAP*, vol. 6, pp. 240–259, 1998.

[45] R. Martin, "Speech enhancement based on minimum square error estimation and supergaussian priors," *IEEE Transactions on Speech and Audio Processing*, vol. 13, no. 5, pp. 845–856, 2005.

[46] I. A. McCowan and H. Bourlard, "Microphone array post-filter based on noise field coherence," *IEEE Transactions on Speech Audio Processing*, vol. 11, pp. 709–716, 2003.
[47] J. McDonough, M. Wölfel, and A. Waibel, "On maximum mutual information speaker–adapted training," *Computer Speech & Language*, December 2007.
[48] J. Meyer and G. W. Elko, "A highly scalable spherical microphone array based on an orthonormal decomposition of the soundfield," in *Proceedings of the ICASSP*, Orlando, FL, May 2002.
[49] J. Meyer and G. W. Elko, "Spherical microphone arrays for 3D sound recording," in *Audio Signal Processing for Next–Generation Multimedia Communication Systems*. Boston: Kluwer Academic, 2004, pp. 67–90.
[50] F. D. Neeser and J. L. Massey, "Proper complex random processes with applications to information theory," *IEEE Trans. on Information Theory*, vol. 39, no. 4, pp. 1293–1302, July 1993.
[51] S. Nordebo, I. Claesson, and S. Nordholm, "Adaptive beamforming: spatial filter designed blocking matrix," *IEEE Journal of Oceanic Engineering*, vol. 19, pp. 583–590, 1994.
[52] S. Nordholm, I. Claesson, and B. Bengtsson, "Adaptive array noise suppression of handsfree speaker input in cars," *IEEE Transactions on Vehicular Technology*, vol. 42, pp. 514–518, 1993.
[53] F. W. J. Olver and L. C. Maximon, "Bessel functions," in *NIST Handbook of Mathematical Functions*, F. W. J. Olver, D. W. Lozier, R. F. Boisvert, and C. W. Clark, Eds. New York: Cambridge University Press, 2010.
[54] M. Omologo and P. Svaizer, "Acoustic event localization using a crosspower–spectrum phase based technique," in *Proceedings of the ICASSP*, vol. II, 1994, pp. 273–6.
[55] A. V. Oppenheim and R. W. Schafer, *Discrete–Time Signal Processing*, 3rd ed. Prentice Hall, Upper Saddle River, NJ, 2010.
[56] A. V. Oppenheim, R. W. Schafer, and J. R. Buck, *Discrete–Time Signal Processing*, 2nd ed. Prentice Hall Inc., Englewood Cliffs, NJ, 2009.
[57] A. Papoulis and S. U. Pillai, *Probability, Random Variables, and Stochastic Processes*, 4th ed. New York: McGraw-Hill, 2002.
[58] R. Roy and T. Kailath, "Esprit–estimation of signal parameters via rotational invariance techniques," *IEEE Transactions on Acoustics, Speech and Signal Processing*, vol. 37, pp. 984–995, 1989.
[59] D. Sharon Gannot, Burshtein, and E. Weinstein, "Signal enhancement using beamforming and nonstationarity withapplications to speech," *IEEE Transactions on SP*, vol. 49, pp. 1614–1626, 2001.
[60] H. Teutsch, *Modal Array Signal Processing: Principles and Applications of Acoustic Wavefield Decomposition*. Heidelberg: Springer, 2007.
[61] L. Uebel and P. Woodland, "Improvements in linear transform based speaker adaptation," in *Proceedings of the ICASSP*, 2001.
[62] P. P. Vaidyanathan, *Multirate Systems and Filter Banks*. Englewood Cliffs: Prentice-Hall, 1993.
[63] H. L. Van Trees, *Optimum Array Processing*. New York: John Wiley & Sons, 2002.
[64] E. Warsitz, A. Krueger, and R. Haeb-Umbach, "Speech enhancement with a new generalized eigenvector blocking matrix for application in a generalized sidelobe canceller," in *Proceedings of the ICASSP*, 2008.
[65] B. Widrow, K. M. Duvall, R. P. Gooch, and W. C. Newman, "Signal cancellation phenomena in adaptive antennas: Causes and cures," *IEEE Transactions on Antennas and Propagation*, vol. AP-30, pp. 469–478, 1982.
[66] E. G. Williams, *Fourier Acoustics*. San Diego, CA, USA: Academic Press, 1999.
[67] M. Wolfel and J. McDonough, *Distant Speech Recognition*. London: John Wiley & Sons, 2009.
[68] S. Yan, H. Sun, U. P. Svensson, X. Ma, and J. M. Hovem, "Optimal modal beamforming for spherical microphone arrays," *IEEE Transactions on Audio Speech Language Processing*, vol. 19, no. 2, pp. 361–371, February 2011.

Part Three

Feature Enhancement

7

From Signals to Speech Features by Digital Signal Processing

Matthias Wölfel
Pforzheim University, Germany

7.1 Introduction

Acoustic classification of speech signals as well as some speech feature-enhancement techniques require that the speech waveform $s(t)$ is processed to get a sequence of feature vectors—the so called *speech features*—of a relative small number of dimensions. This reduction is necessary to not waste resources by representing irrelevant information and to prevent the *curse of dimensionality*[1]. The transformation of the speech waveform into a set of dimension-reduced features is known as *speech feature extraction, acoustic preprocessing*, or *front-end processing*.

The set of transformations has to be carefully chosen such that the resulting features will contain only relevant information to perform the desired task. Feature extraction as applied in *automatic speech recognition* (ASR) systems aims to preserve the information needed to determine the phonetic class while being invariant to other factors including speaker differences such as accent, emotions, fundamental frequency (in the case of nontonal languages), or speaking rate or other distortion such as background noise, channel effects, or reverberation. For other systems, different information might be needed. For example, in speaker verification one is interested in keeping the speaker-specific characteristics. Note that the correct choice of feature transformation and reduction is critical, because if useful information is lost in this step it cannot be recovered in following processing steps.

Experience has proven that feature-extraction techniques based on characteristics of the human auditory system, as described in Chapter 8, are likely to provide classification performance for various audio tasks such as ASR and speaker identification that is superior to

[1] The term *curse of dimensionality* was coined by Bellmann [1] and describes the problem whereby volume increases exponentially as additional dimensions are added to a space.

Techniques for Noise Robustness in Automatic Speech Recognition, First Edition.
Edited by Tuomas Virtanen, Rita Singh, and Bhiksha Raj.

naive or ad hoc techniques. This stems from the fact that the human auditory system evolved concurrently ove millions of years with the human speech-production apparatus, and hence is highly "tuned" to the perception of all aspects of human speech.

7.1.1 *About this Chapter*

This chapter describes methods how to transform speech signals into speech features based on signal processing. While signal processing, in particular digital signal processing, is based on "abstract" mathematical methods, the choice of the processing steps to extract speech features is partly motivated by aspects of the human auditory system. Due to space constrains, it is not possible to cover the broad field of speech feature extraction extensively. The selected topic presented here have been carefully chosen such that a reader with basic knowledge of signal processing is able to follow and that it provides the necessary information for the following chapters about speech enhancement. For a broader coverage of feature extraction, the reader should refer to [8, 17, 33, 40] where more space is devoted on feature extraction and its mathematical background.

7.2 The Speech Signal

Although there are many possible speech sounds which can be produced by humans, the shape of the vocal tract and its mode of excitation are restricted. Without knowing much about its origin, we can observe several basic characteristics of the speech signal by looking at a speech waveform, such as the one depicted in Figure 7.1.

We observe that the speech signal:

- is *time variant;*
- is *quasi-periodic* in some segments (at voiced regions), has a *stochastic spectral character* (at unvoiced regions) or is *paused;*
- is *quasi-stationary* in time intervals of 5–25 ms, which implies that the vocal tract shape and thus its transfer function remain nearly unchanged within this period; and
- is *changing continuously and gradually*, not abruptly.

These characteristics of the speech signal are defined by its generation process originating in the lungs as the speaker exhales. Speech consists of pressure waves created by the airflow

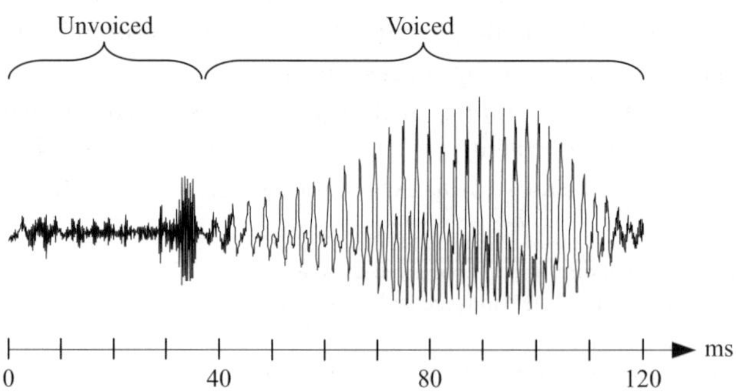

Figure 7.1 A speech waveform of unvoiced and voiced speech.

through the vocal tract. The vocal folds in the larynx can open and close quasi-periodically to interrupt this airflow. The result is *voiced speech*. Vowels are the most prominent examples of this type. It is characterized by its periodicity, where the frequency of the excitation provided by the vocal cords during voiced speech is known as the *fundamental frequency*. The periodicity of voiced speech gives rise to a spectrum containing harmonics lf_0 of the fundamental frequency f_0 for integer $l \geq 1$. These harmonics are known as *partials*. A truly periodic sequence, observed over an infinite interval, will have a discrete-line spectrum, but voiced sounds are only locally quasi-periodic. For *unvoiced* speech the vocal cords do not vibrate. Rather, the excitation is provided by turbulent airflow through a constriction in the vocal tract, imparting to the phonemes falling into this class a noisy characteristic. The positions of the other articulators in the vocal tract serve to filter the noisy excitation, amplifying certain frequencies while attenuating others. The spectra for unvoiced speech range from a flat shape to spectral patterns lacking low-frequency components.

7.3 Spectral Processing

Spectral processing, estimation, and analysis are fundamental for many speech applications including recognition, identification, compression, coding, and voice conversion, to just name a few, because it allows for additional analysis possibilities in contrast to the time signal. These applications impose a variety of requirements on the spectral estimate, including

- spectral resolution;
- nonlinear modeling of the frequency axis;
- variance of the estimated spectra; and
- the capacity to model or suppress the fundamental frequency during voiced speech.

To satisfy these requirements, a broad variety of solutions has been proposed in the literature, all of which can be classified into either

- *nonparametric methods* based on periodograms (e.g., the power spectrum); or
- *parametric methods*, using a small number of parameters estimated from the data (e.g., the spectral envelope).

In this section, we will consider only such spectral estimation techniques which are useful in extracting speech features for further processing or analysis. An overview of spectral estimation approaches which are useful in speech processing is given in Table 7.1. Those methods not treated here (namely warped power spectrum, perceptual linear prediction, warped-twice linear prediction, and warped-twice minimum variance distortionless response) are covered in [40].

7.3.1 Windowing

In order to analyze the speech signal in alternative domains for better analysis, such as the spectral domain (as discussed in the remainder of this section) or cepstral domain (Section 7.4), it is necessary to split its continuous stream into short segments by multiplying the speech signal component wise with an analysis window. Selecting the length of these windows, the

Table 7.1 Overview of spectral estimation methods.

		Properties		
Spectrum	Method	Frequency axis	Frequency resolution	Pitch sensitivity
PS	Nonparametric	Linear	Static	Very high
Warped PS	Nonparametric	Nonlinear	Static	Very high
Mel-filter bank	Nonparametric	Nonlinear	Static	High
LP	Parametric	Linear	Static	Medium
Perceptual LP	Parametric	Nonlinear	Static	Medium
Warped LP	Parametric	Nonlinear	Static	Medium
Warped-twice LP	Parametric	Nonlinear	Adaptive	Medium
MVDR	Parametric	Linear	Static	Low
Warped MVDR	Parametric	Nonlinear	Static	Low
Warped-twice MVDR	Parametric	Nonlinear	Adaptive	Low

PS = power spectrum, LP = linear prediction; MVDR = minimum variance distortionless response.

so called *frame size*, involves a trade off between two conflicting requirements. Each segment must be

- *short enough* to provide the required time resolution, and
- *long enough* to ensure adequate frequency resolution[2] and to be insensitive to its exact position relative to the glottal cycle[3].

The choice of *frame shift*, which defines the step size of the window, and frame size are dependent on the velocity of the articulators, which determines how quickly the vocal tract changes shape. Some speech sounds, such as stop consonants or diphthongs, have sharp spectral transitions with a spectral peak shift of up to 80 Hz/ms [20]. It is common to adjust the frame shift and frame size together; as a shorter frame shift can track more rapid variations of the shape of the vocal tract, the analysis window size should also be shortened to achieve better localization in fast-changing articulator configurations. In general, it can be said that the advantage of a long observation segment is that it smoothes out some of the temporal variations of unvoiced speech while its disadvantage is that it blurs rapid events, such as the release of stop consonants. In speech processing a frame size of 16–32 ms and a frame shift of 5–15 ms are commonly used.

Besides choosing the right frame shift and size, the shape of the analysis window is very important. The *windowing theorem* states that the Fourier transform of the time window is convolved with the short-term spectrum of the actual signal. This means that true spectral characteristics of the signal will be "smeared" with the Fourier transform of the analysis window. The amount of spectral leakage is directly related to the size of the side lobes in the frequency domain. Windowing functions without abrupt discontinuities at their edges in

[2] The *frequency resolution* is defined as the inverse time interval width (Rayleigh frequency resolution).

[3] One glottal cycle can be described in three phases: the *opening phase*, during which the glottis opens, the *returning phase*, the interval when the glottis is closing and the *closed phase*, the time during the glottis is closed and there is no glottal excitation.

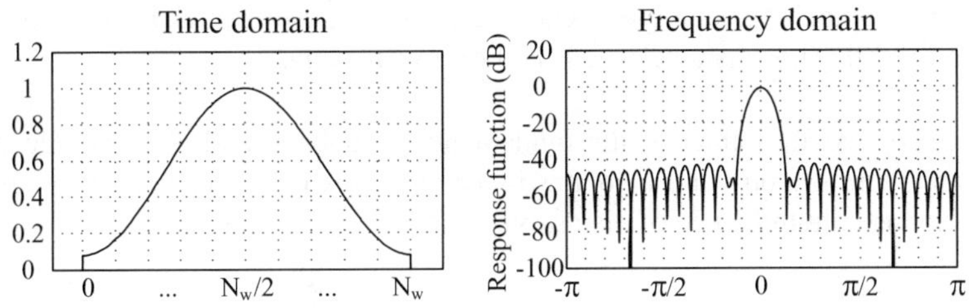

Figure 7.2 The Hamming window in time and frequency domain.

the time domain are known to have a smaller smearing effect. Many types of windowing functions of this kind have been proposed in the literature, including the Hann, Blackman, Kaiser, and Bartlett windows. A detailed description of the different windowing functions can be found, for example, in Oppenheim and Schafer [30, Theorem 7.4.1]. In speech processing, the *Hamming window*, for a window length N_w, is defined by

$$w[n] = \begin{cases} 0.54 - 0.46\cos\left(\frac{2\pi n}{N_w}\right), & \forall\, 0 \le n \le N_w, \\ 0, & \text{otherwise}, \end{cases} \tag{7.1}$$

and used almost exclusively. The Hamming window is illustrated in Figure 7.2. From its frequency response it can be observed that the Hamming window has a wide main lobe[4], but its side lobes[5] are more than 40 dB below the main lobe.

7.3.2 Power Spectrum

A very simple approach to spectral analysis of a signal $x[n]$, given an observation window of *M* samples $n = 0, \ldots, M-1$, is to calculate its *power spectrum*. The power spectrum plots, for each spectral bin, the signal's power falling within a frequency range defined by the sampling frequency and observation window. It can be obtained through the calculation of the *discrete circular autocorrelation*

$$\phi[l] = \sum_{n=0}^{M-1-l} x[n]x[(n+l)\%M], \tag{7.2}$$

where % is the modulo operator. Thereafter, the discrete Fourier transform of the autocorrelation coefficients is calculated, resulting in the *discrete power spectrum*

$$S[m] = \sum_{l=0}^{M-1} \phi[l] e^{-j2\pi lm/M} \;\forall\, m = 0, 1, \ldots, M-1,$$

where m is the discrete angle frequency.

The power spectrum is widely used in speech processing because it can be quickly calculated via the *fast Fourier transform* (FFT) [6]. Nonetheless, it is poorly suited to the estimation of

[4] The main lobe is the lobe containing the maximum power.

[5] Side lobes are the lobes that are not the main lobe. Usually they are not wanted.

speech spectra intended for automatic classification, because it models spectral peaks and valleys equally well. This characteristic is bad for two reasons:

1. It cannot suppress the effect of the fundamental frequency and its harmonics in voiced speech, and therefore provides a poor estimate of the response function of the vocal tract.[6]
2. It cannot reduce the influence of distortions.

7.3.3 Spectral Envelopes

The *spectral envelope* is a plot of power versus frequency representing the resonances of the vocal tract. The spectral envelope is typically subject to certain smoothness criteria, such that the spectral effects of the periodic or noisy excitations, which are provided by the vocal cords or by the turbulent flow of air through a constriction of the vocal tract, respectively, are excluded. The spectral envelope models the spectral peaks of the power spectra nearly exactly, but may devote less precision to modeling the spectral valleys. It may be impossible or undesirable to model every individual peak; for example if a group of peaks is close together. In such cases, the spectral envelope should provide a reasonable approximation. In addition, the method used for estimating the spectral envelope should be stable and applicable to a wide range of signals with very different characteristics. To provide robustness in the presence of distortion, it is desirable that a local change in signal frequency does not affect the intensity of the spectral estimate at frequencies well apart from this point. Moreover, spectral envelope representation should be resilient to distortions in the data. We begin our discussion of the spectral envelope estimation with the most popular method, namely, linear prediction.

7.3.4 LP Envelope

The idea behind *linear prediction* (LP) is to predict the signal $x[n]$ at time n by a weighted linear combination of M immediately preceding samples and some input $u[n]$, such that

$$\hat{x}_M[n] = -\sum_{m=1}^{M} a_m\, x[n-m] + G \cdot u[n],$$

where M is known as the *model order* and G as the *gain*. Hence, it is necessary to determine the values of the weights $\{a\}_{m=1}^{M}$ which are dubbed LP coefficients in case of all-pole modeling. Assuming that $u[n]$ is unknown and thus that $x[n]$ must be predicted solely from a weighted combination of prior samples, the error between $x[n]$ and the prediction $\hat{x}_M[n]$ is given by the *error term*

$$e_M[n] = x[n] + \sum_{m=1}^{M} a_m\, x[n-m]. \tag{7.3}$$

The higher the error term $e_M[n]$, the worse the approximation of the "true" signal representation by the linear LP coefficients. Note that the goal of LP is not always, in particular in spectral

[6] To model the vocal tract is important since it contains the relevant information about uttered phonemes.

estimation, to faithfully represent the "tuned" signal, but to represent the transfer function of the vocal tract. And therefore, a higher model order, which leads to a better approximation of the "tuned" signal representation, does not necessarily lead to a better representation with respect to the goal.

The vector of prediction coefficients $\mathbf{a} = \begin{bmatrix} a_1 & a_2 & \cdots & a_m \end{bmatrix}^T$ is estimated by minimizing the total power of the prediction error as:

$$\hat{\mathbf{a}} = \underset{\mathbf{a}=[a_1,a_2,\cdots,a_M]}{\text{argmin}} \sum_{n=-\infty}^{\infty} \left(x[n] + \sum_{m=1}^{M} a_m \, x[n-m] \right)^2 . \tag{7.4}$$

Three principal methods to solve this minimization problem, yielding slightly different prediction coefficients [19], exist:

- the *autocorrelation method*;
- the *covariance method* which is based on the covariance matrix; and
- the *lattice method*.

The autocorrelation coefficients, in the autocorrelation method, have a very simple and symmetric structure. This structure allows for a recursive solution procedure in which each predictor coefficient may be derived in turn from previous coefficients. Due to this simple procedure, which is known as the *Levinson–Durbin recursion* [19], the autocorrelation method is used almost exclusively in speech processing. The Levinson–Durbin recursion is summarized in Table 7.2 where

$$\phi[m] = \sum_{n=0}^{N} x[n] \, x[n-m] \tag{7.5}$$

is the autocorrelatation sequence of the input signal x of length N. A detailed introduction to LP covering also the covariance and lattice method can be found in Strobach [35].

Table 7.2 The Levinson–Durbin recursion.

Step	Description
1.	Initialize with $a_{0,0} = 1$ and $e_0 = \phi[0]$
2.	For $m = 1, 2, \cdots, M$ $k_m = \frac{-1}{e_{m-1}} \sum_{i=0}^{m-1} \phi[i-m] \, a_{i,m-1}$ with $a_{i,m} = \begin{cases} 1, & i = 0, \\ a_{i,m-1} + k_m \, a^*_{m-i,m-1}, & i = 1, 2, \cdots, m-1, \\ k_m, & i = m \end{cases}$ and $e_m = e_{m-1}(1 - \lvert k_m \rvert^2)$
3.	The final set of linear prediction coefficients is given by $\{a_i = a_{i,M}\}_i$.

Frequency Domain Formulation

So far we have introduced the basic concept of LP from a time domain perspective. By applying the z-transform to (7.3), and summing over z to get the total power of the prediction error, we obtain the formulation in the transform domain:

$$\hat{\mathbf{a}} = \underset{\mathbf{a}=[a_1,a_2,\cdots,a_M]}{\operatorname{argmin}} \sum_{z=-\infty}^{\infty} \left(\left(1 + \sum_{m=1}^{M} a_m \, z^{-m} \right) X(z) \right)^2 .$$

Replacing z by $e^{j\omega}$ and applying Parseval's theorem [30] to replace the infinite summation by a finite integral, we get

$$\hat{\mathbf{a}} = \underset{\mathbf{a}=[a_1,a_2,\cdots a_M]}{\operatorname{argmin}} \frac{1}{2\pi} \int_{-\pi}^{\pi} \left| A\left(e^{j\omega}\right) \cdot X\left(e^{j\omega}\right) \right|^2 \, d\omega, \tag{7.6}$$

where

$$A\left(e^{j\omega}\right) = 1 + \sum_{m=1}^{M} a_m \, e^{-jm\omega}. \tag{7.7}$$

Once the LP coefficients a and the squared prediction error $e_M = G^2$ have been obtained from the Levinson–Durbin recursion, the transfer function of the discrete all-pole model can be expressed as

$$H(z) = \frac{G}{A(z)} = \frac{G}{1 + \sum_{m=1}^{M} a_m \, z^{-m}}, \tag{7.8}$$

where the *gain G* matches the scale of the LP model to the spectrum of the original signal. The all-pole spectral estimate $\hat{S}\left(e^{j\omega}\right)$, henceforth known as the *LP envelope*, is then given by

$$\hat{S}\left(e^{j\omega}\right) = \left| H\left(e^{j\omega}\right) \right|^2 = \frac{e_M}{\left| 1 + \sum_{m=1}^{M} a_m \, e^{-jm\omega} \right|^2}. \tag{7.9}$$

Limitation of LP Envelopes

Figure 7.3 illustrates the limitation of LP spectral estimation. By comparing the LP envelope with the Fourier spectrum in Figure 7.3, it can be observed that the LP envelope overestimates the spectral peak at 4 kHz.

To understand the limitation of LP envelopes from a mathematical standpoint, we can oversimplify voiced speech by modeling only the harmonics [24]. This can be represented as the short-time spectrum of a segment of voiced speech as the overtone series

$$S_{\text{harmonic}}\left(e^{j\omega}\right) = \sum_{l=1}^{L} 2\pi \frac{|b_l|^2}{4} \left[\delta(\omega + \omega_0 l) + \delta(\omega - \omega_0 l) \right], \tag{7.10}$$

where δ is the Dirac delta function, $\omega_0 = 2\pi f_0$ for a fundamental frequency of f_0. In the above, b_l is the amplitude of the lth harmonic and $L = f_s/2f_0$ is the number of harmonics, where f_s is

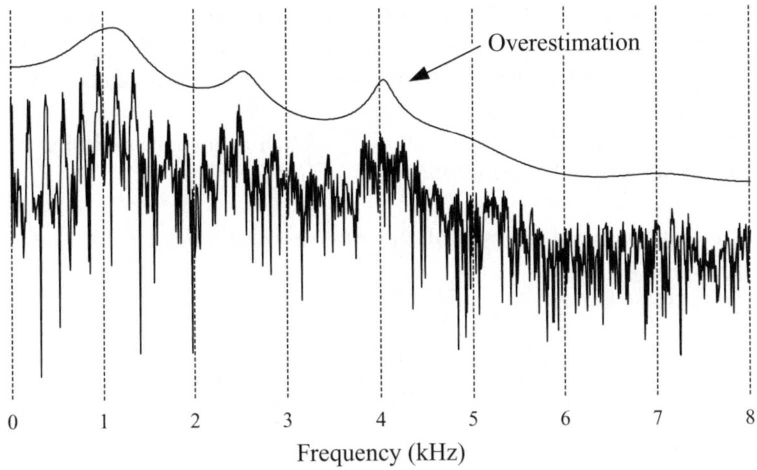

Figure 7.3 Spectral estimation of voiced speech for the linear prediction envelope (top) with model order 15 and the Fourier spectrum (bottom).

the sampling frequency. We can now set $\left|X\left(e^{j\omega}\right)\right|^2 = S_{\text{harmonic}}\left(e^{j\omega}\right)$ and substitute (7.10) into (7.6) to obtain

$$\hat{\mathbf{a}} = \underset{\mathbf{a}=[a_1,a_2,\cdots,a_M]}{\operatorname{argmin}} \frac{1}{2\pi}\int_{-\omega}^{\omega}\left|A\left(\mathrm{e}^{j\omega}\right)\right|^2 \cdot S_{\text{harmonic}}\left(\mathrm{e}^{j\omega}\right)\,d\omega,$$

or, equivalently,

$$\underset{\mathbf{a}=[a_1,a_2,\cdots,a_M]}{\operatorname{argmin}} \sum_{l=1}^{L}\frac{|b_l|^2}{2}\left|A(\mathrm{e}^{jl\omega_0})\right|^2.$$

To achieve the desired minimization of the prediction error, the LP filter (7.7) attempts to null out the harmonics $l\omega_0$ present in the original spectrum. With increasing model order M, the ability of the LP filter to null out these harmonics increases. But in the process, the zeros of the LP filter move ever closer to the unit circle, thereby causing sharper contours in the spectral envelope (7.9) and an overestimation of the spectral power at the harmonics [24]. Such effects are particularly problematic for medium- and high-pitched voices, because the harmonics are more spread as in low-pitched voices. As such, the LP method does not provide spectral envelopes which reliably estimate the power at the harmonic frequencies in voiced speech.

7.3.5 *MVDR Envelope*

Here, we briefly review the *minimum variance distortionless response* (MVDR)[7] as originally introduced by originally introduced by Capon [5]. It has been adopted by Lacoss [18] who demonstrated that this method provides an unbiased minimum variance estimate of the spectral

[7] Also known as Capon's method or the maximum-likelihood method [25].

components, hence the name MVDR. In order to overcome the problems associated with LP, Murthi and Rao [23] proposed the use of the MVDR for all-pole modeling of speech signals. A detailed discussion of speech spectral estimation using the MVDR can be found in Murthi and Rao [24].

MVDR spectral estimation can be posed as a problem in filter bank design, wherein the final filter bank is subject to the *distortionless constraint* [15]:

> *The signal at the* frequency of interest (FOI) ω_{foi} *must pass undistorted with unity gain.*

This can be expressed as

$$H\left(\mathrm{e}^{j\omega_{\text{foi}}}\right) = \sum_{m=0}^{M} h[m]\,\mathrm{e}^{-jm\omega_{\text{foi}}} = 1.$$

This constraint can be rewritten in vector form as:

$$\mathbf{v}^H\left(\mathrm{e}^{j\omega_{\text{foi}}}\right)\mathbf{h} = 1,$$

where $\mathbf{v}\left(\mathrm{e}^{j\omega_{\text{foi}}}\right)$ is the *fixed frequency vector*,

$$\mathbf{v}\left(\mathrm{e}^{j\omega}\right) = \begin{bmatrix}1 & \mathrm{e}^{-j\omega} & \mathrm{e}^{-j2\omega} & \cdots & \mathrm{e}^{-jM\omega}\end{bmatrix}^T,$$

and $\mathbf{h}$ is the *stacked impulse response*,

$$\mathbf{h} = \begin{bmatrix}h[0] & h[1] & \cdots & h[M]\end{bmatrix}^T.$$

The distortionless filter $\mathbf{h}$ can now be obtained by solving for the constrained minimization problem

$$\min_{\mathbf{h}} \mathbf{h}^H \mathbf{\Phi}\mathbf{h} \text{ subject to } \mathbf{v}^H\left(\mathrm{e}^{j\omega_{\text{foi}}}\right)\mathbf{h} = 1, \qquad (7.11)$$

where $\mathbf{\Phi}$ is the $(M+1)\times(M+1)$ Toeplitz autocorrelation matrix with (m,n)th element $\mathbf{\Phi}[m,n] = \phi[m-n]$ of the input signal (7.5). The minimization of the output power in (7.11) guarantees minimum leakage from other frequencies. The solution of the constrained minimization problem can be expressed as [15]:

$$\mathbf{h} = \frac{\mathbf{\Phi}^{-1}\,\mathbf{v}(\mathrm{e}^{j\omega_{\text{foi}}})}{\mathbf{v}^H(\mathrm{e}^{j\omega_{\text{foi}}})\,\mathbf{\Phi}^{-1}\,\mathbf{v}(\mathrm{e}^{j\omega_{\text{foi}}})}.$$

This implies that $\mathbf{h}$ is the impulse response of the distortionless filter for the frequency ω_{foi}. The MVDR envelope of the spectrum $S(\mathrm{e}^{-j\omega})$ at frequency ω_{foi} is then obtained as the output of the optimized constrained filter [15]:

$$S_{\text{MVDR}}(\mathrm{e}^{j\omega_{\text{foi}}}) = \frac{1}{2\pi}\int_{-\pi}^{\pi}\left|H(\mathrm{e}^{j\omega_{\text{foi}}})\right|^2 S(\mathrm{e}^{-j\omega})\,d\omega. \qquad (7.12)$$

Although MVDR spectral estimation was posed as a problem of designing a distortionless filter for a given frequency ω_{foi}, this was only a conceptual device. The MVDR spectral envelope can in fact be represented in parametric form for all frequencies and computed as

$$S_{\text{MVDR}}(\mathrm{e}^{j\omega}) = \frac{1}{\mathbf{v}^H(\mathrm{e}^{j\omega})\,\mathbf{\Phi}^{-1}\mathbf{v}(\mathrm{e}^{j\omega})}.$$

Table 7.3 Fast MVDR spectral envelope calculation.

Step	Description
1.	Compute the LPCs $a_0, a_1, \cdots, a_M$ of order M and the prediction error e_M as defined in Table 7.2.
2.	Correlate the LPCs, as $\mu_m = \begin{cases} \frac{1}{e_M} \sum_{i=0}^{M-m} (M+1-m-2i)\, a_i^{(M)}\, a_{i+m}^{*(M)}, & m = 0, 1, 2, \cdots, M, \\ \mu_{-m}^*, & m = -M, \cdots, -1 \end{cases}$
3.	Compute the *MVDR envelope* $S_{\text{MVDR}}(\mathrm{e}^{j\omega}) = \frac{1}{\sum_{m=-M}^{M} \mu_m \mathrm{e}^{-j\omega m}}$

Under the assumption that matrix $\mathbf{\Phi}$ is positive definite and thus invertible, Musicus [25] derived a fast algorithm to calculate the MVDR spectral envelope from a set of *linear prediction coefficients* (LPC)s, as given in Table 7.3.

The MVDR envelope copes well with the problem of overestimation of the spectral power at the harmonics of voiced speech. To show this, we once more model voiced speech as the sum of harmonics (7.10). Using the frequency form of the MVDR envelope given by (7.12), the spectral estimate at $\omega_l = \omega_0 l \; \forall \, l = 1, 2, \ldots$ is given by

$$S_{\text{MVDR}}(\mathrm{e}^{j\omega_0 l}) = \sum_{l=1}^{L} \frac{|b_l|^2}{4} \left\{ |H(\mathrm{e}^{j\omega_l})|^2 + |H(\mathrm{e}^{-j\omega_l})|^2 \right\},$$

where b_l is the amplitude of the lth harmonic. Thus, the MVDR distortionless filter $\mathbf{h}$ faithfully preserves the input power at $\omega_0 l$ while treating the other $(2L-1)$ exponentials as interference and attempting to minimize their influence on the output of the filter. Hence, the MVDR envelope models the perceptually important speech harmonics very well. However, unlike warped spectra, as introduced in the following section, it does not mimic the human auditory system and does not model the different frequency bands with varying accuracy.

7.3.6 Warping the Frequency Axis

So far we have investigated methods on a linear frequency scale. To better approximate the human auditory analysis and to increase classification[8], it is necessary to map the linear frequency axis ω to a nonlinear frequency axis $\tilde{\omega}$. Two widely used, but in principal very different methods, exist:

- *Critical band filtering* applies an array of overlapping bandpass filters which, set accordingly, are able to approximating the mel or Bark scale.
- *Warping* applies a conformal map to approximate the mel or Bark scale.

[8] The application of nonlinear frequency scale, such as the mel and the Bark scale, has been demonstrated to increase performance of speech recognition as well as speaker identification.

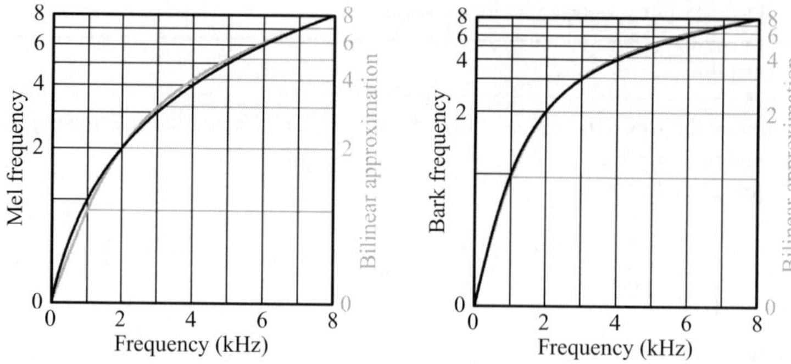

Figure 7.4 Mel frequency (scale shown along left edge of left image) and Bark frequency (scale shown along left edge of right image) can be approximated by a BLT (scale shown along right edges) for a sampling rate of 16 kHz, $\alpha_{\text{mel}} = 0.4595$, $\alpha_{\text{Bark}} = 0.6254$.

While critical band filtering [32] sums frequency bins[9] and therefore always looses spectral resolution, this is not the case for warping. A convenient way to implement warping is through a conformal map, such as a first order all-pass filter [30, Theorem 5.5], which is also known as the *bilinear transform* (BLT) [3, 31], or a Blaschke factor [12, Theorem 9.1]. It is defined in the z-domain as

$$\tilde{z}^{-1} = Q_z = \frac{z^{-1} - \alpha}{1 - \alpha \cdot z^{-1}} \;\forall\; -1 < \alpha < +1, \tag{7.13}$$

where α is the *warp factor*. A particular characteristic of the BLT is that it preserves the unit circle, such that

$$\left|Q\left(\mathrm{e}^{j\omega}\right)\right| = 1 \;\forall\; -\pi < \omega \leq \pi.$$

Indeed, this latter property is the reason behind the designation *all-pass*. The relationship between $\tilde{\omega}$ and ω is nonlinear as indicated by the phase function of the all-pass filter [21]:

$$\tilde{\omega} = \arg\left(Q\left(\mathrm{e}^{-j\omega}\right)\right) = \omega + 2\arctan\left(\frac{\alpha \sin\omega}{1 - \alpha\cos\omega}\right). \tag{7.14}$$

A good approximation of the mel and Bark scale by the BLT is possible if the warp factor is set accordingly. The optimal warp factor depends on the sampling frequency and can be found by different optimization methods [34]. Figure 7.4 compares the mel-scale and the Bark-scale with their approximated counterparts for a sampling frequency of 16 kHz.

Warping can be applied in different domains. Note that the frequency axis is nonlinearly scaled independent of the application of the BLT in the time or frequency domain. However,

9 due to the application of the bandpass filters in the discrete frequency domain

the achieved accuracy of the spectral estimate depends on wether the BLT is applied in the time or frequency domain. This effect can be explained as follows:

- *Warping in the time domain* modifies the values in the autocorrelation matrix and therefore, in the case of autoregressive models more coefficients are used to describe lower frequencies and fewer coefficients to describe higher frequencies.
- *Warping in the frequency or cepstral domain* does not change the spectral resolution as the transform is applied after spectral analysis.

  As indicated by Nocerino *et al.* [27], a general warping transform in the same domain, such as the BLT, is equivalent to a matrix multiplication

$$\mathbf{f}_{\text{warp}} = \mathbf{L}(\alpha)\,\mathbf{f},$$

  where $\mathbf{L}(\alpha)$ denotes the transformation matrix which depends on the warp factor α. It follows that the values $\mathbf{f}_{\text{warp}}$ on the warped scale are a linear interpolation of the values $\mathbf{f}$ on the linear scale and therefore the spectral resolution is not altered.

Figure 7.5 demonstrates the effect of warping, with a warp factor $\alpha > 0$, on the spectral envelope if applied either in the time or in the frequency domain. The figure also compares the warped spectral envelopes with its unwarped counterpart. We observe, that if the BLT is applied in the time domain, that spectral resolution decreases as frequency increases. In comparison to the resolution provided by the linear frequency scale, corresponding to $\alpha = 0$, the warped frequency resolution increases for low frequencies up to the *turning point* (TP) frequency [13]

$$f_{\text{tp}}(\alpha) = \pm\frac{f_{\text{s}}}{2\pi}\arccos(\alpha), \tag{7.15}$$

where f_{s} represents the sampling frequency. At the TP frequency, the spectral resolution is not affected. Above the TP frequency, the frequency resolution decreases in comparison to the resolution provided by the linear frequency scale.

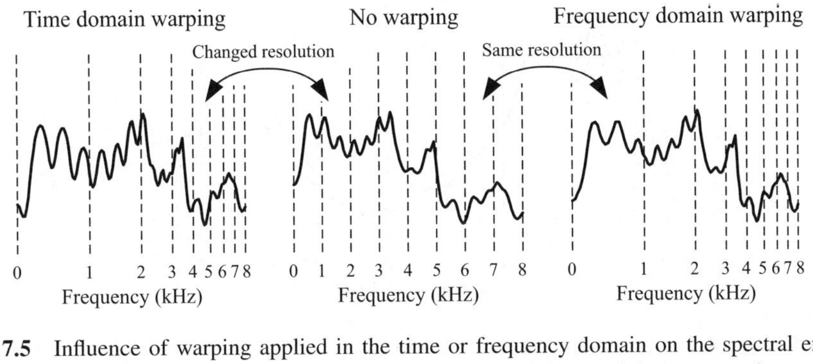

Figure 7.5 Influence of warping applied in the time or frequency domain on the spectral envelope. While warping in the time domain changes the spectral resolution and frequency axis, warping in frequency domain does not alter the spectral resolution but still changes the frequency axis.

As observed by Strube [36], prediction error minimization of the predictors $\tilde{a}_m$ in the warped domain is equivalent to the minimization of the output power of the warped inverse filter

$$\tilde{A}(z) = 1 + \sum_{m=1}^{M} \tilde{a}_m \, \tilde{z}^{-m}(z), \tag{7.16}$$

in the linear domain, where each unit delay element z^{-1} is replaced by a BLT $\tilde{z}^{-1}$. The prediction error is therefore given by

$$E\left(e^{j\omega}\right) = \left|\tilde{A}\left(e^{j\omega}\right)\right|^2 P\left(e^{j\omega}\right), \tag{7.17}$$

where $P\left(e^{j\omega}\right)$ is the power spectrum of the signal. From Parseval's theorem [29] it then follows that the total squared prediction error can be expressed as

$$\sigma^2 = \int_{-\pi}^{\pi} E\left(e^{j\tilde{\omega}}\right) d\tilde{\omega} = \int_{-\pi}^{\pi} E\left(e^{j\omega}\right) W^2\left(e^{j\omega}\right) d\omega, \tag{7.18}$$

where $W(z)$ denotes the weighting filter

$$W(z) = \frac{\sqrt{1-\alpha^2}}{1-\alpha z^{-1}}. \tag{7.19}$$

However, the minimization of the squared prediction error σ^2 does *not* lead to minimization of the power, but the power of the error signal filtered by the weighting filter $W(z)$, which is apparent from the presence of this factor in (7.18). Thus, the BLT introduces an unwanted spectral tilt. To compensate for this negative effect, the inverse weighting function

$$\left|\tilde{W}(\tilde{z}) \cdot \tilde{W}(\tilde{z}^{-1})\right|^{-1} = \frac{\left|1+\alpha\cdot\tilde{z}^{-1}\right|^2}{1-\alpha^2} \tag{7.20}$$

can be applied. The effect of the spectral tilt introduced by the BLT and its correction (7.20) are depicted in Figure 7.6. For a fixed warp parameter α, it can be easily applied in the frequency domain by a precalculated frequency dependent correction term.

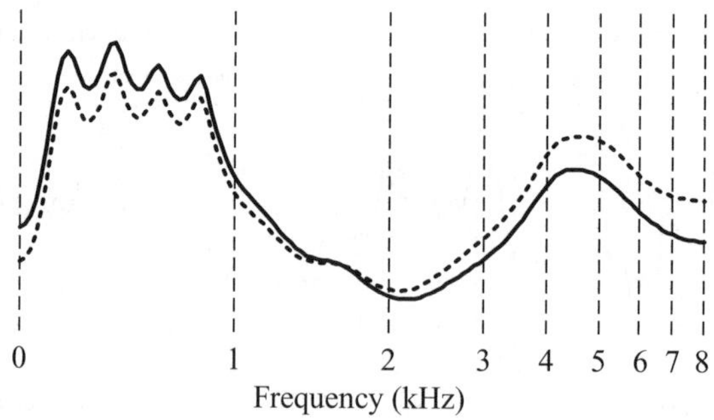

Figure 7.6 The plot of two warped spectral envelopes $\alpha = \alpha_{\mathrm{mel}}$ demonstrates the effect of spectral tilt. While the spectral tilt is not compensated for the dashed line, it is compensated for the solid line. It is clear to see that high frequencies are emphasized for $\alpha > 0$ if no compensation is applied.

7.3.7 Warped LP Envelope

The LP and MVDR all-pole models approximate speech spectra equally well at all frequency bands, while the human auditory system provides frequency resolution which is higher for low frequencies and lower for high frequencies. To eliminate for this inconsistency, two widely used modifications exist for LP:

- *perceptual linear prediction* [16]
  Perceptual linear prediction (PLP) modifies LP spectral analysis by the introduction of the Bark scale and logarithmic amplitude compression (raised to the power of 0.33 to simulate the power law of hearing) prior to the Levinson–Durbin recursion.
- *warped linear prediction* [36]
  An alternative to PLP, for which there is no need to convert between time and frequency domains, is to perform LP analysis on a *warped* frequency axis by replacing the unit delay element $e^{-jk\omega}$ with a cascade of first-order all-pass filters such as were presented in Section 7.3.6.

Similar to LP, warped LP can be estimated with the Levinson–Durbin recursion, Table 7.2, but yet using autocorrelation coefficients derived from a warped frequency scale. Those autocorrelation coefficients are referred to as warped autocorrelation coefficients. Note that applying the BLT to the spectrum of a finite sequence produces a spectrum corresponding to an infinite sequence

$$\tilde{X}(\tilde{z}) = \sum_{n=0}^{\infty} \tilde{x}[n]\, \tilde{z}^{-n} = X(z) = \sum_{n=0}^{N-1} x[n]\, z^{-n}.$$

Thus, the direct calculation of the warped autocorrelation coefficients

$$\tilde{\phi}[m] = \sum_{n=0}^{\infty} \tilde{x}[n]\, \tilde{x}[n-m], \tag{7.21}$$

is not feasible. To overcome this problem, a variety of solutions has been proposed [10, 36, 37]. Here, we give the algorithm proposed by Matsumoto *et al.* [22]. To obtain the warped predictors, we must solve the normal equations

$$\sum_{y=1}^{p} \tilde{\mathbf{\Phi}}[m,n]\, \tilde{a}_{m,n} = -\tilde{\mathbf{\Phi}}[m,0], \;\; \forall\; m = 1, 2, \cdots, p, \tag{7.22}$$

where

$$\tilde{\mathbf{\Phi}}[m,n] = \sum_{l=0}^{\infty} y_m[l]\, y_n[l],$$

and

$$y_m[n] = \alpha \cdot (y_m[n-1] - y_{m-1}[n]) - y_{m-1}[n-1]$$

is the output of the mth order all-pass filter excited by $y_0[n] = x[n]$. The last equation implies that $\tilde{\mathbf{\Phi}}[m,n]$ is a component of the warped autocorrelation function

$$\tilde{\mathbf{\Phi}}[m,n] = \tilde{\phi}[|m-n|]. \tag{7.23}$$

Thus, (7.22) is revealed to be an autocorrelation equation, exactly like the autocorrelation equation found in standard LP analysis. Furthermore, as $\tilde{\mathbf{\Phi}}[m,n]$ depends only on the difference $|m-n|$, we can replace (7.21) by

$$\tilde{\phi}[|m-n|] = \sum_{l=0}^{N-1-|m-n|} x[l]\, y_{|m-n|}[l]. \tag{7.24}$$

Hence, the warped autocorrelation coefficients $\tilde{\mathbf{\Phi}}[m,n]$ can be calculated with a finite sum.

Given the warped-LP coefficients, we can now obtain the transfer function $H_{\text{warped-LP}}(z)$. Thereby, we derive an all-pole spectral estimate in the warped-frequency domain, henceforth referred to as the *warped LP envelope*

$$S_{\text{warped-LP}}(\mathrm{e}^{j\omega}) = \left|H_{\text{warped-LP}}\left(\mathrm{e}^{j\omega}\right)\right|^2 = \frac{\tilde{e}_M}{\left|1+\sum_{m=1}^{M}\tilde{a}_m\,\mathrm{e}^{-jm\omega}\right|^2}, \tag{7.25}$$

where $\tilde{e}$ is the squared prediction error of the warped estimate.

Note that if α is set appropriately, the spectrum (7.25) is already in the mel-warped frequency domain and therefore it is necessary to either

- eliminate the mel spaced triangular filter bank traditionally used in the extraction of mel-frequency cepstral coefficients, or
- replace it by a filter bank of uniform half-overlapping triangular filters to provide feature reduction or additional spectral smoothing.

The warping of the LP envelope addresses the inconsistency between LP spectral estimation and that performed by the human auditory system. Unfortunately, for high-pitched voiced speech the lower harmonics become so sparse that single harmonics appear as spectral poles, which is highly undesirable in all-pole modeling. One proposed approach to overcome this drawback is to weight the warped autocorrelation coefficient $\widetilde{\phi}[m]$ with a lag window [21]. An alternative is to use the warped MVDR envelope as described in Section 7.3.8.

7.3.8 Warped MVDR Envelope

To overcome the problems inherent in LP while emphasizing the perceptually relevant portions of the spectrum, the BLT must be applied prior to MVDR spectral envelope estimation [41]. Let us define the *warped frequency vector* $\tilde{\mathrm{v}}$ as

$$\tilde{\mathrm{v}}(\mathrm{e}^{j\omega}) = \left[1 \quad \frac{\mathrm{e}^{-j\omega}-\alpha}{1-\alpha\cdot\mathrm{e}^{-j\omega}} \quad \frac{\mathrm{e}^{-j2\omega}-\alpha}{1-\alpha\cdot\mathrm{e}^{-j2\omega}} \quad \cdots \quad \frac{\mathrm{e}^{-jM\omega}-\alpha}{1-\alpha\cdot\mathrm{e}^{-jM\omega}}\right]^T.$$

In order to calculate the distortionless filter $\tilde{\mathbf{h}}$ in the warped domain, we must once more solve the constrained minimization problem

$$\min_{\tilde{\mathbf{h}}} \tilde{\mathbf{h}}^H \tilde{\mathbf{\Phi}} \tilde{\mathbf{h}} \text{ subject to } \tilde{\mathbf{v}}^H(\mathrm{e}^{j\omega_{\mathrm{foi}}})\tilde{\mathbf{h}} = 1, \tag{7.26}$$

where $\tilde{\mathbf{\Phi}}$ is the Toeplitz autocorrelation matrix as defined by (7.23). Clearly, this solution is different from MVDR on the linear frequency scale. However, the way to solve for the warped constrained minimization problem is very similar to its unwarped counterpart. The warped MVDR envelope of the spectrum $S(\mathrm{e}^{-j\omega})$ at frequency ω_{foi} can be obtained as the output of the optimal filter

$$S_{\mathrm{warpedMVDR}}(\mathrm{e}^{j\omega_{\mathrm{foi}}}) = \frac{1}{2\pi}\int_{-\pi}^{\pi}\left|\tilde{H}(\mathrm{e}^{j\omega_{\mathrm{foi}}})\right|^2 S(\mathrm{e}^{-j\omega})\mathrm{d}\omega, \tag{7.27}$$

under the same constraint as in MVDR, that the signal at the frequency of interest must pass undistorted with unity gain, but yet on a nonlinear scaled frequency axis

$$\tilde{H}(\mathrm{e}^{j\omega_{\mathrm{foi}}}) = \sum_{m=0}^{M}\tilde{h}(m)\frac{\mathrm{e}^{-jm\omega_{\mathrm{foi}}} - \alpha}{1 - \alpha\cdot\mathrm{e}^{-jm\omega_{\mathrm{foi}}}} = 1.$$

Assuming that the Hermitian Toeplitz correlation matrix $\tilde{\mathbf{\Phi}}$ is positive definite and thus invertible, Musicus' (1985) algorithm, as given in Table 7.3, can be readily applied to compute the warped MVDR spectral envelope. However, the LPCs and the error term in Step 1 of Table 7.3 must be replaced by their warped counterparts from Section 7.3.7. Note that the spectrum (7.27) derived by the modified, fast algorithm has a warped frequency axis and should be handled as suggested in Section 7.3.7.

Warped-Twice MVDR Envelope

If one is aiming for adjusting spectral resolution to lower or higher frequency regions without changing the frequency axis it should be noted that the warping of the frequency axis due to the BLT in the time domain can be compensated through a second BLT in the frequency domain [26, 38]. Due to the application of two warping stages in MVDR spectral estimation, this approach is dubbed *warped-twice MVDR*.

7.3.9 Comparison of Spectral Estimates

Figure 7.7 displays plots of the spectral envelopes derived from the power spectrum as well as the LP and MVDR models on both a linear and nonlinear frequency scale. The warp factor for the warped LP and warped MVDR was set to 0.4595 so as to simulate the mel-frequency for a signal sampled at 16 kHz. Due to its stronger smoothing properties, the model order of the MVDR envelope was set to 30, while that of the LP envelope was set to 15.

The spectral estimates on the nonlinear frequency scale differ from those in the linear frequency scale inasmuch as more parameters are apportioned to describe the lower as compared to the higher frequency regions. Thus, the warped estimates provide a higher spectral resolution in low frequencies and lower spectral resolution in high-frequency regions. Therefore,

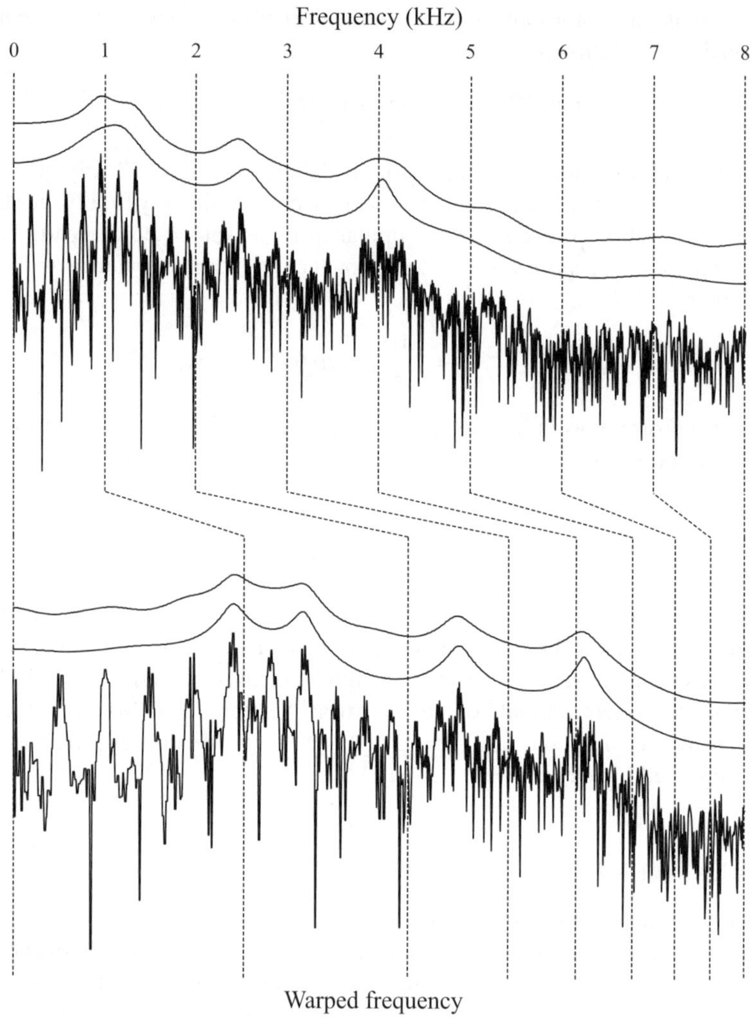

Figure 7.7 Different spectral estimations (logarithmic power vs. frequency, for a better visualization each envelope has a particular energy offset) of voiced speech and the influence of warping. From top to bottom: minimum variance distortionless response envelope with model order 30, linear prediction envelope with model order 15 and Fourier spectrum, and their mel warped, $\alpha = 0.4595$, counterparts with same model order.

warping prior to spectral analysis provides properties which cannot be achieved when the spectral analysis is *followed* by frequency warping.

The MVDR envelope prevents the unwanted overestimation of the harmonic peaks in medium- and high-pitched voiced speech that is seen in the LP envelope. As it is apparent from Figure 7.7, the LP envelope overestimates the spectral peak at 4 kHz, which is apparent upon comparing the LP envelope with the Fourier spectrum. Unlike the LP envelope, the MVDR envelope provides a broad peak which matches the true spectrum better.

Spectral Relationship Between LP and MVDR Envelopes

Burg [4] showed that the MVDR envelope of model order M can also be expressed as the harmonic mean of the LP spectra $S_{\text{LP}}^{(M)}(e^{j\omega})$ of orders 0 through M:

$$S_{\text{MVDR}}^{(M)}(\mathrm{e}^{j\omega}) = \left[\sum_{m=0}^{M} \frac{1}{S_{\text{LP}}^{(m)}(\mathrm{e}^{j\omega})} \right]^{-1}.$$

The given relation, which also holds for the warped counterparts, explains why the (warped) MVDR spectral envelope exhibits a smoother frequency response with decreased variance than the corresponding (warped) LP spectrum if compared for the same model order.

7.3.10 *The Spectrogram*

The *spectrogram* is a graphical representation of the energy density as a function of discrete angular frequency m and discrete time frames t

$$\text{spectrogram}_t(m) = |X[n, m)|^2,$$

where $X[n, m)$ can be some of the previously described spectral estimation methods. A spectrogram is typically displayed in gray scale, such that the higher the energy at a specific frequency range and a given time frame, the darker this region appears in the time-frequency plane. Hence, spectral peaks are shown in black, while spectral valleys are shown in white. Values in between these two extremes have a gray shade. Due to the large dynamic range of human speech, spectrograms are alternatively displayed in a logarithmic scale

$$\text{logarithmic spectrogram}_t(m) = 20 \log_{10} |X[n, m)|.$$

Depending on the window size used we differentiate between:

- *The wide-band spectrogram.* In this case a short duration window of less than a pitch period, typically 10 ms, is used. This provides good time resolution, but smears the harmonic structure, thereby yielding spectra similar to those of spectral envelopes.
- *The narrow-band spectrogram.* In this case, a long duration window of at least the length of two pitch periods is used. The narrow-band spectrogram provides good frequency resolution but poor time resolution. Due to the increased frequency resolution, the harmonics of f_0 can be observed as horizontal striations during segments of voiced speech.

An example of a wide-band as well as the narrow-band spectrogram is given in Figure 7.8.

7.4 Cepstral Processing

Cepstral[10] features were originally invented by Bogert *et al.* [2] to distinguish between earthquakes and underground nuclear explosions. Just one year later, Noll [28] adopted them into speech processing for vocal pitch detection. The cepstrum was introduced to other fields of

[10] *Cepstra* stems from the reversal of the first four letters of *spectra*.

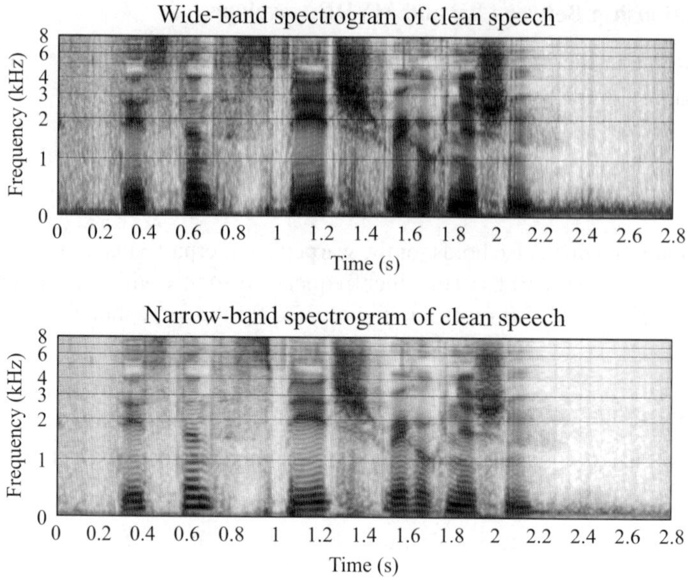

Figure 7.8 Narrow-band and wide-band, mel-scaled, logarithmic spectrogram of clean speech. All spectrograms were produced from the phrase "distant speech recognition" spoken by a male speaker.

speech processing in the early 1980s. Its first use for the purpose of ASR and for speaker verification was reported by Davis and Mermelstein [7] and Furui [11], respectively. Nowadays cepstral features are widely used in a broad variety of speech applications.

7.4.1 Definition and Calculation of Cepstral Coefficients

The cepstrum is defined as the result of taking the Fourier transform of the (warped) logarithmic spectrum. It can thus be interpreted as containing information about rate changes in the different spectrum bands. The cepstrum is in particular useful in speech processing. This is due to the fact that the low-frequency periodic excitation from the vocal cords and the formant filtering of the vocal tract are located in different regions in the cepstral domain. Thus, the influence of the excitation and the shape of the vocal tract in a signal can be easily separated in the cepstral domain.

The *real cepstrum*[11] is defined as the sequence

$$c_x[k] = \frac{1}{2\pi} \int_{-\pi}^{\pi} \log \left| X\left(e^{j\omega}\right) \right| e^{j\omega k} d\omega, \tag{7.28}$$

where $X\left(e^{j\omega}\right)$ can be any of the presented spectral estimate (the mel or the Bark scale is most commonly used) as presented in Section 7.3.

[11] The cepstrum can be defined complex; however, the real cepstrum is exclusively used in speech technology. The reader who is interested in the definition of the complex spectrum should refer to Oppenheim and Schafer [30].

To generate acoustic features, for the purposes of speech recognition or speaker identification/verification, the minimum phase[12] equivalent

$$\hat{x}_{\min}[k] = \begin{cases} 0, & \forall\ k < 0, \\ c_x[0] & k = 0, \\ 2c_x[k] & \forall\ k > 0 \end{cases} \tag{7.29}$$

of the cepstral sequence is almost invariably used. Such features can be calculated by using the inverse discrete Fourier transform to calculate $c_x[k]$ as in (7.28), then using this intermediate result to calculate $\hat{x}_{\min}[k]$ as in (7.29).

Noting that the logarithmic magnitude and power spectra are real symmetric functions [33], (7.29) can also be represented as the Type 2 *discrete cosine transform* (DCT). The transformation can thus be conveniently represented as a simple matrix multiplication to the log-power spectral density vector $\log\left|X\left(e^{j\omega}\right)\right|$, such that

$$\hat{x}_{\min}[k] = \sum_{m=0}^{M-1} \log\left|X\left(e^{j\omega_m}\right)\right| T_{k,m}^{(2)}, \tag{7.30}$$

where $T_{k,m}^{(2)}$ are the individual components of the Type 2 DCT.

7.4.2 Characteristics of Cepstral Sequences

To investigate the characteristics of cepstral sequences, we define a system transfer function with M_i zeros p_m and N_i poles q_m *inside* the unit circle and M_o zeros d_m and N_o poles e_m *outside* the unit circle. Together with the scale term K we can then write the equation of the transfer function as

$$H(z) = K\,\frac{\prod_{m=1}^{M_\text{o}}(1 - d_m\,z)\prod_{m=1}^{M_\text{i}}(1 - p_m\,z^{-1})}{\prod_{m=1}^{N_\text{o}}(1 - e_m\,z)\prod_{m=1}^{N_\text{i}}(1 - q_m\,z^{-1})}. \tag{7.31}$$

Note that, by definition, $|d_m|, |e_m|, |p_m|, |q_m| < 1$ for all poles and zeros.

In order to determine that cepstral sequence $\hat{h}[k]$ of which the transform pair is the transfer function $\hat{H}(z)$, we can make use of the series expansions

$$\log\left(1 - \alpha\,z^{-1}\right) = -\textstyle\sum_{k=1}^{\infty} \frac{\alpha^k}{k}\,z^{-k}\ \forall\ |z| > |\alpha|, \tag{7.32}$$

$$\log\left(1 - \beta\,z\right) = -\textstyle\sum_{k=1}^{\infty} \frac{\beta^k}{k}\,z^{k}\ \forall\ |z| < |\beta|^{-1}. \tag{7.33}$$

Hence, we can express $\hat{h}[k]$ as

$$\hat{h}[k] = \begin{cases} \log|K|, & k = 0, \\ -\sum_{m=1}^{M_\text{i}} \frac{p_m^k}{k} + \sum_{m=1}^{N_i} \frac{q_m^k}{k}, & \forall\ k > 0, \\ \sum_{m=1}^{M_\text{o}} \frac{d_m^{-k}}{k} - \sum_{m=1}^{N_\text{o}} \frac{e_m^{-k}}{k}, & \forall\ k < 0. \end{cases} \tag{7.34}$$

[12] A system is *minimum-phase* if its inverse is causal and stable [30].

Several characteristics of cepstral sequences emerge from (7.34):

- A minimum phase system will have a *causal* sequence of cepstral coefficients, which implies $\hat{h}[k] = 0 \; \forall \; k < 0$.
- For real $h[k]$, which implies that the complex poles and zeros of $H(z)$ occur in complex-conjugate pairs, the cepstral coefficients $\hat{h}[k]$ will also be real as stated at the outset.
- The cepstral coefficients $\hat{h}[k]$ decay at *least* by the factor $1/k$. Hence, it can be concluded, that the lower order coefficients contain *most* of the information about the overall spectral shape of $H\left(e^{j\omega}\right)$.

The low order cepstral coefficients, especially $\hat{x}_{\min}[0]$ and $\hat{x}_{\min}[1]$, can be given a particular intuitive meaning. The initial value $\hat{x}_{\min}[0]$ represents the *average power* of the input signal. The next value $\hat{x}_{\min}[1]$ indicates the *distribution of spectral energy* between low and high frequencies. A positive value indicates a sonorant sound, as the preponderance of the spectral energy will be concentrated in the low-frequency regions. A negative value indicates a fricative, inasmuch as most of the spectral energy will be concentrated at high frequencies [9]. Higher order cepstral coefficients represent ever increasing levels of spectral detail. Note that a finite input sequence results in an infinite number of cepstral coefficients. However, it is a well-established fact that a finite number of coefficients, typically ranging between 12 and 20 depending on the sampling rate and estimation method, is optimal for speech analysis [17].

7.5 Influence of Distortions on Different Speech Features

In this section, we investigate the influence of different kinds of distortions, namely noise, reverberation, and a change in fundamental frequency, both in the logarithmic power spectral domain[13] and cepstral domain. The comparison is performed on the power spectrum as a prototypical representative for nonparametric spectral estimation methods and warped MVDR as a prototypical representation for parametric spectral estimation and spectral envelopes. We have chosen these prototypes because the power spectrum is probably the most common spectral estimation method and warped MVDR, a method, which has been demonstrated superior to a broad range of alternative methods in speech processing [39] for the reasons discussed in Section 7.3. Figure 7.9 shows two block diagrams illustrating all the processing steps required for the calculation of the compared mel-frequency cepstral coefficient (employing the power spectrum) and warped MVDR cepstral coefficient (employing the warped MVDR spectral envelope) front-ends.

But before we can start with the analysis and comparison, we have to first define different useful objective functions.

7.5.1 *Objective Functions*

To evaluate and compare the quality of different feature-extraction methods *objective functions* are required which are able to judge the overall, or particular parts, of the system quality. To

[13] The spectral domain is easier comprehensible by humans than the cepstral domain and thus, it is preferable to do some of the analysis in this domain.

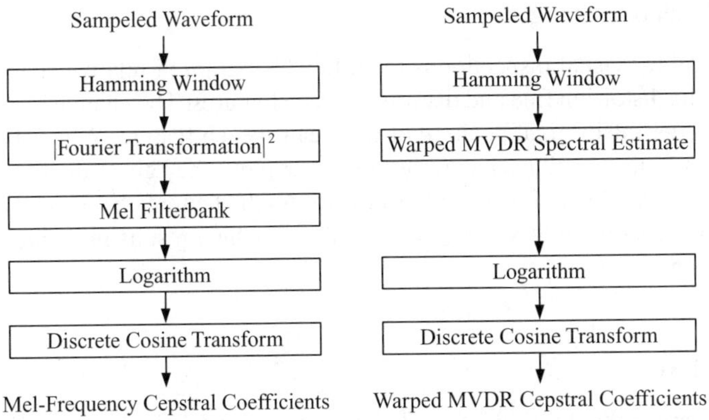

Figure 7.9 Block diagrams illustrating all the processing steps required for the calculation of the compared mel-frequency cepstral coefficient and warped MVDR cepstral coefficients front-ends.

evaluate speech features the overall system quality could be measured by the word error rate in case of speech recognition or acceptance and rejection ratios in case of speaker verification. Evaluating the whole system, which would result in high computational costs and turnaround times, might not only be unhandy for the given reasons, but might also not be a useful tool for analysis. To overcome high computational costs and turnaround times objective functions can be used which correlate strongly to methods judging the overall system performance, but without the need to run the whole system. One such method, to evaluate the quality of speech features, which is strongly correlated to the word error rate, is *class separability*. Class separability measures how good different classes can be distinguished from each other. For analysis, alternative methods which are able to provide additional information exist. We introduce two of such techniques in the next sections which are particular useful to investigate the influence of distortions on speech features.

Scatter Plot of Clean versus Distorted Features

To compare how speech features are changing in respect to a given distortion, one can prepare a scatter plot. These scatter plots can either be derived over all dimension (used in our experiments) or for a subset (e.g., if the influence of a particular frequency range should be investigated). In our scatter plots, the x-axis represents clean speech features, while the y-axis represents their distorted[14] counterparts. Therefore, each point in the scatter plot correlates the position of a clean observation to the corresponding distorted observation. A darker color in the scatter plot reflects more "hits" at a particular position while a lighter color has less "hits." Therefore, a distortion-free signal is reflected by different shades of gray on the line $x = y$. If a signal becomes more and more distorted more points are farther from the line $x = y$. While the scatter plot gives a good indication on "how" and "where" the features are moved, it does not allow to see the amount of relative changes.

[14] Distorted in this respect can be any kind of unwanted change on the feature values.

Histogram of Relative changes

A second method for visual inspection is to build a histogram of relative changes (the difference) between the distortion free and distorted speech features. The change is relative, because it does not consider the original position of the clean speech feature. While those histograms are not able to show the absolute change, as the scatter plots, they give a distribution of change. The distortion free case would have all hits at zero. As the features get increasingly distorted, the distribution is becoming less centered around zero which results in an increased variance of the distribution.

Class Separability

Under the assumption that for each of M classes[15], there exists a set $\{\mathbf{y}_{m,k}\}$ of labeled training samples, class separability can be expressed as a function of two (out of three)[16] scatter matrices. With the given definitions

- K_m is the number of samples in the mth class;
- $\boldsymbol{\mu}_m$ is the mean of all samples in the mth class;
- K is the total number of samples; and
- $\boldsymbol{\mu}$ is the mean of all samples regardless of class;

these matrices can be defined as

- The *within-class scatter matrix*

$$\mathbf{S}_{\mathrm{w}} = \frac{1}{K} \sum_{m=1}^{M} \left[\sum_{k=1}^{K_m} (\mathbf{y}_{m,k} - \boldsymbol{\mu}_m)(\mathbf{y}_{m,k} - \boldsymbol{\mu}_m)^T \right],$$

  which is defined as the expected covariance of each of the classes. It is a measure of "how compact" the classes are.
- The *between-class scatter matrix*

$$\mathbf{S}_{\mathrm{b}} = \frac{1}{K} \sum_{m=1}^{M} K_m (\boldsymbol{\mu}_m - \boldsymbol{\mu})(\boldsymbol{\mu}_m - \boldsymbol{\mu})^T,$$

  which is defined as the covariance of the mean values of each of the classes. It is a measure of "how separated" classes are to each other.
- The *total scatter matrix*

$$\mathbf{S}_{\mathrm{t}} = \mathbf{S}_{\mathrm{w}} + \mathbf{S}_{\mathrm{b}} = \frac{1}{K} \sum_{m=1}^{M} \left[\sum_{k=1}^{K_m} (\mathbf{y}_{m,k} - \boldsymbol{\mu})(\mathbf{y}_{m,k} - \boldsymbol{\mu})^T \right],$$

  which is defined as the expected covariance of the data set whose members are the mean vectors of each class.

[15] In speech recognition, a class can be defined as phoneme or subphoneme.
[16] Any of the scatter matrices can always be derived from the other two.

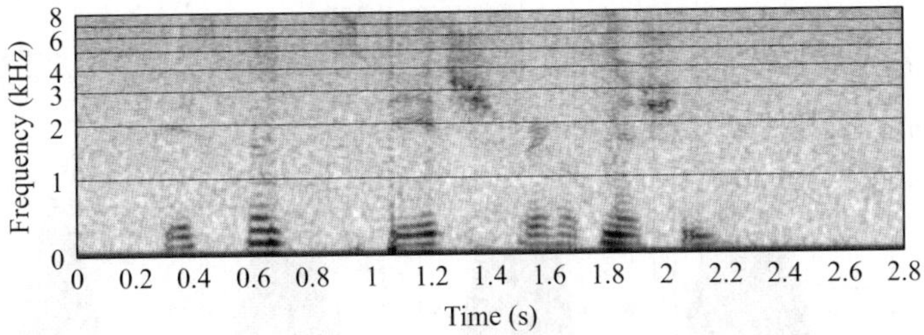

Figure 7.10 Wide-band, mel-scaled, logarithmic spectrogram of noisy speech.

From the above relations, it is clear that for a high class separability all vectors belonging to the same class must be close together and well separated from the feature vectors of other classes. Class separability can be expressed in a single value as

$$d = \operatorname{trace}\left(\mathbf{S}_{\mathrm{w}}^{-1}\mathbf{S}_{\mathrm{b}}\right). \tag{7.35}$$

7.5.2 *Robustness against Noise*

Figure 7.10 shows a wide-band spectrogram of noisy speech. The frequency axis of this spectrogram is scaled according to the mel scale while the energy axis is logarithmic. It is easy to see, compare to Figure 7.8, that the additive noise fills up the regions of the time-frequency plane with low speech energy.

To further investigate this effect on the power spectrum and warped MVDR, we plot scatter matrices of logarithmic power (see Figure 7.11). In the case of additive noise, the lower values are lifted to higher energies; that is the low-energy components are masked by noise and their information is *missing*. This effect is more apparent on the power spectrum as on the warped MVDR envelope. The warped MVDR envelope has less features in low-energy regions and

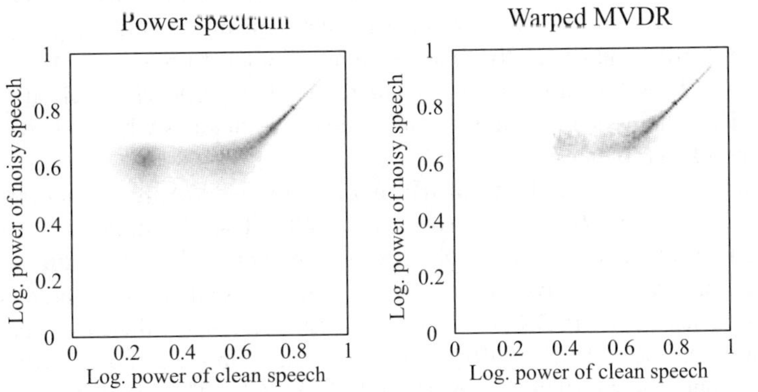

Figure 7.11 Scatter plots showing the influence of noise on speech in the logarithmic power domain. The average signal-to-noise ratio is 10 dB.

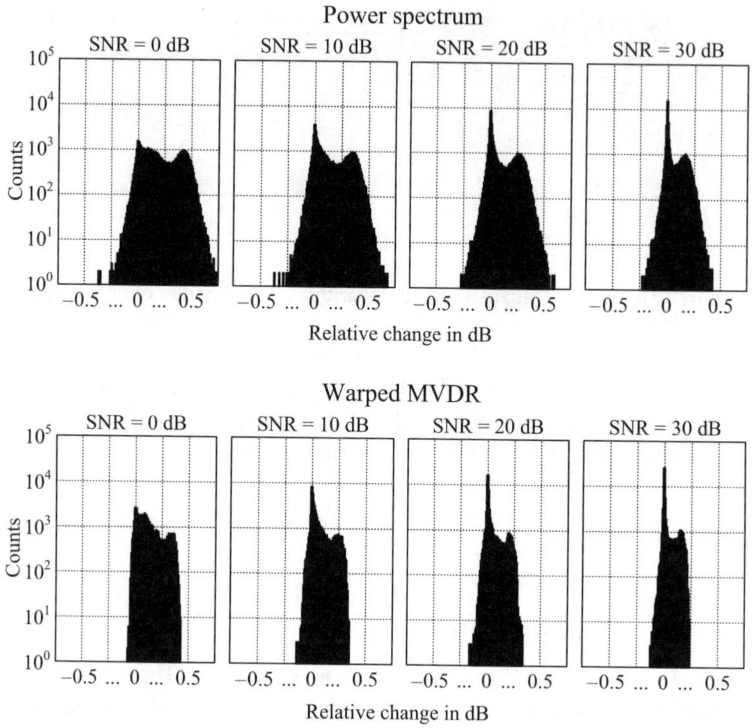

Figure 7.12 Histogram of relative changes between the distortion free and distorted speech features for different signal-to-noise ratios.

thus, the features are closer to the x–y-axis for the case of warped MVDR than the power spectrum.

The robustness to distortions for different features is clearer by plotting the histogram of relative changes. By comparing the histograms of relative changes, for the two spectral estimation methods, in Figure 7.12 for different SNRs we observe that the power spectrum is more spread and therefore has a wider variance than that of the warped MVDR. Also note that the distribution of relative changes is not limited to positive values, but contains negative values. Those negative values are due to the phase between the clean signal and the distortions. This finding is important for enhancement techniques where it is still common to "only" subtract spectral energy.

Comparing the class separability of the two features *mel-frequency cepstral coefficients* (MFCCs) and *warped MVDR cepstral coefficients* (WMVDRCCs) in Figure 7.13, we observe that WMVDRCCs have, on average, a higher class separability. It is interesting to observe that for very high distortions the two features have nearly the same class separability, while for higher SNR values the WMVDRCCs shows its advantages. While the class separability stays nearly equal in the case of WMVDRCCs for SNR values above 20 db, the class separability degrades continuously from higher to lower SNRs for MFCCs. This difference can be explained by the fact of the unequal modeling of high- and low-energy regions by spectral envelopes as already discussed.

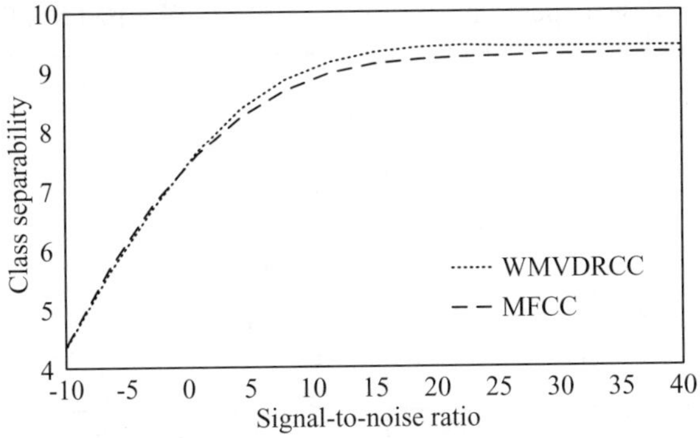

Figure 7.13 Influence of noise on class separability.

7.5.3 *Robustness against Echo and Reverberation*

Figure 7.14 shows a wide-band spectrogram of reverberant speech. In contrast to additive distortions reverberation smears the spectral energy along the time axis.

Comparing the scatter plots of speech distorted by reverberation, Figure 7.15, with those of speech distorted by noise, Figure 7.11, we observe that the distribution of the scatter points is quite different. Reverberation influences spectral features nearly independently of the power of the clean speech (with a tendency to lower variations for higher energies), while in contrast noise distortions influence in particular spectral features where the power of the clean speech signal is low. By comparing the scatter plot of the power spectrum (left image) and the scatter plot of the warped MVDR spectral envelope (right image) in Figure 7.15, it can be observed that—as before—there are not as many low-energy values for the MVDR spectral envelope as compared to the power spectrum.

As before by looking at the histogram of relative change for different reverberation times, Figure 7.16, we once more observe that the power spectrum is more spread, and therefore has a wider variance than the warped MVDR. Thereofore, we can conclude that the warped MVDR spectral envelope is more robust to additive as well as reverberant distortions than the power spectrum. The findings, that the distribution of relative changes is not limited to positive values, but contains negative values, also hold for reverberation.

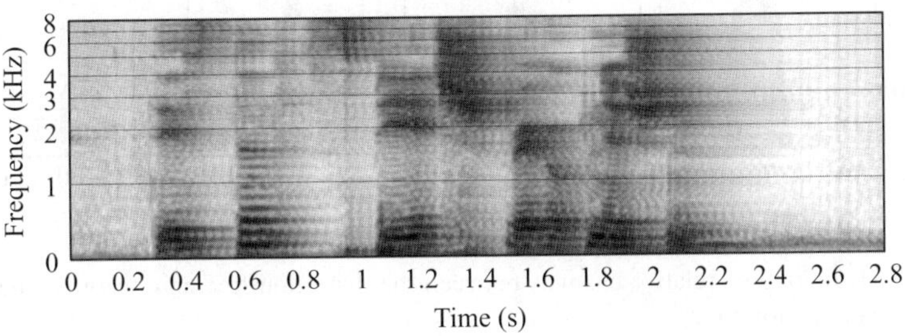

Figure 7.14 Narrow-band, mel-scaled, logarithmic spectrogram of reverberant speech.

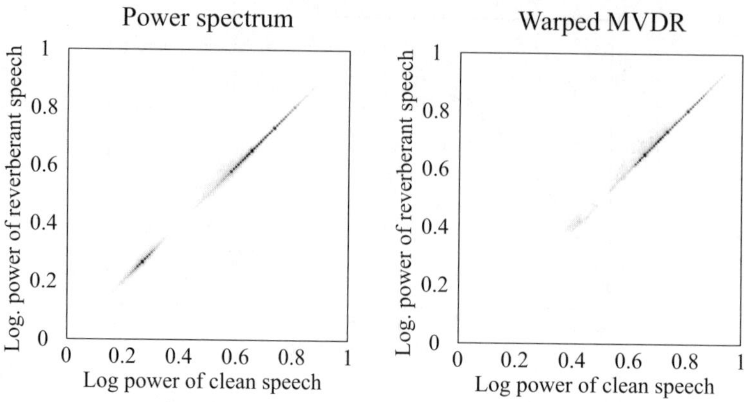

Figure 7.15 Scatter plots showing the influence of reverberation on captured speech in the logarithmic power domain. The average RT_{60} is 2 s.

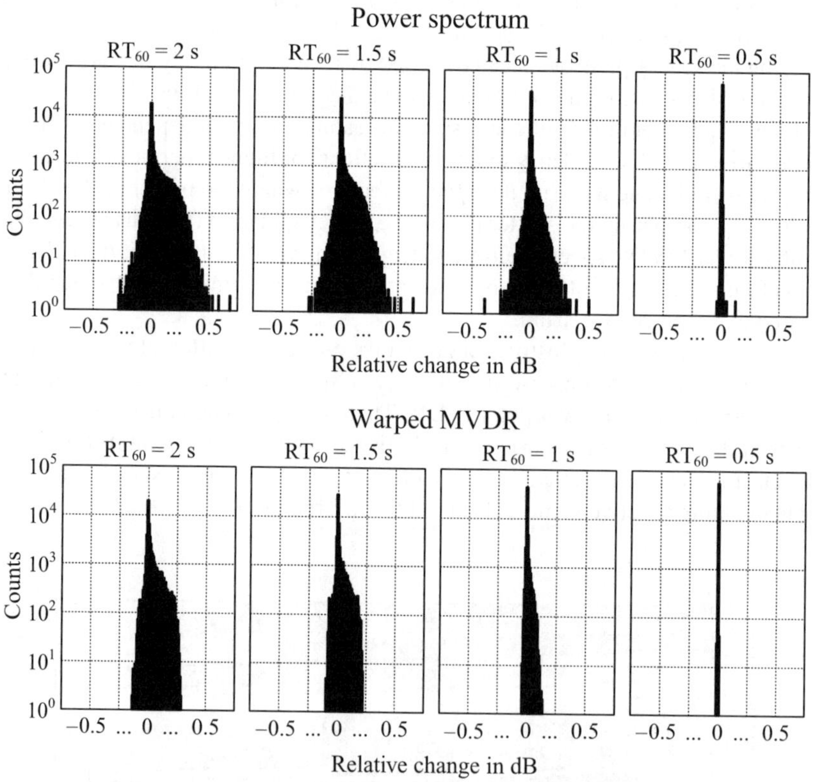

Figure 7.16 Histogram of relative changes between the distortion free and distorted speech features for different reverberation times.

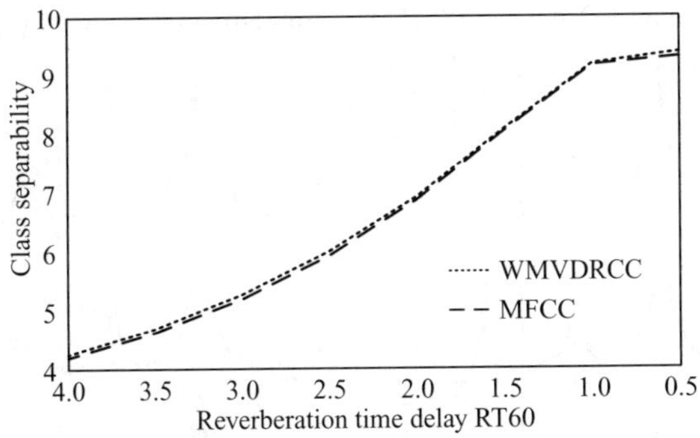

Figure 7.17 Influence of noise on class separability.

Looking at the class separability in Figure 7.17 for different time delay values RT_{60}, we observe that the curve of the degrading effect is differently pronounced than in the case of noise. In contrast to noise, for reverberation the difference between the two features, MFCCs and WMVDRCCs, is not very apparent, but for most values the class separability is slightly higher for WMVDRCCs.

7.5.4 *Robustness against Changes in Fundamental Frequency*

While the field of *robustness*, at least in the case of speech recognition, is dominated by methods against noise and echo/reverberation, it is easily overseen that other distortions such as speaker specific information or emotion (which might be a source of wanted information for other tasks such as speaker identification or verification) might play a significant role in increasing the variance of the extracted speech features, and thus result in an increased mismatch between training and testing. Thereofore, a key role of speech feature-extraction methods for speech recognition is not only to provide robustness against noise and echo/reverberation, but also against other variations such as speaker variation or the emotional state.

Due to the nature of speech production, the previously introduced analysis methods (Section 7.5.1), cannot be applied here. The influence of the model order in spectral envelope estimation is represented on voiced speech in Figure 7.18:

- A *higher model order* shows more detail of the fine structure of the spectrum and represents the fundamental frequency.
- A *low model order* reduces the influence of the excitation and is more or less a representation of the transfer function of the vocal track.

The overall performance of speech recognition and speaker identification/verification accuracy for WMVDRCCs for model orders between 20 and 120 is depicted in Figure 7.19. Comparing the optimal choice of model order for speech recognition—which is around 30[17]—with the optimal choice for speaker identification/verification—where larger model

[17] Note that in those experiments in the literature where a significant higher model order is used for speech recognition, the spectral estimate is followed by a filter bank which adds additional smoothing.

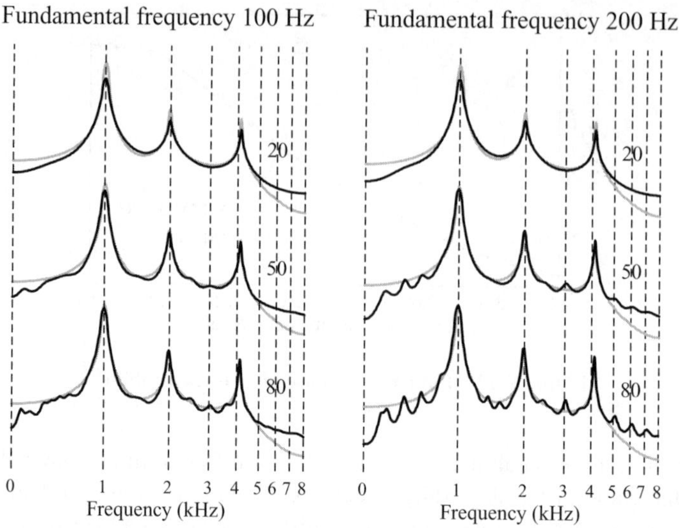

Figure 7.18 Influence of the model order on the spectral envelope of voiced speech for fundamental frequencies of 100 Hz (left image) and 200 Hz (right image). The gray line represents the transfer function of the vocal tract.

orders perform better—it becomes apparent that the optimal choice of model order, for the two tasks, is significantly different. And thus the required or provided information:

- For speaker identification/verification tasks, it is important to keep the information about the fundamental frequency and their harmonics (which is speaker specific).
- For speech recognition (at least for nontonal languages), it is important to suppress the influence of the fundamental frequency and their harmonics.

By suppression this particular classification irrelevant information spectral estimates, and thus speech features, become more robust to pitch variation of the same speaker as well as to speaker specific characteristics.

It is interesting to note that for lower model orders the performance difference—for speech recognition as well as speaker identification/verification—is relative high, while for higher

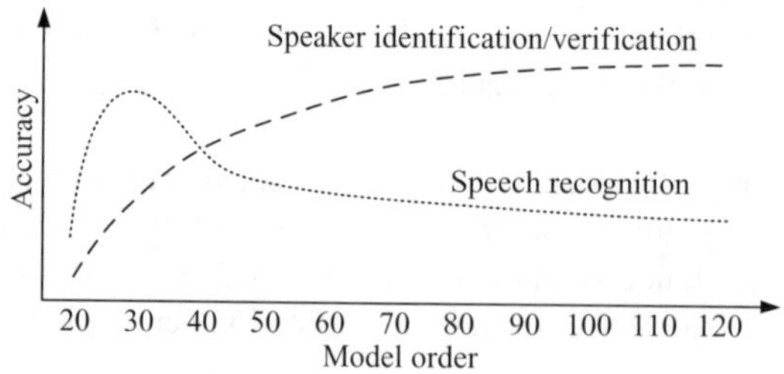

Figure 7.19 Speech-recognition and speaker identification/verification accuracy for warped MVDR cepstral coefficients versus model order.

model orders the performance difference is relative low. This can be explained by the fact that the influence of the overall spectral resolution is higher, for the same step size, in the case of lower model orders in comparison to higher model orders.

7.6 Summary and Further Reading

This chapter has given a brief insight into digital signal processing to extract speech features. We have seen that one contribution to the robust extraction of acoustic features is the appropriate estimation of the spectral representation. The bilinear transform has been introduced as an alternative to nonlinear scaled filter banks to represent nonlinear resolution of the human ear. Additional robustness to speech features is provided by a successive processing chain using cepstral processing and truncation[18] [30]. Human, as well as automatic, phonetic categorization and discrimination is poor for short observation windows. This fact suggest the use of longer observation context. Due to space constrains this has not been considered here, the interested reader should refer to Huang *et al.* [17] or Wölfel and McDonough [40]. Feature reduction is another important step not treated here. Linear discriminant analysis and heteroscedastic linear discriminant analysis which is frequently used in acoustic front-ends is explained in Hastie *et al.* [14].

References

[1] R. Bellman, *Adaptive Control Processes*. Princeton, New Jersey, USA: Princeton University Press, 1961.

[2] B. Bogert, M. Healy, and J. Tukey, "The quefrency alanysis of time series for echoes," *Proceedings of the Symposium on Time Series Analysis*, 1963.

[3] C. Braccini and A. V. Oppenheim, "Unequal bandwidth spectral analysis using digital frequency warping," *IEEE Transactions on ASSP*, vol. 22, pp. 236–244, August 1974.

[4] J. Burg, "The relationship between maximum entropy and maximum likelihood spectra," *Geophysics*, vol. 37, pp. 375–376, April 1972.

[5] J. Capon, "High–resolution frequency–wavenumber spectrum analysis," *Proceedings of the IEEE*, vol. 57, pp. 1408–1418, August 1969.

[6] J. Cooley and J. Tukey, "An algorithm for the machine calculation of complex fourier series," *Mathematics of Computation*, vol. 19 (90), pp. 297–301, 1965.

[7] S. B. Davis and P. Mermelstein, "Comparison of parametric representations for monosyllabic word recognition in continuously spoken sentences," *IEEE Transactions on ASSP*, vol. 28, pp. 357–366, August 1980.

[8] J. Deller, H. J.H.L., and J. Proakis, *Discrete-Time Processing of Speech Signals*. Piscataway, NJ, USA: Wiley-IEEE Press, 1999.

[9] L. Deng and D. O'Shaughnessy, *Speech processing: A Dynamic and Optimization–Oriented Approach*. New York, NY, USA: Marcel Dekker, Inc., 2003.

[10] B. Edler and G. Schuller, "Audio coding using a psychoacoustic pre– and postfilter," in *Proceedings of ICASSP*, vol. 2, 2000, pp. 881–884.

[11] S. Furui, "Cepstral analysis technique for automatic speaker verification," *IEEE Transactions on ASSP*, vol. 29, no. 2, pp. 254–272, April 1981.

[12] R. E. Greene and S. G. Krantz, *Function Theory of One Complex Variable*. New York: John Wiley & Sons, 1997.

[13] A. Härmä and U. Laine, "A comparison of warped and conventional linear predictive coding," *IEEE Transactions on SAP*, vol. 9, no. 5, pp. 579–588, 2001.

[14] T. Hastie, R. Tibshirani, and J. Friedman, *The Elements of Statistical Learning*. Heidelberg, Germany: Springer, 2001.

[18] the use of a finite number of coefficients, typically ranging between 12 and 20

[15] S. Haykin, *Adaptive Filter Theory*, 4th ed. New York: Prentice-Hall, 2002.

[16] H. Hermansky, "Perceptual linear predictive (PLP) analysis of speech," *Journal of ASA*, vol. 87, no. 4, pp. 1738–1752, April 1990.

[17] Huang, X., Acero, A., and Hon, H.W., *Spoken language processing*. Englewood Cliffs, N.J, USA: Prentice-Hall, 2001.

[18] R. Lacoss, "Data adaptive spectral analysis methods," *Geophysics*, vol. 36, no. 4, pp. 661–675, 1971.

[19] J. Makhoul, "Linear prediction: A tutorial review," *Proceedings of the IEEE*, vol. 63, no. 4, pp. 561–580, April 1975.

[20] J. Markel and A. H. J. Gray, *Linear Prediction of Speech*. Heidelberg, Germany: Springer, 1980.

[21] H. Matsumoto and M. Moroto, "Evaluation of Mel–LPC cepstrum in a large vocabulary continuous speech recognition," *Proceedings of ICASSP*, vol. 1, pp. 117–120, 2001.

[22] M. Matsumoto, Y. Nakatoh, and Y. Furuhata, "An efficient mel–LPC analysis method for speech recognition," *Proceedings of ICSLP*, pp. 1051–1054, 1998.

[23] M. Murthi and B. Rao, "Minimum variance distortionless response (MVDR) modelling of voiced speech," *Proceedings of ICASSP*, 1997.

[24] M. Murthi and B. Rao, "All–pole modeling of speech based on the minimum variance distortionless response spectrum," *IEEE Transactions on SAP*, vol. 8, no. 3, pp. 221–239, May 2000.

[25] B. Musicus, "Fast MLM power spectrum estimation from uniformly spaced correlations," *IEEE Transactions on ASSP*, vol. 33, pp. 1333–1335, 1985.

[26] Y. Nakatoh, M. Nishizaki, S. Yoshizawa, and M. Yamada, "An adaptive Mel–LP analysis for speech recognition," *Proceedings of ICSLP*, 2004.

[27] N. Nocerino, F. Soong, L. Rabiner, and D. Klatt, "Comparative study of several distortion measures for speech recognition," in *Proceedings of ICASSP*, 1985.

[28] A. Noll, "Short–time spectrum and 'cepstrum' technique for vocal–pitch detection," *Journal of ASA*, vol. 36, pp. 296–302, 1964.

[29] A. V. Oppenheim and R. W. Schafer, *Discrete-Time Signal Processing*. Englewood Cliffs, New Jersey: Prentice-Hall, 1999.

[30] A. Oppenheim and R. Schafer, *Discrete–Time Signal Processing*. Upper Saddle River, NJ, USA: Prentice-Hall Inc., 1989.

[31] A. Oppenheim, D. Johnson, and K. Steiglitz, "Computation of spectra with unequal resolution using the fast fourier transform," *IEEE Proceedings Letters*, Vol. 59, No. 2, 229–301, February 1971.

[32] R. Patterson, "Auditory filter shapes derived with noise stimuli," *Journal of the Acoustical Society of America*, vol. 59, no. 3, pp. 640–654, 1976.

[33] L. Rabiner and R. Schafer, *Digital Processing of Speech Signals*. Upper Saddle River, NJ, USA: Prentice-Hall, 1978.

[34] J. O. Smith III and J. S. Abel, "Bark and ERB bilinear transforms," *IEEE Transactions on SAP*, vol. 7, no. 6, pp. 697–708, 1999.

[35] P. Strobach, *Linear prediction theory: A mathematical basis for adaptive systems*. Heidelberg, Germany: Springer, 1990.

[36] H. Strube, "Linear prediction on a warped frequency scale," *Journal of ASA*, vol. 68, no. 8, pp. 1071–1076, 1980.

[37] K. Tokuda, T. Kobayashi, and S. Imai, "Adaptive cepstral analysis of speech," *IEEE Transactions on SAP*, vol. 3, no. 6, pp. 481–489, 1995.

[38] M. Wölfel, "Warped–twice minimum variance distortionless response spectral estimation," *Proceedings of EUSIPCO*, 2006.

[39] M. Wölfel, "Signal adaptive spectral envelope estimation for robust speech recognition," *Speech Communication*, vol. 51, pp. 551–561, 2009.

[40] M. Wölfel and J. McDonough, *Distant Speech Recognition*. Chichester, West Sussex, UK: John Wiley & Sons, 2009.

[41] M. Wölfel and J. McDonough, "Minimum variance distortionless response spectral estimation, review and refinements," *IEEE Signal Processing Magazine*, vol. 22, no. 5, pp. 117–126, September 2005.

8

Features Based on Auditory Physiology and Perception

Richard M. Stern[1], Nelson Morgan[2]
[1]*Carnegie Mellon University, USA*
[2]*International Computer Science Institute and the University of California, Berkeley, USA*

8.1 Introduction

It is well known that human speech processing capabilities far surpass the capabilities of current automatic speech recognition and related technologies, despite very intensive research in automated speech technologies in recent decades. Indeed, since the early 1980s, this observation has motivated the development of speech-recognition feature-extraction approaches that are inspired by auditory processing and perception, but it is only relatively recently that these approaches have become effective in their application to computer speech processing. The goal of this chapter is to review some of the major ways in which feature extraction schemes based on auditory processing have facilitated greater speech-recognition accuracy in recent years, as well as to provide some insight into the nature of current trends and future directions in this area.

We begin this chapter with a brief review of some of the major physiological and perceptual phenomena that have motivated feature-extraction algorithms based on auditory processing. We continue with a review and discussion of three seminal "classical" auditory models of the 1980s that have had a major impact on the approaches taken by more recent contributors to this field. Finally, we turn our attention to selected more recent topics of interest in auditory feature analysis, along with some of the feature extraction approaches that have been based on them. We conclude with a discussion of the attributes of auditory models that appear to be most effective in improving speech-recognition accuracy in difficult acoustic environments.

Techniques for Noise Robustness in Automatic Speech Recognition, First Edition.
Edited by Tuomas Virtanen, Rita Singh, and Bhiksha Raj.

8.2 Some Attributes of Auditory Physiology and Perception

In this section, we very briefly review and discuss a selected set of attributes of auditory physiology that historically or currently have been the object of attention by developers of auditory-based features. This discussion has been simplified for clarity's sake at the expense of other interesting phenomena that have received less attention in constructing models, at least to date, and it is far from comprehensive, even with respect to the auditory response to speech sounds. Furthermore, the physiological response to speech sounds is the object of much current attention, so that any report on current progress will inevitably be quickly out of date. The reader is referred to standard texts and reviews in physiology and psychoacoustics (e.g., [77, 88, 117]) for more comprehensive descriptions of general auditory physiology as well as the psychoacoustical response to speech sounds. The physiological response of the auditory system to speech and speech-like sounds is described in [86], among other places.

8.2.1 Peripheral Processing

Mechanical Response to Sound

The peripheral auditory response to sound has been well documented over a period of many years. Very briefly, small increases and decreases in pressure of the air that impinges on the two ears induce small inward and outward motion on the part of the *tympanic membrane* (eardrum). The eardrum is connected mechanically to the three bones in the middle ear, the malleus, incus, and stapes (or, more commonly, the hammer, anvil, and stirrup). The *cochlea* is the organ that converts the mechanical vibrations in the middle ear to neural impulses that can be processed by the brainstem and brain. The cochlea can be thought of as a fluid-filled spiral tube, and the mechanical vibrations of the structures of the middle ear induce wave motion of the fluid in the cochlea. The *basilar membrane* is a structure that runs the length of the cochlea. As one moves from the basal end of the cochlea (closest to the stapes) to the apical end (away from the stapes), the stiffness of the basilar membrane decreases, causing its fundamental resonant frequency to decrease as well. Figure 8.1 illustrates some classical measurements of cochlear motion by Georg von Békésy [112] which were obtained using stroboscopic techniques in the 1940s. These curves show that the membrane responds to high-frequency tones primarily at the basal end, while low-frequency signals elicit maximal vibration at the apical end, although

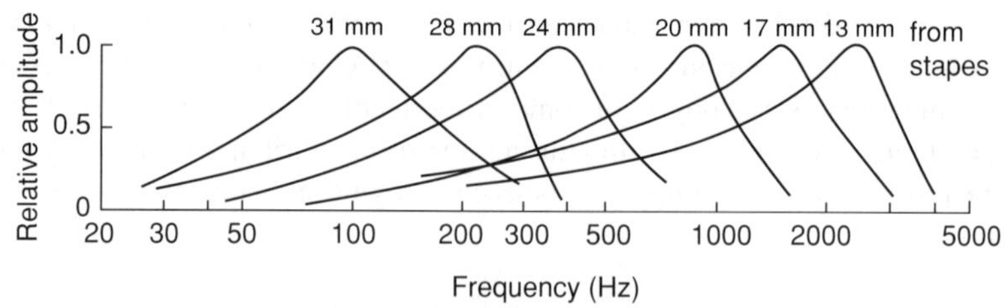

Figure 8.1 The response of the basilar membrane as a function of frequency, measured at six different distances from the stapes. As the frequency axis is plotted on a logarithmic scale, it can be easily seen that the effective bandwidth is proportional to center frequency at higher frequencies; effective bandwidth is roughly constant at lower frequencies (from [112]).

the response to low-frequency sounds is more asymmetric and distributed more broadly along the membrane.

Affixed to the human basilar membrane are about 15 000 hair cells, which enervate about 30 000 individual fibers of the auditory nerve. Through an electrochemical mechanism, the mechanical motion of the hair cells elicits the generation of a brief transient or "spike" in the voltage inside the cell wall. This transient is then propagated along the nerve fibers and beyond to the cochlear nuclei and subsequently to higher centers in the brainstem and the brain. The most important attribute of these spikes is the time at which they take place. Because each nerve fiber takes input from a relatively small number of fibers that in turn move in response to vibration over only a limited length along the basilar membrane, and because a given location along the basilar membrane is most sensitive to a narrow range of frequencies, each fiber of the auditory nerve also only responds to a similar range of frequencies.

It should be borne in mind that the basic description above is highly simplified, ignoring nonlinearities in the cochlea and in the hair-cell response. In addition, the there are actually two different types of hair cells with systematic differences in response. The *inner hair cells* transduce and pass on information from the basilar membrane to higher levels of analysis in the auditory system. The *outer hair cells*, which constitute the larger fraction of the total population, have a response that is affected in part by efferent feedback from higher centers of neural processing. These cells appear to amplify the incoming signals nonlinearly, with low-level inputs amplified more than more intense signal components. This amplification produces a compression in dynamic range, and it is also believed to play an important role in the mechanism underlying lateral suppression, which improves frequency resolution. We describe some of the simple attributes of the auditory-nerve response to simple signals in the sections below, focussing on those attributes that are most commonly incorporated into feature extraction for automatic speech recognition.

Transient Response of Auditory-Nerve Fibers

The spikes that are generated by fibers of the auditory nerve occur at random times, and hence the response of the nerve fibers must be characterized statistically. Figure 8.2 is a "poststimulus-time" (PST) histogram of the rate of firing in response to tone bursts as a

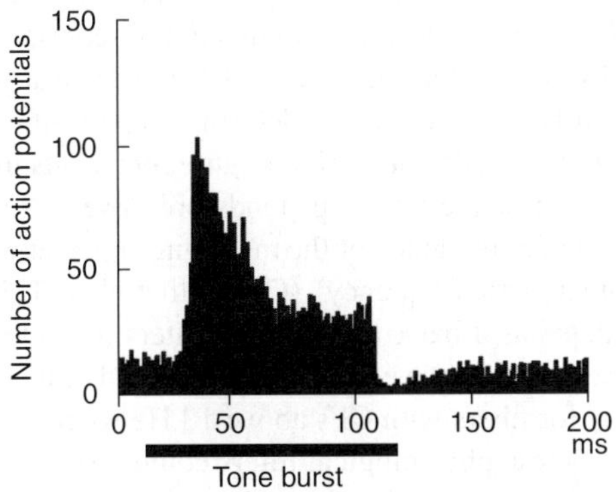

Figure 8.2 PST histograms in response to tone bursts (from [88]).

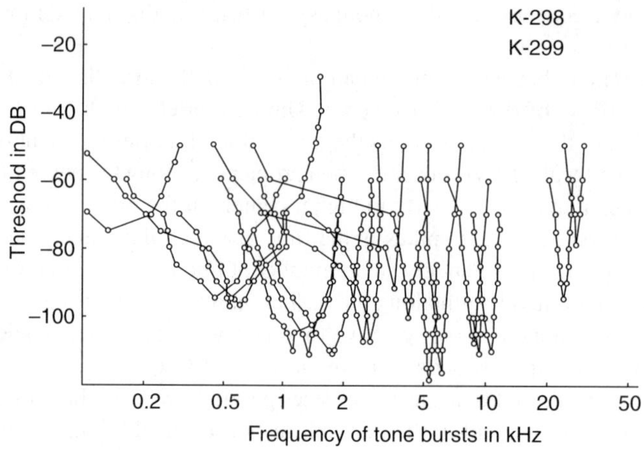

Figure 8.3 Tuning curves indicating the response threshold for pure tones for an assortment of auditory-nerve fibers with different CFs (from [48]).

function of the time after the initiation of the burst, averaged over many presentations of the tone burst. It can be seen there is a low level of spontaneous activity before the tone burst is gated on. When the tone is suddenly turned on, there is an "overshoot" in the amount of activity, which eventually settles down to about 50 spikes per second. Similarly, when the tone is gated off, the response drops below the spontaneous rate before rising to it. These results illustrate the property that the auditory system tends to emphasize transients in the response to the signals, with less response from the steady-state portions.

Frequency Resolution of the Auditory System

As was noted above, different frequency components of an incoming signal elicit maximal vibrations of the basilar membrane at different locations along the membrane. Because each of the roughly 30 000 fibers of the auditory nerve is connected to a particular location along the membrane, the response of each of these fibers is frequency specific as well, as illustrated by the curves of Figure 8.3. Each of the curves in this figure represents the intensity at a given frequency that is needed to cause the mean rate of firing from a particular auditory-nerve fiber in response to a sine tone to increase a predetermined percentage above the spontaneous average firing rate for that fiber. Each curve in the figure represents the response of a different fiber. It can be seen that each of the fibers responds only over a relatively narrow range of frequency and there is a specific frequency of the incoming signal at which the fiber is the most sensitive, called the "characteristic frequency" (CF) of that fiber. This portion of the auditory system is frequently modeled as a bank of bandpass filters (despite the many nonlinearities in the physiological processing), and we note that the "bandwidth" of the filters appears to be approximately constant for fibers with CFs above 1 kHz when plotted as a function of log frequency. This means that these physiological filters could be considered to be "constant-Q" in that the nominal bandwidth is roughly proportional to center frequency. The bandwidth of the filters is roughly constant at lower frequencies, although this is less obvious from the

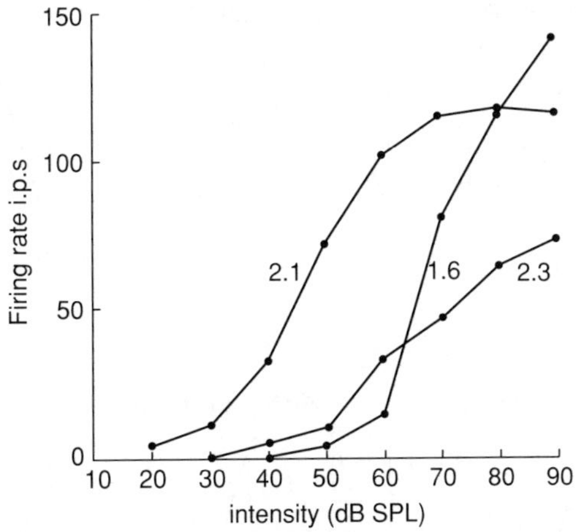

Figure 8.4 Rate of spike discharges as a function of intensity for three auditory-nerve fibers with different CFs (as indicated), after subtracting the spontaneous rate of firing (from [89]).

curves in Figure 8.3. This frequency-specific or "tonotopic" organization of individual parallel channels is generally maintained as we move up from the auditory nerve to higher centers of processing in the brainstem and the auditory cortex.

Rate-Level Responses

We have previously stated that many aspects of auditory processing are nonlinear in nature. This is illustrated rather directly in Figure 8.4, which shows the manner in which the rate of response increases as a function of signal intensity. (The spontaneous rate of firing for each fiber has been subtracted from the curves.) As can be seen, the rate-intensity function is roughly S-shaped, with a relatively flat portion corresponding to intensities below the threshold intensity for the fiber, a limited range of about 20 dB in which the response rate increases in roughly linear proportion to the signal intensity, and a saturation region in which the response is again essentially independent of the incoming signal intensity. (There are some exceptions to this, such as the fiber with CF 1.6 kHz in the figure.) The fact that each individual fiber is limited to approximately 20 dB of active response implies that psychophysical phenomena such as loudness perception must be mediated by the combined response of a number of fibers over a range of frequencies.

Synchronized Response to Low-Frequency Tones

When the excitation for a particular auditory-nerve fiber is below the threshold intensity level for that fiber, spikes will be generated randomly, following a Poisson interval distribution with a refractory interval of no response for about 4 ms after each spike. Figure 8.5 is a "postzero-crossing histogram" (PZC histogram) which describes the firing rate that is observed as a function of the phase of the input signal. We note that the response roughly follows the shape of the input signal at least when the signal amplitude is positive (which actually corresponds

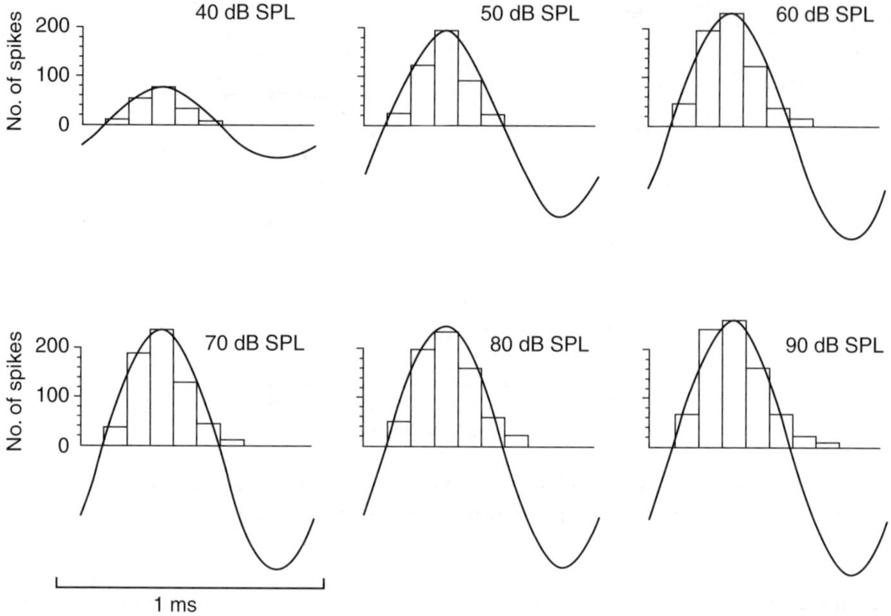

Figure 8.5 Period histograms in response to a 1100-Hz pure tone at various signal intensities (from [94] as redrawn by [88]).

to times at which the instantaneous pressure is lower than the baseline level). This "phase-locking" behavior enables the auditory system to compare arrival times of signals to the two ears at low frequencies, which is the basis for the spatial localization of a sound source at these frequencies. While the auditory system loses the ability to respond in synchrony to the fine structure of higher frequency components of the input signal, its response is synchronized to the *envelopes* of these signal components (e.g., [21]). The frequency at which the auditory system loses its ability to track the fine structure of the incoming signal in this fashion is approximately the frequency at which such timing information becomes useless because that information becomes ambiguous for localization purposes, which strongly suggests that the primary biological role for low-frequency synchrony is indeed sound localization.

While temporal coding is clearly important for binaural sound localization, it may also play a role in the robust interpretation of the signals from each individual ear as well. For example, the upper panel of Figure 8.6 depicts the mean rate of response to a synthetic vowel sound by an ensemble of auditory-nerve fibers as a function of the CF of the fibers, with the intensity of the vowel sound varied over a wide range of input intensities, as described by Sachs and Young [96]. The lower panel of that figure depicts the derived *averaged localized synchronized rate* (or *ALSR*) to the same signals [119], which describes the extent to which the neural response at a given CF is synchronized to the nearest harmonic of the fundamental frequency of the vowel. It can be easily seen that the mean rate of response varies dramatically as the input intensity changes, while the ALSR remains substantially unchanged. These results suggest that the timing information associated with the response to low-frequency components of a signal can be substantially more robust to variations in intensity (and potentially various other

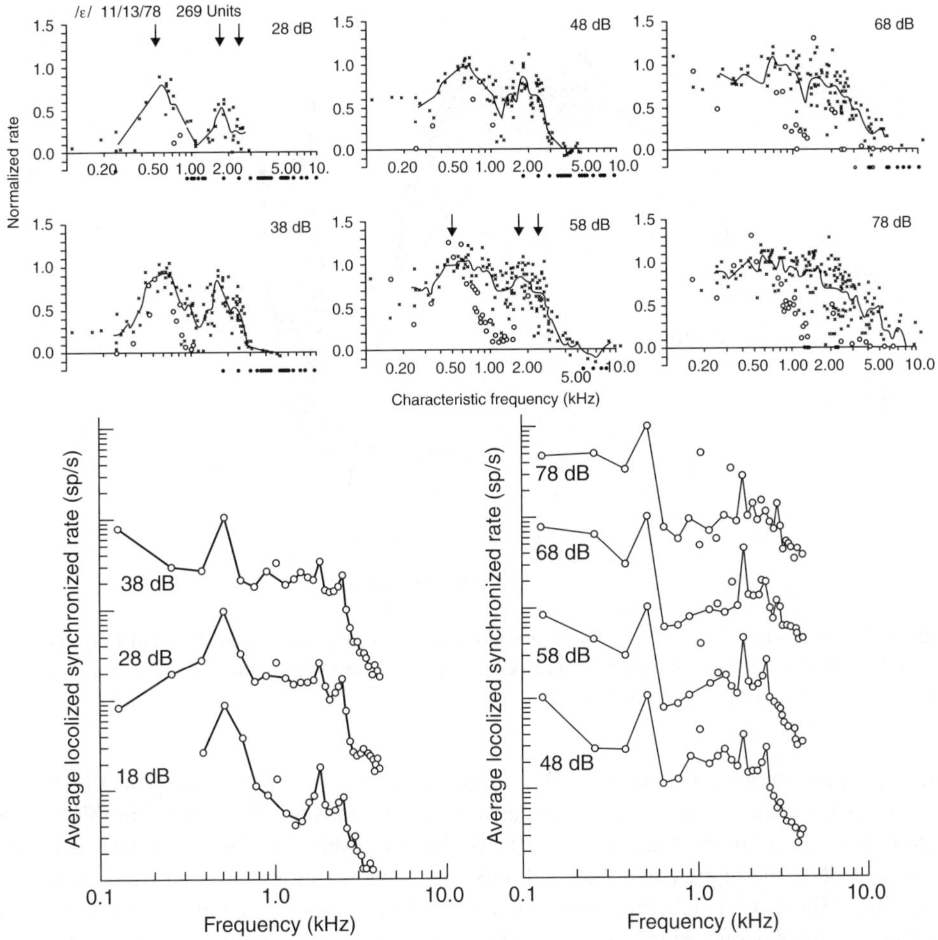

Figure 8.6 Comparison of auditory-nerve responses to a computer-simulated vowel sound at various intensities based on mean rate (upper panel) and synchrony to the signal (see text). For clarity of presentation, the lower panel is drawn with the curves shifted along the vertical axis (redrawn from [96] and [119]).

types of signal variability and/or degradation) than the mean rate of the neural response. Most conventional feature extraction schemes (such as MFCC and PLP coefficients) are based on short-time energy in each frequency band, which is more directly related to mean rate than temporal synchrony in the physiological responses.

Lateral Suppression

The response of auditory-nerve fibers to more complex signals also depends on the nature of the spectral content of the signals, as the response to signals at a given frequency may be

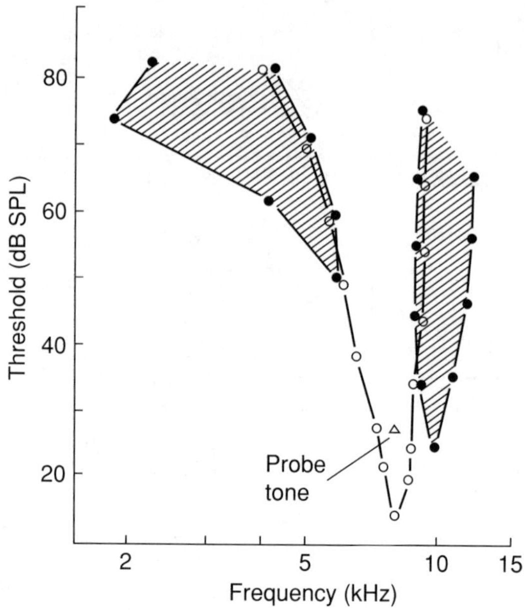

Figure 8.7 The shaded portions of the figure indicate combinations of intensities and frequencies at which the presence of a second tone suppresses the auditory-nerve response to a tone at a fiber's CF presented 10 dB above threshold (from [5]).

suppressed or inhibited by energy at adjacent frequencies (e.g., [5, 95]). For example, Figure 8.7 summarizes some aspects of the response to a pairs of tones. The signal in this case is a pair of tones, a "probe tone" that is 10 dB above threshold at the CF (indicated by the open triangle in the figure) plus a second tone presented at various different frequencies and intensities. The cross-hatched regions indicate the frequencies and intensities for which the response to the two tones combined is less than the response to the probe tone at the CF alone. The open circles outline the tuning curve for the fiber that describes the threshold intensity for the probe tone alone as a function of frequency. It can be seen that the presence of the second tone over a range of frequencies surrounding the CF inhibits the response to the probe tone at CF, even when the second tone is presented at intensities that would be below threshold if it had been presented in isolation. This form of "lateral suppression" has the effect of enhancing the response to changes in the signal content with respect to frequency, just as the overshoots and undershoots in the transient response have the effect of enhancing the response to changes in signal level over time.

8.2.2 Processing at more Central Levels

While the phenomena described above are all observed at the level of the cochlea or the auditory nerve, substantial processing takes place at the level of the precortical centers of the brainstem as well as in the auditory cortex itself. We note here three sets of more central phenomena that also play significant roles in auditory modeling.

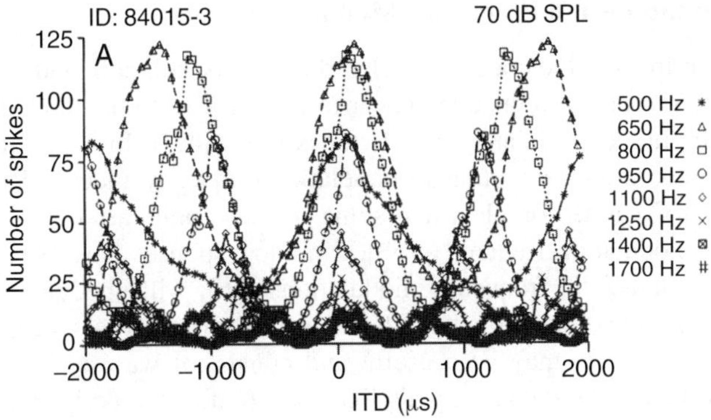

Figure 8.8 Response of a unit in the MSO to pure tones at various frequencies plotted as a function of the stimulus ITD (from [116]).

Sensitivity to Interaural Time Delay and Other Binaural Phenomena

It is well known that two important cues for human localization of the direction of arrival of a sound are the interaural time difference (ITD) and interaural intensity difference (IID) (e.g., [24]). As first noted by Rayleigh [108], ITDs are most useful at low frequencies and IIDs are only present at higher frequencies for reasons related to spatial aliasing and physical diffraction, respectively. Physiologists have observed units in the superior olivary complex and the inferior colliculus that appear to respond maximally to a single "characteristic" ITD (e.g., [93,116]). As an example, Figure 8.8 depicts the response of a unit in the superior olivary complex in the brainstem of a cat which responds in excitatory fashion when signals are presented binaurally. The figure plots the relative number of spikes in response to tones at various frequencies as a function of the ITD with which the signals are presented to the two ears. We note that this unit exhibits a maximum in response when the tones are presented with an ITD of approximately 33 μs for frequencies ranging from 500 to 1700 Hz. In other words, the function of this unit appears to be the detection of a specific ITD, and that ITD of best response is sometimes referred to as the *characteristic delay* (CD) of the unit. An ensemble of such units with a range of CFs and CDs can produce a display that represents the interaural cross-correlation of the signals to the two ears after the frequency-dependent and nonlinear processing of the auditory periphery. Over the years many theories have been developed that describe how a display of this sort can be used to describe and predict a wide range of binaural phenomena as reviewed by Stern and Trahiotis [103] and Stern *et al.* [105]). Units have also been described that appear to record the IIDs of a stimulus (e.g., [93]).

Another important attribute of human binaural hearing is that localization is dominated by the first-arriving components of a complex sound [113]. This phenomenon, which is referred to as the *precedence effect*, is clearly helpful in causing the perceived location of a source in a reverberant environment to remain constant, as it is dominated by the characteristics of the direct field (which arrives straight from the sound source) while suppressing the potential impact of later-arriving reflected components from other directions. In addition to its role in maintaining perceived constancy of direction of arrival in reverberation, the precedence effect is also believed by some to improve speech intelligibility in reverberant environments.

Sensitivity to Amplitude and Frequency Modulation

Physiological recordings in the cochlear nuclei, the inferior colliculi, and the auditory cortex have revealed the presence of units that appear to be sensitive to the modulation frequencies of sinusoidally-amplitude-modulated (SAM) tones (e.g., [47]). In some of these cases, response would be maximum at a particular modulation frequency, independently of the carrier frequency of the SAM tone complex, and some of these units are organized anatomically according to best modulation frequency [58]. Similar responses have been observed in the cochlear nuclei to sinusoidal frequency modulations [76], with modulation frequencies of 50–300 Hz providing maximal response. These results have lead to speculation that the so-called *modulation spectrum* may be a useful and consistent way to describe the dynamic temporal characteristics of complex signals like speech after the peripheral frequency analysis. In one perceptual study Drullman *et al.* [22, 23] conducted a series of experiments that characterized the perception of speech that had been analyzed and resynthesized with modified temporal envelopes in each frequency band, concluding that modulation spectrum components between 4 and 16 Hz are critical for speech intelligibility. Nevertheless, the extent to which the physiological representation of amplitude modulation is preserved and remains invariant at higher levels remains an open issue at present.

Feature Detection at Higher Levels: Spectro-Temporal Receptive Fields

There is a rich variety of feature-detection mechanisms that have been observed at the higher levels of the brainstem and in the auditory cortex as reviewed by Palmer and Shamma [86]. A characterization that has proved useful is that if the *spectro-temporal response field* or *STRF*, which can, in principle, be used to describe sensitivity to amplitude modulation, frequency modulation, as well as a more general sensitivity to sweeps in frequency over time, as might be useful in detecting formant transitions in speech. As an example, researchers at the University of Maryland and elsewhere have used used dynamic "ripple" stimuli, with drifting sinusoidal spectral envelopes, to develop the STRF patterns in the responses of units of the primary auditory cortex (A1) in ferrets [20,56]. They reported units with a variety of types of response patterns, including sensitivity to upward and downward ripples, as well as a range of best frequencies, bandwidths, asymmetries in response with respect to change in frequency, temporal dynamics, and related characteristics. It is frequently convenient to illustrate the observed STRFs as color temperature patterns in the time-frequency plane.

8.2.3 Psychoacoustical Correlates of Physiological Observations

All of the phenomena cited above have perceptual counterparts, which are observed by carefully-designed psychoacoustical experiments. The results of these experiments give us direct insight into the characteristics and limitations of auditory perception, although we must infer the mechanism underlying the experimental results. (In contrast, physiological experiments provide direct measurements of the internal response to sound, but the perceptual significance of a given observation must be inferred.) Interesting auditory phenomena are frequently first revealed through psychoacoustical experimentation, with the probable physiological mechanism underlying the perceptual observation identified at a later date. We briefly discuss four sets of basic psychoacoustic observations that have played a major role in auditory modeling.

The Psychoacoustical Transfer Function

The original psychoacousticians were physicists and philosophers of the nineteenth century who had the goal of developing mathematical functions that related sensation and perception, such as the dependence of the subjective loudness of a sound on its physical intensity. As can be expected, the nature of the relationships will depend on the temporal and spectral properties of the signals, as well as how the scales are constructed. The original psychophysical scales for intensities were based on the empirical observations of Weber [115], who observed that the increment in intensity needed to just barely perceive that a simple sound (such as a tone) was louder than another was a constant fraction of the reference intensity level. This type of dependence of the *just-noticeable difference* or *JND* of intensity on reference intensity is observed in other sensory modalities as well, such as the perception of the brightness of light or the weight of a mass. Fechner [26] proposed that a psychophysical scale that describes perceived intensity as a function of the intensity of the physical stimulus could be constructed by combining Weber's empirical observation with the assumption that JNDs in intensity should represent equal *intervals* on the perceptual scale. It is easy to show that this assumption implies a logarithmic perceptual scale

$$\Psi = C \log(\Phi), \tag{8.1}$$

where Φ in the above equation represents physical intensity and Ψ represents its perceptual correlate (presumably loudness in hearing). The logarithmic scale for intensity perception, of course, motivated the decibel scale, and it is partially supported by the fact that there is typically a linear relation between the neural rate of response and intensity in dB for intermediate intensity levels, as in intermediate range of the curves in Figure 8.4. Many years later Stevens proposed an alternate loudness scale, which implicitly assumes that JNDs in intensity should represent equal *ratios* on the perceptual scale. This gives rise to the power law relationship

$$\Psi = K_1 \Phi^{K_2}, \tag{8.2}$$

where Φ and Ψ are as in Equation (8.1). The Stevens power law is supported by the results of many *magnitude estimation* experiments in which subjects are asked to apply a subjective numerical label to the perceived intensity of a signal. While the value of the exponent K_2 depends to some extent on the nature of the signal and how the experiment is conducted, it is typically on the order of 0.33 when physical intensity is expressed in terms of stimulus amplitude [106]. More extensive discussion of these theories and their derivation are available in texts by Gescheider [31] and Baird and Noma [10], among many other sources.

Auditory Frequency Resolution

As noted above, the individual parallel channels of the auditory system are frequency selective in their response to sound. It is generally assumed that the first stage of auditory processing may be modeled as a bank of bandpass filters, and all modern theories of auditory perception are attentive to the impact of processing by the peripheral auditory system on the representation of sound. For example, the detection of a tonal signal in a broadband masker is commonly assumed to be mediated by the signal-to-noise ratio at the output of the auditory filter that contains the target tone. Auditory frequency resolution was first studied psychophysically in the 1930s by Fletcher and colleagues at Bell Laboratories [28], which preceded Békésy's physiological measurements of cochlear mechanics in the 1940s [112] as well as subsequent descriptions of the frequency-specific physiological response to sound at the level of the fibers of the auditory nerve and at higher centers (e.g., [48]). There are a number of ways of measuring auditory frequency selectivity (*cf.* [78] and Chapter 3 of [77]), and to some extent the estimated bandwidth of the auditory channels (commonly referred to as the "critical band" associated with each channel) depends on the assumed filter shape and the way in which bandwidth is measured. In general, the estimated channel bandwidth increases with increasing center frequency of the channel, and at higher frequencies the filter bandwidth tends to be roughly proportional to center frequency, as was observed in Figure 8.3.

From these experiments, three distinct frequency scales have emerged that describe the bandwidths of the auditory filters and, correspondingly, the center frequencies of the filters that are needed to ensure that the filters are separated by a constant number of critical bands at all frequencies. The *Bark scale* (named after Heinrich Barkhausen), based on estimates of the critical band from traditional masking experiments, was first proposed by Zwicker [123], quantified by Zwicker and Terhardt [124], and subsequently represented in simplified form by Traunmüller [110] as

$$Bark(f) = [26.8/(1 + (1960/f))] - 0.53, \tag{8.3}$$

where the frequency f is in Hz.

The *mel scale* (which refers to the word "melody") was proposed by Stevens *et al.* [107] and is based on pitch comparisons; it is approximated by the formula [85]

$$Mel(f) = 2595 \log_{10}\left(1 + \frac{f}{700}\right). \tag{8.4}$$

The original critical band estimates of Fletcher were based on the simplifying assumption that the auditory filters were rectangular in shape. The shape of the auditory filters has been estimated in several ways, frequently making use of notch-shaped maskers (e.g., [87]). A popular scale proposed by Moore and Glasberg [79] called the *ERB* scale describes the *equivalent rectangular bandwidth* of these filters. The number of ERBs as a function of frequency is approximated by the formula

$$ERB_N(f) = 21.4 \log_{10}(1 + 4.37f/1000), \tag{8.5}$$

where again f is in Hz. For example, at 1 kHz this function is equal to about 130 Hz, which means that an increase of frequency of 130 Hz centered about 1 kHz would constitute one ERB.

Figure 8.9 compares the Bark, Mel, and ERB scales from the equations above after multiplying each curve by a constant that was chosen to minimize the squared difference between the curves. It can be seen that despite the differences in how the frequency scales were formulated, they all look similar, reflecting the fact that the perceptual scale is expanded with respect to frequency at low frequencies and compressed at higher frequencies. All common models of

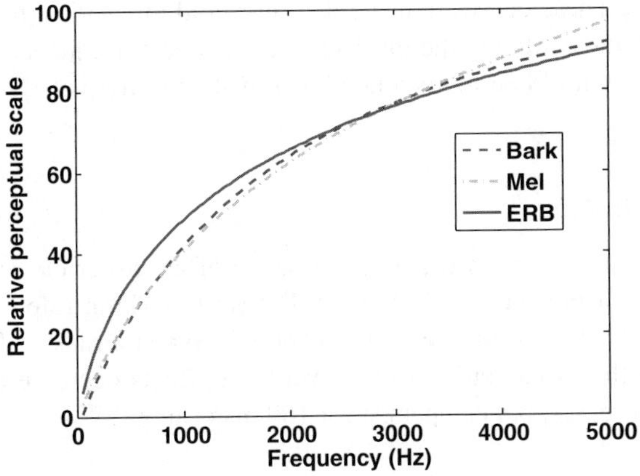

Figure 8.9 Comparison of frequency scales derived from the Bark, mel, and ERB scales.

auditory processing begin with a bank of filters whose center frequencies and bandwidths are based on one of the three frequency scales depicted in this figure.

Loudness Matching and Auditory Thresholds

A final set of results that have had an impact on feature extraction and auditory models is the set of *equal loudness contours* depicted in Figure 8.10 after measurements by Fletcher and Munson [29]. Each curve depicts the intensity of a tone at an arbitrary frequency that matches the loudness of a tone of a specified intensity at 1 kHz, which is defined to be the loudness

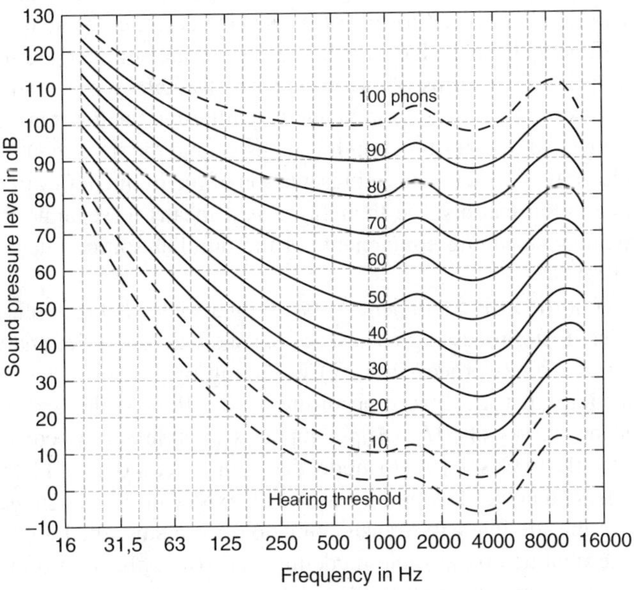

Figure 8.10 Equal-loudness matching curves (from [77] after Fletcher and Munson [29]).

of that tone in *phons*. These curves indicate that threshold intensities (the lowest curve) vary with frequency, with the ear being the most sensitive between frequencies of about 1000 and 4000 Hz. The upper limit of hearing is much less sensitive to frequency.

Nonsimultaneous Masking

Nonsimultaneous masking occurs when the presence of a masker elevates the threshold intensity for a target that precedes or follows it. Forward masking refers to inhibition of the perception of a target after the masker is switched off. When a masker follows the probe in time, the effect is called backward masking. Masking effects decrease as the time between masker and probe increases, but can persist for 100 ms or more [77].

8.2.4 The Impact of Auditory Processing on Conventional Feature Extraction

The overwhelming majority of speech-recognition systems today make use of features that are based on either *Mel-Frequency Cepstral Coefficients* (*MFCCs*) first proposed by Davis and Mermelstein in 1980 [19] or features based on *perceptual linear predictive (PLP)* analysis of speech [36], proposed by Hermansky in 1990. We briefly discuss MFCC and PLP processing in this section. The major functional blocks used in these procedures are summarized in Figure 8.11.

As is well known, MFCC analysis consists of (1) short-time Fourier analysis using Hamming windows, (2) weighting of the short-time magnitude spectrum by a series of triangularly-shaped functions with peaks that are equally spaced in frequency according to the Mel scale, (3) computation of the log of the total energy in the weighted spectrum, and (4) computation of a relatively small number of coefficients of the inverse discrete-cosine transform (DCT) of the log power coefficients from each channel. These steps are summarized in the left column of Figure 8.11. Expressed in terms of the principles of auditory processing, the triangular weighting functions serve as a crude form of auditory filtering, the log transformation mimics Fechner's psychophysical transfer function for intensity, and the inverse DCT can be thought of as providing a lowpass Fourier series representation of the frequency-warped log spectrum. The cepstral computation can also be thought of as a means to separate the effects of the excitation and frequency-shaping components of the familiar source-filter model of speech production (e.g., [90]).

The computation of the PLP coefficients is based on a somewhat different implementation of similar principles. PLP processing consists of (1) short-time Fourier analysis using Hamming windows (as in MFCC processing), (2) weighting of the power spectrum by a set of asymmetrical functions that are spaced according to the Bark scale, and that are based on the auditory masking curves of [97], (3) preemphasis to simulate the equal-loudness curve suggested by Makhoul and Cosell [66] to model the loudness contours of Fletcher and Munson (as in Figure 8.10), (4) a power-law nonlinearity with exponent 0.33 as suggested by Stevens *et al.* [107] to describe the intensity transfer function, (5) a smoothed approximation to the frequency response obtained by all-pole modeling, and (6) application of a linear recursion that converts the coefficients of the all-pole model to cepstral coefficients.

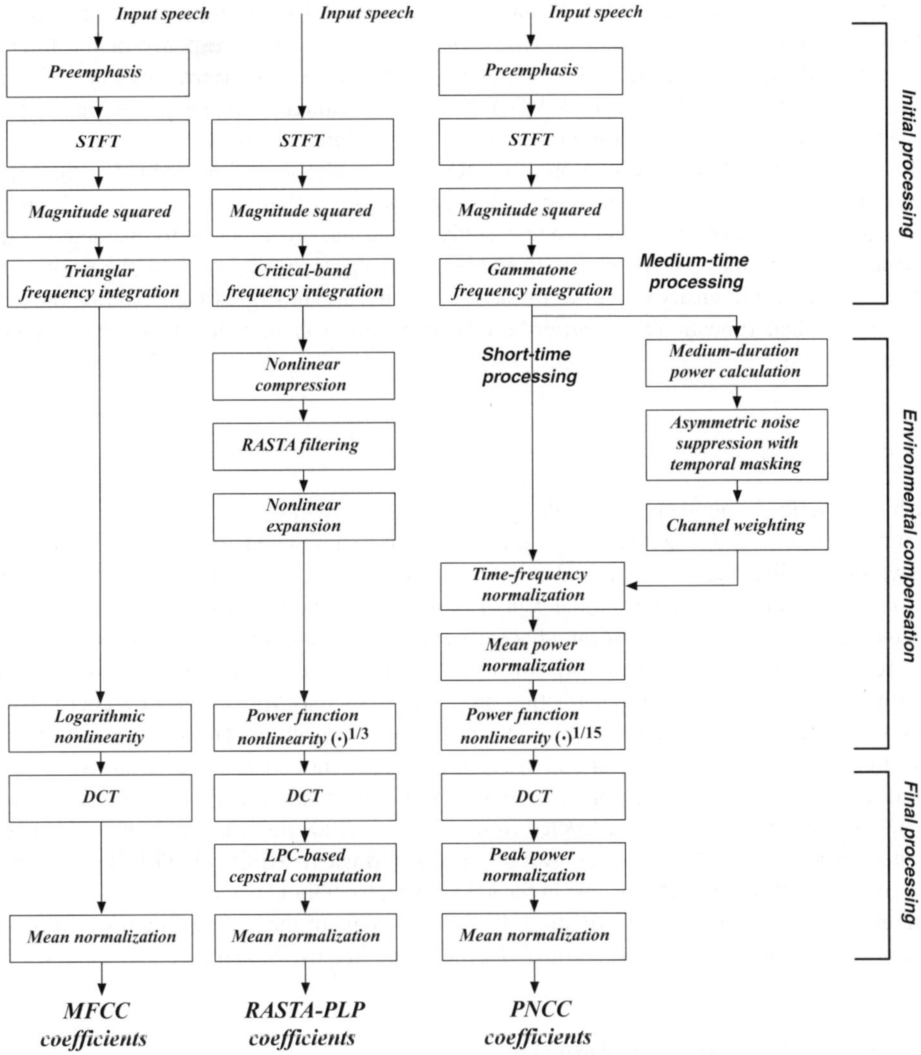

Figure 8.11 Comparison of major functional blocks of the MFCC, PLP-RASTA, and PNCC processing methods. (PNCC processing is discussed in Section 8.4.)

PLP processing is also frequently used in conjunction with Hermansky and Morgan's RASTA algorithm [37], a contraction of *relative spectral analysis*. RASTA processing in effect applies a bandpass filter to the compressed spectral amplitudes that emerge between Steps (3) and (4) of the PLP processing above. RASTA processing also models the tendency of the auditory periphery to emphasize the transient portions of incoming signals, as noted in Section 8.2.1 above. In practice, the bandpass nature of the filter causes the mean values of the spectral coefficients to equal zero, which effects a normalization that is similar to the normalization provided by cepstral mean normalization (CMN) that is commonly used in conjunction with

MFCC processing. Both the RASTA filter and CMN are effective in compensating for the effects of unknown linear filtering in cases for which the impulse response of the filter is shorter than the duration of the analysis window used for processing. Hermansky and Morgan [37] also propose an extension to RASTA processing, called *J-RASTA* processing, which provides similar compensation for additive noise at low signal levels.

In summary, PLP feature extraction is an attempt to model several perceptual attributes of the auditory system more exactly than MFCC processing, including the use of the Zwicker filters to represent peripheral frequency selectivity and the preemphasis to characterize the dependence of loudness on frequency. In addition, it replaces the mel scale by the Bark scale, the log relation for intensity by a power function, and it uses autoregressive modeling of a low order (rather than truncation of a Fourier-based expansion) to obtain a smoothed representation of the spectrum.

8.2.5 Summary

We have described a number of physiological phenomena that have motivated the development of auditory modeling for automatic speech recognition. These phenomena include frequency analysis in parallel channels, a limited dynamic range of response within each channel, preservation of temporal fine structure, enhancement of temporal contrast at signal onsets and offsets, enhancement of spectral contrast (at adjacent frequencies), and preservation of temporal fine structure (at least at low frequencies). Most of these phenomena also have psychoacoustical correlates. Conventional feature extraction procedures (such as MFCC and PLP coefficients) preserve some of these attributes (such as basic frequency selectivity and spectral bandwidth) but omit others (such as temporal and spectral enhancement and detailed timing structure). As an example, Figure 8.12 compares a high-resolution spectrogram in response to a short utterance to a spectrogram reconstructed from MFCC coefficients computed for the same utterance. In addition to the frequency warping that is intrinsic to MFCC (and PLP) processing, it is clear that substantial detail is lost in the MFCC representation, some of which is sacrificed deliberately to remove pitch information. One of the goals of the auditory representations is to restore some of this lost information about the signal in a useful and efficient fashion.

8.3 "Classic" Auditory Representations

The first significant attempts to develop models of the peripheral auditory system for use as front ends to ASR systems occurred in the 1980s with the models of Seneff [99], Ghitza [32], and Lyon [61, 62], which we summarize in this section. The Seneff model, in particular, has been the basis for many subsequent studies, in part because of it was described in great detail and it is easily available in MATLAB form as part of the Auditory Toolbox developed and distributed by [102]. This very useful resource, which also includes the Lyon model, is based on earlier work by Lyon and Slaney using Mathematica.

Seneff's Auditory Model

Seneff's auditory model [99] is summarized in block diagram form in Figure 8.13. The first of three stages of the model consisted of 40 recursive linear filters implemented in cascade form

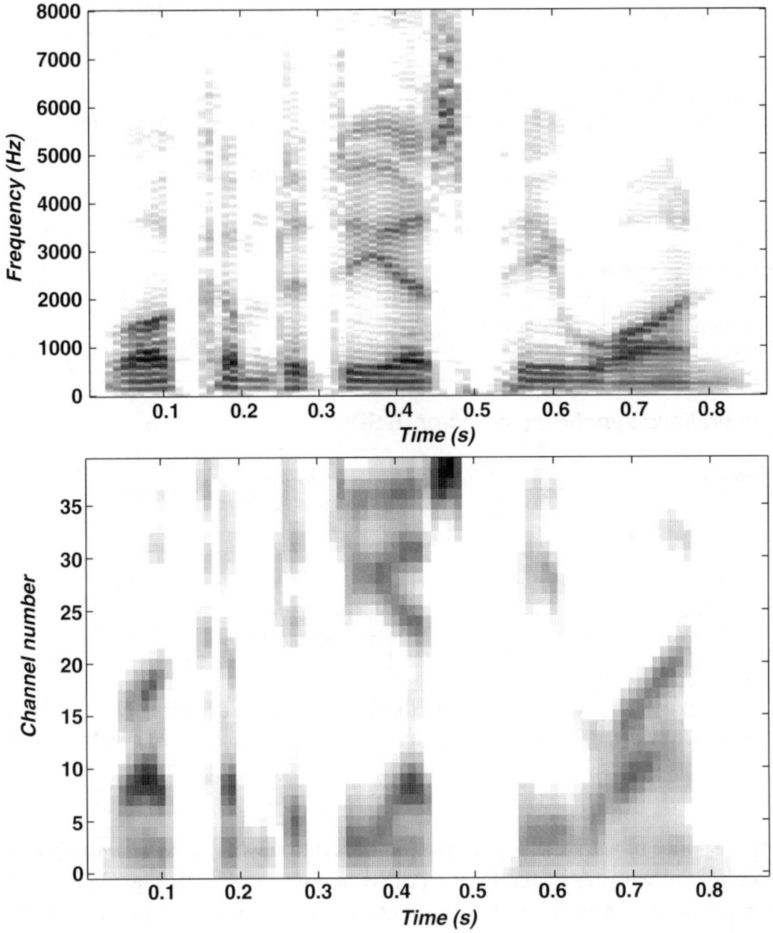

Figure 8.12 *Upper panel*: wide-band spectrogram of a sample utterance. *Lower panel*: reconstruction of the spectrogram after MFCC analysis.

to mimic the nominal auditory-nerve frequency responses as described by Kiang *et al.* [48] and other contemporary physiologists.

Substantial effort was devoted in Stage II to describing the nonlinear transduction from the motion of the basilar membrane to the mean rate of of auditory-nerve spike discharges. As indicated in the central panel of Figure 8.13, this "inner-hair-cell model" included four stages: (1) nonlinear half-wave rectification using an inverse tangent function for positive inputs and an exponential function for negative inputs, (2) short-term adaptation that modeled the release of transmitter in the synapse, (3) a lowpass filter with cut-off frequency of approximately 1 kHz to suppress synchronous response at higher input frequencies, and (4) a rapid automatic gain control (AGC) stage to maintain an approximately-constant response rate at higher input intensities when an auditory-nerve fiber is nominally in saturation.

Stage III consisted of two parallel operations on the hair-cell model outputs. The first of these was an envelope detector, which produced a statistic intended to model the instantaneous

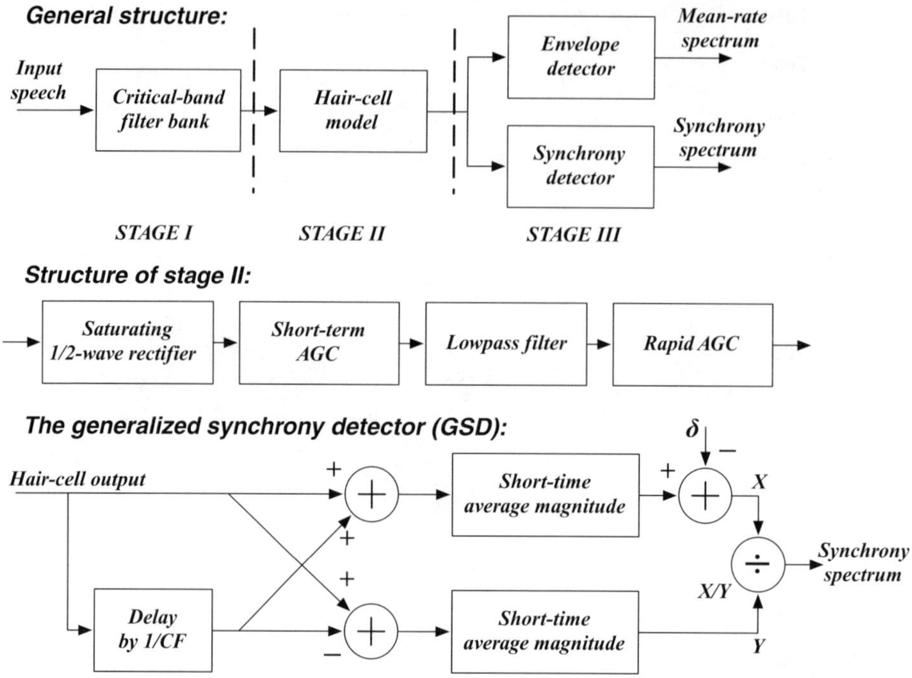

Figure 8.13 *Upper panel*: general structure of the Seneff model. *Central panel*: block diagram of the Seneff hair cell model. *Lower panel*: block diagram of of Seneff's generalized synchrony detector (after [99]).

mean rate of response of a given fiber. The second operation was called a *generalized synchrony detector (GSD)*, and was motivated by the ALSR measure of [119]. The GSD is summarized in the lower panel of Figure 8.13. The hair-cell output is compared to itself delayed by the reciprocal of the center frequency of the filter in each channel, and the short-time averages of the sums and differences of these two functions are divided by one another. The threshold δ is introduced to suppress response to low-intensity signals, and the resulting quotient is passed through a saturating half-wave rectifier to limit the magnitude of the predicted synchrony.

Ghitza's EIH Model

A second classic auditory model developed by Ghitza [32] is called the *Ensemble Interval Histogram* (EIH) model and is summarized in Figure 8.14. Ghitza makes use of the peripheral auditory model proposed by Allen [3] to describe the transformation of sound pressure into the neural rate of firing and focussed on the mechanism used to interpret the neural firing rates. The most interesting aspect of the EIH model is its use of timing information to develop a spectral representation of the incoming sound. Specifically, the EIH model records in each frequency channel the times at which the outputs of the auditory model crosses a set of seven thresholds that are logarithmically spaced over the dynamic range of each channel. Histograms are compiled of the reciprocals of the times between the threshold crossings of each threshold

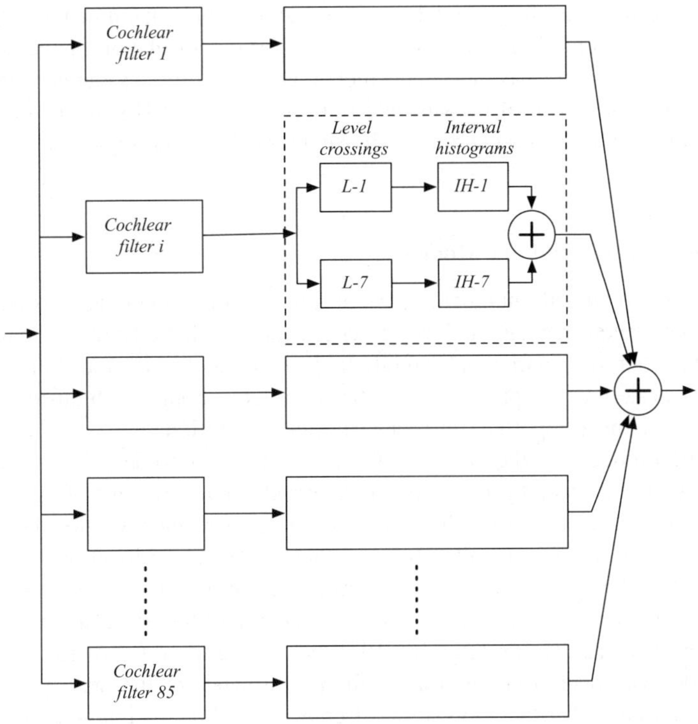

Figure 8.14 General structure of the Ghitza model (after [32]).

in each channel, and these histograms are summed over all thresholds and channels, producing an estimate of the internal spectral response to the incoming sound.

The EIH model was the only one of the three original auditory models for which the developer included speech-recognition evaluations with the original description of the model. Ghitza obtained these results using a contemporary DTW recognizer [32]. He observed that while the use of the auditory model provided no advantage in clean speech (and in some cases degraded performance compared to baseline MFCC processing), improvements were noted in noise and reverberation.

Lyon's Model

The third major model of the 1980s was described initially by Lyon [61,62]. Lyon's model for auditory-nerve activity [61] included many of the same elements as the models of Seneff and Ghitza (such as bandpass filtering, nonlinear rectification, and compression, along with several types of short-time temporal adaptation), as well as a mechanism for lateral suppression, which was unique among the classical models. Lyon was particularly concerned with the nature and shape of the filters used to model peripheral analysis and a longitudinal of his perspective on this subject may be found in [64]. In an extension of this work, Lyon proposed a "correlogram" display [63] that is derived from the short-time autocorrelation of the outputs of each channel that was believed to be useful for pitch detection and timbre identification. In 1983, Lyon described a computational binaural model based on cross-correlation of the corresponding outputs from the monaural processors. This model has the ability to separate signals based on differences in time of arrival of the signals to the two ears, and is similar in concept to the classic mechanism for extracting interaural time delays (ITDs) first suggested by Jeffress [46].

Performance of Early Auditory Models

The classic models included a number of attributes of auditory processing beyond MFCC/PLP feature extraction: more realistic auditory filtering, more realistic auditory nonlinearity, and in some cases synchrony extraction, lateral suppression, and interaural correlation. Unsurprisingly, each system developer had his or her own idea about which attribute of auditory processing was the most important for robust speech recognition.

While the EIH model was the only one of the original three to be evaluated quantitatively for speech-recognition accuracy at the time of its introduction, a number of early studies compared the recognition accuracy of auditory-based front ends with conventional representations (e.g., [32,43,45,70]). It was generally observed that while conventional feature extraction in some cases provided best accuracy when recognizing clean speech, auditory-based processing would provide superior results when speech was degraded by added noise. Early work in the CMU Robust Speech Group (e.g., [84,104]) confirmed these trends for reverberation as well as for additive noise in an analysis of the performance of the Seneff model. We also noted, disappointingly, that the application of conventional engineering-approaches such as *codeword-dependent cepstral normalization* (CDCN, [1]) provided recognition accuracy that was as good as or better than the accuracy obtained using auditory-based features in degraded acoustical environments. In a more recent analysis of the Seneff model, we observed that the saturating half-wave nonlinearity in Stage II of the Seneff model is the functional element that appears to provide the greatest improvement in recognition accuracy compared to baseline MFCC processing [17].

One auditory model of the late 1980s that was successful was developed by Cohen [18], and it exhibited a number of the physiological and psychoacoustical properties of hearing described in Section 8.2. Cohen's model included a bank of filters that modeled critical-band filtering, an empirical intensity scaling to describe equal loudness according to the curves of Fletcher and Munson [29], a cube-root power-law compressive nonlinearity to describe loudness scaling after Stevens [106]. The final stage of the model was a simple differential equation that models the time-varying release of neural transmitter based on the model of Schroeder and Hall [98]. This stage provided the type of transient overshoots observed in Figure 8.2. Feature extraction based on Cohen's auditory model provided consistently better recognition accuracy than features that approximated cepstral coefficients derived from a similar bank of bandpass filters for a variety of speakers and microphones in relatively quiet rooms. On the basis of these results, Cohen's auditory model was adopted as the feature extraction procedure for the IBM Tangora system and was used routinely for about a decade.

Despite the adoption of Cohen's feature extraction in Tangora and interesting demonstrations using the outputs of the models of Seneff, Ghitza, and Lyon, interest in the use of auditory models generally diminished for a period of time around the late 1980s. As noted above, the auditory models generally failed to provide superior performance when processing clean speech, which was the emphasis for much of the research community at this time. In part, this may well have been a consequence of the typical assumption in the speech-recognition systems of the day that the probability densities of the features were normally distributed. In contrast, the actual outputs of the auditory models were distinctly non-Gaussian in nature. For example, Chigier and Leung [16] noted that the accuracy of speech-recognition systems that used features based on the Seneff model was greatly improved when a multilayer perceptron (which learns the shape of the feature distributions without *a priori* assumptions) is used instead of a Bayesian classifier that assumed the use of unimodal Gaussian densities. The classical auditory models fared even worse when computation was taken into account. Ohshima [83], for example, observed that the Seneff model requires about 40 times as many multiplies and 33 times as many additions compared to MFCC or PLP feature extraction. And in all cases, the desire to improve robustness in speech recognition in those years was secondary to the need to resolve far more basic issues in acoustic modeling, large-vocabulary search, etc.

8.4 Current Trends in Auditory Feature Analysis

By the late 1990s, physiologically-motivated and perceptually-motivated feature extraction methods began to flourish once again for several reasons. Computational capabilities had advanced over the decade to a significant degree, and front-end signal processing came to consume a relatively small fraction of the computational demands of large-vocabulary speech recognition compared to score evaluation, graph search, etc. The development of fully continuous hidden Markov models using Gaussian mixture densities as probabilities for the features, along with the development of efficient techniques to train the parameters of these acoustic models, meant that the non-Gaussian form of the output densities of the auditory models was no longer a factor that limited their performance.

In this section, we describe some of these trends in auditory processing that have become important for feature extraction beginning in the 1990s. These trends include closer attention to the details of the physiology, a reconsideration of mechanisms of synchrony extraction, more effective and mature approaches to information fusion, serious attention to the temporal evolution of the outputs of the auditory filters, the development of models based on spectro-temporal response fields, concern for dealing with the effects of room reverberation as well as additive noise, and the use of two or more microphones motivated by binaural processing (which we do not discuss in this chapter).

In the sections below, with some exceptions, we characterize the performance of the systems considered only indirectly. This is because it is almost impossible to meaningfully compare recognition results across different research sites and different experimental paradigms. For example, the baseline level of recognition accuracy will depend on many factors including the types of acoustical models employed and the degree of constraint imposed by language modeling. The type of additive noise used typically affects the degree of improvement to be expected from robust signal processing approaches: for example, it is relatively easy to ameliorate the effects of additive white noise, but effective compensation for the effects of background music is far more difficult to achieve. As the amount of available acoustic training increases, the degree of improvement observed by advanced feature-extraction or signal-processing techniques diminishes because the initial acoustical models become intrinsically more robust. While most of the results in robust speech recognition that are reported in the literature are based on training on clean speech, the amount of improvement provided by signal processing also diminishes when an ASR system is trained in a variety of acoustical environments (multistyle training) or when the acoustical conditions of the training and testing data are matched.

We begin with peripheral phenomena and continue with more central phenomena.

Speech Recognition Based on Detailed Physiological Models

In addition to the "practical" abstractions proposed by speech researchers including the classical representations discussed in Section 8.3, auditory physiologists have also proposed models of their own that describe and predict the functioning of the auditory periphery in detail. For example, the model of Meddis and his colleagues (e.g., [68, 69]) is a relatively early formulation that has been quite influential in speech processing. The Meddis model characterizes the rate of spikes in terms of a mechanism based on the dynamics of the flow of neurotransmitter from inner hair cells into the synaptic cleft, followed by its subsequent uptake once again by the hair cell. Its initial formulation, which has been refined over the years, was able to predict a number of the physiological phenomena described in Section 8.2.1 including the nonlinear rate-intensity curve, the transient behavior of envelopes of tone bursts, synchronous response to low-frequency inputs, the interspike interval histogram, and other phenomena. Hewitt and Meddis reviewed the physiological mechanisms underlying seven contemporary models of auditory transduction, and compared their ability to describe a range of physiological data, concluding that their own formulation described the largest set of physiological phenomena most accurately [42].

The Carney group (e.g., Zhang *et al.* [120]; Heinz *et al.* [35]; Zilarny *et al.* [122]) has also developed a series of physiologically based models of auditory-nerve activity over the years. The original goal of the work of Carney and her colleagues had been to develop a model that can describe the response to more complex signals such as noise-masked signals and speech, primarily through the inclusion into the model of the compressive nonlinearity of the cochlear amplifier in the inner ear. A diagram of most of the functional blocks of the model of Zhang *et al.* is depicted in the left panel of Figure 8.15. As can be seen in the diagram, the model includes a *signal path* that has many of the attributes of the basic phenomenological models introduced in Section 8.3, with a time-varying nonlinear narrow-band peripheral filter that is followed by a linear filter. Both of these filters are based on gammatone filters. The time constant that determines the gain and bandwidth of the nonlinear filter in the signal path is controlled by the output of the wide-band *control path* that is depicted on the right side of the panel. The level-dependent gain and bandwidth of the control path enable the model to describe phenomena such as two-tone suppression within a single auditory-nerve channel, without needing to depend on inhibitory inputs from fiber at adjacent frequencies, as in Lyon's model [61].

A few years ago Kim *et al.* [52] from the CMU group presented some initial speech recognition results that developed simple measures of mean rate and synchrony from the outputs of the model of Zhang *et al.* Figure 8.16 compares the recognition accuracy for speech in white noise using feature-extraction procedures that were based on the putative mean rate of auditory-nerve response [52]. The CMU Sphinx-3 ASR system was trained using clean speech for these experiments. The curves in Figure 8.16 describe the recognition accuracy obtained

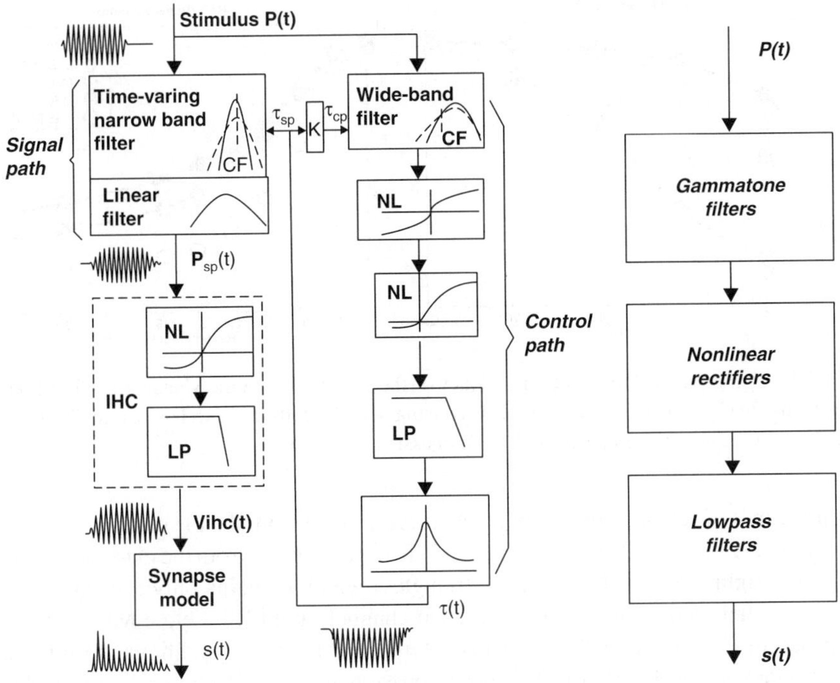

Figure 8.15 *Left panel*: block diagram of the Zhang–Carney model (from Zhang *et al.*, 2001). *Right panel*: block diagram of a much simpler computational model of auditory processing.

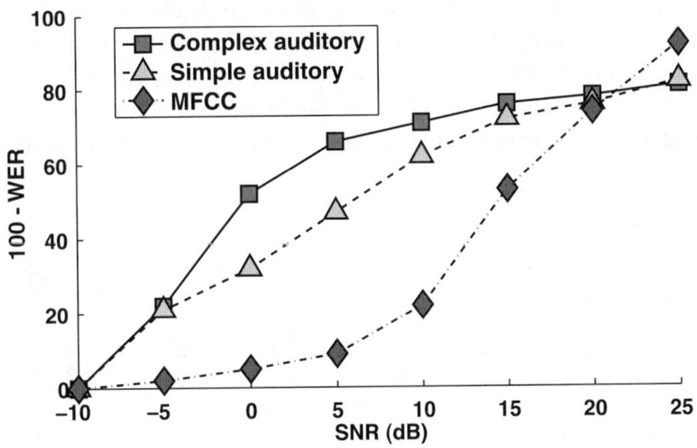

Figure 8.16 Comparison of speech-recognition accuracy obtained using features derived from the Zhang–Carney model (squares), features obtained from the much simpler model in the right panel of Figure 8.15 (triangles), and conventional MFCC coefficients (diamonds). Data were obtained using sentences from the DARPA Resource Management corpus corrupted by additive white noise. The language model is detuned, which increases the absolute word error rate from the best possible value (replotted from [52]).

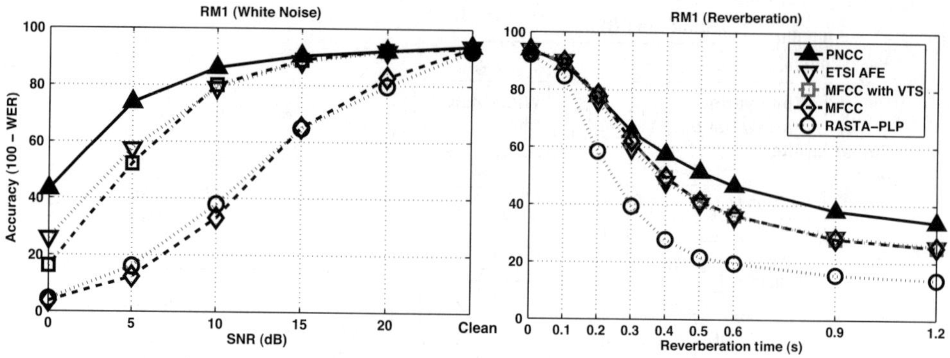

Figure 8.17 Comparison of recognition accuracy on the DARPA Resource Management RM1 database, obtained using PNCC processing with processing using MFCC features, RASTA-PLP features, the ETSI AFE, and MFCC features augmented by VTS processing (from [51]).

using three types of feature extraction: (1) features derived from the mean rate response based on the complete model of Zhang *et al.* [120]; (2) features derived from the extremely simplified model in the right panel of Figure 8.15 (triangles) which contains only a bandpass filter, a nonlinear rectifier, and a lowpass filter in each channel; and (3) baseline MFCC processing as described in [19] (diamonds). As can be seen, for this set of conditions the full auditory model provides about 15 dB of effective improvement in SNR compared to the baseline MFCC processing, while the highly simplified model provides about a 10-dB improvement. Unfortunately, the computational cost of features based on the complete model of Zhang *et al.* is on the order of 250 times the computational cost incurred by the baseline MFCC processing. In contrast, the simplified auditory processing consumes only about twice the computation of the baseline MFCC processing. We note that ASR performance in small tasks including the DARPA Resource Management task used for these comparisons can easily become dominated by the impact of a strong language model. In obtaining the results for this figure, as well as for Figure 8.17, we deliberately manipulated the language weight parameter to reduce the impact

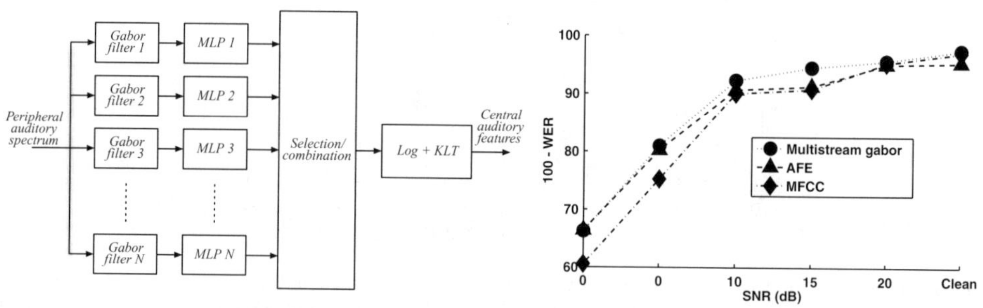

Figure 8.18 (left) Generation of multistream Gabor filter features. See text for details. (Right) ASR word accuracy on the Numbers95 test set in the presence of a range of real-world noise sources using a system trained on clean speech. Results shown use the Gabor-based features to augment an MFCC feature vector, using SRI's Decipher system (from [92]).

of the language model in order to emphasize differences in recognition accuracy that were due to changes in feature-extraction procedures. As a consequence, the absolute recognition accuracy is not as good as it would have been had we optimized all system parameters.

Power-Normalized Cepstral Coefficients (PNCC Processing)

The extreme computational costs associated with the implementation of a complete physiological model such as that of Zhang *et al.* [120] have motivated many researchers to develop simplified models that capture the essentials of auditory processing that are believed to be most relevant for speech perception. The development of *power-normalized cepstral coefficients* (PNCC, [49–51]) is a convenient example of computationally-efficient "pragmatic" physiologically motivated feature extraction. PNCC processing was developed with the goal of obtaining features that incorporate some of the relevant physiological phenomena in a computationally efficient fashion. A summary of the major functional blocks of PNCC processing is provided in the right column of Figure 8.11. PNCC processing includes (1) traditional pre-emphasis and short-time Fourier transformation, (2) integration of the squared energy of the STFT outputs using gammatone frequency weighting, (3) "medium-time" nonlinear processing that suppresses the effects of additive noise and room reverberation, (4) a power-function nonlinearity with exponent 1/15, and (5) generation of cepstral-like coefficients using a discrete cosine transform (DCT) and mean normalization. The power law, rather than the more common logarithmic transformation, was adopted because it provides reduced variability at very low signal intensities, and the exponent of 1/15 was selected because it provides a best fit to the onset portion of the rate-intensity curve developed by the model of Heinz *et al.* [35]. The power-function nonlinearity has the additional advantage of preserving ratios of responses that are independent of input amplitude.

For the most part, noise and reverberation suppression is introduced to PNCC processing through the system blocks labeled "medium-time processing" in the far right column of Figure 8.11 [51]. Medium-time processing operates on segments of the waveform on the order of 50–150 ms duration (as do other waveform-based compensation approaches) in contrast to compensation algorithms such as vector Taylor series (VTS, [80]) that manipulate cepstral coefficients derived from from analysis windows on the order of 20–30 ms in duration.

Figure 8.17 compares the recognition accuracy obtained using PNCC processing with the accuracy obtained using baseline MFCC processing [19], PLP-RASTA processing [37], MFCC with VTS [80], and the "Advanced Front End" (AFE), a newer feature-extraction scheme developed as a standard for the European Telecommunications Standards Institute (ETSI), which also has noise-robustness capabilities [110]. It can be seen from the panels of Figure 8.17 that the recognition accuracy obtained using features derived with PNCC processing is substantially better than baseline processing using either MFCC or RASTA-PLP features, MFCC features augmented by the VTS noise-reduction algorithms, or the ETSI Advanced Front End for speech that had been degraded by additive white noise and simulated reverberation. In considering these comparisons, it must be borne in mind that neither MFCC nor RASTA-PLP coefficients were developed with the goal of robustness with respect to additive noise. A version of RASTA-PLP known as J-RASTA [37] is far more effective in the presence of additive noise. A much more thorough discussion of PNCC processing, including recognition results in the presence of a number of other types of degradations, may be found in [51]. PNCC processing is only about 30% more computationally costly than MFCC processing, and comparable in computational cost to RASTA-PLP. All of these methods require substantially less computation than either the ETSI Advanced Front End or the VTS approach to noise robustness.

Spectral Profiles Based on Synchrony Information

Since the 1980s, the approaches of Seneff and Ghitza for developing a spectral representation from the temporal fine structure of auditory-nerve firings (rather than simply their mean rate) have been elaborated upon, and other techniques have been introduced as well. We summarize some of these approaches in this section.

Ali *et al.* [2] proposed a simple but useful extension of the Seneff GSD model that develops a synchrony spectrum by simply averaging the responses of several GSDs tuned to the same frequency using inputs from bandpass filters with CFs in a small neighborhood about a central frequency. As described by Ali *et al.*, this approach, referred to as *average localized synchrony detection* (ALSD), produces a synchrony spectrum with smaller spurious peaks than are obtained using either Seneff's original GSD detector, mean-rate-based spectral estimates, or the synchrony spectrum produced by the lateral inhibitory network (LIN) of Shamma [100], and it provides the best recognition results of the methods considered for a small vowel-classification task in white noise.

D. Kim *et al.* [53] proposed a type of processing called *zero-crossing peak analysis* (ZCPA) that could be considered to be an elaboration of Ghitza's EIH processing, but without the complication of the multiple thresholds that are part of the EIH model. Positive-going zero crossings are recorded directly from the outputs of each of the auditory filters, and the times of these zero crossings are recorded on a channel by channel basis. A histogram is generated of the reciprocal of the intervals between the zero crossings (a measure of instantaneous frequency), weighted by the amplitude of the peak between the zero crossings. While quantitative analysis of zero crossings of a random process is always difficult, the authors argue that setting the threshold for marking an event to zero will minimize the variance of the observations. Kim *et al.* [53] compared the recognition accuracy in a small isolated word task using ZCPA with similar results obtained using LPC-based features, features from the EIH model, and features obtained using zero crossings without the weighting by the peak amplitude. The ZCPA approach provided the greatest accuracy in all cases, especially at low SNRs. Ghulam *et al.* [33, 34] augmented the ZCPA procedure by adding auditory masking, Wiener filtering, and a weighting of the frequency histograms to emphasize components that are close to harmonics of the fundamental frequency.

C. Kim *et al.* [52] implemented a synchrony-based estimation of spectral contours using a third method: direct Fourier transformation of the phase-locked temporal envelopes of the outputs of the critical-band filters. This produces a high-resolution spectral representation at low frequencies for which the auditory nerve is synchronized to the input up to about 2.2 kHz, and which includes the effects of all of the nonlinearities of the peripheral processing. The use of the synchrony processing at low frequencies provided only a modest improvement compared to the auditory model with mean-rate processing as shown in Figure 8.16, although it was a large improvement compared to baseline MFCC processing.

Multistream Processing

The *articulation index* model of speech perception, which was suggested by Fletcher [28] and French and Steinberg [30], and revived by Allen [4], modeled phonetic speech recognition as arising from independent estimators for critical bands. This initially led to a great deal of interest in the development of *multiband systems* based on this view of independent detectors per critical band that were developed to improve robustness of speech recognition, particularly for narrow-band noise (e.g., [11,40,75]). This approach in turn can be generalized to the consideration of fusion of information from parallel detectors that are presumed to provide complementary information about the incoming speech. This information can be combined at the input (feature) level [81,82], at the level at which the HMM search takes place, which is sometimes referred to as "state combination" [44,65], or at the output level by merging hypothesis lattices [27,67,101]. In a systematic comparison of all of these approaches, Li [59] observed that state combination provides the best recognition accuracy by a small margin.

The *Tandem* approach, first proposed by Hermansky, Ellis, and Sharma [41], has been particularly successful in facilitating the combination of multiple information streams at the feature level. Typically, the Tandem method is applied by expressing the outputs of a multilayer perceptron (MLP) as probabilities, which can be combined linearly or nonlinearly across the streams. These combined probabilities are then in turn (after some simple transformations, such as the logarithm followed by principal components analysis) used as features to a conventional hidden Markov model classifier. If the linear stream weights can be determined dynamically, there is at least the potential for robustness to time-varying environmental conditions. The MLP training is quite robust to the nature of the input distribution, and in particular can easily be used to handle acoustic inputs covering a large temporal context. Over the years, the Tandem approach has proven to be a very useful way of combining rather diverse sets of features.

Long-Time Temporal Evolution

An additional major trend has been the development of features that are based on the temporal evolution of the envelopes of the outputs of the bandpass filters that are part of any description of the auditory system. As noted in Section 8.2.2, some units in the brainstem of various mammals exhibit a sensitivity to amplitude modulation, with maximal responses at a particular modulation frequency independently of the carrier frequency. Psychoacoustical results also indicate that humans are sensitive to modulation frequency [111,118]), with temporal modulation transfer functions indicating greatest sensitivity to temporal modulations at approximately the same frequencies as in the physiological data, despite the obvious species differences.

Initially, this information has been used to implement features based on frequency components of these temporal envelopes, which (as noted in Section 8.2.2) are referred to by Kingsbury and others as the *modulation spectrum* [55]. Specifically, Kingsbury *et al.* [54] obtained lowpass and bandpass representations of the envelopes of the outputs of the critical-band filters by passing the filter outputs through a square-root nonlinearity, followed by a lowpass filter with a 16-Hz cutoff and a bandpass filter with passband from 2 to 16 Hz (in parallel), and two subsequent AGC stages. The modulation spectrum is obtained by expressing these signals as a function of the center frequencies of the critical-band filters. This is a useful representation because speech signals typically exhibit temporal modulations with modulation frequencies in the range that is passed by this processing, while noise components often exhibit frequencies of amplitude modulation outside this range. Tchorz and Kollmeier [109] also developed an influential physiologically-motivated feature extraction system at about the same time that included the usual stages of filtering, rectification, and transient enhancement. They were also concerned about the impact of modulation spectra, noting that their model provided the greatest output for temporal modulations around 6 Hz, and that in general lowpass filtering the envelopes of the outputs of the auditory model in each channel reduced the variability introduced by background noise.

Other researchers have subsequently characterized the temporal patterns more explicitly. In general, these procedures operate on the time-varying envelope or log energy of a long temporal segment that is the output of a single critical-band filter. These representations effectively slice a spectrographic representation into horizontal "slices" rather than the vertical slices isolated by the conventional windowing procedure, which is brief and time and broad and frequency. As an example, Hermansky and Sharma [38] developed the *TRAPS* representation (for *TempoRAl PatternS*), which operates on 1-second segments of the log spectral energies that emerge from each of 15 critical-band filters. In the original implementation, these outputs were classified directly by a multilayer perceptron (MLP). This work was extended by Chen *et al.* [13] who developed *HATS* (for *Hidden Activation TRAPS*), which trains an additional MLP layer at the level of each critical band filter to provide a set of basis functions optimized to maximize the discriminability of the data to be classified.

Athineos and Ellis [6,7,9] have developed *frequency-domain linear prediction*, or *FDLP*. In this process, the temporal envelopes of the outputs of critical band filters are represented by linear prediction. Much as linear-predictive parameters computed from the time-domain signal within a short analysis window (e.g., 25 ms) represent the envelopes of the short-time spectrum within a slice of time, the FDLP parameters represent the Hilbert envelope of the *temporal sequence* within a slice of spectrum. This method was further incorporated into a method called *LP-TRAPs* [8], in which the FDLP-derived Hilbert envelopes were used as input to MLPs that learned phonetically relevant transformations for later use in speech recognition. LP-TRAPS can be considered to be a parametric estimation approach to characterizing the trajectories of the temporal envelopes, while traditional TRAPS is nonparametric in nature.

It is also worth restating that RASTA processing, described in Section 8.2.4, was developed to emphasize the critical temporal modulations, and in so doing RASTA emphasizes transitions, roughly models forward masking, and reduces sensitivity to irrelevant steady-state convolutional factors. More recently, temporal modulation in subbands was normalized to improve ASR in reverberant environments [60].

Joint Feature Representation in Frequency, Rate, and Scale

There has been substantial interest in recent years in developing computational models of speech processing based on the *spectro-temporal receptive fields* (STRFs) that were described in Section 8.2.2. In an influential set of studies, Chi *et al.* [14] conducted a series of psychoacoustical experiments that measured the spectro-temporal modulation transfer functions (MTF) in response to moving ripple signals such as those used to develop physiological STRFs, arguing that the results were consistent the physiologically measured STRFs, and that the spectro-temporal MTFs are separable into the product of a spectral MTF and a temporal MTF. Subsequently, this enabled Chi *et al.* to propose a model of central auditory processing with three independent variables: auditory frequency, "rate" (characterizing temporal modulation), and "scale" (characterizing spectral modulation), with receptive fields of varying extent as would be obtained by successive stages of wavelet processing [15]. The model relates this representation to feature extraction at the level of the brainstem and the cortex, including detectors based on STRFs, incorporating an auditory model similar to those described above that provides the input to the STRF filtering. Chi *et al.* also generated speech from the model outputs and compared the intelligibility of the reconstructed speech to the degree of spectral and temporal modulation in the signal.

A number of researchers have found it convenient to use two-dimensional Gabor filters as a reasonable and computationally-tractable approximation to the STRFs of A1 neurons. This representation was used successfully by Mesgarani *et al.* to implement features for speech/nonspeech discrimination [71] and similar approaches were used to extract features for ASR by multiple researchers (e.g., [39,57,73,121]). In many of these cases, MLPs were used to transform the filter outputs into a form that is more amenable to use by Gaussian mixture-based HMMs, typically using the Tandem approach described above [41]. The filters can either be used as part of a single modulation filter bank that either does or does not incorporate a final MLP, or the filters can be split into multiple streams, each with their own MLP, as described in the multistream section above. The choice of filters can either be data-driven (as in [57]) or chosen to span the space of interest, that is to cover the range of temporal and spectral modulations that are significant components of speech (e.g., [72, 74]).

In one recent study, Ravuri [91] developed a complex model that incorporates hundreds of two-dimensional Gabor filters, each with their own discriminatively-trained neural network to generate noise-insensitive features for ASR. As an example, as shown in Figure 8.18, Ravuri and Morgan [92] describe the recognition accuracy that is obtained by incorporating feature streams developed by modeling a range of STRFs of mel spectra using Gabor filters. The resulting MLP posterior probability outputs are linearly combined across streams (the "Selection/Combination" block in the left panel of Figure 8.18), where each stream is weighted by inverse entropy of the posterior distribution for each stream's MLP. The combined stream is further processed with a log function to roughly Gaussianize it, and with a Karhunen–Loeve transformation to orthogonalize the features; both steps are taken provide a better statistical match of the features to systems based on Gaussian mixtures. The system was trained on the high-SNR Numbers95 database and tested on an independent Numbers95 set with noises added from the RSG-10 database, which comprises a range of noises including speech babble, factory noise, etc. The results labeled "multistream Gabor" were obtained using 174 feature streams, each of which included a single spectro-temporal filter followed by an MLP trained for phonetic discrimination. It can be seen that the use of the multi-stream Gabor filter features improves accuracy at all SNRs, although in some cases by only a small amount.

8.5 Summary

In this chapter, we have reviewed a number of signal processing concepts that have been abstracted from several decades of concentrated study of how the auditory system responds to sound. The results of these studies have provided insight into the development of more environmentally robust approaches to feature extraction for automatic speech recognition. While we have only explicitly included a limited number of experimental ASR results from our own groups, many physiologically motivated feature-extraction procedures have demonstrated recognition accuracy that is as good as or better than the recognition accuracy provided by conventional signal processing, at least in degraded acoustical environments.

Although there remains no universally accepted theory about which aspects of auditory processing are the most relevant to robust speech recognition, we may speculate with some confidence about some of the reasons for the general success of auditory models. The increasing bandwidth of the auditory analysis filters with increasing center frequency enables good spectral resolution at low CFs (which is useful for tracking formant frequencies precisely) and better temporal resolution at higher CFs (which is helpful in marking the precise time structure of consonant bursts). The nonlinear nature of the auditory rate-intensity function tends to suppress feature variability caused by additive low-level noise, and an appropriate shape of the nonlinearity can provide normalization to absolute amplitude as well. The short-time temporal suppression and lateral frequency suppression provides an ongoing enhancement of change with respect to running time and analysis frequency. As has been noted by Wang and Shamma [114] and others, the tendency of the auditory system to enhance local spectro-temporal contrast while averaging the incoming signals over a broader range of time and frequency enables the system to provide a degree of suppression to the effects of noise and reverberation. Bandpass filtering of the modulation spectrum between 2 and 16 Hz will help to separate the responses to speech and noise, as many disturbances produce modulations outside that range. In many respects, the more central representations of speech at the level of the brainstem and the cortex are based primarily on the dynamic aspects of the speech signal, and perceptual results from classical auditory scene analysis [12] confirm the importance of many of these cues in segregating individual sound sources in cluttered acoustical environments.

The good success of relatively simple feature extraction procedures such as PNCC processing, RASTA, and similar approaches suggests that the potential benefits from the use of auditory processing are widespread. While our understanding of how to harness the potential of more central representations such as the spectro-temporal receptive fields is presently in its infancy, we expect that we will be able to continue to improve the robustness and overall utility of our representations for speech as we continue to deepen our understanding of how speech is processed by the auditory system.

Acknowledgments

This research was supported by NSF (Grants IIS-0420866 and IIS-0916918) at CMU, and Cisco, Microsoft, and Intel Corporations and internal funds at ICSI. The authors are grateful to Chanwoo Kim and Yu-Hsiang (Bosco) Chiu for sharing their data, along with Mark Harvilla, Kshitiz Kumar, Bhiksha Raj, and Rita Singh at CMU, as well as Suman Ravuri, Bernd Meyer, and Sherry Zhao at ICSI for many helpful discussions.

References

[1] A. Acero and R. M. Stern, "Environmental robustness in automatic speech recognition," in *Proceedings of the IEEE International Conference on Acoustics, Speech and Signal Processing*, 1990, pp. 849–852.

[2] A. M. A. Ali, J. V. der Spiegel, and P. Mueller, "Robust auditory-based speech processing using the average localized synchrony detection," *IEEE Transactions on Speech and Audio Processing*, vol. 10, pp. 279–292, 1999.

[3] J. B. Allen, "Cochlear modeling," *IEEE ASSP Magazine*, vol. 1, pp. 3–29, 1985.
[4] J. B. Allen, "How do humans process and recognize speech?" *IEEE Transactions on Speech and Audio*, vol. 2, pp. 567–577, 1994.
[5] R. M. Arthur, R. R. Pfeiffer, and N. Suga, "Properties of two-tone inhibition in primary auditory neurons," *Journal of Physiology*, vol. 212, pp. 593–609, 1971.
[6] M. Athineos and D. P. W. Ellis, "Frequency-domain linear prediction for temporal features," in *Proceedings of the IEEE ASRU Workshop*, 2003, pp. 261–266.
[7] M. Athineos and D. P. W. Ellis, "Autoregressive modeling of termporal envelopes," *IEEE Transactions on Signal Processing*, vol. 15, pp. 5237–5245, 2007.
[8] M. Athineos, H. Hermansky, and D. P. W. Ellis, "LP-TRAP: Linear predictive temporal patterns," in *Proceedings of the International Conference on Spoken Language Processing*, 2004, pp. 949–952.
[9] M. Athineos, H. Hermansky, and D. P. W. Ellis, "PLP^2: Autoregressive modeling of auditory-like 2-D spectro-temporal patterns," in *Proceedings of the ISCA Tutorial and Research Workshop on Statistical and Perceptual Audio Processing SAPA-04*, 2004, pp. 25–30.
[10] J. C. Baird and E. Noma, *Fundamentals of Scaling and Psychophysics*. Hoboken, NJ, USA: John Wiley & Sons, 1978.
[11] H. Bourlard, S. Dupont, H. Hermansky, and N. Morgan, "Towards sub-band-based speech recognition," in *Proceedings of the European Signal Processing Conference*, 1996, pp. 1579–1582.
[12] A. S. Bregman, *Auditory scene analysis*. Cambridge, MA: MIT Press, 1990.
[13] B. Chen, S. Chang, and S. Sivadas, "Learning discriminative temporal patterns in speech: Development of novel TRAPS-like classifiers," in *Proceedings of Eurospeech*, 2003.
[14] T. Chi, Y. Gao, M. C. Guyton, P. Ru, and S. A. Shamma, "Spectro-temporal modulation transfer functions and speech intelligibility," *Journal of the Acoustical Society of America*, vol. 106, pp. 719–732, 1999.
[15] T. Chi, R. Ru, and S. A. Shamma, "Multiresolution spectrotemporal analysis of complex sounds," *Journal of the Acoustical Society of America*, vol. 118, no. 2, pp. 887–906, August 2005.
[16] B. Chigier and H. C. Leung, "The effects of signal representations, phonetic classification techniques, and the telephone network," in *Proceedings of the International Conference of Spoken Language Processing*, 1992, pp. 97–100.
[17] Y.-H. Chiu and R. M. Stern, "Analysis of physiologically-motivated signal processing for robust speech recognition," in *Proceedings of Interspeech*, 2008.
[18] J. R. Cohen, "Application of an auditory model to speech recognition," *Journal of the Acoustical Society of America*, vol. 85, pp. 2623–2629, 1989.
[19] S. B. Davis and P. Mermelstein, "Comparison of parametric representations for monosyllabic word recognition in continuously spoken sentences," *IEEE Transactions on Acoustics, Speech and Signal Processing*, vol. 28, pp. 357–366, 1980.
[20] D. A. Depireux, J. Z. Simon, D. J. Klein, and S. A. Shamma, "Spectro-temporal response field characterization with dynamic ripples in ferret primary auditory cortex," *Journal of Neurophysiology*, vol. 85, pp. 1220–1234, 2001.
[21] A. Dreyer and B. Delgutte, "Phase locking of auditory-nerve fibers to the envelopes of high-frequency sounds: implications for sound localization," *Journal of Neurophysiology*, vol. 96, no. 5, pp. 2327–2341, 2006.
[22] R. Drullman, J. M. Festen, and R. Plomp, "Effects of temporal envelope smearing on speech reception," *Journal of the Acoustical Society of America*, vol. 95, pp. 1053–1064, 1994.
[23] R. Drullman, J. M. Festen, and R. Plomp, "Effect of reducing slow temporal modulations on speech reception," *Journal of the Acoustical Society of America*, vol. 95, pp. 2670–2680, 1994.
[24] N. I. Durlach and H. S. Colburn, "Binaural phenomena," in *Hearing*, series on Handbook of Perception, E. C. Carterette and M. P. Friedman, Eds. Academic Press, New York, 1978, vol. IV, Ch. 10, pp. 365–466.
[25] European Telecommunications Standards Institute, "Speech processing, transmission and quality aspects (STQ); Distributed speech recognition; advanced front-end feature extraction algorithm; compression algorithms," Technical Report ETSI ES 202 050, Rev. 1.1.5, January 2007.
[26] G. T. Fechner, *Element der Psychophysik*. Breitkopf & Härterl; (English translation by H. E. Adler, Holt, Rinehart, and Winston, 1966), 1860.
[27] J. Fiscus, "A post-processing system to yield reduced word error rates: recognizer output voting error reduction (ROVER)," in *Proceedings of the IEEE ASRU Workshop*, 1997, pp. 347–354.
[28] H. Fletcher, "Auditory patterns," *Reviews of Modern Physics*, vol. 12, pp. 47–65, 1940.

[29] H. Fletcher and W. A. Munson, "Loudness, its definition, measurement and calculation," *Journal of the Acoustical Society of America*, vol. 5, pp. 82–108, 1933.
[30] N. R. French and J. C. Steinberg, "Factors governing the intelligibility of speech sounds," *Journal of the Acoustical Society of America*, vol. 19, pp. 90–119, 1947.
[31] G. A. Gescheider, *Psychophysics: The Fundamentals*. London, UK: Psychology Press, 1997.
[32] O. Ghitza, "Auditory nerve representation as a front-end for speech recognition in a noisy environment," *Computer Speech & Language*, vol. 1, pp. 109–130, 1986.
[33] M. Ghulam, T. Fukuda, J. Horikawa, and T. Niita, "Pitch-synchronous ZCPA (PS-ZCPA)-based feature extraction with auditory masking," in *Proceedings of the IEEE International Conference on Acoustics, Speech and Signal Processing*, 2005.
[34] M. Ghulam, T. Fukuda, K. Katsurada, J. Horikawa, and T. Niita, "PS-ZCPA based feature extraction with auditoy masking, modulation enhancement and noise reduction," *IECE Transactions on Information and Systems*, vol. E89-D, pp. 1015–1023, 2006.
[35] M. G. Heinz, X. Zhang, I. C. Bruce, and L. H. Carney, "Auditory-nerve model for predicting performance limits of normal and impaired listeners," *Acoustics Research Letters Online*, vol. 2, pp. 91–96, July 2001.
[36] H. Hermansky, "Perceptual linear predictive (PLP) anlysis of speech," *Journal of the Acoustical Society of America*, vol. 87, pp. 1738–1752, 1990.
[37] H. Hermansky and N. Morgan, "RASTA processing of speech," *IEEE Transactions on Speech and Audio Processing*, vol. 2, pp. 578–589, 1994.
[38] H. Hermansky and S. Sharma, "Temporal patterns (TRAPS) in ASR of noisy speech," in *Proceedings of the IEEE International Conference on Acoustics, Speech and Signal Processing*, 1999.
[39] H. Hermansky and F. Valente, "Hierarchical and parallel processing of modulation spectrum for ASR applications," in *Proceedings of the IEEE International Conference on Acoustics, Speech and Signal Processing*, 2008, pp. 4165–4168.
[40] H. Hermansky, S. Tibrewala, and M. Pavel, "Towards asr on partially corrupted speech," in *Proceedings of the International Conference on Spoken Language Processing*, vol. 1, 1996, pp. 462–465.
[41] H. Hermansky, D. P. W. Ellis, and S. Sharma, "Tandem connectionist feature extraction for conventional hmm systems," in *Proceedings of the IEEE ICASSP*, 2000, pp. 1635–1638.
[42] M. J. Hewitt and R. Meddis, "An evaluation of eight computer modles of mammalian inner hair-cell function," *Journal of the Acoustical Society of America*, vol. 90, pp. 904–917, 1991.
[43] M. J. Hunt and C. Lefebvre, "A comparison of several acoustic representations for speech recognition with degraded and undegraded speech," in *Proceedings of the IEEE International Conference on Acoustics, Speech and Signal Processing*, 1989, pp. 262–265.
[44] A. Janin, D. P. W. Ellis, and N. Morgan, "Multi-stream speech recognition: Ready for prime time?" *Proceedings of the Eurospeech*, pp. 591–594, 1999.
[45] C. R. Jankowski, H.-D. H. Vo, and R. P. Lippmann, "A comparison of signal processing front ends for automatic word recognition," *IEEE Transactions on Speech and Audio Processing*, vol. 3, pp. 286–293, 1995.
[46] L. A. Jeffress, "A place theory of sound localization," *Journal of the Comparative & Physiological Psychology*, vol. 41, pp. 35–39, 1948.
[47] P. X. Joris, C. E. Schreiner, and A. Rees, "Neural processing of amplitude-modulated sounds," *Physiological Reviews*, vol. 84, pp. 541–577, 2004.
[48] N. Y.-S. Kiang, T. Watanabe, W. C. Thomas, and L. F. Clark, *Discharge Patterns of Single Fibers in the Cat's Auditory Nerve*. Cambridge, MA, USA: MIT Press, 1966.
[49] C. Kim and R. M. Stern, "Feature extraction for robust speech recognition based on maximizing the sharpness of the power distribution and on power flooring," in *Proceedings of the IEEE International Conference on Acoustics, Speech and Signal Processing*, March 2010, pp. 4574–4577.
[50] C. Kim and R. M. Stern, "Feature extraction for robust speech recognition using a power-law nonlinearity and power-bias subtraction," in *Proceedings of Interspeech*, September 2009, pp. 28–31.
[51] C. Kim and R. M. Stern, "Power-normalized cepstral coefficients (PNCC) for robust speech recognition," *IEEE Transactions on Audio, Speech and Language Processing* (accepted for publication), 2012.
[52] C. Kim, Y.-H. Chiu, and R. M. Stern, "Physiologically-motivated synchrony-based processing for robust automatic speech recognition," in *Proceedings of Interspeech*, 2006, pp. 1975–1978.
[53] D.-S. Kim, S.-Y. Lee, and R. Kil, "Auditory processing of speech signals for robust speech recognition in real world noisy environments," *IEEE Transactions on Speech and Audio Processing*, vol. 7, pp. 55–59, 1999.

[54] B. E. D. Kingsbury, "Perceptually inspired signal-processing strategies for robust speech recognition in reverberant environments," PhD dissertation, University of California, Berkeley, 1998.

[55] B. E. D. Kingsbury, N. Morgan, and S. Greenberg, "Robust speech recognition using the modulation spectrogram," *Speech Communication*, vol. 25, no. 1–3, pp. 117–132, 1998.

[56] D. J. Klein, D. A. Depireux, J. Z. Simon, and S. A. Shamma, "Robust spectro-temporal reverse correlation for the auditory system: Optimizing stimulus design," *Journal of Comparative Neuroscience*, vol. 9, pp. 85–111, 2000.

[57] M. Kleinschmidt, "Localized spectro-temporal features for automatic speech recognition," in *Proceedings of the Eurospeech*, 2003, pp. 2573–2576.

[58] G. Langner and C. E. Schreiner, "Periodicity coding in the inferior colliculus of the cat. I. neuronal mechanisms," *Journal of Neurophysiology*, vol. 60, pp. 1799–1822, 1988.

[59] X. Li, "Combination and generation of parallel feature streams for improved speech recognition," PhD dissertation, Carnegie Mellon University, 2005.

[60] X. Lu, M. Unoki, and S. Nakamura, "Subband temporal modulation spectrum normalization for automatic speech recognition in reverberant envioronments," in *Proceedings of Interspeech*, 2009.

[61] R. F. Lyon, "A computational model of filtering, detection and compression in the cochlea," in *Proceedings of the IEEE International Conference on Acoustics, Speech and Signal Processing*, Paris, May 1982, pp. 1282–1285.

[62] R. F. Lyon, "A computational model of binaural localization and separation," in *Proceedings of the International Conference on Acoustics, Speech and Signal Processing*, 1983, pp. 1148–1151.

[63] R. F. Lyon, "Computational models of neural auditory processing," in *Proceedings of the IEEE International Conference on Acoustics, Speech and Signal Processing*, 1984, pp. 36.1.1–36.1.4.

[64] R. F. Lyon, A. G. Katsiamis, and E. M. Drakakis, "History and future of auditory filter models," in *Proceedings of the IEEE International Symposium on Circuits and Systems*, 2010, pp. 3809–3812.

[65] C. Ma, K.-K. J. Kuo, H. Soltau, X. Cui, U. Chaudhari, L. Mangu, and C.-H. Lee, "A comparative study on system combination schemes for LVCSR," in *Proceedings of the IEEE ICASSP*, 2010, pp. 4394–4397.

[66] J. Makhoul and L. Cosell, "LPCW: An LPC vocoder with linear predictive spectral warping," in *Proceedings of the IEEE International Conference on Acoustics, Speech and Signal Processing*, 1976, pp. 466–469.

[67] L. Mangu, E. Brill, and A. Stolcke, "Finding consensus in speech recognition; word error minimization and other applications of confusion networks," *Computer Speech & Language*, vol. 14, pp. 373–400, 2000.

[68] R. Meddis, "Simulation of mechanical to neural transduction in the auditory receptor," *Journal of the Acoustical Society of America*, vol. 79, pp. 702–711, 1986.

[69] R. Meddis, "Simulation of auditory-neural transduction: further studies," *Journal of the Acoustical Society of America*, vol. 83, pp. 1056–1063, 1988.

[70] H. Meng and V. W. Zue, "A comparative study of acoustic representations of speech for vowel classification using multi-layer perceptrons," in *Proceedings of the International Conference on Spoken Language Processing*, 1990, pp. 1053–1056.

[71] N. Mesgarani, M. Slaney, and S. A. Shamma, "Discrimination of speech from nonspeech based on multiscale spectro-temporal modulations," *IEEE Transactions on Audio, Speech and Language Proc.*, vol. 14, pp. 920–929, 2006.

[72] B. T. Meyer and B. Kollmeier, "Robustness of spectrotemporal features against intrinsic and extrinsic variations in automatic speech recognition," *Speech Communication*, vol. 53, pp. 753–767, 2011.

[73] B. T. Meyer, T. Jürgens, T. Wesker, T. Brand, and B. Kollmeier, "Human phoneme recognition as a function of speech-intrinsic variabilities," *Journal of the Acoustical Society of America*, vol. 128, pp. 3126–3141, 2010.

[74] B. T. Meyer, S. V. Ravuri, M. R. Schaedler, and N. Morgan, "Comparing different flavors of spectro-temporal features for asr," in *Proceedings of Interspeech*, 2011.

[75] N. Mirghafori, "A multi-band approach to automatic speech recognition," PhD dissertation, University of California, Berkeley, Berkeley CA, January 1999.

[76] A. R. Møller, "Coding of amplitude and frequency modulated sounds in the cochlear nucleus," *Acustica*, vol. 31, pp. 292–299, 1974.

[77] B. C. J. Moore, *An Introduction to the Psychology of Hearing*, 5th ed. London: Academic Press, 2003.

[78] B. C. J. Moore, "Frequency analysis and masking," in *Hearing*, 2nd ed., series on Handbook of Perception and Cognition, B. C. J. Moore, Ed. San Diego, CA, USA: Academic Press, 1995, Ch. 5, pp. 161–205.

[79] B. C. J. Moore and B. R. Glasberg, "Suggested formulae for calculating auditory-filter bandwidths and excitation patterns," *Journal of the Acoustical Society of America*, vol. 74, pp. 750–731, 1983.

[80] P. J. Moreno, B. Raj, and R. M. Stern, "A vector Taylor series approach for environment-independent speech recognition," in *Proceedings of the IEEE International Conference on Acoustics, Speech and Signal Processing*, May 1996, pp. 733–736.

[81] N. Morgan, "Deep and wide: Multiple layers in automatic speech recognition," *IEEE Transactions on Audio, Speech and Language Processing*, vol. 20, pp. 7–13, 2012.

[82] N. Morgan, Q. Zhu, A. Stolcke, K. Sonmez, S. Sivadas, T. Shinozaki, M. Ostendorf, P. Jain, H. Hermansky, D. Ellis, G. Doddington, B. Chen, O. Cetin, H. Bourlard, and M. Athineos, "Pushing the envelope—aside," *IEEE Signal Processing Magazine*, vol. 22, pp. 81–88, 2005.

[83] Y. Ohshima, "Environmental robustness in speech recognition using physiological-motivted signal processing," PhD dissertation, Carnegie Mellon University, December 1993.

[84] Y. Ohshima and R. M. Stern, "Environmental robustness in automatic speech recognition using physiologically-motivated signal processing," in *Proceedings of the International Conference of Spoken Language Processing*, 1994.

[85] D. O'Shaughnessy, *Speech Communication: Human and Machine*, 2nd ed. Hoboken, NJ, USA: Wiley-IEEE Press, 2000.

[86] A. Palmer and S. Shamma, "Physiological representations of speech," in *Speech Processing in the Auditory System*, series on Springer Handbook of Auditory Research, S. Greenberg, A. N. Popper, and R. R. Fay, Eds. New York, NY, USA: Springer-Verlag, 2004, Ch. 4.

[87] R. D. Patterson and I. Nimmo-Smith, "Off-frequency listening and auditory filter asymmetry," *Journal of the Acoustical Society of America*, vol. 67, no. 1, pp. 229–245, 1980.

[88] J. O. Pickles, *An Introduction to the Physiology of Hearing*, 3rd ed. Bingley, UK: Academic Press, 2008.

[89] J. O. Pickles, "The neurophysiological basis of frequency selectivity," in *Frequency Selectivity in Hearing*, B. C. J. Moore, Ed. New York: Plenum, 1986.

[90] L. R. Rabiner and R. W. Schafer, *Theory and Applications of Digital Speech Processing*. Englewood Cliffs, NJ, USA: Prentice-Hall, 2010.

[91] S. Ravuri, "On the use of spectro-temporal features in noise-additive speech," Master's thesis, University of California, Berkeley, 2011.

[92] S. Ravuri and N. Morgan, "Easy does it: Many-stream ASR without fine tuning streams," in *Proceedings of the IEEE International Conference on Acoustical, Speech and Signal Processing*, 2012.

[93] J. E. Rose, N. B. Gross, C. D. Geisler, and J. E. Hind, "Some neural mechanisms in the inferior colliculus of the cat which may be relevant to localization of a sound source," *Journal of Neurophysiology*, vol. 29, pp. 288–314, 1966.

[94] J. E. Rose, J. E. Hind, D. J. Anderson, and J. F. Brugge, "Some effects of stimulus intensity on response of auditory nerve fibers in the squirrel monkey," *Journal of Neurophysiology*, vol. 34, pp. 685–699, 1971.

[95] M. B. Sachs and N. Y.-S. Kiang, "Two-tone inhibition in auditory-nerve fibers," *Journal of the Acoustical Society of America*, vol. 43, pp. 1120–1128, 1968.

[96] M. B. Sachs and E. D. Young, "Encoding of steady-state vowels in the auditory nerve: Representation in terms of discharge rate," *Journal of the Acoustical Society of America*, vol. 55, pp. 470–479, 1979.

[97] M. R. Schroeder, "Recognition of complex acoustic signals," in *Life Sciences Research Report 5*, T. H. Bullock, Ed. Berlin: Abakon Verlag, 1977.

[98] M. R. Schroeder and J. L. Hall, "A model for mechanical to neural transwduction in the auditory receptor," *Journal of the Acoustical Society of America*, vol. 55, pp. 1055–1060, 1974.

[99] S. Seneff, "A joint synchrony/mean-rate model of auditory speech processing," *Journal of Phonetics*, vol. 15, pp. 55–76, 1988.

[100] S. A. Shamma, "The acoustic features of speech sounds in a model of auditory processing: Vowels and voiceless fricatives," *Journal of Phonetics*, vol. 16, pp. 77–91, 1988.

[101] R. Singh, M. L. Seltzer, B. Raj, and R. M. Stern, "Speech in noisy environments: robust automatic segmentation, feature extraction, and hypothesis combination," in *Proceedings of the IEEE International Conference on Acoustics, Speech and Signal Processing*, 2001, pp. 273–276.

[102] M. Slaney, *Auditory Toolbox (V.2)*, 1998, http://www.slaney.org/malcolm/pubs.html. [Online]. Available at http://www.slaney.org/malcolm/pubs.html

[103] R. M. Stern and C. Trahiotis, "Models of binaural interaction," in *Hearing*, 2nd ed., series on Handbook of Perception and Cognition, B. C. J. Moore, Ed. New York: Academic, 1995, Ch. 10, pp. 347–386.

[104] R. M. Stern, A. Acero, F.-H. Liu, and Y. Ohshima, "Signal processing for robust speech recognition," in *Automatic Speech and Speaker Recognition*, C.-H. Lee, F. K. Soong and K. K. Paliwal, Eds. Norwell, MA, USA: Kluwer Academic Publishers, 1996, Ch. 14, pp. 351–378.

[105] R. M. Stern, D. Wang, and G. J. Brown, "Binaural sound localization," in *Computational Auditory Scene Analysis*, D. Wang and G. J. Brown, Eds. Hooken, NJ, USA: Wiley-IEEE Press, 2006, Ch. 5.

[106] S. S. Stevens, "On the psychophysical law," *Psychological Review*, vol. 64, pp. 153–181, 1957.

[107] S. S. Stevens, J. Volkman, and E. Newman, "A scale for the measurement of the psychological magnitude pitch," *Journal of the Acoustical Society of America*, vol. 8, pp. 185–190, March 1937.

[108] Strutt JW, Third Baron of Rayleigh, "On our perception of sound direction," *Philosophical Magazine*, vol. 13, pp. 214–232, 1907.

[109] J. Tchorz and B. Kollmeier, "A model of auditory perception as front end for automatic speech recognition," *Journal of the Acoustical Society of America*, vol. 106, no. 4, pp. 2040–2060, October 1999.

[110] H. Traunmüller, "Analytical expressions for the tonotopic sensory scale," *Journal of the Acoustical Society of America*, vol. 88, pp. 97–100, 1990.

[111] N. Viemeister, "Temporal modulation transfer function based on modulation thresholds," *Journal of the Acoustical Society of America*, vol. 66, pp. 1364–1380, 1979.

[112] G. von Békésy, "On the resonance curve and the decay period at various points on the cochlear partition." *Journal of the Acoustical Society of America*, 21, pp. 245–254, 1949.

[113] H. W. Wallach, E. B. Newman, and M. R. Rosenzweig, "The precedence effect in sound localization," *American Journal of Psychology*, vol. 62, pp. 315–337, 1949.

[114] K. Wang and S. A. Shamma, "Self-normalization and noise-robustness in early auditory representations," *IEEE Transactions on Speech and Audio Processing*, vol. 2, pp. 421–435, 1994.

[115] E. H. Weber, *De pulsu, resorpitione, auditu et tactu: Annotations anatomicae et physiologicae*. Leipzig: Koehlor, 1834.

[116] T. C. T. Yin and J. C. K. Chan, "Interaural time sensitivity in medial superior olive of cat," *Journal of Neurophysiology*, vol. 64, pp. 465–474, 1990.

[117] W. A. Yost, *Fundamentals of Hearing: An Introduction*, 5th ed. Bingley, UK: Emerald Group Publishing, 2006.

[118] W. A. Yost and M. J. Moore, "Temporal changes in a complex spectral profile," *Journal of the Acoustical Society of America*, vol. 81, pp. 1896–1905, 1987.

[119] E. D. Young and M. B. Sachs, "Representation of steady-state vowels in the emporal aspects of the discharge patterns of populations of auditory-nerve fibers," *Journal of the Acoustical Society of America*, vol. 66, pp. 1381–1403, 1979.

[120] X. Zhang, M. G. Heinz, I. C. Bruce, and L. H. Carney, "A phenomenologial model for the response of auditory-nerve fibers: I. nonlinear tuning with compression and suppresion," *Journal of the Acoustical Society of America*, vol. 109, pp. 648–670, 2001.

[121] S. Y. Zhao, S. Ravuri, and N. Morgan, "Multi-stream to many-stream: Using spectgro-temporal features in ASR," in *Proceedings of Interspeech*, 2009.

[122] M. S. A. Zilany, I. C. Bruce, P. C. Nelson, and L. H. Carney, "A phenomonological model of the synapse between the inner hair cell and auditory nerve: mlong-term adaptation with power-law dynamics," *Journal of the Acoustical Society of America*, vol. 126, pp. 2390–2412, 2009.

[123] E. Zwicker, "Subdivision of the audible frequency range into critical bands (frequenzgruppen)," *Journal of the Acoustical Society of America*, vol. 33, p. 248, February 1961.

[124] E. Zwicker and E. Terhardt, "Analytical expressions for critical-band rate and critical bandwidth as a function of frequency," *Journal of the Acoustical Society of America*, vol. 68, pp. 1523–1525, November 1980.

9

Feature Compensation

Jasha Droppo
Microsoft Research, USA

9.1 Life in an Ideal World

People convey linguistic messages by generating acoustic speech signals. In an ideal world, we could record that signal and derive acoustic features that contain all of the necessary information to achieve perfect recognition accuracy, and nothing else.

In our world, the acoustic features are computed from acoustic signals recorded by a microphone, and the information we need is obscured by noise and other irrelevant variabilities. To make matters worse, these features often suffer from linear and nonlinear channel effects, reverberation, and a significant amount of additive noise. Even in the absence of these distortions, the speech portion of the signal itself contains more information than what was said, including how it was said and who said it.

Figure 9.1 shows the connection between the ideal speech features that we want, the clean speech features that we may be able to get by carefully controlling the environmental conditions at the time of capture, and the noisy speech that we must often tolerate.

This chapter focuses on feature-enhancement techniques, which strive to remove extraneous information and distortion from a sequence of speech-recognition features, while retaining information about what was said.

9.1.1 Noise Robustness Tasks

When building and testing robust automatic speech-recognition systems, their relative performance often changes with several factors, including the degree of mismatch between the training and testing data, the size of the vocabulary, and the complexity of the acoustic model. Therefore, it is useful to control for these variables by comparing performance on a standard set of data and tasks.

The European Telecommunications Standards Institute's technical committee for Speech Transmission Planning and Quality of Service (ETSI STQ) have generated several such data

Techniques for Noise Robustness in Automatic Speech Recognition, First Edition.
Edited by Tuomas Virtanen, Rita Singh, and Bhiksha Raj.

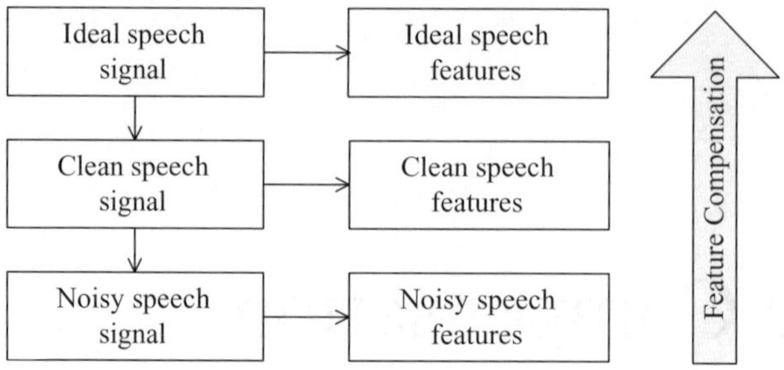

Figure 9.1 The goal of feature compensation is to recover more ideal speech features from observed noisy speech features.

sets. Their AURORA digital speech-recognition working group released a series of tasks for this purpose, system evaluation. Each contains the necessary data and specifications for running an experiment, including data and recipes for building acoustic and language models and several relevant defined training and testing scenarios.

In this chapter, the relative merits of the discussed techniques are demonstrated on two different standard noise robust speech-recognition tasks, summarized in Table 9.1. Both are similar in that they were created by artificially mixing clean speech from a base corpus with noise of various types. The Aurora 2 corpus is based on the TI-Digits corpus and is a small vocabulary system, consisting only of the numbers zero through nine and the word "oh." The Aurora 4 corpus is based on the *Wall Street Journal* corpus, which has a much larger vocabulary.

For both tasks, we follow the standard recipes for building acoustic models. In the case of Aurora 2, this means maximum-likelihood whole-word digit models with 16 emitting states and 20 mixture components per state. For Aurora 4, we build a maximum-likelihood (ML) triphone acoustic model with approximately 1500 shared states and 16 mixture components per state.

9.1.2 Probabilistic Feature Enhancement

The feature compensation methods discussed in this chapter share a common underlying probabilistic framework.

According to Figure 9.1, the clean and corrupted speech signals are both generated from the ideal speech signal. As a result, the features they generate will be correlated, allowing us to build a probabilistic model that relates the two. If the clean speech signal generates features $\mathbf{x}$, and the noisy speech signal generates $\mathbf{y}$, then their joint distribution $p(\mathbf{x}, \mathbf{y})$ contains all of the information we need.

Table 9.1 Standard tasks used in this chapter.

Task name	Base corpus	Vocabulary
Aurora 2	TI-Digits	11
Aurora 4	*Wall Street Journal*	5000

With this joint distribution, we can estimate the speech features in one domain from speech features in another. One popular estimate is the minimum mean squared error (MMSE) estimate of the clean speech feature $\mathbf{x}$ given the noisy speech feature $\mathbf{y}$. It produces an estimate that minimizes the expected squared error to the true value of $\mathbf{x}$, and is well known to be the conditional expectation of $\mathbf{x}$ given $\mathbf{y}$:

$$\hat{\mathbf{x}}_{\mathrm{MMSE}} = E\left[\mathbf{x}|\mathbf{y}\right] = \int \mathrm{d}\,\mathbf{x} p(\mathbf{x}|\mathbf{y})\mathbf{x}.$$

Or, we can estimate the most likely value of $\mathbf{x}$ as the ML estimate:

$$\hat{\mathbf{x}}_{\mathrm{ML}} = \max_{\mathbf{x}} p(\mathbf{x}|\mathbf{y}).$$

Both the MMSE and ML estimates contain the conditional probability $p(\mathbf{x}|\mathbf{y})$. The max or $E[\cdot]$ operators use this conditional probability to transform from one feature representation to another.

To use this form of feature compensation, we need to choose a suitable form for the joint distribution of $\mathbf{x}$ and $\mathbf{y}$, derive the mapping function from the joint distribution, and estimate the parameters for that form from suitable training data.

9.1.3 Gaussian Mixture Models

The most common model for the joint distribution of clean and noisy speech features is a Gaussian mixture model (GMM). The variable k indexes the individual mixture components of the model, each one being a joint Gaussian (normal) distribution over $\mathbf{x}$ and $\mathbf{y}$:

$$p(\mathbf{x},\mathbf{y}) = \sum_k p(k)p(\mathbf{x},\mathbf{y}|k).$$

Each Gaussian component can be written in form where $\mathbf{x}$ and $\mathbf{y}$ are concatenated to form a single vector space. In this form, the component conditional means $\boldsymbol{\mu}_{\mathbf{x}|k}$ and $\boldsymbol{\mu}_{\mathbf{y}|k}$ as well as the variance and covariance matrices $\boldsymbol{\Sigma}_{\mathbf{xx}|k}$, $\boldsymbol{\Sigma}_{\mathbf{yy}|k}$ and $\boldsymbol{\Sigma}_{\mathbf{xy}|k}$ are readily apparent:

$$p(\mathbf{x},\mathbf{y}|k) = \mathcal{N}\left(\begin{bmatrix}\mathbf{x}\\ \mathbf{y}\end{bmatrix}; \begin{bmatrix}\boldsymbol{\mu}_{\mathbf{x}|k}\\ \boldsymbol{\mu}_{\mathbf{y}|k}\end{bmatrix}, \begin{bmatrix}\boldsymbol{\Sigma}_{\mathbf{xx}|k} & \boldsymbol{\Sigma}_{\mathbf{xy}|k}\\ \boldsymbol{\Sigma}^T_{\mathbf{xy}|k} & \boldsymbol{\Sigma}_{\mathbf{yy}|k}\end{bmatrix}\right).$$

Given this parametrization, the conditional probability distribution function of $\mathbf{x}$ given $\mathbf{y}$ and k is also Gaussian:

$$\begin{aligned} p(\mathbf{x}|\mathbf{y},k) &= \mathcal{N}\left(\mathbf{x}; \boldsymbol{\mu}_{\mathbf{x}|k} + \boldsymbol{\Sigma}_{\mathbf{xy}|k}\boldsymbol{\Sigma}^{-1}_{\mathbf{xx}|k}(\mathbf{y}-\boldsymbol{\mu}_{\mathbf{y}|k}), \boldsymbol{\Sigma}_{\mathbf{xx}|k} - (\boldsymbol{\Sigma}^T_{\mathbf{xy}|k})^{-1}\boldsymbol{\Sigma}_{\mathbf{yy}|k}\boldsymbol{\Sigma}^{-1}_{\mathbf{xy}|k}\right)\\ &= \mathcal{N}\left(\mathbf{x}; \boldsymbol{\mu}_{\mathbf{x}|\mathbf{y},k}; \boldsymbol{\Sigma}_{\mathbf{x}|\mathbf{y},k}\right). \end{aligned}$$

Although the formulae for the parameters of $p(\mathbf{x}|\mathbf{y})$ look complicated, they are nothing more than an affine transformation of $\mathbf{y}$ and a residual estimation error $\boldsymbol{\Sigma}_{\mathbf{x}|\mathbf{y},k}$:

$$\boldsymbol{\mu}_{\mathbf{x}|\mathbf{y},k} = \mathbf{A}_k\mathbf{y} + \mathbf{b}_k, \text{ where} \tag{9.1}$$

$$\mathbf{A} = \boldsymbol{\Sigma}_{\mathbf{xy}|k}\boldsymbol{\Sigma}^{-1}_{\mathbf{xx}|k} \text{ and} \tag{9.2}$$

$$\mathbf{b} = \boldsymbol{\mu}_{\mathbf{x}|k} - \boldsymbol{\Sigma}_{\mathbf{xy}|k}\boldsymbol{\Sigma}_{\mathbf{xx}|k}^{-1}\boldsymbol{\mu}_{\mathbf{y}|k}. \tag{9.3}$$

As a result, it is sufficient to estimate the parameters $\mathbf{A}$, $\mathbf{b}$, and $\boldsymbol{\Sigma}$ for each Gaussian component k the model. The final form is

$$p(\mathbf{x}, \mathbf{y}) = \sum_k p(\mathbf{x}|\mathbf{y}, k)p(\mathbf{y}|k)p(k), \tag{9.4}$$

$$p(\mathbf{x}|\mathbf{y}, k) = \mathcal{N}(\mathbf{x}; \mathbf{A}_k\mathbf{y} + \mathbf{b}_k, \boldsymbol{\Sigma}_{\mathbf{x}|\mathbf{y},k}), \tag{9.5}$$

$$p(\mathbf{y}|k) = \mathcal{N}(\mathbf{y}; \boldsymbol{\mu}_{\mathbf{y}|k}, \boldsymbol{\Sigma}_{\mathbf{y}|k}). \tag{9.6}$$

The minimum mean squared estimate of the clean speech feature $\mathbf{x}$ given the noisy speech feature $\mathbf{y}$ is equal to the expected value of $\mathbf{x}$ given the observed noisy speech $\mathbf{y}$:

$$\begin{aligned}
\hat{\mathbf{x}} &= E\left[\mathbf{x}|\mathbf{y}\right] \\
&= \int p(\mathbf{x}|\mathbf{y})\mathbf{x}\, d\mathbf{x} \\
&= \int \sum_k p(\mathbf{x}|\mathbf{y}, k)p(k|\mathbf{y})\mathbf{x}\, d\mathbf{x} \\
&= \sum_k p(k|\mathbf{y})E\left[\mathbf{x}|\mathbf{y}, k\right] \\
&= \sum_k p(k|\mathbf{y})\left(\mathbf{A}_k\mathbf{y} + \mathbf{b}_k\right).
\end{aligned}$$

This general form is nice, because conceptually k is breaking the feature space into K overlapping partitions. Within each partition, the mapping is reduced to a simple affine transformation, which is then interpolated by $p(k|\mathbf{y})$. Sometimes, such as when using feature-based uncertainty decoding (Chapter 17), it is useful to know the variance of the estimate $\hat{\mathbf{x}}$, which is given by the second central moment of $p(\mathbf{x}|\mathbf{y})$:

$$\begin{aligned}
\hat{\boldsymbol{\Sigma}}_{\mathbf{x}} &= E\left[(\mathbf{x} - \hat{\mathbf{x}})(\mathbf{x} - \hat{\mathbf{x}})^T|\mathbf{y}\right] \\
&= \int p(\mathbf{x}|\mathbf{y})(\mathbf{x} - \hat{\mathbf{x}})(\mathbf{x} - \hat{\mathbf{x}})^T\, d\mathbf{x} \\
&= \int \sum_k p(\mathbf{x}|\mathbf{y}, k)p(k|\mathbf{y})(\mathbf{x} - \hat{\mathbf{x}})(\mathbf{x} - \hat{\mathbf{x}})^T\, d\mathbf{x} \\
&= -\hat{\mathbf{x}}\hat{\mathbf{x}}^T + \sum_k p(k|\mathbf{y})\left(\boldsymbol{\Sigma}_{\mathbf{x}|\mathbf{y},k} + \boldsymbol{\mu}_{\mathbf{x}|\mathbf{y},k}\boldsymbol{\mu}_{\mathbf{x}|\mathbf{y},k}^T\right).
\end{aligned}$$

The following sections describe different feature compensation methods that vary depending on the quality and quantity of data available to estimate the transform parameters.

9.2 MMSE-SPLICE

The SPLICE technique was first introduced as a method for overcoming noisy speech in [10]. SPLICE is an acronym of "Stereo piecewise linear compensation for environment." Its purpose is to estimate clean speech features $\mathbf{x}$ from observed speech features $\mathbf{y}$ taken from a noisy environment.[1] It is similar in spirit to the FCDCN of [1] and the RATZ of [18], with the notable difference that the previous methods learn clean speech GMMs, and SPLICE uses noisy speech GMMs.

The MMSE-SPLICE transform $\hat{\mathbf{x}} = f(\mathbf{y};\boldsymbol{\theta})$ is defined as the minimum mean squared estimate of $\mathbf{x}$, given $\mathbf{y}$ and the model parameters $\boldsymbol{\theta}$. The parameters of the transform are learned from pairs of observations $(\mathbf{x},\mathbf{y})$ that contain the exact same speech in both clean and noisy conditions. This data might be generated by a close-talking microphone and far-field microphone in a noisy room, or by artificially mixing noise into an existing clean utterance.

Using the GMM derivation above, the parameters to learn are those of the GMM of noisy speech $\left\{\boldsymbol{\mu}_{\mathbf{y}|k}, \boldsymbol{\Sigma}_{\mathbf{y}|k}\right\}$ and those of the conditional model's affine transformations $\{\mathbf{A}_k, \mathbf{b}_k\}$.

The parameters of the GMM for noisy speech can be estimated with standard expectation-maximization techniques, and then held fixed for the remainder of training.

Depending on the amount of training data available, the model can be scaled to have an appropriate number of parameters to train. Plentiful data one can train a large model with component-specific affine transformations:

$$\mathbf{x} = \mathbf{y} + \sum_i p(i|\mathbf{y})(\mathbf{A}_i\mathbf{y} + \mathbf{b}_i).$$

With a moderate amount of data, a large number of components can still be trained by assuming each $\mathbf{A}_i$ is the identity matrix. As a result, the transform is a component-specific offset:

$$\mathbf{x} = \mathbf{y} + \sum_i p(i|\mathbf{y})\mathbf{b}_i.$$

With a small amount of training data, the model parameters can be shared across all components of the GMM. The resulting SPLICE transformation is mathematically equivalent to the affine transformation used in C-MLLR, ([6], also covered in Chapter 11):

$$\mathbf{x} = \mathbf{A}\mathbf{y} + \mathbf{b}.$$

9.2.1 Parameter Estimation

Holding the parameters of $p(\mathbf{y})$ constant, we find estimates of the $\mathbf{b}_k$ and $\mathbf{A}_k$ parameters. Because $\boldsymbol{\Sigma}_{\mathbf{x}|\mathbf{y},k}$ does not appear in the transformation, it is not covered here. But, it may be useful when doing uncertainty decoding as in [9], which is covered more deeply in Section 17.4.1.

The squared error between $\hat{\mathbf{x}}$ and $\mathbf{x}$ in the entire corpus is given by the following. Training is accomplished by finding model parameters that minimize this squared error

$$\text{Error} = \sum_t \left\| \sum_i p(i|\mathbf{y}_t)(\mathbf{A}_i\mathbf{y}_t + \mathbf{b}_i) - \mathbf{x}_t \right\|^2, \tag{9.7}$$

where t is the index of the pair of clean and noisy speech observations in the training set.

[1] In this section, $\mathbf{y}$ is a traditional feature vector based on static cepstra and its derivatives. But, it is possible to expand $\mathbf{y}$ to include more context information, finer frequency detail, or other nontraditional features.

For the case where the SPLICE parametrization is a set of offsets $\mathbf{b}_i$, the rotations $\mathbf{A}_i$ are fixed, and the optimal set of offsets is easy to find. After making the substitution $\mathbf{A}_i = \mathbf{I}$, the squared error is a quadratic function of $\mathbf{b}_i$, and optimal values can be found by setting the derivative of the error function to zero, and solving for the set of $\mathbf{b}_i$:

$$0 = \frac{\partial}{\partial b_j} \text{Error}|_{\mathbf{A}_i=\mathbf{I}},$$

$$0 = 2\sum_t p(j|\mathbf{y}_t)\left(\mathbf{y}_t - \mathbf{x}_t + \sum_i p(i|\mathbf{y}_t)\mathbf{b}_i\right),$$

$$\sum_i \mathbf{b}_i \sum_t p(i|\mathbf{y}_t)p(j|\mathbf{y}_t) = \sum_t p(j|\mathbf{y}_t)\left(\mathbf{x}_t - \mathbf{y}_t\right).$$

This yields a simultaneous set of K equations, one for each possible value of j. The structure of this set becomes clear when written as a single equation $\mathbf{BU} = \mathbf{V}$, where the matrices are formed by concatenating the individual terms above. The MMSE solution for $\mathbf{B}$ is $\mathbf{B} = \mathbf{VU}^{-1}$.

The columns of the $\mathbf{B}$ matrix are the offsets $\mathbf{b}_j$ that we hope to solve for

$$\mathbf{B} = \begin{bmatrix} \mathbf{b}_1, \mathbf{b}_2, \ldots, \mathbf{b}_K \end{bmatrix}.$$

The $\mathbf{U}$ matrix is formed by taking the sum of the outer product of posterior component likelihoods for each frame of training data:

$$u_{ij} = \sum_t p(i|\mathbf{y}_t)p(j|\mathbf{y}_t).$$

Because this sum contains many more terms than dimensions, the matrix $\mathbf{U}$ is generally full rank. If it is not, then some mixture components in the GMM are highly correlated and should be discarded.

The columns of the $\mathbf{V}$ matrix are component-weighted differences between $\mathbf{x}$ and $\mathbf{y}$:

$$\mathbf{v}_j = \sum_t p(j|\mathbf{y}_t)\left(\mathbf{x}_t - \mathbf{y}_t\right).$$

For the more general case, where both $\mathbf{A}_j$ and $\mathbf{b}_j$ are estimated, Equation (9.7) can be simplified by representing the affine transformation more compactly as

$$\begin{aligned} \mathbf{A}_i\mathbf{y}_t + \mathbf{b}_i &= \bar{\mathbf{A}}_i\bar{\mathbf{y}}_t, \text{ where} \\ \bar{\mathbf{A}}_i &= \begin{bmatrix} \mathbf{A}_i & \mathbf{b}_i \end{bmatrix} \text{ and} \\ \bar{\mathbf{y}}_t &= \begin{bmatrix} \mathbf{y}_t \\ 1 \end{bmatrix}. \end{aligned}$$

Because the error function above is now a quadratic function of $\bar{\mathbf{A}}_i$, an optimal value can be found by setting the derivative of the error function with respect to $\bar{\mathbf{A}}_i$ to zero, and solving for $\bar{\mathbf{A}}_i$:

$$0 = \frac{\partial}{\partial \bar{\mathbf{A}}_j}\text{Error},$$

$$0 = \frac{\partial}{\partial \bar{\mathbf{A}}_j}\sum_t \left\| \mathbf{x}_t - \sum_i p(i|\mathbf{y}_t)\bar{\mathbf{A}}_i\bar{\mathbf{y}}_t \right\|^2,$$

$$0 = \sum_t \frac{\partial}{\partial \bar{\mathbf{A}}_j}\left(\mathbf{x}_t - \sum_i p(i|\mathbf{y}_t)\bar{\mathbf{A}}_i\bar{\mathbf{y}}_t\right)^T \left(\mathbf{x}_t - \sum_i p(i|\mathbf{y}_t)\bar{\mathbf{A}}_i\bar{\mathbf{y}}_t\right),$$

$$0 = 2\sum_t \left(\sum_i p(i|\mathbf{y}_t)\bar{\mathbf{A}}_i\bar{\mathbf{y}}_t - \mathbf{x}_t\right) p(j|\mathbf{y}_t)\left(\bar{\mathbf{y}}_t\right)^T.$$

As before, we collect the terms dependent on $\bar{\mathbf{A}}_i$ to one side of the equation and simplify:

$$\sum_t \left(\sum_i p(i|\mathbf{y}_t)\bar{\mathbf{A}}_i\bar{\mathbf{y}}_t\right) p(j|\mathbf{y}_t)\left(\bar{\mathbf{y}}_t\right)^T = \sum_t \mathbf{x}_t p(j|\mathbf{y}_t)\left(\bar{\mathbf{y}}_t\right)^T,$$

$$\sum_i \bar{\mathbf{A}}_i \sum_t p(i|\mathbf{y}_t)p(j|\mathbf{y}_t)\bar{\mathbf{y}}_t\left(\bar{\mathbf{y}}_t\right)^T = \sum_t p(j|\mathbf{y}_t)\mathbf{x}_t\left(\bar{\mathbf{y}}_t\right)^T. \tag{9.8}$$

Because this equation holds regardless of the value chosen for *j*, we again have a system of linear equations that can be easily solved for an optimal set of $\bar{\mathbf{A}}_j$. The easiest way to see the solution is to define two new variables. The first, $\mathbf{q}_t$, is a super-vector composed by concatenating a number of posterior-scaled values of $\bar{\mathbf{y}}_t$. The second, $\bar{\mathbf{A}}$, is a similar concatenation of the $\bar{\mathbf{A}}_i$ matrices:

$$\mathbf{q}_t = \left[p(i=1|\mathbf{y}_t)\bar{\mathbf{y}}_t^T, p(i=2|\mathbf{y}_t)\bar{\mathbf{y}}_t^T, \ldots p(i=K|\mathbf{y}_t)\bar{\mathbf{y}}_t^T\right]^T,$$

$$\bar{\mathbf{A}} = \left[\bar{\mathbf{A}}_1, \bar{\mathbf{A}}_2, \ldots \bar{\mathbf{A}}_K\right].$$

With these in hand, Equation (9.8) can be reformulated and solved:

$$\bar{\mathbf{A}}\sum_t \mathbf{q}_t\left(\mathbf{q}_t\right)^T = \sum_t \mathbf{x}_t\left(\mathbf{q}_t\right)^T,$$

$$\bar{\mathbf{A}} = \left(\sum_t \mathbf{x}_t\left(\mathbf{q}_t\right)^T\right)\left(\sum_t \mathbf{q}_t\left(\mathbf{q}_t\right)^T\right)^{-1}.$$

It is possible that the necessary matrix inverse can not be computed because it it not of full rank. This can happen if feature space has correlated dimensions or if some Gaussians in the clean speech model are too similar. In the former case, the offending dimensions should be removed. In the latter, the clean speech model is too complex and should be shrunk to fewer components.

Table 9.2 Word error rate on Aurora 2, demonstrating the benefit of using MMSE-SPLICE.

Training data	Feature enhancement	Set A	Set B	Set C	Average
Clean	None	41.42	47.10	31.73	41.75
Clean	MMSE-SPLICE	11.67	12.25	13.67	12.30
Multistyle	None	8.64	10.23	11.96	9.94
Multistyle MMSE-SPLICE	MMSE-SPLICE	6.66	8.99	7.85	7.83

9.2.2 Results

A 256 mixture component offset-only variant of MMSE-SPLICE was trained and applied to the Aurora 2 task. The offsets were trained by using the 8440 utterances meant for clean acoustic model training and their 8440 counterparts meant for multistyle training as a stereo data set. A 256-component GMM was trained on all of the noisy data, and then the stereo data were used as described above to train the offset vectors.

For both the clean and multistyle training data, SPLICE was applied to only the static cepstral coefficients. When processing clean training data, the standard delta and acceleration coefficients were computed from the enhanced static coefficients. When processing noisy training data, the delta and acceleration coefficients were computed from the noisy static coefficients. We have found this setup to be optimal across many algorithms, and unless stated otherwise, will use this convention for the rest of the chapter.

The word accuracy results for MMSE-SPLICE and Aurora 2 are shown in Table 9.2. For each row in the table, here and in the rest of this chapter, the same feature-enhancement processing was applied to both the training and testing data.

For the "Clean" training condition, when no feature enhancement is employed, the average digit accuracy is 58.25%. MMSE-SPLICE, by design, reduces the mismatch between the noisy evaluation data and the clean training data and increases the average digit accuracy to 87.70%. For the multistyle training condition, the baseline system is much more robust to additive noise and the MMSE-SPLICE is only able to reduce the average digit error rate by about 20%.

The same type of transform was also trained and used in our Aurora 4 system. The noisy speech GMM was trained on the 7138 utterances in the "multinoise" set, to which the "clean" set was added to generate the required stereo training data.

Table 9.3 shows the results of adding a 256 mixture component, offset-only variant of SPLICE to the Aurora 4 task. Because Aurora 4 is significantly different from Aurora 1 in several ways, including vocabulary size, signal-to-noise ratio (SNR), and bandwidth, the relative gains of applying MMSE-SPLICE have changed. MMSE-SPLICE provides about

Table 9.3 Word error rate on Aurora 4, demonstrating the benefit of using MMSE-SPLICE.

Training data	Feature enhancement	Test 1 (clean)	Test 2–7 (noisy)
Clean	None	8.4	33.9
Clean	MMSE-SPLICE	8.3	29.9
Multistyle	None	14.0	19.3
Multistyle MMSE-SPLICE	MMSE-SPLICE	13.4	19.2

a 10% reduction in word error rate for the clean training condition, but much less of an improvement for the multistyle training condition.

One place where the technique does provide gains is in the accuracy of recognizing clean speech based on multistyle training data. It is well known that multistyle training data increase performance on noisy test data, while simultaneously decreasing performance on clean test data. For the results in Table 9.3, this is clearly shown by the word error rate rising from 8.4% to 14.0% on clean test data, while falling from 33.9% to 19.3% on noisy test data. Because the MMSE-SPLICE technique maps the multistyle training data to look more like clean training data, some of this performance degradation can be regained.

9.3 Discriminative SPLICE

Even though MMSE-SPLICE is easy to train and deploy, it may not be the best solution. Its output may be optimal in a MMSE sense, but it might not produce the most accurate composite system. Furthermore, if stereo data are unavailable, it is impossible to train.

Discriminative SPLICE addresses both of those shortcomings. It uses examples of noisy speech, together with a speech-recognizer's acoustic model, to learn SPLICE parameters that map the noisy speech into features that the acoustic model is more likely to recognize correctly.

Figure 9.2 illustrates how this is accomplished. The parameters $\boldsymbol{\theta}$ define the SPLICE transformation, and the parameters λ define the acoustic model. The objective function can come from any of a number of well-studied techniques such as minimum classification error (MCE, as in [15, 16]), maximum mutual information (MMI, as in [2, 19], or minimum phone error (MPE, as in [20]). Discriminative splice modifies the SPLICE parameters $\boldsymbol{\theta}$ in such a way as to improve the chosen discriminative objective function.

No explicit constraints are placed on the intermediate feature representation $\mathbf{x}$, other than it comes from a feature space that improves our objective function. This gives the system more freedom than existing methods that define $\mathbf{x}$ as clean speech (e.g. [10]) or phone posteriors (e.g. [13]).

This style of training can produce superior results to the two channel MMSE approach, and can even increase accuracy when the data are relatively noise-free. One disadvantage is that

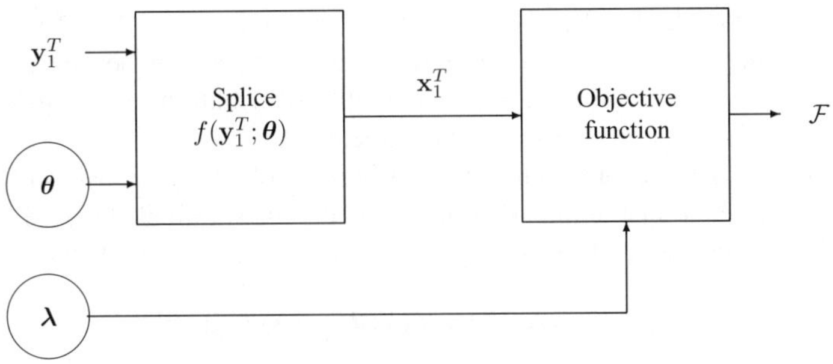

Figure 9.2 The SPLICE transform exists between a sequence of noisy speech features $\mathbf{y}_1^T$ and the objective function calculation. The objective function may be improved through modification of the SPLICE parameters $\boldsymbol{\theta}$ or the acoustic model parameters $\boldsymbol{\lambda}$.

it is easy to overtrain the parameters, leading to an increase in error rate on held out testing data. To mitigate this problem, the standard tools of regularization and careful partitioning of development and test data should be employed.

The recommended form of regularization is to bias the model toward a benign set of SPLICE parameters. Toward that end, it is usually sufficient to add penalty terms to the chosen objective function that increase as the SPLICE transform diverges from the identity transform. Two reasonable penalty terms are as follows, where $||\cdot||_2$ is the L_2 matrix norm, and the scaling constants ϵ_b and ϵ_A should be tuned on the development set:

$$\text{Bias penalty} = \epsilon_b \sum_i \mathbf{b}_i^T \mathbf{b}_i,$$

$$\text{Rotation penalty} = \epsilon_A \sum_i ||\mathbf{A}_i - \mathbf{I}||_2.$$

For the results presented in this chapter, the SPLICE models were small enough that no regularization was necessary.

9.3.1 The MMI Objective Function

This chapter demonstrates discriminative SPLICE using a MMI objective function. MMI maximizes the mutual information between the acoustics and the class (model) labels corresponding to the reference transcriptions. It is defined as the sum of the log conditional probabilities for all correct transcriptions w_r of utterance number r, given the corresponding noisy training data $\mathcal{Y}_r$. Note that, unlike MMSE-SPLICE, there is no need for corresponding clean training data:

$$\mathcal{F} = \sum_{r=0}^{R-1} \mathcal{F}_r = \sum_{r=0}^{R-1} \ln p(w_r|\mathcal{Y}_r). \tag{9.9}$$

This function favors increasing likelihood assigned by the model to the correct words given the acoustics and model parameters. That is, the model parameters are changed so that it is more likely to reproduce the correct transcriptions.

To derive $\ln p(w_r|\mathcal{Y}_r)$, both the front-end and back-end halves of the acoustic processing need to be considered. The front-end feature transformation $\mathcal{X}_r = f(\mathcal{Y}_r; \boldsymbol{\theta})$ is parameterized by $\boldsymbol{\theta}$ and converts the rth input sequence $\mathcal{Y}_r$ into the feature vector sequence $\mathcal{X}_r$. The back-end acoustic score $p_{\mathcal{X}}(\mathcal{X}_r, w; \boldsymbol{\lambda})$ defines a joint probability distribution over transformed feature sequences $\mathcal{X}_r$ and transcriptions w using the parameters $\boldsymbol{\lambda}$.

Assuming the front-end feature transform is differentiable, its composition with the back-end acoustic score is a simple application of the change of variables theorem. If $J_f(\mathcal{Y}; \boldsymbol{\theta})$ is the Jacobian determinant of the transformation $f(\mathcal{Y}; \boldsymbol{\theta})$ evaluated at $\mathcal{Y}$, then

$$p_{\mathcal{Y}}(\mathcal{Y}, w; \boldsymbol{\theta}, \boldsymbol{\lambda}) = p_{\mathcal{X}}(f(\mathcal{Y}; \boldsymbol{\theta}), w; \boldsymbol{\lambda}) J_f(\mathcal{Y}; \boldsymbol{\theta}). \tag{9.10}$$

Then, we use Bayes' rule to derive the desired conditional log probability:

$$\ln p(w_r|\mathcal{Y}_r) = \ln \frac{p_{\mathcal{Y}}(\mathcal{Y}_r, w_r)}{p_{\mathcal{Y}}(\mathcal{Y}_r)} = \ln \frac{p_{\mathcal{Y}}(\mathcal{Y}_r, w_r)}{\sum_w p_{\mathcal{Y}}(\mathcal{Y}_r, w)}. \tag{9.11}$$

The utterance-dependant portion of the objective function is given by combining Equations (9.10) and (9.11):

$$\mathcal{F}_r = \ln \frac{p_{\mathcal{X}}(f(\mathcal{Y}_r;\boldsymbol{\theta}), w_r; \boldsymbol{\lambda}) J_f(\mathcal{Y}_r)}{\sum_w p_{\mathcal{X}}(f(\mathcal{Y}_r;\boldsymbol{\theta}), w; \boldsymbol{\lambda}) J_f(\mathcal{Y}_r)}. \tag{9.12}$$

When the Jacobian determinant is nonzero, it disappears entirely from Equation (9.12). For the remainder of this chapter, we assume that to be the case.

Since exact computation of the denominator of Equation (9.12) would be computationally expensive, the probabilities $p_{\mathcal{X}}(\mathcal{X}_r, w; \boldsymbol{\lambda})$ are approximated on word lattices[2] generated by the baseline ML acoustic model. The numerator is calculated over the best path that corresponds with the correct transcription, and the denominator is calculated over all paths in the lattice.

As is commonly done in lattice-based MMI estimation, the objective function should also be modified to include posterior flattening [20], the time marks in the lattice should be held fixed, and forward-backward performed within each arc to determine arc conditional posterior probabilities.

9.3.2 Training the Front-End Parameters

In this section, we detail the procedure for discriminative training of the SPLICE parameters $\boldsymbol{\theta}$ after the back-end acoustic model parameters have been fully trained. It is also possible to jointly optimize $\boldsymbol{\theta}$ and $\boldsymbol{\lambda}$ using the same objective function. For details, see [7].

Because there is no known closed form solution to find the parameters $\boldsymbol{\theta}$ that maximize Equation (9.12), we resort to gradient-based optimization methods. If $p_{\mathcal{X}}(\mathcal{X}_r, w; \boldsymbol{\lambda})$ and $f(\mathcal{Y}_r;\boldsymbol{\theta})$ are continuous and differentiable, we should be able to compute the gradient of $\mathcal{F}$ with respect to $\boldsymbol{\theta}$.

Every $\mathcal{F}_r$ is a function of many acoustic model state conditional probabilities $p(\mathbf{x}_t^r|s_t^r)$.[3] Each of these are functions of the front-end processed acoustic features x_{it}^r. And, each transformed feature is a function of the front-end parameters $\boldsymbol{\theta}$. This structure allows a simple application of the chain rule:

$$\frac{\partial \mathcal{F}_r}{\partial \boldsymbol{\theta}} = \sum_{t,s,i} \frac{\partial \mathcal{F}_r}{\partial \ln p(\mathbf{x}_t^r|s_t^r = s)} \frac{\partial \ln p(\mathbf{x}_t^r|s_t^r = s)}{\partial x_{it}^r} \frac{\partial x_{it}^r}{\partial \boldsymbol{\theta}}. \tag{9.13}$$

Here, r is an index into a particular utterance in the training data. The tth observation vector in utterance r is identified by $\mathbf{x}_t^r$. The scalar x_{it}^r is the ith dimension of that vector. The back-end acoustic model state at time t in utterance r is s_t^r.

The first term in Equation (9.13) captures the sensitivity of the objective function to individual acoustic likelihoods in the model. It is equal to the difference of the conditional and unconditional posterior, with respect to the correct transcription. These are simply the flattened

[2] A lattice is a mathematical structure that efficiently describes a mapping from sequences of symbols to real numbers, without explicitly enumerating all possible symbol sequences. In our case, the symbols are words, and the real numbers represent the likelihood of the word sequences under the current model.

[3] This derivation assumes one Gaussian mixture component per state of the acoustic model. For the multiple mixture component case, the variable s indexes not state, but the individual mixture components. Nothing else needs to be changed.

numerator and denominator terms that occur in standard lattice-based MMI estimation:

$$\begin{aligned}\frac{\partial \mathcal{F}_r}{\partial \ln p(x_t^r | s_t^r = s)} &= p(s_t^r = s|\mathcal{X}_r, w_r) - p(s_t^r = s|\mathcal{X}_r) \\ &= \gamma_{rts}^{\text{num}} - \gamma_{rts}^{\text{den}}. \end{aligned} \tag{9.14}$$

The second term in Equation (9.13) captures the sensitivity of individual likelihoods in the acoustic model with respect to the front-end transformed features. Because this is a Gaussian likelihood, computing its differential is a simple matter:

$$\frac{\partial \ln p(x_t^r | s_t^r = s)}{\partial x_t^r} = -\boldsymbol{\Sigma}_s^{-1}(x_t^r - \boldsymbol{\mu}_s). \tag{9.15}$$

Here, $\boldsymbol{\mu}_s$ and $\boldsymbol{\Sigma}_s$ are mean and variance parameters from the Gaussian component associated with state *s* in the back end acoustic model.

The final term in Equation (9.13) captures the relationship between the transformed features and the parameters of the front-end. Here, we restrict ourselves to training the offset parameters $\mathbf{b}_m$ only.[4] For the *u*th element of the vector $\mathbf{b}_m$:

$$\begin{aligned}\frac{\partial x_{it}^r}{\partial b_{um}} &= \frac{\partial}{\partial b_{um}}\left(y_{ut}^r + \sum_{m'} b_{im'} p(m'|\mathbf{y}_t^r)\right) \\ &= \delta(i = u) p(m|\mathbf{y}_t^r). \end{aligned} \tag{9.16}$$

Here, we have used the Kronecker delta function $\delta(\cdot)$, which takes the value 1 when its argument is true, and the value 0 otherwise.

Combining Equations (9.9), (9.13), (9.16), the complete gradient with respect to the vector $\mathbf{b}_m$ is

$$\frac{\partial \mathcal{F}}{\partial \mathbf{b}_m} = -\sum_{r,t,s} p(m|\mathbf{y}_t^r)\left(\gamma_{rts}^{\text{num}} - \gamma_{rts}^{\text{den}}\right) \boldsymbol{\Sigma}_s^{-1}(\mathbf{x}_t^r - \boldsymbol{\mu}_s). \tag{9.17}$$

Equation (9.17) can then used to compute the gradient of the objective function with respect to the offset vectors during gradient-based optimization of the objective function.

9.3.3 The Rprop Algorithm

Rprop is a well-known gradient-based algorithm that was originally developed to train neural networks [22]. Here, Rprop is employed to train the transform parameters $\boldsymbol{\theta}$ that improve the MMI objective function, Equation (9.9).

By design, Rprop only needs the sign of the gradient of the objective function with respect to each parameter. Unlike many other gradient ascent methods, the scale of the step size for parameter *i*, Δ_i, is unrelated to the magnitude of the current gradient $\frac{\partial \mathcal{F}}{\partial \theta_i}$. As a result, Rprop is quite easy to implement and is robust to nonuniform scaling across the feature dimensions.

Each iteration *l* of the Rprop training algorithm can be simply described as a loop over parameter values, a computation of the sign of the gradient of the objective function with respect

---

[4] Training the rotations $\mathbf{A}_m$ is also possible, and was derived in [12].

to each parameter, and a rule-based parameter update. The procedure is run independently on each parameter θ_i in $\boldsymbol{\theta}$.

For each parameter θ_i {
  $d \leftarrow \frac{\partial \mathcal{F}}{\partial \theta_i}(l-1) \cdot \frac{\partial \mathcal{F}}{\partial \theta_i}(l)$
  if ($d >= 0$) then {
    if ($d > 0$) then $\Delta_i \leftarrow \min(1.2\,\Delta_i, \Delta_{\max})$
    $\theta_i(l+1) \leftarrow \theta_i(l) + \mathrm{sign}(\frac{\partial \mathcal{F}}{\partial \theta_i}(l)) \cdot \Delta_i$
  } else if ($d < 0$) then \{
    $\Delta_i \leftarrow \max(0.5\,\Delta_i, \Delta_{\min})$
    $\theta_i(l+1) \leftarrow \theta_i(l-1)$
    $\frac{\partial \mathcal{F}}{\partial \theta_i}(l) \leftarrow 0$
  }
}

There are only a handful of parameters to set in this training algorithm. We typically choose $\Delta_{\min} = 10^{-5}$ and $\Delta_{\max} = 0.1$ to bound the step size within a reasonable range. For the first iteration, set the value for the initial step size to be small, such as $\Delta_i = 0.01, \forall i$, and assume the previous gradient was zero.

Analysis of the Rprop algorithm is simple. At each iteration, for every parameter, Rprop does one of three things:

1. If the current and previous gradient are nonzero and in the same direction ($d > 0$), the step size Δ_i is increased and applied in the same direction as the current gradient.
2. If the current and previous gradient are in opposite directions ($d < 0$), it means that a local maximum has been overshot. In this case, the step size is reduced and the parameter is reset to its value before the last update. Also, the memory of the current gradient is set to zero. This serves as a flag for the next iteration of the algorithm.
3. If either the current or previous gradient are zero, then $d = 0$ and the current step size is applied in the direction of the current gradient. This is appropriate whether the current gradient is zero, and Rprop has found a local maximum, or the previous gradient is zero, indicating that the algorithm had overshot and backtracked during the previous iteration.

9.3.4 Results

To demonstrate how MMI-SPLICE affects the accuracy of speech recognition, we apply the technique to the Aurora 4 task. Several systems were trained with a 128 component, offset-only version of MMI-SPLICE. As with MMSE-SPLICE, the noisy SPLICE GMM was trained on the "multinoise" training set. This same "multinoise" data were used in training the SPLICE parameters when evaluating the MMI objective function. The SPLICE offset parameters are initialized to zero so that it will mimic the identity transform.

In Table 9.4, an acoustic model is first trained on the clean Aurora 4 training data. The MMI-SPLICE offset parameters are initialized to zero, and several iterations of Rprop are made. With each iteration, the word error rate of the clean test set increases, and the word error rate of the noisy test sets decrease. Overall, the average word error rate drops by about 23%.

Table 9.5 illustrates how MMI-SPLICE can affect the word error rate for systems trained with multistyle acoustic data. As with MMSE-SPLICE, normalizing the multistyle training

Table 9.4 Word error rate on Aurora 4, demonstrating how increasing the number of Rprop training iterations changes the word error rate for seven test conditions.

Iteration	Clean	Car	Babble	Restaurant	Street	Airport	Train	Average
0	9.0	40.3	57.2	47.6	51.0	50.7	54.3	44.3
20	9.2	34.1	50.1	44.5	46.2	45.1	48.8	39.7
40	10.4	28.2	44.6	43.3	44.7	40.7	47.8	37.1
60	11.6	26.1	41.8	43.1	44.1	39.6	46.6	36.1
80	12.5	24.7	39.1	43.4	41.9	38.0	44.6	34.9
100	13.7	23.6	37.5	43.7	40.3	36.3	42.6	34.0

data toward a more consistent distribution has mitigated some of the expected degradation on clean testing data. Unlike MMSE-SPLICE, MMI-SPLICE is also able to decrease the word error rate result on the noisy test utterances.

9.4 Model-Based Feature Enhancement

The previous sections assumed that the functional form of the mismatch between clean speech and noisy speech was unknown. In that case, we trained general models with many parameters from a large set of training data.

In this section, we derive an example of model-based feature enhancement (MBFE). It consists of a model that represents our belief about how the speech has been corrupted, and a small handful of learnable parameters.

Model-based feature-enhancement algorithms are powerful and can learn their parameters on an utterance by utterance basis, but are limited to modeling those distortions accounted for in the initial model design.

For simplicity, we assume here that the model is built in the static cepstral domain, and that standard delta or acceleration parameters are not used. It is, of course, possible to use them under the same framework, with further approximation, as in [5].

Figure 9.3 illustrates the necessary steps. First, the relationship between clean speech and noisy speech is derived. This additive noise-mixing equation is then combined with a clean speech GMM to produce a joint probability model for the clean and noisy speech. Because inference on this model is difficult, it is then linearized using vector Taylor series. This resulting model can then be used to enhance speech-recognition features, in the same way as SPLICE was used previously.

Table 9.5 Word error rate on Aurora 4, demonstrating that MMI-SPLICE improves upon both the baseline system and MMSE-SPLICE.

Training data	Feature enhancement	Test 1 (clean)	Test 2–7 (noisy)
Multistyle	None	14.0	19.3
Multistyle MMSE-SPLICE	MMSE-SPLICE	13.4	19.2
Multistyle MMI-SPLICE	MMI-SPLICE	13.4	18.8

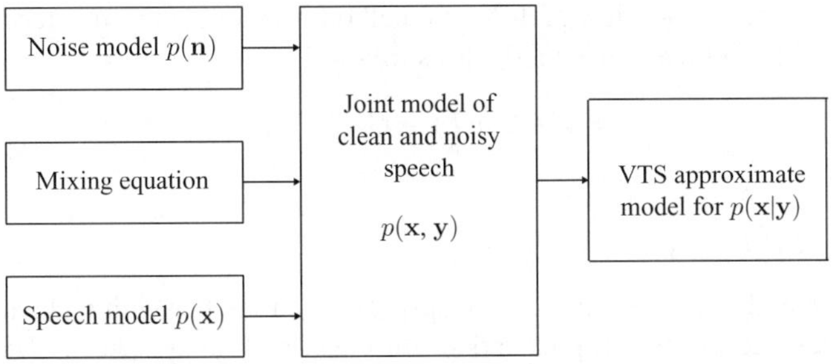

Figure 9.3 Steps in deriving a tractable model for model-based feature enhancement.

9.4.1 *The Additive Noise-Mixing Equation*

To address mismatch due to background noise in the signal, we construct a mismatch function for additive noise. In particular, the noisy time series $y[t]$ is a linear combination of the clean time series $x[t]$ and the noise time series $n[t]$:

$$y[t] = x[t] + n[t].$$

If this noisy signal is passed through a mel-frequency cepstral coefficient (MFCC) front-end, the relationship between noisy speech, clean speech, and the additive noise gets a lot more complicated.

Let $\mathbf{x}$ represent the cepstral features that the clean speech would have generated in the absence of noise, and $\mathbf{n}$ represent the cepstral features that the noise would have generated in the absence of speech. Define the matrix $\mathbf{C}$ as the discrete cosine transform used to transform log mel-frequency energies into the cepstral domain, and $\mathbf{D}$ its right inverse such that $\mathbf{CD} = \mathbf{I}$.

The approximate noise power, $\mathbf{n}^p$, approximate clean speech power, $\mathbf{x}^p$, and approximate noisy speech power, $\mathbf{y}^p$, can be found by inverting the last two steps of the cepstral calculation:

$$\begin{aligned} \mathbf{x}^p &\approx \mathrm{e}^{\mathbf{Dx}}, \\ \mathbf{n}^p &\approx \mathrm{e}^{\mathbf{Dn}}, \\ \mathbf{y}^p &\approx \mathrm{e}^{\mathbf{Dy}}. \end{aligned}$$

Because the power spectra of the noise and speech should mix linearly, we get the following relationship, which is easily manipulated into a more standard form:

$$\begin{aligned} \mathbf{y}^p &= \mathbf{x}^p + \mathbf{n}^p, \\ \mathrm{e}^{\mathbf{Dy}} &= \mathrm{e}^{\mathbf{Dx}} + \mathrm{e}^{\mathbf{Dn}}, \\ \mathbf{Dy} &= \ln\left(\mathrm{e} + \mathbf{Dx} + \mathrm{e}^{\mathbf{Dn}}\right), \\ \mathbf{Dy} &= \mathbf{Dx} + \ln\left(\mathbf{1} + \mathrm{e}^{\mathbf{D}(\mathbf{n}-\mathbf{x})}\right), \\ \mathbf{y} &= \mathbf{x} + \mathbf{C}\ln\left(\mathbf{1} + \mathrm{e}^{\mathbf{D}(\mathbf{n}-\mathbf{x})}\right). \end{aligned}$$

Because of the approximations, namely that the power spectra mix linearly and deterministically, and that the cepstral rotation can be inverted with $\mathbf{D}$, it is customary to include a stochastic error $\mathbf{e}$. For an overview of different methods to model this error term, see [11]. The final formula for the additive noise model is as follows:

$$\mathbf{y} = \mathbf{x} + \mathbf{C}\ln\left(\mathbf{1} + e^{\mathbf{D}(\mathbf{n}-\mathbf{x})}\right) + \mathbf{e}. \tag{9.18}$$

9.4.2 The Joint Probability Model

Here, we use the additive noise model, Equation (9.18), together with a clean speech GMM and a noise model, to develop a joint model for clean and noisy speech. The clean speech model $p(\mathbf{x})$ is pretrained with high signal-to-noise utterances:

$$p(\mathbf{x}) = \sum_s p(\mathbf{x}|s)p(s) = \sum_s \mathcal{N}(\mathbf{x}; \boldsymbol{\mu}_{\mathbf{x}|s}, \boldsymbol{\Sigma}_{\mathbf{x}|s})p(s).$$

The simplest form of noise model, which is what we will use here, is a single Gaussian component trained from known or suspected nonspeech segments of the current utterance. Other common choices are to recursively track nonstationary noises in concert with the model [4], or to use an auxiliary noise tracker such as minimum controller recursive averaging [3]:

$$p(\mathbf{n}) = \mathcal{N}(\mathbf{n}; \boldsymbol{\mu}_{\mathbf{n}}, \boldsymbol{\Sigma}_{\mathbf{n}}).$$

Of course, we can not directly observe the $\mathbf{x}$ or $\mathbf{n}$ once they have been mixed and presented to the system as $\mathbf{y}$. We can only parameterize their stochastic relationship. For simplicity, we assume that the error term $\mathbf{e}$ in Equation (9.18) is deterministically equal to zero. This formulation is known as the zero variance model (ZVM), and similar to several related models, including [14, 17, 24].

The joint PDF, shown in Figure 9.4, is a distribution over the clean speech $\mathbf{x}$, the noise $\mathbf{n}$, the noisy observation $\mathbf{y}$ and the speech state s. It is a generative model in which the speech state affects the clean speech, which mixes with noise to produce the noisy observation:

$$p(\mathbf{y}, \mathbf{x}, \mathbf{n}, s) = p(\mathbf{y}|\mathbf{x}, \mathbf{n})p(\mathbf{x}, s)p(\mathbf{n}).$$

Because we expect the speech and noise estimates to be highly correlated in the posterior $p(\mathbf{x}, \mathbf{n}|\mathbf{y})$, we generally introduce an instantaneous SNR variable $\mathbf{r} = \mathbf{x} - \mathbf{n}$ that can capture this relationship and stabilize the processing:

$$p(\mathbf{y}, \mathbf{r}, \mathbf{x}, \mathbf{n}, s) = p(\mathbf{y}|\mathbf{x}, \mathbf{n})p(\mathbf{r}|\mathbf{x}, \mathbf{n})p(\mathbf{x}, s)p(\mathbf{n}).$$

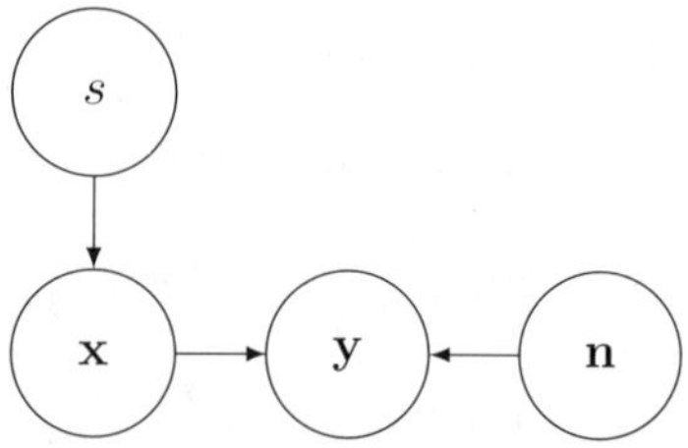

Figure 9.4 Graphical representation of the observation model. The observation $\mathbf{y}$ is a nonlinear function of speech $\mathbf{x}$ and noise $\mathbf{n}$.

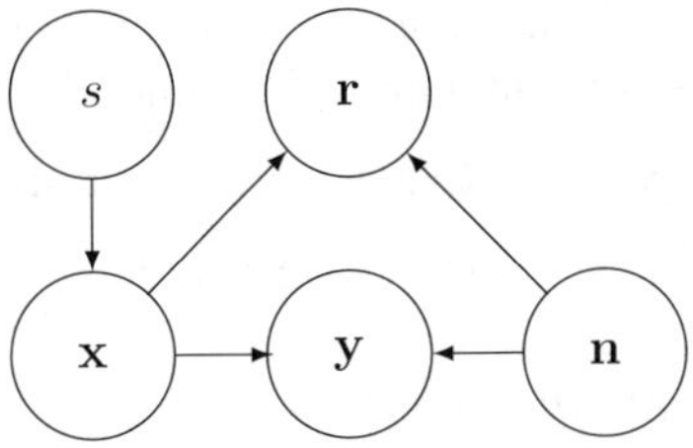

Figure 9.5 Graphical system model.

Figure 9.5 contains a graphical representation of the complete model. Here, $\mathbf{x}$ and $\mathbf{n}$ represent the continuous hidden speech and noise vectors. The observation $\mathbf{y}$ and SNR $\mathbf{r}$ are both deterministic functions of $\mathbf{x}$ and $\mathbf{n}$. As a result, the conditional probabilities $p(\mathbf{y}|\mathbf{x}, \mathbf{n})$ and $p(\mathbf{r}|\mathbf{x}, \mathbf{n})$ can be represented by Dirac delta functions:

$$p(\mathbf{y}|\mathbf{x}, \mathbf{n}) = \delta\left(\mathbf{x} + \mathbf{C}\ln\left(\mathbf{1} + e^{\mathbf{D}(\mathbf{x}-\mathbf{n})}\right) - \mathbf{y}\right), \tag{9.19}$$

$$p(\mathbf{r}|\mathbf{x}, \mathbf{n}) = \delta\left(\mathbf{x} - \mathbf{n} - \mathbf{r}\right), \tag{9.20}$$

where $\mathbf{1}$ is a vector consisting of ones. This allows us to marginalize the continuous variables $\mathbf{x}$ and $\mathbf{n}$, as follows, so the only remaining continuous hidden variable is the instantaneous SNR $\mathbf{r}$:

$$\begin{aligned}
p(\mathbf{y}, \mathbf{r}, s) &= \int d\mathbf{x} \int d\mathbf{n}\, p(\mathbf{y}, \mathbf{r}, \mathbf{x}, \mathbf{n}, s) \\
&= \int d\mathbf{x} \int d\mathbf{n}\, p(\mathbf{y}|\mathbf{x}, \mathbf{n}) p(\mathbf{r}|\mathbf{x}, \mathbf{n}) p(\mathbf{x}, s) p(\mathbf{n}) \\
&= \int d\mathbf{x} \int d\mathbf{n}\, \delta\left(\mathbf{x} + \mathbf{C}\ln\left(\mathbf{1} + e^{\mathbf{D}(\mathbf{n}-\mathbf{x})}\right) - \mathbf{y}\right) \delta\left(\mathbf{x} - \mathbf{n} - \mathbf{r}\right) p(\mathbf{x}, s) p(\mathbf{n}) \\
&= p(\mathbf{x}, s)|_{\mathbf{x}=\mathbf{y}-\mathbf{C}\ln(\mathbf{1}+e^{\mathbf{Dr}})+\mathbf{r}}\; p(\mathbf{n})|_{\mathbf{n}=\mathbf{y}-\mathbf{C}\ln(\mathbf{1}+e^{\mathbf{Dr}})} \\
&= \mathcal{N}(\mathbf{y} - \mathbf{C}\ln\left(\mathbf{1} + e^{\mathbf{Dr}}\right) + \mathbf{r}; \boldsymbol{\mu}_{\mathbf{x}|s}, \boldsymbol{\Sigma}_{\mathbf{x}|s}) p(s) \\
&\quad \mathcal{N}(\mathbf{y} - \mathbf{C}\ln\left(\mathbf{1} + e^{\mathbf{Dr}}\right); \boldsymbol{\mu}_{\mathbf{n}}, \boldsymbol{\Sigma}_{\mathbf{n}}).
\end{aligned} \tag{9.21}$$

The behavior of this joint PDF is intuitive. At high SNR, the elements of $\mathbf{Dr}$ will be much greater than zero, and we have the approximate relationship:

$$\mathbf{C}\ln\left(\mathbf{1} + e^{\mathbf{Dr}}\right)) \approx \mathbf{r}, \text{ so}$$

$$p(\mathbf{y}, \mathbf{r}, s) \approx \mathcal{N}(\mathbf{y}; \boldsymbol{\mu}_{\mathbf{x}|s}, \boldsymbol{\Sigma}_{\mathbf{x}|s}) p(s) \mathcal{N}(\mathbf{y} - \mathbf{r}; \boldsymbol{\mu}_{\mathbf{n}}, \boldsymbol{\Sigma}_{n}).$$

In this high SNR case, the clean speech Gaussian component $\mathcal{N}(\cdot; \boldsymbol{\mu}_{\mathbf{x}|s}, \boldsymbol{\Sigma}_{\mathbf{x}|s})$ is evaluated at the noisy observation $\mathbf{y}$, and the noise Gaussian $\mathcal{N}(\cdot; \boldsymbol{\mu}_{\mathbf{n}}, \boldsymbol{\Sigma}_{n})$ is evaluated at the noise estimate $\mathbf{y} - \mathbf{r}$. That is, the observation $\mathbf{y}$ is assumed to be clean speech, and the noise is at a level $\mathbf{r}$ units below the observation. The converse is true for low SNR, where the elements of $\mathbf{Dr}$ are much less than zero. In this case, the noise Gaussian will be evaluated at the noisy observation $\mathbf{y}$, and the clean speech Gaussian component will be evaluated at the clean speech estimate $\mathbf{y} - \mathbf{r}$.

By performing inference on this new variable, an estimate for the instantaneous SNR can be mapped back into estimates of $\mathbf{x}$ and $\mathbf{n}$ through a nonlinear transformation:

$$\mathbf{x} = \mathbf{y} - \mathbf{C}\ln(\mathbf{1} + e^{\mathbf{Dr}}) + \mathbf{r} \text{ and} \tag{9.22}$$

$$\mathbf{n} = \mathbf{y} - \mathbf{C}\ln(\mathbf{1} + e^{\mathbf{Dr}}). \tag{9.23}$$

9.4.3 Vector Taylor Series Approximation

Unfortunately, even though the outer form of both terms for $p(\mathbf{y}, \mathbf{r}, s)$ in Equation (9.21) are Gaussian, the nonlinearity on $\mathbf{r}$ makes the overall distribution non-Gaussian. Because the difficulty is the nonlinearity, we replace it with a first-order vector Taylor series (VTS) approximation. This approximation will linearize the arguments to the Gaussian components in Equation (9.21) and make learning and inference tractable.

The offending nonlinear function $\mathbf{g}(\mathbf{r})$ is defined and approximated as

$$\mathbf{g}(\mathbf{r}) = \mathbf{C}\ln\left(\mathbf{1} + e^{\mathbf{Dr}}\right) \approx \mathbf{g}(\mathbf{r}') + \mathbf{G}(\mathbf{r}')(\mathbf{r} - \mathbf{r}'),$$

where the matrix-valued function $\mathbf{G}(\mathbf{r}')$ is the Jacobian of $\mathbf{g}(\mathbf{r})$ evaluated at $\mathbf{r}'$. It can be decomposed into a product of the cepstral rotation matrix $\mathbf{C}$, a matrix-valued function $\mathbf{F}(\mathbf{r}')$, and the cepstral rotation pseudo-inverse $\mathbf{D}$:

$$\mathbf{G}(\mathbf{r}') = \mathbf{C}\mathbf{F}(\mathbf{r}')\mathbf{D}.$$

The function $\mathbf{F}$ produces diagonal matrices, whose elements are given by a vector $\mathbf{f}(\mathbf{r}')$, which in turn is

$$\mathbf{f}(\mathbf{r}') = \frac{1}{1 + \exp(\mathbf{D}(\mathbf{r}'))}.$$

The distribution $p(\mathbf{y}, \mathbf{r})$ based on this approximation is derived by substituting the Taylor series approximation for $\mathbf{g}(\mathbf{r})$ into Equation (9.21). Note that the expansion point $\mathbf{r}'_s$ may be state specific:

$$\begin{aligned}
p(\mathbf{y}, \mathbf{r}, s) &\approx \mathcal{N}(\mathbf{y} - \mathbf{g}(\mathbf{r}'_s) - \mathbf{G}(\mathbf{r}'_s)(\mathbf{r} - \mathbf{r}'_s) + \mathbf{r}; \boldsymbol{\mu}_{\mathbf{x}|s}, \boldsymbol{\Sigma}_{\mathbf{x}|s}) \\
&\quad \mathcal{N}(\mathbf{y} - \mathbf{g}(\mathbf{r}'_s) - \mathbf{G}(\mathbf{r}'_s)(\mathbf{r} - \mathbf{r}'_s); \boldsymbol{\mu}_{\mathbf{n}}, \boldsymbol{\Sigma}_{\mathbf{n}})p(s) \\
&= \mathcal{N}(\mathbf{r}; \boldsymbol{\mu}_{\mathbf{r}|s}, \boldsymbol{\Sigma}_{\mathbf{r}|s})\mathcal{N}(\mathbf{a}_s; \mathbf{b}_s, \mathbf{C}_s)p(s) \\
&= p(\mathbf{r}|\mathbf{y}, s)p(\mathbf{y}|s)p(s).
\end{aligned}$$

Standard Gaussian manipulation formulas are used to bring $p(\mathbf{y}, \mathbf{r}, s)$ into this factored form:

$$p(\mathbf{r}|\mathbf{y}, s) = N\left(\mathbf{r}; \boldsymbol{\mu}_{\mathbf{r}|s}, \boldsymbol{\Sigma}_{\mathbf{r}|s}\right), \tag{9.24}$$

$$\boldsymbol{\Sigma}^{-1}_{\mathbf{r}|s} = (\mathbf{G}(\mathbf{r}'_s) - \mathbf{I})^T\boldsymbol{\Sigma}^{-1}_{\mathbf{x}|s}(\mathbf{G}(\mathbf{r}'_s) - \mathbf{I}) + \mathbf{G}^T(\mathbf{r}'_s)(\boldsymbol{\Sigma}_{\mathbf{n}})^{-1}\mathbf{G}(\mathbf{r}'_s), \tag{9.25}$$

$$\boldsymbol{\mu}_{\mathbf{r}|s} = \boldsymbol{\Sigma}_{\mathbf{r}|s}\left((\mathbf{G}(\mathbf{r}'_s) - \mathbf{I})^T\boldsymbol{\Sigma}^{-1}_{\mathbf{x}|s}(\mathbf{a}_s - \boldsymbol{\mu}_{\mathbf{x}|s}) + (\mathbf{G}(\mathbf{r}'_s))^T\boldsymbol{\Sigma}^{-1}_{\mathbf{n}}(\mathbf{a}_s - \boldsymbol{\mu}_{\mathbf{n}})\right), \tag{9.26}$$

$$p(\mathbf{y}|s) = \mathcal{N}(\mathbf{a}_s; \mathbf{b}_s, \mathbf{C}_s), \tag{9.27}$$

$$\mathbf{a}_s = \mathbf{y} - \mathbf{g}(\mathbf{r}'_s) + \mathbf{G}(\mathbf{r}'_s)\mathbf{r}'_s, \tag{9.28}$$

$$\mathbf{b}_s = \boldsymbol{\mu}_{\mathbf{n}} + \mathbf{G}(\mathbf{r}'_s)(\boldsymbol{\mu}_{\mathbf{x}|s} - \boldsymbol{\mu}_{\mathbf{n}}), \tag{9.29}$$

$$\mathbf{C}_s = (\mathbf{G}(\mathbf{r}'_s))^T\boldsymbol{\Sigma}_{\mathbf{x}|s}\mathbf{G}(\mathbf{r}'_s) + (\mathbf{G}(\mathbf{r}'_s) - \mathbf{I})^T\boldsymbol{\Sigma}_{\mathbf{n}}(\mathbf{G}(\mathbf{r}'_s) - \mathbf{I}). \tag{9.30}$$

9.4.4 Estimating Clean Speech

To estimate clean speech using Equations (9.24) and (9.27), a suitable set of expansion points $\mathbf{r}'_s$ must be found. Toward that end, an iterative procedure is performed for each mixture component in the clean speech GMM. Convergence usually occurs within the first two iterations:

1. Initialize the expansion point $\mathbf{r}'_s$ to be the expected SNR for that mixture component, $\boldsymbol{\mu}_{\mathbf{x}|s} - \boldsymbol{\mu}_{\mathbf{n}}$.
2. Use Equations (9.24)–(9.26) to find the expected value of $\mathbf{r}$ under the approximate model, $\boldsymbol{\mu}_{\mathbf{r}|s}$.
3. If the value of $\boldsymbol{\mu}_{\mathbf{r}|s}$ has changed significantly, repeat step 2.

After an appropriate expansion point for each mixture component is found, a minimum mean squared estimate of $\mathbf{r}$ given $\mathbf{y}$ under the model is computed by taking the expectation of our approximate model:

$$\hat{\mathbf{r}}_{\mathrm{MMSE}} = E[\mathbf{r}|\mathbf{y}] = \sum_s p(s|\mathbf{y})E[\mathbf{r}|\mathbf{y}, s].$$

Here, $p(s|\mathbf{y})$ is the posterior probability of state s under the model, which can be found by Bayes' rule:

$$p(s|\mathbf{y}) = \frac{p(\mathbf{y}|s)p(s)}{\sum_{s'} p(\mathbf{y}|s')p(s')}.$$

Finally, $\hat{\mathbf{r}}_{\mathrm{MMSE}}$ is used to estimate $\mathbf{x}$ by undoing the previous mapping using Equation (9.22). Note that, since the transformation is nonlinear, our estimate of clean speech $\hat{\mathbf{x}}$ is not the optimal MMSE estimator for $\mathbf{x}$:

$$\hat{\mathbf{x}} = \mathbf{y} - \mathbf{C}\ln\left(\mathbf{1} + \mathbf{e}^{\mathbf{D}\hat{\mathbf{r}}_{\mathrm{MMSE}}}\right) + \hat{\mathbf{r}}_{\mathrm{MMSE}}.$$

9.4.5 Results

Table 9.6 shows how a feature-enhancing VTS front-end with 32 mixture components improves accuracy on our Aurora 2 system. The clean speech GMM was trained using the "clean" training data provided with the corpus. The "multistyle" training data were only used to train the recognition system's acoustic models. The VTS front-end improves adapts to each utterance independently, allowing it to improve upon the global model learned by either MMSE-SPLICE or MMI-SPLICE.

Table 9.6 Word error rate on Aurora 2, demonstrating the benefits of VTS feature enhancement.

Training data	Feature enhancement	Set A	Set B	Set C	Average
Clean	None	41.42	47.10	31.73	41.75
Clean+VTS	VTS	11.08	10.61	12.72	10.82
Multistyle	None	8.64	10.23	11.96	9.94
Multistyle+VTS	VTS	6.28	6.81	7.21	6.68

Table 9.7 Word error rate on Aurora 4, demonstrating that VTS feature enhancement also works in this harder task.

Training data	Feature enhancement	Test 1 (clean)	Test 2–7 (noisy)
Clean	None	8.4	33.9
Clean	VTS	7.9	23.6
Multistyle	None	14.0	19.3
Multistyle VTS	VTS	12.3	17.6

Table 9.7 shows that the VTS feature enhancement also improves accuracy on the Aurora 4 tests. Again, the VTS feature enhancement proves to be more powerful than either the MMSE-SPLICE or MMI-SPLICE algorithms.

Although VTS feature enhancement is more effective than SPLICE at reducing the effects of additive noise, it comes with two drawbacks: complexity and specificity. Whereas SPLICE is simple to implement and consumes few resources, VTS feature enhancement is difficult to implement and consumes much more processing power. Also, SPLICE can compensate for any systematic corruption in the feature space, but VTS feature enhancement is limited to the additive noise formula at its core.

9.5 Switching Linear Dynamic System

The previous sections have described how frames of noisy speech features can be transformed into estimates of clean speech feature frames. Speech and noise models have consisted of time-independent probability models, and there has been little attention paid to the time evolution of either.

Because speech and noise tend to evolve from one frame to the next, it stands to reason that models which incorporate this evolution should outperform models that do not. This intuition forms the basis for systems that use switching linear dynamic models, first applied to the feature-enhancement problem in [8, 21], and a good survey of which appears in [23].

A switching linear dynamic model (SLDM) describes the time evolution of a vector-valued sequence such as speech or noise cepstra. The model parameters consist of state-conditional rotations $\mathbf{H}_s$ and offsets $\mathbf{h}_s$. These are combined to form a model for the current vector $\mathbf{x}_t$ when the previous vector $\mathbf{x}_{t-1}$ is known:

$$\mathbf{x}_t = \mathbf{H}_{s_t}\mathbf{x}_{t-1} + \mathbf{h}_{s_t} + \mathbf{v}_{s_t}.$$

The vector-valued sequence $\mathbf{v}_{s_t}$ is a zero-mean Gaussian innovation source which drives the system.

Figure 9.6 illustrates the structure of the model for a sequence of length $T = 4$. Every unique state sequence s_1^T generates a different nonstationary linear dynamic model. As a result, it is appropriate for describing a number of time-varying systems, including nonstationary noise processes and the evolution of speech features over time.

The difficulty in replacing the clean speech or noise models with a SLDM is in estimation of the hidden state sequence. An exact solution would need to sum over all possible state

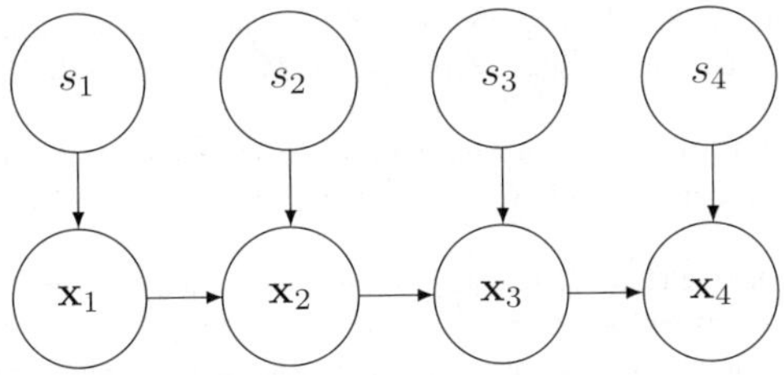

Figure 9.6 Graphical representation of the switching LDM useful for modeling clean speech features.

sequences, a problem that increases exponentially with the length of the sequence. Approximate solutions that work well are the generalized pseudo-Bayes technique used by Droppo and Acero [8] and the particle filtering approach of Raj *et al.* [21].

9.6 Conclusion

Feature-enhancement algorithms live between the front-end of the speech-recognition system, where speech features are generated, and the back-end of the speech-recognition system, where acoustic likelihoods are calculated. Their goal is to reduce any information in the feature stream that might be confusing or unnecessary to the recognition task, and thereby move closer to the ideal features we would like to use.

This chapter has presented three different feature-enhancement techniques. Each one uses a joint probability distribution to generate an estimate of the clean speech features given the noisy speech features.

MMSE-SPLICE is a good choice if you have or can generate parallel examples of noisy and clean speech. It is easy to code and can produce dramatic results against a clean acoustic model. If the acoustic model has been trained with multistyle data, the improvement may be less dramatic.

A discriminative SPLICE, such as MMI-SPLICE is more complicated to implement, but is more effective than MMSE-SPLICE both in clean and noisy test environments.

Model-based feature enhancement, such as iterative VTS feature enhancement, produces the best word error rates of the systems discussed in this chapter, but is very complex to implement and uses a great deal more processing power. It is the preferred method when there is enough processing power available, and when the distortion is expected to match the additive noise mixing equation.

References

[1] A. Acero and R. M. Stern, "Robust speech recognition by normalization of the acoustic space," in *Proceedings of the IEEE ICASSP*, vol. 2, pp. 893–896, 1991.

[2] L. R. Bahl, P. F. Brown, P. V. D. Souza, and R. L. Mercer, "Maximum mutual information estimation of hidden Markov model parameters for speech recognition," in *International Conference on Acoustics, Speech and Signal Processing*, 1997.

[3] I. Cohen and B. Berdugo, "Noise estimation by minima controlled recursive averaging for robust speech enhancement," *IEEE Signal Processing Letters*, vol. 9, no. 1, pp. 12–15, 2002.

[4] L. Deng, J. Droppo, and A. Acero, "Log-domain speech feature enhancement using sequential map noise estimate and a nonlinear model of acoustic environment," in *Proceedings of the ICSLP*, Denver, CO, September 2002.

[5] L. Deng, J. Droppo, and A. Acero, "Estimating cepstrum of speech under the presence of noise using a joint prior of static and dynamic features," *IEEE Transactions on Speech and Audio Processing*, vol. 12, no. 3, pp. 218–233, May 2004.

[6] V. V. Digalakis, D. Rtischev, and L. G. Neumeyer, "Speaker adaptation using constrained estimation of gaussian mixtures," *IEEE Transactions on Speech and Audio Processing*, vol. 3, pp. 357–366, September 1995.

[7] J. Droppo and A. Acero, "Joint discriminative front end and back end training for improved speech recognition accuracy," in *International Conference on Acoustics, Speech and Signal Processing*, Toulouse, France, May 2006.

[8] J. Droppo and A. Acero, "Noise robust speech recognition with a switching linear dynamic model," in *International Conference on Acoustics, Speech and Signal Processing*, Montreal, Canada, May 2004.

[9] J. Droppo, A. Acero, and L. Deng, "Uncertainty decoding with SPLICE for noise robust speech recognition," in *Proceedings of the 2002 ICASSP*, Orlando, Florida, May 2002.

[10] J. Droppo, L. Deng, and A. Acero, "Evaluation of SPLICE on the Aurora 2 and 3 tasks," in *Proceedings of the ICSLP*, pp. 29–32, 2002.

[11] J. Droppo, L. Deng, and A. Acero, "A comparison of three non-linear observation models for noisy speech features," in *Proceedings of the 2003 Eurospeech*, Geneva, Switzerland, September 2003, pp. 681–684.

[12] J. Droppo, M. Mahajan, A. Gunawardana, and A. Acero, "How to train a discriminative front end with stochastic gradient descent and maximum mutual information," in *Proceedings of the IEEE ASRU*, 2005.

[13] D. Ellis and M. Gomez, "Investigations into tandem acoustic modeling for the Aurora task," in *Proceedings of the Eurospeech*, 2001, pp. 189–192.

[14] B. Frey, L. Deng, A. Acero, and T. Kristjansson, "ALGONQUIN: Iterating Laplace's method to remove multiple types of acoustic distortion for robust speech recognition," in *Proceedings of the 2001 Eurospeech*, Aalbork, Denmark, September 2001.

[15] B.-H. Juang, W. Hou, and C.-H. Lee, "Minimum classification error rate methods for speech recognition," *IEEE Transactions on Speech and Audio Processing*, vol. 5, pp. 257–265, 1997.

[16] E. McDermott and S. Katagiri, "Prototype-based minimum classification error/generalized probabilistic descent training for various speech units," *Computer Speech & Language*, vol. 8, pp. 351–368, 1994.

[17] P. Moreno, "Speech recognition in noisy environments," PhD dissertation, Carnegie Mellon University, 1996.

[18] P. J. Moreno, B. Raj, E. Gouvea, and R. M. Stern, "Multivariate-gaussian-based cepstral normalization for robust speech recognition," in *International Conference on Acoustics, Speech and Signal Processing*, 1995, pp. 137–140.

[19] Y. Normandin, "Hidden Markov models, maximum mutual information estimation and the speech recognition problem," PhD dissertation, McGill University, 1991.

[20] D. Povey, "Discriminative training for large vocabulary speech recognition," PhD dissertation, Cambridge University, 2003.

[21] B. Raj, R. Singh, and R. Stern, "On tracking noise with linear dynamical system models," in *International Conference on Acoustics, Speech and Signal Processing*, Montreal, Canada, May 2004.

[22] M. Riedmiller and H. Braun, "A direct adaptive method for faster backpropagation learning: The RPROP algorithm," in *IEEE International Conference on Neural Networks*, vol. 1, 1993, pp. 586–91.

[23] B. Schuller, M. Wollmer, T. Moosmayr, and G. Rigoll, "Recognition of noisy speech: A comparative survey of robust model architecture and feature enhancement," *EURASIP Journal on Audio, Speech and Music Processing*, 2009.

[24] V. Stouten, H. Van hamme, K. Demuynck, and P. Wambacq, "Robust speech recognition using model-based feature enhancement," in *Proceedings of the 2003 Eurospeech*, Geneva, Switzerland, September 2003, pp. 17–20.

10

Reverberant Speech Recognition

Reinhold Haeb-Umbach, Alexander Krueger
University of Paderborn, Germany

10.1 Introduction

From a usage point of view, there are a number of reasons why in many applications of automatic speech recognition (ASR) distant talking microphones are to be preferred over close-talking microphones. The first is convenience: freeing the user from holding a microphone or wearing a headset increases the ease of use, and thus raises the acceptance of appliances or services operated by voice commands. A second reason is safety: there are numerous applications, where the hands are needed for more important tasks than for holding a microphone to capture the user's speech. Examples include the hands-free control of a cellular phone or a car navigation system while driving, or the control of some apparatus by a surgeon while being busy with an operation. Finally, moving the microphone away from the mouth of the speaker is in line with the disappearing computer and the ambient intelligence paradigm, which has been put forward already for several years [1]. It describes the vision of technology that is invisible, embedded in our surroundings while still being present whenever we need it. Interacting with it should be simple and effortless, and speech, as a "remote control" that a user has with him all the time, is the ideal means of interaction.

However, increasing the distance between the speaker and the microphone has dramatic consequences on the quality of the captured speech signal. There is first the signal attenuation due to the propagation from source to sensor. In free space, the value of the signal power is inversely proportional to the squared distance. Elko [11] considered the following example: assume that the microphone is located 2 cm away from the speaker's mouth. Increasing the distance to 10 cm or 1 m would correspond to an attenuation by 14 and 34 dB, respectively. If one wanted to compensate this loss by the directional gain of a microphone array, this would require at least 5 (omnidirectional) microphones for the distance of 10 cm and 50 microphones for 1 m, assuming an acoustic environment that is characterized by diffuse noise.

Second, if the distance between speaker and microphone is increased, it becomes likely that the sensor also captures other acoustic events, in addition to the desired speaker's signal.

Techniques for Noise Robustness in Automatic Speech Recognition, First Edition.
Edited by Tuomas Virtanen, Rita Singh, and Bhiksha Raj.

Such other acoustic signals could be noise produced by equipment (e.g., fan noise), or speech by other speakers in the same room. If microphone arrays are used which direct their beam of sensitivity towards the speaker, directional signal sources from directions of arrival (DoA) that are sufficiently different from that of the target signal can be well suppressed. However, problematic are distortions that originate from the same DoA; also diffuse noise cannot be suppressed very effectively.

Finally, and this is the theme of this contribution, distant talking microphones will capture reverberant speech, if source and sensor are located in a reverberant enclosure. Reverberation refers to the process of multipath propagation of an acoustic signal from its source to the microphone. The received signal generally consists of a direct sound, reflections that arrive shortly (up to about $80 - 100$ ms) after the direct sound, called *early reverberation* or *early reflections*, and reflections that arrive later, termed *late reverberation* or *late reflections* [19]. Reflections are caused by walls, ceiling, floor, and objects, that are located in the source-sensor enclosure.

This contribution is organized as follows. In the next section, we briefly describe the physical phenomenon of reverberation and discuss its impact on human and machine intelligibility. We also derive an expression which shows how feature vectors used for speech recognition are affected by reverberation. In Section 10.3, we present a taxonomy and an overview of the literature on approaches to improve recognition of reverberant speech. These techniques typically require knowledge of certain characteristics of the reverberation. As a complete characterization in terms of the *acoustic impulse response (AIR)* from the speaker to the microphone is usually not available, simplified models whose parameters are easier to estimate than the true AIR are to be preferred. In Section 10.4, we discuss feature domain models of the AIR that have been used in recognition systems. In Section 10.5, we discuss one feature-enhancement method that has been developed by the authors of this chapter in a bit more detail. The following experimental section presents results both on a small and a large vocabulary task. In particular, we will compare the recognition accuracy obtained by matched reverberant training with what is achieved by the presented feature enhancement and by an acoustic model-compensation technique. Finally, some conclusions are drawn in Section 10.7.

10.2 The Effect of Reverberation

10.2.1 What is Reverberation?

As mentioned in the introduction, reverberation denotes the multipath propagation from an acoustic source to the sensor. As a result, the microphone signal is a superposition of multiple replicas of the source signal, each with different delay and attenuation. A system theoretic model for this dispersion is the convolution of the source signal $s(l)$ with an AIR $h_l(p)$:

$$y(l) = \sum_{p=0}^{\infty} h_l(p) s(l-p). \tag{10.1}$$

The AIR $h_l(p)$ from the source to the microphone is in general time variant. Here, the subscript l indicates the time variance whereas $p \in \mathbb{N}_0$ is the lag index.

The time variance of the AIR is due to changes within the source-sensor enclosure, for example, movements of the speaker, movements in the environment or changes in temperature.

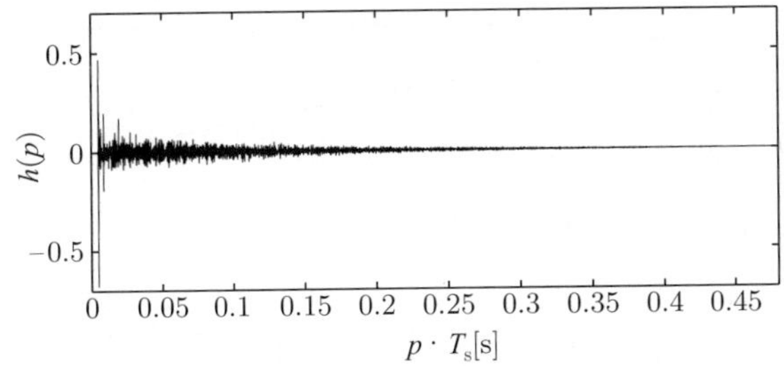

Figure 10.1 Example of a typical acoustic impulse response, measured in a large office room (T_s sampling interval).

To illustrate that even small movements have an effect consider a signal recorded at a sampling rate of $1/T_s = 8$ kHz. The distance traveled during a sampling period is $d = c \cdot T_s = 4.3$ cm, where c denotes the sound velocity of $343\,\mathrm{m/s}$. Therefore, a $4.3\,\mathrm{cm}$ change in the length of an echo path, for example, caused by a slight head movement by the speaker, moves the related impulse by one sampling interval, illustrating the high sensitivity of the impulse response to even smallest movements. Based on a statistical model of room acoustics Radlovic *et al.* [47] demonstrated that even a small movement of the speaker of the order of a tenth of the acoustic wavelength can cause significant changes in the AIR.

However, in the following, we will assume that the AIR is time invariant, that is $h_l(p) = h(p)\ \forall l \in \mathbb{N}_0$. Figure 10.1 shows a typical AIR, that has been measured in a large office room. At the very left of the curve, one can observe the direct signal component, which arrives at the microphone after a short propagation time of below 0.01 s, followed by a few early reflections (the spikes at the beginning of the AIR). The late reflections appear as a noise-like signal with a decaying envelope.

The so-called *energy decay curve* (*EDC*) is insightful for studying the properties of the energy decay of the AIR. The EDC is defined as

$$\mathrm{EDC}_h(l) := \frac{\sum_{p'=l}^{\infty} h^2(p')}{\sum_{p'=0}^{\infty} h^2(p')}. \tag{10.2}$$

Figure 10.2 portrays the EDC corresponding to the AIR of Figure 10.1. The straight line indicates that the energy of the AIR decreases exponentially with time. The deviation from the straight line at very low lags is due to the direct signal component and the early reflections, while the deviation at large lags ($l \cdot T_s > 0.4$ s) is a measurement artefact.

If the distance between the source and the microphone increases, the energy which is related to the direct path decreases, while the combined energy of the early and late reflections is approximately constant. The distance, at which the direct path energy is equal to the combined energy of the early and late reflections is called the *critical distance*. Usually, the microphones are further away than this critical distance, that is the reflective energy is larger than the direct path energy. For example, a room with a volume of $100\,\mathrm{m}^3$ and a reverberation time of $500\,\mathrm{ms}$ has a critical distance of $80\,\mathrm{cm}$, according to Sabine's formula [34].

The convolution of a speech signal with an AIR like the one in Figure 10.1 results in a temporal smearing or dispersion of the input signal. The audible temporal smearing is caused

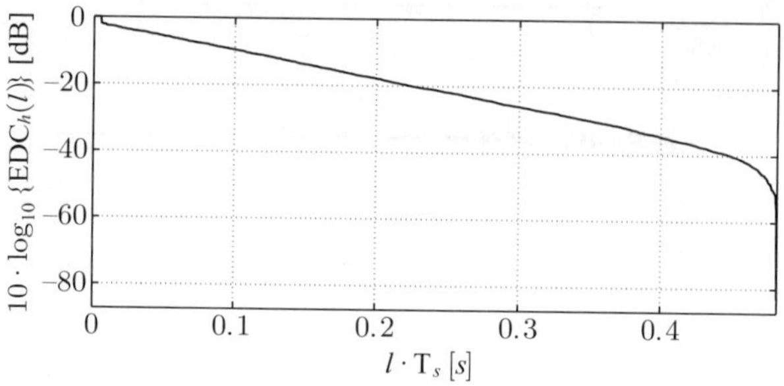

Figure 10.2 Energy decay curve (on a logarithmic ordinate) correspondig to AIR given in Figure 10.1.

by late reverberation, while early reverberation also causes an audible effect called coloration, which is a frequency-dependent amplification or attenuation.

The effect of reverberation on human speech perception has been extensively studied in the literature. Berkley [5] showed that the perception is mainly affected by two factors, coloration and echo. The former is related to the *direct-to-reverberation ratio (DRR)*, which is defined to be the ratio of the energy of the direct sound to the total reflective energy (both early and late reflections). The latter in turn can be quantified by the reverberation time T_{60}, which is defined to be the time required for the energy of the AIR to decay by 60 dB compared to its initial level. Typical reverberation times are in the range from 200 to 1000 ms. Decreasing DRR and increasing T_{60} negatively affect speech intelligibility. The example in Figure 10.1 has a DRR of about 0 dB and $T_{60} \approx 750$ ms.

Still, the intelligibility of reverberant speech by humans is strikingly good, and is often attributed to the so-called *precedence effect*. This psychoacoustic effect states that similar sounds arriving from different locations are solely localized in the direction of the first sound wave arriving at our ears [39]. Early reverberation is actually perceived to reinforce the direct sound and is, therefore, considered useful with regard to speech intelligibility. Late reverberation, on the other hand, impairs speech intelligibility.

10.2.2 The Relationship between Clean and Reverberant Speech Features

In this section, we will derive an analytic expression which relates feature vectors computed from a reverberant speech signal to those of the corresponding nonreverberant signal.

The microphone signal $x(l)$ is modeled to be the superposition of reverberant speech $y(l)$ and noise $n(l)$:

$$x(l) = y(l) + n(l), \tag{10.3}$$

where $y(l)$ is given by Equation (10.1). However, in the following, we will assume a time invariant AIR of finite length, that is

$$h_l(p) = \begin{cases} h(p) & 0 \leq p < L_h \\ 0 & \text{else} \end{cases} \quad \forall l \in \mathbb{N}_0. \tag{10.4}$$

Further note that we assume that only a single microphone is present.

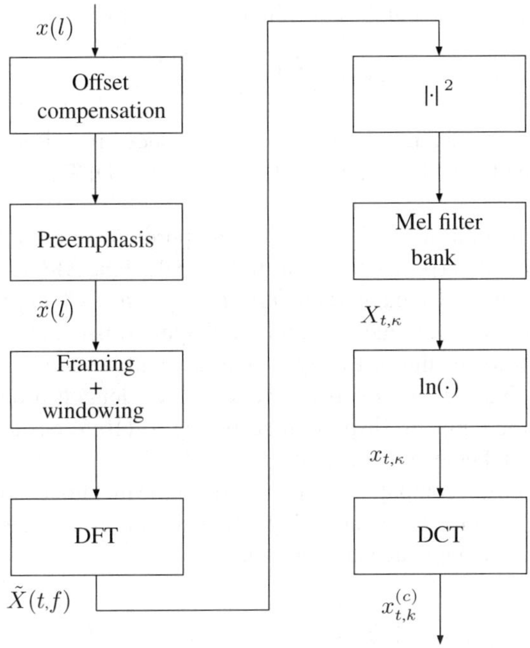

Figure 10.3 Modified version of the feature extraction according to ETSI standard front end (SFE), ES 201 108 [13]. The modification is the replacement of the magnitude spectrum by the power spectrum and the replacement of the logarithmic frame energy by the zeroth cepstral component.

The noise component $n(l)$ includes all reverberant noise signals, which originate from noise sources located in the environment, as well as inherent microphone noise. The two signal components, $s(l)$ and $n(l)$, are modeled as realizations of independent random processes, since they originate from independent sources.

The feature extraction to be studied next follows mostly the ETSI ES 201 108 standard front end (SFE) for the computation of *mel frequency cepstral coefficients (MFCCs)* [13]. Two slight modifications have been applied, which simplify the mathematical analysis: the magnitude spectrum is replaced by the power spectrum and the logarithmic frame energy is replaced by the zeroth cepstral component. Figure 10.3 illustrates the processing stages of the feature extraction and the notation used for intermediate signals.

First, bias removal and preemphasis are applied to the input signal given by Equation (10.3). The resulting signal $\tilde{x}(l)$ is subsequently framed with a frame shift of B samples and windowed with a Hamming window, whose length is denoted by L_w. The windowed signal is transformed into the frequency domain by an F-point *discrete Fourier transform (DFT)*. In the *short-time discrete Fourier transform (STDFT)* domain the relationships of Equations (10.1) and (10.3) turn into the following relationship among the STDFTs $\tilde{X}(t, f)$, $\tilde{S}(t, f)$, and $\tilde{N}(t, f)$ of the noisy reverberant speech signal $\tilde{x}(l)$, the clean speech signal $\tilde{s}(l)$ and the noise signal $\tilde{n}(l)$:

$$\tilde{X}(t, f) \approx \sum_{t'=0}^{L_{\mathcal{H}}} \tilde{S}(t - t', f)\mathcal{H}(t', f) + \tilde{N}(t, f), \tag{10.5}$$

where the number of summands in the convolution is given by:

$$L_{\mathcal{H}} = \left\lfloor \frac{L_h + L_w - 2}{B} \right\rfloor. \tag{10.6}$$

In (10.5), t denotes the frame index, while f is the frequency bin index. $\mathcal{H}(t, f)$ is a frequency domain representation of the AIR, which is different to the STDFT $H(t, f)$ in general (see [31] for details).

Note that the convolution of (10.1) results in an approximate convolution in the STDFT domain, however now with respect to the frame index. Only if the AIR length is short compared to that of the window, the so-called *multiplicative transfer function approximation* (*MTFA*) is valid and the convolution in the time domain is well approximated by a multiplication in the STDFT domain [3]. However, this is usually not the case with reverberation, which is typically in the order of several hundred milliseconds, while window lenghts used in ASR are between 20 and 40 ms. Then the temporal dispersion according to (10.5) occurs, that is, $\tilde{X}(t, f)$ not only depends on $\tilde{S}(t, f)$, but also on the past $\tilde{S}(t - t', f)$, $t' = 1, \ldots, L_{\mathcal{H}}$.

After the power spectrum computation, application of the mel filter bank and transformation into the logarithmic domain, the relationship between nonreverberant and the reverberant features becomes highly complicated and nonlinear [31]:

$$x_{t,\kappa} = \ln\left(\sum_{t'=0}^{L_{\mathcal{H}}} \mathrm{e}^{s_{t-t',\kappa} + \overline{h}_{t',\kappa}} + \mathrm{e}^{n_{t,\kappa}}\right) + v_{t,\kappa}. \tag{10.7}$$

Here, $x_{t,\kappa}$, $s_{t,\kappa}$ and $n_{t,\kappa}$ are the *logarithmic mel power spectral coefficients (LMPSCs)* of the noisy reverberant speech, the nonreverberant speech and of the noise signal, respectively, at frame t and mel filter index κ, $\kappa = 1, \ldots, N_\kappa$, where N_κ is the number of triangular shaped mel filters. The term $v_{t,\kappa}$ captures all errors resulting from the various approximations that had to be introduced to arrive at (10.7). These include, among others, an error term resulting from the averaging operation within a mel filter, the error introduced by the omission of the cross terms between speech and noise STDFTs in the power spectrum computation, and the error resulting from using a simplified model of the AIR (see Section 10.4 further [31]).

The coefficients

$$\overline{h}_{t',\kappa} := \ln\left(\overline{\mathcal{H}}_{t',\kappa}\right) \tag{10.8}$$

can be interpreted as a logarithmic mel power spectral representation of the AIR with

$$\overline{\mathcal{H}}_{t',\kappa} := \left(\frac{1}{F_\kappa^{(\mathrm{up})} - F_\kappa^{(\mathrm{lo})} + 1}\right) \sum_{f=F_\kappa^{(\mathrm{lo})}}^{F_\kappa^{(\mathrm{up})}} |\mathcal{H}(t', f)|^2 \tag{10.9}$$

denoting the averaged AIR power per mel band, where $F_\kappa^{(\mathrm{lo})}$ and $F_\kappa^{(\mathrm{up})}$ are the frequency indices of the band edges of the κth mel filter.

Capturing all coefficients $x_{t,\kappa}$, $\kappa = 1, \ldots, N_\kappa$ in a vector $\mathbf{x}_t$ finally results in the following relationship between the LMPSCs of clean speech, noise and noisy reverberant speech signals:

$$\begin{aligned} \mathbf{x}_t &= \ln\left(\sum_{t'=0}^{L_{\mathcal{H}}} \mathrm{e}^{\mathbf{s}_{t-t'} + \overline{\mathbf{h}}_{t'}} + \mathrm{e}^{\mathbf{n}_t}\right) + \mathbf{v}_t \\ &= f\left(\mathbf{s}_{t-L_{\mathcal{H}}:t}, \overline{\mathbf{h}}_{0:L_{\mathcal{H}}}, \mathbf{n}_t\right) + \mathbf{v}_t, \end{aligned} \tag{10.10}$$

where

$$f\left(\mathbf{s}_{t-L_{\mathcal{H}}:t}, \overline{\mathbf{h}}_{0:L_{\mathcal{H}}}, \mathbf{n}_t\right) := \ln\left(\sum_{t'=0}^{L_{\mathcal{H}}} \mathrm{e}^{\mathbf{s}_{t-t'}+\overline{\mathbf{h}}_{t'}} + \mathrm{e}^{\mathbf{n}_t}\right), \tag{10.11}$$

and where

$$\mathbf{s}_{t-L_{\mathcal{H}}:t} := \mathbf{s}_{t-L_{\mathcal{H}}}, \ldots, \mathbf{s}_t \tag{10.12}$$

$$\overline{\mathbf{h}}_{0:L_{\mathcal{H}}} := \overline{\mathbf{h}}_0, ..., \overline{\mathbf{h}}_{L_{\mathcal{H}}} \tag{10.13}$$

denote the sequence of the vectors for the logarithmic mel power spectrum of clean speech and the representation of the AIR.

The last step in the feature extraction according to Figure 10.3 is the application of the *discrete cosine transform (DCT)* to $\mathbf{x}_t$ to obtain the MFCCs $x_{t,k}^{(c)}$, where $k = 0, \ldots, K-1$ denotes the cepstral index.

Figure 10.4 shows an example of the logarithmic mel power spectrum of a nonreverberant and a reverberant speech signal, respectively, which corresponds to the connected digits utterance (“one”, “one”, “six”, “eight”, “five”, “two”, “two”). From the figure, a temporal dispersion can be observed in the logarithmic mel power spectrum of the reverberant signal.

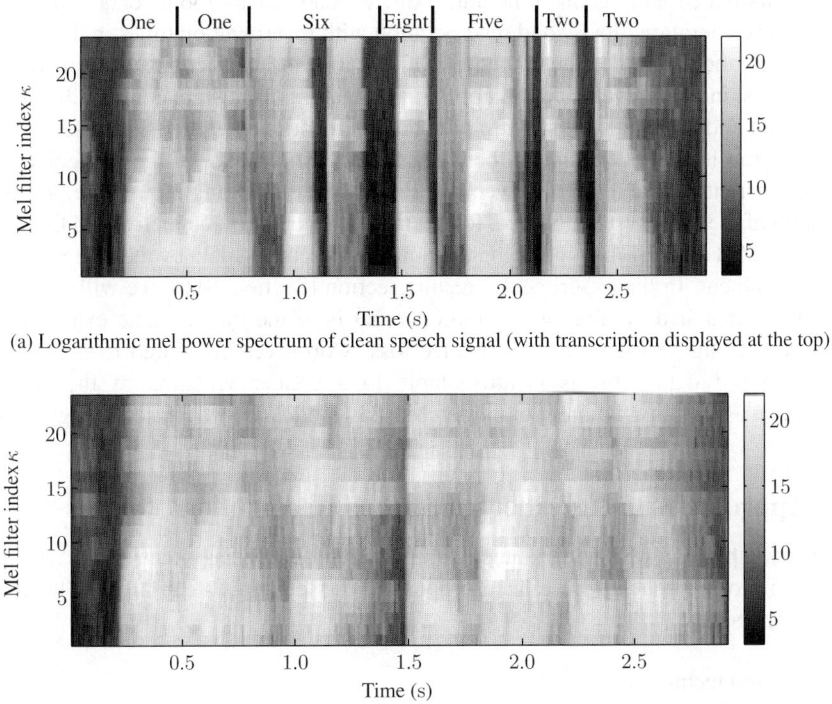

(a) Logarithmic mel power spectrum of clean speech signal (with transcription displayed at the top)

(b) Logarithmic mel power spectrum of corresponding reverberant speech signal

Figure 10.4 Logarithmic mel power spectra of exemplary nonreverberant and reverberant speech signal (reverberation time $T_{60} \approx 450$ ms) corresponding to the connected digits utterance (“one”, “one”, “six”, “eight”, “five”, “two”, “two”).

Table 10.1 Word error rate [%] on the Aurora 5 and the reverberated Aurora 4 database using the ETSI standard front end.

	Nonreverberant	Office	Living room
Aurora 5	0.64	6.32	14.94
Aurora 4-rev	14.00	47.47	73.44

Consider, for example, the glottal stop during the pronunciation of the digit "six" at about 1.15 s, which is clearly visible in the LMPSC trajectory of the nonreverberant signal in Figure 10.4a, while it has completely disappeared in the reverberant case (see Figure 10.4b).

10.2.3 The Effect of Reverberation on ASR Performance

The effect of reverberation on the ASR performance is illustrated in Table 10.1, which presents word error rates on the Aurora 5 database, which is related to a connected digit recognition task, and a reverberated version of the Aurora 4 database, here referred to as Aurora 4-rev, which is related to a large vocabulary recognition task (see Section 10.6.1 for a description of the databases). While acoustic model training is always performed on nonreverberant data, the test set is subdivided into "nonreverberant," "office," and "living room" data, where the first exhibits no reverberation, the second reverberation with reverberation times in the range from 300 to 400 ms, and the third in the range from 400 to 500 ms. The recognizer employed MFCC feature vectors, including velocity and acceleration features, that were computed by the ETSI standard front end [13] with the slight modifications mentioned earlier. The dispersive effect of reverberation leads to a mismatch between the acoustic model of the recognizer trained on nonreverberant data and the observed reverberant data, resulting in a severe performance degradation of the ASR system. Notice, for example, the factor of ten increase in the error rate on the Aurora 5 database when replacing the nonreverberant test data by the data related to the office environment. In the experimental results section (Section 10.6), we will see that even in the case of matched training-test conditions, that is, if the training data exhibit the same reverberation as the test data, a performance loss is observed compared to nonreverberant training and test data. This is is attributable to a greater violation of the conditional independence assumption inherent in hidden Markov model (HMM)-based recognizers.

10.3 Approaches to Reverberant Speech Recognition

Research on improving ASR performance on reverberant speech has intensified in recent years, and a number of approaches have been proposed. The techniques can broadly be classified into three categories:

1. Signal-based techniques
2. Front-end techniques
3. Back-end techniques

depending on where in the processing chain reverberation is addressed.

10.3.1 Signal-Based Techniques

Signal-based techniques aim at an enhancement of the time domain signal by speech dereverberation. This is a complicated task if dereverberation is to be performed blindly, since the source signal and the AIR, which usually consists of several hundred coefficients, are both unknown to the receiver. Problems are exacerbated by the fact that the z-transform of the AIR, the *acoustic transfer function (ATF)*, is nonminimum phase and often has zeros close to the unit circle, which cause significant noise amplification if equalized. We already mentioned the high sensitivity of the AIR toward small speaker movements which means that in a practical setting the AIR has to be considered time variant, further complicating the estimation task.

Two classes of approaches to dereverberation may be discerned: reverberation cancellation and reverberation suppression. Reverberation cancellation aims at equalizing the AIR, while reverberation suppression targets the reduction of the effect of reverberation.

A more detailed review of techniques for dereverberating the acoustic signal can be found in [19, 24, 45] and Chapter 3.

Reverberation Cancellation

Methods falling in the first class are theoretically more appealing as they try to remove the cause of reverberation rather than its effect. A pioneering work in blind acoustic channel equalization was done by Miyoshi and Kaneda [42]. They developed a multichannel technique, known as *multiple input/output inverse theorem* (*MINT*), by which theoretically perfect inverse filtering can be achieved if the AIRs from the source to the microphones are known, even in the case of mixed-phase AIRs. The existence of the inverse filter is guaranteed if the number of microphones exceeds the number of active sources and if the ATFs in the individual channels do not share common zeros. However, even a moderate channel estimation error causes significant spectral distortions in the output signal.

In most applications, one cannot expect the AIRs to be time invariant and known in advance. The estimation of the AIRs poses a significant challenge on its own right. Even theoretically, the channel is only identifiable under two conditions: first, the ATFs from the source to the sensors are not allowed to have common zeros in the z-plane, and second, the autocorrelation matrix of the source signal must be full rank [25]. But even then the ATFs can only identified up to a multiplicative constant, and the issue remains how to separate the transfer function of the vocal tract from the ATFs. For an identifiable system, Huang and Benesty [25] proposed two multichannel adaptive approaches, based on the construction of an error signal based on the cross correlations between different channels, whose power is then minimized using *least mean square (LMS)* and Newton algorithms.

If the source signal samples are assumed to be independent and identically distributed (i.i.d.), equalization can be achieved without an explicit estimation of the AIR, that is, rather than estimating the AIR and inverting it, an attempt is made to directly estimate the inverse by whitening the input signal. It is based on an application of linear prediction for blind equalization. However, the hypothesis of i.i.d. samples does not hold for speech signals, and whitening the input signal destroys its correlation structure. To compensate for that effect, the transfer function of the vocal tract needs to be estimated and accounted for in the inverse filter [7]. As the impulse response of the vocal tract is considerably shorter than the AIR, an alternative is to conduct long-term multistep linear prediction, whereby only the higher lag correlations due to reverberation are removed. Thereby a sample is predicted not by its immediate predecessors but by samples further in the past [30].

Reverberation Suppression

Techniques falling in the second class do not equalize the AIRs but reduce the effect of reverberation instead. One of the earliest approaches to reverberation suppression is the enhancement of the *linear prediction (LP)* residuals by Allen [2]. According to the well-known source-filter model, the speech production is described in terms of an excitation sequence (the so-called LP residual) driving a time variant all-pole filter. The excitation signal is either random noise for unvoiced speech or a quasi-periodic pulse train for voiced speech. The effect of reverberation on voiced speech is seen in the LP residual by extraneous peaks caused by multipath reflections. Consequently dereverberation is achieved by attenuating these peaks and synthesizing the enhanced speech waveform using the modified LP residual and the time varying all-pole filter with coefficients calculated from the reverberant speech. The assumption that only the LP residual is affected by reverberation, while the *linear predictive coding (LPC)* coefficients characterizing the vocal tract remain unaffected, was later dropped. A detailed analysis of the effect of reverberation on the poles of the LP model can be found in [16].

Reverberation is known to reduce the modulation index, that is the degree of amplitude modulation of the speech signal. Temporal envelope filtering aims at restoring the depth of the amplitude modulation by applying a temporal filtering operation to the time trajectory of the output of subband filters [35, 57].

Another approach to reverberation cancellation that has been widely studied is a spectral-enhancement technique that borrows ideas from spectral subtraction known from the enhancement of noisy speech. Here, the late reverberation is considered to be uncorrelated with the direct sound component, an assumption motivated by the randomness of late reflections. Then the spectrum of the late reverberation is estimated and subtracted from that of the reverberant signal to obtain an estimate of the spectrum of the nonreverberant signal [19, 36].

Unfortunately, speech enhancement resulting in less reverberant speech from the viewpoint of a human listener does not necessarily translate into an improved ASR accuracy. A likely explanation is that the processing may introduce artefacts, which, even if inaudible, have an unpredictable effect on the subsequent feature extraction leading to a degradation of the recognition performance. An observation made by Delcroix *et al.* [8] supports this conjecture. They have noticed that speech dereverberation by long-term multistep linear prediction increases the variance of the acoustic features.

10.3.2 Front-End Techniques

Front-end methods aim at computing acoustic features for speech recognition which are insensitive to reverberation. As the human speech perception is much less sensitive to reverberation than today's speech recognition systems are, perceptually motivated processing stages have been included in the acoustic feature extraction. Among those are an analysis of the modulation spectrum of speech with the aim of enhancing the sensitivity towards amplitude modulations in the range of 2–16 Hz with a maximum sensitivity at 4 Hz, the syllable rate of speech [29], the use of a gammatone filterbank replacing the mel filters [41], and nonlinear filtering of the power spectral coefficients of each frequency band to mimick the precedence effect [27].

Other front-end approaches aim at normalizing the features such that they are insensitive to reverberation or removing the effect of reverberation on the computed features, as will be described next.

Feature Normalization

The most well-known feature normalization technique is probably *cepstral mean normalization (CMN)*, also called *cepstral mean subtraction (CMS)*. It is based on the following idea: the *discrete-time Fourier transform (DTFT)* $\tilde{Y}(e^{j\theta})$ of the reverberated speech signal $y(l)$ can be written as the product of the clean speech signal DTFT $\tilde{S}(e^{j\theta})$ and the DTFT of the AIR $H(e^{j\theta})$:

$$\tilde{Y}(e^{j\theta}) = \tilde{S}(e^{j\theta})H(e^{j\theta}).$$

If a STDFT is applied instead of the DTFT, the multiplicative relationship still approximately holds if the analysis window is larger than the duration of the AIR:

$$\tilde{Y}(t, f) \approx \tilde{S}(t, f)H(0, f). \tag{10.14}$$

Applying the log-operation on the power spectrum then gives

$$\ln |\tilde{Y}(t, f)|^2 \approx \ln |\tilde{S}(t, f)|^2 + \ln |H(0, f)|^2,$$

and the relationship remains additive when going to the cepstral domain. Note that $\ln |H(0, f)|^2$ is independent of the frame index t. Thus, if the temporal mean is subtracted from $\ln |\tilde{Y}(t, f))|^2$, or from its cepstral representation, the result is independent of the ATF H. The MFCC feature extraction described in Section 10.2.2 additionally includes the mel filterbank, but the above argumentation still approximately holds.

CMN has been successfully applied to MFCC feature vectors to make a speech recognizer insensitive to different microphone characteristics. However, its usefulness for reverberant speech recognition is limited, because the duration of the AIR is usually much larger than typical window sizes used in speech recognition. Then the MTFA does not hold and the simple multiplicative relationship of Equation (10.14) has to be replaced by a convolution (see Equation (10.5)). As a result, classical CMN and related techniques, such as the *RelAtive SpecTrAl* method known by the acronym RASTA [20], perform only poorly on reverberant speech [28]. As a remedy, Avendano *et al.* proposed to use an analysis window of the order of 1–2 s [4]. After this so-called *long-term cepstral mean normalization* a transformation is applied to obtain features with a time-frequency resolution typically used in speech recognition. Alternatively, the time domain signal can be reconstructed after long-term CMN followed by ordinary ASR feature extraction. By doing so, a significant reduction of the word error rate in the presence of reverberation could be achieved compared to the performance of the ETSI SFE. However, the technique introduces a latency on the order of several seconds [17].

In CMN, each feature vector component is processed separately. Alternatively, a full transformation matrix can be applied to the feature vector as a whole. In *constrained maximum likelihood linear regression* (*CMLLR*), also called *feature space maximum likelihood linear regression* (*FMLLR*), the transformation matrix is chosen such as to maximize the likelihood on adaptation data [15]. However, this affine transformation approach suffers from the same deficiencies as ordinary CMN: since the MTFA does not hold for reverberant speech due to the small analysis windows used in speech recognition, the gains by the frame-by-frame transformation of CMLLR are limited. If the feature vector to be transformed also contains dynamic features, the vector captures the temporal smearing introduced by reverberation to some extent, and some improvements in word accuracy by CMLLR/FMLLR can be observed [31]. Model adaptation techniques such as CMLLR are discussed in Chapter 11 of this book.

Feature Enhancement

Feature enhancement techniques aim at estimating the feature vectors of the underlying clean speech signal, given the feature vectors computed from the corresponding reverberant speech signal. If a Bayesian framework is used, the problem can be stated as follows: given an *a priori* model of clean speech features and an observation model that describes the relationship between clean and reverberant features, estimate the posterior probability of the clean features, given the observed reverberant features.

The Bayesian approach was taken in [59] and in own prior work [31, 32]. The observation model used by Wölfel [59] modeled reverberation as an additive distortion in the mel power spectral domain, as is suggested by the late reflection model of [36], which was already mentioned in Section 10.3.1. Its parameters were estimated by multistep linear prediction [30]. In contrast, Krueger and Haeb-Umbach [31] used the observation model derived in Section 10.2.2, which accounts for the convolutive effect of reverberation in the feature domain. The error term $\{\mathbf{v}_t\}_{t\in\mathbb{N}}$ in Equation (10.10) was assumed to be a realization of a white Gaussian stochastic process with mean vector $\boldsymbol{\mu}_{\mathrm{v}}$ and covariance matrix $\boldsymbol{\Sigma}_{\mathrm{v}}$. Options for estimating these parameters are discussed in [32].

Other differences between the two works are in the used *a priori* model and the inference algorithm. Wölfel [59] employed higher order autoregressive processes as *a priori* model and computationally expensive particle filters for tracking. On the other hand, Krueger and Haeb-Umbach [31] employed a *switching linear dynamic model* [26] as an *a priori* model of the trajectory of clean speech feature vectors and a bank of Kalman filters for inference. More details of this approach can be found in Section 10.5.

10.3.3 Back-End Techniques

The third category of approaches to reverberant speech recognition are concerned with modifying the acoustic models and the decoding rule of the recognizer.

Modification of the Acoustic Model

A straight-forward approach to obtain an acoustic model appropriate for the recognition of reverberant speech is to train the recognizer with reverberant data recorded in the target environment. If the exact test conditions are not known at training time, a pool of models for different reverberation conditions can be trained in advance, and the one of the trained models that is deemed appropriate is selected during recognition [6]. To avoid the effort of recording a database in the target environment, alternatively only the AIR can be measured and the reverberant training data is generated by convolving the signals of the nonreverberant database with the measured AIR [54]. However, this does not account for the time variance of the AIR and thus may lead to less effective acoustic models. To even save the effort of an AIR measurement, artificially generated AIRs, for example, generated by the image method [2], can be employed, if their parameters, such as the reverberation time, approximately match those of the target environment.

Alternatively, an acoustic model that is trained on nonreverberant data, can be adapted to the test reverberation conditions. This can be done either statically, that is prior to recognition, or dynamically and continuously alongside recognition.

The aforementioned *maximum likelihood linear regression* (*MLLR*) [37], an example of static adaptation, can also be applied to the acoustic models rather than to the feature vectors. The advantage is that different transformation matrices can be used for different regions of the acoustic space, while CMLLR is restricted to a single global transformation. The number of transformation matrices can be chosen according to the amount of adaptation data available. Originally meant to adapt an acoustic model to new speakers, it was employed for the recognition of reverberant speech by Toh *et al.* [56]. However, the same restrictions apply that were mentioned earlier in the context of CMLLR: MLLR assumes that the distortion of the acoustic features can be described by affine transforms applied to individual feature vectors. However, this is an inadequate assumption for reverberation, even if the acoustic feature vector includes velocity and acceleration features which are computed from an interval of several successive frames, as reverberation often extends well beyond this interval.

A more exact model must account for the temporal convolution in the spectral domain, as is seen from Equations (10.5) and (10.7). Thus, the modifications of the emission probabilities of the acoustic model necessary to account for reverberation greatly depend on the preceding context. One way to do so is to split an HMM state into substates to take into account the energy dispersion due to reverberation [49]. The number of substates is chosen according to the average state occupancy within an HMM state.

Hirsch and Finster [22] proposed to adapt the HMM means, originally trained on nonreverberant data, to the reverberant test data prior to recognition in a way that is reminiscent of the well-known *parallel model combination (PMC)* method [14]: the mean of the emission probability of a certain HMM state can be adapted to the reverberant environment by retransforming it into the power spectral domain, adding the contribution of the preceding states and transforming it back to the feature, that is cepstral, domain. In [22], the contribution of the preceding HMM states to a current state caused by reverberation is assumed to decay exponentially with the temporal distance from the current state. With this and an assumption about the average state duration, the modified mean of the HMM state emission probability was computed from the emission probabilities of the preceding states. Obviously, this requires knowledge of the preceding states. If whole-word HMMs are assumed, this information is available, at least within a word. However, a dispersion of energy across word boundaries could not be accounted for. In a triphone-based recognizer the left context that can be accounted for by this kind of static adaptation is even smaller (on the order of 100 ms) and thus much smaller than typical reverberation times encountered in offices or living rooms.

A better treatment of the energy dispersion is possible with dynamic adaptation, where the adaptation of the models is carried out in parallel to the decoding of the word sequence. During decoding different hypotheses about the left context of a current HMM state are evaluated, and for each context considered its specific impact on the emission probability of the current state caused by reverberation can be evaluated [53, 55]. However, this much more precise modeling is bought at the price of a greatly increased computational complexity. Therefore, acoustic model adaptation techniques have been mostly used in the context of small vocabulary recognition tasks.

Modification of the Decoder

In the method proposed by Sehr *et al.* [50], the effect of reverberation is accounted for during decoding by modifying the acoustic likelihood computation. The likelihood computation is concerned with the evaluation of the emission probability of an HMM state q_t at the given observation $\mathbf{x}_t$. By introducing the clean feature vector $\mathbf{s}_t$ and the representation of the AIR $\overline{\mathbf{h}}_{0:L_\mathcal{H}}$ as hidden variables the acoustic likelihood of an observation $\mathbf{x}_t$ for a given HMM state q_t can be expressed as

$$\begin{aligned} p(\mathbf{x}_t|q_t) &= \int\int p(\mathbf{x}_t, \mathbf{s}_{t-L_\mathcal{H}:t}, \overline{\mathbf{h}}_{0:L_\mathcal{H}}|q_t)\mathrm{d}\mathbf{s}_{t-L_\mathcal{H}:t}\mathrm{d}\overline{\mathbf{h}}_{0:L_\mathcal{H}} \\ &= \int\int p(\mathbf{x}_t|\mathbf{s}_{t-L_\mathcal{H}:t}, \overline{\mathbf{h}}_{0:L_\mathcal{H}})p(\mathbf{s}_{t-L_\mathcal{H}:t}|q_t)p(\overline{\mathbf{h}}_{0:L_\mathcal{H}})\mathrm{d}\mathbf{s}_{t-L_\mathcal{H}:t}\mathrm{d}\overline{\mathbf{h}}_{0:L_\mathcal{H}}. \quad (10.15) \end{aligned}$$

Here, the AIR representation $\overline{\mathbf{h}}_{0:L_\mathcal{H}}$ is modeled as a realization of a random variable, which can be assumed to be statistically independent of the clean speech feature vector. Further, we made use of the fact that $\mathbf{x}_t$ can be considered independent of q_t, if $\mathbf{s}_{t-L_\mathcal{H}:t}$ is given.

Sehr *et al.* [50] assumed that the error term $\mathbf{v}_t$ in (10.10) is zero and that additive noise is absent ($\mathbf{n}_t = 0$). Then $\mathbf{x}_t$ is a deterministic function of the clean speech features and the AIR representation:

$$p(\mathbf{x}_t|\mathbf{s}_{t-L_\mathcal{H}:t}, \overline{\mathbf{h}}_{0:L_\mathcal{H}}) = \delta\left(\mathbf{x}_t - f\left(\mathbf{s}_{t-L_\mathcal{H}:t}, \overline{\mathbf{h}}_{0:L_\mathcal{H}}\right)\right) \quad (10.16)$$

and (10.15) becomes

$$\begin{aligned} &p(\mathbf{x}_t|q_t) = \\ &\int\int \delta\left(\mathbf{x}_t - f\left(\mathbf{s}_{t-L_\mathcal{H}:t}, \overline{\mathbf{h}}_{0:L_\mathcal{H}}\right)\right) p(\mathbf{s}_{t-L_\mathcal{H}:t}|q_t)p(\overline{\mathbf{h}}_{0:L_\mathcal{H}})\mathrm{d}\mathbf{s}_{t-L_\mathcal{H}:t}\mathrm{d}\overline{\mathbf{h}}_{0:L_\mathcal{H}}. \quad (10.17) \end{aligned}$$

Here, $\delta(\cdot)$ denotes the Dirac delta function. This expression states that the integration has to be carried out over all possible clean speech and AIR realizations. However, only those combinations deliver a contribution, for which $f\left(\mathbf{s}_{t-L_\mathcal{H}:t}, \overline{\mathbf{h}}_{0:L_\mathcal{H}}\right)$ is equal to the observed $\mathbf{x}_t$. Sehr *et al.* [50] have approximated this integral by the integrand that delivers the largest contribution resulting in the following acoustic likelihood:

$$\begin{aligned} p(\mathbf{x}_t|q_t) := \max_{\mathbf{s}_{t-L_\mathcal{H}:t}, \overline{\mathbf{h}}_{0:L_\mathcal{H}}} &\left\{p(\mathbf{s}_{t-L_\mathcal{H}:t}|q_t)p(\overline{\mathbf{h}}_{0:L_\mathcal{H}})\right\} \\ \text{subject to } \mathbf{x}_t &= f\left(\mathbf{s}_{t-L_\mathcal{H}:t}, \overline{\mathbf{h}}_{0:L_\mathcal{H}}\right). \quad (10.18) \end{aligned}$$

Thus, each acoustic likelihood computation requires an optimization to be carried out, which significantly increases the decoding complexity. So far, this concept could only be applied to mel or log mel power spectral representations of the speech signal, single Gaussian emission probabilities, and noise-free input data [51]. An efficient algorithm for the widely used MFCCs and *Gaussian mixture models (GMMs)* has yet to be developed.

Another variant of accounting for the impact of reverberation in the decoder is the use of uncertainty information. Thereby the contribution of those features to the overall decision on the word sequence is deemphasized which are deemed to be strongly affected by reverberation and thus to be unreliable. In the so-called *missing-data* technique unreliable features are even completely excluded from consideration [18].

Finally, it should be mentioned that different techniques for reverberant speech recognition can also be combined. For example, speech dereverberation may be followed by a modification of the decoder in the spirit of uncertainty decoding, where those features are assigned an increased variance which are considered to be highly distorted [8].

10.3.4 Concluding Remarks

Speech dereverberation techniques have the advantage that they deliver a dereverberated time domain signal, that may be used for human-to-human communication. On the other hand, the previous discussion showed that signal-based techniques tend to be less robust than feature-based methods: the estimation the the AIR or of the inverse filter is very challenging, in particular, in time variant and noisy acoustic environments. Further, signal-based techniques typically require multiple microphones, while the other methods that are closer to the recognizer usually operate on single-channel input, which may be considered an important practical advantage.

Acoustic model and decoder-based techniques tend to be computationally expensive compared to feature-based methods, and therefore seem to be applicable only for recognition tasks with a small vocabulary size. A further advantage of feature enhancement methods is that they basically leave the feature extraction and the back-end untouched, as the enhancement is carried out after feature extraction and prior to decoding. Thus, it is conjectured that their integration into different recognition systems should be less challenging than the integration of model-based or decoder-based methods.

All techniques require more or less exact knowledge about the reverberation conditions. Signal-based techniques aim at estimating the AIR or an inverse filter, which is still a widely unsolved task in the presence of time variant acoustic conditions. Many front-end and back-end methods employ simplified models of the AIR. This is often sufficient, as feature extraction conducts a decimation in time and frequency anyway: the frame rate is much smaller than the sampling rate (e.g., 100 vs. 8000 Hz) and the number of mel channels is smaller than the DFT size (e.g., 23 channels vs. 256 DFT bins). Different options for a feature domain AIR model are discussed next.

10.4 Feature Domain Model of the Acoustic Impulse Response

The feature domain representation of the AIR can be either a deterministic or a stochastic model. While the first has been adopted by Hirsch and Finster [22] and by Krueger and Haeb-Umbach [31], a stochastic model is employed in [50, 51]. The stochastic model is a way to account for the time variance of the AIR, but Sehr *et al.* [51] observed that the stochastic model is beneficial, even if the AIR is time invariant. The reason probably is that in their model the error term $\mathbf{v}_t$ in Equation (10.10) has been neglected, and the fluctuations due to this term then have to be captured by the AIR model.

Sehr *et al.* [50] considered a matrix $\overline{\mathbf{H}}$ as mel power spectral domain AIR model, whose elements $\overline{H}_{t',\kappa}$, where t' is the lag (row) index and κ the mel filter (column) index, are modeled as independent random variables. The sequence of matrices $\overline{\mathbf{H}}_t$, where t is the frame index, is assumed to be an i.i.d. matrix-valued stationary Gaussian random process. Different methods have been proposed to estimate the means and variances of the matrix elements.

The most straight-forward way is to conduct a number of AIR measurements with different loudspeaker and microphone positions in the target room, compute the mel power spectral domain representation of each AIR and estimate the mean and the variance for each mel band and lag by computing the empirical mean and variance of these representations [50]. The model can also be derived from a few calibration utterances, whose transcriptions are known and which are recorded in the target environment. Maximum likelihood estimates of the means and variances are derived in [52]. A blind-estimation technique has been proposed by Wen *et al.* [58].

Hirsch and Finster [22] used a much coarser model of the AIR. In an idealized model the effect of multiple reflections of sound in a room can be described by an exponential decay of the acoustic energy [34]. Since the authors were only interested in an average representation they replaced the square of the AIR by its envelope which they modeled as a decaying exponential of finite length:

$$h^2(l) \approx \sigma_h^2 \cdot u(l) \cdot \mathrm{e}^{-2l/\tau}, \tag{10.19}$$

where

$$\tau = \frac{T_{60}}{3\ln(10) \cdot T_s} \tag{10.20}$$

(T_s: sampling interval). The AIR's support indicator function

$$u(l) := \begin{cases} 1 & \text{for } 0 \leq l < L_h \\ 0 & \text{else} \end{cases} \tag{10.21}$$

enforces the AIR to be causal having a finite length L_h. Further, σ_h is a normalizing scalar which determines the AIR's overall energy.

The authors of this chapter used a more sophisticated AIR model [31], which was originally proposed by Polack [46]. Here, the AIR is modeled as a realization of a zero-mean white Gaussian process $\zeta(l)$ of finite length with an exponentially decaying envelope

$$h(l) \approx \sigma_h \cdot u(l) \cdot \zeta(l) \cdot \mathrm{e}^{-\frac{l}{\tau}}. \tag{10.22}$$

In Monte Carlo simulations it was observed that the distributions of the feature domain AIR coefficients $\overline{h}_{t',\kappa}$ defined in Equation (10.8), under the stochastic AIR model (10.22) can be well approximated by Gaussians.

If the distributions of the log mel power spectral coefficients are Gaussians, the coefficients $\overline{\mathcal{H}}_{t',\kappa}$ defined in Equation (10.8), are lognormally distributed. From the mean and variance of the lognormal distribution for $\overline{\mathcal{H}}_{t',\kappa}$ the mean of the normal density of $\overline{h}_{t',\kappa}$ is obtained as follows:

$$E\left[\overline{h}_{t',\kappa}\right] = \frac{1}{2}\ln\left(\frac{\mu_{t',\kappa}^{4(\mathcal{H})}}{\sigma_{t',\kappa}^{2(\mathcal{H})} + \mu_{t',\kappa}^{2(\mathcal{H})}}\right), \tag{10.23}$$

where $E[\cdot]$ denotes the expectation. Here, $\mu_{t',\kappa}^{(\mathcal{H})}$ and $\sigma_{t',\kappa}^{2(\mathcal{H})}$ denote the mean and variance of the averaged AIR power in the κth mel band, see (10.9), under the model (10.22). Their values can be computed from the model (10.22) [31].

The unknown AIR coefficients $\overline{h}_{t',\kappa}$ in the observation model (10.7) are then replaced by their mean

$$\hat{\overline{h}}_{t',\kappa} := E[\overline{h}_{t',\kappa}]; \quad t' = 0, \ldots, L_{\mathcal{H}}; \; \kappa = 1, \ldots, N_\kappa. \tag{10.24}$$

Note that, despite the presence of the noise term $\zeta(l)$ in (10.22), the logarithmic mel power spectral representation $\hat{\overline{h}}_{t',\kappa}$ is not a stochastic random variable, but is a deterministic parameter instead.

Obviously the models by Hirsch and Finster [22] and our own [31] are much coarser than the stochastic model by Sehr *et al.* [50] described first. This can be viewed both as an advantage and a disadvantage. The great advantage over more sophisticated models is that the estimation of the model is much simpler, as it is determined by just two parameters.

The first, σ_h^2, is closely related to the energy of the AIR. Often it is assumed that the AIR has unit energy ($E[\sum_{l=0}^{L_h-1} h^2(l)] = 1$). This holds for many artificially generated databases, where the reverberant signal is obtained by the convolution of clean data with an AIR, while it will usually not hold for true reverberant recordings. Then, a normalization has to be carried out such that the average power of the speech in the training data equals that of the speech in the test data. This can be achieved with techniques like CMN.

The second parameter is the room reverberation time T_{60}. There are a number of techniques how this parameter can be estimated, where the blind approaches that estimate it from the reverberant speech signal, rather than from the AIR, are practically more relevant. A maximum likelihood technique has been proposed by Ratnam *et al.* [48]. Wen *et al.* [58] estimated the parameters of a two-slope model, while Löllmann and Vary [40] treated the estimation of T_{60} in the presence of additive noise. Hirsch and Finster [22] proposed to select this parameter based on the likelihood computed by the ASR decoder: T_{60} is estimated after multiple recognitions of an utterance with HMMs adapted under different hypotheses for the value of T_{60}. They decided then on that value of T_{60}, whose corresponding set of HMMs achieved the highest likelihood in a forced alignment of the feature vector sequence with the recognized sentence.

Thus, the simplified models of the AIR, Equations (10.19) and (10.22), do not require AIR measurements in the target room, nor do they ask for calibration sentences with known transcription. This does not only make the approach less complicated, it also increases its flexibility, as a change of the room can be easily accounted for, since only two parameters have to be reestimated. These parameters can be estimated blindly without the need for any calibration sentences, measurements, or offline processing. However, it is likely that the coarser models will lead to some performance loss in the speech recognizer.

Figure 10.5 depicts the logarithmic mel power spectral representations of two AIRs used for the compilation of the Aurora 5 database for the office and living room environment, respectively, together with their approximations (10.24) obtained from the simplified AIR model (10.22). A critical approximation is probably the poor modeling of the direct signal component and the early reflections. Further, the simplified model assumes no frequency dependence of the reverberation time.

10.5 Bayesian Feature Enhancement

In this section we will give some more information about the Bayesian feature enhancement approach to reverberant speech recognition mentioned earlier. We will describe the principle

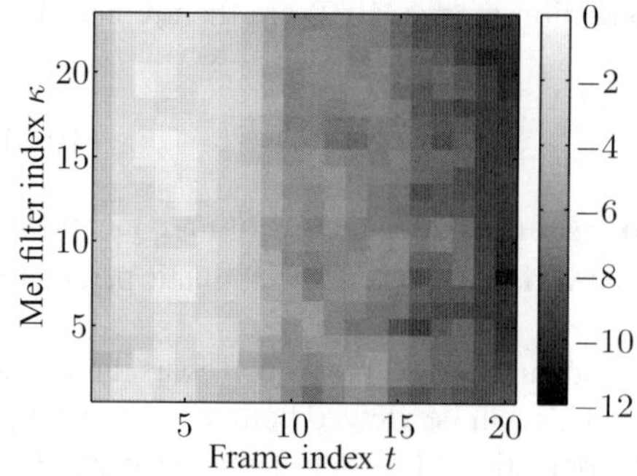

(a) Logarithmic mel power spectral representation $\overline{h}_{t,\kappa}$ of true AIR for the office environment ($T_{60} \approx 350\,\text{ms}$).

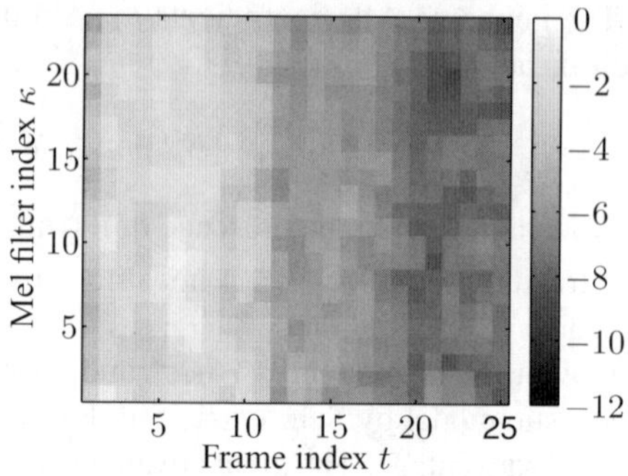

(b) Logarithmic mel power spectral representation $\overline{h}_{t,\kappa}$ of true AIR for the living room environment ($T_{60} \approx 450\,\text{ms}$).

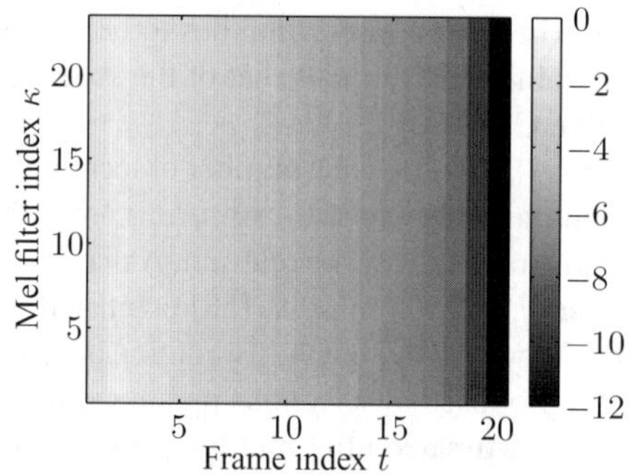

(c) Approximate logarithmic mel power spectral representation $\hat{\overline{h}}_{t,\kappa}$ of AIR for the office environment according to (10.24) ($T_{60} \approx 350\,\text{ms}$).

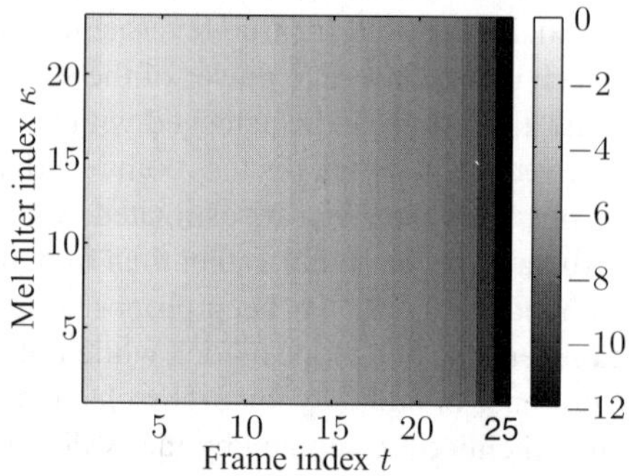

(d) Approximate logarithmic mel power spectral representation $\hat{\overline{h}}_{t,\kappa}$ of AIR for the living room environment according to (10.24) ($T_{60} \approx 450\,\text{ms}$).

Figure 10.5 Logarithmic mel power spectral representations of two AIRs used for the compilation of the Aurora 5 database together with their approximations.

behind it and the modeling assumptions. For a full treatment, the interested reader is referred to [31] and [32].

10.5.1 Basic Approach

The well-known *vector Taylor series* (*VTS*) approach was probably one of the earliest attempts to apply the Bayesian principle to robust ASR [43]. For the case of feature enhancement in the presence of noise and reverberation, the goal is to estimate the posterior density $p(\mathbf{s}_t|\mathbf{x}_{1:t})$ of the sought-after clean speech feature vector $\mathbf{s}_t$, given the observed noisy reverberant feature vectors $\mathbf{x}_{1:t}$.

From the derivation of Equation (10.10), it is clear that this density depends on the noise $\mathbf{n}_t$ and the representation $\overline{\mathbf{h}}_{0:L_\mathcal{H}}$ of the AIR. The two quantities can be either modeled as unknown (deterministic) parameters or as stochastic random variables. In [32] the deterministic model of (10.24) was used for the reverberation, while the log mel power spectrum of noise was modeled as a random process which is tracked along with the clean speech. The goal is thus to estimate the joint posterior distribution $p\left(\mathbf{d}_t|\mathbf{x}_{1:t}\right)$ of speech and noise, where

$$\mathbf{d}_t := \left(\mathbf{s}_t^T, \mathbf{n}_t^T\right)^T. \tag{10.25}$$

Here, $(\cdot)^T$ denotes vector transposition. From this, the posterior of the clean speech feature vector can be obtained by marginalization.

The posterior $p\left(\mathbf{d}_t|\mathbf{x}_{1:t}\right)$ is estimated recursively employing two steps, measurement update and time update, as will be explained next.

10.5.2 *Measurement Update*

In the *measurement update* step, the desired posterior distribution $p\left(\mathbf{d}_t|\mathbf{x}_{1:t}\right)$ is obtained from the so-called predictive distribution $p\left(\mathbf{d}_t|\mathbf{x}_{1:t-1}\right)$ by

$$p\left(\mathbf{d}_t|\mathbf{x}_{1:t}\right) = \frac{p\left(\mathbf{x}_t|\mathbf{d}_t, \mathbf{x}_{1:t-1}\right) p\left(\mathbf{d}_t|\mathbf{x}_{1:t-1}\right)}{\int p(\mathbf{x}_t|\tilde{\mathbf{d}}_t, \mathbf{x}_{1:t-1}) p(\tilde{\mathbf{d}}_t|\mathbf{x}_{1:t-1}) d\tilde{\mathbf{d}}_t}, \tag{10.26}$$

employing the observation $\mathbf{x}_t$ and an observation distribution $p\left(\mathbf{x}_t|\mathbf{d}_t, \mathbf{x}_{1:t-1}\right)$.

For the enhancement of noisy speech feature vectors the dependence of $\mathbf{x}_t$ on the previous observations $\mathbf{x}_{1:t-1}$ is usually neglected [9,43]:

$$p\left(\mathbf{x}_t|\mathbf{d}_t, \mathbf{x}_{1:t-1}\right) \approx p\left(\mathbf{x}_t|\mathbf{d}_t\right). \tag{10.27}$$

However, this approximation may be too coarse in the case of reverberant speech recognition, since successive reverberant feature vectors are strongly correlated with each other due to the dispersive effect of reverberation. To partly compensate for the error introduced by this approximation Krueger and Haeb-Umbach [31] proposed to use an extended vector

$$\breve{\mathbf{s}}_t := \left(\mathbf{s}_t^T, \mathbf{s}_{t-1}^T, \ldots, \mathbf{s}_{t-L_C+1}^T\right)^T \tag{10.28}$$

consisting of the current and an appropriate number of $L_C - 1$ previous clean speech feature vectors, and the vector $\mathbf{d}_t$ is replaced by the overall state vector

$$\mathbf{z}_t := \left(\breve{\mathbf{s}}_t^T, \mathbf{n}_t^T\right)^T. \tag{10.29}$$

In the following, we are referring to this variable as state vector, as is common practice in recursive Bayesian estimation using Kalman filtering techniques.

With this extended state vector, it is better justified to neglect the dependence on $\mathbf{x}_{1:t-1}$ and to approximate the observation distribution by

$$p\left(\mathbf{x}_t|\mathbf{d}_t, \mathbf{x}_{1:t-1}\right) \approx p\left(\mathbf{x}_t|\mathbf{z}_t\right). \tag{10.30}$$

The relationship between $\mathbf{x}_t$ and $\mathbf{z}_t$ is highly nonlinear, see Equation (10.10), and an exact analytic expression for the observation probability, Equation (10.30), is not known, even for the less complicated case of absence of reverberation and distortions only by noise.

A common solution is to approximate (10.11) by a Taylor series, which is truncated after the linear term. Recently, a more exact approximation has been proposed, where the Gaussian for $\mathbf{s}_t$ was approximated by a Gaussian mixture [38]. Since each component density of the mixture has a smaller variance than the original Gaussian, the Taylor series approximation of each Gaussian results overall in a smaller linearization error. Alternatively, the unscented transform [9] or Monte Carlo techniques [59] can be used to approximate $p(\mathbf{x}_t|\mathbf{z}_t)$.

10.5.3 Time Update

In the *time update* step, a predictive distribution for the extended feature vector $\mathbf{z}_t$ based on the previous reverberant noisy observations $\mathbf{x}_{1:t-1}$ is computed as:

$$p(\mathbf{z}_t|\mathbf{x}_{1:t-1}) = \int p(\mathbf{z}_t|\mathbf{z}_{t-1}, \mathbf{x}_{1:t-1})\, p(\mathbf{z}_{t-1}|\mathbf{x}_{1:t-1})\, d\mathbf{z}_{t-1} \tag{10.31}$$

requiring a predictive distribution $p(\mathbf{z}_t|\mathbf{z}_{t-1}, \mathbf{x}_{1:t-1})$ for the clean speech and noise feature vectors.

With Equation (10.29) the predictive density can be factorized due to the independence of speech and noise

$$p(\mathbf{z}_t|\mathbf{z}_{t-1}, \mathbf{x}_{1:t-1}) = p(\breve{\mathbf{s}}_t|\breve{\mathbf{s}}_{t-1}, \mathbf{x}_{1:t-1})\, p(\mathbf{n}_t|\mathbf{n}_{t-1}, \mathbf{x}_{1:t-1}). \tag{10.32}$$

The predictive distribution of noise may be usually approximated by a single Gaussian distribution

$$p(\mathbf{n}_t|\mathbf{n}_{t-1}, \mathbf{x}_{1:t-1}) \approx p(\mathbf{n}_t) = \mathcal{N}(\mathbf{n}_t; \boldsymbol{\mu}_\mathbf{n}, \boldsymbol{\Sigma}_\mathbf{n}) \tag{10.33}$$

whose parameters $\boldsymbol{\mu}_\mathbf{n}$ and $\boldsymbol{\Sigma}_\mathbf{n}$ are assumed to be constant for the duration of an utterance (e.g., [43]).

For the predictive distribution of speech, it is advantageous to take a model which accounts for the correlation between successive speech feature vectors. This can be achieved by a linear dynamic model. However, it has been observed that a single dynamic model is inappropriate to model the complicated dynamics of speech, and an interaction of multiple dynamic models, a *switching linear dynamic model (SLDM)* [26], describes the dynamics of speech features much better:

$$p(\breve{\mathbf{s}}_t|\breve{\mathbf{s}}_{t-1}, \mathbf{x}_{1:t-1}) = \sum_{i=1}^{I} p(\breve{\mathbf{s}}_t|\breve{\mathbf{s}}_{t-1}, \mathbf{x}_{1:t-1}, \gamma_t = i)\, P(\gamma_t = i|\breve{\mathbf{s}}_{t-1}, \mathbf{x}_{1:t-1}), \tag{10.34}$$

where $\gamma_t \in \{1, ..., I\}$ is a realization of a hidden regime variable indicating the active model at frame t. The distribution $p(\breve{\mathbf{s}}_t|\breve{\mathbf{s}}_{t-1}, \mathbf{x}_{1:t-1}, \gamma_t = i)$ is completely determined by $p(\mathbf{s}_t|\breve{\mathbf{s}}_{t-1}, \mathbf{x}_{1:t-1}, \gamma_t = i)$ which in turn is approximated by a linear, autoregressive prediction model as:

$$p(\mathbf{s}_t|\breve{\mathbf{s}}_{t-1}, \mathbf{x}_{1:t-1}, \gamma_t = i) \approx p(\mathbf{s}_t|\mathbf{s}_{t-1}, \gamma_t = i) \tag{10.35}$$

$$\approx \begin{cases} \mathcal{N}\left(\mathbf{s}_1; \boldsymbol{\mu}_\mathbf{s}^{(i)}, \boldsymbol{\Sigma}_\mathbf{s}^{(i)}\right) & \text{for } t = 1 \\ \mathcal{N}\left(\mathbf{s}_t; \mathbf{A}^{(i)}\mathbf{s}_{t-1} + \mathbf{b}^{(i)}, \mathbf{V}^{(i)}\right) & \text{for } t > 1. \end{cases} \tag{10.36}$$

According to this model, the first clean speech feature vector is modeled by a GMM with mean vectors $\boldsymbol{\mu}_{\mathrm{s}}^{(i)}$ and covariance matrices $\boldsymbol{\Sigma}_{\mathrm{s}}^{(i)}$, $i = 1, \ldots, I$. All successive clean feature vectors are predicted from their predecessors by an affine transformation described by the state transition matrix $\mathbf{A}^{(i)}$ and the bias vector $\mathbf{b}^{(i)}$. The prediction error is assumed to be zero mean having a Gaussian distribution with the covariance matrix $\mathbf{V}^{(i)}$.

The mixing probabilities may be approximated by

$$P\left(\gamma_t = i | \breve{\mathbf{s}}_{t-1}, \mathbf{x}_{1:t-1}\right) = \begin{cases} \alpha^{(i)} & \text{for } t = 1 \\ \sum_{j=1}^{I} a_{ij} P\left(\gamma_{t-1} = j | \mathbf{x}_{1:t-1}\right) & \text{for } t > 1 \end{cases} \tag{10.37}$$

with time invariant state transition probabilities

$$a_{ij} := P\left(\gamma_t = i | \gamma_{t-1} = j\right) \tag{10.38}$$

and model probabilities for the first frame $t = 1$

$$\alpha^{(i)} := P\left(\gamma_1 = i\right). \tag{10.39}$$

This kind of *a priori* model explicitly considers correlations between successive speech feature vectors, which are due to the speech production process on the one hand and the feature extraction process on the other. A variant of this model has been used successfully for noise robust speech recognition by Droppo and Acero [10]. The SLDM model parameters can be trained on clean speech training data using the *expectation maximization (EM)* algorithm [44].

10.5.4 Inference

Assuming all involved distributions to be Gaussians or mixtures of those, the inference simplifies to recursively computing the mean vectors and covariance matrices of the posterior distribution $p\left(\mathbf{z}_t | \mathbf{x}_{1:t}\right)$

$$\mathbf{z}_{t|t} := E\left[\mathbf{z}_t | \mathbf{x}_{1:t}\right], \tag{10.40}$$

$$\boldsymbol{\Sigma}_{\mathbf{z}_{t|t}} := E\left[\left(\mathbf{z}_t - \mathbf{z}_{t|t}\right)\left(\mathbf{z}_t - \mathbf{z}_{t|t}\right)^T \middle| \mathbf{x}_{1:t}\right]. \tag{10.41}$$

The knowledge about the posterior distribution allows the computation of different kinds of point estimates for the clean speech feature vector $\mathbf{s}_t$, such as the *minimum mean squared error (MMSE)* estimate

$$\hat{\mathbf{s}}_t^{(\mathrm{MMSE})} := E\left[\mathbf{s}_t | \mathbf{x}_{1:t}\right]. \tag{10.42}$$

Further, the posterior covariance matrix

$$\hat{\boldsymbol{\Sigma}}_{\mathbf{s}_t}^{(\mathrm{MMSE})} := E\left[\left(\mathbf{s}_t - \hat{\mathbf{s}}_t^{(\mathrm{MMSE})}\right)\left(\mathbf{s}_t - \hat{\mathbf{s}}_t^{(\mathrm{MMSE})}\right)^T \middle| \mathbf{x}_{1:t}\right] \tag{10.43}$$

can be regarded as a measure of the degree of uncertainty in the point estimate $\hat{\mathbf{s}}_t^{(\mathrm{MMSE})}$.

To obtain a first qualitative impression of the power of Bayesian feature enhancement, consider Figure 10.6. It depicts the spectrogram of the same utterance as in Figure 10.4, this time after the feature enhancement stage. One can clearly see that Bayesian feature enhancement is able to undo some of the effects of reverberation. For example, the glottal stop at about 1.15 s becomes visible again.

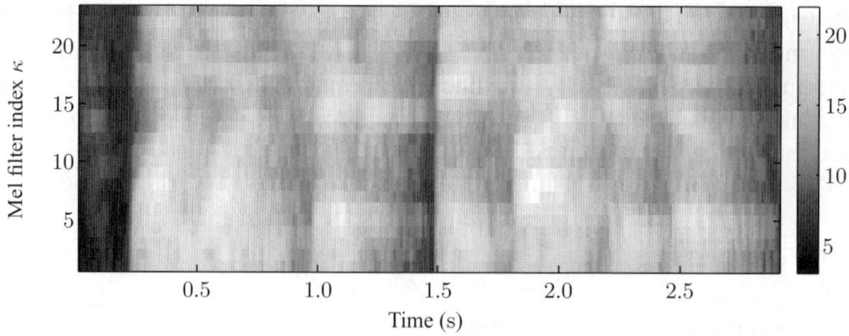

(a) Logarithmic mel power spectrum of reverberant speech signal (repeated from Figure 10.4b for comparison purposes).

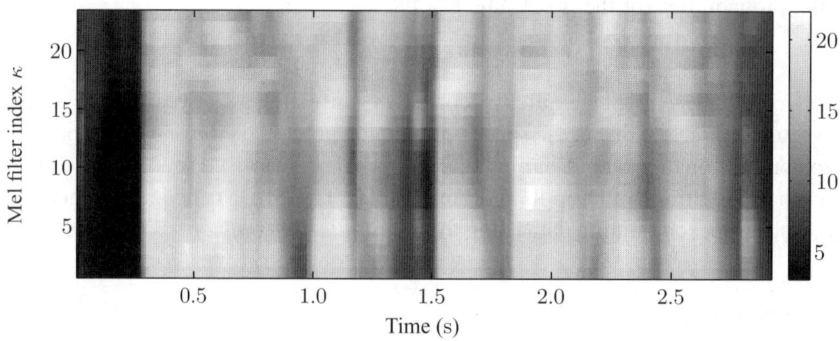

(b) Enhanced logarithmic mel power spectrum.

Figure 10.6 Logarithmic mel power spectrogram of reverberant speech signal and that enhanced by Bayesian feature enhancement. Same utterance as in Figure 10.4.

10.6 Experimental Results

10.6.1 Databases

Most publications on reverberant speech recognition consider small vocabulary tasks only. However, here we are going to present experimental results on both a small vocabulary and a large vocabulary task.

For the small vocabulary task we consider the Aurora 5 database [23], which contains utterances of connected digits sampled at a rate of $1/T_{\mathrm{s}} = 8$ kHz. The database comprises 8623 clean nonreverberant training utterances, which were employed for the training of the acoustic model of the speech recognizer and for the training of the SLDM parameters.

The test set consisted of reverberant and noisy reverberant utterances under different reverberation conditions and different *signal-to-noise ratios (SNRs)*. The considered reverberant environments comprise an office and a living room. The individual reverberant utterances of the database had been created by convolving the corresponding clean utterances with artificial AIRs [23], where the reverberation time T_{60} was varied in the range between 300 and 400 ms for the office, and in the range between 400 and 500 ms for the living room. The noisy reverberant utterances were created by adding interior noise at different SNR levels, where the noise signals in part contained nonstationary segments. For each condition there were 8700 test utterances.

For the large vocabulary task, we consider the Aurora 4 database, which comprises the Wallstreet Journal Nov'92 5k evaluation test set to which noise at varying SNR levels and varying types has been added [21]. In our experiments, we used the downsampled version of the data with a sampling rate of 8 kHz. The Aurora 4 test set consists of seven versions of a set of 166 utterances with 2715 words per set. One version is the clean data while the six other versions have been obtained from the clean data by artificially adding noise of different characteristics at a randomly chosen SNR between 5 and 15 dB. As the Aurora 4 test database does not contain reverberant speech, we modified the test set as follows. The set of clean data was artificially convolved with the same AIRs as those used for the generation of the Aurora 5 test set, resulting in the noise-free *office* and *living room* test sets. To these data, we then added the noise that was also used on the Aurora 5 database at SNRs of 15, 10, 5, and 0 dB. We employed the noise types of Aurora 5 rather than those of Aurora 4, as many of the Aurora 4 noise types (e.g., train, airport, car, and street) are not representative of reverberant environments. This modified test database is called Aurora 4-rev in the following.

10.6.2 Overview of the Tested Methods

As a baseline, we consider the performance obtained with the ETSI standards ES 201 108 (standard front end (SFE)) [13] and ES 202 050 (advanced front end (AFE)) [12], to which we applied the two small modifications that have been discussed in Section 10.2.2. The AFE, which is also discussed in Section 4.5.4 of this book, is in its essence an extension of the SFE by a two-stage Wiener filter and a blind equalization to compensate for noise and acoustic mismatch caused by different transducer characteristics. The MFCC feature vector is 39-dimensional, resulting from the static MFCCs as well as their corresponding velocity and acceleration features. Velocity features were obtained from the slope of a linear regression line over the static features in a window of ± 4 frames around the current frame. For the acceleration features, the window extended over ± 2 frames. The parameters of the front ends are summarized in Table 10.2.

As another reference, we will consider matched training and test conditions, with the SFE used for feature extraction. Please note that we used exactly the same set of AIRs to reverberate the training data as was used to generate the reverberant test data. While this may not be realistic in practice, it nevertheless provides an upper bound of what is achievable with matched reverberant training.

Results will also be presented for the *Bayesian feature enhancement (BFE)* outlined in Section 10.5. For the Bayesian feature enhancement, we employed an SLDM consisting of $I = 4$ linear dynamic models as the *a priori* model of speech. The SLDM was trained on the same clean speech training data as the acoustic model. For the initialization of the four dynamic

Table 10.2 Parameters for feature extraction according to the ETSI standards.

Frame shift B	Window size L_w	DFT length F	#mel filters N_κ	#ceptral coeffs. K	Window type
80	200	256	23	13	Hamming

models a "k-means++"-like algorithm [33] was used. The models were then refined using four iterations of the EM algorithm similar to [44]. In the extended state vector, we considered $L_C = 6$ successive feature vectors (see Equation (10.28)). For the feature domain AIR model, we assumed a constant reverberation time of $T_{60} = 350$ ms for the office and $T_{60} = 450$ ms for the living room environment. The length of the AIR model, $L_{\mathcal{H}}$, was set to 20 and 25 for office and living room, respectively. This resulted in less than 0.1% error between the energy of the nontruncated AIR model and the model truncated to L_h samples (see Equation (10.22)).

For the digit recognition task, the acoustic model consisted of speaker and gender independent word based HMMs with 16 states per word and four Gaussian mixture components per state. Simple left-to-right models without skips over states were used. For the large vocabulary task, training has been carried out on the WSJ0 training database. The HMMs were three-state triphone models with ten Gaussians per mixture. The training of the HMM parameters and Viterbi decoding for the recognition were carried out using the hidden Markov model toolkit (HTK) software [60]. For the Aurora 4-rev experiments a bigram language model was applied.

We will also present results with a system which was trained on the reverberant training data, after Bayesian feature enhancement has been applied to it. The motivation is that these processed training data contain the artefacts that the feature enhancement may introduce. Recognition was then carried out with the same Bayesian feature enhancement that was applied to the training data. In the tables, this condition is denoted as "BFE (trn+recog)."

To complement the comparison, we also cite the results of Hirsch and Finster [22], who employed an acoustic model based approach to reverberant speech recognition, where the acoustic emission probabilities were modified by incorporating the influence of the preceding frames, as was briefly discussed in Section 10.3.3. Note that Hirsch and Finster had a somewhat different setup: they used gender dependent HMMs with two Gaussians per mixture in their digit recognition experiments. For the *Wall Street Journal* task, Hirsch and Finster employed triphone HMMs with four Gaussians per mixture. They experimented with the 16 kHz version of the database whereas the 8 kHz version was taken for the other approaches. With their settings, they achieved a word error rate of 48.8% on the reverberant office data set of Aurora 4-rev using the SFE, whereas we obtained 47.37% with the SFE and our settings. As these two results are fairly close, we conclude that despite the somewhat different setup a comparison of our experimental results with the ones reported in [22] appears to be valid.

10.6.3 Recognition Results on Reverberant Speech

In this section, we compare the recognition performance on noise-free reverberant speech. Tables 10.3 and 10.4 present the word error rates obtained on noise-free test sets of the Aurora 5 and the reverberant Aurora 4 database.

The drastic increase in error rate of the SFE from nonreverberant to reverberant test data has already been mentioned in Section 10.1. In the results presented here, it can be seen that also the AFE is obviously also unable to cope with reverberant speech: the performance on reverberant speech is as poor as that of the SFE.

This does not only hold for the small vocabulary task related to the Aurora 5 database, but also for the large vocabulary task related to the Aurora 4-rev database. On Aurora 4 the error rates of the SFE and AFE on the nonreverberant noise-free test set had been 14.00% and 14.40%, respectively.

Table 10.3 Word error rates [%] on Aurora 5 (noise-free reverberant test data) for the ETSI SFE and AFE, Bayesian feature enhancement (BFE), BFE applied to reverberant training data and to recognition data (BFE trn+recog), acoustic model adaptation (AMA) (cited from [22]), and reverberant training and recognition without any enhancement.

	Office	Living room
SFE	6.32	14.94
AFE	6.11	14.53
BFE	1.97	3.61
BFE (trn+recog)	2.00	3.35
AMA [22]	3.30	8.00
Reverberant training	1.29	2.61

One can observe that the *Bayesian feature enhancement (BFE)* greatly improves the word accuracy compared to the SFE and AFE. For example, on Aurora 5 the word error rate is reduced from 6.32% for the office and 14.94% for the living room environment with the SFE to 1.97% and 3.61%, respectively, which corresponds to a recovery of approximately 75% of the errors caused by reverberation. The *acoustic model adaptation (AMA)* technique is not quite as effective: only about 50% of the errors caused by reverberation can be recovered on the Aurora 5 database. However, both techniques are outperformed by matched reverberant training, which achieves 1.29% and 2.61%, respectively.

A large improvement by Bayesian feature enhancement is also observed on the reverberant Aurora 4 data, where the error rate decreased from 47.37% (SFE, office) and 73.44% (SFE, living room) to 27.77% and 40.44%, respectively. If reverberant training data are used and BFE is applied on them as well, then word error rates are achieved, which equal or outperform matched reverberant training: 24.38% versus 24.24% and 32.97% versus 36.32%.

Unfortunately, for the reverberated Aurora 4 database Hirsch and Finster [22] presented only results for the office environment and, in particular, they did not report any details about the kind of errors. However, for the office environment the word error rate could only be reduced to 39.80%.

Table 10.4 Word error rates (WER) [%] on Aurora 4-rev (noise-free reverberant test data) for the ETSI SFE and AFE, Bayesian feature enhancement (BFE), BFE applied to reverberant training data and to recognition data (BFE trn+recog), acoustic model adaptation (AMA) (cited from [22]), and reverberant training and recognition without any enhancement. Where available, the decomposition of the WER into substitutions (Sub), deletions (Del), and insertions (Ins) is shown.

	Office				Living room			
	Sub	Del	Ins	WER	Sub	Del	Ins	WER
SFE	34.84	5.52	7.00	**47.37**	56.06	10.87	6.52	**73.44**
AFE	34.03	6.48	6.08	**46.59**	55.99	11.16	5.45	**72.60**
BFE	20.07	2.54	5.16	**27.77**	29.10	3.61	7.73	**40.44**
BFE (trn+recog)	17.46	2.43	4.49	**24.38**	23.72	3.61	5.64	**32.97**
AMA [22]	–	–	–	**39.80**	–	–	–	–
Reverberant training	18.01	3.06	3.17	**24.24**	26.26	6.08	3.98	**36.32**

10.6.4 Recognition Results on Noisy Reverberant Speech

In this section, we are going to evaluate the performance on test data to which noise was artificially added, in addition to the reverberation. We present results for SNRs of 15, 10, 5, and 0 dB on Aurora 5 and 15 and 10 dB on Aurora 4-rev. The error rates on the large vocabulary task at 10 dB were already so high that we decided not to include the results on lower SNRs. Note that no change was made to the training conditions, that is training has always been carried out on noise-free data.

From Table 10.5, it can be observed that the AFE manages to provide a high degree of robustness against noise on Aurora 5. For the low SNR ranges, where the noise is the dominant error source, the AFE clearly outperforms the matched reverberant training (which was carried out on noise-free training data).

The Bayesian feature enhancement manages to keep up with the AFE for moderate noise levels, while the AFE is superior in low SNR conditions.

The PMC-like model adaptation of [22] shows a great degree of noise robustness and achieves the best recognition rates on the noisy Aurora 5 database. This is due to the adaptation of the means of the acoustic emission probabilities to both, the reverberation and the noise seen on the test data. As the noise is reestimated on a per frame basis, the method is able to follow nonstationary noise, however, at the expense of a considerable computational complexity.

Table 10.5 Word error rates [%] on Aurora 5 (noisy reverberant test data) for the ETSI SFE and AFE, Bayesian feature enhancement (BFE), acoustic model adaptation (AMA) (cited from [22]) and reverberant training.

SNR [dB]		Office	Living room
15	SFE	19.93	35.58
	AFE	10.92	21.31
	BFE	7.47	12.21
	AMA [22]	6.20	9.20
	Reverberant training	15.44	14.58
10	SFE	44.75	57.38
	AFE	17.26	29.17
	BFE	16.83	24.04
	AMA [22]	11.50	16.90
	Reverberant training	38.31	51.19
5	SFE	71.73	79.01
	AFE	30.09	43.06
	BFE	35.13	44.33
	AMA [22]	24.30	32.00
	Reverberant training	67.81	77.88
0	SFE	88.10	89.72
	AFE	51.41	62.65
	BFE	62.44	69.51
	AMA [22]	49.20	60.00
	Reverberant training	87.63	91.88

Table 10.6 Word error rates [%] on Aurora 4-rev with SNRs of 15 and 10 dB for the ETSI SFE and AFE, Bayesian feature enhancement and reverberant training.

		Office				Living room			
SNR [dB]		Sub	Del	Ins	WER	Sub	Del	Ins	WER
15	SFE	49.13	9.80	8.58	**67.51**	58.64	16.80	7.55	**82.98**
	AFE	35.14	7.07	7.40	**49.61**	50.68	8.14	9.10	**67.92**
	BFE	31.09	4.53	10.64	**46.26**	43.76	5.45	11.57	**60.77**
Reverberant training		28.40	4.90	12.97	**46.26**	36.13	6.85	11.57	**54.55**
10	SFE	57.90	20.77	7.11	**85.78**	55.58	32.15	4.05	**91.79**
	AFE	44.01	9.47	9.43	**62.91**	59.08	10.72	9.21	**79.01**
	BFE	46.15	7.29	12.63	**66.08**	55.32	10.28	14.11	**79.71**
Reverberant training		43.09	9.54	13.33	**65.97**	48.25	12.78	10.20	**71.23**

For the Bayesian feature enhancement, the parameters of the noise model, Equation (10.33), are estimated on the first and last fifteen (Aurora 5) and fifty (Aurora 4-rev) frames of each utterance and are then kept constant for the duration of the utterance, resulting in the time invariant noise model of Equation (10.33). However, this is a poor modeling assumption for the nonstationary noise types encountered in the databases.

On Aurora 4-rev, all methods perform fairly poor (see Table 10.6). The word error rates achieved seem to be too high to be of practical interest. It seems that substantial progress needs to be made until large vocabulary speech recognizers work reliably on noisy reverberant speech.

10.7 Conclusions

We have presented an overview of approaches to reverberant speech recognition and discussed signal-based, feature-based, and acoustic model or decoder-based methods. As in practice, distant microphones also capture an increased amount of additive noise compared to close-talking microphones we also considered the case of noisy reverberant speech. The relationship between the clean and the noisy reverberant log mel power spectral feature vectors was derived, and it was pointed out that it is this highly nonlinear relationship and the convolutive nature of the distortion introduced by reverberation which makes the development of ASR algorithms robust to reverberation a tough problem, even if the AIR from the source to the sensor is known.

As the AIR is usually not available and furthermore time variant, there is an interest in developing simplified models of the AIR in the feature domain, whose parameters can be estimated more easily than the AIR itself. Some of these models have been discussed in this chapter.

A Bayesian feature enhancement approach was discussed in somewhat more detail. Its performance was evaluated as well as that of an acoustic model adaptation approach that employed the PMC principle, and three reference systems: the ETSI SFE, and AFE, and a recognizer that was trained on reverberant data.

It was observed that the ETSI AFE performs equally poor as the SFE on the noise-free reverberant data, while its increased noise robustness became apparent on the noisy reverberant test sets. Bayesian feature enhancement and acoustic model adaptation are both able to greatly improve the recognition performance over the SFE: on the noise-free subset of Aurora 5 feature enhancement recovered up to 75% of all errors introduced by reverberation, while the acoustic model adaptation recovered 50%. While the acoustic model adaptation does not require a complete new training when used in an enclosure with a different AIR, it still asks for the modification of all acoustic model parameters, which is computationally rather expensive. Feature enhancement, on the other hand, entails low computational complexity which furthermore is independent of the acoustic model size and thus the size of the vocabulary.

Matched reverberant training delivers a reverberation robust system, that is hardly beaten by the other approaches. While being conceptually simple, it may still not be very practical, as it assumes that an acoustic model specific to the target environment can be trained. Moving to a room with a different AIR would require a whole new training.

Finally, one should bear in mind that the test sets were artificial in the sense that nonreverberant data were convolved with a constant AIR. But even in this case the word error rates on a noisy reverberant 5000-word vocabulary task were too high to be of practical interest. In any practical setup, a speaker will move, and we have discussed that even slight head movements can have already a large impact on the AIR. It seems that significant progress is required until the problem of automatic recognition of noisy and reverberant speech is solved.

Acknowledgment

We are very grateful to Hans-Günter Hirsch for providing the AIRs as well as the interior noise recordings used for the compilation of the reverberant Aurora 4 database.

References

[1] E. Aarts and S. Marzano, *The New Everyday: Views on Ambient Intelligence*. Rotterdam, The Netherlands: 010 Uitgeverij, 2004.

[2] J. Allen, "Synthesis of pure speech from a reverberant signal," US Patent 3 786 188, 1974.

[3] Y. Avargel and I. Cohen, "On multiplicative transfer function approximation in the short-time Fourier transform domain," *IEEE Signal Processing Letters*, vol. 14, no. 5, pp. 337–340, May 2007.

[4] C. Avendano, S. Tibrewala, and H. Hermansky, "Multiresolution channel normalization for ASR in reverberant environments," in *Proceedings of the Eurospeech*, 1997, pp. 1107–1110.

[5] D. A. Berkley, "Normal listeners in typical rooms—reverberation perception, simulation, and reduction," in *Acoustical Factors Affecting Hearing Aid Performance*, G. A. Studebaker and I. Hochberg, Eds. Baltimore: University Park Press, 1980, pp. 3–24.

[6] L. Couvreur and C. Couvreur, "Blind model selection for automatic speech recognition in reverberant environments," *Journal of VLSI Signal Processing*, vol. 36, no. 2/3, pp. 189–203, 2004.

[7] M. Delcroix, T. Hikichi, and M. Miyoshi, "Precise dereverberation using multichannel linear prediction," *IEEE Transactions on Audio, Speech and Language Processing*, vol. 15, no. 2, pp. 430–440, Feb. 2007.

[8] M. Delcroix, T. Nakatani, and S. Watanabe, "Static and dynamic variance compensation for recognition of reverberant speech with dereverberation preprocessing," *IEEE Transactions on Audio, Speech and Language Processing*, vol. 17, no. 2, pp. 324–334, February 2009.

[9] J. Deng, M. Bouchard, and T.H. Yeap, "Feature enhancement for noisy speech recognition with a time-variant linear predictive HMM structure," *IEEE Transactions on Audio, Speech, and Language Processing*, vol. 16, no. 5, July 2008, pp. 891–899.

[10] J. Droppo and A. Acero, "Noise robust speech recognition with a switching linear dynamic model," in *Proceedings of the IEEE International Conference on Acoustics, Speech and Signal Processing (ICASSP)*, vol. 1, 2004, pp. I–953–6.

[11] G. W. Elko, "Microphone arrays," in *Proceedings of the International Workhop on Hands-free Speech Communication*, Kyoto, Japan, April 2001.
[12] ETSI, "ETSI standard document, speech processing, transmission and quality aspects (stq); distributed speech recognition; advanced front-end feature extraction algorithm; compression algorithms, v1.1.5," Technical Report, ETSI ES 202 050, 2007.
[13] ETSI, "ETSI standard document, speech processing, transmission and quality aspects (stq); distributed speech recognition; front-end feature extraction algorithm; compression algorithms, ETSI ES 201 108 v1.1.3," Technical Report, 2003.
[14] M. Gales, "Model-based techniques for noise robust speech recognition," PhD dissertation, Cambridge University, 1995.
[15] M. Gales, "Maximum likelihood linear transformations for HMM-based speech recognition," *Computer Speech & Language*, vol. 12, pp. 75–98, 1998.
[16] N. D. Gaubitch, D. B. Ward, and P. A. Naylor, "Statistical analysis of the autoregressive modeling of reverberant speech," *Journal of the Acoustical Society of America*, vol. 120, no. 6, pp. 4031–4039, 2006.
[17] D. Gelbart and N. Morgan, "Evaluating long-term spectral subtraction for reverberant ASR," in *IEEE Workshop on Automatic Speech Recognition and Understanding*, Madonna di Campiglio, Trento, Italy, 2001.
[18] J. Gemmeke, M. Van Segbroeck, Y. Wang, B. Cranen, and H. Van hamme, "Automatic speech recognition using missing data techniques: Handling of real-world data," in *Robust Speech Recognition of Uncertain or Missing Data*, R. Haeb-Umbach and D. Kolossa, Eds. Berlin, Heidelberg: Springer, 2011.
[19] E. Habets, "Single- and multimicrophone speech dereverberation using spectral enhancement," PhD dissertation, Technische Universiteit Eindhoven, 2007.
[20] H. Hermansky and N. Morgan, "Rasta processing of speech," *IEEE Transactions on Speech and Audio Processing*, vol. 2, no. 4, pp. 578–589, October 1994.
[21] H.-G. Hirsch, "Experimental framework for the performance evaluation of speech recognition front-ends on a large vocabulary task," Technical Report, STQ AURORA DSR WORKING GROUP, November 2002.
[22] H.-G. Hirsch and H. Finster, "A new approach for the adaptation of HMMs to reverberation and background noise," *Speech Commununication*, vol. 50, no. 3, pp. 244–263, 2008.
[23] H. Hirsch, "Aurora-5 experimental framework for the performance evaluation of speech recognition in case of a hands-free speech input in noisy environments," Technical Report, Niederrhein University of Applied Sciences, 2007.
[24] Y. Huang, J. Benesty, and J. Chen, "Dereverberation," in *Springer Handbook of Speech Processing*, J. Benesty, M. Sondhi, and Y. Huang, Eds. Berlin, Heidelberg: Springer, 2008.
[25] Y. A. Huang and J. Benesty, "Adaptive multichannel least mean square and Newton algorithms for blind channel identification," *Signal Processing*, vol. 82, pp. 1127–1138, August 2002.
[26] C.-J. Kim, "Dynamic linear models with Markov-switching," York (Canada)—Department of Economics, Papers 91–8, 1991. Available at http://ideas.repec.org/p/fth/yorkca/91-8.html.
[27] C. Kim and R. M. Stern, "Nonlinear enhancement of onset for robust speech recognition," in *Proceedings of the Annual Conference of the International Speech Communication Association (Interspeech)*, 2010.
[28] B. E. D. Kingsbury and N. Morgan, "Recognizing reverberant speech with RASTA-PLP," in *Proceedings of the IEEE International Conference on Acoustics, Speech and Signal Processing (ICASSP)*, vol. 2, April 21–24, 1997, pp. 1259–1262.
[29] B. E. D. Kingsbury, N. Morgan, and S. Greenberg, "Robust speech recognition using the modulation spectrogram," *Speech Communication*, vol. 25, pp. 117–132, August 1998.
[30] K. Kinoshita, M. Delcroix, T. Nakatani, and M. Miyoshi, "Suppression of late reverberation effect on speech signal using long-term multiple-step linear prediction," *IEEE Transactions on Audio, Speech and Language Processing*, vol. 17, no. 4, pp. 534–545, May 2009.
[31] A. Krueger and R. Haeb-Umbach, "Model-based feature enhancement for reverberant speech recognition," *IEEE Transactions on Audio, Speech and Language Processing*, vol. 18, no. 7, pp. 1692–1707, September 2010.
[32] A. Krueger and R. Haeb-Umbach, "A model-based approach to joint compensation of noise and reverberation for speech recognition," in *Robust Speech Recognition of Uncertain or Missing Data*, R. Haeb-Umbach and D. Kolossa, Eds. Berlin, Heidelberg: Springer, 2011.
[33] A. Krueger, V. Leutnant, R. Haeb-Umbach, A. Marcel, and J. Bloemer, "On the initialisation of dynamic models for speech features," in *Proceedings of the ITG Fachtagung Sprachkommunikation*, 2010.
[34] H. Kuttruff, *Room Acoustics*. London: Taylor & Francis Group, 2009.

[35] T. Langhans and H. Strube, "Speech enhancement by nonlinear multiband envelope filtering," in *Proceedings of the IEEE International Conference on Acoustics, Speech and Signal Processing (ICASSP)*, vol. 7, May 1982, pp. 156–159.
[36] K. Lebart, J. Boucher, and P. Denbigh, "A new method based on spectral subtraction for speech dereverberation," *Acta Acustica united with Acustica*, vol. 87, pp. 359–366(8), 2001.
[37] C. J. Leggetter and P. C. Woodland, "Maximum likelihood linear regression for speaker adaptation of continuous density hidden Markov models," *Computer Speech & Language*, vol. 9, no. 2, pp. 171–185, 1995.
[38] V. Leutnant, A. Krueger, and R. Haeb-Umbach, "A versatile Gaussian splitting approach to nonlinear state estimation and its application to noise-robust ASR," in *Proceedings of the Annual Conference of the International Speech Communication Association (Interspeech)*, Florence, Italy, August 2011.
[39] R. Litovsky, H. Colburn, W. Yost, and S. Guzman, "The precedence effect," *Journal of the Acoustical Society of America*, vol. 106, no. 4, 1999.
[40] H. Löllmann and P. Vary, "Estimation of the reverberation time in noisy environments," in *Proceedings of International Workshop on Acoustic Echo and Noise Control (IWAENC)*, Seattle, Washington, September 2008.
[41] H. K. Maganti and M. Matassoni, "An auditory based modulation spectral feature for reverberant speech recognition," in *Proceedings of the Annual Conference of the International Speech Communication Association (Interspeech)*, Makuhari, Japan, 2010.
[42] M. Miyoshi and Y. Kaneda, "Inverse filtering of room acoustics," *IEEE Transactions on Acoustics, Speech and Signal Processing*, vol. 36, no. 2, pp. 145–152, February 1988.
[43] P. Moreno, B. Raj, and R. Stern, "A vector Taylor series approach for environment-independent speech recognition," in *Proceedings of the IEEE International Conference on Acoustics, Speech and Signal Processing (ICASSP)*, vol. 2, May 1996, pp. 733–736.
[44] K. Murphy, "Switching Kalman filters," Technical Report, U.C. Berkeley, 1998.
[45] P. Naylor and N. Gaubitch, "Speech dereverberation". Berlin, Heidelberg: Springer, 2010.
[46] J. Polack, "La transmission de l'énergie sonore dans les salles," Dissertation, Université du Maine, 1988.
[47] B. Radlovic, R. Williamson, and R. Kennedy, "On the poor robustness of sound equalization in reverberant environments," in *Proceedings of the IEEE International Conference on Acoustics, Speech and Signal Processing (ICASSP)*, vol. 2, Mar. 1999, pp. 881–884.
[48] R. Ratnam, D. L. Jones, B. C. Wheeler, J. William D. O'Brien, C. R. Lansing, and A. S. Feng, "Blind estimation of reverberation time," *Journal of the Acoustical Society of America*, vol. 114, no. 5, pp. 2877–2892, 2003.
[49] C. Raut, T. Nishimoto, and S. Sagayama, "Model adaptation for long convolutional distortion by maximum likelihood based state filtering approach," in *Proceedings of the IEEE International Conference on Acoustics, Speech and Signal Processing (ICASSP)*, vol. 1, 14–19 May 2006, pp. I–I.
[50] A. Sehr, M. Zeller, and W. Kellermann, "Distant-talking continuous speech recognition based on a novel reverberation model in the feature domain," in *Proceedings of the Annual Conference of the International Speech Communication Association (Interspeech)*, Pittsburgh, PA, September 2006.
[51] A. Sehr, R. Maas, and W. Kellermann, "Reverberation model-based decoding in the logmelspec domain for robust distant-talking speech recognition," *IEEE Transactions on Audio, Speech and Language Processing*, vol. 18, no. 7, pp. 1676–1691, September 2010.
[52] A. Sehr, Y. Zheng, E. Nöth, and W. Kellermann, "Maximum likelihood estimation of a reverberation model for robust distant-talking speech recognition," in *Proceedings of the European Signal Processing Conference (EUSIPCO)*, 2007, pp. 1299–1303.
[53] A. Sehr, R. Maas, and W. Kellermann, "Frame-wise HMM adaptation using state-dependent reverberation estimates," in *Proceedings of the IEEE International Conference on Acoustics, Speech and Signal Processing (ICASSP)*, May 2011, pp. 5484–5487.
[54] V. Stahl, A. Fischer, and R. Bippus, "Acoustic synthesis of training data for speech recognition in living room environments," in *Proceedings of the IEEE International Conference on Acoustics, Speech and Signal Processing (ICASSP)*, vol. 1, 2001, pp. 21–24.
[55] T. Takiguchi and M. Nishimura, "Acoustic model adaptation using first order prediction for reverberant speech," in *Proceedings of the IEEE International Conference on Acoustics, Speech and Signal Processing (ICASSP)*, vol. 1, May 2004, pp. 869–72.
[56] A. Toh, R. Togneri, and S. Nordholm, "Combining MLLR adaptation and feature extraction for robust speech recognition in reverberant environments," in *Proceedings of the Internation Conference on Speech Science and Technology (SST)*, Auckland, New Zealand, December 2006.

[57] M. Unoki, K. Sakata, M. Furukawa, and M. Akagi, "A speech dereverberation method based on the MTF concept in power envelope restoration," *Acoustical Science and Technology*, vol. 25, no. 4, pp. 243–254, 2004.

[58] J. Y. C. Wen, A. Sehr, P. A. Naylor, and W. Kellermann, "Blind estimation of a feature-domain reverberation model in nondiffuse environments with variance adjustment," in *Proceedings of the European Signal Processing Conference (EUSIPCO)*, 2009, pp. 175–179.

[59] M. Wölfel, "Enhanced speech features by single-channel joint compensation of noise and reverberation," *IEEE Transactions on Audio, Speech and Language Processing*, vol. 17, no. 2, pp. 312–323, February 2009.

[60] S. J. Young, G. Evermann, M. J. F. Gales, T. Hain, D. Kershaw, G. Moore, J. Odell, D. Ollason, D. Povey, V. Valtchev, and P. C. Woodland, *The HTK Book, version 3.4*. Cambridge, UK: Cambridge University Engineering Department, 2006.

Part Four

Model Enhancement

11

Adaptation and Discriminative Training of Acoustic Models

Yannick Estève, Paul Deléglise
University of Le Mans, France

11.1 Introduction

The main weakness of automatic speech-recognition (ASR) systems resides in their lack of robustness to variability. All the knowledge bases used in an ASR system are affected by this problem: the dictionary – that is the list of the words recognizable by the system, along with their pronunciation variants – the language models as well as the acoustic models. Those knowledge bases – most particularly language and acoustic models, of probabilistic essence – are very dependent on the data used to estimate their various parameters. The problem posed by this dependence of probabilistic models on their training corpora is made more significant by the high cost of building such corpora. As a result of that cost, in practice, it is common for probabilistic models to be used in application contexts that differ considerably from the context of their training data.

Such mismatch between training data and application context causes the models to lose some of their precision and predictive power, in turn degrading the quality of speech recognition. This is a well-known problem, which has led to the development of many techniques aiming at lessening its impact. Model adaptation consists in reducing the mismatch between probabilistic models and the data against which they are used.

Noise is a cause of mismatch: it constitutes a variable phenomenon with potentially numerous, unexpected origins; it may appear suddenly, or be constantly present, or during an indeterminate time. It is a phenomenon that perturbs acoustic models. The perturbation is less noticeable in the case of a constant noise that is already present in the data used for the training of the acoustic models – an example of such constant noise (known as stationary additive noise) is the hum of a video projector during a meeting.

Acoustic model-adaptation techniques, initially developed with a focus on speaker adaptation, have seen their scope broaden to cover stationary additive noise, which is a kind of

Techniques for Noise Robustness in Automatic Speech Recognition, First Edition.
Edited by Tuomas Virtanen, Rita Singh, and Bhiksha Raj.

acoustic mismatch that they can handle very well (along with similar mismatches such as changes of transmission channels: microphone and telephone). As a consequence, acoustic model adaptation is now an essential tool as a means to lessen the impact of noise on the performance of ASR systems.

Some other kinds of noise, in particular nonstationary additive noise, are found almost impossible to compensate for using acoustic model adaptation. Yet such adaptation can still play a role in this case: other techniques presented in this book are able to reduce these kinds of noise, but usually at the price of transforming the signal itself; acoustic model adaptation can then be useful in order to minimize the side effects induced by these processes.

In addition to acoustic model adaptation, discriminative training techniques applied to HMM are really relevant to make acoustic models more robust to noise. They can be applied conjointly with some adaptation techniques.

The objective of this chapter is to explain to the reader the fundamentals of the most widely used techniques for acoustic model adaptation. In order to illustrate these theoretical aspects, algorithms used to implement the acoustic model-adaptation techniques are also described. This chapter is organized as follows: a brief overview of HMM (Hidden Markov Model) estimation used for acoustic modeling is first presented, followed by a general discussion on acoustic model adaptation for noise robustness, including some recent results. Maximum *A Posteriori* (MAP) reestimation is then described, followed by the Maximum Likelihood Linear Regression (MLLR) approach, including Constrained MLLR (CMLLR). Two discriminative training methods are also presented: Maximum Mutual Information Estimation (MMIE) and Minimum Phone Error (MPE) training. The last part of this chapter proposes a discussion on the use of these different techniques in ASR systems.

11.1.1 Acoustic Models

As described in Chapter 2, acoustic models are based on first-order HMMs, each of which models a phonetic unit, usually in context, such as triphones. In an acoustic model, each HMM is composed of a set $\mathcal{Q}$ of states, whose cardinality is set by hand; of a set A of discrete transition probabilities $a_{i,j} = P(q_{t+1} = j|q_t = i)$ from state i to state j $(i, j \in \mathcal{Q})$; of a set B of emission probability distributions – with one state-dependent output probability distribution for each state of the HMM; and of a set Π of initial probabilities over states π_q, giving the probability for each state $q \in \mathcal{Q}$ of being the initial state of the HMM. Thus, a HMM is defined by the four-tuple $(\mathcal{Q}, A, B, \Pi)$. Emission probabilities $b \in B$ are either discrete distributions, or mixtures of continuous density functions. In this chapter, we will focus on adaptation of Continuous Density HMM (CDHMM) acoustic models, which are the most widely used models: it will be assumed that the state-dependent output probability distributions are multi-Gaussian, continuous probability density functions, represented by Gaussian Mixture Models (GMMs). The likelihood of emission x at time t for state q of an HMM is then expressed as

$$b_q(\mathbf{x}_t) = \sum_{i=1}^{I} w_{q,i} \mathcal{N}(\mathbf{x}_t; \boldsymbol{\mu}_{q,i}, \boldsymbol{\Theta}_{q,i}), \tag{11.1}$$

where

- $\mathbf{x}_t$ is the output at time t.
- I is the number of mixture components.

- $w_{q,i}$ is the mixing coefficient for the ith mixture of the qth state.
- $\mathcal{N}$ is a Gaussian multivariate density function:

$$\mathcal{N}(\mathbf{x}_t; \boldsymbol{\mu}_{q,i}, \boldsymbol{\Theta}_{q,i}) = \frac{1}{\sqrt{(2\pi)^D |\boldsymbol{\Theta}_{q,i}|}} \exp -\frac{1}{2}(\mathbf{x}_t - \boldsymbol{\mu}_{q,i})^\mathsf{T} \boldsymbol{\Theta}_{q,i}^{-1} (\mathbf{x}_t - \boldsymbol{\mu}_{q,i}).$$

- $\boldsymbol{\mu}_{q,i}$ is the mean vector for the ith mixture of the qth state.
- $\boldsymbol{\Theta}_{q,i}$ is the covariance matrix for the ith mixture of the qth state.
- D is the dimension of the vector space.
- $|\boldsymbol{\Theta}_{q,i}|$ is the determinant of $\boldsymbol{\Theta}_{q,i}$.

From this point on, the covariance matrices will be assumed diagonal, except when this will be specified. This diagonal matrix will be represented as a vector, denoted $\boldsymbol{\theta}$, composed by its diagonal entries. This assumption makes calculations on Gaussians easier, in terms of both feasibility and computation time. For this reason, this assumption is usually found in practice in ASR systems.

11.1.2 Maximum Likelihood Estimation

Training acoustic models consists in estimating parameters $(\mathcal{Q}, A, B, \Pi)$ for each of the HMMs that model a phonetic unit. The structure of the HMMs, that is $\mathcal{Q}$, is set by hand and does not change. The initial probabilities over states (π_q) are clearly defined in the case of the left-right HMMs used for speech recognition. The parameters $\Lambda = (A, B)$ have to be estimated using the training data; the most widely used technique to do so is maximum likelihood estimation (MLE). Given representation X of the training corpus as a sequence of observation vectors ($X = \mathbf{x}_1, \mathbf{x}_2, \ldots, \mathbf{x}_m$), MLE consists in maximizing the likelihood $L(\Lambda|X)$, finding $\hat{\Lambda}$ such as

$$\hat{\Lambda} = \underset{\Lambda}{\operatorname{argmax}}\, L(\Lambda|X) = \underset{\Lambda}{\operatorname{argmax}} \sum_{Y \in \mathcal{Y}} \prod_{t=1}^{m} b_{y(t)}(\mathbf{x}_t) a_{y(t),y(t+1)}, \tag{11.2}$$

where $\mathcal{Y}$ is the set of all the admissible paths in HMMs and $Y = y(1) \ldots y(m)$ is a sequence of HMM states.

The Baum–Welch algorithm, presented in [3], is the most used to compute the value of $\hat{\Lambda}$. It constitutes a particular case of the EM algorithm described in [5]. We will assume here that the reader knows the Baum–Welch algorithm.

The training corpus is usually seen as a set $X = X^1, X^2, \ldots, X^R$ of sequences $X^r = \mathbf{x}_1^r, \mathbf{x}_2^r, \ldots, \mathbf{x}_{M_r}^r$ of observation vectors $\mathbf{x}_m^r$. Usually, each sequence X^r corresponds to an utterance from the training corpus. A rough initialization of parameters Λ for these sequences is usually done and training acoustic models consists in reestimating these parameters to match Equation (11.2). In the case of MLE using the Baum–Welch algorithm, the values of transition probabilities $\hat{a}_{i,j}$ can be obtained using the equation below:

$$\hat{a}_{i,j} = \frac{\sum_{r=1}^{R} \frac{1}{L_r} \sum_{t=1}^{M_r - 1} \alpha_i^r(t) a_{i,j} b_j(\mathbf{x}_{t+1}^r) \beta_j^r(t+1)}{\sum_{r=1}^{R} \frac{1}{L_r} \sum_{t=1}^{T_r - 1} \alpha_i^r(t) a_{i,j} \beta_i^r(t)}, \tag{11.3}$$

where

- $a_{i,j}$ and b_j are the initial parameters, pre-reestimation; they are the parameters used during the forward-backward process involved in the Baum–Welch algorithm.
- R is the number of observation sequences.
- M_r is the number of observation vectors in the rth sequence (X^r).
- $\alpha_i^r(t)$ is the forward probability of state i of the HMM – as obtained with the forward-backward algorithm – at time t of observation sequence X^r.
- $\beta_i^r(t)$ is the backward probability of state i of the HMM at time t of observation sequence X^r.
- L_r is the total likelihood of observation sequence X^r, obtained through the forward-backward algorithm[1]; actually, $L_r = \alpha_N^r(M_r) = \beta_1^r(1)$ with N being the number of states in the HMM. L_r can be considered as the sum over likelihoods of all the paths starting from $t = 0$ and ending to $t = R$ consuming each observation vector.

Estimating the emission probability function for state $b_q(\mathbf{x}_t)$ consists in estimating parameters $\Gamma_q = (w_{q,i}, \boldsymbol{\mu}_{q,i}, \boldsymbol{\Theta}_{q,i})$ of the corresponding GMM. The use of the Baum–Welch algorithm for MLE yields the following equations:

$$w_{q,i} = \frac{\sum_{r=1}^{R}\sum_{t=1}^{T_r}\gamma_{q,i}^r(t)}{\sum_{r=1}^{R}\sum_{t=1}^{T_r}\gamma_q^r(t)}, \tag{11.4}$$

$$\hat{\boldsymbol{\mu}}_{q,i} = \frac{\sum_{r=1}^{R}\sum_{t=1}^{T_r}\gamma_{q,i}^r(t)\mathbf{x}_t^r}{\sum_{r=1}^{R}\sum_{t=1}^{T_r}\gamma_{q,i}^r(t)}, \tag{11.5}$$

$$\hat{\boldsymbol{\Theta}}_{q,i} = \frac{\sum_{r=1}^{R}\sum_{t=1}^{T_r}\gamma_{q,i}^r(t)(\mathbf{x}_t^r - \hat{\boldsymbol{\mu}}_{q,i})(\mathbf{x}_t^r - \hat{\boldsymbol{\mu}}_{q,i})^{\mathsf{T}}}{\sum_{r=1}^{R}\sum_{t=1}^{T_r}\gamma_{q,i}^r(t)}, \tag{11.6}$$

$\gamma_{q,i}^r(t)$ being the probability of state occupancy for state q of the ith component of the Gaussian mixture at time t, and $\gamma_q^r(t)$ being the probability of state occupancy for state q at time t defined as: $\gamma_q^r(t) = \frac{1}{L_r}\alpha_q(t)\beta_q(t)$. Reestimation of the whole set of parameters Λ is repeated as long as $L(\Lambda|X)$ increases enough at each iteration of the process.

11.2 Acoustic Model Adaptation and Noise Robustness

In the next sections, the most well-known techniques for acoustic model adaptation will be presented. These techniques were not initially developed specifically to adapt acoustic models to noise, but they undeniably increase the accuracy of speech-recognition systems in noisy environments. These adaptation techniques may be used either *supervised* or *unsupervised*. In supervised mode, manual transcriptions of recordings in noisy environments are available. These data are ordinarily put to use in a static way: all the data are used at once to adapt preexisting acoustic models, before the speech-recognition system is started. In unsupervised mode, no manual transcriptions are available: such transcriptions must be generated by a speech-recognition system. Contrary to supervised mode, an important issue

[1] It is assumed that there is only one allowable initial state and one allowable final state.

in unsupervised mode consists in dealing with transcription errors. For instance, this can be done by using confidence measures provided by the ASR system in order to filter automatic transcripts. Adaptation of acoustic models using on these automatic transcriptions of audio recordings may then be carried out either statically or, more usually, dynamically–in the latter case, adaptation is performed online, while processing applicative data. Static and dynamic terms were used in [11] or [37] which also used respectively off-line (or batch) and on-line (or incremental) terms.

11.2.1 Static (or Offline) Adaptation

When the type of noise present in the audio signal is known in advance and in-domain recordings, with the same acoustic environment, are available, the acoustic models should be adapted before processing that signal–on condition that the initial acoustic models be more efficient than whatever acoustic models that could be estimated from scratch using only the available in-domain data.

Be it in supervised or in unsupervised mode, any of the adaptation techniques described in the present chapter can be applied for static adaptation using available in-domain data. The experiments carried out in [4] show that on the Aurora 2 corpus [14]–which is composed of recordings in noisy environments–using MLLR adaptation in static, unsupervised mode allows to increase word recognition accuracy. For example, for a signal-to-noise ratio (SNR) of 15 dB, word recognition accuracy goes from 87.5% with no adaptation to 93.6% with adaptation; while for a SNR of 5 dB, word recognition accuracy jumps from 39.5% to 72.8%. [4] presents various results, depending on the size of the adaptation corpus and on the amount or nature of the noise. In all cases, a larger adaptation corpus yields better results.

In unsupervised mode, the main risk is to reinforce the biases found in the unadapted acoustic models, which yield errors during the generation of automatic transcriptions of the in-domain data. In [32], the authors propose an unsupervised cross-validation technique, coupled with an aggregated adaptation technique, in order to reduce that overfitting problem. These techniques rely on MAP for unsupervised domain adaptation, together with unsupervised MLLR speaker adaptation. Their experiments in noisy environment on the Spontaneous Japanese corpus described in [17] show that associating MAP and MLLR techniques allows a decrease in word error rate, from 35% with no adaptation down to 29% with a classical unsupervised adaptation approach. When used aggregated as proposed in the article, the same techniques allow a word error rate of 25.8% on that same corpus.

11.2.2 Dynamic (or Online) Adaptation

In the case of unexpected noise in the speech signal, it is difficult to have the correct adaptation corpus beforehand. The unexpectedness of that noise usually implies that it will be present for a limited time only–therefore, the system should be able to adapt the acoustic models to the current acoustic conditions. Unsupervised dynamic adaptation fits that case, particularly when using MLLR or CMLLR adaptation techniques. Indeed, these techniques are relevant when the amount of adaptation data is small, and they exhibit relatively good robustness to transcription errors. In that case, the ASR system must be a multipass system: the first pass yields automatic transcriptions (using generic acoustic models), used to compute linear

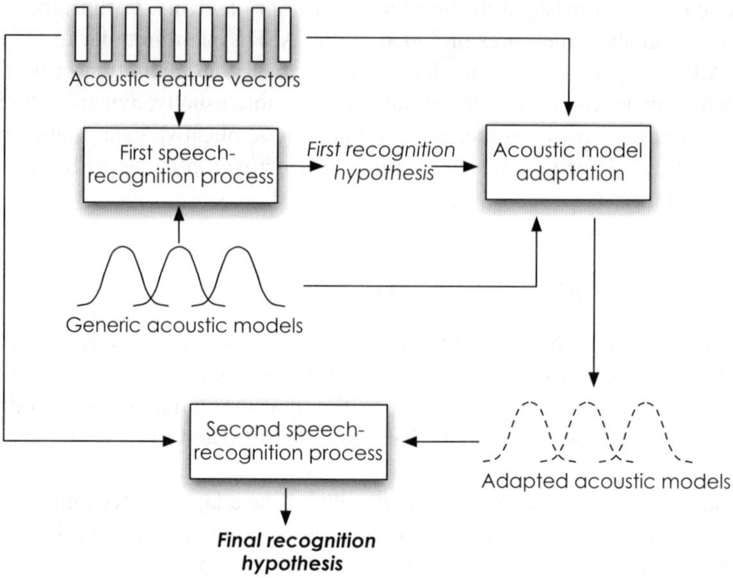

Figure 11.1 Dynamic adaptation of acoustic models: two speech-recognition passes are necessary to process an utterance. The first one produces an automatic transcription by applying generic acoustic models. This automatic transcription is then used to adapt acoustic models. The adapted acoustic models are used to process the utterance again.

transformation matrices for MLLR or CMLLR; the next pass then applies those transformation matrices, thus achieving unsupervised, dynamic adaptation of the acoustic models. Figure 11.1 illustrates dynamic adaptation of acoustic models[2].

In [29], the authors apply CMLLR-based unsupervised, dynamic adaptation to noisy data from the SPEECON database [15]. For a 14 dB SNR, the resulting word error rate is 43.8% with no adaptation and goes down to 28.8% when using CMLLR adaptation. For an 8 dB SNR, WER goes from 67.7% with no adaptation to 44.9% with CMLLR adaptation. In [18], the authors present an extension to CMLLR designed more specifically for adaptation to noise. The technique is named Noisy CMLLR (NCMLLR) and is presented in detail in Chapter 17. Results are given for phone number recognition in a car driven on a highway (*Toshiba In-Car Task*). In that situation, the average SNR is 18 dB. Without adaptation, but with MPE models trained on in-domain data, the word error rate is 5%. With NCMLLR adaptation in unsupervised, dynamic mode, the word error rate can go as low as 1.4%.

11.3 Maximum *A Posteriori* Reestimation

(MAP), described in [10], is used for model adaptation in cases where there is a significant mismatch between training data and application data, for instance when application data

[2] Figure 11.1 refers to MLLR acoustic model adaptation. CMLLR adaptation usually transforms acoustic features vectors instead of acoustic models, contrary to what is shown in the figure: this point will be explained later in this chapter.

have different noise or channel conditions from the training data, and a sufficient amount of adaptation data are available. MAP adaptation relies on the availability of an *a priori* distribution $p_0(\Lambda)$ of the parameters to be adapted. Where in the case of MLE, the parameters $\hat{\Lambda}$ of an HMM are chosen in order to maximize $L(\Lambda|X)$, in the case of MAP the *a priori* distribution is involved in the estimation of $\hat{\Lambda}$:

$$\hat{\Lambda}_{\mathsf{map}} = \underset{\Lambda}{\operatorname{argmax}}\, L(\Lambda|X) p_0(\Lambda). \tag{11.7}$$

In practice, $p_0(\Lambda)$ corresponds to the distribution of parameters Λ as it was computed during acoustic model training. Since training is usually done with large amounts of data, taking distribution $p_0(\Lambda)$ into account allows robust reestimation of parameters Λ on a comparatively smaller amount of data. It has to be noted that if distribution $p_0(\Lambda)$ is uniform, it does not provide any information and MAP reestimation then amounts to MLE.

MAP reestimation of the mean $\hat{\mu}$ of a given Gaussian, starting with prior mean μ_0, can be described as:

$$\hat{\boldsymbol{\mu}}_{q,i} = \frac{\tau_{q,i}\boldsymbol{\mu}_0 + \sum_{r=1}^{R}\sum_{t=1}^{T_r}\gamma_{q,i}^{r}(t)\mathbf{x}_t^r}{\tau_{q,i} + \sum_{r=1}^{R}\sum_{t=1}^{T_r}\gamma_{q,i}^{r}(t)}, \tag{11.8}$$

where $\tau_{q,i}$ is a meta-parameter, usually set empirically to a value between 2 and 20, identically for each pair (q, i). $\tau_{q,i}$ allows to set the weight of the prior mean relative to the weight of the mean computed through maximum likelihood estimation. A low value for $\tau_{q,i}$ results in a low influence of the prior mean in the MAP reestimation. Equation (11.8) comes from [10], which also proposes similar equations to adapt the covariance matrix $\Theta_{q,i}$ of each Gaussian, as well as its weight $w_{q,i}$ within the mixture.

Algorithm 1 proposes an implementation of the reestimation of all these parameters. Figure 11.2 illustrates the fact that by using MAP adaptation each Gaussian in the original acoustic models is updated individually. In this figure, each Gaussian is located using its isodensity contour in a two-dimensional acoustic feature space, this isodensity being defined for example as its 85% peak likelihood contour. For a Gaussian, the importance of the update depends to the number of occurrences of this Gaussian in the adaptation corpus.

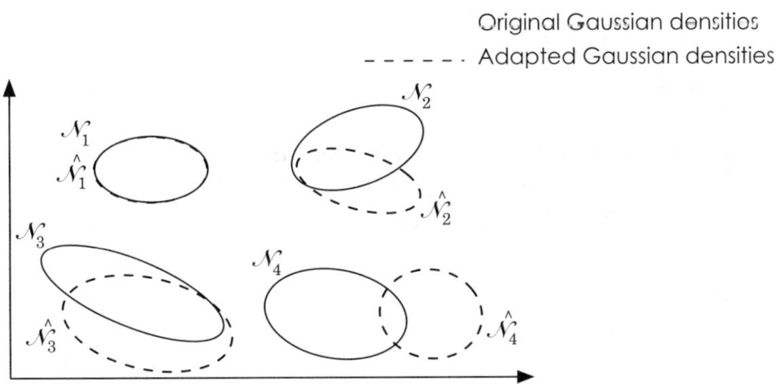

Figure 11.2 Schema on the effects of MAP reestimation in a two-dimensional feature space: each Gaussian is updated individually.

Algorithm 1: Example of MAP adaptation implementation

Input: τ value, prior means $\boldsymbol{\mu}_{\text{prior}}$, variances $\boldsymbol{\theta}_{\text{prior}}$, and mixture weights w_{prior}
Output: MAP adapted means $\boldsymbol{\mu}_{\text{map}}$, variances $\boldsymbol{\theta}_{\text{map}}$, and mixture weights w_{map}
Data: adaptation data X

$L_{\text{old}} = 0$;
repeat
  MLE reestimation on adaptation data using Equations (11.4), (11.5), and (11.6) to compute the w_{mle}, $\boldsymbol{\mu}_{\text{mle}}$, *and* $\boldsymbol{\theta}_{\text{mle}}$ *values.After the first iteration,* w_{map}, $\boldsymbol{\mu}_{\text{map}}$, *and* $\boldsymbol{\theta}_{\text{map}}$ *– which are values computed during the previous iteration – are used as parameters of initial acoustic models for the MLE reestimation process.*
  if $(L(\Lambda_{\text{mle}}|X) - L_{\text{old}}) < \Delta$ **then**
    `/* Stop condition of the main loop */`
    break ;
  end
  $L_{\text{old}} = L(\Lambda_{\text{mle}}|X)$;
  foreach *state* q **do**
    foreach *gaussian density* i **do**
      w = w_{mle}[q,i];
      w_{map}[q,i] = w + τ;
      if $w_{\text{mle}}[q,i] > \tau/10$ **then**
        `/* only if enough observations */`
        foreach *acoustic vector component* c **do**
          m = ($\tau \times \mu_{\text{prior}}$[q,i,c] + w $\times$ μ_{mle}[q,i,c]) / (τ+ w);
          v = $\tau \times \theta_{\text{prior}}$[q,i,c]
            + w$\times$(θ_{mle}[q,i,c] + μ_{mle}[q,i,c]$\times\mu_{\text{mle}}$[q,i,c]
                - 2$\times\mu_{\text{mle}}$[q,i,c]$\times$m + m$\times$m)
            + $\tau\times$(μ_{prior}[q,i,c] - m)$\times$(μ_{prior}[q,i,c] - m);
          if $\theta_{\text{mle}}[q,i,c] > \epsilon$ **then**
            `/* only if the variance is large enough */`
            `/*` ϵ `has to be empirically estimated */`
            v = v/(τ + w);
          else
            v=θ_{prior}[q,i,c] ;
          end
          μ_{map}[q,i,c] =m;
          θ_{map}[q,i,c] =v;
        end
      else
        foreach *acoustic vector component* c **do**
          μ_{map}[q,i,c] = μ_{prior}[q,i,c];
          θ_{map}[q,i,c] = θ_{prior}[q,i,c];
        end
      end
    end
  end
until $false$;

11.4 Maximum Likelihood Linear Regression

When, as part of speaker adaptation of acoustic models, transformations are only applied to the means of the Gaussians, as described in [20], or to both the means and covariance matrices as in [8], they can be seen as repositioning acoustic classes from the acoustic space of the initial models to the acoustic space of a new speaker, or new recording environment. MLLR is the main linear-transformation technique for adapting only the means, and sometimes variances, of a Gaussian mixture associated with an HMM, without updating the weight of the mixture components. When using this linear regression approach, the Gaussian mean parameters are updated as follows:

$$\hat{\boldsymbol{\mu}}_{q,i} = \mathbf{A}_c \boldsymbol{\mu}_{q,i} + \mathbf{b}_c, \tag{11.9}$$

where $\mathbf{A}_c$ is an $D \times D$ regression matrix and $\mathbf{b}_c$ is an additive D-dimensional vector (D being the dimensionality of the observations). $\mathbf{A}_c$ and $\mathbf{b}_c$ are associated with acoustic class c which can be a set of shared HMM states or a class of close phonemes. The same transformation can be applied to all the Gaussians that are part of one same given acoustic class. This makes the MLLR approach interesting when only a small amount of adaptation data is available, since it is possible to set the number of classes – and hence the number of transformations to define – as a function of that amount.

Equation (11.9) is usually rewritten as:

$$\hat{\boldsymbol{\mu}}_{q,i} = \mathbf{W}_c \boldsymbol{\xi}_{q,i}, \tag{11.10}$$

where $\mathbf{W}_c$ is a $D \times (D+1)$ regression matrix and $\boldsymbol{\xi}_{q,i}$ is the extended mean vector, with $\mathbf{W}_c = [\mathbf{b}_c, \mathbf{A}_c]$ and $\boldsymbol{\xi}_{q,i} = [1, \boldsymbol{\mu}_{q,i}^{\mathsf{T}}]^{\mathsf{T}}$. Computing the matrices $\mathbf{W}_c$ is the central point of acoustic model adaptation through MLLR. The matrices $\mathbf{W}_c$ are estimated so that the likelihood of the resulting adapted model is maximized on the adaptation data. As usual, the Expectation-Maximization (EM) algorithm is well suited to the resolution of this problem. Maximization of the auxiliary function used in the EM algorithm, of which the details are given in [20], yields the following equation for computation of the elements of $\mathbf{W}_c$:

$$\mathbf{w}_{c,r} = \mathbf{k}_c^{(r)} \mathbf{G}_c^{(r)-1}, \tag{11.11}$$

where $\mathbf{w}_{c,r}$ is the rth row of transformation matrix $\mathbf{W}_c$ corresponding to acoustic class c, and

$$\mathbf{G}_c^{(r)} = \sum_{i_c=1}^{I_c} \frac{1}{\theta_{i_c r}^2} \boldsymbol{\xi}_{i_c} \boldsymbol{\xi}_{i_c}^{\mathsf{T}} \sum_{t=1}^{T} \gamma_{i_c}(t), \tag{11.12}$$

$$\mathbf{k}_c^{(r)} = \sum_{i_c=1}^{I_c} \sum_{t=1}^{T} \gamma_{i_c}(t) \frac{1}{\theta_{i_c r}^2} \mathbf{x}_r(t) \boldsymbol{\xi}_{i_c}^{\mathsf{T}}, \tag{11.13}$$

where i_c is a mixture component of regression class c, $\theta_{i_c r}$ is the rth element on the diagonal in the covariance matrice $\boldsymbol{\Theta}_{i_c}$ of this component, γ_{i_c} is the occupancy probability for i_c at time t, and $\boldsymbol{\xi}_{i_c}$ is the extended mean vector for the mixture component i_c.

Mean adaptation is where MLLR has the greatest impact, but variance adaptation may also bring an extra, slight decrease in word error rate. To adapt variances, the transformation matrix $\mathbf{H}_c$ can be computed from the means adapted by using the transformation matrix $\mathbf{W}_c$.

Variances are adapted as follows:

$$\hat{\Theta}_{q,i} = \mathbf{C}^{-1^{\mathsf{T}}} \mathbf{H}_c \mathbf{C}^{-1}, \tag{11.14}$$

where $\mathbf{C}^{-1}$ is the inverse of the Choleski factor of $\Theta_{q,i}$, that is $\Theta_{q,i}^{-1} = \mathbf{C}\mathbf{C}^{\mathsf{T}}$. Estimating $\mathbf{H}_c$ then amounts to

$$\mathbf{H}_c = \frac{\sum_{i_c=1}^{I_c} \mathbf{C}_{i_c}^{\mathsf{T}} \left[\gamma_{i_c}(t)(\mathbf{x}(t) - \hat{\boldsymbol{\mu}}_{i_c})(\mathbf{x}(t) - \hat{\boldsymbol{\mu}}_{i_c})^{\mathsf{T}} \right] \mathbf{C}_{i_c}}{\gamma_{i_c}(t)}. \tag{11.15}$$

The equation above is only for the case of diagonal covariance matrices, which, as explained earlier, is the most frequently used configuration for reasons of simplicity and computation time. An implementation of the estimation of MLLR matrices for mean transformation is presented in Algorithm 2. Figure 11.3 illustrates the fact that by using MLLR adaptation all the Gaussians group to the same MLLR classes in the original acoustic models are updated in one block, contrary to MAP adaptation illustrated in Figure 11.2.

11.4.1 Class Regression Tree

The number of linear transformation matrices that it is be possible to estimate depends on the available amount of adaptation data. Each transformation corresponds to an acoustic class. The definition of such a class is variable. An acoustic class may be manually defined, in a static way. For example, an acoustic class may be the set of phonemes belonging to the same predefined broad phoneme class: sonorant, stop, fricative, closure, silence, vowel, etc. The transformation for this class will be applied to all the Gaussian components taking part in the corresponding HMMs.

Acoustic classes are more usually generated automatically and in a dynamic way, which allows to adapt the number of acoustic classes to the amount of available adaptation data, and

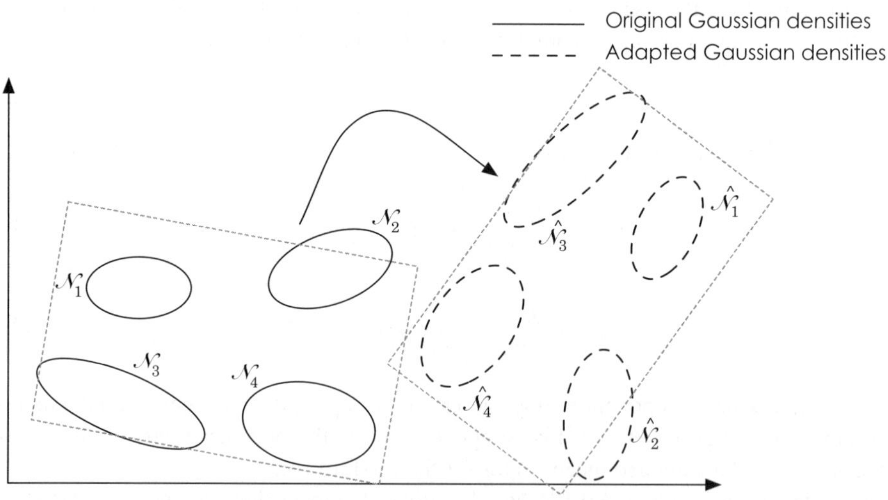

Figure 11.3 Schema on the effects of MLLR adaptation in a two-dimensional feature space: all the Gaussians grouped in the same MLLR classes are updated with the same transformation.

Algorithm 2: Estimation of MLLR matrices for mean transformation

Input: Initial acoustic models estimated on the training corpus: means $\boldsymbol{\mu}_{i_c}$ and variances $\boldsymbol{\theta}_{i_c}$
Output: MLLR mean transformation matrices: $\mathbf{W}_c$
Data: Adaptation data: $\mathbf{x}_t$

```
/* First step. Modified Baum-Welch algorithm to compute G_c et k_c
   during the backward pass for each time t                                */
foreach regression class c do
    foreach gaussian i_c included to the class c do
        for p : 0 → (D − 1) do
            for q : p → (D − 1) do
                /* Computing ξ_{i_c} ξ_{i_c}^T represented by mProd              */
                mProd[p][q] = μ_{i_c}[p] × μ_{i_c}[q] ;
            end
        end
        for r : 0 → (D − 1) do
            temp1 = γ_{i_c}(t) / θ_{i_c}[r] ;
            for p : 0 → (D − 1) do
                for q : p → (D − 1) do
                    G_c[r][p][q] = temp1 × mProd[p][q];
                end
                temp2 = temp1 × μ_{i_c}[p]
                G_c[r][p][D] += temp2 ;
                k_c[r][p] + temp2 × x_t[r] ;
            end
            G_c[r][D][D] += temp2 ;
            k_c[r][D] += temp1 × x_t[r] ;
        end
    end
end
/* Second step. Resolving Equation (11.11) to obtain the MLLR
   transformation matrix W_c                                               */
foreach regression class c do
    for j : 0 → (D − 1) do
        if G_c is invertible then
            Resolve G_c W_c = k_c /* classical AX = B linear equation      */
        end
    end
end
```

makes MLLR adaptation more robust. Clustering into acoustic classes is done directly at the Gaussian component level, using a class regression tree.

The class regression tree is built using the initial acoustic models and does not depend on the adaptation data. It is a binary tree, with each node grouping Gaussian components which are close in the acoustic space. The deeper a node is in the tree, the closer its components are to each other. During MLLR adaptation, each node may be attached to a transformation matrix, depending on the amount of adaptation available to estimate that matrix. The advantage of grouping Gaussian components lies in that is allows applying one same transformation to a whole set of components, thus also adapting distributions which were not observed in the adaptation data.

In the ideal case where the amount of adaptation data is large enough, the matrices that will be used will be the ones attached to the leaf nodes of the class regression tree. The Gaussian component clusters attached to the leaves are named *base regression classes*. In practice it is often the case that adaptation data are not available in amounts large enough to allow matrix estimation for the base regression classes. The solution is then to use lower ranked nodes, going all the way down to the root node in the worst case. In the latter case, the one same transformation, known as the global transformation, will be applied to every Gaussian component.

Acoustic proximity is usually measured for two Gaussians by comparing their means. In order to build the class regression tree, a centroid-splitting algorithm using a Euclidean distance measure is usually applied to the whole set of Gaussian components. Algorithm 3 shows a brief summary of class regression tree building.

Algorithm 3: Example of an algorithm building a class regression tree

Input: HMM that were estimated on the training data; number of expected base regression classes

Output: Class regression tree (hierarchized nodes)

Compute the mean $\boldsymbol{\mu}_0$ and the variance $\boldsymbol{\Theta}_0$ from the mean values of Gaussians associated with the root node n_0 containing all the Gaussians, weighted according to their occupancy in the training data;

repeat

  Select the node n_p to split: for example the one with the largest Euclidean distance between a mean vector $\boldsymbol{\mu}_i$ of a Gaussian i in this node and $\boldsymbol{\mu}_p$;

  Insert two new children n_l and n_r and initialize their means $\boldsymbol{\mu}_l$ and $\boldsymbol{\mu}_r$ with their parent's mean $\boldsymbol{\mu}_p$, by translating them to opposite directions: $\boldsymbol{\mu}_l = \boldsymbol{\mu}_p - \lambda\sqrt{diag(\boldsymbol{\Theta}_p)}$ and $\boldsymbol{\mu}_r = \boldsymbol{\mu}_p + \lambda\sqrt{diag(\boldsymbol{\Theta}_p)}$ where $\sqrt{diag(\boldsymbol{\Theta}_p)}$ is the component-wise square root of a vector formed from the diagonal elements of $\boldsymbol{\Theta}_p$.**foreach** *gaussian component in node* n_p **do**

    Assign this component to the closest child by using the Euclidean distance measure;

  end

  Recompute the new mean $\boldsymbol{\mu}_l$ (resp. $\boldsymbol{\mu}_r$) and the variance $\boldsymbol{\Theta}_l$ (resp. $\boldsymbol{\Theta}_r$) from the mean values of Gaussians associated with node n_l (resp. n_r), weighted by their occupancy in the training data;

until *the number of expected final nodes is reached*;

11.4.2 Constrained Maximum Likelihood Linear Regression

MLLR proposes a method to estimate transformations for the Gaussian means and variances of acoustic models. Estimation of the transformation matrices is done independently for the means and for the variances. CMLLR proposes an approach where the transformations are constrained as follows:

$$\hat{\boldsymbol{\mu}}_{q,i} = \mathbf{A}_c \boldsymbol{\mu}_{q,i} - \mathbf{b}_c, \tag{11.16}$$

$$\hat{\boldsymbol{\Theta}}_{q,i} = \mathbf{A}_c^{\mathsf{T}} \boldsymbol{\Theta}_{q,i} \mathbf{A}_c, \tag{11.17}$$

where $\mathbf{A}_c$ is still an $D \times D$ regression matrix and $\mathbf{b}_c$ is an additive D-dimensional vector (D being the dimensionality of the observations). This approach was first presented in [6] for use with diagonal transformation matrices, and later extended in [7] to cover full transformation matrices. In practice, CMLLR is used more as an adaptation technique for acoustic features rather than to adapt models. Indeed, applying the constraints defined in Equations (11.16) and (11.17) in order to maximize likelihood for the acoustic models on the adaptation data amounts to modifying the value of the observation vectors as follows:

$$\hat{\mathbf{x}}_t = \mathbf{A}_c^{-1} \mathbf{x}_t + \mathbf{A}_c^{-1} \mathbf{b}_c. \tag{11.18}$$

The concept of acoustic classes is not really relevant in the case where the acoustic features themselves are transformed, because of the difficulty of determining a direct relationship between an observation and a class of Gaussians. Therefore, only one CMLLR transformation matrix is usually applied to the acoustic features before those features are matched against the acoustic models as part of the speech-recognition process.

11.4.3 CMLLR Implementation

For CMLLR implementation, the problem will be different from the one encountered for MLLR. We will try to find $\mathbf{A}$ and $\mathbf{b}$ such as

$$\hat{\mathbf{x}}_t = \mathbf{A}_c \mathbf{x}_t + \mathbf{b}_c. \tag{11.19}$$

By defining

$$\mathbf{W}_c = [\mathbf{A}_c\ \mathbf{b}_c], \tag{11.20}$$

it can be rewritten as

$$\hat{\mathbf{x}}_t = \mathbf{W}_c \boldsymbol{\zeta}_t, \tag{11.21}$$

$$\boldsymbol{\zeta}_t = [\mathbf{x}_t^{\mathsf{T}}\ 1]^{\mathsf{T}}. \tag{11.22}$$

In the same way as in the MLLR approach, computing the matrices $\mathbf{W}_c$ is a crucial point of acoustic model adaptation through CMLLR. Matrices $\mathbf{W}_c$ have to be estimated so as to maximize the likelihood of the resulting adapted acoustic models on the adaptation data. Maximizing the auxiliary function amounts to maximizing the following expression:

$$\mathcal{Q} = \beta_c \log(\det(\mathbf{A}_c)) - \frac{1}{2} \sum_{r=0}^{D-1} \left(\mathbf{w}_{c,r} \mathbf{G}_c^{(r)} \mathbf{w}_{c,r}^{\mathsf{T}} - 2\mathbf{w}_{c,r} \mathbf{k}_c^{(j)\mathsf{T}} \right), \tag{11.23}$$

where $\mathbf{w}_{c,r}$ is the rth row of transformation matrix $\mathbf{W}_c$ corresponding to acoustic class c and

$$\beta_c = \sum_{t=1}^{T} \sum_{i_c=1}^{l_c} \gamma_{i_c}(t), \tag{11.24}$$

$$\mathbf{G}_c^{(r)} = \sum_{i_c=1}^{l_c} \frac{1}{\theta_{i_c r}} \sum_{t=1}^{T} \gamma_{i_c}(t) \boldsymbol{\zeta}_t \boldsymbol{\zeta}_t^{\mathsf{T}}, \tag{11.25}$$

$$\mathbf{k}_c^{(r)} = \sum_{i_c=1}^{l_c} \frac{1}{\theta_{i_c,r}} \boldsymbol{\mu}_{i_c,r} \sum_{t=1}^{T} \gamma_{i_c}(t) \boldsymbol{\zeta}_t^{\mathsf{T}}. \tag{11.26}$$

In order to get the derivative of the expression with respect to w_r, it can be noticed that

$$\det(\mathbf{A}) = \mathrm{cofact}(\mathbf{A})_r \mathbf{a}_r^{\mathsf{T}}, \tag{11.27}$$

where $\mathrm{cofact}(\mathbf{A})_r$ is the rth row of the matrix of cofactors of $\mathbf{A}$, and $\mathbf{a}_r^{\mathsf{T}}$ is the transposed of the rth row of matrix $\mathbf{A}$. By writing $\mathbf{q}_{c,r} = [\mathrm{cofact}(\mathbf{A}_c)_r \; 0]$, we get

$$\det(\mathbf{A}_c) = \mathbf{q}_{c,r} \mathbf{w}_{c,r}^{\mathsf{T}}. \tag{11.28}$$

Differentiating $\mathcal{Q}$ presented in Equation (11.23) with respect to $\mathbf{w}_{c,r}$ yields

$$\frac{d\mathcal{Q}}{d\mathbf{w}_{c,r}} = \beta_c \frac{\mathbf{q}_{c,r}}{\mathbf{q}_{c,r}\mathbf{w}_{c,r}^{\mathsf{T}}} - \mathbf{w}_{c,r}\mathbf{G}_c^{(r)} + \mathbf{k}_c^{(r)}. \tag{11.29}$$

Since CMLLR is usually done with one class, the sums over i_c become sums of $i \in \mathcal{G}$, with $\mathcal{G}$ being the set of Gaussians that exist in the model. In this case, index c disappears and β is equal to the total number of frames taken into account in the algorithm.

Practical realization is done in two steps. First, accumulators (β, $\mathbf{G}^{(r)}_{\{0 \le r < D\}}$, $\mathbf{k}^{(r)}_{\{0 \le r < D\}}$) have to be computed over the whole adaptation corpus. Then, Equation (11.29) has to be solved through an iterative process described in [7].

11.4.4 Speaker Adaptive Training

In order to reduce variation due to speaker, channel, or acoustic condition, [1] has proposed the speaker adaptive training (SAT) approach, which uses MLLR or CMLLR transformations during model training. This approach allows to prepare the initial acoustic models in view of MLLR or CMLLR adaptation at the time they get used in a dynamic way[3] during the speech-recognition process.

SAT consists in grouping the training data of acoustic models by speaker, after the initial models have been estimated. For each speaker, MLLR or CMLLR transformations are computed using the training data corresponding to the speaker. In the case of MLLR, the transformations for each speaker are applied to the initial models, which are then reestimated

[3] *Cf.* Section 11.2 in this chapter.

on the whole training corpus. The resulting models have a higher likelihood on the training corpus, as well as smaller variances. This approach yields a significant decrease in word error rate, as was shown in [1,28].

The use of CMLLR seems more beneficial than that of MLLR: applying the transformations to the acoustic features is far easier than applying them to the model means; and, on top of that, CMLLR/SAT training can be used conjointly with a discriminative training method, with additive benefits.

The success of SAT which makes ASR systems more robust to speaker variability was the main motivation to develop the noise adaptive training (NAT) approach, described in Chapter 13.

11.5 Discriminative Training

Speech recognition these days sees more and more widespread use of discriminative training techniques applied to HMMs. However, it should be noted that those techniques, which are in competition against generative training approaches such as MLE seen in Section 11.1.2, have been studied by the speech-recognition community for decades. The first works, based on information theory, resulted in the MMIE approach, presented in [2]. But, for a long time, discriminative training was lagging behind generative training in terms of results, except for a few simple tasks such as digit recognition as was presented in [22]. Except for [35], it was after 2000, and particularly following the work presented in [25], that discriminative training started being interesting for very large vocabulary speech-recognition systems.

These days, using discriminative training allows to decrease the word error rate for most applications of automatic speech recognition. Discriminative training may be seen as an adaptation technique in the same way as the MAP approach: it can not be used dynamically – as CMLLR can – but only statically, like MAP. Discriminative training is commonly used to adapt acoustic models to the targeted task.

In fact, discriminative training improves the robustness of the acoustic models. Reference [38] presents a study on such robustness according to the size of the margin of a model: this margin is defined as *"the desired minimum distance between any training sample to the decision boundary of the model in a separable classification case."* Noise distortion reduces the relevance of the decision boundary of a model, especially when this model was trained on clean data. Figure 11.4 shows the impact of noise distortion, while Figure 11.5 illustrates the definition of the margin of a model. The both were inspired from [38]. Discriminative training implicitly increases the margin of a model by choosing utterances near the decision boundary to modify the model: larger margins make acoustic models more robust to noise distortions.

In [38], authors compare acoustic models trained with MLE, acoustic models estimated with discriminating training, and acoustic models trained with a margin-based method, called soft-margin estimation (SMEs). In their experiments, acoustic models trained with discriminative training and SME outperform very significantly the ones trained with MLE on noisy data on the Aurora-2 corpus, while SME allows to reach slightly better results than discriminative training in these experimental conditions. For example, in one of their experiments, acoustic models trained with MLE reach 85.58% of word accuracy for a signal-to-noise ratio of 15dB,

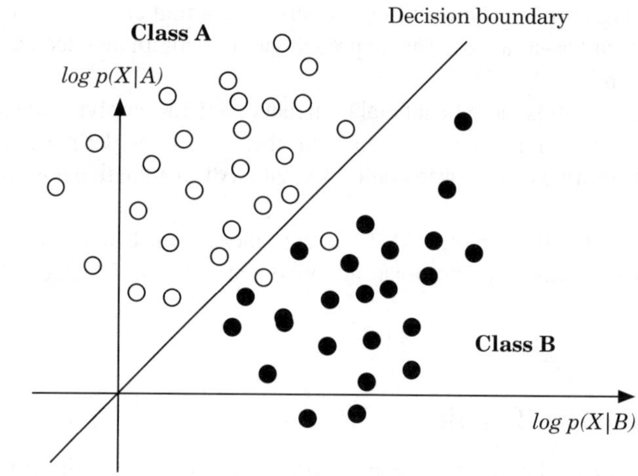

Figure 11.4 Noise robustness illustrated by a two-classification problem in log likelihood domain. Elements in class A are represented by white circles and elements in class B by black ones. Noise distortion reduces the relevance of the decision boundary of a model: noisy samples may cross the decision boundary and may be wrongly classified.

while the ones trained with discriminative training reach 92.27% and the ones trained with SME reach 92.95%.

In this section, we will present the fundamentals of discriminative training with MMIE or minimum phone error (MPE), as well as the bases of an implementation of MPE. Readers wishing to learn more about discriminative training will find helpful material in the tutorial presented in [13]. A very complete overview of that domain can also be found in [16].

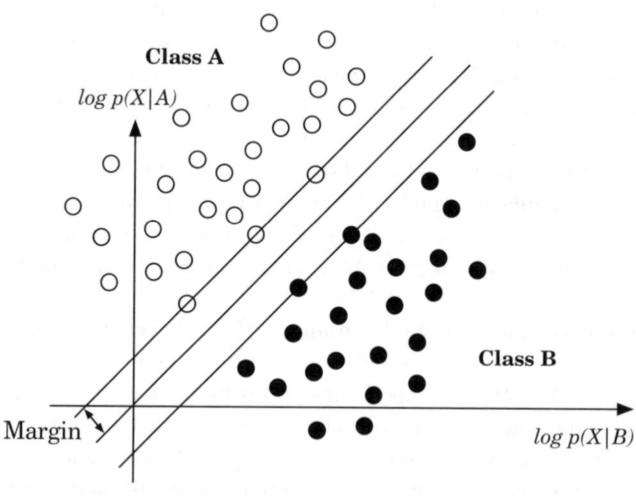

Figure 11.5 Increasing the margin makes the model more robust. Elements in class A are represented by white circles and elements in class B by black ones.

11.5.1 MMI Discriminative Training Criterion

Each training criterion used to estimate the Λ parameters of HMMs can be represented as an objective function $\mathcal{F}(\Lambda, X^1 ... X^R)$ that should be either maximized or minimized on the training data $X^1 \ldots X^R$. For example, the objective function associated with MLE, to be maximized (*cf.* Section 11.1.2), can be written as: $\mathcal{F}_{\mathsf{MLE}} = \sum_{r=1}^{R} \log P_\Lambda(X^r|s^r)$, where s^r is the correct transcription of X^r, the rth sequence of acoustic data.

The MMI objective function used for MMIE is written as follows [2,30]:

$$\mathcal{F}_{\mathsf{MMI}}(\Lambda) = \sum_{r=1}^{R} \log \frac{P_\Lambda(X^r|s^r)^{\mathcal{K}} P(s^r)^{\mathcal{K}}}{\sum_s P_\Lambda(X^r|s)^{\mathcal{K}} P(s)^{\mathcal{K}}}, \tag{11.30}$$

where $P(s)$ is the language model probability, including linguistic weights and word insertion penalties for sentence s; $P_\Lambda(X^r|s^r)$ is the likelihood provided by acoustic models for the correct sentence s^r; and $\mathcal{K}$ is a probability scale optimized to get better results on test data [30].

In fact, the numerator in Equation (11.30) is the likelihood of the correct transcription combined with its linguistic probability, while the denominator is the total likelihood for all the word sequences combined with linguistic probabilities: $\mathcal{F}_{\mathsf{MMI}}(\Lambda)$ equals posterior probability of correct sentences given data and acoustic models with Λ parameters. Maximizing $\mathcal{F}_{\mathsf{MMI}}(\Lambda)$ corresponds to maximizing the ratio of the MLE training criterion by the denominator. That way the MMI criterion should be more correlated to the word error rate than the ML criterion is. This is true in the case of the training corpus; we will see later that it will be necessary to take into account the generalization of the MMIE approach.

The extended Baum–Welch algorithm presented in [12,21] is generally used in order to optimize $\mathcal{F}_{\mathsf{MMI}}(\Lambda)$. It then yields the equations below for the estimation of means and variances obtained through MMIE:

$$\hat{\boldsymbol{\mu}}_{q,i} = \frac{\left\{\varsigma_{q,i}^{\mathsf{num}}(X) - \varsigma_{q,i}^{\mathsf{den}}(X)\right\} + \mathcal{D}_{q,i}\boldsymbol{\mu}_{q,i}}{\left\{\gamma_{q,i}^{\mathsf{num}} - \gamma_{q,i}^{\mathsf{den}}\right\} + \mathcal{D}_{q,i}}, \tag{11.31}$$

$$\hat{\boldsymbol{\theta}}_{q,i} = \frac{\left\{\varsigma_{q,i}^{\mathsf{num}}(X^2) - \varsigma_{q,i}^{\mathsf{den}}(X^2)\right\} + \mathcal{D}_{q,i}(\boldsymbol{\theta}_{q,i} + \boldsymbol{\mu}_{q,i}^2)}{\left\{\gamma_{q,i}^{\mathsf{num}} - \gamma_{q,i}^{\mathsf{den}}\right\} + \mathcal{D}_{q,i}} - \hat{\boldsymbol{\mu}}_{q,i}^2, \tag{11.32}$$

where

- $\varsigma_{q,i}(X)$ and $\varsigma_{q,i}(X^2)$ are sums of data and squared data, respectively, weighted by their probability for Gaussian i of state q, data being represented by acoustic observation vectors.
- $\gamma_{q,i}$ are Gaussian occupancies summed over time.
- num corresponds to the correct word sequence, and den corresponds to the recognition model computed from all the possible word sequences: in practice all the possible recognition hypotheses are approximated by using word lattices[4].

[4] A word lattice is a directed acyclic graph with a set of nodes (including a starting node) and arcs (or transitions) between these nodes. Each hypothesis word is associated to a node or to an arc: this depends to the representation chosen in the ASR system. Different scores (acoustic, linguistic, . . .) can also label nodes or arcs.

- $\boldsymbol{\mu}_{q,i}$ and $\boldsymbol{\theta}_{q,i}$ are the prior means and variances, $\boldsymbol{\mu}_{q,i}^2$ denoting element-wise squaring, with $\boldsymbol{\mu}_{q,i}^2 = \boldsymbol{\mu}_{q,i} \otimes \boldsymbol{\mu}_{q,i}$.
- $\mathcal{D}_{q,i}$ are positive smoothing constants for each Gaussian i of state q: a low value of $\mathcal{D}_{q,i}$ increases optimization speed, but $\mathcal{D}_{q,i}$ must be large enough to ensure positive variances.

The process of discriminative training with MMIE is summarized as follows:

1. Process training (or adaptation) data in order to generate word lattices using MLE acoustic models with a weak language model, for example unigram or bigram.
2. Align HMMs with the word lattices in order to specify HMM models boundaries.
3. While word error rate on development data increases
   (a) compute $\mathcal{D}_{q,i}$ values which ensure that the variances stay positive; and
   (b) estimate $\hat{\boldsymbol{\mu}}_{q,i}$ and $\hat{\boldsymbol{\theta}}_{q,i}$ following, respectively, Equations (11.31) and (11.32).

11.5.2 MPE Discriminative Training Criterion

The goal of MMIE was to propose an estimation of HMM parameters resulting in acoustic models that would be more discriminant than those obtained through generative training of the MLE kind. To reach this objective, estimation of acoustic models is based on getting sharper distinction between wrong and good recognition hypotheses. The purpose of MPE is the same, the difference being in the discriminative criterion, which in this case is based on an approximation of the phone error rate [24, 25]. MPE generally yields better results than MMIE, for a similar training complexity.

While $\mathcal{F}_{\mathsf{MMI}}(\hat{\Lambda})$ is the posterior probability of the correct utterance given the speech data, the MPE objective function, $\mathcal{F}_{\mathsf{MPE}}(\hat{\Lambda})$, is a weighted average of a measure of phone accuracy, weighted by sentence likelihood, as described in [25]:

$$\mathcal{F}_{\mathsf{MPE}}(\Lambda) = \sum_{r=1}^{R} \frac{\sum_s P_\Lambda(X^r|s)^{\mathcal{K}} P(s)^{\mathcal{K}} \text{RawPhoneAccuracy}(s)}{\sum_s P_\Lambda(X^r|s)^{\mathcal{K}} P(s)^{\mathcal{K}}}, \tag{11.33}$$

where most of the terms are the same as for the MMI objective function shown in Equation (11.30), with the addition of function RawPhoneAccuracy which is a measure of the number of phones correctly transcribed in sentence s and is equal to the sum of PhoneAcc(p) for all phones p in sentence s, with

$$\text{PhoneAcc}(p) = \begin{cases} 1 & \text{if correct phone} \\ 0 & \text{if substitution} \\ -1 & \text{if insertion.} \end{cases} \tag{11.34}$$

Mean and variance reestimation with MPE relies on the same equations as for MMIE – that is, Equations (11.31) and (11.32), respectively. The difference is found in the way the terms composing the equations are computed: γ^{num}, $\varsigma^{\mathsf{num}}(X)$, $\varsigma^{\mathsf{num}}(X^2)$, γ^{den}, $\varsigma^{\mathsf{den}}(X)$, and $\varsigma^{\mathsf{den}}(X^2)$. Computation of those terms in the MPE case is explained below.

In the same way as for MMIE, implementation of MPE estimation uses a lattice framework: for each sentence of the training corpus, speech recognition is done using the initial acoustic models – usually estimated through MLE – in order to build a lattice of phones, analogous to a classical lattice of words used in speech recognition, each phone hypothesis being associated to one arc. That lattice will be used, along with the phones of the corresponding reference transcription, to carry out MPE training.

The most important accumulator, which holds statistics gathered on the training corpus (or on the adaptation corpus) for later use, is γ_a^{MPE}:

$$\gamma_a^{\mathsf{MPE}} = \frac{1}{\mathcal{K}} \frac{\delta \mathcal{F}_{\mathsf{MPE}}}{\delta \log \mathcal{L}(a)}. \tag{11.35}$$

That accumulator is a value computed for each arc a of the phone lattice, and is the differential of the objective function $\mathcal{F}_{\mathsf{MPE}}$ with respect to the log likelihood $\mathcal{L}(a)$ of arc a. In practice, to evaluate γ_a^{MPE}, some approximations can be made, as presented in Equation (11.47) in Section 11.5.4. For MPE, $\mathcal{K}$ is usually equal to the inverse of the language model weight used in the speech-recognition process that lead to the lattice. The terms used to update means and variances according to Equations (11.31) and (11.32) can then be written as follows:

$$\gamma_{q,i}^{\mathsf{num}} = \sum_{a=1}^{\mathcal{A}} \sum_{t=s_a}^{e_a} \gamma_{a,q,i}(t) \max(0, \gamma_a^{\mathsf{MPE}}), \tag{11.36}$$

$$\varsigma_{q,i}^{\mathsf{num}}(X) = \sum_{a=1}^{\mathcal{A}} \sum_{t=s_a}^{e_a} \gamma_{a,q,i}(t) \max(0, \gamma_a^{\mathsf{MPE}})\mathbf{x}(t), \tag{11.37}$$

$$\varsigma_{q,i}^{\mathsf{num}}(X^2) = \sum_{a=1}^{\mathcal{A}} \sum_{t=s_a}^{e_a} \gamma_{a,q,i}(t) \max(0, \gamma_a^{\mathsf{MPE}})\mathbf{x}^2(t), \tag{11.38}$$

$$\tag{11.39}$$

$$\gamma_{q,i}^{\mathsf{den}} = \sum_{a=1}^{\mathcal{A}} \sum_{t=s_a}^{e_a} \gamma_{a,q,i}(t) \max(0, -\gamma_a^{\mathsf{MPE}}), \tag{11.40}$$

$$\varsigma_{q,i}^{\mathsf{den}}(X) = \sum_{a=1}^{\mathcal{A}} \sum_{t=s_a}^{e_a} \gamma_{a,q,i}(t) \max(0, -\gamma_a^{\mathsf{MPE}})\mathbf{x}(t), \tag{11.41}$$

$$\varsigma_{q,i}^{\mathsf{den}}(X^2) = \sum_{a=1}^{\mathcal{A}} \sum_{t=s_a}^{e_a} \gamma_{a,q,i}(t) \max(0, -\gamma_a^{\mathsf{MPE}})\mathbf{x}^2(t), \tag{11.42}$$

where $\mathcal{A}$ is the number of arcs in the phone lattice, $\gamma_{a,q,i}(t)$ is the occupation probability at time t for Gaussian i of state q for the arc a, s_a is the starting time of arc a, e_a its ending time, and $\mathbf{x}^2(t)$ denotes element-wise squaring of $\mathbf{x}(t)$, with $\mathbf{x}^2(t) = \mathbf{x}(t) \otimes \mathbf{x}(t)$.

In order to increase the generalization power of MPE estimation, in turn leading to better results, those terms are usually modified by applying the I-smoothing technique.

11.5.3 I-smoothing

I-smoothing is a technique that was developed to avoid overtraining caused by the use of discriminative criteria such as MMI or MPE. It consists in smoothing parameter estimation by integrating a prior distribution, for example, based on MLE, in a spirit similar to that of the MAP approach. That technique is crucial for acoustic model adaptation, particularly when in presence of a small amount of adaptation data. In the case of MPE, I-smoothing is a requisite for good results. It implies the following updates:

$$\gamma_{q,i}^{\prime\mathsf{num}} = \gamma_{q,i}^{\mathsf{num}} + \tau^{I}, \tag{11.43}$$

$$\varsigma_{q,i}^{\prime\mathsf{num}}(X) = \varsigma_{q,i}^{\mathsf{num}}(X) + \frac{\tau^{I}}{\gamma_{q,i}^{\mathsf{mle}}}\varsigma_{q,i}^{\mathsf{mle}}(X), \tag{11.44}$$

$$\varsigma_{q,i}^{\prime\mathsf{num}}(X^2) = \varsigma_{q,i}^{\mathsf{num}}(X^2) + \frac{\tau^{I}}{\gamma_{q,i}^{\mathsf{mle}}}\varsigma_{q,i}^{\mathsf{mle}}(X^2), \tag{11.45}$$

where

- $\varsigma_{q,i}(X)^{\mathsf{mle}}$ and $\varsigma_{q,i}(X^2)^{\mathsf{mle}}$ are sums of data and squared data respectively, weighted by their probability for prior Gaussian i of state q estimated using MLE.
- $\gamma_{q,i}^{\mathsf{mle}}$ are prior Gaussian occupancies summed over time.
- τ^{I} is an empirically tuned value which is usually equal to 50.

Thus, when using I-smoothing, those modified terms will be used in Equations (11.31) and (11.32).

11.5.4 MPE Implementation

This section presents the key points of MPE implementation; for a more thorough description of the implementation, the reader can refer to the PhD dissertation of [23]. The general principle of MPE training is similar to that of MMIE; the main difference resides in the use of a phone recognition lattice for MPE instead of a word lattice for MMIE. Hence, the implementations of MMIE and MPE differ mostly by how the various statistics required to optimize the discriminative criterion are computed.

Approximate Alignment

It should first be noted that computing RawPhoneAccuracy(s) is complicated and computationally intensive since it requires a phonetic alignment of each recognition hypothesis in the lattice. In order to simplify the matter, it is common to use an approximation of PhoneAcc(p), based on time-alignment information. In this case, PhoneAcc(p) is computed as follows to integrate alignment approximations:

$$\text{PhoneAcc}(p) = \max_{z} \begin{cases} -1 + 2e(p,z) & \text{if } p \text{ and } z \text{ are same phone} \\ -1 + e(p,z) & \text{if } p \text{ and } z \text{ are different phones,} \end{cases} \tag{11.46}$$

where p is a hypothesis phone, z is a phone in the reference transcript, and $e(p,z)$ is the proportion of the length of phone z overlapping with phone p. It results in much easier computations than going using strict alignment would. Despite its imperfection – coming from the choice of the reference phone z, on which the hypothesis phone p is aligned, being done locally – that approximation gives very good results, and is well suited to a lattice context.

Updating Means and Variances

The terms γ_a^{MPE}, defined for each arc a of the phone recognition lattice, are essential to the computation of the elements defined in Equations (11.36)–(11.45), and are directly used to update the means and variances in Equations (11.31) and (11.32). By using the approximate alignment presented above, computation of accumulator γ_a^{MPE}, presented in Equation (11.35), can be done as follows:

$$\gamma_a^{\text{MPE}} = \gamma_a(c(a) - c_{\text{avg}}), \tag{11.47}$$

where γ_a is the occupancy probability of the arc a computed from the forward-backward algorithm below, $c(a)$ is the average RawPhoneAccuracy of sentences passing through the arc a, and c_{avg} is the average RawPhoneAccuracy of all the sentences in the recognition lattice.

Algorithm 4 shows how to compute γ_a^{MPE}. This algorithm is a variant of the one given in [23]: it can easily be optimized in order to limit the number of operations.

Updating weights

For MPE training, as well as for MMIE, updating the HMM - of iterations by applying the equation below:

$$w_{q,i}^{(\mathcal{I}+1)} = \frac{\gamma_{q,i}^{\text{num}} + w_{q,i}^{(\mathcal{I})} k_{q,i}^{(\mathcal{I})}}{\sum_i \gamma_{q,i}^{\text{num}} + w_{q,i}^{(\mathcal{I})} k_{q,i}^{(\mathcal{I})}} \tag{11.48}$$

with

$$k_{q,i}^{(\mathcal{I})} = \left(\max_m \frac{\gamma_{q,i}^{\text{den}}}{w_{q,i}^{\text{orig}}} \left(\frac{w_{q,i}^{(\mathcal{I})}}{w_{q,i}^{\text{orig}}} \right)^{C-1} \right) - \frac{\gamma_{q,i}^{\text{den}}}{w_{q,i}^{\text{orig}}} \left(\frac{w_{q,i}^{(\mathcal{I})}}{w_{q,i}^{\text{orig}}} \right)^{C-1}, \tag{11.49}$$

where $w_{q,i}^{\text{orig}}$ is the initial model weight before MPE, and $w_{q,i}^{(0)} - w_{q,i}^{\text{orig}}$. $(\mathcal{I})$ is the iteration rank used to compute the weight $w_{q,i}^{(\mathcal{I}+1)}$, following the recurrence relation existing between $w_{q,i}^{(\mathcal{I}+1)}$ and $w_{q,i}^{(\mathcal{I})}$, and described in Equation (11.48). C is a positive smoothing constant which allows to regulate convergence (the higher the value of C, the faster convergence will be reached). Interestingly, the value of C is often set to 1 (resulting in a considerably simpler equation). Usually, 100 iterations of Equation (11.48) are done; hence, the value of the updated weights come from the terms $w_{q,i}^{(100)}$.

Computing $\mathcal{D}_{q,i}$ Values

In order to update the means and variances of acoustic models, Equations (11.31) and (11.32) use the smoothing constants $\mathcal{D}_{q,i}$, which ensure that the variances stay positive. Several means of computing the values $\mathcal{D}_{q,i}$ have been proposed for MMIE [31,34]. In his work on the MPE

Algorithm 4: Algorithm for γ_a^{MPE} values computation, which are the most important accumulators for MPE training; γ_a^{MPE} is defined in Equation (11.35)

Input: Phone recognition lattice with $\mathcal{A}$ arcs a sorted chronologically, $\mathcal{K}$ is the inverse of the linguistic weight used during the initial recognition process, $\mathcal{L}(a)$ is the log likelihood of arc a computed by initial MLE acoustic models, $\mathcal{T}_{ba}$ is the lattice transition probability between b and a derived from the language model.

Output: γ_a^{MPE} for each arc a

```
for a : 1 → A do
    if a is a starting arc then
        α_a = L(a)^K;
        α'_a = PhoneAcc(a);
    else
        α_a = 0;
        tmpSum1 = tmpSum2 = 0;
        foreach arc b preceding a do
            α_a = α_a + α_b × T_ba^K × L(a)^K;
            tmpSum1 = tmpSum1 + α'_b × α_b × T_ba^K;
            tmpSum2 = tmpSum2 + α_b × T_ba^K;
        end
        α'_a = PhoneAcc(a) + tmpSum1/tmpSum2;
    end
end
for a : A → 1 do
    if a is an ending arc then
        β_a = 1;
        β'_a = 0;
    else
        β = 0;
        tmpSum = 0;
        foreach arc f following a do
            β_a = β_a + T_af^K × L(f)^K × β_f;
            tmpSum = tmpSum + T_af^K × L(f)^K × β_f × (β'_f + PhoneAcc(f));
        end
        β'_a = tmpSum/β_a;
    end
end
tmpSum1 = tmpSum2 = 0;
foreach ending arc a do
    tmpSum1 = tmpSum1 + α'_a × α_a;
    tmpSum2 = tmpSum2 + α_a;
end
c_avg = tmpSum1/tmpSum2;
foreach arc a do
    γ_a = (α_a × β_a)/tmpSum2;
    c(a) = α'_a + β'_a;
    γ_a^MPE = γ_a × (c(a) − c_avg);
end
```

approach, presented in [23], Povey proposes to choose $\mathcal{D}_{q,i}$ such as

$$\mathcal{D}_{q,i} = \max\left(2 \times \mathcal{D}_{q,i}^{\mathsf{min}}, E \times \gamma_{q,i}^{\mathsf{den}}\right), \tag{11.50}$$

where $\mathcal{D}_{q,i}^{\mathsf{min}}$ is the value that guarantees that variances are positive for each dimension of the ith Gaussian of state q; and E is a constant equal to 1 or 2. Experiments carried out in [23] show that the best value for E depends on the corpus and on the number of Gaussians per state in the HMM set. With data for which an ASR system has a low error rate, and for which a large number of Gaussians is available, $E = 1$ yields the best results. For the other cases, $E = 2$ is usually the best choice.

11.6 Conclusion

Adaptation techniques for acoustic models are in use in the best speech-recognition systems, such as the ones presented in [9,19,36]. There exist variants of the acoustic model-adaptation techniques presented in this chapter, such as structural MAP, lattice-based MLLR, MAP-MMIE, or MAP-MPE, MCE (close to MMI), MWE (which is the same as MPE but focusing on word errors instead of phone errors), fMPE, presented in [26] which extends MPE to acoustic features instead of acoustic models, but the fundamentals are the same.

The ASR systems that yield the best results these days make use of all the techniques described here, as they are usually complementary. For example, Figure 11.6 illustrates a possible processing sequence combining all of those techniques in order to enhance the accuracy of a speech-recognition system. As a first step, the acoustic models are estimated over a training corpus by using the classical maximum likelihood criterion. A first adaptation can then be done using MAP, for example to build gender-dependent acoustic models by splitting the training corpus into mono-gender corpora. That gender-based specialization of the initial models can then be enhanced through a decrease of interspeaker distance – from an acoustic model perspective—within each of the gender-dependent corpora. By applying the CMLLR/SAT method to the acoustic features of the corpora, it is then possible to obtain more precise acoustic models. In particular, the use of CMLLR transformations when processing the test data will benefit from the use of CMLLR/SAT during the training phase. Finally, since CMLLR/SAT only applies to acoustic features and not directly to the parameters of the HMMs themselves, its use can easily be completed with a reestimation of those parameters through discriminative training—for example using the MPE criterion. To make the ASR

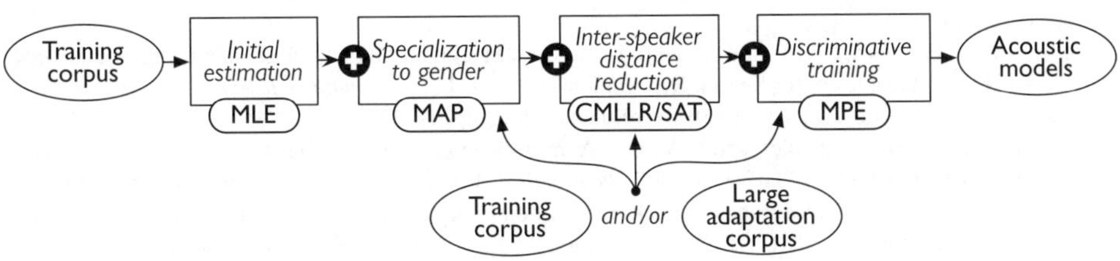

Figure 11.6 Example of a sequential process for acoustic model estimation combining ML estimation and adaptation techniques presented in this chapter.

system significantly more robust to noise, this static acoustic model adaptation is usually combined with dynamic adaptation, as presented in Section 11.2.

Other combinations are of course possible, using a different order, or using variants of the adaptation techniques presented here. For example, some ASR systems use the vocal tract length normalization (VTLN) approach in addition to MLLR or CMLLR. Reference [33] showed that VTLN was an interesting complement to MLLR. However, it also concluded that VTLN did not bring any enhancement as an addition to CMLLR with a high enough number of iterations in the estimation of transformations matrices for the latter.

Some strategies consist in generating a set of different acoustic models through reestimation using different adaptation techniques, in order to merge the results obtained by the various models, as presented in [9]. More generally, the cross-adaptation approach, consisting in using several ASR systems and adapting the acoustic models of each of them by using the outputs of the other ones, offers very interesting prospects [27, 36].

Quantifying precisely the benefits of each of the techniques presented in this paper is difficult. The gain depends on the applicative context, which can present many variations in terms of task, language, training corpus size, adaptation corpus size, expected response time for the ASR system, *etc*. Moreover, it is now rare to find one of those techniques used alone, and the latest published results are based on combinations of many or all of those techniques. It can be noticed that the benefits brought by acoustic model adaptation are crucial to the development of a state-of-the-art ASR system these days: [9] showed that acoustic model adaptation alone allowed to decrease word error rate from 17.2% down to 12.3% in the context of broadcast news transcription in English. In [27], in the context of conversational telephone in English, the use of cross-adaptation allows to bring the transcription error rate from 21% down to 16.7%.

In conclusion, it has been seen that several acoustic model-adaptation techniques exist in order to make an ASR system more robust. The major part of these techniques were not initially proposed specifically to adapt acoustic models to noise, but they allow a significant decrease of the word error rate in presence of noise, as shown in results presented in Section 11.2 Among these techniques, CMLLR, and its variants applied dynamically to acoustic features in an unsupervised mode, seems to be one of the most interesting ones, especially because it can be used in conjunction with other adaptation techniques.

References

[1] T. Anastasakos, J. McDonough, R. Schwartz, and J. Makhoul, "A Compact Model for Speaker Adaptive Training," in *International Conference on Spoken Language Processing (ICSLP'1996)*, Philadelphia, USA, 1996, pp. 1137–1140.

[2] L. R. Bahl, P. F. Brown, P. V. de Souza, and R. L. Mercer, "Maximum Mutual Information Estimation of Hidden Markov Model Parameters for Speech Recognition," in *IEEE International Conference on Acoustics, Speech and Signal Processing (ICASSP'1986)*, Tokyo, Japan, 1986, pp. 49–52.

[3] L. E. Baum, T. Petrie, G. Soules, and N. Weiss, "A maximization technique occurring in the statistical analysis of probabilistic functions in Markov chains," *The Annals of Mathematical Statistics*, pp. 164–171, 1970.

[4] X. Cui and A. Alwan, "Noise Robust Speech Recognition Using Feature Compensation Based on Polynomial Regression of Utterance SNR," *IEEE Transactions on Speech and Audio Processing*, vol. 13, no. 6, pp. 1161–1172, 2005.

[5] A. P. Dempster, N. M. Laird, and D. B. Rubin, "Maximum likelihood from incomplete data via the EM algorithm," *Journal of the Royal Statistical Society, Series B*, vol. 34, pp. 1–38, 1977.

[6] V. Digalakis, D. Rtischev, and L. Neumeyer, "Speaker Adaptation Using Constrained Estimation of Gaussian Mixtures," *IEEE Transactions on Speech and Audio Processing*, vol. 3, pp. 357–366, 1995.

[7] M. J. F. Gales, "Maximum Likelihood Linear Transformations for HMM-based Speech Recognition," *Computer Speech & Language*, vol. 12, pp. 75–98, 1998.

[8] M. J. F. Gales and P. C. Woodland, "Mean and variance adaptation within the MLLR framework," *Computer Speech & Language*, vol. 10, no. 4, pp. 249–264, 1996.

[9] M. J. F. Gales, D. Y. Kim, P. C. Woodland, H. Y. Chan, D. Mrva, R. Sinha, and S. E. Tranter, "Progress in the CU-HTK Broadcast News Transcription System," *IEEE Transactions on Speech and Audio Processing*, vol. 3, pp. 357–366, 2006.

[10] J.-L. Gauvain and C.-H. Lee, "Maximum A Posteriori Estimation for Multivariate Gaussian Mixture Observations of Markov Chcains," *IEEE Transactions on Speech and Audio Processing*, vol. 2, pp. 291–298, 1994.

[11] D. Giuliani and R. De Mori, "Speaker adaptation," in *Spoken dialogues with computers*, R. De Mori, Ed. Academic Press Inc, 1998, ch. 11, pp. 363–404.

[12] P. Gopalakrishnan, D. Kanevsky, A. Nádas, and D. Nahamoo, "An inequality for rational functions with applications to some statistical estimation problems," *IEEE Transactions on Information Theory*, vol. 37, no. 1, 1991.

[13] X. He, L. Deng, and W. Chou, "Discriminative learning in sequential pattern recognition: a unifying review for optimization-based speech recognition," *IEEE Signal Processing Magazine*, pp. 14–36, 2008.

[14] H. G. Hirsch and D. Pearce, "The AURORA experimental framework for the performance evaluations of speech recognition systems under noisy conditions," in *ISCA ITRW ASR 2000 Automatic Speech Recognition: Challenges for the Next Millennium*, Paris, France, 2000, pp. 181–188.

[15] D. Iskra, B. Grosskopf, K. Marasek, H. van den Heuvel, F. Diehl, and A. Kiessling, "SPEECON - speech databases for consumer devices: Database specification and validation," in *International Conference on Language Resources and Evaluation (LREC 2002)*, Las Palmas, Canary Islands, Spain, 2002, pp. 329–333.

[16] H. Jiang, "Discriminative training of HMMs for automatic speech recognition: A survey," *Computer Speech & Language*, vol. 24, pp. 589–608, 2010.

[17] T. Kawahara, H. Nanjo, T. Shinozaki, and S. Furui, "Benchmark test for speech recognition using the Corpus of spontaneous Japanese," in *ISCA & IEEE Workshop on Spontaneous Speech Processing and Recognition (SSPR 2003)*, Tokyo, Japan, 2003, pp. 135–138.

[18] D. K. Kim and M. J. F. Gales, "Noisy Constrained Maximum-Likelihood Linear Regression for Noise-Robust Speech Recognition," *IEEE Transactions on Speech and Audio Processing*, vol. 19, no. 2, pp. 315–325, 2011.

[19] V.-B. Le, L. Lamel, and J.-L. Gauvain, "Multistyle MLP Features for BN Transcription," in *IEEE International Conference on Acoustics, Speech and Signal Processing (ICASSP'2010)*, Dallas, Texas, USA, 2010, pp. 4866–4869.

[20] C. J. Leggetter and P. C. Woodland, "Maximum likelihood linear regression for speaker adaptation of continuous density hidden Markov models," *Computer Speech & Language*, vol. 9, no. 2, pp. 171–185, 1995.

[21] Y. Normandin and S. D. Morgera, "An improved MMIE training algorithm for speaker-independent, small vocabulary, continuous speech recognition," in *IEEE International Conference on Acoustics, Speech and Signal Processing (ICASSP'1991)*, vol. 1, Toronto, Ontario, Canada, 1991, pp. 537–540.

[22] Y. Normandin, R. Cardin, and R. De Mori, "High-performance connected digit recognition using maximum mutual information estimation," *IEEE Transactions on Speech and Audio Processing*, vol. 2, no. 2, 1994, pp. 229–311.

[23] D. Povey, "Discriminative Training for Large Vocabulary Speech Recognition," PhD dissertation, Cambridge University Engineering Department, 2003.

[24] D. Povey and B. Kingsbury, "Evaluation of Proposed Modifications to MPE for Large Scale Discriminative Training," in *IEEE International Conference on Acoustics, Speech and Signal Processing (ICASSP'2007)*, vol. 4, Honolulu, Hawaii, USA, 2007, pp. 321–324.

[25] D. Povey and P. C. Woodland, "Minimum Phone Error and I-Smoothing for Improved Discriminative Training," in *IEEE International Conference on Acoustics, Speech and Signal Processing (ICASSP'2002)*, vol. 1, Orlando, Florida, USA, 2002, pp. 105–108.

[26] D. Povey, B. Kingsbury, L. Mangu, G. Saon, H. Soltau, and G. Zweig, "fMPE: Discriminatively trained features for speech recognition," in *IEEE International Conference on Acoustics, Speech and Signal Processing (ICASSP'2005)*, vol. 1, Philadelphia, Pennsylvania, USA, 2005, pp. 961–964.

[27] R. Prasad, S. Matsoukas, C.-L. Kao, J. Ma, D.-X. Xu, T. Colthurst, G. Thattai, O. Kimball, R. Schwartz, J.-L. Gauvain, L. Lamel, H. Schwenk, G. Adda, and F. Lefevre, "The 2004 BBN/LIMSI 20xRT English Conversational Telephone Speech System," in *DARPA RT04*, Palisades, NY, USA, 2004.

[28] D. Pye and P. C. Woodland, "Experiments in Speaker Normalisation and Adaptation for Large Vocabulary Speech Recognition," in *IEEE International Conference on Acoustics, Speech and Signal Processing ICASSP'1997)*, Munich, Germany, 1997, pp. 1047–1050.

[29] U. Remes, K. J. Palomäki, and M. Kurimo, "Robust automatic speech recognition using acoustic model adaptation prior to missing feature reconstruction," in *17th European Signal Processing Conference (EUSIPCO 2009)*, Glasgow, Scotland, 2009.

[30] R. Schlüter and W. Macherey, "Comparison of Discriminative Training Criteria," in *IEEE International Conference on Acoustics, Speech and Signal Processing (ICASSP'1998)*, Seattle, USA, 1998, pp. 493–496.

[31] R. Schlüter, W. Macherey, S. Kanthak, and H. Ney, "Comparison of Optimization Methods for Discriminative Training," in *European Conference on Speech Communication and Technology (EUROSPEECH'1997)*, Rhodes, Greece, 1997, pp. 15–18.

[32] T. Shinozaki, Y. Kubota, and S. Furui, "Unsupervised Acoustic Model Adaptation Based on Ensemble Methods," *IEEE Journal of Selected Topics in Signal Processing*, vol. 4, no. 6, pp. 1007–1015, 2010.

[33] L. F. Uebel and P. C. Woodland, "An investigation into vocal tract length normalisation," in *6th European Conference on Speech Communication and Technology (Eurospeech'1999)*, vol. 6, 1999, pp. 2527–2530.

[34] V. Valtchev, P. C. Woodland, and S. J. Young, "Lattice-based discriminative training for large vocabulary speech recognition system," in *IEEE International Conference on Acoustics, Speech and Signal Processing (ICASSP'1996)*, Atlanta, Georgia, USA, 1996, pp. 605–608.

[35] V. Valtchev, J. J. Odell, P. C. Woodland, and S. J. Young, "MMIE training of large vocabulary speech recognition system," *Speech Communication*, vol. 22, pp. 303–314, 1997.

[36] D. Vergyri, A. Mandal, W. Wang, A. Stolcke, J. Zheng, M. Graciarena, D. Rybach, C. Gollan, R. Schlüter, K. Kirchhoff, A. Faria, and N. Morgan, "Development of the SRI/nightingale Arabic ASR system," in *9th Annual Conference of the International Speech Communication Association (Interspeech'2008)*, 2008, pp. 1437–1440.

[37] P. C. Woodland, "Speaker Adaptation for Continuous Density HMMs: A Review," in *ITRW on Adaptation Methods for Speech Recognitionn*, Sophia Antipolis, France, 2001, pp. 11–19.

[38] X. Xiao, J. Li, E. S. Chng, H. Li, and C. H. Lee, "A Study on the Generalization Capability of Acoustic Models for Robust Speech Recognition," *IEEE Transactions on Audio, Speech and Language Processing*, vol. 22, no. 6, pp. 1158–1169, 2010.

12

Factorial Models for Noise Robust Speech Recognition

John R. Hershey[1], Steven J. Rennie[2], Jonathan Le Roux[1]
[1]*Mitsubishi Electric Research Laboratories, USA*
[2]*IBM Thomas J. Watson Research Center, USA*

12.1 Introduction

Noise compensation techniques for robust automatic speech recognition (ASR) attempt to improve system performance in the presence of acoustic interference. In *feature-based* noise compensation, which includes *speech enhancement* approaches, the acoustic features that are sent to the recognizer are first processed to remove the effects of noise (see Chapter 9). *Model compensation* approaches, in contrast, are concerned with modifying and even extending the acoustic model of speech to account for the effects of noise. A taxonomy of the different approaches to noise compensation is depicted in Figure 12.1, which serves as a road map for the present discussion.

The two main strategies used for model compensation approaches are *model adaptation* and *model-based noise compensation*. Model adaptation approaches implicitly account for noise by adjusting the parameters of the acoustic model of speech, whereas model-based noise compensation approaches explicitly model the noise and its effect on the noisy speech features. Common adaptation approaches include *maximum likelihood linear regression* (MLLR) [55], maximum *a posteriori* (MAP) adaptation [32], and their generalizations [17, 29, 47]. These approaches, which are discussed in Chapter 11, alter the speech acoustic model in a completely data-driven way given additional training data or test data. Adaptation methods are somewhat more general than model-based approaches in that they may handle effects on the signal that are difficult to explicitly model, such as nonlinear distortion and changes in the voice in reaction to noise (the Lombard effect [53]). However, in the presence of additive noise, failing to take into account the known interactions between speech and noise can be detrimental to performance.

Techniques for Noise Robustness in Automatic Speech Recognition, First Edition.
Edited by Tuomas Virtanen, Rita Singh, and Bhiksha Raj.

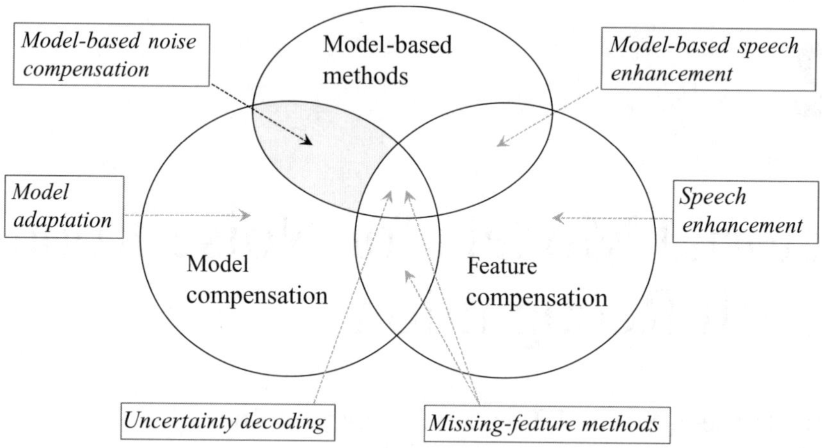

Figure 12.1 Noise compensation methods in a Venn diagram. The shaded region represents model-based noise compensation, the subject of this chapter. Note that the term "model" in "model compensation" refers to the recognizer's acoustic model, whereas in "model-based noise compensation," it refers to the models of additive noise.

Model-based noise compensation approaches, in contrast to adaptation approaches, explicitly model the different factors present in the acoustic environment: the speech, the various sources of acoustic interference, and how they interact to form the noisy speech signal. By modeling the noise separately from the speech, these factorial models can generalize to combinations of speech and noise sounds not seen during training, and can explicitly represent the dynamics of individual noise sources. A significant advantage of this approach is that the compensated speech features and recognition result are jointly inferred, unlike in feature-based approaches. The recognizer's model of speech dynamics can be directly employed to better infer the acoustic states and parameters of the interference model. Similarly, the model of acoustic interference and its dynamics can be utilized to more accurately estimate the sequence of states of the speech model. Performing these inference processes jointly allows the recognizer to consider different possible combinations of speech and interference.

Approaches that lie somewhere between feature-based and model-based noise compensation include uncertainty decoding [19,57], which is discussed in Chapter 17, and missing-feature methods [68, 83], which are discussed in Chapters 14–16. These methods involve additional communication from the feature enhancement algorithm to the recognizer about the uncertainty associated with the enhanced features being estimated. Model-based compensation approaches can be seen as taking the idea of uncertainty decoding to its logical conclusion: by placing the enhancement model inside the recognizer, the information about uncertainty is considered jointly in terms of the noise model and the full speech model of the recognizer.

Difficult obstacles must be overcome in order to realize the full benefit of the model-based approach. A primary challenge is the complexity of inference: if implemented naively, joint inference in factorial models requires performing computations for each combination of the states of the models. Because of the potential combinatorial explosion, this is prohibitively expensive for many real applications. Alleviating these problems continues to be a core challenge in the field, and therefore efficient inference is a central theme in this chapter.

Another challenge is the dilemma of feature domains. In feature domains where the interaction between speech and noise is additive, isolating the phonetic content of the speech signal can be difficult. This is because phonetic content is imparted to speech by the filtering effect of the vocal tract, which is approximately multiplicative in the power spectrum. However, in the log spectrum domain the vocal tract filter is additive. Speech recognizers exploit this by using features that are linear transforms of the log spectrum domain. In such domains, the effect of noise is nonlinear, and compensating for it becomes difficult. As such, a major focus of research has been to derive tractable inference algorithms by approximating the interaction between speech and noise in the log spectrum domain.

This chapter presents the fundamental concepts and current state of the art in model-based compensation, while hinting along the way at potential future directions.[1] First, the general framework of the model-based approach is introduced. This is followed by a review of the feature domains commonly used for representing signals, focusing on the way in which additive signals interact deterministically in each domain. A probabilistic perspective on these interaction functions and their approximations is then presented. Following this, several commonly used inference methods which utilize these approximate interaction functions are described in detail. Because computational complexity is of paramount importance in speech processing, we also describe an array of methods which can be used to alleviate the complexity of evaluating factorial models of noisy speech. The chapter concludes with a discussion of many promising research directions in this exciting and rapidly evolving area, with a focus on how complex and highly structured models of noise can be utilized for robust speech recognition.

12.2 The Model-Based Approach

Model-based approaches start with probabilistic models of the features of speech and the noise, and combine them using an *interaction model*, which describes the distribution of the observed noisy speech given the speech and noise. To make this explicit we will need some notation: $p(x)$ denotes a probability distribution. In the case that x is a discrete random variable, p denotes a *probability mass function*, and if x is a continuous random variable, it denotes a *probability density function* (pdf). To simplify notation, we shall specify the random variable considered as a subscript, for example, $p_{\mathsf{x}}(x)$, only when required to avoid confusion. Assume that we have probabilistic models for the features of the clean speech, $\boldsymbol{x}_t$, and the noise $\boldsymbol{n}_t$ at time t: $p(\boldsymbol{x}_t|s_t^{\mathsf{x}})$ and $p(\boldsymbol{n}_t|s_t^{\mathsf{n}})$, which depend on some states s_t^{x} and s_t^{n}. In the context of speech recognition, the clean speech model is typically a hidden Markov model (HMM), which describes the dynamical properties of speech via transition probabilities over the unobserved states s_t^{x}. The interaction model then describes the conditional probability of the noisy speech given the clean speech and the noise, $p(\boldsymbol{y}_t|\boldsymbol{x}_t, \boldsymbol{n}_t)$. Inference in the model-based method involves computing one or more of the following basic quantities: the *state likelihood* $p(\boldsymbol{y}_t|s_t^{\mathsf{x}}, s_t^{\mathsf{n}})$, the *joint clean speech and noise posterior* $p(\boldsymbol{x}_t, \boldsymbol{n}_t|\boldsymbol{y}_t, s_t^{\mathsf{x}}, s_t^{\mathsf{n}})$, and the *clean speech estimate* $\mathrm{E}(\boldsymbol{x}_t|\boldsymbol{y}_t, s_t^{\mathsf{x}}, s_t^{\mathsf{n}})$ for a given hypothesis of the speech and noise states s_t^{x} and s_t^{n}. The state likelihood, which is

[1] Additional perspectives and background material may be found in recent reviews on this topic [13, 30].

needed in speech recognition to compute the posterior probability of state sequences, involves the integral

$$p(\boldsymbol{y}_t|s_t^{\mathsf{x}}, s_t^{\mathsf{n}}) = \int p(\boldsymbol{y}_t|\boldsymbol{x}_t, \boldsymbol{n}_t)p(\boldsymbol{n}_t|s_t^{\mathsf{n}})p(\boldsymbol{x}_t|s_t^{\mathsf{x}})\, \mathrm{d}\boldsymbol{x}_t\, \mathrm{d}\boldsymbol{n}_t. \tag{12.1}$$

The joint posterior of the speech and noise features can be computed using the above integral:

$$p(\boldsymbol{x}_t, \boldsymbol{n}_t|\boldsymbol{y}_t, s_t^{\mathsf{x}}, s_t^{\mathsf{n}}) = \frac{p(\boldsymbol{y}_t|\boldsymbol{x}_t, \boldsymbol{n}_t)p(\boldsymbol{n}_t|s_t^{\mathsf{n}})p(\boldsymbol{x}_t|s_t^{\mathsf{x}})}{p(\boldsymbol{y}_t|s_t^{\mathsf{x}}, s_t^{\mathsf{n}})}. \tag{12.2}$$

The expected value of the speech features, used in feature-based compensation, can then be obtained as follows:

$$\mathrm{E}(\boldsymbol{x}_t|s_t^{\mathsf{x}}, s_t^{\mathsf{n}}) = \int \boldsymbol{x}_t\; p(\boldsymbol{x}_t, \boldsymbol{n}_t|\boldsymbol{y}_t, s_t^{\mathsf{x}}, s_t^{\mathsf{n}})\, \mathrm{d}\boldsymbol{x}_t\, \mathrm{d}\boldsymbol{n}_t. \tag{12.3}$$

For uncertainty-decoding approaches, a measure of uncertainty such as the posterior variance, $\mathrm{Var}(\boldsymbol{x}_t|s_t^{\mathsf{x}}, s_t^{\mathsf{n}})$, would also need to be computed (see Chapter 17 for more details). Note that there are typically mixture components for each state, so that $p(\boldsymbol{x}_t|s_t^{\mathsf{x}}) = \sum_{c_t^{\mathsf{x}}} p(\boldsymbol{x}_t|c_t^{\mathsf{x}})p(c_t^{\mathsf{x}}|s_t^{\mathsf{x}})$. In the rest of this chapter, we neglect mixture components to avoid clutter, as introducing them is straightforward and irrelevant to the main problem of computing the above integrals.

Given this general framework, what remains is to show how the above integrals can be accurately and efficiently estimated in the feature domains commonly used in speech modeling. To that end, we turn to the interaction functions that result from analysis of signals in different feature domains.

12.3 Signal Feature Domains

We shall present here the different representations of a signal commonly involved in automatic speech recognition, introduce the corresponding notations, and describe the interaction functions between clean speech and noise in each domain. Due to the complexity of these interactions, and in particular due to the nonlinear transformations involved, approximations are often required. We shall point them out as we proceed, and mention the conditions under which they can be considered to be justified.

We assume that the observed signal is a degraded version of the clean signal, where the degradation is classically modeled as the combination of linear channel distortion and additive noise [1]. The flow chart of the basic front-end signal processing is shown in Figure 12.2. Denoting by $y[t]$ the observed speech, $x[t]$ the clean speech, $n[t]$ the noise signal, and $h[t]$ the impulse response of the linear channel-distortion filter, we obtain the following relationship in the time domain, where $*$ denotes convolution:

$$y[t] = (h * x)[t] + n[t]. \tag{12.4}$$

The frequency content of the observed signal is then generally analyzed using the short-term *discrete Fourier transform* (DFT): overlapping frames of the signal are windowed and the DFT is computed, leading to the complex short-term spectrum. Let us denote by $Y_{t,f}$ (respectively, $X_{t,f}$ and $N_{t,f}$) the spectrum of the observed speech (respectively, the clean speech and the noise) at time frame t and frequency bin f, and by H_f the DFT of h (assumed shorter than the

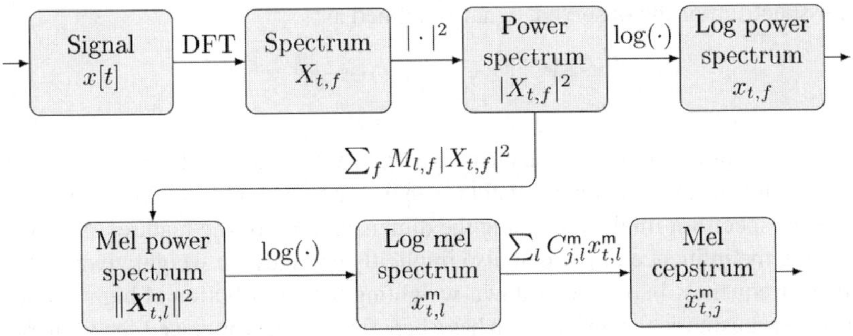

Figure 12.2 Basic front-end signal processing showing the notation used throughout the chapter for different feature domains.

window length). Under the so-called narrowband approximation, the relationship between the complex short-term spectra can be written as

$$Y_{t,f} \approx H_f X_{t,f} + N_{t,f}. \tag{12.5}$$

Note that this approximation can only be justified for a short channel-distortion filter and a smooth window function (i.e., whose Fourier transform is concentrated at low frequencies) [48]. This approximation is extremely common in frequency-domain source separation [42, 87].

We are now ready to transform (12.5) to the power spectrum domain:

$$|Y_{t,f}|^2 = |H_f|^2|X_{t,f}|^2 + |N_{t,f}|^2 + 2|H_f||X_{t,f}||N_{t,f}|\cos(\phi_{t,f}), \tag{12.6}$$

where $\phi_{t,f}$ is the phase difference between $H_f X_{t,f}$ and $N_{t,f}$. The third term is often assumed to be zero, leading to the following approximate interaction:

$$|Y_{t,f}|^2 \approx |H_f|^2|X_{t,f}|^2 + |N_{t,f}|^2. \tag{12.7}$$

This approximation is commonly justified by noticing that the expected value of the cross-term $|H_f||X_{t,f}||N_{t,f}|\cos(\phi_{t,f})$ is zero if x and n are assumed statistically independent. However, the expected value being equal to zero does not tell us much about the particular value taken at a given time-frequency bin. A slightly stronger argument to justify the above approximation is that of the sparsity of audio signals: if the speech and noise signals are sparse in the time-frequency domain, their cross-term is likely to be very small most of the time. Nonetheless, this term is not equal to zero in general, and we will see that the influence of the cross-term is actually very complex.

In order to reduce the influence of pitch (and thus reduce the within-class variance relative to the between-class variance when recognizing phonemes or sub-phonemes), the power spectrum is converted to the so-called mel power spectrum. The mel power spectrum is obtained by filtering the power spectrum using a small number L (typically 20 to 24 at a sampling rate of 8 kHz, 40 at 16 kHz) of overlapping triangular filters with both center frequencies and bandwidths equally spaced on the mel scale, believed to well approximate the human perception of frequency. Denoting by $M_{l,f}$ the response of filter l in frequency f, the mel power spectrum of the observed signal is defined as:

$$\|\boldsymbol{Y}^{\mathsf{m}}_{t,l}\|^2 = \sum_f M_{l,f}|Y_{t,f}|^2 \tag{12.8}$$

with similar definitions for that of the clean speech, $\|\boldsymbol{X}^{\mathsf{m}}\|^2$, and of the noise, $\|\boldsymbol{N}^{\mathsf{m}}\|^2$. As the number L of filters is typically much smaller than the number F of frequency bins, considering the mel power spectrum implies reducing the dimensionality of the features. Moreover, apart from reducing the influence of pitch, it also implicitly changes the weight given to the data as a function of frequency, in particular down-weighting the contribution of high frequencies. In terms of noise robustness, the mel domain has a beneficial effect for voiced speech in broadband noise: it gives preferential weight to the peaks of the spectrum, which are likely to correspond to the harmonics of speech, where the signal-to-noise ratio is greatest. This is easy to see in the log mel domain, since $\log \sum_f M_{l,f}|Y_{t,f}|^2 \approx \max_f \left(\log(|Y_{t,f}|^2) + \log M_{l,f}\right)$. Finally, we shall see that, as a side effect, it also leads to greater accuracy in the log-sum approximation, which is introduced in Section 12.4.3.

We can now obtain an analog of (12.6) on the mel spectra:

$$\|\boldsymbol{Y}^{\mathsf{m}}_{t,l}\|^2 = \|\boldsymbol{H}^{\mathsf{m}}_{t,l}\|^2\|\boldsymbol{X}^{\mathsf{m}}_{t,l}\|^2 + \|\boldsymbol{N}^{\mathsf{m}}_{t,l}\|^2 + 2\sqrt{\|\boldsymbol{H}^{\mathsf{m}}_{t,l}\|^2\|\boldsymbol{X}^{\mathsf{m}}_{t,l}\|^2\|\boldsymbol{N}^{\mathsf{m}}_{t,l}\|^2}\,\alpha^{\mathsf{m}}_{t,l}, \tag{12.9}$$

where the two newly introduced quantities

$$\|\boldsymbol{H}^{\mathsf{m}}_{t,l}\|^2 = \frac{\sum_f M_{l,f}|H_f|^2|X_{t,f}|^2}{\|\boldsymbol{X}^{\mathsf{m}}_{t,l}\|^2}, \tag{12.10}$$

$$\alpha^{\mathsf{m}}_{t,l} = \frac{\sum_f M_{l,f}|H_f||X_{t,f}||N_{t,f}|\cos(\phi_{t,f})}{\sqrt{\|\boldsymbol{H}^{\mathsf{m}}_{t,l}\|^2\|\boldsymbol{X}^{\mathsf{m}}_{t,l}\|^2\|\boldsymbol{N}^{\mathsf{m}}_{t,l}\|^2}} \tag{12.11}$$

incorporate complex interactions between the various terms. These terms are typically not handled using the above formulae but through approximate models, as shown later in this chapter.

In order to deal with the very wide dynamic range of speech, and motivated by considerations on the roughly logarithmic perception of loudness by humans, the power spectrum and mel power spectrum are often converted to the log domain. We define the log power spectrum of the observed signal as

$$y_{t,f} = \log(|Y_{t,f}|^2) \tag{12.12}$$

with analogous definitions for the log power spectra of the clean speech $x_{t,f}$, the noise $n_{t,f}$, and the channel distortion h_f. This leads to the following interaction function in the log power domain:

$$y_{t,f} = \log\left(\mathrm{e}^{h_f + x_{t,f}} + \mathrm{e}^{n_{t,f}} + 2\mathrm{e}^{\frac{h_f + x_{t,f} + n_{t,f}}{2}}\cos(\phi_{t,f})\right). \tag{12.13}$$

Similarly to the log power spectrum, we define the log mel power spectrum of the noisy speech as

$$y^{\mathsf{m}}_{t,l} = \log(\|\boldsymbol{Y}^{\mathsf{m}}_{t,l}\|^2) \tag{12.14}$$

and analogously for the log power spectra of the clean speech, $x^{\mathrm{m}}_{t,l}$, the noise, $n^{\mathrm{m}}_{t,l}$, and the channel distortion, $h^{\mathrm{m}}_{t,l}$. The interaction function in the log mel power domain becomes

$$y^{\mathrm{m}}_{t,l} = \log\left(\mathrm{e}^{h^{\mathrm{m}}_{t,l}+x^{\mathrm{m}}_{t,l}} + \mathrm{e}^{n^{\mathrm{m}}_{t,l}} + 2\mathrm{e}^{\frac{h^{\mathrm{m}}_{t,l}+x^{\mathrm{m}}_{t,l}+n^{\mathrm{m}}_{t,l}}{2}} \alpha^{\mathrm{m}}_{t,l}\right). \tag{12.15}$$

To decorrelate the features, and focus on the envelope characteristics of the log mel power spectrum, which are likely to be related to the characteristics of the vocal tract, most speech recognition systems further compute the so-called mel cepstrum, which consists of the low-frequency components of the discrete cosine transform (DCT) of the log mel power spectrum. Introducing the DCT matrix $\mathbf{C}^{\mathrm{m}}$ of size $K \times L$, where K is the number of mel cepstral coefficients (typically around 13), the mel cepstrum of the noisy signal is defined as

$$\tilde{\boldsymbol{y}}^{\mathrm{m}}_t = \mathbf{C}^{\mathrm{m}} \boldsymbol{y}^{\mathrm{m}}_t \tag{12.16}$$

with similar definitions of the mel cepstra of the clean speech, $\tilde{\boldsymbol{x}}^{\mathrm{m}}_t$, the noise, $\tilde{\boldsymbol{n}}^{\mathrm{m}}_t$, and the channel distortion, $\tilde{\boldsymbol{h}}^{\mathrm{m}}_t$. As the matrix $\mathbf{C}^{\mathrm{m}}$ is typically not invertible, the interaction function in the mel cepstrum domain is generally approximated by

$$\tilde{\boldsymbol{y}}^{\mathrm{m}}_t = \mathbf{C}^{\mathrm{m}} \log\left(\mathrm{e}^{\mathbf{D}^{\mathrm{m}}(\tilde{\boldsymbol{h}}^{\mathrm{m}}_t+\tilde{\boldsymbol{x}}^{\mathrm{m}}_t)} + \mathrm{e}^{\mathbf{D}^{\mathrm{m}}\tilde{\boldsymbol{n}}^{\mathrm{m}}_t} + 2\mathrm{e}^{\mathbf{D}^{\mathrm{m}}\frac{\tilde{\boldsymbol{h}}^{\mathrm{m}}_t+\tilde{\boldsymbol{x}}^{\mathrm{m}}_t+\tilde{\boldsymbol{n}}^{\mathrm{m}}_t}{2}} \circ \boldsymbol{\alpha}^{\mathrm{m}}_t\right), \tag{12.17}$$

where $\mathbf{D}^{\mathrm{m}}$ is a pseudoinverse of $\mathbf{C}^{\mathrm{m}}$, such as the Moore–Penrose pseudoinverse, and $\circ$ denotes the element-wise product.

Notice that the interaction becomes more and more complicated and nonlinear as we move closer to the features that are used in modern speech recognizers. When we consider using probabilistic models of the speech and noise in the feature domain, the more complicated the interaction function is, the less tractable inference becomes. To make matters worse, state-of-the-art systems do not stop at the mel cepstrum, but introduce further transformations that encompass multiple frames. These include linear transformations of several frames of features, such as the so-called *delta* and *delta-delta* features and *linear discriminant analysis* (LDA), as well as nonlinear transformations such as *feature-based minimum phone error* (fMPE). Except for the simplest cases, model-based noise compensation with such features has not yet been addressed. We shall thus limit our presentation mainly to *static* (i.e., single frame) features.

12.4 Interaction Models

For each of the feature domains introduced above, we have shown that a domain-specific interaction function describes how noisy features relate to those of the clean speech and the additive noise. In feature domains such as the complex spectrum, which contain complete information about the underlying signals, the interaction function is deterministic. However, in feature domains that omit some information, the unknown information leads to uncertainty about the interaction, which in the model-based approach is described using a probabilistic interaction function. In general, the model-based approach thus requires a distribution $p(\boldsymbol{y}_t|\boldsymbol{x}_t, \boldsymbol{n}_t)$ over the observed noisy features $\boldsymbol{y}_t$ given the speech features $\boldsymbol{x}_t$ and the noise features $\boldsymbol{n}_t$. The definition of this function varies depending on the feature domain. In the feature domains most used for modeling speech, approximations are generally required to make inference tractable. In this section, we review interaction models for log spectrum features, as well as some of their extensions to the mel spectrum domain and the mel cepstrum. From here on we omit time subscripts to simplify notation, bearing in mind that we are modeling the interaction in a particular time frame t.

12.4.1 Exact Interaction Model

We consider for now the modeling of speech and noise energy in the log power spectrum domain. In (12.13), the unknown phase and channel are a source of uncertainty in the relationship between the power spectra of the speech and noise. In the log power spectrum, the effect of the acoustic channel is well approximated as an additive constant, for stationary reverberation with an impulse response of length less than a frame. We can thus model the channel implicitly as part of the speech feature, to simplify our discussion, with little loss in generality. See [13] for a review in which it is explicitly included in the interaction model. The difference in phase ϕ_f between the speech and noise signals is a remaining source of uncertainty

$$p(y_f|x_f, n_f, \phi_f) = \delta\left(y_f - \log\left(\mathrm{e}^{x_f} + \mathrm{e}^{n_f} + 2\mathrm{e}^{\frac{x_f+n_f}{2}}\cos(\phi_f)\right)\right). \tag{12.18}$$

We need to compute $\int_{-\pi}^{\pi} p(y_f|x_f, n_f, \phi_f)p(\phi_f)\,\mathrm{d}\phi_f$. We define $\alpha_f = \cos(\phi_f)$ and derive $p_{\alpha_f}(\alpha_f)$ from $p_{\phi_f}(\phi_f)$, noting that $\cos(\phi_f) = \cos(-\phi_f)$, so that for $\phi_f \in (-\pi, \pi)$, we have two solutions to $|\phi_f| = \cos^{-1}(\alpha_f)$:

$$p_{\alpha_f}(\alpha_f) = \frac{p_{\phi_f}(\phi_f) + p_{\phi_f}(-\phi_f)}{\left|\frac{\partial \cos(\phi_f)}{\partial \phi_f}\right|} = \frac{p_{\phi_f}(\cos^{-1}(\alpha_f)) + p_{\phi_f}(-\cos^{-1}(\alpha_f))}{\sqrt{1-\alpha_f^2}}. \tag{12.19}$$

Given a distribution over α_f, the log spectrum interaction model can be written generally as

$$p(y_f|x_f, n_f) = p_{\alpha_f}(\alpha_f)\left|\frac{\partial y_f}{\partial \alpha_f}\right|^{-1} \tag{12.20}$$

$$= p_{\alpha_f}\left(\frac{1}{2}\left(\mathrm{e}^{y_f - \frac{x_f+n_f}{2}} - \mathrm{e}^{\frac{x_f-n_f}{2}} - \mathrm{e}^{\frac{n_f-x_f}{2}}\right)\right)\frac{1}{2}\mathrm{e}^{y_f - \frac{x_f+n_f}{2}}, \tag{12.21}$$

where α_f is obtained as a function of y_f, x_f and n_f from (12.13).

If we assume that the phase difference between speech and noise ϕ_f is uniformly distributed, $p_{\phi_f}(\phi_f) = \frac{1}{2\pi}$, then a change of variables leads to

$$p_{\alpha_f}(\alpha_f) = \frac{1}{\pi\sqrt{1-\alpha^2{}_f}}. \tag{12.22}$$

This is a shifted *beta distribution*: $p_{\alpha_f}(\alpha_f) = \frac{1}{2}\mathrm{Beta}\left(\frac{\alpha_f+1}{2}; a = \frac{1}{2}, b = \frac{1}{2}\right)$ where $\mathrm{Beta}(x; a, b) = x^{a-1}(1-x)^{b-1}\frac{\Gamma(a+b)}{\Gamma(a)\Gamma(b)}$ [23]. Indicating uniform phase by $\mathcal{U}(\phi)$, we

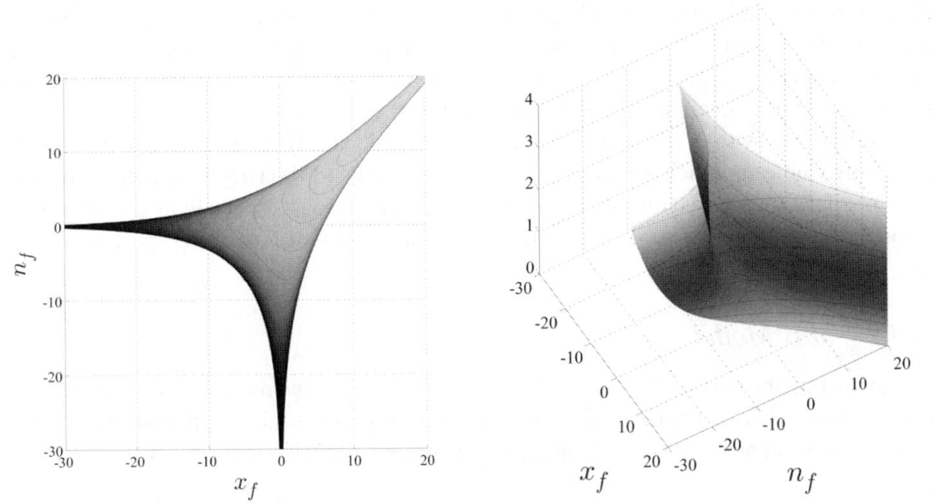

(a) Contour plot of exact interaction model

(b) Surface plot of exact interaction model

Figure 12.3 (a) Contour plot of the exact density in the log spectrum $p_{\mathcal{U}(\phi)}(y_f = 0|x_f, n_f)$ and (b) surface plot of the same, showing how the function is unbounded at the edges. Contour line spacing is logarithmic and the function has been truncated to fit in the plot box.

substitute (12.22) yielding

$$p_{\mathcal{U}(\phi)}(y_f|x_f,n_f) = \frac{\frac{1}{2\pi}\mathrm{e}^{y_f - \frac{x_f+n_f}{2}}}{\sqrt{1-\frac{1}{4}\left(\mathrm{e}^{y_f-\frac{x_f+n_f}{2}} - \mathrm{e}^{\frac{x_f-n_f}{2}} - \mathrm{e}^{\frac{n_f-x_f}{2}}\right)^2}} \tag{12.23}$$

as shown in [37], where it is called the *devil function* after its tortuous shape. Note that interesting alternate expressions for the same quantity can be obtained after some algebraic manipulations:

$$p_{\mathcal{U}(\phi)}(y_f|x_f,n_f)$$

$$= \frac{\frac{1}{\pi}\mathrm{e}^{y_f}}{\sqrt{\left(\mathrm{e}^{\frac{y_f}{2}}+\mathrm{e}^{\frac{x_f}{2}}+\mathrm{e}^{\frac{n_f}{2}}\right)\left(-\mathrm{e}^{\frac{y_f}{2}}+\mathrm{e}^{\frac{x_f}{2}}+\mathrm{e}^{\frac{n_f}{2}}\right)\left(\mathrm{e}^{\frac{y_f}{2}}-\mathrm{e}^{\frac{x_f}{2}}+\mathrm{e}^{\frac{n_f}{2}}\right)\left(\mathrm{e}^{\frac{y_f}{2}}+\mathrm{e}^{\frac{x_f}{2}}-\mathrm{e}^{\frac{n_f}{2}}\right)}} \tag{12.24}$$

$$= \frac{\frac{1}{\pi}\mathrm{e}^{y_f}}{\sqrt{\left(\mathrm{e}^{y_f}+\mathrm{e}^{x_f}+\mathrm{e}^{n_f}\right)^2 - 2\left(\mathrm{e}^{2y_f}+\mathrm{e}^{2x_f}+\mathrm{e}^{2n_f}\right)}}. \tag{12.25}$$

Later, we discuss application of similar derivations to the mel domain considered in [90]. In the amplitude domain, a similar distribution is known in the wireless communications literature as the *two-wave envelope* pdf [21].

Figures 12.3(a) and 12.3(b) show the exact interaction density function (12.23). The interaction density is highly nonlinear, and diverges to infinity along the edges of the feasible region.

The edge toward the bottom left of Figure 12.3(a), where $x_f < 0$ and $n_f < 0$, results from cases where the phase difference is zero and the signal amplitudes add up to the observation. The two other edges, where $x_f > 0$ or $n_f > 0$, result from cases where the signals have opposing phase and cancel to generate the observed signal.

Unfortunately, with (12.23), the integral in (12.1) is generally intractable, leaving sampling as the only viable approach for inference (see, for example, [37]). Therefore, there have been a series of approaches based on approximate interaction functions, especially in the mel domain, to which we will turn after discussing more basic approximations in the log spectrum domain.

12.4.2 Max Model

Approximating the sum of two signals in a frequency band as the maximum of the two signals is an intuitive idea that roughly follows our knowledge of masking phenomena in human hearing[2], and can be justified mathematically. Expressing (12.13) in the form

$$y_f = \max(x_f, n_f) + \log\left(1 + \mathrm{e}^{-|x_f - n_f|} + 2\mathrm{e}^{-\frac{|x_f - n_f|}{2}} \cos(\phi_f)\right), \qquad (12.26)$$

we can see that when one signal dominates the other, the second term approaches zero, taking the effect of phase along with it. This motivates the *max approximation*:

$$y_f \approx \max(x_f, n_f), \qquad (12.27)$$

which can be interpreted probabilistically using a Dirac delta:

$$p_{\max}(y_f | x_f, n_f) \overset{\text{def}}{=} \delta\left(y_f - \max(x_f, n_f)\right). \qquad (12.28)$$

Note that more general models based on the max approximation could be defined by additionally modeling the uncertainty associated with the approximation. For example, the approximation error could be modeled as Gaussian, and, optionally, made dependent on SNR. Such modeling has been thoroughly investigated for the log-sum approximation, as described below, but, to the best of our knowledge, has not yet been investigated for the max approximation.

Remarkably, the max approximation is the mean of the exact interaction function (12.23) [66][3]:

$$\mathrm{E}(y_f | x_f, n_f) = \int y_f \, p_{\mathcal{U}(\phi)}(y_f | x_f, n_f) \, \mathrm{d}y_f = \max(x_f, n_f). \qquad (12.29)$$

[2] A high-intensity signal at a given frequency affects the human hearing threshold for other signals at that frequency (signals roughly 6 dB below the dominant signal are not heard), and nearby frequencies, with diminishing effect, as a function of frequency difference. Consult [59] for details.

[3] While [66] reverts to an integration table to complete the proof of (12.29), it can be shown from (12.26) by noticing that $\forall \eta \in [0, 1)$, $\int \log(1 + \eta^2 + 2\eta\cos(\theta)) \, \mathrm{d}\theta = \int \log|1 + \eta \mathrm{e}^{\mathrm{i}\theta}|^2 \, \mathrm{d}\theta = 2\mathrm{Re}(\int \log(1 + \eta \mathrm{e}^{\mathrm{i}\theta}) \, \mathrm{d}\theta) = 0$, where θ is integrated over $[0, 2\pi)$. After a change of variable $z = \eta \mathrm{e}^{\mathrm{i}\theta}$, this can be obtained using Cauchy's integral formula $f(a) = \frac{1}{2\pi\mathrm{i}} \oint_\gamma \frac{f(z)}{z-a} \, \mathrm{d}z$ applied to the holomorphic function $f : z \mapsto \log(1 + z)$ defined in the open disk $\{z \in \mathbb{C} : |z| < 1\}$ of the complex plane, with $a = 0$ and on the circle γ of center 0 and radius η. The case $\eta = 1$ results from simple computations and amounts to showing $\int_0^\pi \log(\sin(\theta)) \, \mathrm{d}\theta = -\pi \log(2)$.

The max approximation was first used for noise compensation in [62]. Shortly thereafter, in [91], it was used to compute joint state likelihoods of speech and noise and find their optimal state sequence using a factorial hidden Markov model.

Inference in the max model is generally intractable when $p(\boldsymbol{x}|s^{\mathsf{x}})$ or $p(\boldsymbol{n}|s^{\mathsf{n}})$ have dependencies across frequency, as do, for example, full-covariance Gaussians. However, for conditionally independent models of the form $p(\boldsymbol{x}|s^{\mathsf{x}}) = \prod_f p(x_f|s^{\mathsf{x}})$, the state likelihoods and the posterior of $(\boldsymbol{x}, \boldsymbol{n})$ given the states can be readily computed, as shown below. Moreover, the max model is also highly amenable to approximate inference when explicitly evaluating all state combinations is computationally intractable, as described in Section 12.6.

12.4.3 Log-Sum Model

The *log-sum model*, used in [27,60] based on the additivity assumption in the power domain [7], uses the log of the expected value in the power domain to define an interaction function:

$$y_f \approx \log \mathrm{E}(\mathrm{e}^{y_f}|x_f, n_f) = \log(\mathrm{e}^{x_f} + \mathrm{e}^{n_f}), \tag{12.30}$$

which can then be interpreted probabilistically using

$$p_{\text{logsum}}(y_f|x_f, n_f) \stackrel{\text{def}}{=} \mathcal{N}(y_f; \log(\mathrm{e}^{x_f} + \mathrm{e}^{n_f}), \psi_f), \tag{12.31}$$

where $\boldsymbol{\Psi}$ is a variance intended to compensate for the effects of phase. In the limit as $\psi_f \to 0$, $p_{\text{logsum}}(y_f|x_f, n_f)$ becomes a Dirac delta function, leading to the model investigated in [20].

In the case of the log mel spectrum, which is closer to the features used by a recognizer, matters are made worse by the lack of a closed form expression for $p(y_l^{\mathsf{m}}|x_l^{\mathsf{m}}, n_l^{\mathsf{m}})$. This situation arises because the mel quantities are averages across frequency, but the signal interaction involves the whole frequency domain, as can be seen for example in (12.11). On the other hand, since the mel frequency domain averages together multiple bins, the effect of phase averages out. In this case, the log-sum approximation becomes more accurate, as shown in Figure 12.4(b).

However, the log sum approximation does not account for the changing variance of y_f^{m} as a function of the SNR stemming from the complicated phase term in (12.9). Various approximations have been proposed to handle this [16,49,85,86,90].

12.4.4 Mel Interaction Model

Although directly integrating out phase in the mel spectrum interaction (12.15) is intractable, a frequently used approximation is to assume that the term α_l^{m} in (12.15) has a known distribution, $\tilde{p}(\alpha_l^{\mathsf{m}})$, that is independent of x_l^{m} and n_l^{m}. Using this approximation, we can directly use (12.21):

$$p_{\text{mel}}(y_l^{\mathsf{m}}|x_l^{\mathsf{m}}, n_l^{\mathsf{m}}) \approx \tilde{p}_{\alpha_l^{\mathsf{m}}}\left(\tfrac{1}{2}\left(\mathrm{e}^{y_l^{\mathsf{m}} - \frac{x_l^{\mathsf{m}} + n_l^{\mathsf{m}}}{2}} - \mathrm{e}^{\frac{x_l^{\mathsf{m}} - n_l^{\mathsf{m}}}{2}} - \mathrm{e}^{\frac{n_l^{\mathsf{m}} - x_l^{\mathsf{m}}}{2}}\right)\right) \tfrac{1}{2}\mathrm{e}^{y_l^{\mathsf{m}} - \frac{x_l^{\mathsf{m}} + n_l^{\mathsf{m}}}{2}}. \tag{12.32}$$

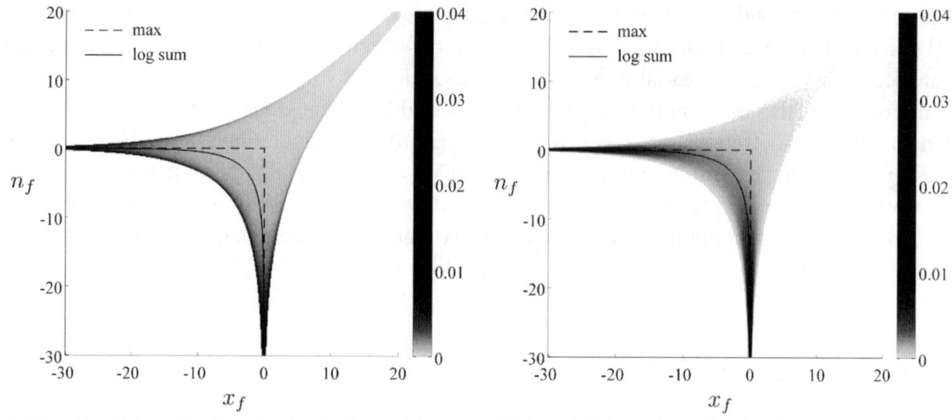

(a) Empirical interaction function for the log spectrum.

(b) Empirical interaction function for an average of five power spectrum bins.

Figure 12.4 Histograms representing empirical measurements of the interaction function for (a) the log spectrum domain and (b) an average of five power spectrum bins typical of the mel spectrum domain.

Unfortunately, it is still intractable to perform exact inference in this model. Hence, in [90], the integrals in (12.1) are computed by Monte Carlo, using a truncated Gaussian approximation to $\tilde{p}(\alpha_l^{\mathsf{m}})$. The shifted beta distribution mentioned earlier also has the feature that it can approximate a Gaussian for parameters a and b such that $ab \gg 1$, so perhaps it could be used as a unifying distribution, with empirically trained parameters, to handle the full range of cases. Approximate inference methods are discussed in Section 12.5.

12.5 Inference Methods

We have defined a number of interaction models, and now turn to inference methods for these interaction models. The main quantity of interest for speech recognition is the state likelihood $p(\boldsymbol{y}|s^{\mathsf{x}}, s^{\mathsf{n}})$ defined in (12.1). The posterior distribution of speech and noise $p(\boldsymbol{x}, \boldsymbol{n}|\boldsymbol{y}, s^{\mathsf{x}}, s^{\mathsf{n}})$ defined in (12.2) is also important but can often be computed using the same approximation methods as the likelihood. Figures 12.6 and 12.7 show how different the likelihoods can be for the various approximate inference methods described in this section.

12.5.1 Max Model Inference

The likelihood of the speech and noise features, $\boldsymbol{x}$ and $\boldsymbol{n}$, under the max model is

$$
\begin{aligned}
p_{\max}(\boldsymbol{y}|\boldsymbol{x}, \boldsymbol{n}) &= \prod_f p_{\max}(y_f|x_f, n_f) \\
&= \prod_f \delta(y_f - \max(x_f, n_f)).
\end{aligned}
\tag{12.33}
$$

For models of the form $p(\boldsymbol{x}|s^{\mathsf{x}}) = \prod_f p(x_f|s^{\mathsf{x}})$ with conditionally independent features given the states (e.g., diagonal-covariance Gaussians), the state likelihoods and the posterior of $\boldsymbol{x}$

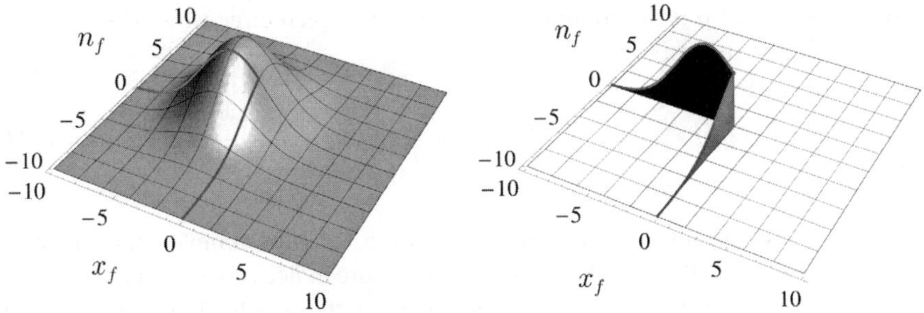

(a) Prior density (b) Posterior density

Figure 12.5 Inference under the max interaction model for clean speech x_f and noise n_f, for a single combination of states. In (a) the conditional prior $p(x_f, n_f|s^{\mathsf{x}}, s^{\mathsf{n}}) = p(x_f|s^{\mathsf{x}})p(n_f|s^{\mathsf{n}})$ is shown for a single feature dimension. The support of the likelihood function $p_{\max}(y_f|x_f, n_f) = \delta\left(y_f - \max(x_f, n_f)\right)$, for $y_f = 0$, is represented by a thick contour. The state likelihood $p(y_f = 0|s^{\mathsf{x}}, s^{\mathsf{n}})$ is the integral along this contour. The feature posterior $p(x_f, n_f|s^{\mathsf{x}}, s^{\mathsf{n}}, y_f = 0)$, which is proportional to the product of the prior and likelihood functions, is shown in (b).

given the states can be readily computed. Define the probability density of the event $\mathsf{x}_f = y_f$ given state s^{x} as $p_{\mathsf{x}_f}(y_f|s^{\mathsf{x}})$, and the probability of the event that $\mathsf{x}_f \leq y_f$ as $\Phi_{\mathsf{x}_f}(y_f|s^{\mathsf{x}}) \stackrel{\text{def}}{=} p(\mathsf{x}_f \leq y_f) = \int_{-\infty}^{y_f} p(x_f|s^{\mathsf{x}})\mathrm{d}x_f$, which is the cumulative distribution function (cdf) of x_f given s^{x} evaluated at y_f.

For a given combination of states $s^{\mathsf{x}}, s^{\mathsf{n}}$, the cdf of y_f under the model factors for independent sources [62, 75]:

$$\begin{aligned} p(\mathsf{y}_f \leq y_f|s^{\mathsf{x}}, s^{\mathsf{n}}) &= p(\max(\mathsf{x}_f, \mathsf{n}_f) \leq y_f|s^{\mathsf{x}}, s^{\mathsf{n}}) \\ &= p(\mathsf{x}_f \leq y_f, \mathsf{n}_f \leq y_f|s^{\mathsf{x}}, s^{\mathsf{n}}) \\ &= \Phi_{\mathsf{x}_f}(y_f|s^{\mathsf{x}})\Phi_{\mathsf{n}_f}(y_f|s^{\mathsf{n}}). \end{aligned} \tag{12.34}$$

The density of y_f is obtained by differentiating the cdf:

$$\begin{aligned} p(y_f|s^{\mathsf{x}}, s^{\mathsf{n}}) &= \frac{\mathrm{d}}{\mathrm{d}y_f}\left(\Phi_{\mathsf{x}_f}(y_f|s^{\mathsf{x}})\Phi_{\mathsf{n}_f}(y_f|s^{\mathsf{n}})\right) \\ &= p_{\mathsf{x}_f}(y_f|s^{\mathsf{x}})\Phi_{\mathsf{n}_f}(y_f|s^{\mathsf{n}}) + p_{\mathsf{n}_f}(y_f|s^{\mathsf{n}})\Phi_{\mathsf{x}_f}(y_f|s^{\mathsf{x}}). \end{aligned} \tag{12.35}$$

The density of $\boldsymbol{y}$ then is

$$p(\boldsymbol{y}|s^{\mathsf{x}}, s^{\mathsf{n}}) = \prod_f \left(p_{\mathsf{x}_f}(y_f|s^{\mathsf{x}})\Phi_{\mathsf{n}_f}(y_f|s^{\mathsf{n}}) + p_{\mathsf{n}_f}(y_f|s^{\mathsf{n}})\Phi_{\mathsf{x}_f}(y_f|s^{\mathsf{x}})\right). \tag{12.36}$$

Inference under this model is illustrated in Figure 12.5, and compared to other methods in Figures 12.6 and 12.7.

In the case that the $p(\boldsymbol{x}|s^{\mathsf{x}})$ or $p(\boldsymbol{n}|s^{\mathsf{n}})$ have dependencies across frequency, such as with full-covariance Gaussians, inference in the max model is generally intractable. When the conditional joint cdf of y is differentiated with respect to each dimension of y, we obtain an expression having 2^F terms:

$$p(y|s^{\mathsf{x}}, s^{\mathsf{n}}) = \sum_{\mathcal{F} \in \mathcal{P}([1..F])} \frac{\partial \Phi_{\mathsf{x}}(y|s^{\mathsf{x}})}{\partial \{y_f\}_{f \in \mathcal{F}}} \frac{\partial \Phi_{\mathsf{n}}(y|s^{\mathsf{n}})}{\partial \{y_{f'}\}_{f' \in \bar{\mathcal{F}}}}, \tag{12.37}$$

where $\mathcal{F} \subset [1..F]$ is any subset of the feature dimensions, $\bar{\mathcal{F}}$ is its complement, and the power set $\mathcal{P}([1..F])$ is the set of all such subsets. A set $\mathcal{F}$ of feature indices corresponds to a hypothesis that $\mathsf{x}_f > \mathsf{n}_f, f \in \mathcal{F}$, or in other words that x dominates in the selected frequency bands. When computing these quantities we would typically start with the joint pdf for each source, and integrate to obtain the term of interest:

$$\frac{\partial \Phi_{\mathsf{x}}(y|s^{\mathsf{x}})}{\partial \{y_f\}_{f \in \mathcal{F}}} = \int_{\mathcal{R}_{\bar{\mathcal{F}}}} p_{\mathsf{x}}(y_{\mathcal{F}}, y_{\bar{\mathcal{F}}}|s^{\mathsf{x}}) dy_{\bar{\mathcal{F}}}, \tag{12.38}$$

where we denote a subset of the variables indexed by set $\mathcal{F}$ as $y_{\mathcal{F}} = \{y_f\}_{f \in \mathcal{F}}$, and the region of integration is the negative half-space of $y_{\bar{\mathcal{F}}}$ defined by $\mathcal{R}_{\bar{\mathcal{F}}} = \bigotimes_{f \in \bar{\mathcal{F}}} (-\infty, y_f]$. These integrals are intractable, in general, for conditionally dependent models. Such integrals are also used in the marginalization approach to missing data methods as discussed in Chapter 14, and are a source of difficulty in applying these methods in the cepstral domain.

The equations above can be directly generalized to the case of multiple independent sources, as shown in [75]. In the general case of conditionally dependent features, there are then K^F terms in the conditional pdf of y, where K is the number of source signals. In the case of conditionally independent features, the model factorizes over frequency, and only univariate forms of the integrals above have to be computed. However, there remains an exponential number of combinations of the states of each source that need to be considered. Approximate techniques for addressing this computational issue are discussed below in the section on efficient inference methods.

12.5.2 Parallel Model Combination

In an approach known as *parallel model combination* (PMC), [28] makes use of the log-sum approximation, and assumes that the conditional probability $p_{\text{pmc}}(y_f|s^{\mathsf{x}}, s^{\mathsf{n}})$ is a normal distribution in the log spectrum or log mel spectrum domain. Moment-matching is then used in the power domain to estimate the parameters of $p_{\text{pmc}}(y_f|s^{\mathsf{x}}, s^{\mathsf{n}})$. To avoid clutter, we omit conditioning on the states and simply write $p_{\text{pmc}}(y_f)$ in this section. For simplicity, we present the method using diagonal-covariance models. The method is straightforward to extend to the case where the models are full-covariance [28], or are defined in a transformed domain such as the mel cepstrum, although at considerable additional computational cost. PMC defines $p_{\text{pmc}}(y_f) = \mathcal{N}(y_f; \hat{\mu}_{\mathsf{y}_f}, \hat{\sigma}_{\mathsf{y}_f})$, and chooses the mean $\hat{\mu}_{\mathsf{y}_f}$ and the variance $\hat{\sigma}_{\mathsf{y}_f}$ so that

$$\begin{aligned} \mathrm{E}_{\mathcal{N}(y_f; \hat{\mu}_{\mathsf{y}_f}, \hat{\sigma}_{\mathsf{y}_f})}(|Y_f|^2) &= \mathrm{E}(|X_f|^2) + \mathrm{E}(|N_f|^2), \\ \mathrm{Var}_{\mathcal{N}(y_f; \hat{\mu}_{\mathsf{y}_f}, \hat{\sigma}_{\mathsf{y}_f})}(|Y_f|^2) &= \mathrm{Var}(|X_f|^2) + \mathrm{Var}(|N_f|^2). \end{aligned} \tag{12.39}$$

As $x_f \sim \mathcal{N}(x_f; \mu_{\mathsf{x}_f}, \sigma_{\mathsf{x}_f})$, the following identities hold:

$$\begin{aligned} \mathrm{E}(|X_f|^2) &= \mathrm{E}(\mathrm{e}^{x_f}) = \mathrm{e}^{\mu_{\mathsf{x}_f} + \frac{1}{2}\sigma_{\mathsf{x}_f}}, \\ \mathrm{Var}(|X_f|^2) &= \mathrm{Var}(\mathrm{e}^{x_f}) = (\mathrm{e}^{\sigma_{\mathsf{x}_f}} - 1)\mathrm{e}^{2\mu_{\mathsf{x}_f} + \sigma_{\mathsf{x}_f}}, \end{aligned} \tag{12.40}$$

and similarly for n_f. These identities can be inverted for y_f to yield

$$\begin{aligned} \hat{\mu}_{\mathsf{y}_f} &= \log \mathrm{E}_{\mathcal{N}(y_f; \hat{\mu}_{\mathsf{y}_f}, \hat{\sigma}_{\mathsf{y}_f})}(|Y_f|^2) - \frac{1}{2}\hat{\sigma}_{\mathsf{y}_f}, \\ \hat{\sigma}_{\mathsf{y}_f} &= \log\left(1 + \frac{\mathrm{Var}_{\mathcal{N}(y_f; \hat{\mu}_{\mathsf{y}_f}, \hat{\sigma}_{\mathsf{y}_f})}(|Y_f|^2)}{(\mathrm{E}_{\mathcal{N}(y_f; \hat{\mu}_{\mathsf{y}_f}, \hat{\sigma}_{\mathsf{y}_f})}(|Y_f|^2))^2}\right). \end{aligned} \tag{12.41}$$

Substituting (12.39) and then (12.40) into (12.41) yields

$$\begin{aligned} \hat{\mu}_{\mathsf{y}_f} &= \log \frac{\mathrm{e}^{\mu_{\mathsf{x}_f} + \frac{1}{2}\sigma_{\mathsf{x}_f}} + \mathrm{e}^{\mu_{\mathsf{n}_f} + \frac{1}{2}\sigma_{\mathsf{n}_f}}}{\mathrm{e}^{\frac{1}{2}\hat{\sigma}_{\mathsf{y}_f}}}, \\ \hat{\sigma}_{\mathsf{y}_f} &= \log\left(1 + \frac{(\mathrm{e}^{\sigma_{\mathsf{x}_f}} - 1)\mathrm{e}^{2\mu_{\mathsf{x}_f} + \sigma_{\mathsf{x}_f}} + (\mathrm{e}^{\sigma_{\mathsf{n}_f}} - 1)\mathrm{e}^{2\mu_{\mathsf{n}_f} + \sigma_{\mathsf{n}_f}}}{(\mathrm{e}^{\mu_{\mathsf{x}_f} + \frac{1}{2}\sigma_{\mathsf{x}_f}} + \mathrm{e}^{\mu_{\mathsf{n}_f} + \frac{1}{2}\sigma_{\mathsf{n}_f}})^2}\right). \end{aligned} \tag{12.42}$$

In other words, PMC assumes that the distributions of the clean speech and the noise are log-normal, and approximates the sum of two log-normal distribution as another log-normal distribution whose parameters are estimated by moment-matching in the power domain. This method is known as the Fenton–Wilkinson method [24]. Returning to writing state-conditional models, and with the parameters of $p_{\mathrm{pmc}}(y_f|s^{\mathsf{x}}, s^{\mathsf{n}})$ in hand, the state likelihood can now be evaluated. Note that this method does not supply an estimate of the speech features given the noisy features.

PMC is the result of three approximations: the log-sum approximation, the assumption that $p(y_f|s^{\mathsf{x}}, s^{\mathsf{n}})$ is Gaussian in the log domain, and the Fenton–Wilkinson approximation, which uses moment-matching in the power domain instead of moment-matching in the log domain. The latter is problematic because the mean and variance in the power domain are not sufficient statistics of a log-normal distribution. Because of this, the mean and variance of the true conditional distribution $p(y_f|s^{\mathsf{x}}, s^{\mathsf{n}})$ in the log domain are generally different from those estimated by the Fenton–Wilkinson method, as can be seen in Figure 12.6, where the Fenton–Wilkinson approximation is compared to Monte-Carlo approximations of the true conditional distribution. A Monte-Carlo method known as *data-driven PMC* was developed in [28, 30] to address this problem. Data-driven PMC estimates the mean of y by sampling from the prior distributions of speech and noise, and computing the empirical mean of the noisy speech under the log-sum approximation. Other log-normal approximation methods for the sum of independent log-normal distributions have been proposed which instead directly estimate the sufficient statistics in the log domain [69,82,95]. Section 12.5.3 concerns another method that in some cases abandons the assumption that $p(y_f|s^{\mathsf{x}}, s^{\mathsf{n}})$ is log-normal altogether.

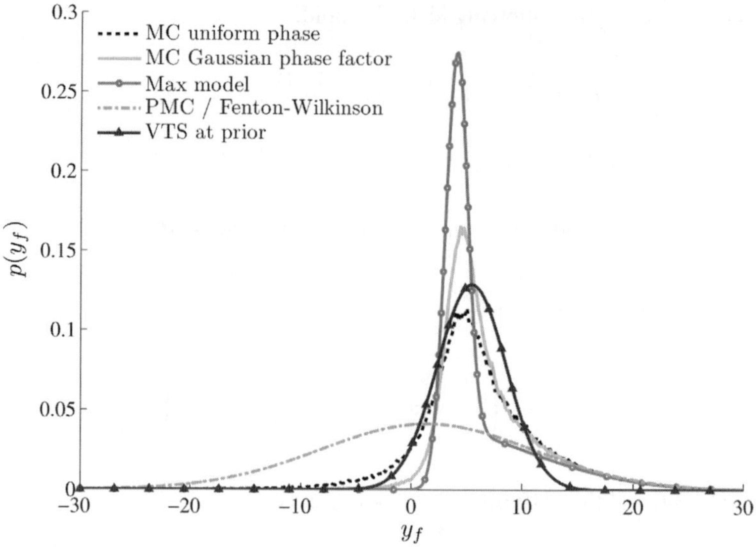

Figure 12.6 Comparison of the probability distribution $p(y_f|s^{\mathsf{x}}, s^{\mathsf{n}})$ under Gaussian priors for the speech and noise for different interaction models and inference methods. In all cases, the speech prior has mean 4 dB, and standard deviation 1 dB, and the noise prior has mean 0 dB and standard deviation 10 dB. The *MC uniform phase* is a Monte-Carlo approximation to the exact interaction model in the log spectral domain with uniform phase (the devil function), (12.21), using the max-model control variate method of [37]. *MC Gaussian phase factor* is the Monte-Carlo approximation to (12.53) with α variance 0.2, using a similar control variate approach. Both Monte-Carlo estimates are here computed with 10000 samples per value of y_f. The *max model* and *PMC (Fenton–Wilkinson)* approaches are straightforward, whereas VTS approaches depend upon the expansion point and iteration. Here, we show VTS expanded at the prior mean.

12.5.3 Vector Taylor Series Approaches

Unlike PMC, the vector Taylor series (VTS)-based approaches do not assume that the conditional probability distribution $p(y_f|s^{\mathsf{x}}, s^{\mathsf{n}})$ is Gaussian in the log spectrum or log mel spectrum domain. Instead, they linearize the log-sum interaction function (12.31) about an expansion point that is optimized for each observed y_f. The resulting conditional probability distribution is non-Gaussian and performs better in general than the PMC approximation. As an added benefit, the method yields estimates of the clean speech and is amenable to feature-based and model-based methods. The early VTS work of [60] was further developed by completing the probabilistic framework and introducing iterations on the expansion point in an algorithm known as Algonquin [26,50], which we describe here. Although the original algorithms included reverberation of the speech in the framework, we here relegate these channel components to the speech model for simplicity.

Here, we present the algorithm for general full-covariance models, and omit the dependency on states of each model for simplicity of notation. Note that for diagonal-covariance models, the features decouple and can be handled using the formula below independently for each feature. To handle the joint posterior, we concatenate $\boldsymbol{x}$ and $\boldsymbol{n}$ to form the joint vector $\boldsymbol{z} = [\boldsymbol{x}^\top \boldsymbol{n}^\top]^\top$ and use the function $g(\boldsymbol{z}) = \log(\mathrm{e}^{\boldsymbol{x}} + \mathrm{e}^{\boldsymbol{n}})$, where the logarithm and exponents operate

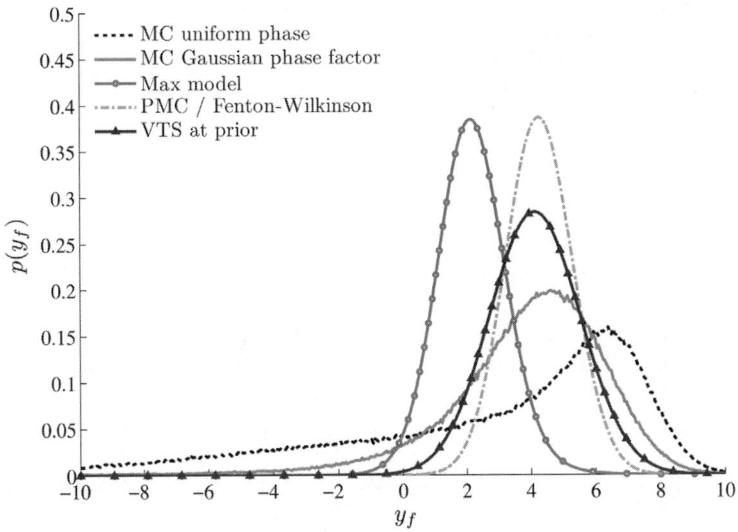

Figure 12.7 Comparison of the same probability distributions as Figure 12.6, but with different Gaussian priors for the speech and noise. Here, the speech prior has mean 2 dB, and standard deviation 1 dB, and the noise prior has mean 0 dB and standard deviation 2 dB. In this case, the prior is close to the point where the max model is less accurate (0 dB SNR). PMC, on the other hand, appears to do better because the variances of speech and noise are closer to each other. VTS also is more accurate because the expansion point at the prior is closer to the posterior mode.

element-wise on $\boldsymbol{x}$ and $\boldsymbol{n}$. Using a first-order Taylor series expansion at the point $\boldsymbol{z}_0$, the conditional distribution $p_{\text{logsum}}(\boldsymbol{y}|\boldsymbol{x},\boldsymbol{n})$ introduced in (12.31) is approximated as

$$p_{\text{logsum}}(\boldsymbol{y}|\boldsymbol{z}) \approx p_{\text{linear}}(\boldsymbol{y}|\boldsymbol{z};\boldsymbol{z}_0) = \mathcal{N}(\boldsymbol{y}; g(\boldsymbol{z}_0) + \mathbf{J}_g(\boldsymbol{z}_0)(\boldsymbol{z}-\boldsymbol{z}_0), \boldsymbol{\Psi}), \tag{12.43}$$

where $\boldsymbol{\Psi} = (\psi_f)_f$ and $\mathbf{J}_g(\boldsymbol{z}_0)$ is the Jacobian matrix of g, evaluated at $\boldsymbol{z}_0$:

$$\begin{aligned}\mathbf{J}_g(\boldsymbol{z}_0) = \left.\frac{\partial g}{\partial \boldsymbol{z}}\right|_{\boldsymbol{z}_0} &= \left[\operatorname{diag}\left(\frac{\partial g}{\partial \boldsymbol{x}}\right) \quad \operatorname{diag}\left(\frac{\partial g}{\partial \boldsymbol{n}}\right)\right]\Bigg|_{\boldsymbol{x}_0,\boldsymbol{n}_0} \\ &= \left[\operatorname{diag}\left(\frac{1}{1+e^{\boldsymbol{n}_0-\boldsymbol{x}_0}}\right) \quad \operatorname{diag}\left(\frac{1}{1+e^{\boldsymbol{x}_0-\boldsymbol{n}_0}}\right)\right].\end{aligned} \tag{12.44}$$

We assume that $\boldsymbol{x}$ and $\boldsymbol{n}$ are independent and Gaussian distributed when conditioning on the corresponding speech and noise states (which we here omit):

$$p(\boldsymbol{x}) = \mathcal{N}(\boldsymbol{x};\boldsymbol{\mu}_{\mathsf{x}},\boldsymbol{\Sigma}_{\mathsf{x}}), \qquad p(\boldsymbol{n}) = \mathcal{N}(\boldsymbol{n};\boldsymbol{\mu}_{\mathsf{n}},\boldsymbol{\Sigma}_{\mathsf{n}}). \tag{12.45}$$

Hence, $\boldsymbol{z}$ is Gaussian distributed with mean and covariance

$$\boldsymbol{\mu}_{\mathsf{z}} = \begin{bmatrix}\boldsymbol{\mu}_{\mathsf{x}} \\ \boldsymbol{\mu}_{\mathsf{n}}\end{bmatrix}, \qquad \boldsymbol{\Sigma}_{\mathsf{z}} = \begin{bmatrix}\boldsymbol{\Sigma}_{\mathsf{x}} & 0 \\ 0 & \boldsymbol{\Sigma}_{\mathsf{n}}\end{bmatrix}. \tag{12.46}$$

This leads to a simple linear Gaussian model with a Gaussian prior $p(\boldsymbol{z})$ and a Gaussian conditional distribution $p_{\text{linear}}(\boldsymbol{y}|\boldsymbol{z};\boldsymbol{z}_0)$ whose mean is a linear function of $\boldsymbol{z}$ and whose covariance is independent of $\boldsymbol{z}$. It is an easy and classical result in Bayesian theory that both $p_{\text{linear}}(\boldsymbol{y};\boldsymbol{z}_0)$ and the posterior $p_{\text{linear}}(\boldsymbol{z}|\boldsymbol{y};\boldsymbol{z}_0)$ are then Gaussian, and that their mean and

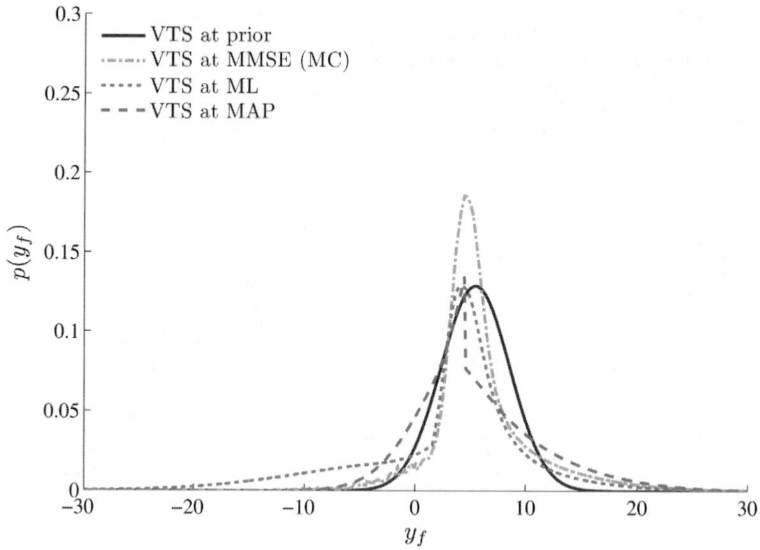

Figure 12.8 Comparison of the probability distribution $p(y_f|s^{\mathsf{x}}, s^{\mathsf{n}})$ for VTS computed with different expansion points, using the same priors as Figure 12.6. Expansion points are (a) the prior mean (*prior*) which is the most commonly used expansion point, (b) the posterior mean, (*MMSE*) estimated using the *MC Gaussian phase factor* method shown in Figure 12.6, (c) the point having maximum likelihood (*ML*) under the linearization, and (d) the mode of the posterior distribution (*MAP*) on the log-sum approximation curve, computed by grid search. Note that the latter is discontinuous because it switches from one mode of the posterior to another which has greater posterior density, but less likelihood when integrated under the linearization.

covariance can be easily computed from those of the prior and the conditional distribution. In particular, we can obtain the mean and covariance of the posterior by completing the square with respect to $\boldsymbol{z}$ in the exponent of $p_{\text{linear}}(\boldsymbol{y}|\boldsymbol{z}; \boldsymbol{z}_0)p(\boldsymbol{z})$. The covariance $\boldsymbol{\Sigma}_{\mathsf{z}|\boldsymbol{y}}$ turns out to be independent of $\boldsymbol{y}$:

$$\boldsymbol{\Sigma}_{\mathsf{z}|\boldsymbol{y}} = \left[\boldsymbol{\Sigma}_{\mathsf{z}}^{-1} + \mathbf{J}_g(\boldsymbol{z}_0)^\top \boldsymbol{\Psi}^{-1} \mathbf{J}_g(\boldsymbol{z}_0)\right]^{-1}, \tag{12.47}$$

while the mean is given by

$$\boldsymbol{\mu}_{\mathsf{z}|\boldsymbol{y}} = \boldsymbol{\Sigma}_{\mathsf{z}|\boldsymbol{y}} \left[\boldsymbol{\Sigma}_{\mathsf{z}}^{-1}\boldsymbol{\mu}_{\mathsf{z}} + \mathbf{J}_g(\boldsymbol{z}_0)^\top \boldsymbol{\Psi}^{-1}(\boldsymbol{y} - g(\boldsymbol{z}_0) + \mathbf{J}_g(\boldsymbol{z}_0)\boldsymbol{z}_0)\right] \stackrel{\text{def}}{=} \begin{bmatrix} \boldsymbol{\mu}_{\mathsf{x}|\boldsymbol{y}} \\ \boldsymbol{\mu}_{\mathsf{n}|\boldsymbol{y}} \end{bmatrix}. \tag{12.48}$$

By further integrating out $\boldsymbol{z}$ in $p_{\text{linear}}(\boldsymbol{y}|\boldsymbol{z}; \boldsymbol{z}0)p(\boldsymbol{z})$, we obtain the mean and covariance of $p(\boldsymbol{y}; \boldsymbol{z}_0)$:

$$\boldsymbol{\mu}_{\mathsf{y}} = g(\boldsymbol{z}_0) + \mathbf{J}_g(\boldsymbol{z}_0)(\boldsymbol{\mu}_{\mathsf{z}} - \boldsymbol{z}_0), \tag{12.49}$$

$$\boldsymbol{\Sigma}_{\mathsf{y}} = \boldsymbol{\Psi} + \mathbf{J}_g(\boldsymbol{z}_0)\boldsymbol{\Sigma}_{\mathsf{z}}\mathbf{J}_g(\boldsymbol{z}_0)^\top. \tag{12.50}$$

Note that although $p_{\text{linear}}(\boldsymbol{y}; \boldsymbol{z}_0)$ is Gaussian for a given expansion point, the value of $\boldsymbol{z}_0$ is the result of optimization and depends on $\boldsymbol{y}$ in a nonlinear way, so that the state likelihood is non-Gaussian as a function of $\boldsymbol{y}$.

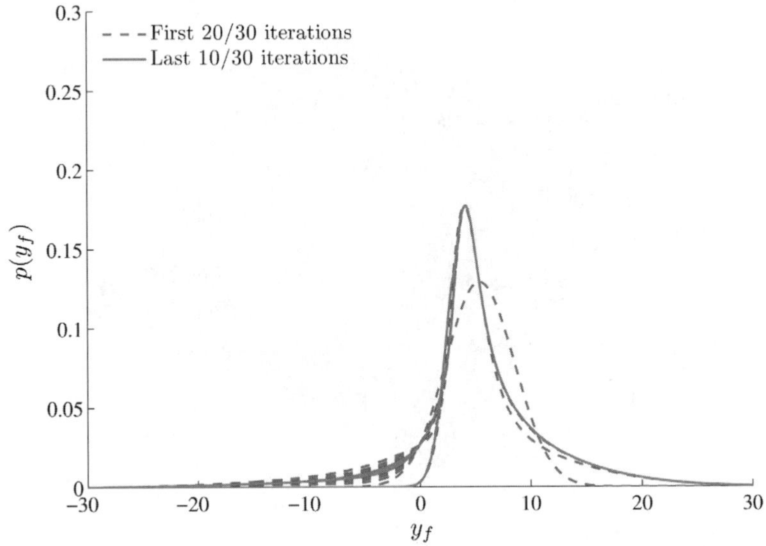

Figure 12.9 Comparison of the probability distribution $p(y_f|s^{\mathrm{x}}, s^{\mathrm{n}})$ for iterative VTS at different iterations, using the same priors as Figure 12.6. The convergence properties of iterative VTS are shown by plotting each of the first 20 iterations, followed by each of the last 10 iterations for a total of 30. The fact that these last iterations still differ on the left-hand tail of the distribution indicates that the algorithm is oscillating between different solutions. Here, the iterations are started at the prior mean, but other expansion points lead to similar behavior. It is interesting to note that in this case, the minimum of the last several iterations of VTS makes a nice approximation of the probability distribution given by *MC Gaussian phase factor* shown in Figure 12.6.

We shall note as well that the posterior mean can be rewritten in a simpler and more intuitive way using the above covariance:

$$\boldsymbol{\mu}_{\mathrm{z}|\boldsymbol{y}} = \boldsymbol{\mu}_{\mathrm{z}} + \boldsymbol{\Sigma}_{\mathrm{z}}\mathbf{J}_g(\boldsymbol{z}_0)^\top \boldsymbol{\Sigma}_{\mathrm{y}}^{-1} (\boldsymbol{y} - \boldsymbol{\mu}_{\mathrm{y}}) . \tag{12.51}$$

The posterior mean is thus obtained as the sum of the prior mean and a renormalized version of the bias between the observed noisy speech and the predicted value of the noisy speech at the prior mean given a linearization of the interaction function at $\boldsymbol{z}_0$.

The linearization point is important to the accuracy of the algorithm, as can be seen in Figure 12.8, and theoretically should be near the mode of the "true" posterior obtained using $p_{\mathrm{logsum}}(\boldsymbol{y}|\boldsymbol{x}, \boldsymbol{n})$ as the conditional probability. Therefore, whereas the initial linearization point is at the prior mean, in each iteration the estimated posterior mean is used to obtain a new expansion point $\boldsymbol{z}_0 = \boldsymbol{\mu}_{\mathrm{z}|\boldsymbol{y}}$. Because the interaction function is shift invariant, in the sense that $\boldsymbol{y} + \boldsymbol{v} = g(\boldsymbol{x} + \boldsymbol{v}, \boldsymbol{n} + \boldsymbol{v})$ for any $\boldsymbol{v}$, the linearization at $\boldsymbol{z}_0 = [\boldsymbol{x}_0; \boldsymbol{n}_0]$ is a plane tangent to g along the line in $\boldsymbol{x}$, $\boldsymbol{n}$ defined by $\boldsymbol{x} - \boldsymbol{x}_0 = \boldsymbol{n} - \boldsymbol{n}_0$. Since $\boldsymbol{y}$ is observed, this is equivalent, as illustrated in Figure 12.10, to linearizing at a point on the curve determined by the observation $\boldsymbol{y} = g(\boldsymbol{x}_0', \boldsymbol{n}_0')$, defined by $\boldsymbol{x}_0' = \boldsymbol{\mu}_{\mathrm{x}|\boldsymbol{y}} + \boldsymbol{v}$, $\boldsymbol{n}_0' = \boldsymbol{\mu}_{\mathrm{n}|\boldsymbol{y}} + \boldsymbol{v}$, where $\boldsymbol{v} = \boldsymbol{y} - g(\boldsymbol{\mu}_{\mathrm{x}|\boldsymbol{y}}, \boldsymbol{\mu}_{\mathrm{n}|\boldsymbol{y}})$. This point is not necessarily at the posterior mode along the curve, so the expansion can be a source of trouble for the algorithm. Most notably, it does not guarantee the convergence of the likelihood estimate which is known to fail in many conditions [49], as illustrated in Figure 12.9. It may be better to pose the problem in terms of finding the mode of the posterior distribution

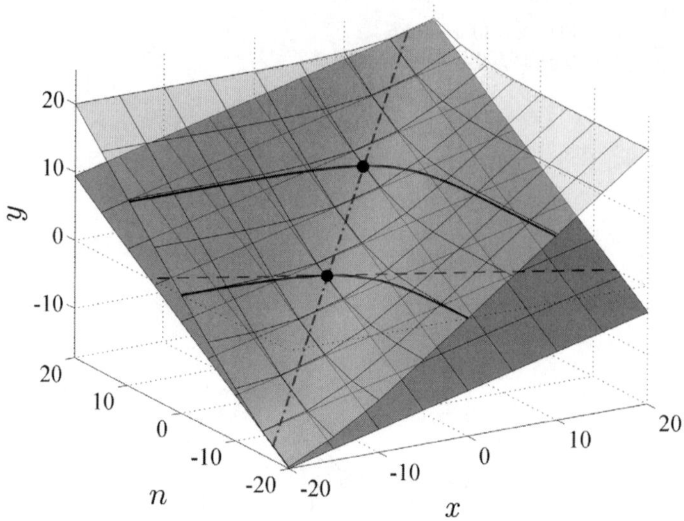

Figure 12.10 Illustration of the linearization procedure in VTS for a single frequency. The transparent surface on top represents the log-sum interaction $y = \log(e^x + e^n)$, while the plane below it is the linearization of g, which is tangent to that surface at $(x_0, n_0, g(x_0, n_0))$, for $(x_0, n_0) = (5, 8)$. This plane is also tangent along the dash-dotted line, because g has the property that $g(x + v, n + v) = y + v$ for any v. The two solid curves represent $y = \log(e^x + e^n)$ for $y = 0$ and $y = g(x_0, n_0)$. The dashed line is the tangent to $0 = \log(e^x + e^n)$ at (x_0, n_0) in the $y = 0$ plane.

directly. Optimization methods such as quasi-Newton methods involve differentiating the log posterior, and thus compute differentials of $g(x)$, but can step toward the optimum in a smoother and faster way [51].

Our discussion of VTS approaches above has assumed the use of source models based in the log (mel) power spectral domain, rather than cepstral domain, and neglected the explicit modeling of channel effects. Both circumstances can be readily handled in the VTS framework, assuming that an invertible DCT matrix $\mathbf{C}$ is used to transform to the cepstral domain. However, often the cepstra are generated by eliminating higher-order coefficients, in order to minimize the influence of pitch, and the Moore–Penrose pseudoinverse is commonly used. A more principled approach would be to supply a model of the upper cepstra so that the transformation is invertible.

In general, recognizers model features of multiple frames rather than a single one. This creates a model in which inference at the current frame is dependent upon previous frames. In [15], models of both *static* (i.e., single frame) and *dynamic* (i.e., differences across frame) features are used as priors for Algonquin. Although exact inference in such a model is generally intractable, [15] made the expedient approximation of using point estimates of the clean speech of previous frames to compute the priors of the current frame.

Unfortunately, as mentioned earlier, state-of-the-art speech recognizers use more complex and non-invertible transformations of multiple frames, such as LDA or fMPE transforms. Because of the nonlinearity and dimensionality reduction, further approximations would be necessary to perform model-based noise compensation with such models. In general, as previously mentioned, fleshing out the models to provide some distributions of the dimensions that are normally discarded is one avenue of attack.

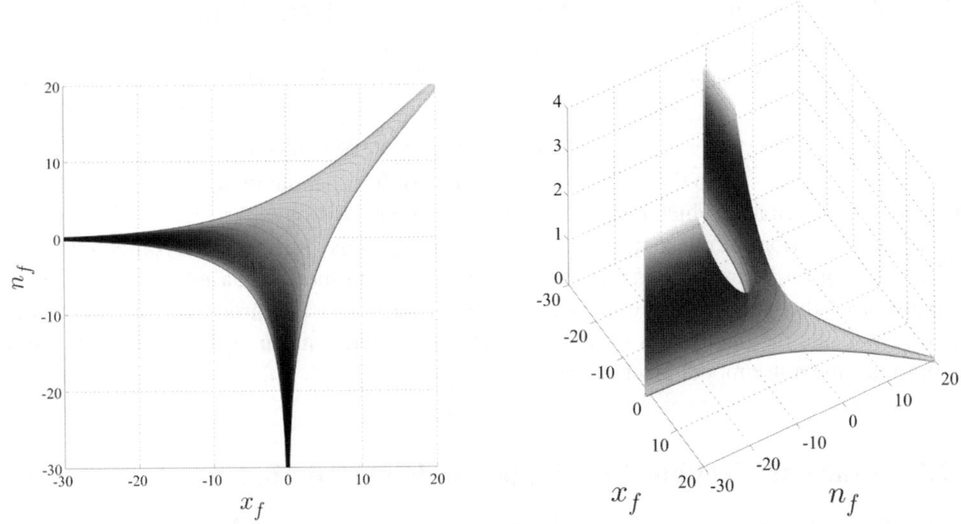

(a) Contour plot of phase-factor model (b) Surface plot of phase-factor model

Figure 12.11 (a) Contour plot of phase factor approximation and (b) surface plot of the same. Contour line spacing is logarithmic and the function has been truncated to fit in the plot box.

12.5.4 SNR-Dependent Approaches

SNR-dependent approaches [16, 20, 49], also known as "phase-sensitive" approaches, are similar to the basic VTS model except that a Gaussian model is used for the *phase factor* $\boldsymbol{\alpha}$ in (12.9), rather than for the entire phase term. Thus, neglecting the channel effects, the model is

$$\boldsymbol{y} = \log(\mathrm{e}^{\boldsymbol{x}} + \mathrm{e}^{\boldsymbol{n}} + 2\mathrm{e}^{\frac{\boldsymbol{x}+\boldsymbol{n}}{2}} \circ \boldsymbol{\alpha}). \tag{12.52}$$

The phase factor $\boldsymbol{\alpha} \in [-1, 1]^F$ is modeled as a zero mean Gaussian $p(\boldsymbol{\alpha}) = \mathcal{N}(\boldsymbol{\alpha}; 0, \boldsymbol{\Sigma}_\alpha)$ truncated to the interval $[-1, 1]$ in each dimension. The variance $\boldsymbol{\Sigma}_\alpha$ is usually assumed to be diagonal. Using (12.21), we then have

$$p_{\mathrm{snrdep}}(\boldsymbol{y}|\boldsymbol{x}, \boldsymbol{n}) = \mathcal{N}_\alpha\left(\frac{1}{2}(\mathrm{e}^{\boldsymbol{y}-\frac{\boldsymbol{x}+\boldsymbol{n}}{2}} - \mathrm{e}^{\frac{\boldsymbol{x}-\boldsymbol{n}}{2}} - \mathrm{e}^{\frac{\boldsymbol{n}-\boldsymbol{x}}{2}}); 0, \boldsymbol{\Sigma}_\alpha\right)\left|\mathrm{diag}\left(\frac{1}{2}\mathrm{e}^{\boldsymbol{y}-\frac{\boldsymbol{x}+\boldsymbol{n}}{2}}\right)\right|. \tag{12.53}$$

This distribution is illustrated in Figure 12.11. It is especially appropriate in the log mel domain and corresponds closely to the empirical distribution shown in Figure 12.4(b). Although the variance of α does not change as a function of SNR, the uncertainty of $\boldsymbol{y}$ given x and $\boldsymbol{n}$ becomes a function of SNR due to the nonlinearity of the interaction. In [49], in addition to modeling α as a Gaussian, the interaction was also approximated using

$$\begin{aligned}\boldsymbol{y} &= \log(\mathrm{e}^{\boldsymbol{x}} + \mathrm{e}^{\boldsymbol{n}}) + \log\left(1 + \frac{2}{\mathrm{e}^{\frac{\boldsymbol{x}-\boldsymbol{n}}{2}} + \mathrm{e}^{\frac{\boldsymbol{n}-\boldsymbol{x}}{2}}} \circ \boldsymbol{\alpha}\right) \\ &\approx \log(\mathrm{e}^{\boldsymbol{x}} + \mathrm{e}^{\boldsymbol{n}}) + \frac{2}{\mathrm{e}^{\frac{\boldsymbol{x}-\boldsymbol{n}}{2}} + \mathrm{e}^{\frac{\boldsymbol{n}-\boldsymbol{x}}{2}}} \circ \boldsymbol{\alpha}.\end{aligned} \tag{12.54}$$

Using this interaction function in (12.20) leads to a conditionally Gaussian likelihood function:

$$p_{\text{snrdepvar}}(\boldsymbol{y}|\boldsymbol{x},\boldsymbol{n}) = \mathcal{N}\left(\boldsymbol{y}; \log(\mathrm{e}^{\boldsymbol{x}} + \mathrm{e}^{\boldsymbol{n}}), \mathbf{A}^{\top}\boldsymbol{\Sigma}_{\alpha}\mathbf{A}\right). \tag{12.55}$$

The matrix $\mathbf{A} \stackrel{\text{def}}{=} \operatorname{diag}\left(2/(\mathrm{e}^{\frac{\boldsymbol{x}-\boldsymbol{n}}{2}} + \mathrm{e}^{\frac{\boldsymbol{n}-\boldsymbol{x}}{2}})\right)$, where division is defined element-wise, is a function of the SNR, $\boldsymbol{x} - \boldsymbol{n}$. In this case, the dependency of the uncertainty upon the SNR clearly appears in the variance, which reaches a maximum for an SNR of zero.

In [16], posteriors of clean speech and likelihoods of noisy speech were computed using (12.53), using an improved version of the VTS/Algonquin method, based on second-order expansion of the joint distribution, $p_{\text{snrdep}}(\boldsymbol{y}|\boldsymbol{x},\boldsymbol{n})p(\boldsymbol{x})p(\boldsymbol{n})$. The proposed algorithm was used to estimate the likelihoods of noisy speech, the posterior mean of the clean speech and to optimize the noise model given noisy speech.

12.6 Efficient Likelihood Evaluation in Factorial Models

Exact inference methods for robust ASR using factorial models require computing the joint state likelihood $p(\boldsymbol{y}|s^{\mathsf{x}}, s^{\mathsf{n}}) \equiv p(\boldsymbol{y}|s^{\mathsf{z}})$, introduced in (12.1), for all combinations of speech and noise states. Therefore, exact inference generally becomes computationally intractable when the number of state combinations is large. Efficient approximate inference naturally involves either reducing the amount of computation required to estimate $p(\boldsymbol{y}|s^{\mathsf{x}}, s^{\mathsf{n}})$, reducing the number of state combinations that are evaluated, or both.

12.6.1 Efficient Inference using the Max Model

In Section 12.5.1, we showed that if the conditional prior distributions of speech and noise have no statistical dependencies between features, the joint likelihood of a given combination of speech and noise states under the max interaction model is given by:

$$p_{\max}(\boldsymbol{y}|s^{\mathsf{x}}, s^{\mathsf{n}}) = \prod_f \left(p_{\mathsf{x}_f}(y_f|s^{\mathsf{x}})\Phi_{\mathsf{n}_f}(y_f|s^{\mathsf{n}}) + p_{\mathsf{n}_f}(y_f|s^{\mathsf{n}})\Phi_{\mathsf{x}_f}(y_f|s^{\mathsf{x}})\right). \tag{12.56}$$

For K explicitly modeled acoustic sources, the result becomes

$$p(\boldsymbol{y}|\{s^k\}) = \prod_f \sum_k p_{\mathsf{x}_f^k}\left(y_f|s^k\right) \prod_{j\neq k} \Phi_{\mathsf{x}_f^j}\left(y_f|s^j\right), \tag{12.57}$$

where s^k denotes the acoustic state of the kth source x^k, and $\{s^k\}$ denotes $\{s^k\}_{k=1}^K = \{s^1, s^2, \ldots, s^K\}$, a particular configuration of the state variables of each source.

An advantageous property of this likelihood function is that it is composed of terms with factors that depend on the state of a single acoustic source. Therefore, the cost of computing these factors scales linearly with the number of acoustic sources that are explicitly modeled. However, exact inference using the max model requires that the product of sums in (12.36) be computed for every combination of states, which scales exponentially with the number of sources. This is true even for models in which the feature dimensions are conditionally independent given the states.

The joint likelihood (12.57) is often approximated to depend only on the acoustic model of a single source, for example, as done in [38], where $p(y_f|\{s^k\}) \approx p_{x_f^i}(y_f|s^i), i = \arg\max_k \mu_{s^k}$. This averts the cost of computing the cumulative distribution functions and the additions and multiplications in (12.57), but inference still scales exponentially with K, since the resulting likelihood function is, in general, different for every combination of states. In the case that all Gaussians in all acoustic models share the same variance at each dimension, the branch-and-bound algorithm in [80] can be applied to do an approximate search for the MAP state configuration, but this approach also has exponential worst-case complexity, and is not well suited for approximating the likelihoods of the states, because the upper bounds produced during the search are very loose.

Recently, a new variational framework for the max model was introduced [76–78]. The framework hinges on the observation that in each feature dimension, a latent hidden variable, corresponding to the identity of the source that explains the data in that dimension, is being integrated out in the sum in (12.57). Denoting the *mask variable* for feature f by d_f, and a particular choice of mask values for all of the features by $\{d_f\} \stackrel{\text{def}}{=} \{d_f\}_{f=1}^F$, we have

$$p(\boldsymbol{y}, \{d_f\}|\{s^k\}) = \prod_f p(y_f, \{d_f\}|\{s^k\}) \tag{12.58}$$

$$= \prod_f p_{x_f^{d_f}}(y_f|s^{d_f}) \prod_{j \neq d_f} \Phi_{x_f^j}(y_f|s^j), \tag{12.59}$$

where $x_f^{d_f}$ and s^{d_f} denote the feature and state of the source that explains feature f. Note that $p(y_f, \{d_f\}|\{s^k\})$ is simply the product of the probability that source d_f explains the data, and all other source features have values less than the data. This *lifted max model* is derived more rigorously in [78], and explicitly models which source explains each feature dimension. The lifted max model has the special property that $p(\boldsymbol{y}, \{d_f\}|\{s^k\})$ factors over the acoustic sources, which immediately implies that if the mask values $\{d_f\}$ are known, inference of the acoustic sources decouples. Since $p(y_f, \{d_f\}|\{s^k\})$ factors over frequency, it also follows that if the state combination is known, then the inference of each d_f decouples from the others. In general, it is intractable to compute all possible acoustic masks (2^F), or all possible state combinations ($\prod_k |s^k|$), but these properties can be exploited using variational methods.

By Jensen's inequality, the log probability of the data under the lifted max model can be lower-bounded as follows:

$$\log p(\boldsymbol{y}) = \log \sum_{\{s^k\},\{d_f\}} p(\boldsymbol{y}, \{s^k\}, \{d_f\}) \tag{12.60}$$

$$\geq \sum_{\{s^k\},\{d_f\}} q(\{s^k\}, \{d_f\}) \log \frac{p(\boldsymbol{y}, \{s^k\}, \{d_f\})}{q(\{s^k\}, \{d_f\})} \stackrel{\text{def}}{=} \mathcal{L} \tag{12.61}$$

for any probability distribution q on the states $\{s^k\}$ and masks $\{d_f\}$. The difference between (12.60) and (12.61) is the Kullback–Leibler (KL) divergence between the exact posterior under the model, $p(\{s^k\}, \{d_f\}|\boldsymbol{y})$, and $q(\{s^k\}, \{d_f\})$ [44]:

$$D(q_{\{s_k\},\{d_f\}} \,||\, p_{\{s_k\},\{d_f\}|\boldsymbol{y}}) = \log p(\boldsymbol{y}) - \mathcal{L}, \tag{12.62}$$

where we use the random variable notation for s_k and d_f to indicate that the divergence is only a function of their distribution and not their values. When $q(\{s^k\}, \{d_f\}) = p(\{s^k\}, \{d_f\}|\boldsymbol{y})$, the bound is tight. By optimizing the variational parameters of $q(\{s^k\}, \{d_f\})$ to maximize the lower bound $\mathcal{L}$ in (12.61), we at the same time minimize (12.62). The resulting q distribution can be utilized as a surrogate for the true posterior $p(\{s^k\}, \{d_f\}|\boldsymbol{y})$, and used to make predictions. Because the joint distribution $p(\boldsymbol{y}, \{s^k\}, \{d_f\})$ factors, any form of $q(\{s^k\}, \{d_f\})$ that factors over both $\{s^k\}$ and $\{d_f\}$ makes optimizing the bound $\mathcal{L}$ in (12.61) *linear* in the number of sources K, the number of features F, and number of states $\sum_k |s^k|$. For example, if $q(\{s^k\}, \{d_f\}) = \prod_f q(d_f) \prod_k q(s^k)$, the bound $\mathcal{L}$ becomes

$$\begin{aligned}
\mathcal{L} &= \sum_{\{s^k\},\{d_f\}} \prod_f q(d_f) \prod_k q(s^k) \log \frac{\prod_f p_{\mathsf{x}_f^{d_f}}(y_f|s^{d_f}) \prod_{j \neq d_f} \Phi_{\mathsf{x}_f^j}(y_f|s^j) \prod_k p(s^k)}{\prod_f q(d_f) \prod_k q(s^k)} \\
&= \sum_{f,k} \Big(q_{\mathsf{d}_f}(k) \sum_{s^k} q(s^k) \log p_{\mathsf{x}_f^k}(y_f|s^k) + (1 - q_{\mathsf{d}_f}(k)) \sum_{s^k} q(s^k) \log \Phi_{\mathsf{x}_f^k}(y_f|s^k) \Big) \\
&\quad + \sum_f H(q_{\mathsf{d}_f}) - \sum_k D(q_{\mathsf{s}_k} \,||\, p_{\mathsf{s}_k})
\end{aligned} \tag{12.63}$$

as shown in [76], where $H(q_{\mathsf{d}_f}) = -\sum_{d_f} q(d_f) \log q(d_f)$ denotes the entropy of q_{d_f}. Clearly the bound can be computed without considering combinations of source states, or combinations of feature mask configurations, and so scales linearly with the number of sources, states per source, and feature dimension. Importantly, this implies that the chosen q distribution can also be iteratively inferred in time linear in these variables. As described in Section 12.7.2, these variational approximations and their extensions have been explored in the context of multi-talker speech recognition.

12.6.2 Efficient Vector-Taylor Series Approaches

To make inference in VTS-based systems more efficient, the following approximations are typically made:

- The noise is modeled by a single Gaussian to reduce the number of joint states to the number of states in the speech model, so that $|s^{\mathsf{z}}| = |s^{\mathsf{x}}|$, where $|s|$ denotes the number of discrete values that the state s can take.
- The data likelihood $p_{\text{linear}}(\boldsymbol{y}|s^{\mathsf{z}}; \boldsymbol{z}_0)$ is assumed to have a diagonal covariance matrix:

$$\begin{aligned}
&p_{\text{linear}}(\boldsymbol{y}|s^{\mathsf{z}}; \boldsymbol{z}_0) = \mathcal{N}(\boldsymbol{y}; \boldsymbol{\mu}_{\mathsf{y}|s^{\mathsf{z}}}, \boldsymbol{\Sigma}_{\mathsf{y}|s^{\mathsf{z}}}) \approx \mathcal{N}(\boldsymbol{y}; \boldsymbol{\mu}_{\mathsf{y}|s^{\mathsf{z}}}, \text{diag}(\boldsymbol{\Sigma}_{\mathsf{y}|s^{\mathsf{z}}})), \\
&\boldsymbol{\mu}_{\mathsf{y}|s^{\mathsf{z}}} = g(\boldsymbol{z}_0) + \mathbf{J}_g(\boldsymbol{z}_0)(\boldsymbol{\mu}_{\mathsf{z}|s^{\mathsf{z}}} - \boldsymbol{z}_0), \\
&\boldsymbol{\Sigma}_{\mathsf{y}|s^{\mathsf{z}}} = \boldsymbol{\Psi} + \mathbf{J}_g(\boldsymbol{z}_0)\boldsymbol{\Sigma}_{\mathsf{z}|s^{\mathsf{z}}}\mathbf{J}_g(\boldsymbol{z}_0)^\top.
\end{aligned}$$

This reduces the cost of evaluating $p_{\text{linear}}(\boldsymbol{y}|s^{\mathsf{z}}; \boldsymbol{z}_0)$ by a factor of F, the dimension of $\boldsymbol{y}$.

- The approximation of the conditional likelihood $p_{\text{logsum}}(\boldsymbol{y}|\boldsymbol{x},\boldsymbol{n})$ as a Gaussian with mean linear in $\boldsymbol{x}$ and $\boldsymbol{n}$ is shared by sufficiently similar speech states

$$p_{\text{logsum}}(\boldsymbol{y}|\boldsymbol{z}) \approx p_{\text{linear}}(\boldsymbol{y}|\boldsymbol{z}, s^{\mathsf{z}}) \approx p_{\text{linear}}(\boldsymbol{y}|\boldsymbol{z}, s^{\mathsf{r_z}}, s^{\mathsf{z}}), \tag{12.64}$$

where the state $s^{\mathsf{r_z}}$ is a "low-resolution" surrogate for the joint state s^{z}, and $|s^{\mathsf{r_z}}| \ll |s^{\mathsf{z}}|$. $s^{\mathsf{r_z}}$ is often referred to in the speech literature as a "regression class variable" [31]. Similarly, hierarchical acoustic models, which consist of multiple acoustic models trained at different model resolutions in terms of number of components can be used to compute surrogate likelihoods using VTS-methods while "searching" for probable state combinations.

The amount of computational savings brought by (12.64) depends on the specific approximations made, and several have been proposed [30]. Techniques such as joint uncertainty decoding (JUD) and VTS-JUD [57, 96], introduced in more detail in Chapter 17, have the advantage that only $|s^{\mathsf{r_z}}|$ sets of "compensation" parameters need to be computed, but the parameters of all $|s^{\mathsf{z}}|$ states of the acoustic model need to be transformed. Predictive CMLLR (PCMLLR) [31], conversely, implements model compensation via a feature transformation:

$$p_{\text{linear}}(\boldsymbol{y}|s^{\mathsf{z}}) \approx p_{\text{cmllr}}(\boldsymbol{y}|s^{\mathsf{z}}, s^{\mathsf{r_z}}) = |\mathbf{A}_{s^{\mathsf{r_z}}}|\mathcal{N}(\mathbf{A}_{s^{\mathsf{r_z}}}\boldsymbol{y} + \boldsymbol{b}_{s^{\mathsf{r_z}}}; \boldsymbol{\mu}_{\mathsf{x}|s^{\mathsf{x}}}, \boldsymbol{\Sigma}_{\mathsf{x}|s^{\mathsf{x}}}), \tag{12.65}$$

where $\mathbf{A}_{s^{\mathsf{r_z}}}$ and $\boldsymbol{b}_{s^{\mathsf{r_z}}}$ are estimated to minimize the KL divergence of $p_{\text{cmllr}}(\boldsymbol{y}|s^{\mathsf{z}}, s^{\mathsf{r_z}})$ from $p_{\text{linear}}(\boldsymbol{y}|s^{\mathsf{z}})$, and the Jacobian determinant $|\mathbf{A}_{s^{\mathsf{r_z}}}|$ ensures that the distribution in the right-hand side normalizes over $\boldsymbol{y}$. Note that the parameters of the speech model are not modified. Compared to model transformation methods that utilize diagonal-covariance approximations of $\boldsymbol{\Sigma}_{\mathsf{y}|s^{\mathsf{z}}}$, PCMLLR has the advantage that correlation changes in the feature vector can be modeled via $\mathbf{A}_{s^{\mathsf{r_z}}}$. Such modeling has been shown to improve ASR performance [30]. Another important advantage is that the PCMLLR model can be adapted in a straightforward manner like CMLLR [30].

The computational burden of computing likelihoods for all combinations of states in VTS models can also be alleviated using variational methods. A variational form of Algonquin was first discussed in [27], and is described in detail for the assumption of Gaussian posteriors for $\boldsymbol{x}$ and $\boldsymbol{n}$ in [49]. These algorithms iterate between computing linear approximation(s) of the log-sum function given the current estimate(s) of the speech and noise, and optimizing a variational lower bound on the resulting approximation to the probability of the data to update the speech and noise estimate(s) and acoustic likelihoods. The idea of conditioning the variational posterior on auxiliary state variables to control the number of masks that are inferred when doing inference in the max model [77, 78] could be similarly applied in the Algonquin (or VTS) framework to control the number of Gaussians used to approximate the posterior distribution of the features.

12.6.3 Band Quantization

Band quantization (BQ) is a technique that can be used to reduce the number of likelihoods that need to be computed per dimension for models with conditionally independent features. A *band-quantized Gaussian mixture model* (BQGMM) is a diagonal-covariance GMM that is constrained as follows. At each feature dimension f, an additional discrete random variable a_f^{x}

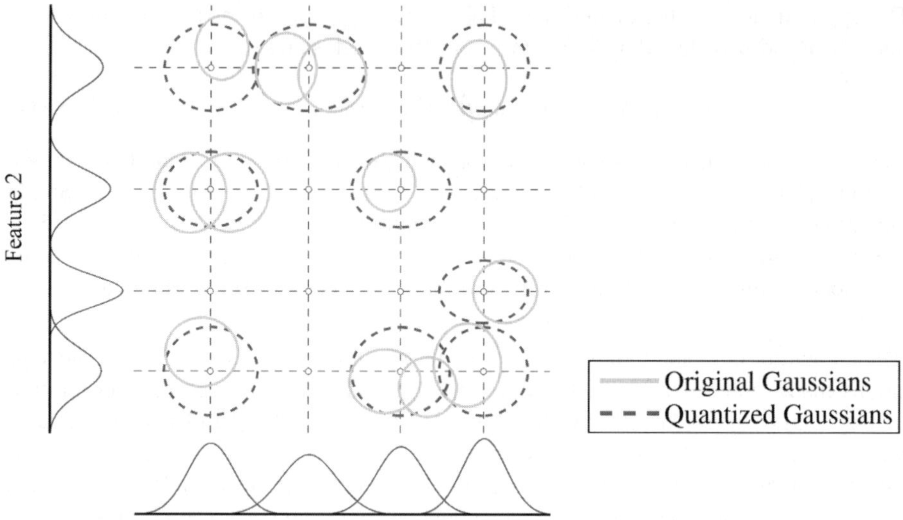

Figure 12.12 In band quantization, a large set of multidimensional Gaussians is represented using a small set of shared one-dimensional Gaussians optimized to best fit the original set of Gaussians. Here, we illustrate 12 two-dimensional Gaussians (solid ellipses). In each dimension, we quantize these to a pool of four shared one-dimensional Gaussians (density plots on axes). The means of these are drawn as a grid (dashed lines), on which the quantized two-dimensional Gaussians (dashed ellipses) can occur only at the intersections. Each quantized two-dimensional Gaussian is constructed from the corresponding pair of one-dimensional Gaussians, one for each feature dimension. In this example, we represent 24 means and variances (12 Gaussians $\times$ 2 dimensions), using 8 means and variances (4 Gaussians $\times$ 2 dimensions).

is introduced, and the feature distribution is assumed to be Gaussian given $a_f^\times$. The mapping from GMM states $c^\times$ to *atoms* $a_f^\times$ is usually constrained to be deterministic:

$$p(\boldsymbol{x}) = \sum_{c^\times} \pi_{c^\times} \mathcal{N}\left(\boldsymbol{x}; \boldsymbol{\mu}_{c^\times}, \boldsymbol{\sigma}^2_{c^\times}\right) \tag{12.66}$$

$$\approx \sum_{c^\times} \pi_{c^\times} \prod_f \mathcal{N}\left(x_f; \mu_{a_f^\times(c^\times)}, \sigma^2_{a_f^\times(c^\times)}\right). \tag{12.67}$$

By design $|a_f^\times| \ll |c^\times|$, so the number of Gaussians per dimension is vastly reduced. Figure 12.12 illustrates the idea. This concept was pioneered in early speech recognizers to reduce the computational load and promote generalization from small training sets [5, 6, 40, 41].

Despite the relatively small number of components $|a_f^\times|$ in each band, taken across bands, BQGMMs are capable of expressing $|a_f^\times|^F$ distinct patterns in an F-dimensional feature space. The computation and storage requirements of a BQGMM relative to its corresponding diagonal-covariance GMM are reduced by approximately a factor of $\frac{|c^\times|}{|a_f^\times|}$. For speech models, this factor is generally on the order of 100 for negligible loss in ASR performance. The computational savings can be even more significant when using factorial models, which in general scale exponentially with the number of acoustic sources that are distinctly modeled. For example, in [38], BQGMMs are used to separate and recognize two simultaneously speaking talkers, and their use speeds up the cost of a full evaluation of the likelihoods by over three orders of magnitude. Importantly, band quantization can be applied to hierarchical acoustic models to reduce their memory footprint, and, depending on the search parameters used, deliver significant additional computational savings [4].

BQGMMs are generally estimated from an existing GMM, by clustering the Gaussians in each dimension using K-means clustering, with the KL-divergence between Gaussians as the distance metric [6]. More generally, BQGMMs can be identified by minimizing the KL-divergence between the BQGMM and an existing GMM. This objective cannot be analytically optimized. An analytic algorithm that uses variational techniques to approximate the KL divergence between GMMs is presented in [36], and was used to construct speech BQGMMs in [38].

12.7 Current Directions

We have reviewed some approaches to handling the problems of intractability in model-based approaches, both at the mathematical level, due to the nonlinearity of feature transforms, and at the computational level, due to the multiplicative number of state combinations for factorial models. We now discuss a few interesting current research directions in model-based robust ASR.

The foregoing has focused on attempts to model signal interaction in feature domains that are known to work well for speech recognition. An alternative is to investigate speech recognition using feature domains in which signal interaction is easily modeled. Approaches to enhancement based on basis decomposition of power spectra attempt to model speech and noise directly in the power spectrum domain [43, 67, 74, 81].

Another direction is to investigate better modeling of speech and noise. The ability to model the noise dynamics is one of the more promising aspects of the model-based compensation framework. We discuss a model with simple linear dynamics on the noise levels that shows strong potential for use within a model-based noise compensation scheme. For more complex noise sources, such as an interfering speaker, noise compensation would be hopeless without complex models of the dynamics of both the target and interfering signals. However, some recent work on factorial HMMs shows that super-human speech recognition is possible and can be performed with far less computation than originally thought [78].

Speech recognition has a history that began with recognition of clean speech, and hence feature optimization has focused on extracting the filtering effects of the vocal tract and eliminating sources of variance that were thought irrelevant to recognition. The voiced parts of speech contain harmonics determined by the pitch, which carry the vocal tract information. However, in non-tonal languages the pitch is largely independent of the words being said. In noise, the situation changes: the harmonics are precisely the frequencies where the SNR is greatest, and so it may be profitable to model the dynamics of pitch along with the vocal tract information, in order to help extract the vocal tract information. Source-filter models also allow the interaction model to operate in the full spectrum, while allowing the recognition part of the model to operate in the filter domain. This type of model has been attempted for speech separation in [34, 46, 56], and for music separation in, for example, [33], and is also used in HMM-based speech synthesis [97].

In the rest of this section, we discuss some of these ideas in more detail. In particular, we discuss dynamic noise models, speech separation with factorial HMMs, and non-negative subspace approaches to signal separation, and their potential use within a speech recognition system.

12.7.1 Dynamic Noise Models for Robust ASR

A fundamental problem in robust ASR (and classification in general) is handling mismatch between training and testing conditions in a highly efficient manner. Maximum Likelihood Linear Regression (MLLR) techniques such as fMLLR, Maximum *a Posteriori* Linear Regression (MAPLR), feature space MAPLR, etc. [10,28,29] are relatively simple, efficient and generally effective approaches to speaker and environmental compensation, and are used (in one form or another) by essentially all state-of-the-art ASR systems today.

However, as we have explained in detail in this chapter, additive noise has a highly nonlinear effect in the log frequency domain. Factorial models of speech and noise can exploit this relationship to learn efficient and representative models of the available training data. The benefits of explicitly modeling canonical variables such as noise are much more pronounced when mismatched data is encountered. Often very little adaptation data is available or very rapid adaptation is preferred. Naturally, an efficient and accurate parameterization of the data can be adapted much more rapidly and can be far more effective than brute-force methods.

The rapid adaptation of a noise model under a factorial representation of noisy speech is an idea with roots tracing back over four decades to early work on front-end denoising using spectral subtraction and Wiener filtering [7, 22]. Speech recognition systems are composed of loosely connected modules: a speech detector, a noise estimator that operates on blocks of data identified as speech-free, and a noise removal system, that produces a speech feature estimate given an estimate of the noise. Ongoing research aims to develop more accurate models of speech, noise, and their interaction, and jointly inferring their configuration under the resulting probability model of the data. More recent, significant work on rapid noise adaptation includes investigations on dynamic forgetting factor algorithms for noise parameter adaptation [2], stochastic online noise parameter adaptation [14], and dynamic noise adaptation (DNA) [71,79].

A distinguishing feature of DNA in this context is that noise is modeled as a random variable with simple dynamics. DNA maintains an approximation to the posterior distribution of the noise rather than a point estimate, which leads to better decisions about what frequency bands are explained by speech versus noise. A limitation of these rapid noise-adaptation techniques is that they generally utilize very simple models of noise that are estimated in online fashion, and maintain no long-term statistics about previously seen data. The use of pre-trained models of noise to detect and reset the DNA noise tracker has been investigated to an extent [79], as has condition detection (CD): the automatic detection of when explicit noise modeling is not beneficial [72]. The latter approach allows for the use of DNA with multi-condition models for the speech model and back-end acoustic models, as is, without any system re-training, and improves the performance of state-of-the-art ASR systems significantly.

Factorial switching models with pre-trained (conditionally) linear dynamical models for speech and noise have also been investigated [18], and are described briefly in Chapter 9. Future work on dynamic noise modeling should focus on efficiently leveraging stronger noise models that incorporate proven adaptation techniques, and incorporating/improving algorithms that have recently been investigated for multi-talker speech recognition (as described directly below), so that more structured acoustic interference, such as secondary speech and music, can be accurately compensated.

12.7.2 Multi-Talker Speech Recognition using Graphical Models

A hallmark of human perception is our ability to solve the auditory cocktail party problem: even when restricted to a single channel, we can direct our attention to a chosen speaker in the presence of interfering speech, and, more often than not, understand what was said remarkably well. A truly exciting direction of current research in factorial modeling for robust ASR has been the use of graphical models to realize super-human speech recognition performance. These techniques have so far utilized HMMs to model each explicitly represented speaker, and combined them with one or more of the interaction models described in this chapter to realize multi-talker speech separation and recognition systems.

A fundamental challenge of multi-talker speech recognition is computational complexity. As discussed in Section 12.6, in general, exact inference involves computing the likelihood of all combinations of the states of the speakers. Exact inference also entails searching the joint (dynamic) state space of the decoders, which also scales exponentially with the number of speakers. In [35, 38, 52], the two-talker system used to outperform human listeners on the PASCAL monaural speech separation and recognition task [11], utilized band quantization (described in Section 12.6.3) to reduce the cost of acoustic labeling by an exponential factor, and joint-state pruning, which, for this well-constrained task, was very effective at controlling the complexity of the joint decoder. In [75], the idea of using loopy belief propagation to iteratively decode the speakers was introduced. This technique reduces the complexity of decoding from exponential to linear in the number of speakers, with negligible loss in recognition performance. Shortly thereafter in [76], the new variational framework for the max model described in Section 12.6.1 was introduced, and used to make inference linear in the number of speakers. Later in [77, 78], this framework was extended so that the complexity of inference could be precisely controlled. The resulting system was able to separate and recognize the speech of up to five speakers talking simultaneously and mixed in a single channel: a remarkable result, considering that the models necessary to describe the data involve trillions of state combinations for each frame.

These recent advances in multi-talker speech recognition are significant, but several important and exciting problems remain. First and foremost, it is important to emphasize that existing algorithms have so far only been tested in reasonably well-constrained scenarios, and artificially mixed data. The enhancement of these techniques to make them suitable for multi-talker recognition of real data streams with significant background noise, channel distortion, and less-constrained speaker vocabularies involves solving many interesting and challenging problems, some of which we discuss briefly below.

For example, to the best of our knowledge, algorithms that select which and how many speakers (or more generally acoustic sources) to explicitly model have yet to be investigated for more than two concurrently active sources. For the case of two sources, a simple method to detect clean conditions is described in [38]. This work, and existing work on speaker segmentation (e.g., [8]) could be used as a starting point for future investigations. Another important direction of future work is to develop representative models of the acoustic background that extract canonical acoustic components that can be composed to explain new, previously unseen test data, and yet do not over-generalize. Current studies include matrix factorization approaches, as described further below, and factorial models based on graphical models with a distributed state representations, such as deep belief networks (DBNs) of restricted Boltzmann machines (RBMs) for ASR [58], and factorial hidden DBNs of RBMs for robust ASR [73].

In addition, relatively little work has been done on probabilistic models for speech separation and recognition that employ multiple channels in a coherent model [3, 12, 70, 84]. With the availability of two or more channels in speech enabled devices rapidly becoming the rule, rather than the exception, it seems inevitable that the best ASR systems will be those that have multi-channel processing capabilities integrated directly into the acoustic scorer and decoder.

12.7.3 Noise Robust ASR using Non-Negative Basis Representations

We have so far shown how tremendous efforts need to be made in order to bring interaction modeling in the domain of the speech recognizer. Another approach to the problem is to try to perform recognition in a domain where the interaction can be conveniently modeled, such as the magnitude or power domain. A promising angle of attack in this direction is to use techniques based on non-negative matrix factorization (NMF) [54]. In the context of audio signal processing, NMF is generally applied to the magnitude or power spectrogram of the signal, with the hope that the non-negative low-rank decomposition thus obtained will extract relevant parts [88].

In NMF approaches, the model for each source in a given frame (a small window of speech, of approximately 40 ms) is defined by a set of weighted non-negative basis functions in the power spectrum (or similar feature space). Inference involves concatenating the basis sets for different sources into a single basis, and solving in parallel for the weights of all sources that best reconstruct the signal. There is also work to include phase explicitly as a parameter [45], which would allow for exact inference of the complete signal.

This type of approach has the advantage of speed because it avoids considering all combinations of basis functions across speakers. It has proven extremely successful, particularly for music signal transcription and source separation [25, 92], as described in more details in Chapter 5 of this book. The original framework has been extended in many directions. One has been to integrate better constraints, such as temporal continuity, into the models while retaining their computational advantages [92, 94]. Another has been to reformulate NMF in a probabilistic framework [9, 93], which enables posterior probabilities and likelihoods to be computed. This also enables NMF to be used as a component in a graphical model such as a speech HMM.

A recent trend of research has, like the speech separation approaches of the previous section, focused on modeling multiple non-stationary sources through factorial models [61,63,65,89]. An exciting new direction takes this idea even further by using nonparametric Bayesian methods to define factorial models with an unbounded number of factors [39, 64]. The beauty of these methods is that, despite their apparent complexity, they are able to acquire models of all of the components of an acoustic scene. This makes them ideally suited to the task of modeling complex and unknown signals. Applying this kind of approach to the speech recognition problem has, to the best of our knowledge, not yet been attempted, but we think that it is a very promising direction for future research.

References

[1] A. Acero, "Acoustical and environmental robustness in automatic speech recognition," PhD dissertation, Carnegie Mellon University, 1990.

[2] M. Afify and O. Siohan, "Sequential noise estimation with optimal forgetting for robust speech recognition," in *Proceedings of the IEEE International Conference on Acoustics, Speech, and Signal Processing (ICASSP)*, vol. 1, May 2001, pp. 229–232.

[3] H. Attias, "New EM algorithms for source separation and deconvolution with a microphone array," in *Proceedings of the IEEE International Conference on Acoustics, Speech, and Signal Processing (ICASSP)*, vol. 5, April 2003, pp. 297–300.

[4] R. Bakis, D. Nahamoo, M. A. Picheny, and J. Sedivy, "Hierarchical labeler in a speech recognition system," US Patent 6023673, 2000.

[5] J. R. Bellegarda and D. Nahamoo, "Tied mixture continuous parameter models for large vocabulary isolated speech recognition," in *Proceedings of the IEEE International Conference on Acoustics, Speech, and Signal Processing (ICASSP)*, vol. 1, May 1989, pp. 13–16.

[6] E. Bocchieri, "Vector quantization for the efficient computation of continuous density likelihoods," in *Proceedings of the IEEE International Conference on Acoustics, Speech, and Signal Processing (ICASSP)*, vol. 2, April 1993, pp. 692–695.

[7] S. F. Boll, "Suppression of acousic noise in speech using spectral subtraction," *IEEE Transactions on Acoustics, Speech, and Signal Processing*, vol. 27, no. 3, pp. 113–120, April 1979.

[8] F. Castaldo, D. Colibro, E. Dalmasso, P. Laface, and C. Vair, "Stream-based speaker segmentation using speaker factors and eigenvoices," in *Proceedings of the IEEE International Conference on Acoustics, Speech, and Signal Processing (ICASSP)*, April 2008, pp. 4133–4136.

[9] A. T. Cemgil, "Bayesian inference in non-negative matrix factorisation models," *Technical Report* CUED/F-INFENG/TR.609, July 2008, University of Cambridge.

[10] C. Chesta, O. Siohan, and C.-H. Lee, "Maximum a posteriori linear regression for HMM adaptation," in *Proceedings of the Interspeech ISCA European Conference on Speech Communication and Technology (Eurospeech)*, 1999.

[11] M. Cooke, J. R. Hershey, and S. J. Rennie, "Monaural speech separation and recognition challenge," *Computer Speech & Language*, vol. 24, no. 1, pp. 1–15, January 2010.

[12] M. Delcroix, K. Kinoshita, T. Nakatani, S. Araki, A. Ogawa, T. Hori, S. Watanabe, M. Fujimoto, T. Yoshioka, T. Oba, Y. Kubo, M. Souden, S.-J. Hahm, and N. A., "Speech recognition in the presence of highly non-stationary noise based on spatial, spectral and temporal speech/noise modeling combined with dynamic variance adaptation," in *Proceedings of the International Workshop on Machine Listening in Multisource Environments (CHiME)*, September 2011.

[13] L. Deng, "Front-end, back-end, and hybrid techniques to noise-robust speech recognition," in *Robust Speech Recognition of Uncertain or Missing Data*, D. Kolossa and R. Haeb-Umbach, Eds. Berlin Heidelberg: Springer-Verlag, 2011, ch. 4, pp. 67–99.

[14] L. Deng, J. Droppo, and A. Acero, "Recursive estimation of nonstationary noise using iterative stochastic approximation for robust speech recognition," *IEEE Transactions on Speech and Audio Processing*, vol. 11, no. 6, pp. 568–580, 2003.

[15] L. Deng, J. Droppo, and A. Acero, "Estimating cepstrum of speech under the presence of noise using a joint prior of static and dynamic features," *IEEE Transactions on Speech and Audio Processing*, vol. 12, no. 3, pp. 218–233, 2004.

[16] L. Deng, J. Droppo, and A. Acero, "Enhancement of log mel power spectra of speech using a phase-sensitive model of the acoustic environment and sequential estimation of the corrupting noise," *IEEE Transactions on Speech and Audio Processing*, vol. 12, no. 2, pp. 133–143, 2004.

[17] V. V. Digalakis, D. Rtischev, and L. G. Neumeyer, "Speaker adaptation using constrained estimation of Gaussian mixtures," *IEEE Transactions on Speech and Audio Processing*, vol. 3, no. 5, pp. 357–366, 1995.

[18] J. Droppo and A. Acero, "Noise robust speech recognition with a switching linear dynamic model," in *Proceedings of the IEEE International Conference on Acoustics, Speech, and Signal Processing (ICASSP)*, vol. 1, May 2004, pp. 953–956.

[19] J. Droppo, A. Acero, and L. Deng, "Uncertainty decoding with SPLICE for noise robust speech recognition," in *Proceedings of the IEEE International Conference on Acoustics, Speech, and Signal Processing (ICASSP)*, vol. 1, May 2002, pp. 57–60.

[20] J. Droppo, L. Deng, and A. Acero, "A comparison of three non-linear observation models for noisy speech features," in *Proceedings of the Interspeech ISCA European Conference on Speech Communication and Technology (Eurospeech)*, 2003, pp. 681–684.

[21] G. D. Durgin, T. S. Rappaport, and D. A. D. Wolf, "New analytical models and probability density functions for fading in wireless communications," *IEEE Transactions on Communications*, vol. 50, no. 6, pp. 1005–1015, 2002.

[22] ETSI, "Speech processing, transmission and quality aspects (STQ); distributed speech recognition; advanced front-end feature extraction algorithm; compression algorithms," *ETSI TR ES 202 050 VERSION 1.1.3*, 2003.

[23] M. Evans, N. Hastings, and B. Peacock, *Statistical distributions*, 3rd ed. New York: Wiley-Interscience, 2000.

[24] L. F. Fenton, "The sum of lognormal probability distributions in scatter transmission systems," *IRE Transactions on Communication Systems*, vol. CS-8, pp. 57–67, 1960.

[25] C. Févotte, "Itakura-Saito non-negative factorizations of the power spectrogram for music signal decomposition," in *Machine Audition: Principles, Algorithms and Systems*, W. Wang, Ed. IGI Global Press, August 2010, Ch. 11.

[26] B. J. Frey, L. Deng, A. Acero, and T. T. Kristjansson, "ALGONQUIN: Iterating Laplace's method to remove multiple types of acoustic distortion for robust speech recognition," in *Proceedings of the Interspeech ISCA European Conference on Speech Communication and Technology (Eurospeech)*, September 2001, pp. 901–904.

[27] B. J. Frey, T. T. Kristjansson, L. Deng, and A. Acero, "ALGONQUIN - learning dynamic noise models from noisy speech for robust speech recognition," in *Advances in Neural Information Processing Systems 14 (NIPS 2001)*, T. G. Dietterich, S. Becker, and Z. Ghahramani, Eds. Cambridge, Massachusetts: MIT Press, 2002, pp. 1165–1171.

[28] M. J. F. Gales, "Model-based techniques for noise robust speech recognition," PhD dissertation, University of Cambridge, September 1995.

[29] M. J. F. Gales, "Maximum likelihood linear transformations for HMM-based speech recognition," *Computer Speech & Language*, vol. 12, pp. 75–98, January 1998.

[30] M. J. F. Gales, "Model-based approaches to handling uncertainty," in *Robust Speech Recognition of Uncertain or Missing Data*, D. Kolossa and R. Haeb-Umbach, Eds. Berlin Heidelberg:Springer Verlag, 2011, Ch. 5, pp. 101–125.

[31] M. J. F. Gales and R. C. van Dalen, "Predictive linear transforms for noise robust speech recognition," in *Proceedings of the IEEE Workshop on Automatic Speech Recognition and Understanding (ASRU)*, 2007, pp. 59–64.

[32] J.-L. Gauvain and C.-H. Lee, "Maximum a posteriori estimation for multivariate Gaussian mixture observations of Markov chains," *IEEE Transactions on Speech and Audio Processing*, vol. 2, no. 2, pp. 291–298, April 1994.

[33] T. Heittola, A. Klapuri, and T. Virtanen, "Musical instrument recognition in polyphonic audio using source-filter model for sound separation," in *Proceedings of the International Society for Music Information Retrieval Conference (ISMIR)*, November 2009.

[34] J. R. Hershey and M. Casey, "Audio-visual sound separation via hidden Markov models." in *Advances in Neural Information Processing Systems 14 (NIPS 2001)*, T. G. Dietterich, S. Becker, and Z. Ghahramani, Eds. Cambridge, Massachusetts: MIT Press, 2001, pp. 1173–1180.

[35] J. R. Hershey, T. T. Kristjansson, S. J. Rennie, and P. A. Olsen, "Single channel speech separation using factorial dynamics," in *Advances in Neural Information Processing Systems 19 (NIPS 2006)*, B. Schölkopf, J. Platt, and T. Hoffman, Eds. Cambridge, Massachusetts: MIT Press, 2007, pp. 593–600.

[36] J. R. Hershey and P. A. Olsen, "Approximating the Kullback Leibler divergence between Gaussian mixture models," in *Proceedings of the IEEE International Conference on Acoustics, Speech, and Signal Processing (ICASSP)*, April 2007.

[37] J. R. Hershey, P. A. Olsen, and S. J. Rennie, "Signal interaction and the devil function," in *Proceedings of the Interspeech ISCA International Conference on Spoken Language Processing (ICSLP)*, September 2010.

[38] J. R. Hershey, S. J. Rennie, P. A. Olsen, and T. T. Kristjansson, "Super-human multi-talker speech recognition: a graphical modeling approach," *Computer Speech & Language*, vol. 24, no. 1, pp. 45–66, January 2010.

[39] M. D. Hoffman, D. M. Blei, and P. R. Cook, "Bayesian nonparametric matrix factorization for recorded music," in *Proceedings of the International Conference on Machine Learning (ICML)*, 2010.

[40] X. Huang and M. Jack, "Semi-continuous hidden Markov models for speech signals," *Computer Speech & Language*, vol. 3, no. 3, pp. 239–251, 1989.

[41] M.-Y. Hwang, “Sub-phonetic acoustic modeling for speaker-independent continuous speech recognition,” PhD dissertation, Carnegie Mellon University, December 1993.
[42] S. Ikeda and N. Murata, "A method of ICA in time-frequency domain," in *Proceedings of the International Conference on Independent Component Analysis and Blind Source Separation (ICA)*, 1999, pp. 365–370.
[43] G.-J. Jang and T.-W. Lee, “A maximum likelihood approach to single-channel source separation,” *Journal of Machine Learning Research*, vol. 4, pp. 1365–1392, 2003.
[44] M. I. Jordan, Z. Ghahramani, T. S. Jaakkola, and L. K. Saul, “An introduction to variational methods for graphical models,” *Machine Learning*, vol. 37, no. 2, pp. 183–233, November 1999.
[45] H. Kameoka, N. Ono, K. Kashino, and S. Sagayama, “Complex NMF: a new sparse representation for acoustic signals,” in *Proceedings of the IEEE International Conference on Acoustics, Speech, and Signal Processing (ICASSP)*, April 2009, pp. 3437–3440.
[46] H. Kameoka, N. Ono, and S. Sagayama, “Speech spectrum modeling for joint estimation of spectral envelope and fundamental frequency,” *IEEE Transactions on Audio, Speech and Language Processing*, vol. 18, no. 6, pp. 1507–1516, August 2010.
[47] D. K. Kim and M. J. F. Gales, “Noisy constrained maximum likelihood linear regression for noise robust speech recognition,” *IEEE Transactions on Audio, Speech, and Language Processing*, vol. 19, no. 2, pp. 315–325, February 2011.
[48] M. Kowalski, E. Vincent, and R. Gribonval, “Beyond the narrowband approximation: wideband convex methods for under-determined reverberant audio source separation,” *IEEE Transactions on Audio, Speech and Language Processing*, vol. 18, no. 7, pp. 1818–1829, September 2010.
[49] T. T. Kristjansson, “Speech recognition in adverse environments,” PhD dissertation, University of Waterloo, 2002.
[50] T. T. Kristjansson, H. Attias, and J. R. Hershey, “Single microphone source separation using high resolution signal reconstruction,” in *Proceedings of the IEEE International Conference on Acoustics, Speech, and Signal Processing (ICASSP)*, May 2004, pp. 817–820.
[51] T. T. Kristjansson and R. Gopinath, “Cepstrum domain Laplace denoising,” 2005. Available at: http://www.research.ibm.com/people/r/rameshg/kristjansson-icassp2005.pdf.
[52] T. T. Kristjansson, J. R. Hershey, P. A. Olsen, S. J. Rennie, and R. Gopinath, “Super-human multi-talker speech recognition: The IBM 2006 speech separation challenge system,” in *Proceedings of the Interspeech ISCA International Conference on Spoken Language Processing (ICSLP)*, 2006.
[53] H. Lane and B. Tranel, “The Lombard sign and the role of hearing in speech,” *Journal of Speech and Hearing Research*, vol. 14, no. 4, p. 677, 1971.
[54] D. D. Lee and H. S. Seung, "Learning the parts of objects by non-negative matrix factorization," *Nature*, vol. 401, pp. 788–791, October 1999.
[55] C. J. Leggetter and P. C. Woodland, “Maximum likelihood linear regression for speaker adaptation of HMMs,” *Computer Speech and Language*, vol. 9, pp. 171–186, 1995.
[56] J. Le Roux, H. Kameoka, N. Ono, A. de Cheveigné, and S. Sagayama, “Single channel speech and background segregation through harmonic-temporal clustering,” in *Proceedings of the IEEE Workshop on Applications of Signal Processing to Audio and Acoustics (WASPAA)*, October 2007, pp. 279–282.
[57] H. Liao and M. J. F. Gales, “Joint uncertainty decoding for noise robust speech recognition,” in *Proceedings of the Interspeech ISCA International Conference on Spoken Language Processing (ICSLP)*, 2005.
[58] A. Mohamed, G.E. Dahl, and G. Hinton, “Acoustic modeling using deep belief networks,” *IEEE Transactions on Audio, Speech and Language Processing*, vol. 20, no. 1, pp. 14–22, January 2012.
[59] B. C. J. Moore, *An Introduction to the Psychology of Hearing*, 5th ed. London: Academic Press, 2003.
[60] P. J. Moreno, B. Raj, and R. M. Stern, “A vector Taylor series approach for environment-independent speech recognition,” in *Proceedings of the IEEE International Conference on Acoustics, Speech, and Signal Processing (ICASSP)*, vol. 2, May 1996, pp. 733–736.
[61] G. J. Mysore, P. Smaragdis, and B. Raj, “Non-negative hidden Markov modeling of audio with application to source separation,” in *Proceedings of the International Conference on Latent Variable Analysis and Signal Separation (LVA/ICA)*, 2010, pp. 140–148.
[62] A. Nádas, D. Nahamoo, and M. A. Picheny, “Speech recognition using noise-adaptive prototypes,” *IEEE Transactions on Acoustics, Speech, and Signal Processing*, vol. 37, no. 10, pp. 1495–1503, October 1989.
[63] M. Nakano, J. Le Roux, H. Kameoka, N. Ono, and S. Sagayama, “Infinite-state spectrum model for music signal analysis,” in *Proceedings of the IEEE International Conference on Acoustics, Speech, and Signal Processing (ICASSP)*, May 2011, pp. 1972–1975.

[64] M. Nakano, J. Le Roux, H. Kameoka, T. Nakamura, N. Ono, and S. Sagayama, "Bayesian nonparametric spectrogram modeling based on infinite factorial infinite hidden Markov model," in *Proceedings of the IEEE Workshop on Applications of Signal Processing to Audio and Acoustics (WASPAA)*, October 2011.

[65] A. Ozerov, C. Févotte, and M. Charbit, "Factorial scaled hidden Markov model for polyphonic audio representation and source separation," in *Proceedings of the IEEE Workshop on Applications of Signal Processing to Audio and Acoustics (WASPAA)*, October 2009.

[66] M. H. Radfar, A. H. Banihashemi, R. M. Dansereau, and A. Sayadiyan, "Nonlinear minimum mean square error estimator for mixture-maximisation approximation," *Electronics Letters*, vol. 42, no. 12, pp. 724–725, June 2006.

[67] B. Raj and P. Smaragdis, "Latent variable decomposition of spectrograms for single channel speaker separation," in *Proceedings of the IEEE Workshop on Applications of Signal Processing to Audio and Acoustics (WASPAA)*, October 2005, pp. 17–20.

[68] B. Raj and R. M. Stern, "Missing-feature approaches in speech recognition," *IEEE Signal Processing Magazine*, vol. 22, no. 5, pp. 101–116, September 2005.

[69] C. K. Raut, T. Nishimoto, and S. Sagayama, "Model composition by Lagrange polynomial approximation for robust speech recognition in noisy environment," in *Proceedings of the Interspeech ISCA International Conference on Spoken Language Processing (ICSLP)*, 2004, pp. 2809–2812.

[70] S. J. Rennie, P. Aarabi, T. T. Kristjansson, B. Frey, and K. Achan, "Robust variational speech separation using fewer microphones than speakers," in *Proceedings of the IEEE International Conference on Acoustics, Speech, and Signal Processing (ICASSP)*, vol. 1, April 2003, pp. 88–91.

[71] S. J. Rennie, P. Dognin, and P. Fousek, "Robust speech recognition using dynamic noise adaptation," in *Proceedings of the IEEE International Conference on Acoustics, Speech, and Signal Processing (ICASSP)*, May 2011.

[72] S. J. Rennie, P. Fousek, and P. Dognin, "Matched-condition robust dynamic noise adaptation," in *Proceedings of the IEEE Workshop on Automatic Speech Recognition and Understanding (ASRU)*, December 2011.

[73] S. J. Rennie, P. Fousek, and P. Dognin, "Factorial hidden restricted Boltzmann machines for robust ASR," in *Proceedings of the IEEE International Conference on Acoustics, Speech, and Signal Processing (ICASSP)* March 2012, pp. 4297–4300.

[74] S. J. Rennie, J. R. Hershey, and P. A. Olsen, "Efficient model-based speech separation and denoising using non-negative subspace analysis," in *Proceedings of the IEEE International Conference on Acoustics, Speech, and Signal Processing (ICASSP)*, April 2008, pp. 1833–1836.

[75] S. J. Rennie, J. R. Hershey, and P. A. Olsen, "Single-channel speech separation and recognition using loopy belief propagation," in *Proceedings of the IEEE International Conference on Acoustics, Speech, and Signal Processing (ICASSP)*, April 2009, pp. 3845–3848.

[76] S. J. Rennie, J. R. Hershey, and P. A. Olsen, "Variational loopy belief propagation for multi-talker speech recognition," in *Proceedings of the Interspeech ISCA European Conference on Speech Communication and Technology (Eurospeech)*, September 2009, pp. 1331–1334.

[77] S. J. Rennie, J. R. Hershey, and P. A. Olsen, "Hierarchical variational loopy belief propagation for multi-talker speech recognition," in *Proceedings of the IEEE Workshop on Automatic Speech Recognition and Understanding (ASRU)*, December 2009, pp. 176–181.

[78] S. J. Rennie, J. R. Hershey, and P. A. Olsen, "Single-channel multitalker speech recognition," *IEEE Signal Processing Magazine*, vol. 27, no. 6, pp. 66–80, 2010.

[79] S. J. Rennie, T. T. Kristjansson, P. A. Olsen, and R. Gopinath, "Dynamic noise adaptation," in *Proceedings of the IEEE International Conference on Acoustics, Speech, and Signal Processing (ICASSP)*, vol. 1, 2006, pp. 1197–1200.

[80] S. T. Roweis, "Factorial models and refiltering for speech separation and denoising," in *Proceedings of the Interspeech ISCA European Conference on Speech Communication and Technology (Eurospeech)*, September 2003, pp. 1009–1012.

[81] M. N. Schmidt and R. K. Olsson, "Single-channel speech separation using sparse non-negative matrix factorization," in *Proceedings of the Interspeech ISCA International Conference on Spoken Language Processing (ICSLP)*, September 2006, pp. 2614–2617.

[82] S. Schwartz and Y. Yeh, "On the distribution function and moments of power sums with lognormal components," *Bell System Technical Journal*, vol. 61, pp. 1441–1462, 1982.

[83] M. L. Seltzer, B. Raj, and R. M. Stern, "A Bayesian framework for spectrographic mask estimation for missing feature speech recognition," *Speech Communication*, vol. 43, no. 4, pp. 370–393, 2004.

[84] M. L. Seltzer, "Microphone array processing for robust speech recognition," PhD dissertation, Carnegie Mellon University, 2003.

[85] M. L. Seltzer, A. Acero, and K. Kalgaonkar, "Acoustic model adaptation via linear spline interpolation for robust speech recognition," in *Proceedings of the IEEE International Conference on Acoustics, Speech, and Signal Processing (ICASSP)*, 2010, pp. 4550–4553.

[86] Y. Shinohara and M. Akamine, "Bayesian feature enhancement using a mixture of unscented transformation for uncertainty decoding of noisy speech," in *Proceedings of the IEEE International Conference on Acoustics, Speech, and Signal Processing (ICASSP)*, 2009, pp. 4569–4572.

[87] P. Smaragdis, "Blind separation of convolved mixtures in the frequency domain," *Neurocomputing*, vol. 22, no. 1, pp. 21–34, 1998.

[88] P. Smaragdis and J. C. Brown, "Non-negative matrix factorization for polyphonic music transcription," in *Proceedings of the IEEE Workshop on Applications of Signal Processing to Audio and Acoustics (WASPAA)*, 2003, pp. 177–180.

[89] P. Smaragdis and B. Raj, "The Markov selection model for concurrent speech recognition," in *Proceedings of the IEEE International Workshop on Machine Learning for Signal Processing (MLSP)*, 2010, pp. 214–219.

[90] R. C. van Dalen and M. J. F. Gales, "Asymptotically exact noise-corrupted speech likelihoods," in *Proceedings of the Interspeech ISCA International Conference on Spoken Language Processing (ICSLP)*, 2010.

[91] A. P. Varga and R. K. Moore, "Hidden Markov model decomposition of speech and noise," in *Proceedings of the IEEE International Conference on Acoustics, Speech, and Signal Processing (ICASSP)*, April 1990, pp. 845–848.

[92] T. Virtanen, "Monaural sound source separation by non-negative matrix factorization with temporal continuity and sparseness criteria," *IEEE Transactions on Audio, Speech and Language Processing*, vol. 15, no. 3, pp. 1066–1074, March 2007.

[93] T. Virtanen, A. T. Cemgil, and S. Godsill, "Bayesian extensions to non-negative matrix factorisation for audio signal modelling," in *Proceedings of the IEEE International Conference on Acoustics, Speech, and Signal Processing (ICASSP)*, April 2008, pp. 1825–1828.

[94] K. W. Wilson, B. Raj, and P. Smaragdis, "Regularized non-negative matrix factorization with temporal dependencies for speech denoising," in *Proceedings of the Interspeech ISCA International Conference on Spoken Language Processing (ICSLP)*, September 2008, pp. 411–414.

[95] J. Wu, N. B. Mehta, and J. Zhang, "A flexible lognormal sum approximation method," in *Proceedings of the IEEE Global Telecommunications Conference (Globecom)*, December 2005, pp. 3413–3417.

[96] M. Xu, M. J. F. Gales, and K. K. Chin, "Improving joint uncertainty decoding performance by predictive methods for noise robust speech recognition," in *Proceedings of the IEEE Workshop on Automatic Speech Recognition and Understanding (ASRU)*, 2009.

[97] H. Zen, K. Tokuda, and A. Black, "Statistical parametric speech synthesis," *Speech Communication*, vol. 51, no. 11, pp. 1039–1064, 2009.

13

Acoustic Model Training for Robust Speech Recognition

Michael L. Seltzer
Microsoft Research, USA

13.1 Introduction

Traditionally, researchers working on the field of noise robustness have focused their efforts on two areas: front-end enhancement and model compensation. Front-end enhancement encompasses a variety of signal and feature processing methods, such as those discussed in Chapters 4 and 9, that are designed to remove distortions in the speech caused by the acoustic environment [10, 30, 37]. On the other hand, model compensation, described in Chapters 11 and 12, alters the parameters of the speech recognizer's acoustic models to better match the characteristics of the current environment [13, 17, 32]. There is a rich literature in both of these areas that has led to improvements in speech-recognition performance over the years [14].

While all of this effort is focused on noise compensation at runtime, relatively little attention has been paid to the manner in which the speech-recognition systems are trained. Almost all of the robustness algorithms assume, either implicitly or explicitly, that the recognizer has been trained from clean speech, and the job of a noise-robustness technique is to reduce the mismatch between the clean acoustic models and the noisy speech. As a result, performance is determined by how well the captured speech is denoised or how well the clean acoustic models adapt to the environment of the test utterance. However, there are many reasons why this is suboptimal. First, from a theoretical point of view, speech recognizers perform best when the speech used to train the recognizer is close to that seen in deployment. Using clean acoustic models creates the most severe mismatch between training and testing environments. In addition, it is simply impractical to require a system to be trained on clean speech data. Collecting such data requires careful recordings which are time consuming and expensive. Even more importantly, it means that data collected in the field cannot be used to train the system.

Techniques for Noise Robustness in Automatic Speech Recognition, First Edition.
Edited by Tuomas Virtanen, Rita Singh, and Bhiksha Raj.

Alternatively, acoustic models can be trained on on noisy speech data. Because training models using matched data is usually impractical, systems are often trained with data collected in a variety of environmental conditions. While these multicondition systems generally have better performance than a system trained from clean speech, there are drawbacks here as well. First, by using such a diverse set of acoustic data, additional variability is introduced into the model that may not be pertinent to the speech-recognition task. In addition, because the model now represents an average of noisy conditions, it is suboptimal for any one particular environment, and may not generalize well to noises unseen in training.

Recently, a new adaptive training approach for robust speech recognition has been proposed that enables the use of multicondition training data in combination with front-end enhancement [5] or model compensation [18,19,28]. This training paradigm is called noise adaptive training (NAT) and produces acoustic models that have less unwanted variability than multistyle models and do a better job generalizing to unseen data. NAT is inspired by the success of speaker adaptive training (SAT). The main principle behind SAT is that the same speaker adaptation algorithm used at runtime is also applied to the training data. The resulting acoustic models have less interspeaker variability and are therefore optimal (in the maximum likelihood sense) for that particular adaptation strategy. SAT has consistently shown improvements over conventional speaker-independent training when speaker adaptation is used at runtime and is a standard part of many large vocabulary-recognition systems.

This chapter is organized as follows. In Section 13.2, we first review the traditional methods for training speech-recognition systems for noisy environments and compare their performance. We then briefly review SAT in Section 13.3, as it provides many of the foundational elements helpful to understanding NAT. In Section 13.4, we introduce feature-space NAT and evaluate its performance using two different front-end compensation algorithms. NAT in the model domain is then presented in Section 13.5 and a detailed look at its implementation using vector Taylor series (VTS) adaptation is shown in Section 13.6. The performance of NAT with both front-end and model-based compensation is evaluated in a series of experiments on simulated and actual noisy speech. The advantages and disadvantages of the methods described in this chapter are discussed in Section 13.7 highlighting additional related work and open challenges. Finally, the main ideas of the chapter are summarized in Section 13.8.

13.2 Traditional Training Methods for Robust Speech Recognition

There are three basic paradigms for data selection to train a speech-recognition system. The first and most common is to train the system with clean speech signals. These signals are typically recorded in a quiet environment with a close-talking microphone and contain minimal additive noise or reverberation. The motivation for this style of data is that it represents the ideal speech signal and that noisy speech can be cleaned in some manner. This approach results in the most severe mismatch between the acoustic models and the observed noisy speech and places a strong emphasis on the noise-compensation algorithms.

Another approach to training uses so-called matched data that originates from the same environment as the test data. Because of the sheer variety of environmental factors including additive noise, reverberation, speaker-to-microphone distance, signal-to-noise ratio, and microphone type, matched training is impractical to use in a real system. However, such systems are valuable for experimental purposes, as they demonstrate the upper bound in performance in a given environment. It is frequently convenient to train a model for matched conditions

Table 13.1 Word error rates for speech from the WSJ corpus corrupted by white noise using clean, multicondition, and matched training data. Enhancement using spectral subtraction (SS) only provides modest improvements when the models are trained with clean speech.

Training data	5 dB	10 dB	15 dB	20 dB	Clean	Avg
Clean	87.11	55.06	19.76	10.02	4.87	35.36
Clean, SS-test only	75.30	33.79	13.29	8.05	4.65	27.02
Noisy multicondition	28.91	14.84	10.45	7.53	6.09	13.56
Noisy matched	25.41	14.03	8.94	7.05	4.87	12.06

using *single pass retraining* which is a method for mapping an acoustic model trained with one parameterization of the speech signal into another acoustic model based on a different parameterization [36]. In the context of matched training, this technique can be used to map a clean acoustic model to a noisy acoustic model given clean training data and a parallel corpus of noisy training data.

Because matched condition training is impractical, a more general approach called multistyle or multicondition training has been proposed [29]. Multistyle training was first proposed to improve the recognizer's robustness to different speaking styles and was later proposed as means of improved noise robustness [1]. In the same way that training data from many different speakers is pooled to create a speaker independent recognizer, multicondition systems pool data from many different environments in an attempt to create an environment-independent system. For example, for an automotive application, a multistyle model may consist of speech captured at a variety of speeds, in a variety of cars, and with various dashboard instruments turned on. All of this data is pooled to create an acoustic model that captures the average statistics of these environments. They key benefit of multistyle training over clean training is that there is far less mismatch between the acoustic model and the noisy speech seen in deployment. In addition, multicondition systems can be trained from speech collected from the real world, where clean conditions almost never exist.

The performance of these three training methods is shown Table 13.1. The table shows the word error rate (WER) obtained on a noise-corrupted version of the *Wall Street Journal* (WSJ) corpus [33]. In these experiments, the WSJ test set was corrupted by white noise at different SNRs. For the multicondition training, the clean WSJ training set was corrupted by white noise at a variety of SNRs. For matched condition training, a separate acoustic model was created for each test SNR. The table also shows the performance of the system trained from clean speech when the test data is enhanced using spectral subtraction (SS) [4], described in Section 4.5.1. As the results indicate, the performance of the mismatched clean system really degrades at low SNRs. When the test data is denoised, significant gains are obtained. However, much larger gains are obtained using multicondition training, even in the absence of any noise compensation. Note that multicondition training degrades the performance on clean speech, which is to be expected given that the model contains substantially more environmental variability. Of course, the best performance is obtained with matched training.

13.3 A Brief Overview of Speaker Adaptive Training

Much of the motivation for NAT comes from the success of SAT, which creates acoustic models with less interspeaker variability [3]. In traditional speaker-independent acoustic model training, speech data from many speakers are pooled to create a single model that maximizes the likelihood on the training data. However, it is clear that while the pooled-speaker model may be a good model for some average speaker, it is suboptimal for any one particular speaker. This can be interpreted as a version of multistyle training where the primary source of variability comes from the speakers themselves. To combat this unwanted variability at runtime, adaptation techniques such as MLLR can be applied to the acoustic models to better match the user's speech, as described in Chapter 11. The idea of SAT is to employ the same speaker adaptation to the training data so that the speaker variability in the training data can be accounted for by the adaptation transforms rather than the model parameters. This produces a more compact model that captures more of the desired phonetic variation and less interspeaker variability. SAT was originally proposed using maximum likelihood linear regression (MLLR) for mean adaptation [22]. In MLLR, the adapted mean vector can be written as

$$\boldsymbol{\mu}_{qi}^{(r)} = \mathbf{A}^{(r)}\boldsymbol{\mu}_{qi} + \mathbf{b}^{(r)}, \tag{13.1}$$

where $\boldsymbol{\mu}_{qi}$ is the mean vector associated with the Gaussian i in HMM state q, $\{\mathbf{A}^{(r)}, \mathbf{b}^{(r)}\}$ define the MLLR transform for speaker r, and $\boldsymbol{\mu}_{qi}^{(r)}$ is the speaker-adapted mean. Under this model, the auxiliary function used to derive the expectation maximization (EM) update formulae for hidden Markov model (HMM) training must be augmented with the speaker transforms as follows:

$$Q(\Phi_W, \Lambda_S, \bar{\Phi}_W, \bar{\Lambda}_S) = \sum_{r,t,q,i} \gamma_{tqi}^{(r)} \log(\mathcal{N}(\mathbf{s}_t^{(r)}; \mathbf{A}^{(r)}\boldsymbol{\mu}_{qi} + \mathbf{b}^{(r)}, \boldsymbol{\Theta}_{qi})), \tag{13.2}$$

where Λ_S represents the complete set of HMM parameters, Φ_W represents the set of MLLR transforms for all speakers, t, r, q, and i represent the frame, speaker, state, and Gaussian index, respectively, and $\gamma_{tqi}^{(r)}$ is the posterior probability of Gaussian i in state q at frame t from speaker r. The feature vector at frame t for speaker r is represented by $\mathbf{s}_t^{(r)}$ and the variance of Gaussian i in state q is given $\boldsymbol{\Theta}_{qi}$. Throughout this chapter, the current value of a variable to be estimated in an iterative algorithm is indicated with a bar, for example $\bar{\Lambda}_S$.

Using this objective function, the parameters of the HMM and the speaker transforms are optimized in an iterative fashion; first the speaker transforms are updated while the HMM parameters are fixed, then the HMM parameters are updated while the speaker transforms are fixed. This process continues until the likelihood of the training set converges. For example, for a fixed set of speaker transforms, the mean update formula under SAT is

$$\boldsymbol{\mu}_{qi} = \left(\sum_{r,t} \gamma_{tqi}^{(r)} \mathbf{A}^{(r)T} \boldsymbol{\Theta}_{qi}^{-1} \mathbf{A}^{(r)}\right)^{-1} \left(\sum_{r,t} \gamma_{tqi}^{(r)} \mathbf{A}^{(r)T} \boldsymbol{\Theta}_{qi}^{-1} (\mathbf{s}_t^{(r)} - \mathbf{b}^{(r)})\right). \tag{13.3}$$

While SAT was originally proposed for MLLR in the model domain, most current systems used constrained MLLR (CMLLR) which can be implemented as a feature-space transform [8, 12]. SAT experiments reported in [3] showed gains of about 20% reduction in word error rate by performing model adaptation on models trained using SAT compared to conventional speaker-independent models. Clearly, training the model in a way that enables it to be aware of the processing that will be applied at test time provides significant gains. In the remainder of this chapter, we discuss how these ideas can be exploited for training models that are robust to environmental distortions.

13.4 Feature-Space Noise Adaptive Training

There is a large literature of techniques designed to remove distortions from the speech waveform, spectrum, or features. The vast majority of these methods are concerned with additive noise and linear filtering. In typical operation, one of these methods is used to enhance the incoming speech which is then passed to the recognizer that has been trained on clean speech for decoding. The implicit assumption made by this approach is that the methods performed perfect enhancement and there is no residual mismatch in between the enhanced test data and the clean acoustic models. Of course, this is not true. All algorithms have some level of distortion, artifacts, musical nose, or residual noise in the output.

If we put aside the mathematics for a moment, the basic principle of SAT is to apply the same adaptation algorithm during training that will be applied at test time in order to remove the same sources of variability from both the training and test data. For recognizing noisy speech, the obvious analog to speaker adaptation is noise compensation. The different speakers and the variability they introduce are replaced by different environments and the distortions they cause. Therefore, whatever enhancement algorithm will be used during deployment should be run at training time. Of course, in the same way that practical speaker-independent systems are obtained from training data with a variety of speakers, NAT is best performed with noise-corrupted speech that includes the types of distortions and levels expected to be seen in deployment. Of course, if you know that the system will only be used in specific environments, for example in a car, then the types of data used for training can reflect this.

Thus, the basic recipe for *feature-space noise adaptive training* (fNAT) is to select an enhancement algorithm that you believe will be effective in the deployed environment, and apply this algorithm to a collection of multicondition training data that contains much of the same environmental distortions [5], as shown in Figure 13.1. By doing so, the variability caused by the environmental distortion will be removed from the the training data in exactly the same was as the test data. This match between training and test is critical because no enhancement algorithm can generate perfectly clean speech. Any errors or artifacts caused by the algorithm will be generated in the training data, and therefore modeled by the HMMs.

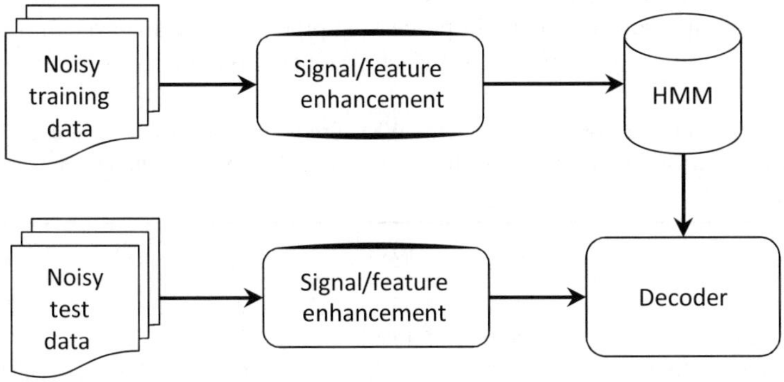

Figure 13.1 Flowchart of feature-space noise-adaptive training (fNAT).

13.4.1 Experiments using fNAT

The performance of fNAT was evaluated on the same noisy WSJ task described in Section 13.2. As before, the acoustic models were trained with clean, multicondition, or matched data. However, in these experiments the noisy training data in the multicondition and matched experiments were also processed using spectral subtraction [4]. As you can see, applying fNAT results a significant gain in performance compared to using clean training data. In addition, if we compare the results in Table 13.1 to those in Table 13.2, we can see that using fNAT even outperforms matched training.

In these experiments, the multicondition training data only varied the level of the noise but not the type. While this assumption may be realistic in some scenarios, there are far more cases where both the type *and* level of noise are unknown. In addition, it is perhaps also reasonable to ask whether these results hold when a more state-of-the-art enhancement algorithm is used. To address these questions, experiments were performed using the Aurora 2 corpus. Aurora 2 consists of clean speech recordings of connected digits degraded with eight types of noise artificially added at signal-to-noise ratios (SNR) varying from −5 to 20 dB and channel distortion [16]. Three test sets provided with the task are contaminated with noise types seen in the training data (Set A), unseen in the training data (Set B), and with additive noise plus channel distortion (Set C). The acoustic models used in these experiments were trained using the standard "complex back-end" Aurora 2 recipe [34]. An HMM with 16 states per digit and a mixture of 20 Gaussians per state is created for each digit as a whole word. In addition, a three state silence model with a mixture of 36 Gaussians per state and a one state short pause model which is tied to the middle state of silence model are used.

In the fNAT experiments, feature enhancement was performed using the ETSI Advanced Front-End (AFE) [30] described in Section 4.5.4. The AFE is a good representation of the state-of-the-art in feature enhancement on this task. The results for clean training and multistyle training are shown in Table 13.3. As the results indicate, applying the AFE to the test data reduces the mismatch to the clean trained models, and significantly improves performance. However, here too, this system is outperformed by simple multicondition training without any additional signal or feature processing. If we used the fNAT approach and apply the AFE to the multicondition training data, a significant gain in performance is obtained. These experiments demonstrate that fNAT in conjunction with multicondition training data has significant advantages over a conventional approach where clean speech is used.

It should be noted that the fNAT can include not just the speech enhancement algorithm but any and all components of the audio-processing pipeline used in a system. In a deployed interactive system, this may include acoustic echo cancellation, microphone array processing,

Table 13.2 Word error rates for speech from the WSJ corpus corrupted by white noise obtained by enhancing the test data using spectral subtraction (SS). For fNAT, spectral subtraction was also applied to the training data.

Training data	5 dB	10 dB	15 dB	20 dB	Clean	Avg
Clean, SS-test	75.30	33.79	13.29	8.05	4.65	27.02
Noisy multicondition	28.91	14.84	10.45	7.53	6.09	13.56
Noisy multicondition, fNAT-SS	20.90	12.22	8.86	7.35	6.57	11.18
Noisy matched, fNAT-SS	21.94	11.74	8.60	6.76	5.02	10.81

Table 13.3 Word accuracy for Aurora 2 using clean and multicondition training sets. The test data was enhanced using the ETSI advanced front-end (AFE). The AFE was also applied to the training data in the fNAT experiment.

Training data	Set A	Set B	Set C	Avg
Clean	60.43	55.85	69.01	60.31
Clean, AFE-test	89.27	87.92	88.53	88.58
Noisy multicondition	91.68	89.74	88.91	90.35
Noisy multicondition, fNAT-AFE	93.74	93.26	92.21	93.24

noise suppression, double-talk detection, and barge-in. For example, in acoustic cancellation, the undesired loudspeaker signal often leaks into the microphone signal when the echo path changes, for example when the talker moves significantly. The harm to recognition accuracy caused by these artifacts can be mitigated if the recognizer is trained on such data.

While fNAT is appealing because it is easy to implement and has been shown to be very effective, it has the drawback that there is little theoretical foundation for it. In traditional SAT, both the adaptation parameters and model parameters are optimized jointly under a common objective function using a maximum likelihood criterion. In fNAT, there is no such framework. Another disadvantage of fNAT is that it relies on point estimates of the clean speech features made by the enhancement algorithm. Using point-estimates implicitly assumes that all training data is equally informative. There is no ability for the front-end to communicate to the model training process any notion of uncertainty about a particular feature. While feature-enhancement methods that generate estimates of uncertainty have been proposed [9], problems with this approach at low SNRs have been discovered [27]. More details about these *uncertainty decoding* methods are discussed in Chapter 17. In Section 13.5, we introduce a model-domain version of NAT that addresses these issues.

13.5 Model-Space Noise Adaptive Training

In NAT in the model space, we seek to jointly optimize the parameters of the HMM and the parameters of a noise compensation algorithm using a set of noise-corrupted training data. Let us assume that there are J utterances in the multicondition training set $\mathcal{X} = \{X^{(j)}\}_{j-1}^{J}$, where $X^{(j)} = \{\mathbf{x}_t^{(j)}\}_{t=1}^{T_j}$ is a sequence of T_j observations corresponding to the jth utterance. We further assume that each utterance $X^{(j)}$ in the training set has an associated distortion model with parameters $\phi^{(j)}$ that represent hidden variables that describe the environment, for example the additive noise and the channel.

In traditional maximum likelihood training, the HMM parameters are estimated such that the resulting generic model Λ_X maximizes the likelihood of the training data. In contrast, the NAT algorithm seeks both the distortion model parameters for all utterances $\Phi = \{\phi^{(j)}\}_{j=1}^{J}$, and the underlying "pseudoclean" HMM parameters Λ_S that jointly maximize the likelihood of the multicondition data when the model Λ_S is transformed to the adapted HMM of $\Lambda_X^{(j)}$. This can be written as

$$(\Lambda_S, \Phi) = \underset{(\Lambda_S, \Phi)}{\operatorname{argmax}} \prod_{j=1}^{J} \mathcal{L}(X^{(j)}; \Lambda_X^{(j)}), \tag{13.4}$$

where

$$\Lambda_X^{(j)} = \mathcal{A}(\Lambda_S, \phi^{(j)}) \tag{13.5}$$

is the adapted HMM and $\mathcal{A}$ represents the specific algorithm used for adaption. The term "pseudoclean" is used to indicate that the model defined by Λ_S is not necessarily equivalent to models trained with clean speech, but rather the models that maximize the likelihood of the multicondition training data when processed by the adaptation scheme.

To jointly learn the distortion model parameters and the pseudoclean speech model parameters, we start with the following EM auxiliary function:

$$Q(\Lambda_S, \Phi, \overline{\Lambda}_S, \overline{\Phi}) = \sum_{j=1}^{J} \sum_{t,q,i} \gamma_{tqi}^{(j)} \log(p(\mathbf{x}_t^{(j)}|q, i, \Lambda_S, \phi^{(j)})), \tag{13.6}$$

where j, t, q, i represent the indices for utterance, frame, state, and Gaussian, respectively, and $\gamma_{tqi}^{(j)}$ is the posterior probability of Gaussian i in the HMM state q for frame t of utterance j

$$\gamma_{tqi}^{(j)} = p(q_t = q, i_t = i|X^{(j)}, \overline{\Lambda}_S, \overline{\phi}^{(j)}). \tag{13.7}$$

This posterior probability is computed as:

$$\gamma_{tqi}^{(j)} = \frac{\alpha_{tq}^{(j)} \beta_{tq}^{(j)}}{\sum_{q'} \alpha_{tq'}^{(j)} \beta_{tq'}^{(j)}} \frac{c_{qi} p(\mathbf{x}_t^{(j)}|q, i, \overline{\Lambda}_S, \overline{\phi}^{(j)})}{\sum_{i'} c_{qi'} p(\mathbf{x}_t^{(j)}|q, i', \overline{\Lambda}_S, \overline{\phi}^{(j)})}, \tag{13.8}$$

where $\alpha_{tq}^{(j)}$ and $\beta_{tq}^{(j)}$ are the conventional forward and backward variables used in the Baum–Welch training algorithm [35], c_{qi} is the mixture weight of the ith Gaussian in state q. In Equations (13.6) and (13.8)

$$p(\mathbf{x}_t^{(j)}|q, i, \overline{\Lambda}_S, \overline{\phi}^{(j)}) \sim \mathcal{N}(\mathbf{x}_t^{(j)}; \boldsymbol{\nu}_{qi}^{(j)}, \boldsymbol{\Psi}_{qi}^{(j)}), \tag{13.9}$$

where $\boldsymbol{\nu}_{qi}^{(j)}, \boldsymbol{\Psi}_{qi}^{(j)}$ are the adapted Gaussian mean and covariance, respectively. Note that these parameters are utterance dependent since they are functions of distortion parameters for that utterance $\phi^{(j)}$. The same parameters with slightly different notation are also used in the regular training of HMMs that is explained in Section 2.3.3.

To perform NAT, the auxiliary function in Equation (13.6) is iteratively maximized with respect to the pseudoclean HMM parameters Λ_X and the distortion parameters Φ. Similarly to SAT, the HMM parameters are updated while the distortion parameters are fixed and then the distortion parameters are fixed while HMM parameters are updated. This process is repeated until the likelihood of the training set converges. Once the pseudoclean model parameters are learned, the distortion parameters Φ are discarded and the HMM parameters Λ_X are ready to be used at runtime along with the adaptation algorithm $\mathcal{A}$. A depiction of this process is shown in Figure 13.2.

In order to apply NAT, an adaptation strategy needs to be chosen. In the following section, an implementation of NAT that uses vector Taylor series (VTS) adaptation is presented.

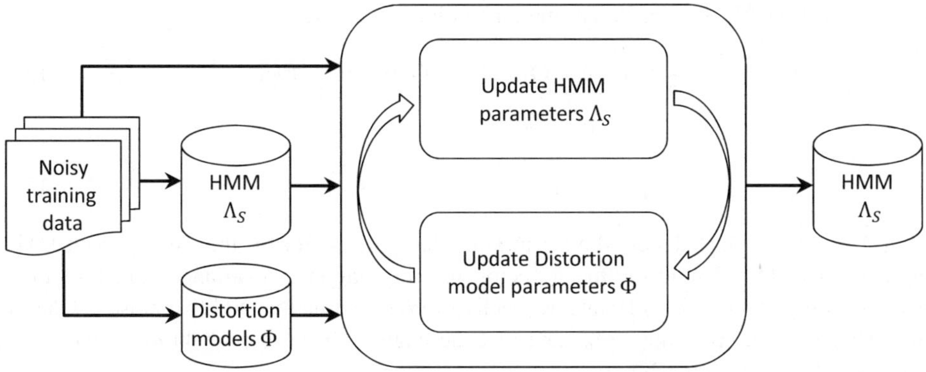

Figure 13.2 Flowchart of model-space noise adaptive training.

13.6 Noise Adaptive Training using VTS Adaptation

We will review VTS adaptation before examining the manner in which it is used for NAT. There have been several implementations of VTS proposed in the literature [2, 21, 23, 32]. While the details about VTS adaptation can be found in Chapter 12, we briefly summarize the algorithm proposed in [23]. This approach which adapts the means and variances of the static, delta, and delta-delta parameters using a generalized EM approach.

13.6.1 Vector Taylor Series HMM Adaptation

In the cepstral domain, the relationship between clean and distorted speech can be expressed as

$$\mathbf{x} = \mathbf{s} + \mathbf{h} + \mathbf{g}(\mathbf{n} - \mathbf{s} - \mathbf{h}), \tag{13.10}$$

where $\mathbf{x}, \mathbf{s}, \mathbf{h}, \mathbf{n}$, are the cepstral vectors corresponding to distorted speech, clean speech, channel, and noise, respectively. In Equation (13.10), the nonlinear function $\mathbf{g}(\mathbf{z})$ is

$$\mathbf{g}(\mathbf{z}) = \mathbf{C}\log(1 + \exp(\mathbf{C}^{\dagger}(\mathbf{z})), \tag{13.11}$$

where $\mathbf{C}$ is the discrete cosine transform (DCT) matrix and $\mathbf{C}^{\dagger}$ is its pseudoinverse. It can be shown that the Jacobian of Equation (13.10) with respect to $\mathbf{s}$ and $\mathbf{h}$ evaluated at a fixed point $(\bar{\mathbf{s}}, \bar{\mathbf{h}}, \bar{\mathbf{n}})$ is

$$\mathbf{G} = \mathbf{C} \cdot \mathrm{diag}\left(\frac{1}{1 + \exp(\mathbf{C}^{\dagger}(\bar{\mathbf{n}} - \bar{\mathbf{s}} - \bar{\mathbf{h}}))}\right) \cdot \mathbf{C}^{\dagger}, \tag{13.12}$$

where $\mathrm{diag}(.)$ represents the diagonal matrix whose elements equal to the value of the vector in the argument. Similarly, the Jacobian of Equation (13.10) with respect to $\mathbf{n}$ can be expressed as $\mathbf{F} = \mathbf{I} - \mathbf{G}$. Then, the nonlinear relationship between the distorted speech, clean speech, and environment parameters (noise and channel) in Equation (13.10) can be approximated by using a first-order VTS expansion around the point $(\bar{\mathbf{s}}, \bar{\mathbf{h}}, \bar{\mathbf{n}})$ as

$$\mathbf{x} \approx \bar{\mathbf{s}} + \bar{\mathbf{h}} + \bar{\mathbf{g}} + \mathbf{G}(\mathbf{s} - \bar{\mathbf{s}}) + \mathbf{G}(\mathbf{h} - \bar{\mathbf{h}}) + \mathbf{F}(\mathbf{n} - \bar{\mathbf{n}}), \tag{13.13}$$

where

$$\bar{\mathbf{g}} = \mathbf{C}\log(1 + \exp(\mathbf{C}^{\dagger}(\bar{\mathbf{n}} - \bar{\mathbf{s}} - \bar{\mathbf{h}}))). \tag{13.14}$$

Let $\Lambda_X = \{\boldsymbol{\mu}_{qi}, \boldsymbol{\Theta}_{qi}\}$ denote the set of Gaussian parameters for the clean speech HMMs where $\boldsymbol{\mu}_{qi}$ and $\boldsymbol{\Theta}_{qi}$ denote the mean vector and the diagonal covariance matrix of the ith Gaussian component in the qth state, respectively. We assume that additive noise is Gaussian with mean $\boldsymbol{\mu}_{\mathbf{n}}$ and covariance $\boldsymbol{\Theta}_{\mathbf{n}}$, and that the channel $\mathbf{h}$ has a probability density of the Kronecker delta function $\delta(\mathbf{h} - \boldsymbol{\mu}_{\mathbf{h}})$.

It is assumed that the environmental distortion does not change the alignment between a speech frame and the corresponding Gaussian component of the HMM. As a result, only the mean vector and covariance matrix for each Gaussian of the HMM will be affected. Using Equation (13.13), we can compute the mean $\boldsymbol{\nu}_{qi}$ and variance $\boldsymbol{\Psi}_{qi}$ of the adapted model Λ_X as

$$\begin{aligned}\boldsymbol{\nu}_{qi} \approx\ & \bar{\boldsymbol{\mu}}_{qi} + \bar{\boldsymbol{\mu}}_{\mathbf{h}} + \bar{\mathbf{g}}_{qi} + \mathbf{G}_{qi}(\boldsymbol{\mu}_{qi} - \bar{\boldsymbol{\mu}}_{qi}) \\ & + \mathbf{G}_{qi}(\boldsymbol{\mu}_{\mathbf{h}} - \bar{\boldsymbol{\mu}}_{\mathbf{h}}) + \mathbf{F}_{qi}(\boldsymbol{\mu}_{\mathbf{n}} - \bar{\boldsymbol{\mu}}_{\mathbf{n}}), \end{aligned} \tag{13.15}$$

$$\boldsymbol{\Psi}_{qi} \approx \mathbf{G}_{qi}\boldsymbol{\Theta}_{qi}\mathbf{G}_{qi}^T + \mathbf{F}_{qi}\boldsymbol{\Theta}_{\mathbf{n}}\mathbf{F}_{qi}^T. \tag{13.16}$$

where $\mathbf{G}, \mathbf{F}$, and $\bar{\mathbf{g}}$ carry the subscript *qi* to emphasize that they are functions of the mean of Gaussian *i* in state *q* of the clean-speech HMM. It can be concluded from Equation (13.16) that even if $\boldsymbol{\Theta}_{qi}$ and $\boldsymbol{\Theta}_{\mathbf{n}}$ are diagonal, $\boldsymbol{\Psi}_{qi}$ is no longer diagonal. However, for compatibility with traditional ASR decoders that have been optimized for diagonal covariances, only diagonal elements of $\boldsymbol{\Psi}_{qi}$ are used.

The means and variances of the delta parameters are typically updated using the continuous-time approximation proposed in [15]. This results in the following mean adaptation formula:

$$\boldsymbol{\nu}_{\Delta qi} \approx \mathbf{G}_{qi}\boldsymbol{\mu}_{\Delta qi}, \tag{13.17}$$

where a Δ in the variable subscript indicates the delta parameters of that variable. Because we assume that the noise is stationary, $\boldsymbol{\mu}_{\Delta\mathbf{n}} = 0$, for all utterances. Similarly, the covariance matrices for the delta features are adapted according to

$$\boldsymbol{\Psi}_{\Delta qi} \approx \mathbf{G}_{qi}\boldsymbol{\Theta}_{\Delta qi}\mathbf{G}_{qi}^T + \mathbf{F}_{qi}\boldsymbol{\Theta}_{\Delta\mathbf{n}}\mathbf{F}_{qi}^T. \tag{13.18}$$

The means and covariance matrices of the delta-delta features are adapted in a similar way to Equations (13.17) and (13.18) by replacing delta parameters with delta-delta parameters.

Adapting the clean-speech model parameters using Equations (13.15)–(13.18) requires estimates of the environment distortion (noise and channel) parameters. As only the noisy speech is observed, these parameters are hidden variables and must be estimated. General purpose noise estimation methods as described in Chapter 4 can be used, but the best performance is typically obtained using integrated EM-based maximum likelihood parameter estimation methods as will be described in Section 13.6.3. The key of NAT is jointly estimating the distortion model parameters with the acoustic model parameters.

13.6.2 Updating the Acoustic Model Parameters

NAT is an iterative procedure based on the generalized EM algorithm that alternately updates the acoustic model parameters and the distortion model parameters. For each iteration, the E-step is performed by accumulating the required sufficient statistics computed from the current estimates of the acoustic model parameters and the utterance-specific distortion model parameters. The M-step then uses these statistics to generate new estimates for these parameters.

We will first derive the update formulae for the acoustic model parameters for a fixed set of distortion model parameters. To compute the pseudoclean model parameters of the ith Gaussian in HMM state q, we can rewrite the auxiliary function given in Equation (13.6) by ignoring the terms constant with respect to the model parameters $\boldsymbol{\mu}_{qi}$ and $\boldsymbol{\Theta}_{qi}$ as follows:

$$Q = \sum_{j,t} \gamma_{tqi}^{(j)} \times \left\{ -\frac{1}{2} \log |\boldsymbol{\Psi}_{qi}^{(j)}| - \frac{1}{2} (\mathbf{x}_t^{(j)} - \boldsymbol{\nu}_{qi}^{(j)})^T \boldsymbol{\Psi}_{qi}^{(j)-1} (\mathbf{x}_t^{(j)} - \boldsymbol{\nu}_{qi}^{(j)}) \right\}. \tag{13.19}$$

where $\gamma_{tqi}^{(j)}$ is the posterior probability defined in Equation (13.8) of the noisy observation $\mathbf{x}_t^{(j)}$ under the utterance-specific VTS-adapted model $\Lambda_X^{(j)}$, and $\{\boldsymbol{\nu}_{qi}^{(j)}, \boldsymbol{\Psi}_{qi}^{(j)}\}$ are the adapted mean and variance as defined in Equations (13.15)–(13.18). Thus, these parameters are utterance specific as well. As in conventional model training, the summation in the auxiliary function is over all frames of all utterances available in the training set.

Updating the Means

To update $\boldsymbol{\mu}_{qi}$, we take the derivative of Equation (13.19) with respect to $\boldsymbol{\mu}_{qi}$, and set the result to zero. This leads to following expression:

$$\sum_{j,t} \gamma_{tqi}^{(j)} \mathbf{G}_{qi}^{(j)T} \boldsymbol{\Psi}_{qi}^{(j)-1} (\mathbf{x}_t^{(j)} - \boldsymbol{\nu}_{qi}^{(j)}) = 0. \tag{13.20}$$

Then, substituting Equation (13.15) into Equation (13.20) produces the following equation:

$$\sum_{j,t} \gamma_{tqi}^{(j)} \mathbf{G}_{qi}^{(j)T} \boldsymbol{\Psi}_{qi}^{(j)-1} \mathbf{G}_{qi}^{(j)} (\boldsymbol{\mu}_{qi} - \bar{\boldsymbol{\mu}}_{qi}) = \sum_{j,t} \gamma_{tqi}^{(j)} \mathbf{G}_{qi}^{(j)T} \boldsymbol{\Psi}_{qi}^{(j)-1} (\mathbf{x}_t^{(j)} - \bar{\boldsymbol{\nu}}_{qi}^{(j)}), \tag{13.21}$$

which can be solved for $\boldsymbol{\mu}_{qi}$ to obtain the following update formula:

$$\boldsymbol{\mu}_{qi} = \bar{\boldsymbol{\mu}}_{qi} + \left(\sum_{j,t} \gamma_{tqi}^{(j)} \mathbf{G}_{qi}^{(j)T} \boldsymbol{\Psi}_{qi}^{(j)-1} \mathbf{G}_{qi}^{(j)} \right)^{-1} \times \left(\sum_{j,t} \gamma_{tqi}^{(j)} \mathbf{G}_{qi}^{(j)T} \boldsymbol{\Psi}_{qi}^{(j)-1} \left(\mathbf{x}_t^{(j)} - \bar{\boldsymbol{\nu}}_{qi}^{(j)} \right) \right). \tag{13.22}$$

The update formulae for the delta and delta-delta mean parameters can be similarly derived, substituting $\Delta \mathbf{x}_t^{(j)}$ and $\boldsymbol{\nu}_{\Delta qi}$ for $\mathbf{x}_t^{(j)}$ and $\boldsymbol{\nu}_{qi}$, respectively, in Equation (13.20), and recalling the definition of $\boldsymbol{\nu}_{\Delta qi}$ in Equation (13.17).

Updating the Variances

There is no closed form solution for covariance matrices of the HMM distributions. As a result, the covariances are optimized using a gradient-based approach. We describe a solution using Newton's method, a second order approach. According to Newton's method, the covariance update equation is

$$\boldsymbol{\Theta}_{qi} = \bar{\boldsymbol{\Theta}}_{qi} - \left[\left(\frac{\partial^2 Q}{\partial^2 \boldsymbol{\Theta}_{qi}} \right)^{-1} \left(\frac{\partial Q}{\partial \boldsymbol{\Theta}_{qi}} \right) \right]_{\boldsymbol{\Theta}_{qi} = \bar{\boldsymbol{\Theta}}_{qi}}, \tag{13.23}$$

where $\bar{\boldsymbol{\Theta}}_{qi}$ is the current estimate of the covariance matrix for Gaussian i and state q.

We assume that the covariance matrices $\boldsymbol{\Psi}_{qi}^{(j)}, \boldsymbol{\Theta}_{qi}, \boldsymbol{\Theta}_{\mathbf{n}}^{(j)}$ are all diagonal. Then, we can write each of these covariance matrices as vectors:

$$\boldsymbol{\psi}_{qi}^{(j)} = [\psi_{qi,1}^{(j)}, \psi_{qi,2}^{(j)}, \ldots, \psi_{qi,D}^{(j)}], \tag{13.24}$$

$$\boldsymbol{\theta}_{qi} = [\theta_{qi,1}, \theta_{qi,2}, \ldots, \theta_{qi,D}], \tag{13.25}$$

$$\boldsymbol{\theta}_{\mathbf{n}}^{(j)} = [\theta_{n,1}^{(j)}, \theta_{n,2}^{(j)}, \ldots, \theta_{n,D}^{(j)}], \tag{13.26}$$

where D is the dimension of the feature vector and $\psi_{qi,d}^{(j)}$, $\theta_{qi,d}$, and $\theta_{n,d}^{(j)}$ are the adapted variance, the acoustic model variance and the noise variance, respectively, for component d of the feature vector. Typically 13-dimensional cepstra are used in the traditional automatic speech recognizers. Then, we can rewrite the auxiliary function in Equation (13.19) as

$$Q = \sum_{j,t} -\frac{1}{2}\gamma_{tqi}^{(j)} \times \left\{ \sum_{d=1}^{D} \left(\log \psi_{qi,d}^{(j)} + \frac{(x_{t,d}^{(j)} - \nu_{qi,d}^{(j)})^2}{\psi_{qi,d}^{(j)}} \right) \right\}, \tag{13.27}$$

where $x_{t,d}^{(j)}$ is the dth component of the feature vector $\mathbf{x}_t^{(j)}$, and $\nu_{qi,d}^{(j)}$ is the dth component of the VTS adapted mean vector $\boldsymbol{\nu}_{qi}^{(j)}$.

To compute the first and second derivatives of the Q function in Equation (13.19) with respect to the clean speech variance $\theta_{qi,d}$, we can expand $\mathbf{G}_{qi}^{(j)}$ and $\mathbf{F}_{qi}^{(j)}$ matrices in Equation (13.19) as

$$\mathbf{G}_{qi}^{(j)} = \begin{bmatrix} g_{11} & g_{12} & \cdots & g_{1D} \\ g_{21} & g_{22} & \cdots & g_{2D} \\ \vdots & \vdots & \vdots & \vdots \\ g_{D1} & g_{D2} & \cdots & g_{DD} \end{bmatrix}, \tag{13.28}$$

$$\mathbf{F}_{qi}^{(j)} = \begin{bmatrix} f_{11} & f_{12} & \cdots & f_{1D} \\ f_{21} & f_{22} & \cdots & f_{2D} \\ \vdots & \vdots & \vdots & \vdots \\ f_{D1} & f_{D2} & \cdots & f_{DD} \end{bmatrix}, \tag{13.29}$$

where indices q, i, and j are omitted to simplify the notation. Then, we can write the formula for the diagonal elements of the covariance matrix $\Psi_{qi}^{(j)}$ given in Equation (13.16) explicitly as

$$\psi_{qi,d}^{(j)} = \sum_{k=1}^{D} g_{dk}^2 \theta_{qi,k} + f_{dk}^2 \theta_{n,k}^{(j)} \quad d = 1, \ldots, D. \tag{13.30}$$

The first derivative of the Q function given in Equation (13.27) with respect to the pth element of the speech variance can be obtained by applying the chain rule as follows:

$$\frac{\partial Q}{\partial \theta_{qi,p}^{(j)}} = \sum_d \frac{\partial Q}{\partial \psi_{qi,d}^{(j)}} \frac{\partial \psi_{qi,d}^{(j)}}{\partial \theta_{qi,p}}, \tag{13.31}$$

where

$$\frac{\partial Q}{\partial \psi_{qi,d}^{(j)}} = -\frac{1}{2} \sum_{j,t} \gamma_{tqi}^{(j)} \left(\frac{1}{\psi_{qi,d}^{(j)}} \left(1 - \frac{(x_{t,d}^{(j)} - \nu_{qi,d}^{(j)})^2}{\psi_{qi,d}^{(j)}} \right) \right) \tag{13.32}$$

and

$$\frac{\partial \psi_{qi,d}^{(j)}}{\partial \theta_{qi,p}^2} = g_{dp}^2. \tag{13.33}$$

Using Equations (13.31)–(13.33), we can write the first derivative of the Q function with respect to the speech variance as

$$\frac{\partial Q}{\partial \theta_{qi,p}^2} = -\frac{1}{2} \sum_{j,t} \gamma_{tqi}^{(j)} \left(\sum_d \frac{g_{dp}^2}{\psi_{qi,d}^{(j)}} \left(1 - \frac{(x_{t,d}^{(j)} - \nu_{qi,d}^{(j)})^2}{\psi_{qi,d}^{(j)}} \right) \right). \tag{13.34}$$

We can continue with the second-derivative of the Q function with respect to the speech variance as follows:

$$\frac{\partial Q^2}{\partial \theta_{qi,p} \partial \theta_{qi,l}} = \frac{\partial}{\partial \theta_{qi,l}} \left(\frac{\partial Q}{\partial \theta_{qi,p}} \right), \tag{13.35}$$

which can be expressed as

$$\frac{\partial Q^2}{\partial \theta_{qi,p} \partial \theta_{qi,l}} = \frac{1}{2} \sum_{j,t} \gamma_{tqi}^{(j)} \left\{ \sum_d \frac{g_{dp}^2 g_{dl}^2}{\psi_{qi,d}^{(j)}} \left(1 - 2\frac{(x_{t,d}^{(j)} - \nu_{qi,d}^{(j)})^2}{\psi_{qi,d}^{(j)}} \right) \right\}. \tag{13.36}$$

As with the mean updates, the covariance matrices for the dynamic features of the pseudo-clean model can be similarly derived by replacing the static parameters and features with the dynamic parameters and features. In order to ensure that the variances remain positive during training, the logarithm of the variances are optimized in practice. This is discussed in more detail in Section 13.6.4.

Updating the Mixture Weights and Transition Probabilities

The transition probabilities, the initial probabilities, and the mixture weights for the pseudoclean model are computed in the same way as traditional ML training of the HMMs (as explained in Section 2.3.3) but using the adapted acoustic model to compute the posterior probabilities.

13.6.3 Updating the Environmental Parameters

Once the HMM parameters are updated, they are fixed and the distortion model parameters are updated. To update the distortion parameters for the *j*th utterance, $\phi^{(j)}$, the auxiliary function is rewritten as

$$Q = \sum_{t,q,i} \gamma_{tqi}^{(j)} \left\{ -\frac{1}{2} \log |\boldsymbol{\Psi}_{qi}^{(j)}| - \frac{1}{2} (\mathbf{x}_t^{(j)} - \boldsymbol{\nu}_{qi}^{(j)})^T \boldsymbol{\Psi}_{qi}^{(j)-1} (\mathbf{x}_t^{(j)} - \boldsymbol{\nu}_{qi}^{(j)}) \right\}. \tag{13.37}$$

As before, the update equations for the noise and channel means are derived by taking the derivative of Equation (13.37) with respect to each of the parameters and setting the result to zero. This leads to the following update equations for the noise and channel means:

$$\boldsymbol{\mu}_{\mathbf{n}}^{(j)} = \bar{\boldsymbol{\mu}}_{\mathbf{n}}^{(j)} + \left(\sum_{t,s,m} \gamma_{tqi}^{(j)} \mathbf{F}_{qi}^{(j)T} \boldsymbol{\Psi}_{qi}^{(j)-1} (\mathbf{F}_{qi}^{(j)}) \right)^{-1} \times \left(\sum_{t,s,m} \gamma_{tqi}^{(j)} \mathbf{F}_{qi}^{(j)T} \boldsymbol{\Psi}_{qi}^{(j)-1} (\mathbf{x}_t^{(j)} - \bar{\boldsymbol{\nu}}_{qi}^{(j)}) \right), \tag{13.38}$$

$$\boldsymbol{\mu}_{\mathbf{h}}^{(j)} = \bar{\boldsymbol{\mu}}_{\mathbf{h}}^{(j)} + \left(\sum_{t,s,m} \gamma_{tqi}^{(j)} \mathbf{G}_{qi}^{(j)T} \boldsymbol{\Psi}_{qi}^{(j)-1} (\mathbf{G}_{qi}^{(j)}) \right)^{-1} \times \left(\sum_{t,s,m} \gamma_{tqi}^{(j)} \mathbf{G}_{qi}^{(j)T} \boldsymbol{\Psi}_{qi}^{(j)-1} (\mathbf{x}_t^{(j)} - \bar{\boldsymbol{\nu}}_{qi}^{(j)}) \right). \tag{13.39}$$

As with the speech variance, the noise variance requires a gradient-based update. We again use Newton's method with the following update expression:

$$\boldsymbol{\Theta}_{\mathbf{n}}^{(j)} = \bar{\boldsymbol{\Theta}}_{\mathbf{n}}^{(j)} - \left[\left(\frac{\partial^2 Q}{\partial^2 \boldsymbol{\Theta}_{\mathbf{n}}^{(j)}} \right)^{-1} \left(\frac{\partial Q}{\partial \boldsymbol{\Theta}_{\mathbf{n}}^{(j)}} \right) \right]_{\boldsymbol{\Theta}_{\mathbf{n}}^{(j)} = \bar{\boldsymbol{\Theta}}_{\mathbf{n}}^{(j)}}. \tag{13.40}$$

The terms in Equation (13.40) can be derived in a similar manner to the terms in the speech variance described previously. The detailed derivation can be found in [20].

13.6.4 Implementation Details

The variances of both the noise and pseudoclean speech are optimized iteratively using Newton's method since there is no closed-form solution. To ensure that the variances remains positive, a change of variable is made such that $\tilde{\boldsymbol{\Theta}} = \log(\boldsymbol{\Theta})$. The optimization is performed on $\tilde{\boldsymbol{\Theta}}$, and then the exponential function is applied to obtain the actual covariance $\boldsymbol{\Theta} = \exp(\tilde{\boldsymbol{\Theta}})$. Maximizing the auxiliary function with respect to the log of the covariance requires changes to the update expression. Complete derivations for the log-covariance update are omitted for space considerations but are given in [20].

There are some well-known numerical issues with Newton's method. If the Hessian matrix is close to singular, its inverse may be unstable. Also, to ensure that the updates converge to a local maximum, the Hessian matrix must be negative definite. A diagonal-loading technique [7] was used to fulfill these constraints as

$$\tilde{\boldsymbol{\Theta}}_{qi} = \tilde{\bar{\boldsymbol{\Theta}}}_{qi} - \left[\left(\frac{\partial^2 Q}{\partial^2 \tilde{\boldsymbol{\Theta}}_{qi}} - \varepsilon I\right)^{-1} \left(\frac{\partial Q}{\partial \tilde{\boldsymbol{\Theta}}_{qi}}\right)\right]_{\tilde{\boldsymbol{\Theta}}_{qi} = \tilde{\bar{\boldsymbol{\Theta}}}_{qi}}, \tag{13.41}$$

where $\varepsilon = 1$ was empirically found to be useful to stabilize the optimization. Also, to ensure the stability, the change of variance was limited such as

$$\tilde{\boldsymbol{\Theta}}_{qi} = \min\left(\max\left(\tilde{\boldsymbol{\Theta}}_{qi}, \tilde{\bar{\boldsymbol{\Theta}}}_{qi} - \varsigma\right), \tilde{\bar{\boldsymbol{\Theta}}}_{qi} + \varsigma\right), \tag{13.42}$$

which in turn limits the change of the original variance $\boldsymbol{\Theta}_{qi}$ by a factor of $\exp(\varsigma)$. In the experiments, ς was set to 1. The noise covariance matrix was also optimized iteratively in the same way.

Finally, because of the approximations in the Taylor series expansion and the gradient-based updates used to learn the covariances, it is theoretically possible for decreases in likelihood to occur during the Generalized EM optimization. To compensate for this, a back-off strategy was proposed where the new parameter estimates are interpolated with their previous values until no decreases in likelihood are observed [26].

The overall algorithm for NAT using VTS adaptation is shown in Algorithm 1.

13.6.5 Experiments using NAT

The effectiveness of the VTS-based NAT algorithm was evaluated through a series of experiments on the Aurora 2 corpus described in Section 13.4 and Aurora 3 corpus, which consists of digit strings in four different languages recorded in an actual car environment.

In these experiments, the distortion parameters for each utterance in the training set were initialized such that the channel mean was set to zero, and the noise mean and covariance were estimated from the first and last 20 frames of each utterance. At each iteration, the static noise mean, static channel mean, and static and dynamic noise variances were reestimated. The dynamic channel and noise means remained set to zero, reflecting the assumptions that the the channel is deterministic and the noise is stochastic but stationary.

Table 13.4 shows the accuracy obtained by several well-known algorithms including cepstral mean normalization (CMN), cepstral mean and variance normalization (CMVN), and the AFE used with fNAT training (fNAT-AFE). All systems were trained using multicondition data, and as before, the complex back-end training recipe was used. The proposed NAT method outperforms all other methods, and provides relative reductions in word error rate of 11.97% relative improvement over CMN, 3.85% over CMVN, 7.54% over fNAT-AFE, and 18.83% over VTS adaptation. Note that NAT-VTS and VTS are exactly the same algorithm at runtime and differ only in the manner in which the acoustic models are trained. The substantial improvement of NAT-VTS over VTS (18.8%) highlights the value of the adaptive training

Algorithm 1: Noise adaptive training using VTS

Input: Initial HMM parameters Λ_S trained from multicondition data, initial distortion parameters Φ for each utterance, multicondition training data $\mathcal{X}$
Output: NAT-VTS-trained HMM model parameters Λ_S

```
repeat
  // Update HMM parameters
  Load HMM parameters Λ_S
  for j = 1 to J do
    Load distortion model parameters φ^(j) for utterance j
    Adapt HMM: Λ_X^(j) ← VTS(Λ_S, φ^(j))
    Compute posterior probabilities γ_tqi using (13.7)
    for all μ_qi, Θ_qi (static and dynamic) do
      Accumulate matrix and vector terms in (13.22)–(13.23)
    end for
  end for
  Update Λ_S and write to file
  // Update distortion model parameters
  Load HMM parameters Λ_S
  for j = 1 to J do
    Load distortion model parameters φ^(j) for utterance j
    Adapt HMM: Λ_X^(j) ← VTS(Λ_S, φ^(j))
    Compute posterior probabilities γ_tqi using (13.7)
    for μ_n, μ_h, Θ_n (static and dynamic) do
      Accumulate matrix and vector terms in (13.38)–(13.40)
    end for
    Update φ^(j) and write to file
  end for
until Likelihood converges
```

approach when multicondition data is used for training. Explain acronyms CMN and CMVN and provide a reference. We will shortly discuss them in the intro as well, so keep this as a placeholder for a reference to there.

We also applied NAT to the acoustic models trained using clean speech. These results are presented in Table 13.5 for Aurora 2. NAT provides a small improvement over the VTS model

Table 13.4 Word accuracy for each set of Aurora 2 using models trained on multicondition data.

Method	Set A	Set B	Set C	Avg
Baseline	91.68	89.74	88.91	90.35
CMN	92.97	92.62	93.32	92.90
CMVN	93.80	93.09	93.70	93.50
fNAT-AFE	93.74	93.26	92.21	93.24
VTS	92.20	91.87	93.37	92.30
NAT-VTS	93.66	93.77	93.89	93.75

Table 13.5 Word accuracy for each set of Aurora 2 using models trained on clean data.

Method	Set A	Set B	Set C	Avg
Baseline	60.43	55.85	69.01	60.31
CMN	68.65	73.71	69.69	70.88
CMVN	84.46	85.55	84.84	84.97
AFE	89.27	87.92	88.53	88.58
VTS	92.61	92.87	92.76	92.75
NAT-VTS	92.79	93.26	92.59	92.94

adaptation (92.75% vs. 92.94%) showing that even clean models have unwanted variability that may be attributed to factors such as microphone characteristics and positioning, instrumental noise, and speaker differences. It is also interesting to note that when the acoustic models are trained with clean data, the model-based techniques VTS and NAT-VTS, perform substantially better than front-end feature enhancement using the AFE.

Because Aurora 2 consists of noisy speech synthetically generated by adding noise to clean speech, we wanted to validate the performance of NAT on real data actually collected in a noisy environment. Aurora 3 consists of connected digit strings recorded in realistic car environments [31]. Each utterance is recorded using either a close-talking or hands-free far field microphone and labeled as coming from either a high, medium, or low noise condition. There are four languages (Finnish, Spanish, German, and Danish) and three experimental conditions (well matched, medium matched, and highly mismatched). The acoustic models were trained using the standard "simple back-end" recipe [16]. An HMM with 16 states per digit and mixture of three Gaussians per state is created for each digit as a whole word. A three state silence model with 6 Gaussian mixtures per state and a one state short pause model which is tied to the middle stage of silence model are included.

In Table 13.6, the results obtained with Aurora 3 are presented for the same set of algorithms compared previously. Note that in Aurora 3, there is no clean data available for training and the acoustic models are trained using the speech provided with the database for each experimental condition. When the ML trained models are adapted at runtime using the VTS algorithm, the average word recognition accuracy is 86.26% for the Aurora 3 task. Using NAT-VTS improves the accuracy to 90.66% (a 32% relative improvement) and outperforms the other methods. As before, the next best performing method is fNAT-AFE.

Table 13.6 Word accuracy for the Aurora 3 experimental conditions.

Method	Well	Mid	High	Avg
Baseline	91.34	78.4	55.84	77.94
CMN	92.97	84.43	71.57	84.63
CMVN	94.22	87.92	83.40	89.31
fNAT-AFE	95.30	86.79	87.25	90.31
VTS	91.33	80.25	86.57	86.26
NAT-VTS	94.44	87.55	88.98	90.66

13.7 Discussion

13.7.1 Comparison of Training Algorithms

We have presented several different training strategies in this chapter and empirically evaluated their performance. It is instructive to directly compare the update equations. In Table 13.7, the mean update equations are shown for four different training strategies for a simple Gaussian mixture model. The Gaussian posterior probabilities are shown directly as $p(i|\mathbf{s}_t)$ rather than the conventional γ_{it} notation, in order to make explicit which features are being used to computer posterior probabilities. As the table shows, clean and multicondition training are identical except in the features used. Obviously, multicondition training will model the noisy data $\mathbf{x}_t$ better than the clean-trained models. However, these models have a lot more unwanted variability from the environmental distortion.

In fNAT, point estimates of the clean speech features $\hat{\mathbf{s}}_t$ generated by an enhancement algorithm are used for training. As we have shown experimentally, this is quite effective at removing much of the unwanted environmental variability in the model. Uncertainty in the enhancement process is captured by the model *implicitly* through the additional variance introduced by errors made by the enhancement algorithm over a large training corpus.

Finally, in the NAT algorithm, the posterior probability is computed directly on the noisy data using the adapted model, but the observations are replaced by the mean of the state-conditional posterior distribution of clean speech. This enables NAT to use a different estimate of the clean speech for each Gaussian. Uncertainty in the adaptation process is captured *explicitly* through the presence of the adapted model variance Ψ_i in the accumulation of the sufficient statistics. Components with high variance will contribute less to the estimate of the updated mean.

13.7.2 Comparison to Speaker Adaptive Training

It is also interesting to compare the the update equations of SAT and NAT, since it was our original motivation for this algorithm. Because the variance update is unchanged in SAT, we only focus on the comparison of the mean update equations here. The VTS mean adaptation formula in Equation (13.15) can be written in the form of MLLR transformation as follows:

$$\boldsymbol{\nu}_{qi}^{(j)} = \mathbf{A}_{qi}^{(j)}\boldsymbol{\mu}_{qi} + \mathbf{b}_{qi}^{(j)}, \tag{13.43}$$

where

$$\mathbf{A}_{qi}^{(j)} = \mathbf{G}_{qi}^{(j)} \tag{13.44}$$

Table 13.7 Comparison of GMM mean update formulae for different training strategies.

Training method	Mean update equation
Clean training	$\boldsymbol{\mu}_i = \left(\sum_t p(i\|\mathbf{s}_t)\right)^{-1}\sum_t p(i\|\mathbf{s}_t)\mathbf{s}_t$
Multi training	$\boldsymbol{\mu}_i = \left(\sum_t p(i\|\mathbf{x}_t)\right)^{-1}\sum_t p(i\|\mathbf{x}_t)\mathbf{x}_t$
fNAT	$\boldsymbol{\mu}_i = \left(\sum_t p(i\|\hat{\mathbf{s}}_t)\right)^{-1}\sum_t p(i\|\hat{\mathbf{s}}_t)\hat{\mathbf{s}}_t$
NAT	$\boldsymbol{\mu}_i = \left(\sum_t p(i\|\mathbf{x}_t)\mathbf{G}_i^T\Psi_i^{-1}\mathbf{G}_i\right)^{-1}\sum_t p(i\|\mathbf{x}_t)\mathbf{G}_i^T\Psi_i^{-1}\boldsymbol{\mu}_{s\|x_t,i}$

and

$$\mathbf{b}_{qi}^{(j)} = \bar{\boldsymbol{\mu}}_{qi} + \bar{\boldsymbol{\mu}}_{\mathbf{h}}^{(j)} + \bar{\mathbf{g}}_{qi}^{(j)} - \mathbf{G}_{qi}^{(j)}\bar{\boldsymbol{\mu}}_{qi}, \tag{13.45}$$

when the VTS expansion point is $\bar{\boldsymbol{\mu}}_{\mathbf{h}}^{(j)} = \boldsymbol{\mu}_{\mathbf{h}}^{(j)}$ and $\bar{\boldsymbol{\mu}}_{\mathbf{n}}^{(j)} = \boldsymbol{\mu}_{\mathbf{n}}^{(j)}$. When written in this form, the mean update equation for NAT can be rederived and expressed as

$$\boldsymbol{\mu}_{qi} = \left(\sum_{j,t} \gamma_{tqi}^{(j)} \mathbf{A}_{qi}^{(j)T} \boldsymbol{\Psi}_{qi}^{(j)-1} \mathbf{A}_{qi}^{(j)}\right)^{-1} \left(\sum_{j,t} \gamma_{tqi}^{(j)} \mathbf{A}_{qi}^{(j)T} \boldsymbol{\Psi}_{qi}^{(j)-1} (\mathbf{x}_t^{(j)} - \mathbf{b}_{qi}^{(j)})\right). \tag{13.46}$$

By comparing Equation (13.46) with Equation (13.3), it is clear the the algorithms are very similar with the following key differences. First, SAT typically uses a global transformation for each speaker r while NAT uses a separate transformation for each each Gaussian in the HMM for each utterance. Second, the parameters of the transformation in SAT are estimated from training or adaptation data, whereas the parameters of the transformations in NAT are fully specified by the utterance-specific distortion parameters, that is the noise and channel model parameters.

13.7.3 Related Adaptive Training Methods

In Section 13.6, we described a specific implementation of NAT that used VTS adaptation for noise compensation. In this, work, the VTS implementation used followed closely that of [23]. However, alternative compensation algorithms have been proposed in the literature and can potentially be incorporated into NAT. For example, improved performance on Aurora 2 was obtained using a phase-sensitive model of the distortion function Equation (13.10). In this model, a variable α is introduced to represent the phase asynchrony between the clean speech and the noise [6]. Although α is theoretically a random variable, it was treated as a tunable parameter in [24], whose optimal value resulted in an accuracy of 93.32% with clean trained acoustic models. If a NAT formulation is used that includes this phase-sensitive VTS distortion model, the accuracy using multicondition training data increases from 93.75% to 94.14%.

In addition to the NAT algorithm presented in this chapter, there are other adaptive training algorithms for compensating for environmental distortion in the literature. For example, the irrelevant variability normalization (IVN) training algorithm [18] uses a different version of the VTS algorithm [21] which enables the optimization to be performed using EM, rather than a gradient-based generalized EM. Joint adaptive training (JAT) [28] uses joint uncertainty decoding as its companion model adaptation scheme [25]. Unlike most adaptation algorithms for environmental distortion which transform every Gaussian in the system individually, JUD computes transformations for a set of regression classes, in a similar manner to MLLR. In addition, JAT uses a second-order method for jointly optimizing both the mean and the variance that results in a different form of the Hessian matrix. Typically, the use of regression classes results in computational savings at the expense of reduced accuracy, as discussed in Chapter 17.

While the material in this chapter has been restricted to training models using a maximum likelihood criterion, discriminative training can also be used in an adaptive training framework [11]. To do so, maximum likelihood NAT is performed until the likelihood of the training data converges. Then, the distortion model parameters are fixed and discriminative training is applied to further optimize the HMM parameters. Using this approach with minimum phone error (MPE) training resulted in further improvements in accuracy for both VTS and JUD compensation algorithms [11].

13.8 Conclusion

In this chapter, we have described different methods to train acoustic models for noise robust speech recognition. We have shown that acoustic models trained from clean speech typically have the highest degree of mismatch to the observed noisy test data and thus, generally have the poorest performance, even after front-end compensation. In addition, relying on acoustic models trained from clean speech typically means that data collected from a deployed application in the field cannot be used to retrain and improve the speech-recognition system. Multicondition training substantially reduces the mismatch between the training and test data but may not generalize well to unseen environmental conditions. Additionally, multicondition-trained models may have weak discriminative power if the environmental distortion in the training data introduces excessive variability.

NAT provides a means of training acoustic models with noisy speech in a manner that removes the unwanted environmental variability from the acoustic models. By following the same methodology used to develop SAT, NAT learns a pseudoclean acoustic model that maximizes the likelihood of noisy training data after the model has been adapted to the noisy environment. A simplified version of NAT can be implemented using front-end enhancement. By simply processing the noisy training data through the same front-end enhancement pipeline as the test data, much of the performance benefits of model-based NAT can be obtained.

References

[1] A. Acero, *Acoustical and environmental robustness in automatic speech recognition*. Boston, MA: Kluwer Academic Publishers, 1993.

[2] A. Acero, L. Deng, T. Kristjansson, and J. Zhang, "HMM adaptation using vector Taylor series for noisy speech recognition," in *Proceedings of ICSLP*, Beijing, China, 2000.

[3] T. Anastasakos, J. McDonough, R. Schwartz, and J. Makhoul, "A compact model for speaker-adaptive training," in *Proceedings of ICSLP*, Philadelphia, PA, 1996.

[4] S. Boll, "Suppression of acoustic noise in speech using spectral subtraction," *IEEE Transactions on Acoustics, Speech and Signal Processing*, vol. 27, no. 2, pp. 113–120, April 1979.

[5] L. Deng, A. Acero, M. Plumpe, and X. Huang, "Large-vocabulary speech recognition under adverse acoustic environments," in *Proceedings of ICSLP*, Beijing, China, 2000.

[6] L. Deng, J. Droppo, and A. Acero, "Enhancement of log mel power spectra of speech using a phase-sensitive model of the acoustic environment and sequential estimation of the corrupting noise," *IEEE Transactions on Speech and Audio Processing*, vol. 12, no. 2, pp. 133–143, March 2004.

[7] J. E. Dennis and R. B. Schnabel, *Numerical methods for unconstrained optimization and nonlinear equations*. Englewood Cliffs, NJ: Prentice-Hall, 1983.

[8] V. Digalakis, D. Rtischev, L. Neumeyer, and E. Sa, "Speaker adaptation using constrained estimation of gaussian mixtures," *IEEE Transactions on Speech and Audio Processing*, vol. 3, pp. 357–366, 1995.

[9] J. Droppo, A. Acero, and L. Deng, "Uncertainty decoding with SPLICE for noise robust speech recognition," in *Proceedings of ICASSP*, Orlando, FL, 2002.

[10] Y. Ephraim and D. Malah, "Speech enhancement using a minimum mean square error log-spectral amplitude estimator," *IEEE Transactions on Acoustics, Speech Signal Processing*, vol. ASSP-33, no. 2, pp. 443–445, April 1985.

[11] F. Flego and M. J. F. Gales, "Discriminative adaptive training with VTS and JUD," in *Proceedings of ASRU*, Merano, Italy, 2009.
[12] M. J. F. Gales, "Maximum likelihood linear transformations for HMM-based speech recognition," *Computer Speech & Language*, vol. 12, pp. 75–98, 1998.
[13] M. Gales, S. Young, and S. J. Young, "Robust continuous speech recognition using parallel model combination," *IEEE Transactions on Speech and Audio Processing*, vol. 4, pp. 352–359, 1996.
[14] Y. Gong, "Speech recognition in noisy environments: a survey," *Speech Communication*, vol. 16, no. 3, pp. 261–291, 1995.
[15] R. A. Gopinath, M. Gales, P. S. Gopalakrishnan, S. Balakrishnan-Aiyer, and M. A. Picheny, "Robust speech recognition in noise: performance of the IBM continuous speech recognizer on the ARPA noise spoke task," in *Proceedings of the ARPA workshop on spoken language systems technology*, 1995.
[16] H. Hirsch and D. Pearce, "The AURORA experimental framework for the performance evaluation of speech recognition systems under noisy conditions," in *Proceedings of ISCA ITRW ASR*, Paris, France, September 2000.
[17] Y. Hu and Q. Huo, "An HMM compensation approach using unscented transformation for noisy speech recognition," in *Proc. ISCSLP*, Nov. 2006, pp. 346–357.
[18] Y. Hu and Q. Huo, "Irrelevant variability normalization based HMM training using VTS approximation of an explicit model of enviromental distortions," in *Proceedings of Interspeech*, Antwerp, Belgium, 2007.
[19] O. Kalinli, M. L. Seltzer, and A. Acero, "Noise adaptive training using a vector Taylor series approach for noise robust automatic speech recognition," in *Proceedings of ICASSP*, Taipei, Taiwan, 2009.
[20] O. Kalinli, M. L. Seltzer, J. Droppo, and A. Acero, "Noise adaptive training for robust automatic speech recognition," *IEEE Transactions on Audio, Speech and Language Processing*, vol. 18, no. 8, pp. 1889–1901, Nov. 2010.
[21] D. Y. Kim, C. K. Un, and N. S. Kim, "Speech recognition in noisy environments using first-order vector Taylor series," *Speech Communication*, vol. 24, no. 1, pp. 39–49, 1998.
[22] C. J. Leggetter and P. C. Woodland, "Maximum likelihood linear regression for speaker adaptation of continuous density hidden Markov models," *Computer Speech & Language*, vol. 9, no. 2, pp. 171–185, 1995.
[23] J. Li, L. Deng, D. Yu, Y. Gong, and A. Acero, "High-performance HMM adaptation with joint compensation of additive and convolutive distortions via vector Taylor series," in *Proceedings of ASRU*, Kyoto, Japan, 2007.
[24] J. Li, L. Deng, D. Yu, Y. Gong, and A. Acero, "HMM adaptation using a phase-sensitive acoustic distortion model for environment-robust speech recognition," in *Proceedings of ICASSP*, Las Vegas, NV, 2008.
[25] H. Liao and M. J. F. Gales, "Joint uncertainty decoding for noise robust speech recognition," in *Proceedings of Interspeech*, Lisbon, Portugal, 2005.
[26] H. Liao and M. J. F. Gales, "Joint uncertainty decoding for robust large vocabulary speech recognition," Technical Report CUED/F-INFENG/TR-522, Cambridge University, Cambridge, England, November 2006.
[27] H. Liao and M. J. F. Gales, "Issues with uncertainty decoding for noise robust speech recognition," in *Proceedings of Interspeech*, Pittsburgh, PA, 2006.
[28] H. Liao and M. J. F. Gales, "Adaptive training with joint uncertainty decoding for robust recognition of noisy data," in *Proceedings of ICASSP*, Honolulu, Hawaii, 2007.
[29] R. P. Lippmann, E. A. Martin, and D. B. Paul, "Multistyle training for robust isolated-word speech recognition," in *Proceedings of ICASSP*, Dallas, TX, 1987.
[30] D. Macho, L. Mauuary, B. Noé, Y. Cheng, D. Ealey, D. Jouvet, H. Kelleher, D. Pearce, and F. Saadoun, "Evaluation of a noise-robust DSR front-end on Aurora databases," in *Proceedings of ICSLP*, Denver, Colorado, 2002.
[31] A. Moreno, B. Lindberg, C. Draxler, G. Richard, K. Choukri, J. Allen, and S. Euler, "Speechdat-car: a large speech database for automotive environments," in *Proc. of LREC*, Athens, Greece, 2000.
[32] P. J. Moreno, B. Raj, and R. M. Stern, "A vector Taylor series approach for environment-independent speech recognition," in *Proceedings of ICASSP*, vol. 2, 1996.
[33] D. B. Paul and J. M. Baker, "The design of the Wall Street Journal-based CSR corpus," in *Proceedings of the ARPA Speech and Natural Language Workshop*, Harriman, NY, February 1992, pp. 357–362.
[34] D. Pierce and A. Gunawardana, "Aurora 2.0 speech recognition in noise: update 2.complex backend definition for Aurora 2.0." Available at: http://icslp2002.colorado.edu/special_sessions/aurora, 2002.
[35] L. R. Rabiner, "A tutorial on hidden Markov models and selected applications in speech recognition," *Proceedings of the IEEE*, vol. 77, no. 2, pp. 257–286, 1989.

[36] P. C. Woodland, M. J. F. Gales, and D. Pye, "Improving environmental robustness in large vocabulary speech recognition," in *Proceedings of the IEEE International Conference on Acoustics, Speech and Signal Processing*, vol. 1, Atlanta, GA, May 1996, pp. 65–69.

[37] D. Yu, L. Deng, J. Droppo, J. Wu, Y. Gong, and A. Acero, "A Minimum-mean-square-error noise reduction algorithm on mel-frequency cepstra for robust speech recognition," in *Proceedings of ICASSP*, Las Vegas, NV, 2008.

Part Five

Compensation for Information Loss

14

Missing-Data Techniques: Recognition with Incomplete Spectrograms

Jon Barker
University of Sheffield, UK

14.1 Introduction

In Part Four of this book, the mismatch between the statistics of noisy observations and those of noise-free speech was presented as the fundamental problem facing robust ASR systems. Techniques were described that aimed to improve performance by reducing this mismatch. This section of the book takes a rather different perspective that emphasises *information loss* rather than *model mismatch*. The difference between these perspective can be illustrated by the visual analogy presented in Figure 14.1.

The top panel of the figure shows a word written is a familiar font that has been partially distorted: the lower half of the word has been passed through a ripple effect. It is clear that the distorted image will be poorly matched to models that have been trained on undistorted characters. However, it is also clear that as long as the parameters of the distortion are known, and as long as the ripple effect is invertible, no information has been lost. Armed with knowledge about how the image had been distorted it would be possible to recover the undistorted word. Inverting a known or estimated distortion to reduce model mismatch is the principle behind many robust speech-recognition technologies, for example, including MLLR (Chapter 11), dereverberation (Chapter 10) and feature compensation methods (Chapter 9).

Contrast the top panel with the situation appearing in the bottom panel. Again, a word written in a familiar font has been distorted and the distortion has been applied to exactly the same region of the image. In this case, the distortion is an occlusion – part of the image has been masked by a black rectangle. This distortion differs from the ripple effect in that it cannot be described by a one-to-one mapping and is therefore clearly not invertible. So, even

Techniques for Noise Robustness in Automatic Speech Recognition, First Edition.
Edited by Tuomas Virtanen, Rita Singh, and Bhiksha Raj.

Figure 14.1 A visual analogy comparing two views of the robust ASR problem: noise as a source of model mismatch versus noise as a source of information loss. The distortion in the top panel is invertible and the original signal could theoretically be recovered if the model for the distortion was known. The occlusion in the bottom panel is not invertible and information has been genuinely lost.

with complete knowledge of the process producing the distortion, it is impossible to recover the original undistorted image. *Information has been lost.* As it happens the word can still be recognised because language (whether encoded in text or as speech) is highly redundant and, in this case, there remains sufficient information in the top half of the image to decode the word. However, note that although this decoding process may involve identifying which regions of the image are affected by the noise, it does not involve restoring the original image by 'removing the noise' from the lower half.

The approaches discussed in this chapter view noise robust speech recognition as a problem of recognising speech on the basis of incomplete spectrograms, and are analogous to the problem of attempting to read a word given an incomplete image shown in the lower half of the figure. The noise is considered to behave like an occluding object that has obstructed the view of the noise-free speech spectrogram. This may seem a strange perspective to take, because whereas it is clear that when a visual object is occluded, light from the object is blocked and *replaced* by uninformative light from the occluder, it is not clear that acoustic sources *occlude* each other in this way. In any spectro-temporal region, energy from the two sources combines in an additive way and, in principle, if the noise source was known it could be subtracted to recover the target speech. However, in practice, in local time-frequency regions the noise may be at a level many decibels above that of the speech, and the variance in the noise estimate may be much larger than the speech energy meaning that subtracting the noise with a useful degree of reliability becomes impossible. In such spectro-temporal regions the information about the underlying speech signal is effectively lost and the spectro-temporal region can be considered just as missing as if it had been obliterated by an opaque masker. The effective opaqueness of noise maskers is in fact consistent with our perception of signals in noise. For example, if white noise is added to a tone, as the level of the noise is increased there will come a point where the tone is no longer detectable and no tone is perceived. At this point, we would describe the tone as being *masked by the noise*. However, it is worth noting that the masking threshold is often far below 0 dB signal-to-noise ratio (SNR).

So, although acoustic objects are not opaque, and acoustic masking is not as absolute as visual occlusion, it can often be treated as though it is. A noisy speech spectrogram can then be seen to consist of separated regions where the speech masks the noise, and other regions where the noise masks the speech. In this view, the regions masked by the noise are said to contain *missing data* and the regions dominated by the speech are said to be *present data*. This chapter will take this seemingly rather simplistic binary approximation of acoustic source mixing and show how it leads to a set of simple but surprisingly effective approaches to robust ASR.

The chapter starts by discussing the problem of classification with incomplete data that is central to all the ASR approaches that will be discussed. The classification problem will be motivated by a trivial non-speech example which will help define some of the key concepts and intuitions on which the formalism will be built. The chapter will then proceed to discuss how these concepts can be applied to the case of masked speech recognition. The basic missing-data ASR approach will be introduced and the compatibility of missing-data ASR and techniques employed in conventional ASR systems will be discussed.

The first half of the chapter makes the assumption that there is prior knowledge that informs us with certainty which spectro-temporal regions are missing. In practice, this will not be the case. In general, our knowledge of the missing-data pattern (or the *missing-data mask*) is itself incomplete. The second half of the chapter will consider how this missing-data mask uncertainty, the *meta-missing data*, can be accommodated. We will see that the basic approach is to represent the missing-data mask as a distribution and sum over missing-data mask hypotheses. Various ways of representing mask distributions will be considered.

The chapter has been written in an attempt to present various missing-data approaches under a common theoretical framework and to provide pointers to the extensive literature. Practical details, such as performance and computational cost will be briefly discussed in the final section, but the chapter is not intended as a comparative review of missing-data recognition performance. There have been no comprehensive comparisons of missing-data approaches but broad conclusions can be drawn from the recent review by Raj and Stern [28].

The techniques in this chapter attempt to accommodate for missing information during the classification stage. An alternative approach is to reconstruct the missing information during a pre-processing stage and then to process the reconstructed feature vectors using conventional speech recognition techniques – this approach, commonly referred to as *missing-data imputation*, is discussed further in Chapter 15 of this book. Note also that this chapter assumes that some approximation of the missing-data mask is available *a priori*. For a discussion of the problem of mask estimation itself the reader is referred to Chapter 16.

14.2 Classification with Incomplete Data

The problem of classification with incomplete data arises in a huge variety of contexts, with data being lost through a wide range of mechanisms including sensor malfunction, censoring, summarisation, compression, or corruption due to noise or transmission error. Despite the importance of handling these situations and the ubiquity of the problem, it was not until relatively recent years that missing data received much attention. Modern *missing data theory* was pioneered in the late 1970s by the seminal work of Rubin [30] and the field was given added relevance by the advent of efficient iterative solutions to learning with incomplete data made practical by modern computers [12]. Missing-data theory is now commonly employed and there are several comprehensive textbook treatments of the subject including Schafer [31], Williams *et al.* [37].

Missing-data problems involve some unexpected subtleties that are easily overlooked when dealing with unfamiliar data sets such as abstract speech feature vectors, so, in order to develop our understanding, we will start by considering a trivial two class classification problem given a pair of discrete observations. Solutions to this simple problem can be written down through appeal to common sense and intuition; however, the section will proceed to develop a formal perspective with which we can validate our initial intuitions and extend the ideas to more complex scenarios. Finally, we will return to the problem of speech recognition with incomplete spectra to see how the general ideas apply to this specific case.

14.2.1 A Simple Missing Data Scenario

Consider a set of survey data showing the height and weight of a set of 100 men and 100 women. For convenience the data has been discretised into five bands for weight which, running from lightest to heaviest, are labeled, `--W`, `-W`, `W`, `+W` and `++W` and five bands for height which are similarly labeled, `--H`, `-H`, `H`, `+H` and `++H`. The width of each band has been tuned to represent 20 percentiles of the complete data set. The data is displayed in Table 14.1 which has entries representing every combination of H and W values, written in the form $(F|M)$ where F is the count for the number of female subjects and M is the count for male subjects. So for example, 20 of the 100 women and 1 of the 100 men are jointly in the height band, `--H`, and weight band `--W`.

Let us now consider that after having collected this data, we are asked to guess the gender of a set of previously unseen subjects having only been told their weight and height bands. Given that the data has been collected from an equal number of male and female subjects, and given that we have no prior reason to expect a random subject to be more likely to be one gender than another, then our best guess will be the gender that has most often been seen to have the feature $\{H, W\}$ in the training data. For example, if a subject is observed to be {`H`, `+W`} then the table records four female and six male subjects with this feature, so our best guess for the gender would be *male*. What happens though if we receive a survey form in which the height is recorded as `H` but the box on the form that records the weight feature is seen to be empty? Interestingly, the correct classification may now depend on information about *why* the

Table 14.1 The distribution of weights and heights of 100 men and 100 women. The weights and heights have been discretised into five 20 percentile band. Table entries are of the form F|M where F is the count for females and M the count for males. The final row shows the counts for each height band summed over weights, and the final column shows the counts for each weight band summed over all heights.

	−−H	−H	H	+H	++H	$\Sigma(H)$
−−W	20|1	12|0	4|1	0|1	0|1	36|4
−W	11|0	11|1	6|6	0|3	0|2	28|12
W	7|0	10|2	8|1	2|8	0|2	27|13
+W	1|0	4|0	4|6	0|14	0|11	9|31
++W	0|0	0|0	0|4	0|12	0|24	0|40
$\Sigma(W)$	39|1	37|3	22|18	2|38	0|40	

weight feature happens to be unrecorded. The importance of understanding the mechanism that caused the missingness is made clear by considering the three following scenarios.

In the first situation, suppose that next to the empty height box we see an annotation, 'The scales were broken', and flicking through the pile of new arriving surveys we find the weight box is empty in them all. Now, in this case, since the scales were not working we truly have no idea what the weight might have been. We know the subject to be in height band `H` and we might argue that according to the training data, people with height `H` are most commonly in weight band `-W`. So we guess the missing weight value to be `-W` – we can say that the missing value has been *imputed*. In which case the subject is most likely to be {`H,-W`} and hence is equally likely to be male or female as the table records (6|6) for this condition. However, basing the decision on the most likely weight does not take into account our uncertainty about the missing data. The more intuitively correct thing to do, which we will see later is also theoretically correct in this case, is to imagine that the weight measurement never existed (neither during collection of the *training data* nor when making a classification) and base the classification on the statistics of the height alone. To do this, we simply look at the *marginal* distribution of the heights of the male and female subjects. This marginal distribution can be computed from the *joint distribution* provided by the training data by summing each height column across all weights. The result is shown in the table by the row labelled $\Sigma(W)$. For $\Sigma(W)$ in the `H` column we see that there are 22 females and 18 males and so the best guess for the gender is female. Note that in this case the result arrived at by marginalisation is different from that achieved by imputation.

In a second situation, we receive a survey form with a blank weight measure and again the form notes that the scales are broken, but looking through the pile of forms we see that the weights are only missing for some subjects. Specifically, we note that weights of `--W`, `-W` and `W` have been recorded but there are none recorded as `+W` and `++W`. We then hypothesise that the scales were systematically failing to work for the heaviest people and we confirm this hypothesis by inquiry. In this case, the data is missing but we know something about the missing value. Specifically, we know that the missing value must either have been `+W` or `++W`. It is no longer appropriate to use the marginal height distribution, that is completely ignoring the weight. The missing value still contains some information. In this case, the correct approach would be to form a marginal distribution for height by summing over only the allowable weights, that is summing the `+W` and `++W` rows in the table. Computing this *bounded marginal* and looking at its value for height, `H`, shows the female to male ratio to be (4|10), so in this case we would guess the gender as male rather than female.

In the previous case, it was clear how to proceed because we had access to exact knowledge of the conditions under which the data had gone missing (the *missing-data model*). However, a good missing-data model is not always so readily available. Consider a third situation in which the procedure for collecting the data has been changed. Rather than the height and weight being recorded by a technician, subjects are asked to make their own measurements and then to record the data themselves on the survey form. Further, the form makes it clear that height and weight data is an optional part of the survey and survey subjects have the right to simply omit the data if they so chose. In this case, if the weight information is missing, after checking that the scales were working, it is most likely that the subject has simply withheld the information. Treating this missing information is now no longer straightforward. We might initially assume that the data omissions are independent of their value and of the gender of the subject, that is any person regardless of gender or weight has an equal probability of choosing not to report their weight. In this case, it might seem natural to proceed as if the scales had simply been broken and base the classification on a simple marginal probability. However, consider that perhaps the omissions are not random. It is possible, for example, that participants in the +W and ++W categories are embarrassed about their weight and hence more likely not to record it. In this case, we would want to treat this situation like the second scenario where the scales had been selectively broken. Alternatively, perhaps women – or men – are on average more private about their weight. It is clear that in this case the missingness in itself tells us something directly about the class. However, in general, we choose to proceed, the classification can only be performed after stating and justifying our beliefs about the missing-data model.

14.2.2 Missing Data Theory

Let us now formalise the missing-data classification problem that we discussed in Section 14.2.1. Let $\mathbf{x}$ be a vector in $\mathcal{R}^N$ representing the complete set of observed features (i.e. the height and weight in the previous example). In the example above N equals two, but the dimensionality of the problem is arbitrary. Let $\mathbf{x}_p \in \mathcal{R}^P$ where $P \leq N$ be a subvector of $\mathbf{x}$ constructed from the elements of $\mathbf{x}$ that are directly observed, that is the *present features*. Likewise $\mathbf{x}_m \in \mathcal{R}^M$, where $M + P = N$, is a vector representing the *missing features*. We will also consider the *missing-data pattern*, $\mathbf{s}$ which is a vector of N binary indicator variables which selects which of the features in $\mathbf{x}$ happen to be present and which happen to be missing. In the missing-data ASR literature, this missing-data pattern is also known as the *missing-data mask* or as the *segmentation*. (In this chapter, it is denoted $\mathbf{s}$ – as in *segmentation* – rather than $\mathbf{m}$ to avoid confusion with the usage of m to refer exclusively to missing features.)

In our initial description both the present features $\mathbf{x}_p$ and the missing-data pattern $\mathbf{s}$ are directly observed and we wish to select the class label, which will be denoted q, which is most probable given these observation (we will later generalise to the case where the missing-data pattern itself is not directly observed). So, the missing-data classification problem can be stated as

$$q' = \underset{q}{\operatorname{argmax}}\, P(q|\mathbf{x}_p, \mathbf{s}). \tag{14.1}$$

Before considering this problem, consider how we would proceed if $\mathbf{x}_p$ was the complete data set, that is, the missing features were not missing but had never existed, then the problem would be written as

$$q' = \underset{q}{\operatorname{argmax}}\, P(q|\mathbf{x}_p) \tag{14.2}$$

which, rearranging using Bayes' rule and ignoring the constant denominator, can be written as

$$q' = \underset{q}{\operatorname{argmax}}\, P(\mathbf{x}_p|q)P(q).$$

Note, $p(\mathbf{x}_p|q)$ can be computed as a marginal distribution of $p(\mathbf{x}|q)$. In the examples of Section 14.2.1, the gender was chosen using the marginal when the scales were totally broken. However, this solution resulted from Equation (14.2) rather than the correct, Equation (14.1). So, when are we justified in ignoring the conditioning on the missingness pattern, $\mathbf{s}$?

To clearly see the relation between Equation (14.2) and Equation (14.1) we will introduce the missing features, $\mathbf{x}_m$, and perform some simple algebra

$$\begin{aligned} q' &= \underset{q}{\operatorname{argmax}}\, P(q|\mathbf{x}_p, \mathbf{s}) \\ &= \underset{q}{\operatorname{argmax}} \sum_{\mathbf{x}_m} P(q, \mathbf{x}_m|\mathbf{x}_p, \mathbf{s}) \\ &= \underset{q}{\operatorname{argmax}} \sum_{\mathbf{x}_m} \frac{P(q, \mathbf{x}, \mathbf{s})}{P(\mathbf{x}_p, \mathbf{s})} \\ &= \underset{q}{\operatorname{argmax}} \sum_{\mathbf{x}_m} \left\{ \frac{P(q|\mathbf{x}, \mathbf{s})P(\mathbf{x}_p, \mathbf{x}_m, \mathbf{s})}{P(\mathbf{x}_p, \mathbf{s})} \right\} \\ &= \underset{q}{\operatorname{argmax}} \sum_{\mathbf{x}_m} \{P(q|\mathbf{x}, \mathbf{s})P(\mathbf{x}_m|\mathbf{x}_p, \mathbf{s})\} \\ &= \underset{q}{\operatorname{argmax}} \sum_{\mathbf{x}_m} \left\{ P(q|\mathbf{x}, \mathbf{s}) \frac{P(\mathbf{s}|\mathbf{x}_p, \mathbf{x}_m)P(\mathbf{x}_m|\mathbf{x}_p)}{P(\mathbf{s}|\mathbf{x}_p)} \right\}. \end{aligned} \tag{14.3}$$

Now $P(q|\mathbf{x}, \mathbf{s}) = P(q|\mathbf{x})$ because then the missing-data pattern tells us nothing extra about the class once we are given the full feature vector. (Note that this remains true only as long as we assume that the missing-data mechanism is not dependent on the class q.)

$$q' = \underset{q}{\operatorname{argmax}} \sum_{\mathbf{x}_m} \left\{ P(q|\mathbf{x})P(\mathbf{x}_m|\mathbf{x}_p) \frac{P(\mathbf{s}|\mathbf{x}_p, \mathbf{x}_m)}{P(\mathbf{s}|\mathbf{x}_p)} \right\}.$$

Now if we assume that

$$P(\mathbf{s}|\mathbf{x}_p, \mathbf{x}_m) = P(\mathbf{s}|\mathbf{x}_p) \tag{14.4}$$

the third term cancels and the $\mathbf{x}_m$ in the first term integrates out, to lead us back to Equation (14.2).

$$\begin{aligned} q' &= \underset{q}{\operatorname{argmax}} \sum_{\mathbf{x}_m} \left\{ P(q|\mathbf{x})P(\mathbf{x}_m|\mathbf{x}_p) \frac{P(s|\mathbf{x}_p, \mathbf{x}_m)}{P(s|\mathbf{x}_p)} \right\} \\ &= \underset{q}{\operatorname{argmax}} \sum_{\mathbf{x}_m} \left\{ P(q, \mathbf{x}_m|\mathbf{x}_p) \frac{P(s|\mathbf{x}_p))}{P(s|\mathbf{x}_p)} \right\} \\ &= \underset{q}{\operatorname{argmax}} \sum_{\mathbf{x}_m} P(q|\mathbf{x}_p). \end{aligned}$$

So, in summary, we are justified in using the marginal distribution directly in cases where we can assume Equation (14.4) holds, that is, in situations where, given the present features, the missingness pattern is independent of the values of the missing features. This condition is known as Missing At Random (MAR) [16].

Considering again our survey scenario. In the first case where the scales were totally broken the data was always missing. So the missingness pattern depended neither on the present or missing values

$$P(\mathbf{s}|\mathbf{x}_p, \mathbf{x}_m) = P(\mathbf{s}). \tag{14.5}$$

This more stringent condition is described in the literature as Missing Completely At Random (MCAR). Data that are MCAR are clearly also MAR, hence, the use of the marginal was justified.

A survey situation that is MAR without being MCAR could be contrived. For example, imagine subjects have their height measured and sent to separate rooms to be weighed with the room depending on their height. Now imagine that the scales in just the room for the tallest people is broken. Now tall people generally weigh more, so more heavy weights will be missing from the survey but conditioned on the observed data, the height, the missingness pattern is independent of weight. The data is still MAR and marginalisation could be validly applied. This is worth bearing in mind because the technical meaning of 'missing at random' does not fit precisely with intuitive ideas of randomness.

In the second and third scenarios, the fact that the data was missing depended, or potentially depended, on the weight (the missing value) itself even when conditioned on the observed height. So now the missing data is Not Missing At Random (NMAR). In this case, the model for the missing-data pattern needs to be explicitly stated:

$$\begin{aligned} q' &= \underset{q}{\operatorname{argmax}} \sum_{\mathbf{x}_m} \left\{ P(q|\mathbf{x}) P(\mathbf{x}_m|\mathbf{x}_p) \frac{P(s|\mathbf{x}_p, \mathbf{x}_m)}{P(s|\mathbf{x}_p)} \right\} \\ &= \underset{q}{\operatorname{argmax}} \sum_{\mathbf{x}_m} \left\{ P(q, \mathbf{x}_m|\mathbf{x}_p) P(s|\mathbf{x}_p, \mathbf{x}_m) \right\}. \end{aligned}$$

14.2.3 Validity of the MAR Assumption

We have seen that in situations where missing data is Missing At Random (MAR) there is a straightforward approach to classification: we take the present features to be complete feature vectors and classify them according to their marginal distributions. Good classification performance can be achieved if there is sufficient redundancy in the full feature vector that the marginal distributions retain some discriminative power.

Before considering how marginalisation is applied to acoustic models in ASR, it is important to consider whether the MAR assumption is ever valid for masked speech. Missing-data ASR was largely inspired by earlier work by Ahmad and Tresp [1] who demonstrated that missing feature theory could be successfully applied to the problem of occlusion in visual object classification. The case for the MAR approximation in vision is clear: it is saying that the chances of an object being occluded depend little on the appearance of the object. There are many visual scenes for which this is a reasonable statement. MAR is less readily justifiable in speech recognition. Indeed, very early demonstrations of missing-data ASR sidestepped the issue completely by randomly deleting spectro-temporal elements directly in the feature domain thus ensuring the data was MAR [8].

The validity of the MAR assumption in speech processing depends to a large extent on the circumstances under which the data has been lost. There are certain common situations where the assumption can be clearly justified. First, consider information that has been lost due to filtering. For example, speech that has been passed through a band-limited communication channel may be matched against models trained on the full-band signal as long as the filtered components are treated as missing data. In this case, the missing-data pattern is fixed and so clearly does not depend on the value of the missing elements. Marginalisation may also be an appropriate technique for dealing with information lost due to transmission error in distributed automatic speech recognition [27,33]. In this case, multiple speech features or complete temporal segments of the signal may be missing. Again, although there may be external factors that are predictive of a transmission error, the chances of error do not usually depend on the information being transmitted, and so the MAR condition holds.

However, missing-data techniques are most commonly applied to deal with data that has been lost due to energetic masking [9]. Is such data MAR? In this case, the obvious answer is no, because any particular channel is more likely to be masked – and hence missing – if the speech signal in that channel has low rather than high energy, that is, the missingness pattern depends on the value of the missing components, so the data is NMAR and a specific missing-data model needs to be introduced. Although this is true for masking in general, for certain noise types, MAR may still be a good approximation. In particular, imagine a noise signal that is spectro-temporally sparse but sufficiently intense that in regions where noise energy is present it consistently masks the speech regardless of the speech energy. Extremely abrupt impulsive noises such as hammer blows, or narrow-band tonal noises such as sirens may approach these conditions. In such situations, simple marginalisation may prove effective.

14.2.4 Marginalising Acoustic Models

In situations where the speech observations are missing at random, classification requires evaluation of the marginal likelihood. Our discussion so far has employed discrete observations, but we will now be dealing with continuous quantities, namely spectral energy observations. However, the formalism remains unaltered except that probability mass functions of discrete observations, $P()$ are replaced with probability density functions, $p()$, and summations over discrete probabilities are replaced with integrals over continuous densities, hence the marginal likelihood is written as

$$p(\mathbf{x}_p|q) = \int_{\mathbf{x}_m} p(\mathbf{x}_p, \mathbf{x}_m|q)\, \mathrm{d}\mathbf{x}_m .$$

The ease with which the marginal can be computed will depend on the form of the p.d.f. used to model the acoustic feature vectors. When modelling speech in the spectral domain it is necessary to account for the dependence between features, that is features in adjacent frequency bands are highly correlated. One possibility is to use a multivariate normal distribution with full covariance

$$p(\mathbf{x}|q) = \frac{1}{K} \exp\left\{ -\frac{1}{2}(\mathbf{x} - \boldsymbol{\mu}_q)^T \boldsymbol{\Sigma}_q^{-1} (\mathbf{x} - \boldsymbol{\mu}_q) \right\},$$

where $\boldsymbol{\mu}_q$ is the mean vector and $\boldsymbol{\Sigma}_q$ is a covariance matrix for class q and K is the normalisation constant. In this case, it can be shown [26] that the marginal is also a multivariate normal distribution

$$p(\mathbf{x}_p|q) = \frac{1}{K'} \exp\left\{-\frac{1}{2}(\mathbf{x}_p - \boldsymbol{\mu}_{q,p})^T \boldsymbol{\Sigma}_{q,pp}^{-1}(\mathbf{x}_p - \boldsymbol{\mu}_{q,p})\right\}, \tag{14.6}$$

where $\boldsymbol{\mu}_{q,p}$ is a vector constructed from the original mean vector, $\boldsymbol{\mu}_q$ by striking out the elements corresponding to the missing values in x, and likewise, $\boldsymbol{\Sigma}_{q,pp}$ is constructed from the covariance matrix $\boldsymbol{\Sigma}_q$ by striking out the rows and columns corresponding to the missing elements. Note that the normalisation constant for the marginal, K', does not have the same value as the constant needed to normalise the full distribution, K. (Henceforth, the class subscript, q, is dropped from $\vec{\mu}$ and $\boldsymbol{\Sigma}$ for the sake of compactness).

Unfortunately, although the marginal has a lower dimensionality that the full distribution, it cannot be computed with the same efficiency: in a standard full-covariance ASR system the precision matrices $\boldsymbol{\Sigma}^{-1}$ can be pre-computed and stored. However, in a missing-data system, the matrices $\boldsymbol{\Sigma}_{pp}^{-1}$ are a function of the missing-data pattern which is generally changing at each frame. Given the exponentially large number of possible missing-data patterns, pre-computation is not a practical option and the matrix inverses have to be computed at run time. For systems with a large number of states, this can be a significant computational burden.

The more common alternative for modelling feature interdependence is to use a Gaussian mixture model (GMM) composed of mixture components having a diagonal covariance matrix. The p.d.f. for a diagonal covariance GMM is given as

$$p(\mathbf{x}|q) = \sum_{m=1}^{M} P(m|q)p(\mathbf{x}|m,q) = \sum_{m=1}^{M} P(m|q)\prod_{i=1}^{N} p(x_i|m,q),$$

where M is the number of mixture components, N is the number of elements in the feature vector x and $p(x_i|m,q)$ is a univariate Gaussian distribution

$$p(x_i|m,q) = \frac{1}{\sqrt{2\pi\sigma_{i,m}^2}} \exp\left\{-\frac{1}{2}\left(\frac{x_i - \mu_{i,m}}{\sigma_{i,m}}\right)^2\right\}.$$

Note, that for a diagonal covariance GMM there is dependency between features so $p(\mathbf{x}_m, \mathbf{x}_p|q) \neq p(\mathbf{x}_m|q)p(\mathbf{x}_p|q)$; however, features are independent within each mixture, $p(\mathbf{x}_m, \mathbf{x}_p|q,m) = p(\mathbf{x}_m|q,m)p(\mathbf{x}_p|q,m)$. This factorisation greatly simplifies the form of the marginal distribution

$$\begin{aligned} p(\mathbf{x}_p|q) &= \int_{\mathbf{x}_m} p(\mathbf{x}_m, \mathbf{x}_p|q) d\mathbf{x}_m \\ &= \int_{\mathbf{x}_m} \sum_{m=1}^{M} P(m|q)p(\mathbf{x}_m, \mathbf{x}_p|m,q) d\mathbf{x}_m \\ &= \sum_{m=1}^{M} P(m|q)p(\mathbf{x}_p|m,q) \underbrace{\int_{\mathbf{x}_m} p(\mathbf{x}_m|m,q)\, d\mathbf{x}_m}_{1} \\ &= \sum_{m=1}^{M} P(m|q) \prod_{i\in\mathcal{P}} p(x_i|m,q), \end{aligned} \tag{14.7}$$

where $\mathcal{P}$ is the set of indices of the present features, that is, each mixture is evaluated as a product of univariate Gaussians, one for each feature that is marked as present.

A missing-data ASR system then can be implemented as a relatively minor modification to a conventional hidden Markov model (HMM)-based automatic speech recognition (ASR) system: HMMs with, typically, diagonal covariance GMM emission distributions are trained on noise free speech using a spectral representation and the usual expectation maximisation (EM) algorithm. At recognition time, the usual Viterbi algorithm is employed but the decoder is also given access to a missing-data pattern. The missing-data pattern and the observed noisy spectra are passed to each state in the model. Then the usual state likelihood computation is replaced with an evaluation of Equation (14.7) – or one of the more sophisticated missing-data likelihood equations that we shall see later. So essentially a missing-data ASR system is akin to a conventional HMM just with a different calculation for the state likelihood. This holds true for all the systems discussed in this chapter other than the speech fragment-decoding technique introduced in Section 14.4.3.

14.3 Energetic Masking

Section 14.2 introduced the missing-data formalism and demonstrated that when the data is Missing at Random (MAR) classification can proceed simply by replacing the probability distributions for the complete observation vector with the marginal distribution of the present data components. A few cases where the MAR approximation holds were discussed and examples of the marginal computation for multivariate gaussian distributions and diagonal covariance Gaussian mixture models have been provided. However, in general, when spectro-temporal data has been lost due to noise masking – the situation where missing-data techniques are most commonly applied – the data is far from being MAR. This section will demonstrate how an explicit missing-data model can be introduced to correct for the lack of randomness. Different missing models will be discussed.

14.3.1 The Max Approximation

Figure 14.2 shows a log energy domain representation of a speech spectrum (dashed) in the presence of a masking noise source whose spectrum is shown by the dotted trace. The spectrum of the combined signals is given by the solid line.

Imagine that some process has provided us with the missingness pattern, s, which identifies the frequency channels in which the speech signal dominates (i.e. present data) and those in which it is masked (i.e. missing data). In the figure the missingness pattern, s, is indicated by the horizontal line that lies over the frequency regions in which the data is present. As before, let the speech features x be partitioned into those that are directly observable $\mathbf{x}_p$ and those that are masked by the interferer and are hence not observed, $\mathbf{x}_m$.

Are the missing values likely to be missing at random? MAR implies that

$$P(\mathbf{s}|\mathbf{x}_p, \mathbf{x}_m) = P(\mathbf{s}|\mathbf{x}_p).$$

This assumption clearly does not hold for the simple fact that low-energy regions of the speech signal are more readily masked than high-energy regions. In fact, if we take y to be the observed spectra, then it is the case that, in the regions where the speech is known to be masked, the unknown speech energy must be lower than the observed masking energy.

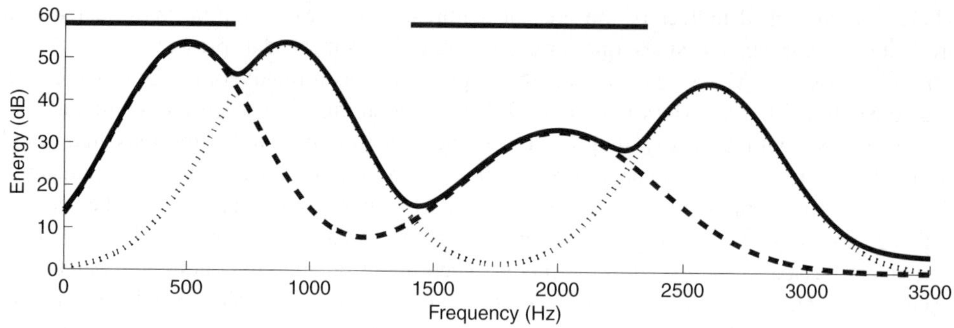

Figure 14.2 A pair of energy spectra (dashed lines) and their summation (solid line) in the log domain. Note that at each frequency point the summation is well approximated by the maximum of the two signals being combined. The maximum error in this approximation is about 3 dB which occurs when the signals being combined have equal energy.

The missing regions are themselves informative. The fact that the speech is masked in these regions provides counter-evidence against acoustic model states that would have produced observations that are more energetic than the masking level [9].

Although the data is NMAR the missing data can be modelled using a simple approximation to describe the addition of the two signals. Operating in the log energy domain let $\mathbf{y}$ represent the spectrum that results from the combination of a log domain speech spectra, $\mathbf{x}$ and a noise spectra, $\mathbf{n}$. Assuming phase independence of the noise and speech source then for each spectral component y_i

$$y_i = \log(\exp(x_i) + \exp(n_i)).$$

Due to the sparsity of the speech signal, in most frequency channels either $\exp(x_i) \gg \exp(n_i)$ or $\exp(n_i) \gg \exp(x_i)$. As such the combination of the speech and noise in the log domain can be modelled using the *max approximation*

$$y_i \approx \max(x_i, n_i).$$

The accuracy of this approximation is illustrated in the figure by the fact that the solid line representing the combined spectra is nearly always close to the bigger of either the dashed or dotted lines representing the component spectra. The maximum deviation is about 3 dB which occurs when the two sources have equal energy. This error of 3 dB is typically small compared to the variability of the spectra across frequency. Ignoring the error in the max approximation, it holds that in regions where the speech dominates it corresponds to the observed signal, $\mathbf{x}_p = \mathbf{y}_p$, and in the masked regions we know that the speech energy lies somewhere between 0 and the observed energy, $\mathbf{0} \leq \mathbf{x}_m \leq \mathbf{y}_m$. This is known as the *missing-data bounds constraint* [9].

14.3.2 Bounded Marginalisation

So far missing data has been discussed in terms of observations that are present or missing and we have performed classification with respect to the present features. The discussion in Section 14.3.1 motivates a subtly different view. A noise-corrupted speech spectrum $\mathbf{y}$ is observed and our interpretation of $\mathbf{y}$ is molded by our belief about the missing-data pattern s. In the masked regions, the noisy observation, $\mathbf{y}$, acts as an upper limit on the possible value of the unobserved speech, whereas in the unmasked regions the 'unobserved' speech energy is equal to the noisy observation $\mathbf{y}$. In this formulation the speech features are all treated as potentially missing, and the noisy spectrum $\mathbf{y}$ and the missingness pattern s are given hence the classification problem is written as

$$q' = \underset{q}{\operatorname{argmax}}\, P(q|\mathbf{s}, \mathbf{y}). \tag{14.8}$$

Now, following Equation (14.3) we arrive at

$$\begin{aligned} q' &= \underset{q}{\operatorname{argmax}} \int P(q, \mathbf{x}|\mathbf{s}, \mathbf{y})\, d\mathbf{x} \\ &= \underset{q}{\operatorname{argmax}} \int P(q|\mathbf{x}) p(\mathbf{x}|\mathbf{s}, \mathbf{y})\, d\mathbf{x} \\ &= \underset{q}{\operatorname{argmax}} \int p(\mathbf{x}|q) \frac{p(\mathbf{x}|\mathbf{s}, \mathbf{y})}{p(\mathbf{x})} d\mathbf{x} P(q) \\ &= \underset{q}{\operatorname{argmax}} \int p(\mathbf{x}|q) W_{s,y}(\mathbf{x})\, d\mathbf{x} P(q). \end{aligned} \tag{14.9}$$

So now the likelihood $p(\mathbf{x}|q)$ used when the speech data is directly observed is replaced by a weighted integral of $p(\mathbf{x}|q)$ computed over all possible speech observations, $\mathbf{x}$. The weighting is dictated by the function $W_{s,y}(\mathbf{x})$. From the previous discussion on the max approximation, $p(\mathbf{x}|\mathbf{s}, \mathbf{y}) \neq 0$ only if $x_i = y_i$ for all i where $s_i = 1$ and $x_i < y_i$ for all i where $s_i = 0$. For values of $\mathbf{x}$ that obey these constraints, we would expect $p(\mathbf{x}|\mathbf{s}, \mathbf{y})$ to be proportional to $p(\mathbf{x})$ (as long as we ignore the fact that speakers adapt their speech in the presence of noise in order to reduce the effects of masking [7, 17]). So W takes a value that is either 0 or constant. Since the scale of the constant will not influence the outcome of the argmax operation, we can take it to be 1 and then we can conveniently factorise $W_{s,y}(\mathbf{x})$ as

$$W_{s,y}(\mathbf{x}) = \prod_{i=1}^{N} W_{s,y}(x_i), \tag{14.10}$$

where

$$W_{s,y,i}(x_i) = \begin{cases} 1 & : \quad s_i = 1 \quad \text{and} \quad x_i = y_i \\ 1 & : \quad s_i = 0 \quad \text{and} \quad x_i < y_i \\ 0 & : \quad \text{else} \end{cases} \tag{14.11}$$

Now if the speech observation distribution is modelled using a diagonal covariance Gaussian mixture model then the integral in Equation (14.9) can be evaluated as follows,

$$\begin{aligned} \int p(\mathbf{x}|q)W_{s,y}(\mathbf{x})\,\mathbf{dx} &= \int \sum_{m=1}^{M} P(m|q) \prod_{i=1}^{N} W_{s,y,i}(x_i)p(x_i|m,q)\mathrm{d}x_i \\ &= \sum_{m=1}^{M} P(m|q) \prod_{i=1}^{N} \left\{ \int W_{s,y,i}(x_i)p(x_i|m,q)\mathrm{d}x_i \right\} \\ &= \sum_{m=1}^{M} P(m|q) \prod_{i\in\mathcal{P}} p(x_i = y_i|m,q) \prod_{i\in\mathcal{M}} \int_{-\infty}^{y_i} p(x_i|m,q)\mathrm{d}x_i, \end{aligned} \tag{14.12}$$

where $\mathcal{P}$ is the set of indices for which $s_i = 1$ (i.e. the present data) and $\mathcal{M}$ is the set of indices for which $s_i = 0$ (i.e. the missing data).

In the missing-data ASR literature Equation (14.12) is typically referred to as the *bounded marginalisation* approach [9]. An integration is being performed similar to that performed in the computation of the marginal distribution, but in this case the integration is only over the permissible values of the missing features whose maximum value is bounded by the observed energy y_i, hence the appearance of the integration bounds. Note that if the masking energy were to be infinite then the bounded marginal Equation (14.12) simplifies to the evaluation of the probability of $\mathbf{x}_p$ using the marginal distribution, Equation (14.7).

14.3.3 Missing Data ASR in the Cepstral Domain

The missing-data techniques that have been discussed so far all operate in the spectral domain, that is the acoustic models are constructed from the statistics of speech energies measured across a range of frequency bands. Modelling in the spectral domain is an obvious choice because it is in the spectral domain that noise masking can be observed to have a *local effect*. Additive noise will obstruct some spectral features, rendering them as missing, while leaving others intact. However, spectral features do not lend themselves well to statistical acoustic modelling. For example, there exist complicated correlations between features which mean many extra parameters have to be estimated. Further, the spectral representation is not invariant to changes in energy level such as might occur if a speaker moves closer to a microphone, nor are they easily normalised for channel variability, such as changes in microphone or changes in room reverberation. Large vocabulary speech-recognition systems have long solved these problems by developing sophisticated chains of feature preprocessing steps [15]. The most significant of these steps is typically the use of linear transforms to decorrelate the features, for example as employed in the construction of mel-frequency cepstral coefficients [19].

Can missing-data techniques be employed with acoustic models trained on cepstral representations? One approach for interfacing spectral missing-data approaches and cepstral speech models is to employ spectral imputation [29]. Imputation-based techniques operate by reconstructing the missing spectral regions based on a best guess of the noise-free spectrum. After reconstruction, it is then straightforward to apply the necessary linear transform to the estimated spectrum to compute an estimated cepstral observation vector (see Chapter 15 for a full account). Although the practical advantages of being able to interface with existing large vocabulary ASR infrastructure should not be underestimated, imputation-based techniques remain sub-optimal from a theoretical point of view because the imputed spectra discard information about the uncertainty. Recall from the simple example described in Section 14.2.1 how the imputation based approach led to an incorrect result. Techniques to incorporate estimates of the imputation uncertainty into the decoding can address these problems, but the uncertainty itself can be difficult to estimate or to model [13, 25].

There have been attempts to directly incorporate cepstral models into the theoretically well-motivated marginalisation approach; however, to do so successfully is problematic for two main reasons.

Firstly, noise corruption that is typically local in the spectral domain becomes spread across all cepstral parameters. Consider a spectral feature vector with one parameter that is missing in an unbound sense, that is the missing feature is free to take any value. Imagine there are 32 frequency channels. If one feature is missing we can compute the necessary $p(\mathbf{x}_p|q)$ using the remaining 31 features and the marginal distribution obtained by integrating over the missing dimension (see Section 14.2.4). Given that the spectral representation contains a lot of redundancy it would be expected that the remaining 31 channels were sufficiently informative to afford a high level of classification performance. Now consider computing a linear transform of the original 32 features to form a set of up to 32 cepstral features. It is easy to see that the uncertainty of the one missing feature is spread across all 32 cepstral features. If we are infinitely uncertain about the value of the one missing spectral feature, we become infinitely uncertain about all 32 cepstral features. In practice however we are never infinitely uncertain about the spectral parameter rather we acknowledge that even if a spectral feature were to be fully missing (e.g. filtered out) we are able to fall back on prior knowledge, that is spectral features have a well defined energy distribution $p(x)$ that can be learned from the training data.

A second more intransigent difficulty with the cepstral domain arises from a computational problem. In the spectral domain, our uncertainty about the missing data as represented by the bounds constraint presents itself as a set of integrations over hyper-cuboid regions which are aligned with the axes. When the distributions are factorised, we have seen how these become simple 1-d integrals. Consider the situation when we perform a linear transform to decorrelate the features. Figure 14.3 illustrates the distribution of two unobserved spectral features. The

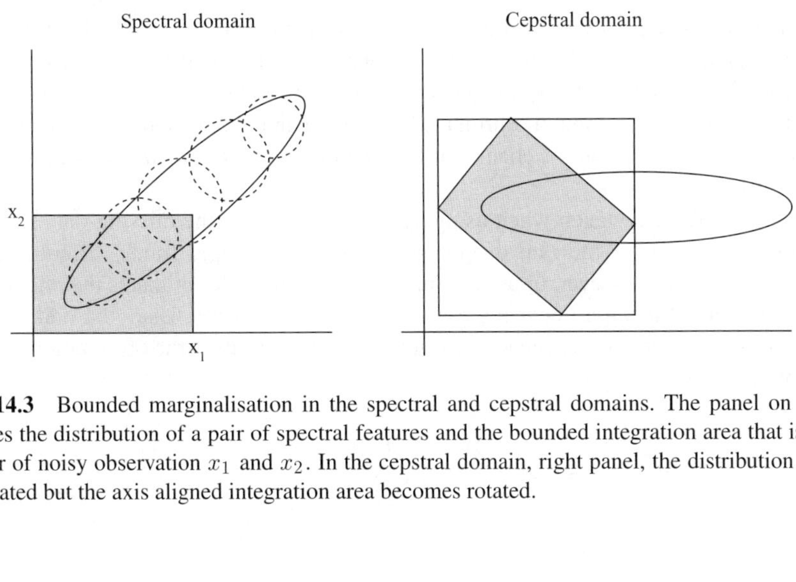

Figure 14.3 Bounded marginalisation in the spectral and cepstral domains. The panel on the right illustrates the distribution of a pair of spectral features and the bounded integration area that is implied by a pair of noisy observation x_1 and x_2. In the cepstral domain, right panel, the distribution becomes decorrelated but the axis aligned integration area becomes rotated.

speech energy is not directly observed but known to be less than the observed masking noise levels and hence the vector $\mathbf{x}$ can lie anywhere in the shaded region. If operating in the log energy domain there is no effective lower bound; however, missing-data systems often work with cube root compressed energies in which case the unobserved speech features are bound below by 0 [9]. Although there is no closed form solution for integrating a full covariance Gaussian distribution over this region, the distribution is typically approximated by a diagonal covariance GMM (as illustrated by the dotted circles in the figure) so the integration becomes the sum of easily computed integrals over mixture components. The panel on the right shows the situation after a diagonalising linear transform has been applied. The features are now decorrelated and hence the ellipse is axis aligned, but now the region of possible speech observations has become a rectangle that is not aligned with the axis. This integral does not have a closed form solution.

One solution to approximating the integral in the cepstral domain is to estimate it with another integral that can be easily computed. For each cepstral feature, it is possible to *independently* compute the minimum and maximum values that can be obtained by considering appropriate extreme choices for the missing *spectral* values, that is points lying on the vertices of the cuboid integration region on the left. Then these minimum and maximum values are used as independent integration bounds for the cepstral features. Geometrically this is equivalent to integrating over the smallest axis-aligned hyper-cuboid that encloses the correct non-axis aligned hyper-cuboid integration region, that is the white box surrounding the grey region in the panel on the right of the figure. However, as the figure makes clear, this technique hugely overestimates the true uncertainty. The significance of the additional corner areas becomes exponentially greater as the number of dimensions increase. Ad-hoc adjustments to the bounds may improve the situation but the underlying problem remains. Further, if operating with log energies, which could in theory be infinitely small, then no suitable axis aligned approximation can be computed. However, ideas similar to these have been applied to good effect in recent missing-data-imputaton systems [14].

14.3.4 Missing Data ASR with Dynamic Features

It is common practice in ASR systems to include the rate of change of the observed acoustic features as an additional set of features that are appended to the feature vector. These dynamic features (often referred to as *delta features* or *velocities*) are equally important whether working in the cepstral or spectral domain. Further information can be extracted by applying this idea twice and computing the rate of change of the delta features, that is *delta delta features*, or *accelerations*.

Some care needs to be taken when using dynamic features in the missing-data framework. The simplest estimate of the rate of change would be a simple frame difference, that is $\Delta x_{i,t} = x_{i,t} - x_{i,t-1}$. However, in order to produce more reliable estimates the rate of change is usually computed using a linear regression over several frames. Using an odd number of frames $(2N + 1)$ centered on x_t the least means squared linear fit to the observations produces a gradient of

$$\Delta x_{i,t} = \frac{\sum_{j=1}^{N} j(x_{i,t+j} - x_{i,t-j})}{\sum_{j=1}^{N} 2j^2}.$$

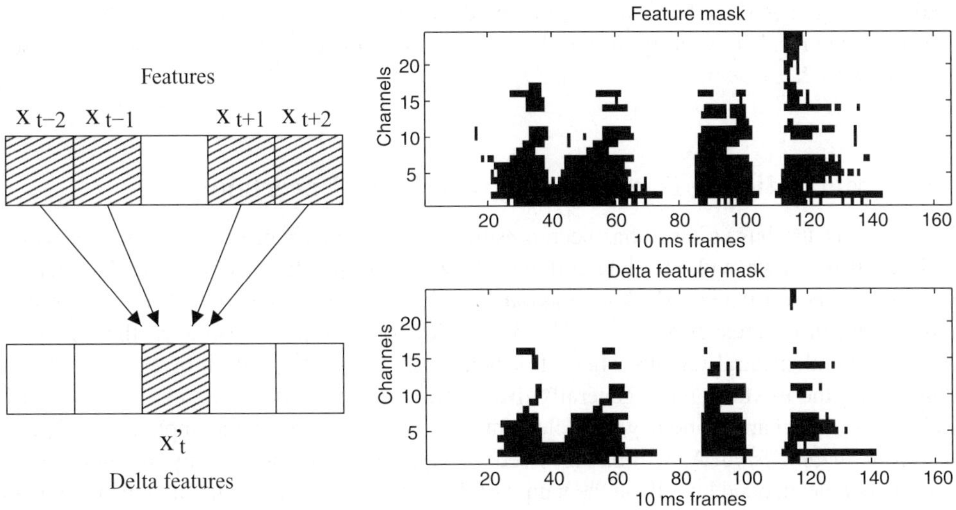

Figure 14.4 The construction of a missing-data pattern for delta features. The panel on the left illustrates the four static features that contribute to a delta feature when employing five frame linear regression. The panel on the right shows a missing-data mask for static features (black indicates present) and the corresponding delta mask in which delta features are only marked present if they can be estimated from present static features.

The panel of the left of Figure 14.4 illustrates the set of static features that are employed in the estimation of a delta feature when using a window size of five, that is $N = 2$. If any of the static features that contributes to the computation of the delta feature is marked as missing then the delta feature cannot be directly computed and should itself be marked as missing [2, 34]. The panel on the right shows an example of a missing-data pattern for a set of spectro-temporal features (top) and the corresponding missing-data pattern for the delta features if this rule is applied (bottom).

Bounds on the missing delta features can be computed, using a similar logic to that discussed in Section 14.3.3, by considering extreme values of the possible missing static features, that is x_{min} and the observed masking energy. The maximum delta would be achieved by setting the missing observations on the left of the window to x_{min} and setting the missing values on the right to be the observed masking energy, and vice versa for the minimum delta. For example, if all static features were missing, and x_{min} was assumed to be 0 – as would be appropriate if using, say, cube root energy features – then the delta bounds would be $\left[-0.2y_{i,t-2} - 0.1y_{i,t-1},\quad 0.1y_{i,t+1} + 0.2y_{i,t+2}\right]$. In practice, however, these bounds are so wide that using them provides little advantage over considering the delta feature to be fully missing. For the sake of simplicity, missing-data systems typically apply full marginalisation to delta and delta delta features [2].

Previously, it was explained that the techniques described in the chapter are based on the assumption that the data is *missing at random* (MAR), which – loosely stated – means that the probability of a feature being missing must not depend on the value of the feature. Since the missingness pattern for the dynamic features depends only on the missing pattern of the static features, then if the MAR assumption holds for the static features it will also hold for the dynamic features. The validity of the MAR assumption for the static features was discussed in Section 14.2.3.

14.4 Meta-Missing Data: Dealing with Mask Uncertainty

In all the discussions so far, it has been presumed that the missingness pattern s is provided and that it is known with absolute certainty. However, in practice, the missing-data pattern itself has to be estimated. Mask estimation is a challenging problem which is discussed in more detail in Chapter 16 (see also [11, 32, 35, 36] for examples). Although there may be some temporal spectral regions where it is clear whether it is the speech or the noise that is dominating the mixture, more generally there will be uncertainty in the missing-data mask estimate – information about which elements are missing is itself incomplete or missing, therefore *meta-missing*. As will be discussed, the appropriate model for missing-data pattern uncertainty can depend largely on the nature of the noise and in the manner in which the mask estimate has been formed. This section will discuss approaches to modelling *mask uncertainty*, and how mask uncertainty can be properly accounted for during missing-data classification.

14.4.1 Missing Data with Soft Masks

So far it has been assumed that there is direct access to the exact missing-data pattern, s. However, in real applications, the missing-data pattern is estimated from an observation of the noisy signal, $\mathbf{y}$. For example, we might choose to employ the single missing-data pattern s' that we believe to be the most likely

$$q' = \underset{q}{\operatorname{argmax}}\, P(q|\mathrm{s}', \mathbf{y}),$$

where

$$\mathrm{s}' = \underset{\mathrm{s}}{\operatorname{argmax}}\, P(\mathrm{s}|\mathbf{y}).$$

However, using a point estimate of the mask does not represent the true uncertainty of our beliefs.This uncertainty can be quite large as there is often no clear evidence on which to base the decision of missingness. The correct thing to do is to sum over all possible masks weighted by their probability, that is

$$\begin{aligned} P(q|\mathbf{y}) &= \sum_{\mathrm{s}} \left\{ \int_{\mathrm{x}} p(q, \mathrm{x}, \mathrm{s}|\mathbf{y})\, d\mathrm{x} \right\} \\ &= \sum_{\mathrm{s}} \left\{ \int_{\mathrm{x}} p(q|\mathrm{x}, \mathrm{s}, \mathbf{y}) p(\mathrm{x}|\mathrm{s}, \mathbf{y}) P(\mathrm{s}|\mathbf{y})\, d\mathrm{x} \right\} \\ &= \sum_{\mathrm{s}} \left\{ \int_{\mathrm{x}} p(\mathrm{x}|q) \frac{p(\mathrm{x}|\mathrm{s}, \mathbf{y})}{p(\mathrm{x})} d\mathrm{x} P(\mathrm{s}|\mathbf{y}) \right\} P(q), \end{aligned}$$

where the last step above is performed by first using the fact that $p(q|\mathbf{x}, \mathbf{s}, \mathbf{y}) = p(q|\mathbf{x})$ and then applying Bayes' rule.

The missing-data pattern distribution, $p(\mathbf{s}|\mathbf{y})$, in the above summation is a distribution over all possible binary vectors. A possible simplifying assumption would be to consider the missingness state of each component of the spectra to be independent given the noisy observation, $\mathbf{y}$

$$P(\mathbf{s}|\mathbf{y}) = \prod_{i=1}^{N} P(s_i|\mathbf{y}).$$

The distribution for a single element can be modelled as a Bernoulli distribution with a single parameter, λ_i, which by convention is taken to represent the probability of the element being *present*. So the missingness pattern s is the outcome of *N* independent Bernoulli trials each with a separate parameter, λ_i. This vector of parameters is often described as a *soft missing-data mask* because it represents the probability of each spectral element being present [2]. In the case where the parameters are all either 0 or 1 then the mask elements are correspondingly either 0 or 1 with certainty and the situation is equivalent to that discussed in earlier sections where certain knowledge of the missing-data pattern is assumed. The soft missing-data mask is typically estimated deterministically from the noisy data, that is $\lambda = f(\mathbf{y})$ so we will sometimes denote $P(\mathbf{s}|\mathbf{y})$ as $P_\lambda(\mathbf{s})$.

Given the soft missing-data mask λ the probability distribution for the missing-data masks, $P_\lambda(\mathbf{s})$ is given by

$$P(\mathbf{s}|\mathbf{y}) = P_\lambda(\mathbf{s}) = \prod_{i \in \mathcal{P}} \lambda_i \prod_{i \in \mathcal{M}} (1 - \lambda_i).$$

As before let us assume a diagonal covariance Gaussian mixture model for $p(\mathbf{x}|q)$ then $P(q|\mathbf{y})$ becomes

$$P(q|\mathbf{y}) = \sum_{\mathbf{s}} \int_{\mathbf{x}} \sum_{m=1}^{M} P(m|q) \prod_{i=1}^{N} p(x_i|m, q) \frac{p(\mathbf{x}|\mathbf{s}, \mathbf{y})}{p(\mathbf{x})} d\mathbf{x} \prod_{i \in \mathcal{P}_s} \lambda_i \prod_{i \in \mathcal{M}_s} (1 - \lambda_i) P(q). \quad (14.13)$$

The direct summation over missing-data patterns would involve 2^N terms and is generally not practicable. The computation would become feasible if, within each mixture component, each of the *N* spectral features made an independent contribution, allowing the equations to be expressed as a product over *N* terms. In order to see how this might be achieved consider the term

$$W_{\mathbf{s},\mathbf{y}}(\mathbf{x}) = \frac{p(\mathbf{x}|\mathbf{s}, \mathbf{y})}{p(\mathbf{x})}.$$

Earlier it was argued that because $p(\mathbf{x}|\mathbf{s}, \mathbf{y})$ is either proportional to $p(\mathbf{x})$ or 0, then *W* is constant everywhere that it is not 0, and therefore *W* can be factorised as shown previously is Equations (14.10) and (14.11) where the constant has been arbitrarily scaled to unity without effect on the argmax operation. However, the constant depends on s and so we cannot arbitrarily scale it to unity (i.e. there is a different constant for each term in the sum over missing-data patterns) and so this term cannot generally be factorised in a useful way. In order to proceed we need to assume that $p(\mathbf{x})$ and $p(\mathbf{x}|\mathbf{s}, \mathbf{y})$ can themselves be factorised into a product of univariate distributions for each spectral feature

$$p(\mathbf{x}) = \prod_{i=1}^{N} p(x_i),$$

$$p(\mathbf{x}|\mathbf{s},\mathbf{y}) = \prod_{i=1}^{N} p(x_i|\mathbf{s},\mathbf{y}).$$

Then there are two cases, either s_i indicates that the spectral feature is dominated by speech energy and so x_i must be equal to y_i

$$p(x_i|s_i = 1, y_i) = \delta(x_i - y_i).$$

Alternatively s_i indicates the feature is masked in which case

$$p(x_i|s_i = 0, y_i) = \begin{cases} F_i.p(x_i) & : \; x_i \leq y_i \\ 0 & : \; xi > y_i \end{cases},$$

where F_i is a normalisation constant need to ensure that the truncated distribution integrates to unity and remains a true pdf, that is

$$F_i = \frac{1}{\int_{-\infty}^{y_i} p(x_i)\mathrm{d}x_i}.$$

Substituting the factored form of *W* into Equation (14.3) produces

$$\begin{aligned} P(q|\mathbf{y}) &= \sum_{\mathbf{s}} \int_{\mathbf{x}} \sum_{m=1}^{M} P(m|q) \prod_{i=1}^{N} \left\{ p(x_i|m,q) \frac{p(x_i|s_i,y_i)}{p(x_i)} \right\} \mathrm{d}x \prod_{i \in \mathcal{P}_s} \lambda_i \prod_{i \in \mathcal{M}_s} (1-\lambda_i) P(q) \\ &= \sum_{m=1}^{M} P(m|q) \sum_{\mathbf{s}} \prod_{i=1}^{N} \left\{ \int_{x_i} p(x_i|m,q) \frac{p(x_i|s_i,y_i)}{p(x_i)} \mathrm{d}x_i \right\} \prod_{i \in \mathcal{P}_s} \lambda_i \prod_{i \in \mathcal{M}_s} (1-\lambda_i) P(q), \end{aligned} \tag{14.14}$$

which, substituting in the missing and present data forms for $p(x_i|m, q)$ and using the notation $p(x = y)$ to represent the value of $p(x)$ evaluated at *y*, can be rewritten as

$$P(q|\mathbf{y}) = \sum_{m=1}^{M} P(m|q) \sum_{\mathbf{s}} \prod_{i \in \mathcal{P}_s} \lambda_i \frac{p(x_i = y_i|m,q)}{p(x_i = y_i)} \prod_{i \in \mathcal{M}_s} (1-\lambda_i) F_i \int_{-\infty}^{y_i} p(x_i|m,q) \mathrm{d}x_i,$$

which can then be rearranged to allow the sum over all possible masks within each of the *M* mixtures to be computed efficiently as a product of *N* terms

$$P(q|\mathbf{y}) = \sum_{m=1}^{M} P(m|q) \prod_{i=1}^{N} \left\{ \lambda_i \frac{p(x_i = y_i|m,q)}{p(x_i = y_i)} + (1-\lambda_i) F_i \int_{-\infty}^{y_i} p(x_i|m,q) \mathrm{d}x_i \right\}.$$

As we are usually searching for the q that maximises $P(q|\mathbf{y})$, and hence constant scaling does not affect the result, the soft missing-data equation is usually seen written as

$$q' = \underset{q}{\text{argmax}} \sum_{m=1}^{M} P(m|q) \prod_{i=1}^{N} \left\{ \lambda_i p(x_i = y_i|m,q) + (1-\lambda_i)\frac{1}{k_i}\int_{-\infty}^{y_i} p(x_i|m,q)\mathrm{d}x_i \right\}.$$

Compared with the discrete mask case, where within each mixture each element of the spectral feature vector either contributes a present-data term or a missing-data term, now each element contributes a weighted sum of the present and missing-data terms where the soft missing-data mask dictates the weighting. The scaling constant k_i is effectively converting the probability into a probability density so that the present and missing-data terms are commensurate. The constant k_i can be calculated as

$$k_i = \frac{1}{F_i p(x_i = y_i)} = \frac{\int_{-\infty}^{y_i} p(x_i)\mathrm{d}x_i}{p(x_i = y_i)}. \tag{14.15}$$

If $p(x_i)$ is assumed to be a uniform distribution between 0 and any arbitrary maximum value then evaluation of Equation (14.15) gives $k_i = y_i$, that is the missing-data integral term is normalised by division by the observed masking energy, y_i. This is exactly the equation presented in the original soft missing-data paper [2]; however, in this original paper the result is derived using an intuitive argument without any formal justification.

14.4.2 Sub-band Combination Approaches

Sub-band combination approaches (e.g. multi-band combination [4, 24] and the probabilistic union model [20–22]) consider the speech spectrum as being split into a small number of frequency sub-bands (typically three or four). Statistical models are then built for features extracted from each band and then some heuristic is employed at recognition time to combine separate band observation probabilities in a way that accommodates the potential for one or more bands to be corrupt. The bands may be modelled directly in the spectral domain, but are more commonly independently transformed using a technique typically applied to the full spectrum, for example transformation to the cepstral domain.

The *multi-band combination* approach replaces the observed data probability $p(\mathbf{x}|q)$ with a score computed as a product over bands

$$f(\mathbf{x}|q) = \prod_{b=1}^{B} p(\mathbf{x}_b|q)^{\lambda_b},$$

where $\mathbf{x}_b$ are the sub-band feature vectors and λ_b are sub-band weighting factors that control the dynamic range of the likelihoods. In the simplest case all weights are equal; however, if there is evidence that a band has low SNR the weight can be reduced, or, in severe noise conditions, set to 0 so that the band is effectively ignored. The effectiveness of the technique depends on how well sub-band SNRs can be estimated. An obvious drawback is that by modelling the sub-bands independently, it is not possible to take advantage of the correlations that exist between widely separated frequency regions and which can be captured by full-band Gaussian mixture modelling.

The *probabilistic union model* notes that the unweighted multi-band approach is equivalent to combining bands with an 'AND' operator, that is the data provides evidence for a class to the extent that *all* bands provide support for that class, whereas in noisy situations it would seem more appropriate to combine bands using an 'OR' operator [21]. The model can be expressed as

$$P(\mathbf{x}|q) = P(\bigvee_{n_1,n_2,\ldots n_{N-M}} \mathbf{x}_{n_1}\mathbf{x}_{n_2}\ldots\mathbf{x}_{n_{N-M}}|q),$$

where $\mathbf{x}_{n_1}\mathbf{x}_{n_2}\ldots\mathbf{x}_{n_{N-M}}$ is a sub-set of $N - M$ of the total N available sub-bands. Sub-bands within each sub-set are combined with an 'AND' operator, whereas $\bigvee$ is the probabilistic 'OR' operation that is operating over all $\binom{N}{N-M}$ sub-sets. The parameter M is called the *order* of the model and represents the number of bands that are believed to be corrupted. So for the case of $N = 3$, the form of $P(\mathbf{x}|w)$ for each possible order would be as follows:

$$M = 0 \Rightarrow P_{M=0}(\mathbf{x}|q) = p(\mathbf{x}_1\mathbf{x}_2\mathbf{x}_3|q) = p(\mathbf{x}_1|q)p(\mathbf{x}_2|q)p(\mathbf{x}_3|q), \tag{14.16}$$

$$\begin{aligned} M = 1 \Rightarrow P_{M=1}(\mathbf{x}|q) &= p(\mathbf{x}_1\mathbf{x}_2 \vee \mathbf{x}_1\mathbf{x}_3 \vee \mathbf{x}_1\mathbf{x}_3|q) \\ &= p(\mathbf{x}_1|q)(\mathbf{x}_2|q) + p(\mathbf{x}_1|q)p(\mathbf{x}_3|q) + p(\mathbf{x}_2|q)p(\mathbf{x}_3|q), \end{aligned} \tag{14.17}$$

$$M = 2 \Rightarrow P_{M=2}(\mathbf{x}|q) = p(\mathbf{x}_1 \vee \mathbf{x}_2 \vee \mathbf{x}_3|q) = p(\mathbf{x}_1) + p(\mathbf{x}_2) + p(\mathbf{x}_3). \tag{14.18}$$

In the above, note that the 'OR'-ed probabilities can be simply added because they are mutually exclusive, and note that the 'AND'-ed terms are being simply multiplied which is only valid under the assumption that the sub-bands are independent given the state (which may be a poor approximation). Unfortunately, the scores from the models of different orders are not directly comparable, so it is not possible to sum over model orders, rather the model order is chosen *a priori* based on some idea of how many bands might be corrupted. A later version of the model termed the *posterior union model* [23] solves this problem by computing the *a posteriori* union probability for each state q given by

$$P_M(q_k|\mathbf{x}) = \frac{P_M(\mathbf{x}|q_k)P(q_k)}{\sum_{i=1}^{Q} P_M(\mathbf{x}|q_i)P(q_i)}. \tag{14.19}$$

However, this model does not resolve the problem of selecting the model order, M. A solution would be to assign a prior distribution over the values of M and sum over all possible models.

It is instructive to compare the probabilistic union model with missing-data with an uncertain missing-data pattern, as discussed in Section 14.4.1:

$$\begin{aligned} P(q|\mathbf{y}) &= \sum_{\mathbf{s}} \int_{\mathbf{x}} p(q, \mathbf{x}, \mathbf{s}|\mathbf{y})d\mathbf{x} \\ &= \sum_{s} \left\{ \int_{\mathbf{x}} p(\mathbf{x}|q) \frac{p(\mathbf{x}|\mathbf{s}, \mathbf{y})}{p(\mathbf{x})} d\mathbf{x} P(\mathbf{s}|\mathbf{y}) \right\} P(q). \end{aligned} \tag{14.20}$$

Consider a case where all missing-data patterns are considered equally probable, that is $P(\mathbf{s}|\mathbf{y})$ is constant, and also consider there to be no bounds constraints on missing values so that $p(\mathbf{x}|\mathbf{s}, \mathbf{y})$ is 0 if present elements of $\mathbf{x}$ do not match the observed signal, $\mathbf{y}$, but otherwise proportional to $p(\mathbf{x})$, which can be expressed compactly as

$$p(\mathbf{x}|\mathbf{s}, \mathbf{y}) = kp(\mathbf{x})\delta(\mathbf{x}_{p_s} - \mathbf{y}_{p_s}), \tag{14.21}$$

where k is a constant set to ensure the distribution integrates to unity and $\mathbf{x}_{p_s}$ is the sub-vector formed from the components of $\mathbf{x}$ marked as present according to s, and $\mathbf{y}_{p_s}$ are the corresponding elements of $\mathbf{y}$. The value of k can be computed by integrating the distribution

$$k = \frac{1}{\int_{\mathbf{x}_{m_s}} p(\mathbf{x}_{p_s} = \mathbf{y}_{p_s}, \mathbf{x}_{m_s}) d\mathbf{x}_{m_s}}. \tag{14.22}$$

Substituting Equation (14.21) into Equation (14.20) and integrating over $\mathbf{x}_{p_s}$ we arrive at

$$p(q|\mathbf{y}) \propto \sum_{\mathbf{s}} \left\{ \frac{\int_{\mathbf{x}_{m_s}} p(\mathbf{x}_{p_s} = \mathbf{y}_{p_s}, \mathbf{x}_{m_s}|q) d\mathbf{x}_{m_s}}{\int_{\mathbf{x}_{m_s}} p(\mathbf{x}_{p_s} = \mathbf{y}_{p_s}, \mathbf{x}_{m_s}) d\mathbf{x}_{m_s}} \right\} P(q). \tag{14.23}$$

In the above the sum is assumed to be over all possible masking patterns. This sum could be approximated by considering masking patterns that accepted or rejected clusters of frequency channels arranged as sub-bands. In this case, an equivalence can be seen between Equation (14.23) and *a posterior* union model (Equation 14.19) in which sub-band probabilities have been computed via missing-data marginalisation and a sum has been taken over all model orders, M. This formalism appears to avoid the model selection problem that is inherent in the posterior union model as presented in Ming and Smith [23].

14.4.3 Speech Fragment Decoding

In the previous sections, we have seen how a single mask estimate can be replaced with a distribution over possible masks. In Section 14.4.1, this distribution is formed by each time-frequency element being independently considered as either present or missing with a probability estimated from the noisy data. This soft mask model may or may not be appropriate. It is successfully used, for example, in situations where the additive noise is estimated to within a simple noise term. In this case, a single missing-data pattern could be computed using the noise mean, and the softness of the mask is in proportion to the variance in the noise estimate. However, this kind of model is not particularly appropriate for everyday listening situations where the noise is typically being generated by multiple sources with unpredictably changing levels of activity. For example, imagine trying to recognise speech in a noisy cafeteria, in a busy street or even in a domestic living room shared with a television set. The end of Section 14.4.2 presented an alternative that summed over all masks that could be formed from a set of sub-bands. Again though, in practice, noise seldom fits neatly into pre-determined frequency bands.

In complex acoustic scenes, direct estimation of the missing-data pattern is extremely challenging. It is usually not possible to say with any certainty whether any particular time frequency element, taken in isolation, is masked or not (i.e. whether s_i is a 0 or a 1). However, it is often possible to make strong predictions about the relation between pairs or groups of time frequency elements. In Chapter 16, it was discussed how certain properties of a sound source can be used to cue grouping rules that bind elements of a source across frequency and time. For example, if a sound is periodic it will have harmonics spread across the entire spectrum but, although appearing in separated frequency bands, these harmonics will all be related by having a common fundamental frequency. Using this cue the vocal resonances of a vowel spoken by one talker can readily be grouped together and seen to be separate from simultaneously occurring vocal resonances of a competing speaker talking with a different

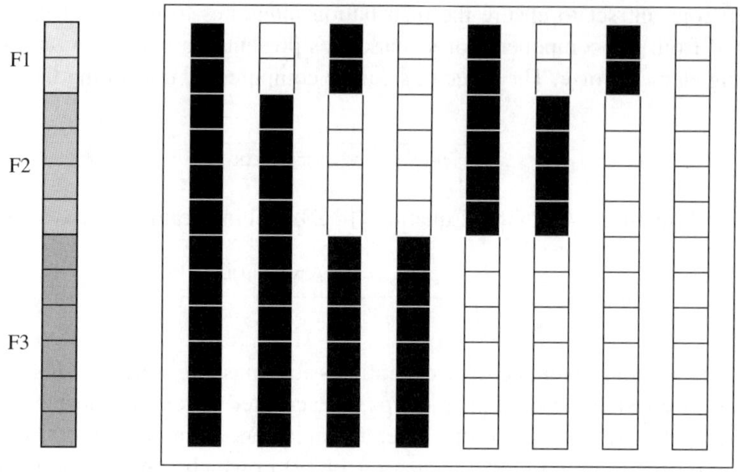

Figure 14.5 On the left, 12 frequency channels have been grouped into three fragments, $F1$, $F2$ and $F3$. On the right are shown the 8 ($2 \times 2 \times 2$) masking patterns (i.e. missing-data patterns) that can be validly constructed from the three fragments.

pitch [18]. Likewise, timing differences between signals arriving at the two ears which indicate the direction of arrival of a sound can be used to separate parts of the spectrum dominated by sounds arriving from one direction from those arriving from another [6]. Using such cues, it is possible to identify *fragments* of sound energy which extend over frequency and time and which appear to be due to a single environmental source. It might be unclear whether this fragment is dominated by the target speech source or by a noise source, but because the fragment is all from one source time-frequency elements in the fragment region must all share a common value in the missingness pattern (i.e. either 0 or 1).

The above discussion of fragment analysis suggests an alternative model for $p(\mathbf{s}|\mathbf{y})$. For the sake of continuity with earlier sections, we will consider initially a single frame of the signal, $\mathbf{y}$. Within this frame, cross-frequency grouping rules have clustered the frequency elements into a small number of mutually exclusive spectral fragments. Each fragment is either a fragment of speech and hence would be represented in $\mathbf{s}$ by 1's or a fragment of noise masker and hence would be represented by 0's. Only missing-data patterns that are consistent with these rules will be permissible, for all others $p(\mathbf{s}|\mathbf{y}) = 0$. We might decide that all permissible missing-data patterns have equal probability. Figure 14.5 shows an example of this model in which three fragments are shown along with the eight possible missing-data patterns that they might generate. $P(q|\mathbf{y})$ is computed just as in Section 14.4.1 by summing over possible segmentations, which in this case gives

$$P(q|\mathbf{y}) = \sum_{m=1}^{M} P(m|q) \sum_{\mathbf{s} \in \mathcal{S}} \left\{ \prod_{i \in \mathcal{P}_s} \frac{p(x_i = y_i|m, q)}{p(x_i = y_i)} \prod_{i \in \mathcal{M}_s} F_i \int_{-\infty}^{y_i} p(x_i|m, q) \mathrm{d}x_i \right\}, \quad (14.24)$$

where $\mathcal{S}$ represents the sets of segmentations permissible given the fragments.

However, this single frame fragment model is not used in practice because the true power of the fragment based approach comes from allowing fragments to extend across time. By tracking source signal properties over time it is possible to construct fragments with a spectro-temporal extent. Let $\mathbf{X} = \{\mathbf{x}_1, ..., \mathbf{x}_T\}$, $\mathbf{Y} = \{\mathbf{y}_1, ..., \mathbf{y}_T\}$, $\mathbf{S} = \{\mathbf{s}_1, ..., \mathbf{s}_T\}$ and Q be the sequence of states $Q = \{q_1, ..., q_T\}$ where T is the number of frames in the utterance. The ASR problem of finding the optimal state sequence can then be posed as

$$Q' = \underset{Q}{\operatorname{argmax}}\, p(Q|\mathbf{Y}) = \underset{Q}{\operatorname{argmax}} \sum_{S} \left\{ \int_{\mathbf{X}} p(Q, \mathbf{X}, \mathbf{S}|\mathbf{Y})\, \mathrm{d}\mathbf{X} \right\}.$$

In general this problem is intractable because of the large number of segmentations that need to be considered; however, the best test sequence, Q', can be estimated by searching for the combined state sequence and single segmentation that jointly maximise $P(Q, \mathbf{S}|\mathbf{Y})$

$$Q', \mathbf{S}' = \underset{Q,\mathbf{S}}{\operatorname{argmax}}\, p(Q, \mathbf{S}|\mathbf{Y}) = \underset{Q,\mathbf{S}}{\operatorname{argmax}} \left\{ \int_{\mathbf{X}} p(Q, \mathbf{X}, \mathbf{S}|\mathbf{Y})\, \mathrm{d}\mathbf{X} \right\}.$$

By making a set of frame-independence assumptions, similar to those made in all HMM-based ASR systems, the problem can be solved by an extension to the standard ASR Viterbi decoding algorithm which includes a parallel search over possible segmentations [3]. For a given segmentation hypothesis, at each time frame, and for each acoustic model state q, the frame-based posterior probabilities $p(q, \mathbf{s}|\mathbf{y})$ need to be computed. These can be computed using Equation (14.24) but with the summation over all s replaced by a single term representing the missing-data pattern hypothesis currently under consideration. The search over full segmentations hypotheses, $\mathbf{S}$ can be achieved efficiently through a graph structure which spawns new hypotheses at the frame where a fragment begins and merges pairs of hypotheses when a fragment ends (see [3] for details).

14.5 Some Perspectives on Performance

It is now over 10 years since the publication of the paper of Cooke *et al.* [9] that first popularised the marginalisation-based missing-data approach to robust speech recognition. Despite the wealth of research that has been conducted in the intervening years, it remains very difficult to say anything very concrete about the performance of missing data techniques. There are strong interactions between techniques for mask estimation, mask uncertainty modelling and acoustic modelling. Achieving a good result requires very careful application and depends to a large extent on the complexity of the recognition task in terms of both the speech content and the noise background.

In the early years of missing-data ASR there was considerable optimism. Much of this optimism was driven by results on small vocabulary tasks using so-called *a priori* masks [9]. In these experiments knowledge of pre-mixed speech and noise signals would be used to generate a missing-data mask by selecting reliable points on the basis of the *true* local SNR. Using such idealised masks, it was shown that recognition performance could remain robust down to global SNRs that would challenge even human perception. However, papers presenting such results seldom point out that *a priori* mask experiments would produce good results even if the real speech and noise mixture contained no cue by which the signals could be separated. It is rather like observing that white text can be read on a white background as long as someone shows you the outline of the letters – but there is clearly no process that can find the outlines given only a blank sheet of paper as input.

Despite the oversold and ambiguous interpretation of *a priori* mask results, the performance gap between these results and genuine missing-data ASR results initially drove researchers to concentrate on improving mask estimation. Estimating missing-data masks is a different problem than recognising words in noisy signals – but it is not clear that it is any more tractable. Despite much research in this direction, and appeals made to auditory scene analysis [5] and its computational models (see Chapter 16), there remains no general purpose algorithm for mask estimation. There are many partial techniques. However, the success of these techniques largely depends on specifics of the application and, in particular, the strength of the assumptions that can be made about the noise, for example, the noise may be at a different spatial location, or it may have a predictable temporal or spectral structure, or it may be sufficiently stationary to allow it to be estimated during non-speech regions. The paradox is that in situations where there is a strong noise model it may be better to use other techniques, for example model combination or noise subtraction (see Chapter 12 and Chapter 4).

The difficulties inherent in estimating accurate missing-data masks have driven the search for techniques that accommodate uncertainty in the mask estimates. Early experiments with soft missing-data masks showed it was possible to gain substantial performance improvements in a wide range of noise conditions even while using simple and ad hocly drawn uncertainty measures [2]. However, in highly non-stationary noise background, it can become difficult to estimate uncertainty in the mask. Errors in the mask pixels are often correlated over local spectro-temporal regions as large fragments of the noise background or the speech background are misallocated. A typical such case would be attempting to recognise speech against a background of competing speech [10]. Soft masks do not offer much in such conditions.

Fragment-decoding techniques provide a potential solution to the general case of speech in unpredictable noise backgrounds [3]. However, even fragment decoding systems are constrained by the quality of the front-end signal processing. Although it is no longer necessary to estimate a complete mask prior to recognition, the front-end still needs to locate locally coherent sound source fragments. Again, whereas experiments with '*a priori* fragments' have shown encouraging results, finding sufficiently coherent fragments to approach this performance using real signal processing has remained an elusive goals.

Over the 10 years since the introduction of missing-data systems there has been a trend towards increasingly complex joined-up systems. The original concept of being able to achieve robust recognition results in a real system by applying a simple missing-or-present masking model while making no assumption about the noise has developed into something more subtle. It has been realised that – apart from in a few exceptional circumstances – the speech data in seldom completely missing but is only more or less uncertain. Missing data has become uncertain data and implicit assumptions about noise sources have become more explicit. The boundaries between missing-data systems, model combination techniques and uncertainty decoding become increasingly blurred. It can be expected that future techniques will benefit by drawing on insights that have been made in each of these areas.

References

[1] S. Ahmad and V. Tresp, "Some solutions to the missing feature problem in vision," *Advances in Neural Information Processing Systems*, vol. 5, pp. 393–400, 1993.

[2] J. Barker, L. Josifovski, M. Cooke, and P. Green, "Soft decisions in missing data techniques for robust automatic speech recognition," in *Sixth International Conference on Spoken Language Processing*, 2000.

[3] J. Barker, M. Cooke, and D. Ellis, "Decoding speech in the presence of other sources," *Speech Communication*, vol. 45, no. 1, pp. 5–25, 2005.

[4] H. Bourlard and S. Dupont, "Subband-based speech recognition," in *IEEE International Conference on Acoustics, Speech and Signal Processing (ICASSP-97)*, vol. 2. IEEE, 1997, pp. 1251–1254.

[5] A. Bregman, *Auditory scene analysis*. Cambridge, MA: MIT press, 1990.

[6] H. Christensen, N. Ma, S. Wrigley, and J. Barker, "A speech fragment approach to localising multiple speakers in reverberant environments," in *IEEE International Conference on Acoustics, Speech and Signal Processing (ICASSP 2009)*, IEEE, 2009, pp. 4593–4596.

[7] M. Cooke and Y. Lu, "Spectral and temporal changes to speech produced in the presence of energetic and informational maskers," *The Journal of the Acoustical Society of America*, vol. 128, pp. 2059–2069, 2010.

[8] M. Cooke, P. Green, and M. Crawford, "Handling missing data in speech recognition," in *Proceeding of the International Conference on Spoken Language Processing*, 1994, pp. 1555–1558.

[9] M. Cooke, P. Green, L. Josifovski, and A. Vizinho, "Robust automatic speech recognition with missing and unreliable acoustic data," *Speech communication*, vol. 34, no. 3, pp. 267–285, 2001.

[10] M. Cooke, J. Hershey, and S. Rennie, "Monaural speech separation and recognition challenge," *Computer Speech & Language*, vol. 24, pp. 1–15, 2010.

[11] S. Demange, C. Cerisara, and J.-P. Haton, "A speech fragment approach to localising multiple speakers in reverberant environments," in *IEEE International Conference on Acoustics, Speech and Signal Processing (ICASSP 2006)*, IEEE, 2006.

[12] A. Dempster *et al.*, "Maximum likelihood from incomplete data via the EM algorithm," *Journal of the Royal Statistical Society. Series B (Methodological)*, vol. 39, no. 1, pp. 1–38, 1977.

[13] J. Droppo, A. Acero, and L. Deng, "Uncertainty decoding with splice for noise robust speech recognition," in *IEEE International Conference on Acoustics Speech and Signal Processing*, vol. 1. IEEE; 1999, 2002.

[14] F. Faubel, J. McDonough, and D. Klakow, "Bounded conditional mean imputation with Gaussian mixture models: a reconstruction approach to partly occluded features," in *IEEE International Conference on Acoustics, Speech and Signal Processing (ICASSP 2009)*, IEEE, 2009, pp. 3869–3872.

[15] B. Gold, N. Morgan, and D. Ellis, *Speech and Audio Signal Processing: Processing and Perception of Speech and Music*, 2nd ed. Hoboken, NJ, USA: John Wiley & Sons, Inc., 2011.

[16] D. Heitjan and S. Basu, "Distinguishing 'Missing At Random' and 'Missing Completely At Random'," *The American Statistician*, vol. 50, no. 3, pp. 207–213, 1996.

[17] E. Lombard, "Le signe de lelevation de la voix," *Annales Des Maladies Oreille, Larynx, Nez, Pharynx*, vol. 37, no. 101–119, p. 25, 1911.

[18] N. Ma, P. Green, J. Barker, and A. Coy, "Exploiting correlogram structure for robust speech recognition with multiple speech sources," *Speech Communication*, vol. 49, no. 12, pp. 874–891, 2007.

[19] P. Mermelstein and S. Davis, "Comparison of parametric representations for monosyllabic word recognition in continuously spoken sentences," *IEEE Transactions on Acoustics, Speech and Signal Processing*, vol. 28, no. 4, p. 357, 1980.

[20] J. Ming and F. Jack Smith, "Union: a model for partial temporal corruption of speech," *Computer Speech & Language*, vol. 15, no. 3, pp. 217–231, 2001.

[21] J. Ming and F. Jack Smith, "Speech recognition with unknown partial feature corruption-a review of the union model," *Computer Speech & Language*, vol. 17, no. 2-3, pp. 287–305, 2003.

[22] J. Ming and F. Smith, "Union: A new approach for combining sub-band observations for noisy speech recognition," *Speech Communication*, vol. 34, no. 1–2, pp. 41–55, 2001.

[23] J. Ming and F. Smith, "A posterior union model for improved robust speech recognition in nonstationary noise," in *IEEE International Conference on Acoustics, Speech and Signal Processing, (ICASSP'03)*, vol. 1. IEEE, 2003.

[24] A. Morris, A. Hagen, and H. Bourlard, "The full combination sub-bands approach to noise robust HMM/ANN based ASR," in *Sixth European Conference on Speech Communication and Technology*, 1999.

[25] A. Morris, J. Barker, and H. Bourlard, "From missing data to maybe useful data: soft data modelling for noise robust ASR," in *Proceedings of WISP*. Stratford-upon-Avon, UK: Citeseer, 2001.

[26] D. Morrison, *Multivariate Statistical Methods*, 3rd ed. McGraw-Hill, 1990.

[27] A. Potamianos and V. Weerackody, "Soft-feature decoding for speech recognition over wireless channels," in *IEEE International Conference on Acoustics, Speech and Signal Processing (ICASSP'01)*, vol. 1. IEEE, 2002, pp. 269–272.

[28] B. Raj and R. Stern, "Missing-feature approaches in speech recognition," *IEEE Signal Processing Magazine*, vol. 22, pp. 101–116, 2005.

[29] B. Raj, M. Seltzer, and R. Stern, "Reconstruction of missing features for robust speech recognition," *Speech Communication*, vol. 43, no. 4, pp. 275–296, 2004.

[30] D. Rubin, "Inference and missing data," *Biometrika*, vol. 63, no. 3, p. 581, 1976.

[31] J. Schafer, *Analysis of incomplete multivariate data*. Boca Raton, FL, USA: Chapman & Hall/CRC, 1997.

[32] M. Seltzer, B. Raj, and R. Stern, "A Bayesian classifier for spec- trographic mask estimation for missing feature speech recognition," *Speech Communication*, vol. 43, no. 4, pp. 379–393, 2004.

[33] Z. Tan, P. Dalsgaard, and B. Lindberg, "Automatic speech recognition over error-prone wireless networks," *Speech Communication*, vol. 47, no. 1–2, pp. 220–242, 2005.

[34] H. van Hamme, "Handling time-derivative features in a missing data framework for robust automatic speech recognition," in *IEEE International Conference on Acoustics, Speech and Signal Processing (ICASSP 2006)*, vol. 1. IEEE, 2006.

[35] M. van Segbroeck and H. van Hamme, "Vector-quantization based mask estimation for missing data automatic speech recognition," in *Proceedings of Interspeech*, Antwerp, Belgium, 2007.

[36] R. Weiss and D. Ellis, "Estimating single-channel source separation masks: Relevance vector machine classifiers vs. pitch-based masking," in *Proceedings of the ISCA Tutorial and Research Workshop on Statistical Perceptual Audition (SAPA)*, Pittsburgh, PA, 2006, pp. 31–36.

[37] D. Williams, X. Liao, Y. Xue, L. Carin, and B. Krishnapuram, "On classification with incomplete data," *IEEE Transactions on Pattern Analysis and Machine Intelligence*, pp. 427–436, 2007.

15

Missing-Data Techniques: Feature Reconstruction

Jort Florent Gemmeke[1], Ulpu Remes[2]
[1]*KU Leuven, Belgium*
[2]*Aalto University School of Science, Finland*

15.1 Introduction

Automatic speech recognition (ASR) performance degrades rapidly when speech is corrupted with increasing levels of noise. Missing-data techniques are a family of methods which tackle noise-robust speech recognition based on the so-called missing-data assumption proposed in [12]. The methods assume that (i) the noisy speech signal can be divided in speech-dominated (reliable) and noise-dominated (unreliable) spectro-temporal components prior to decoding and (ii) the unreliable elements do not retain any information about the corresponding clean speech values. This means that the clean speech values corresponding to the noise-dominated components are effectively missing, and speech recognition must proceed with partially observed data.

Techniques for speech recognition with missing features divide in roughly two categories, marginalization and feature reconstruction. The marginalization approach, discussed in Chapter 14, is based on disregarding the missing components when calculating acoustic model likelihoods: The likelihoods that correspond to the missing components are calculated by integrating over the full range of possible missing-feature values [11]. In this chapter, we focus on the reconstruction approach, where the missing values are substituted (imputed) with clean speech estimates prior to calculating the acoustic model likelihoods [10,43,45]. Since the reconstructed features no longer contain missing data, likelihood calculation does not need to be modified.

In general, all missing-feature imputation methods employ a model of the clean speech to estimate the missing values. Such models range from simple smoothness assumptions [45] to advanced statistical models and exemplar-based approaches. Given the clean speech model

Techniques for Noise Robustness in Automatic Speech Recognition, First Edition.
Edited by Tuomas Virtanen, Rita Singh, and Bhiksha Raj.

and a noisy observation, the missing features are imputed with the values that best match the assumptions of clean speech components at the missing locations.

Most imputation techniques are front-end based, which means they operate independently of the speech recognizer. Front-end imputation methods are attractive for two reasons. First, once the missing features have been replaced with clean speech estimates, any recognizer developed for clean speech conditions can be deployed without further modifications. In addition, the reconstructed features may be subjected to normalization and, for example, converted to cepstra. This is advantageous since the cepstral features are less correlated than spectral features and better suited for processing with ASR systems based on hidden Markov model (HMM) techniques [13]. Using marginalization or imputation techniques directly in the cepstral domain is more difficult since the cepstral transformation spreads the corrupting noise over all cepstral bands.

In this chapter, we primarily discuss four imputation methods that are well known and have been evaluated in various speech recognition tasks. First, we introduce *correlation-based imputation* [43] and *cluster-based imputation* [45] which use statistical models to calculate clean speech estimates using bounded maximum *a posteriori* (MAP) estimation. Correlation-based imputation employs a model that represents the sequence of speech frames as the output of a wide-sense stationary Gaussian process whereas in cluster-based imputation, clean speech is represented with a Gaussian mixture model (GMM). In cluster-based imputation, the bounded MAP estimate is approximated as a weighted sum of estimates calculated for each Gaussian.

The third method we discuss is *class-conditioned imputation* [33,57], where a specific clean speech estimate is calculated for each acoustic model state or Gaussian that is evaluated. This requires classifier modification, and thus, the imputation process is not strictly front-end based as in correlation or cluster-based imputation. Moreover, while class-conditioned imputation otherwise resembles cluster-based imputation, the speech model used for missing-feature estimation must have the same states or the same Gaussian components as the acoustic model used for speech recognition. Traditionally, the acoustic model itself is used.

The last method treated in depth is *sparse imputation* [22,27], which is an exemplar-based method. In contrast to the methods described above, speech is modeled nonparametrically as a linear combination of clean speech example spectrograms spanning multiple time frames. The missing-feature estimates are then determined based on the sparsest possible linear combination of example spectrograms that accurately represents the reliable features of the noisy speech.

For each method, we describe the basic concept and assumptions, discuss practical issues for implementation, and conclude with a description of possible advances on the basic technique proposed in literature. Common advances are combinations with soft missing-data masks or observation uncertainties. In addition to the four well-known imputation methods, we give short overviews of a few other methods that can be used for reconstructing missing features in speech recognition. The methods include reconstruction based on Markov random fields [49], nonlinear state-space models [42], matrix-factorization techniques [35,51], and discrete HMMs [5].

The remainder of this chapter is organized as follows. In Section 15.2, we introduce the concept of feature reconstruction and describe the notation used in the chapter. In Sections 15.3–15.6, we describe the four well-known reconstruction methods, and in Section 15.7, we present short overviews of other methods. In Section 15.8, we discuss results obtained with the various missing-feature reconstruction methods, and finally, we conclude with a discussion in Section 15.9.

15.2 Missing-Data Techniques

In ASR, speech is normally represented as a spectro-temporal distribution of acoustic power, a spectrogram. In noise-free conditions, the value of each time-frequency cell in this two-dimensional matrix depends only on the speech signal power, whereas in noisy conditions, the value of each cell represents both speech and background noise power. Assuming noise is additive and uncorrelated with speech, the power spectrogram of noisy speech can be approximately described as the sum of the individual power spectrograms of clean speech and noise.

The spectrographic features are often mapped to the mel-frequency scale and compressed with a logarithmic function to mimic human hearing. The logarithmic compression of a two-term sum can be approximated by the logarithm of the larger of the two terms [37]. Therefore, it holds for noisy speech features that

$$\mathbf{X} \approx \max(\mathbf{S}, \mathbf{N}) \tag{15.1}$$

with the (mel-frequency) log-power spectrogram matrix $\mathbf{X}$ denoting the noisy speech, $\mathbf{S}$ the clean speech, and $\mathbf{N}$ the background noise spectrogram in the (mel-frequency) log-power domain. The $\max$ operator denotes an element-wise maximum. Based on Equation (15.1), we assume that the features $X(t, f)$, where the time index $1 \leq t \leq T$ and the frequency index $1 \leq f \leq F$, which are dominated by speech energy approximately represent the uncorrupted clean speech signal whereas features dominated by noise energy represent only the noise signal.

The speech-dominated, reliable features $X_r(t, f)$ can be directly used as an estimate of the corresponding clean speech values, $S_r(t, f) = X_r(t, f)$, whereas the noise-dominated, unreliable features $X_u(t, f)$ provide only an upper bound for the unobserved or *missing* clean speech components, $S_u(t, f) \leq X_u(t, f)$. The labels that denote whether a time-frequency component $X(t, f)$ is reliable or unreliable are referred to as a missing-data mask or spectrographic mask $\mathbf{M}$. The mask denotes reliable components as $M(t, f) = 1$ and unreliable components as $M(t, f) = 0$. The missing-feature problem is illustrated in Figure 15.1.

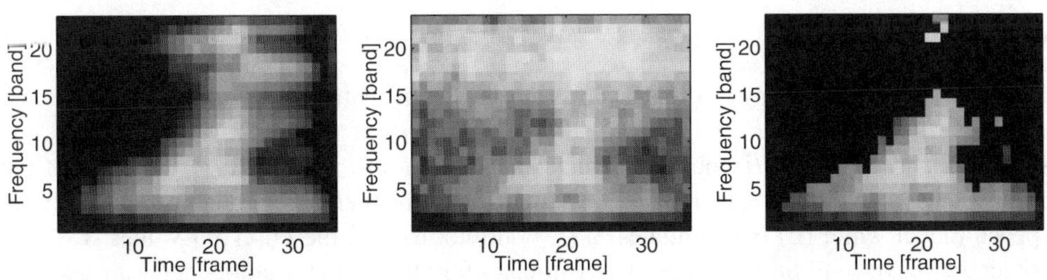

(a) Clean speech (b) Noisy speech (c) Overlayed with mask

Figure 15.1 Figure (a) is a spectro-temporal representation of the digit "one". The horizontal axis represents time, the vertical axis represents frequency, and the intensity represents the acoustic energy. The noisy spectrogram (b) represents the clean speech after it has been artificially corrupted by suburban train noise at SNR $= -5$ dB. In figure (c), the unreliable features of the noisy speech are marked black. We see that a substantial part of the data is missing.

If the clean speech and noise signals or their spectral representations are available so that we know the speech and noise power in each time-frequency cell, a so-called *oracle mask* may be constructed. However, in realistic situations, the location of reliable and unreliable components needs to be estimated. This results in an *estimated mask*. For methods of mask estimation, we refer the reader to the discussion in Chapter 16 or the overview in [9]. Instead of hard labels, assessing the probability that a component is reliable or unreliable has also been proposed. The *probabilistic* or *soft mask* M assigns each time-frequency component a continuous value between 0 and 1 to describe the probability of the component being reliable [3].

15.3 Correlation-Based Imputation

15.3.1 Fundamentals

In correlation-based imputation [43], the missing values are estimated based on their statistical dependencies with reliable observations in the current and neighboring frames. In the following sections, we introduce the clean speech model used in correlation-based imputation and derive the maximum *a posteriori* (MAP) estimates for the missing values based on the model. A correlation-based rule is applied to restrict the set of reliable components to make the estimation procedure computationally feasible.

Clean Speech Model

In correlation-based imputation [43], the matrix $\mathbf{S}$ that represents a sequence of clean speech vectors in the log-power domain is modeled as an output of a wide-sense stationary Gaussian process. Wide-sense stationarity means that the first and second order statistics of the feature vectors $S(t)$ do not vary in time. Therefore, the expected value is constant, $E[S(t)] = \bar{S}$ for all t, and the covariance between feature vectors in any two frames depends only on their relative time difference, $E[S(t)S(t-l)^{\mathsf{T}}] = \mathbf{Q}(l)$ for all t. Moreover, since the process is assumed Gaussian, the joint probability distribution of any time-frequency components $S(t, f)$ in $\mathbf{S}$ is a Gaussian whose parameters are derived from the expectation value vector $\bar{S}$ and the covariance matrices $\mathbf{Q}(l)$ as discussed in Section 15.3.2.

Feature Reconstruction

Given the clean speech model and a noisy spectrogram matrix $\mathbf{X}$ divided in reliable and unreliable features, $\{X_r(t, f)\}$ and $\{X_u(t, f)\}$, estimates for the missing values $\{S_u(t, f)\}$ should be chosen so that (i) the reconstructed spectrogram $\hat{\mathbf{S}}$, where $\hat{S}(t, f) = S_r(t, f)$ if the components is reliable and $\hat{S}(t, f) = \hat{S}_u(t, f)$ if the component is unreliable, fits the clean speech model while (ii) the estimates $\hat{S}_u(t, f)$ do not exceed the observed values $X_u(t, f)$. Under the above conditions, clean speech estimates for the unreliable components are given as

$$\{\hat{S}_u(t, f)\} = \underset{\xi_u}{\operatorname{argmax}}\, P(\xi_u | \{\xi_r(t, f) = X_r(t, f)\}, \{\xi_u(t, f) \leq X_u(t, f)\}, \Lambda), \tag{15.2}$$

where $\xi_r = \{\xi_r(t, f)\}$ and $\xi_u = \{\xi_u(t, f)\}$ denote the random variables corresponding to the reliable and unreliable elements of the clean speech spectrogram matrix $\mathbf{S}$, and Λ denotes the model parameters. The estimates $\hat{S}_u(t, f)$ are referred to as bounded MAP estimates. If the observed spectrograms $\mathbf{X}$ represent, for example, single words, the set $\{\hat{S}_u(t, f)\}$ can be solved using the methods discussed in Box 1. Given a longer utterance, calculations may become computationally infeasible.

Frame-Based Reconstruction

In order to reduce the computational load, Raj [43] proposed using a frame-based approach where the missing values in each time frame are estimated independently. We assume the feature vector $X(t)$ in frame t is divided in reliable and unreliable components that may be rearranged in vectors $X_r(t)$ and $X_u(t)$, respectively. That the missing values in each frame are estimated independently means that $\xi(t, f)$ and $\xi(t', k)$, which correspond to missing components $S_u(t, f)$ and $S_u(t', k)$, are assumed to be uncorrelated if $t' \neq t$. Additionally, because the correlation between any two components $\xi(t, f)$ and $\xi(t', k)$ decreases rapidly when the distance between (t, f) and (t', k) grows, many reliable components $S_r(t', k)$ do not contribute in the clean speech estimate vector $\hat{S}_u(t)$ and can be discarded from Equation (15.2) without a significant effect. The components $\xi(t', k)$ that (i) correspond to reliable observations and (ii) have a correlation greater than certain threshold α with the components of the random variable ξ_u that corresponds to the vector $S_u(t)$ are collected in a vector ξ_n (see Figure 15.2). This vector contains the reliable components that will be used in estimating the missing values in correlation-based imputation.

To summarize, in correlation-based imputation the random variable ξ_u, which corresponds to the vector $S_u(t)$ whose components are the missing clean speech features in frame t, is assumed to be statistically independent of all feature components except the vector ξ_n that contains a subset of reliable features from $\mathbf{S}$. Applying the independence assumption to Equation (15.2),

$$\hat{S}_u(t) = \underset{\xi_u}{\operatorname{argmax}}\, P(\xi_u | \xi_n = X_n(t), \xi_u \leq X_u(t), \Lambda), \tag{15.3}$$

where $X_n(t)$ is a noisy observation vector whose components correspond to the components of ξ_n. Since the clean speech spectrogram matrix $\mathbf{S}$ is modeled as an output of a Gaussian

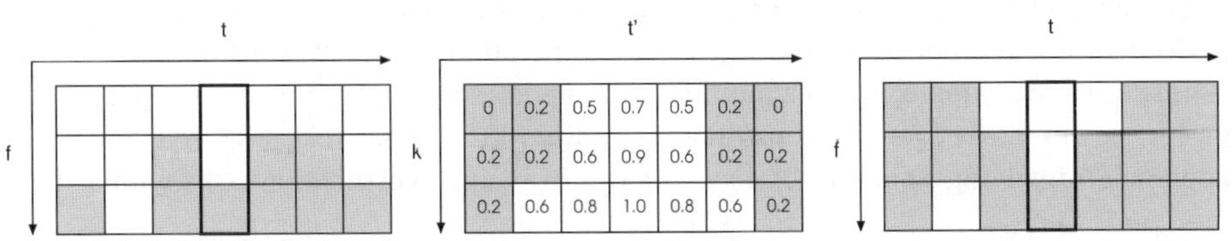

(a) missing data mask (b) correlation (c) reliable features

Figure 15.2 Assume (a) a missing-data mask $\mathbf{M}$ that divides the log-spectral time-frequency components $X(t, f)$ into reliable (white) and unreliable (gray) components. When $t = 4$, the vector constructed from the unreliable components $\xi_u = [\xi(4, 3)]$. The reliable components used in correlation-based imputation are determined based on the mask $\mathbf{M}$ and (b) the correlation between the missing component $\xi(4, 3)$ and the other components $\xi(t', k)$. Components with correlation greater than $\alpha = 0.5$ (white) are assumed to contribute to the clean speech estimate $\hat{S}(4, 3)$ whereas the other components (gray) are assumed independent of $\xi(4, 3)$. Hence, (c) the reliable components used to reconstruct the feature vector $S(4)$ (white) in correlation-based imputation are a subset of all the reliable components in the spectrogram $\mathbf{S}$. The reliable components used for reconstruction are organized in a vector $\xi_n = [\xi(2, 3), \xi(3, 1), \xi(4, 1), \xi(4, 2), \xi(5, 1)]^{\mathsf{T}}$.

process, the joint distribution of ξ_u and ξ_n is Gaussian, and $\hat{S}_u(t)$ can be calculated using the bounded MAP estimation methods discussed in Box 1. The estimation procedure is applied independently in each time frame to obtain the reconstructed spectrogram.

Box 1: Bounded estimation

Assume we are given two random variables ξ_r and ξ_u that are jointly Gaussian and an observation vector X whose components may be arranged in the reliable and unreliable component vectors X_r and X_u so that $\xi_r = X_r$ and $\xi_u \leq X_u$. The bounded MAP estimate for ξ_u is given as:

$$\hat{S}_u(t) = \underset{\xi_u}{\text{argmax}}\{P(\xi_u|\xi_r = X_r, \xi_u \leq X_u, \boldsymbol{\mu}, \boldsymbol{\Theta}\}, \tag{A.1}$$

where $\boldsymbol{\mu}$ and $\boldsymbol{\Theta}$ are the mean and covariance of the vector $\xi = [\xi_r^{\mathsf{T}} \xi_u^{\mathsf{T}}]^{\mathsf{T}}$. If the covariance matrix is diagonal and the components of the observation vector X are rearranged so that $X = [X_r^{\mathsf{T}} X_u^{\mathsf{T}}]^{\mathsf{T}}$, the bounded MAP estimate of each vector component $\xi(f)$ is the minimum of the observed upper bound $X(f)$ and the expected value $E[\xi(f)] = \mu(f)$, where $\mu(f)$ is the fth component of the Gaussian mean $\boldsymbol{\mu}$ in Equation (A.1). If full covariances are used, the bounded estimates cannot be solved in closed form.

The bounded MAP estimation problem with Gaussian variables may be formulated as a constrained optimization task

$$\min_{\xi}\{\frac{1}{2}(\xi - \boldsymbol{\mu})^{\mathsf{T}}\boldsymbol{\Theta}^{-1}(\xi - \boldsymbol{\mu})\} \text{ subject to } \xi_r = X_r \text{ and } \xi_u \leq X_u \ , \tag{A.2}$$

where $\xi = [\xi_r^{\mathsf{T}} \xi_u^{\mathsf{T}}]^{\mathsf{T}}$. The feature vector $\hat{\xi}$ that minimizes the cost function in (A.2) maximizes the conditional probability in Equation (A.1). Finding $\hat{\xi}$ is a general quadratic optimization problem which can be solved using iterative methods such as sequential quadratic programming (SQP). Van hamme [57] compared using a gradient descent method to solve the optimization problem (A.2) with using a multiplicative updates method to solve a non-negative least squares (NNLSQ) problem derived from problem (A.2). Speech recognition performance was reported to converge after 2–5 iterations with either method. In [27], the estimates were computed using a multiplicative updates method for quadratic optimization problems with non-negativity constraints [50]. Raj [43] alternatively proposed an iterative approach for solving Equation (A.1) as a series of bounded estimation tasks with a single missing feature.

15.3.2 Implementation

Correlation-based imputation needs an estimate for the clean speech model parameters and a rule for choosing the reliable components used in reconstructing the tth partially observed feature vector $S(t)$. The speech data may be processed in spectrograms or in fixed-length windows centered around the current frame t. Since this does not affect estimation, we simply assume the missing values are calculated from an $F \times T$ noisy speech segment $\mathbf{X}$. The calculations described in the estimation section below are repeated for every time frame in the observed noisy spectrogram.

Clean Speech Model

All spectrograms $\mathbf{S}_j$ in the clean speech training data are assumed independent observations of the same wide-sense stationary Gaussian process. The maximum likelihood estimate (MLE) for the feature mean is a vector calculated as the sample average

$$\bar{S} = \frac{1}{N} \sum_j \sum_t S_j(t), \tag{15.4}$$

where N is the number of samples in the clean speech training data and $S_j(t)$ denotes the feature vector in the tth frame of the jth spectrogram matrix in the training data. MLEs for the feature covariances are similarly calculated as sample covariances between the $S(t)$ and $S(t-l)$,

$$\mathbf{Q}(l) = \frac{1}{N(l)} \sum_j \sum_t (S_j(t) - \bar{S})(S_j(t-l) - \bar{S})^{\mathsf{T}}, \tag{15.5}$$

where $N(l)$ is the number of samples available for estimating the lth covariance matrix and $\bar{S}$ is the estimated mean vector from Equation (15.4). Considering missing-feature reconstruction, it is noteworthy that the mean and covariance parameters define the distribution of any subset of clean speech features in a spectrogram. More precisely, a sequence of T consecutive clean speech feature vectors concatenated into a single vector s follows a Gaussian distribution with mean and covariance given as

$$\boldsymbol{\mu} = \begin{bmatrix} \bar{S} \\ \vdots \\ \bar{S} \end{bmatrix} \quad \boldsymbol{\Theta} = \begin{bmatrix} \mathbf{Q}(0) & \dots & \mathbf{Q}(T-1) \\ \vdots & & \vdots \\ \mathbf{Q}(T-1) & \dots & \mathbf{Q}(0) \end{bmatrix}, \tag{15.6}$$

where $\boldsymbol{\mu}$ is an FT-dimensional vector constructed by repeating the F-dimensional sample mean T times and $\boldsymbol{\Theta}$ is an $FT \times FT$ matrix constructed from the sample covariances $\mathbf{Q}(l)$.

Reliable Components

In correlation-based imputation, estimates for the missing clean speech values in frame t are calculated from the joint distribution of ξ_n and ξ_u, where ξ_n is a vector constructed from the clean speech components that are used for reconstruction of the vector $S(t)$ in frame t and ξ_u corresponds to the vector $S_u(t)$ whose components are the missing clean speech features in frame t. The components of vector ξ_n are the elements of matrix $\mathbf{S}$ that (i) correspond to the reliable observations $S_r(t',k)$ and (ii) according to the clean speech model, have a correlation greater than a given threshold α with at least one of the components of vector ξ_u. The correlation between $S(t',k)$ and $S(t,f)$ is calculated as

$$r(t-t',f,k) = \frac{q(t-t',f,k)}{\sqrt{q(0,f,f)q(0,k,k)}}, \tag{15.7}$$

where $q(l,f,k)$ denotes the fth row of the kth column of the lth covariance matrix $\mathbf{Q}(l)$ from Equation (15.5).

If the speech data is processed in windows rather than full spectrograms, the window width T should be set so that the correlation between any two components more than $T/2$

frames apart does not exceed the given threshold. Raj [43] reports that the correlation between two feature components $\xi(t, f)$ and $\xi(t', k)$ falls below the proposed threshold $\alpha = 0.5$ when $|t - t'| > 5$. Note that this result depends on the frame rate and feature representation used to compute the spectrograms, so the minimum window width should be determined for each system separately.

Estimation

Bounded MAP estimation (Box 1) is used for computing the clean speech estimate $\hat{S}_u(t)$ based on the extended observation vector $\tilde{X}(t) = [X_n(t)^\mathsf{T} X_u(t)^\mathsf{T}]^\mathsf{T}$ and the mean and covariance of the corresponding clean speech distribution. To form the extended observation vector, let us reshape the $F \times T$ noisy speech segment $\mathbf{X}$ into a single FT-dimensional vector $\mathbf{x}$ by concatenating the feature vectors in subsequent time frames. Given the threshold α for choosing correlated reliable components, we can construct a binary matrix $\mathbf{U}(t)$ that extracts the components of $\tilde{X}(t)$ from $\mathbf{x}$, $\tilde{X}(t) = \mathbf{U}(t)\mathbf{x}$. Similarly, we can construct an extended mask to divide $\tilde{X}(t)$ in reliable and unreliable components as $\tilde{M}(t) = \mathbf{U}(t)\mathbf{m}$, where $\mathbf{m}$ is constructed from the missing-data mask $\mathbf{M}$ by concatenating the mask vectors in subsequent time frames. The corresponding clean speech distribution parameters are given as

$$\tilde{\boldsymbol{\mu}}(t) = \mathbf{U}(t)\boldsymbol{\mu}, \ \tilde{\boldsymbol{\Theta}}(t) = \mathbf{U}(t)\boldsymbol{\Theta}\mathbf{U}(t)^\mathsf{T}, \tag{15.8}$$

where $\boldsymbol{\mu}$ and $\boldsymbol{\Theta}$ are the mean and covariance from Equation (15.6). Calculating the distribution parameters is an $O(d(FT)^2)$ operation, where d denotes the number of components of $\tilde{X}(t)$.

15.4 Cluster-Based Imputation

15.4.1 Fundamentals

In cluster-based imputation [45], clean speech is represented with a Gaussian mixture model (GMM) and the missing values in each frame are estimated based on their statistical relationship with the reliable observations in the current frame. In the following sections, we introduce the clean speech model used in cluster-based imputation and derive the MAP estimates for the missing values based on the model. The estimates are approximated as a weighted sum of cluster-conditional estimates.

Clean Speech Model

In cluster-based imputation [45], the log-domain clean speech feature vectors $S = S(t)$ are modeled as independent and identically distributed (*i.i.d.*) random variables sampled from a mixture of Gaussians:

$$P(S) = \sum_i w_i \mathcal{N}(S\,; \boldsymbol{\mu}_i, \boldsymbol{\Theta}_i), \tag{15.9}$$

where w_i is the weight of the ith mixture component and $\boldsymbol{\mu}_i$ and $\boldsymbol{\Theta}_i$ are the mean vector and covariance matrix of the ith component.

Feature Reconstruction

Considering the noisy observation feature vector $X = X(t)$, we distinguish between its reliable and unreliable component using the vectors X_r and X_u. Accordingly, we denote the corresponding reliable and unreliable (missing) elements of the clean speech vector $S = S(t)$ as S_r and S_u, respectively. Given the clean speech model, the estimate $\hat{S}_u$ for S_u should be chosen so that (i) the reconstructed feature vector $\hat{S}$ fits the clean speech distribution model while (ii) the estimate $\hat{S}_u$ does not exceed the observed values in X_u. The estimate $\hat{S}_u$ under the above conditions is given as

$$\hat{S}_u = \underset{\xi_u}{\operatorname{argmax}}\, P(\xi_u|\xi_r = X_r, \xi_u \leq X_u, \Lambda), \tag{15.10}$$

where ξ_u and ξ_r denote the random variables corresponding to S_u and S_r and Λ denotes the parameters of the clean speech distribution model in Equation (15.9). For GMM-distributed variables, Equation (15.10) can be written as a maximization over a weighted sum of cluster-conditional posterior probabilities,

$$\hat{S}_u = \underset{\xi_u}{\operatorname{argmax}}\{\sum_i P(i|X, \Lambda)P(\xi_u|\xi_r = X_r, \xi_u \leq X_u, \Lambda_i)\}, \tag{15.11}$$

where $P(i|X, \Lambda)$ is the posterior probability for the ith Gaussian component given the noisy observations X and $P(\xi_u|\xi_r = X_r, \xi_u \leq X_u, \Lambda_i)$ is the cluster-conditional posterior probability distribution for the unreliable features. In cluster-based imputation, Equation (15.11) is approximated as

$$\hat{S}_u = \sum_i P(i|X, \Lambda)\, \underset{\xi_u}{\operatorname{argmax}}\{P(\xi_u|\xi_r = X_r, \xi_u \leq X_u, \Lambda_i)\}, \tag{15.12}$$

which is a weighted sum of cluster-conditional bounded MAP estimates for the missing-feature vector S_u. The cluster-conditional estimates can be calculated using the methods discussed in Box 1 and the posterior probabilities for clusters are calculated from the component weights and cluster-conditional observation probabilities as described below.

Cluster Posterior Probabilities

The posterior probability of the clean speech vector S being associated with the ith Gaussian component of the clean speech model is calculated as

$$P(i|X, \Lambda) = \frac{w_i P(X|\Lambda_i)}{\sum_{j=1}^{I} w_j P(X|\Lambda_j)}, \tag{15.13}$$

where w_i is the weight of the ith component from Equation (15.9), $P(X|\Lambda_i)$ the cluster-conditional observation probability, and $\Lambda_i = \{\boldsymbol{\mu}_i, \boldsymbol{\Theta}_i\}$ denotes the mean and covariance of the ith component. Although the clean speech model in Equation (15.9) is assumed to have full covariance matrices, only their diagonal components are considered when calculating the observation probabilities [45]. The probabilities are calculated as

$$P(X|\Lambda_i) = \prod_{f \in f_r} P(\xi(f) = X(f)|\Lambda_i) \prod_{f \in f_u} P(\xi(f) \leq X(f)|\Lambda_i), \tag{15.14}$$

where the sets f_r and f_u denote the reliable and unreliable frequency bands of X. Here, the first term $P(\xi(f) = X(f)|\Lambda_i)$ corresponds to the Gaussian distribution function evaluated at the elements of the reliable observation vector X_r and the second term $P(\xi(f) \leq X(f)|\Lambda_i)$ corresponds to the Gaussian cumulative distribution function evaluated at the elements of the unreliable feature vector X_u.

15.4.2 Implementation

In cluster-based imputation, the statistical dependencies between spectral channels $S(f)$ are modeled as a mixture of Gaussians. In the following sections, we discuss training the model with an appropriate number of components and review the feature reconstruction procedure discussed in Section 15.4.1 from a practical point of view. The calculations described in the estimation section below are repeated for every time frame.

Clean Speech Model

The clean speech GMM may be constructed by clustering clean speech training data in I clusters and modeling each cluster as a Gaussian. This is a simple approach that allows using any available clustering method such as k-means. Alternatively, one can use the expectation-maximization (EM) algorithm to calculate maximum likelihood estimates (MLE) for the GMM parameters. It alternates between calculating cluster membership probabilities for the data given the current parameter estimates (E-step) and calculating estimates for the distribution parameters given the current membership probabilities (M-step). EM-based GMM training has been implemented in, for example, the GMMBAYES Toolbox[1] for MATLAB and the *scikits.learn* module[2] for Python. The training data may also be partitioned based on voicing characteristics, for example, with a separate GMM trained for voiced and unvoiced features, as proposed in [38].

The number of clusters I in the model must be such that the feature vectors in each cluster can be approximately modeled as a Gaussian. This depends on the feature representation and training data. The optimal number for missing value estimation is likely to depend on factors such as the complexity of the noisy speech recognition task and the accuracy of the missing-data mask, and should be determined based on speech recognition experiments. Sometimes a small number is preferred simply because the computational complexity of cluster-based imputation grows in proportion to the number of clusters I and the performance gain from increasing the number of clusters is often small compared to the initial gain from using cluster-based imputation.

The models used in previous experiments with cluster-based imputation have varied in size and training method. A model with 512 components trained using k-means was used in [27] and a model with five components trained using the EM-algorithm in [28]. Since the covariances are assumed diagonal in calculating the cluster posterior probabilities, Raj *et al.* [45] recommend training the GMM with diagonal covariances and estimating the full covariance structure in the final pass of the EM-algorithm. A model with 128 components with full covariances estimated in the final pass was used in [19].

[1] Publicly available at http://www.it.lut.fi/project/gmmbayes/.
[2] Publicly available at http://scikit-learn.sourceforge.net/.

Estimation

Estimates for the missing clean speech vectors S_u are calculated from Equation (15.12). This corresponds to a weighted sum

$$\hat{S}_u = \sum_i \omega_i \hat{S}_u^{(i)}, \tag{15.15}$$

where ω_i are the cluster posterior probabilities and $\hat{S}_u^{(i)}$ the cluster-conditional estimates. The ith cluster-conditional estimate $\hat{S}_u^{(i)}$ is calculated based on the noisy observation vector X and the mean and covariance parameters of the ith Gaussian component as discussed in Box 1. The weights ω_i are calculated as a product of reliable component likelihoods and bounded marginal likelihoods associated with the unreliable components (Equation 15.14). The likelihoods are calculated from the Gaussian distribution function

$$P(\xi(f) = X(f)|\Lambda_i) = \frac{1}{\sqrt{2\pi\theta_i(f)}} \exp\left(\frac{-(X(f) - \mu_i(f))^2}{2\theta_i(f)}\right), \tag{15.16}$$

where $\Lambda_i = \{\boldsymbol{\mu}_i, \boldsymbol{\Theta}_i\}$ denotes the mean and covariance of the ith Gaussian and $\theta_i(f)$ is the fth diagonal component of the ith covariance matrix $\boldsymbol{\Theta}_i$. The bounded marginal likelihoods are calculated from the Gaussian cumulative distribution. They can be solved using the error function as

$$P(\xi(f) \leq X(f)|\Lambda_i) = 0.5 + 0.5 \,\mathrm{erf}\left(\frac{X(f) - \mu_i(f)}{\sqrt{2\theta_i(f)}}\right), \tag{15.17}$$

where $\mathrm{erf}(a)$ denotes the error function evaluated at a. The error function is implemented in all major programming languages.

15.4.3 Advances

In cluster-based imputation, estimates for the missing values have typically been calculated as a weighted sum of cluster-conditional bounded MAP estimates (Equation 15.12) as proposed in [45]. In the following sections, we discuss using soft masks and estimating observation uncertainties, which both require introducing a different estimation criteria for cluster-based imputation. Other recent advances include using multiple prior models [38] to exploit extra information such as voicedness and window-based processing [47] to introduce additional time context in the frame-based reconstruction.

Soft Masks

Using a soft missing-data mask corresponds to assuming that each component of the observed feature vector X is reliable with probability $M(f)$ and unreliable with probability $1 - M(f)$. The use of soft masks in the cluster-based imputation framework was proposed in [44]. When soft masks are used, the clean speech features are reconstructed using minimum mean square error (MMSE) estimation. The MMSE estimate for the fth spectral component of the underlying clean speech vector $S(f)$ is given as

$$\hat{S}(f) = M(f)X(f) + (1 - M(f))E[\xi(f)|\xi \leq X, \Lambda]. \tag{15.18}$$

This may be understood as follows: With probability $M(f)$, the observation $X(f)$ is reliable and the MMSE estimate for $S(f)$ is $X(f)$. With probability $1 - M(f)$, the observations is unreliable and the estimate for $S(f)$ must be calculated as the expectation value of $\xi(f)$ given the observed upper bound $X(f)$ and the clean speech model parameters Λ. The MMSE estimate for $S(f)$ is calculated as

$$E[\xi(f)|\xi \leq X, \Lambda] = \sum_i P(i|X, \Lambda) E[\xi(f)|\xi \leq X, \Lambda_i], \tag{15.19}$$

where $P(i|X, \Lambda)$ is the posterior probability of the ith cluster (Equation 15.13) and the expected value conditioned on Λ_i and the reliable and unreliable observations is the ith cluster-conditional bounded MMSE estimate. Note that all the components are assumed missing, that is $f_u = \{f\}$ and $f_r = \emptyset$, when calculating the observation probabilities used in Equation (15.13) from Equation (15.14).

If noise is assumed to have a uniform distribution in the log-spectral domain, the cluster-conditional estimates $E[\xi(f)|\xi \leq X, \Lambda_i]$ are solved as the expected value of a box-truncated Gaussian distribution [19,44]. Furthermore, if diagonal covariances are used, the ith cluster-conditional bounded MMSE estimate for the fth spectral component $S(f)$ in S_u is given as

$$E[\xi(f)|\xi \leq X, \Lambda_i] = \mu_i(f) - \theta_i(f) \frac{P(\xi(f) = X(f)|\Lambda_i)}{P(\xi(f) \leq X(f)|\Lambda_i)}, \tag{15.20}$$

where $\mu_i(f)$ is the fth component of the mean and $\theta_i(f)$ the fth diagonal component of the covariance matrix of the ith Gaussian component. The likelihood $P(\xi(f) = X(f)|\Lambda_i)$ and the bounded marginal likelihood $P(\xi(f) \leq X(f)|\Lambda_i)$ at the observed value are calculated using Equations (15.16) and (15.17). If full covariances are used, the bounded MMSE estimates must be solved iteratively or the bounded full covariance solution approximated as proposed in [19].

Observation Uncertainty

To improve speech recognition performance after front-end feature enhancement or missing-feature reconstruction, the clean speech vectors S may be associated with a full posterior rather than a point estimate $\hat{S}$. We may then calculate the expected value of the state likelihoods with respect to the clean speech posterior, as discussed in Section 17.2 of this book. The use of a full posterior allows the decoder to consider the uncertainty in each reconstructed feature and recover from errors made in reconstruction.

The clean speech posterior constrained on both the reliable and unreliable observations (Equation 15.11) does not have an analytical solution and is approximated with Equation (15.12) in cluster-based imputation. Therefore, Srinivasan and Wang [52] propose to discard the unreliable observations to calculate the full posterior for the missing clean speech features and use the posterior covariances as a basis for uncertainty estimation. The uncertainty measure proposed in [52] is described in Section 16.5.1.

If the speech signal is decoded in the domain where missing-feature reconstruction is applied, the estimated uncertainties can be directly used as the variance bias Θ_b as described in Section 17.2. Otherwise the uncertainties are first propagated to the acoustic model domain. Uncertainty propagation through the front-end is discussed in, for example [20]. An alternative approach, proposed in [52], uses clean and noisy speech training data to estimate a mapping from the estimated uncertainties in the spectral domain to oracle uncertainties in the cepstral domain. These oracle uncertainties were calculated as the expected squared error between the reconstructed and clean speech features, and the mapping was implemented using multilayer perceptrons (MLP) in [52] and regression trees in [53]. Reconstructed features in the acoustic model domain were given as context information for the frame-based MLP or regression tree mapping.

15.5 Class-Conditioned Imputation

15.5.1 Fundamentals

The class-conditioned imputation approaches employ the same conditional mean imputation principle as correlation and cluster-based imputation. However, instead of using a separate clean speech model, the estimates for the missing features are calculated from the acoustic models, and a separate clean speech estimate is calculated for each acoustic model state [10,33] or distribution component [57]. In the following sections, the acoustic models are assumed to use the same log-power feature representation used in calculating the missing-data mask and the acoustic model states are assumed to have been modeled as GMM distributions. The state and Gaussian-conditioned clean speech estimates are calculated using MAP estimation.

State-Conditioned Imputation

Front-end methods such as correlation and cluster-based imputation replace the observed feature vector $X = X(t)$ with the reconstructed vector $\hat{S}$ in calculating the acoustic model likelihoods. In state-conditioned imputation [10,33], the likelihood for each state q is calculated based on a reconstructed feature whose missing values have been estimated using the GMM distribution associated with the same state,

$$P(X|q) = \sum_i w_{iq}\, P(\hat{S}^{(q)}|\Lambda_{iq}), \tag{15.21}$$

where X is the observed noisy feature vector in the log-power domain and w_{iq} the weight of the ith mixture component in the qth state. The parameters of the qth state distribution and the ith component of the qth state distribution are denoted as Λ_q and Λ_{iq}, respectively, and $\hat{S}^{(q)}$ denotes the reconstructed clean speech vector estimated using the qth state distribution.

We distinguish the reliable and unreliable component of X using the vectors X_r and X_u. Likewise, we denote the corresponding reliable and unreliable (missing) elements of the clean speech vector S as vectors S_r and S_u, respectively. The state-conditioned bounded MAP estimates for the unreliable feature vector S_u are calculated from Equation (15.12) as

$$\hat{S}_u^{(q)} = \sum_i P(i|X, \Lambda_q)\hat{S}_u^{(iq)}, \tag{15.22}$$

where $P(i|X, \Lambda_q)$ is the posterior probability for the ith Gaussian component in the qth acoustic model state and $\hat{S}_u^{(iq)}$ the ith Gaussian-conditioned bounded MAP estimate for the unreliable components. The posterior probabilities are calculated from Equation (15.13) and the Gaussian-conditioned estimates as described in Section 15.5.2.

Gaussian-Conditioned Imputation

Van hamme [57] observed that it is computationally more efficient, and that the speech-recognition performance improves, if missing values are estimated specifically for each Gaussian component rather than each GMM state in the acoustic model. Given a noisy observation X, the state likelihoods are calculated as

$$P(X|q) = \sum_i w_{iq}\, P(\hat{S}^{(iq)}|\Lambda_{iq}), \tag{15.23}$$

where the Gaussian-conditioned bounded MAP estimates $\hat{S}^{(iq)}$ for the reliable components are the observed values X_r, and for the unreliable components, estimates $\hat{S}_u^{(iq)}$ are calculated as described in Section 15.5.2.

15.5.2 Implementation

Class-conditioned imputation differs from cluster-based imputation in that the clean speech estimates are calculated using the acoustic model state distributions. Estimation differs from cluster-based imputation only if the state distributions use diagonal covariances. Note that class-conditioned imputation does not produce a single reconstructed spectrogram $\hat{\mathbf{S}}$ but a set of class-conditioned spectrograms.

Estimation

Assuming the acoustic model states are modeled as GMMs with diagonal covariances, the iqth Gaussian-conditioned bounded MAP estimates for the unreliable components in S_u may be calculated as the minimum of the observed upper bound $X(f)$ and the unbounded MAP estimate. Since the latter coincides with expected value of the missing component in the ith Gaussian in the qth state, the estimates $\hat{S}_u^{(iq)}$ are given as

$$\hat{S}^{(iq)}(f) = \min\{X(f), \mu_{iq}(f)\} \qquad \forall f \in f_u, \tag{15.24}$$

where the set f_u contains the frequency bands corresponding to the unreliable elements in X and where $\boldsymbol{\mu}_{iq}$ denotes the mean of the ith Gaussian in the qth acoustic model state.

The state-conditioned bounded MAP estimates defined in Equation (15.22) are calculated as a weighted sum of the Gaussian-conditioned estimates. The weights are Gaussian posterior probabilities calculated as a product of reliable component likelihoods and bounded marginal likelihoods associated with the unreliable components (Equation 15.13). The likelihoods are calculated as in Equations (15.16)–(15.17) in Section 15.4.2.

Classifier Modification

In principle, the way likelihoods are calculated does not need to be modified even when class-conditioned imputation is used, but since the likelihoods are calculated based on either a state or Gaussian-dependent estimate or the reliable observation, it is necessary to implement some modifications in the likelihood calculating module of the speech recognizer being used. In practice, since modifications are in any case necessary, the most computationally efficient solution may be to slightly modify the likelihood calculation. For example, for

Gaussian-conditioned imputation, the likelihood calculation for the unreliable feature components, $X(f)$ with $f \in f_u$, can be written as:

$$P(\xi(f) = X(f)) = \frac{1}{\sqrt{2\pi\theta_{iq}(f)}} \exp\left(\frac{-(\min\{X(f), \mu_{iq}(f)\} - \mu_{iq}(f))^2}{2\theta_{iq}(f)}\right), \quad (15.25)$$

where $\theta_{iq}(f)$ is the fth diagonal component of the covariance matrix of the ith Gaussian in the qth acoustic model state.

15.5.3 Advances

The log-spectral features $S(t)$ are considered an unattractive feature representation for ASR because their correlatedness requires modeling the feature distribution with full covariances. Since class-conditioned imputation is not a front-end based method, the features cannot be decorrelated after reconstruction. Therefore, the more advanced class-conditioned imputation methods reconstruct the clean speech features directly in a decorrelated feature domain.

Cepstral Domain Imputation

The log-spectral features are normally decorrelated with a linear transformation such as the discrete cosine transformation (DCT). A method for Gaussian-conditioned imputation in the cepstral domain was proposed in [57]. Cepstral feature vectors are calculated as

$$\mathbf{s}(t) = \mathbf{A}S(t), \quad (15.26)$$

where matrix $\mathbf{A}$ denotes the discrete cosine transformation. The bounded MAP estimate for cepstral features $\mathbf{s} = \mathbf{s}(t)$ given a noisy observation is calculated from the distribution of cepstral features as

$$\hat{\mathbf{s}} = \underset{\xi}{\operatorname{argmax}}\, P(\mathbf{A}\xi | \xi_r = X_r, \xi_u \leq X_u, \Lambda_{iq}), \quad (15.27)$$

where $X = X(t)$ is the observed noisy speech feature in log-spectral domain, and its reliable and unreliable elements are described using the vectors X_r and X_u. ξ is the random variable corresponding to the log-spectral feature vector of clean speech, and we distinguish between its reliable and unreliable elements using the vectors ξ_r and ξ_u. The model parameters $\Lambda_{iq} = \{\boldsymbol{\mu}_{iq}, \boldsymbol{\Theta}_{iq}\}$ define the distribution of the cepstral features $\mathbf{s}$ in the ith Gaussian of the qth acoustic model state.

Equation (15.27) cannot be solved in closed form, but it may be formulated as a constrained optimization task

$$\min_{\xi} \{\frac{1}{2}(\mathbf{A}\xi - \boldsymbol{\mu}_{iq})^{\mathsf{T}} \boldsymbol{\Theta}_{iq}^{-1} (\mathbf{A}\xi - \boldsymbol{\mu}_{iq})\} \text{ subject to } \xi_r = X_r \text{ and } \xi_u \leq X_u. \quad (15.28)$$

The clean speech features $\hat{S}$ that minimize the cost function in (15.28) maximize the probability in Equation (15.27), but solving (15.28) requires regularization. Therefore, it is more convenient to express this as an optimization task in the spectral domain,

$$\min_{\xi} \{\frac{1}{2}(\xi - \boldsymbol{\mu}_S)^{\mathsf{T}} \mathbf{P} (\xi - \boldsymbol{\mu}_S)\} \text{ subject to } \xi_r = X_r \text{ and } \xi_u \leq X_u, \quad (15.29)$$

where $\boldsymbol{\mu}_S$ is the log-spectral domain mean of the Gaussian with cepstral mean $\boldsymbol{\mu}_{iq}$ and the precision matrix $\mathbf{P}$ is constructed as

$$\mathbf{P} = \mathbf{A}^\mathsf{T}\boldsymbol{\Theta}_C^{-1}\mathbf{A} + \lambda\boldsymbol{\Theta}_S^{-1}, \tag{15.30}$$

where $\boldsymbol{\Theta}_C = \boldsymbol{\Theta}_{iq}$ is the diagonal covariance matrix in cepstral domain, $\boldsymbol{\Theta}_S$ the diagonal covariance in the log-spectral domain, and λ a regularization parameter. Note that using the formulation (15.29) requires an existence of a spectral and a cepstral acoustic model so that each spectral Gaussian is mapped to a specific cepstral Gaussian. Such spectral model can be obtained, for example, through forced alignment. The optimization problem in (15.29) can be solved using the techniques described in Box 1. A good starting point for minimizing (15.29) is the spectral domain reconstruction from Equation (15.24).

Regularization with $\lambda\boldsymbol{\Theta}_S^{-1}$ is necessary when using the cepstral transformation because the matrix $\mathbf{A}^\mathsf{T}\boldsymbol{\Theta}_C^{-1}\mathbf{A}$ is rank-deficient. In [59], an alternative linear transformation was proposed, the ProsPect transformation. The ProsPect transformation is a low-order approximation of the cepstral transformation, and like cepstral transformation, it largely decorrelates the spectral features. However, the resulting precision matrix $\mathbf{P}$ is full rank, and no regularization is required. In practice, using ProsPect features is more computationally efficient.

Soft Masks

In [60], Gaussian-conditioned imputation was modified to use soft missing-data masks. The soft mask $M(t)$ estimated for the tth frame is represented as a diagonal matrix $\mathbf{W}$ with the mask elements on the diagonal, diag($\mathbf{W}$) $= M(t)$. For soft mask imputation in the cepstral domain, the optimization task in (15.29) becomes

$$\min_{\xi} \{\frac{1}{2}(\xi - \boldsymbol{\mu}_S)^\mathsf{T}\mathbf{V}(\xi - \boldsymbol{\mu}_S) + \frac{1}{2}(\xi - \boldsymbol{\mu}_S)^\mathsf{T}\mathbf{W}(\xi - \boldsymbol{\mu}_S)\} \text{ subject to } \xi \leq X, \tag{15.31}$$

where the matrix $\mathbf{V}$ is given as

$$\mathbf{V} = (\mathbf{I} - \mathbf{W})^{\frac{1}{2}}\mathbf{P}(\mathbf{I} - \mathbf{W})^{\frac{1}{2}}, \tag{15.32}$$

where $\mathbf{I}$ is an identity matrix of the same dimensions as $\mathbf{W}$. In the formulation (15.31), the first term ensures that the optimal point gets as close to the Gaussian mean as permitted by the constraint $\xi \leq X$. The second term that did not exist in the previous formulation (15.29) ensures that if the mask value is 1 and the features are reliable, the optimal point approaches the observed feature value. Note that the formulation (15.31) is not limited to the cepstral domain but is equally valid in the ProsPect or log-spectral domain. For the log-spectral domain, a closed-form solution of (15.31) was presented in [60].

15.6 Sparse Imputation

15.6.1 Fundamentals

The feature-reconstruction methods described in the previous sections are parametric methods that rely on a statistical description of the clean speech characteristics. The front-end based sparse imputation method described in this section, on the other hand, is an exemplar-based feature-reconstruction method. Exemplar-based methods model speech using a collection of actual speech samples, *exemplars*. The exemplars typically span several frames, which allows the estimation to benefit from temporal correlations. The sparse imputation method, first proposed in [22], works by first finding a small subset of clean speech exemplars that sparsely represent the reliable features of the observed noisy speech. This sparse representation is then used to make an estimate of the unreliable features of the noisy speech.

A Sparse Representation of Clean Speech

Consider an utterance that contains only clean speech. The log-power spectrogram matrix of clean speech, $\mathbf{S}$, is reshaped to a single vector $\mathbf{s}$ of dimension $L = F \cdot T$ by concatenating the T subsequent F-dimensional time frames. The assumption is that $\mathbf{s}$ can be represented exactly, or approximated with sufficient accuracy, by a linear, non-negative, combination of exemplar spectrograms $\mathbf{d}_n$, where n denotes a specific exemplar $(1 \leq n \leq N)$ in the dictionary which contains the N available exemplars:

$$\mathbf{s} = \sum_{n=1}^{N} y_n \mathbf{d}_n = \mathbf{D}\mathbf{y} \quad \text{subject to} \quad \mathbf{y} \geq 0, \tag{15.33}$$

where $\mathbf{y}$ is an N-dimensional activation vector. The activation vector is referred to as a sparse representation of $\mathbf{s}$ because the majority of the activations in $\mathbf{y}$ can be zero. The matrix $\mathbf{D}$ denotes the exemplar dictionary $\mathbf{D} = [\mathbf{d}_1\ \mathbf{d}_2 \ldots \mathbf{d}_N]$ with dimensions $L \times N$ and with $N \gg L$.

If the dictionary is large, the system of linear equations in Equation (15.33) typically has no unique solution. However, research in the field of compressive sensing [7,14,15] has shown that if a sparse representation exists, $\mathbf{y}$ can be recovered uniquely by enforcing sparsity. Conceptually, sparsity is important because it forces the selected exemplars to be closer to the underlying, lower dimensional manifolds on which the various speech classes are located [27]. To obtain a sparse solution, we may, for example, regularize the exemplar activations using L1-minimization as proposed in [55]. The sparse activation vector is obtained as

$$\mathbf{y} = \underset{\tilde{\mathbf{y}} \in \mathbb{R}^N}{\operatorname{argmin}} \{ \, \|\mathbf{D}\tilde{\mathbf{y}} - \mathbf{s}\|_2 + \lambda \|\tilde{\mathbf{y}}\|_1 \, \} \quad \text{subject to} \quad \tilde{\mathbf{y}} \geq 0, \tag{15.34}$$

where λ denotes the regularization parameter.

Feature Reconstruction

Sparse imputation is based on finding a sparse representation for partially observed data. First, the subsequent time frames of the spectrographic mask matrix $\mathbf{M}$ are concatenated to form a mask vector $\mathbf{m}$ and the subsequent frames of $\mathbf{X}$ are concatenated to a form a noisy observation vector $\mathbf{x}$. The reliable and unreliable elements of the noisy speech are denoted $\mathbf{x}_r$ and $\mathbf{x}_u$, respectively. The reliable elements $\mathbf{x}_r$ are used as an approximation for the corresponding elements of the now unknown $\mathbf{s}$, so problem (15.34) becomes

$$\mathbf{y} = \underset{\tilde{\mathbf{y}} \in \mathbb{R}^N}{\operatorname{argmin}} \{ \, \|\mathbf{D}_r\tilde{\mathbf{y}} - \mathbf{x}_r\|_2 + \lambda \|\tilde{\mathbf{y}}\|_1 \, \} \quad \text{subject to} \quad \tilde{\mathbf{y}} \geq 0, \tag{15.35}$$

where $\mathbf{D}_r$ denotes the rows of $\mathbf{D}$ that correspond to the elements of $\mathbf{m}$ that are equal to one. The sparse representation $\mathbf{y}$ can be used to estimate the clean observation vector as $\hat{\mathbf{s}} = \mathbf{D}\mathbf{y}$ as illustrated in Figure 15.3.

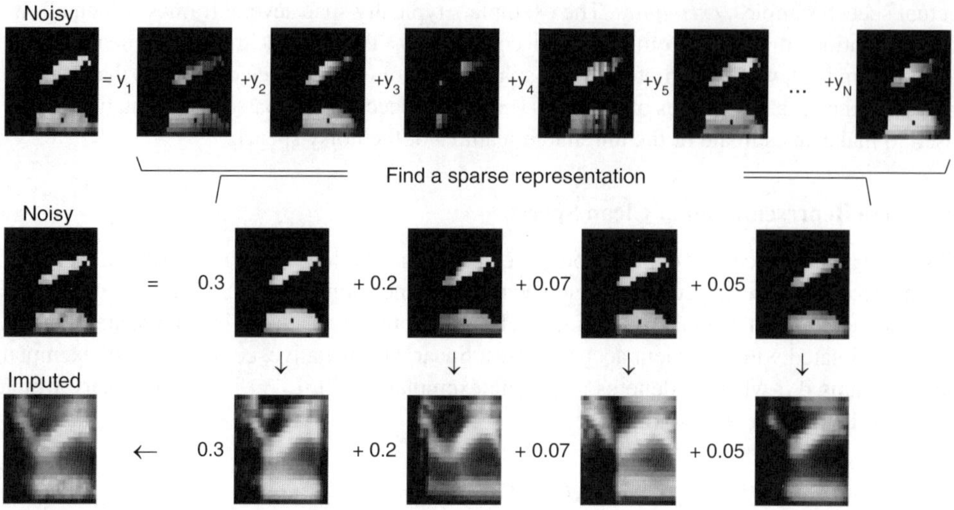

Figure 15.3 Schematic representation of sparse imputation of the noisy digit "three." The top row shows the spectrogram of the noisy digit at SNR -5 dB as a linear combination of masked clean speech exemplars. The missing features as indicated by an oracle mask are shown in black. The middle row shows the largest four nonzero exemplar activations of the sparse representation that describes the masked noisy digit using the masked clean speech exemplars. In the bottom row, imputation is done by linear combination of the unmasked clean speech exemplars.

In practice, the clean speech estimates calculated as $\hat{\mathbf{s}} = \mathbf{D}\mathbf{y}$ have some reconstruction error, so better results are obtained if only the unreliable elements are imputed. Additionally, the constraint that the clean speech estimates should not exceed the noisy feature values can be used: this is referred to as bounded imputation. Approximating this constraint, the clean speech estimates are given as

$$\hat{\mathbf{s}} = \begin{cases} \hat{\mathbf{s}}_r &= \mathbf{x}_r \\ \hat{\mathbf{s}}_u &= \min\left(\mathbf{D}_u\mathbf{y}, \mathbf{x}_u\right), \end{cases} \tag{15.36}$$

where $\hat{\mathbf{s}}_u$ and $\mathbf{D}_u$ are the rows of $\hat{\mathbf{s}}$ and $\mathbf{D}$ that correspond to zero elements in $\mathbf{m}$ and the min operator denotes an element-wise minimum. The reconstructed clean speech spectrogram matrix $\hat{\mathbf{S}}$ is obtained by reshaping $\hat{\mathbf{s}}$ into an $F \times T$ matrix.

15.6.2 Implementation

Continuous Speech

The sparse imputation framework described in Section 15.6.1 is suitable for imputation of noisy speech tokens that can be adequately represented by a fixed number of time frames T. Since arbitrary length utterances clearly do not satisfy this constraint, it is necessary to modify the method. A practical solution was proposed in [24] in the form of a sliding window approach. Here, the utterance is divided in several overlapping windows by sliding a window of length T through the noisy utterance with shifts of Δ, $1 \leq \Delta \leq T$ frames (see Figure 15.4).

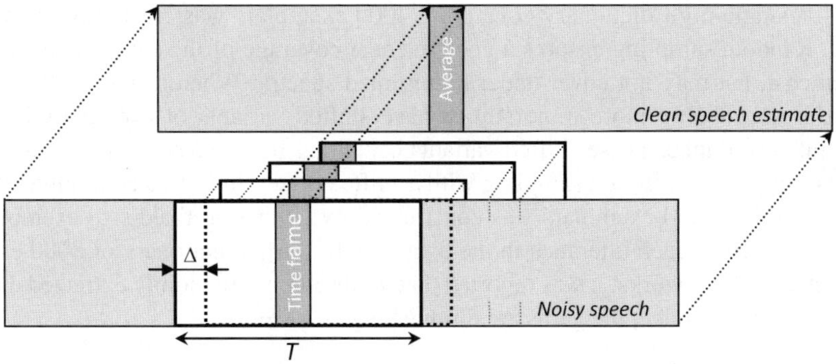

Figure 15.4 Schematic diagram of the sliding window approach for imputation. The dark shaded time frame in the noisy utterance is processed in several fixed-length imputation windows, of which we have shown four. Within each window, the given frame takes a different position due to the window shift Δ. The corresponding time frame in the clean speech estimate is the combination of these individual window-based imputations.

Each window is reconstructed separately using sparse imputation as described in Section 15.6.1.

Assuming the window shift $\Delta < T$, sparse imputation with the sliding window approach results in multiple clean speech estimates for each time frame. The estimates from overlapping windows may be recombined, for example through averaging or by taking the median value. Care must be taken that only clean speech estimates that originate from windows with a nonzero number of reliable elements are used. If for a certain frame, none of the underlying windows contained any reliable features, the sparse imputation method cannot provide a clean speech estimate. If this happens, the clean speech estimate should either not be provided (frame dropping) or it should be calculated based on a different approach such as inserting silence or interpolation.

In [24], it was shown that using larger step sizes Δ reduces computational effort but can decrease imputation accuracy. In most subsequent work on sparse imputation, $\Delta = 1$ has been used. The optimal value of the window length T, which translates directly to the length of the exemplars in the dictionary, depends on the task and should be tuned for each database. For AURORA-2, a connected digit database, $T = 35$ frames (350 ms) were found optimal in [24], whereas for the large-vocabulary SPEECON database, an optimal value of $T \approx 20$ frames (160 ms) was reported in [28].

Creating a Dictionary

Sparse imputation models clean speech as a collection of exemplars, the exemplar dictionary. In the case of small, restricted databases such as those available for small-vocabulary isolated word recognition, the dictionary can be formed by using all the time-normalized training tokens. When the size of the training database increases, the size of such a dictionary could become impractical. In this scenario, the exemplars should be subsampled from the complete database, for which several options exist. These include clustering, self-organizing maps, and random sampling.

In [27], a single-digit dictionary containing 4000 exemplars was created through random sampling. Random sampling ensures a good average coverage of the database and low computational cost, but may not cover under represented spectra. When using a sliding window approach, it is probably more important to have shifted variants of exemplars in order to provide shift-invariance. These shifted variants can either be obtained by artificially shifting extracted exemplars or through sampling with a shifted offset. In [24], a continuous-digit dictionary containing 4000 exemplars was constructed by extracting fixed-size exemplars with a random offset from each utterance in the database. In [28], a dictionary of 8000 exemplars was extracted. In both works, it was reported that while a larger randomly extracted dictionary may improve performance, the gains are diminishing.

Finding a Sparse Representation

The computational and algorithmic complexity of sparse imputation is mainly carried by the minimization (15.35). As minimization using a sparsity constraint has gained considerable interest over the past decade, many off-the-shelf implementations exist. We refer the reader to [8] for an overview and discussion of implementations in various programming languages.

To date, sparse imputation has been used with two different solvers: the basis pursuit interior-point method `l1_ls_nonneg`[3] [34] and the greedy `SolveLasso`[4] solver. For `l1_ls_nonneg`, good results have been obtained using the default settings, for example using the the utility function `find_lambdamax_l1_ls_nonneg` to determine the regularization parameter λ and using a duality gap of 0.01 as a stopping criterion [23]. The `SolveLasso` solver may yield solutions better suited for sparse imputation when terminated after a fixed number of iterations [54]. The number of iterations that is optimal depends on the used dictionary and should be empirically tuned; in [28], 30 iterations were used.

15.6.3 Advances

Soft Masks

In [25], an extension to sparse imputation was proposed that allows the use of a soft missing-data mask. When using a soft missing-data mask M, formulation (15.35) becomes a weighted minimization task

$$\mathbf{y} = \underset{\tilde{\mathbf{y}} \in \mathbb{R}^N}{\operatorname{argmin}} \{ \|\mathbf{W}\mathbf{D}\tilde{\mathbf{y}} - \mathbf{W}\mathbf{x}\|_2 + \lambda \|\tilde{\mathbf{y}}\|_1 \} \quad \text{subject to} \quad \tilde{\mathbf{y}} \geq 0, \tag{15.37}$$

where $\mathbf{W}$ is a diagonal matrix the elements of which are determined by the soft missing-data mask M. The weights on the diagonal are given as $\text{diag}(\mathbf{W}) = \mathbf{m}$, where $\mathbf{m}$ is the mask vector constructed from M. After obtaining the sparse representation $\mathbf{y}$, imputation is done as

$$\hat{\mathbf{s}} = \min(\mathbf{D}\mathbf{y}, \mathbf{x}). \tag{15.38}$$

[3] This solver is publicly available from http://www.stanford.edu/~boyd/l1_ls/.

[4] This solver is implemented as part of the `SparseLab` toolbox which is publicly available from http://www.sparselab.stanford.edu.

In the new formulation (15.37), using a binary mask would be equivalent to using $\mathbf{W}$ as a row selector picking only the rows of $\mathbf{D}$ and $\mathbf{x}$ that are assumed to contain reliable data. In the case of a soft mask, the weights on the diagonal influence the impact of each spectrographic element on the calculation of the reconstruction error.

Observation Uncertainty

For cluster-based imputation and other statistical approaches, the observation uncertainty matrix $\mathbf{\Theta}_\mathrm{b}$ can be estimated using the variance of the clean speech posterior as discussed in Section 15.4.3. Since the sparse imputation method does not support such variance estimation, various heuristic measures were proposed to characterize the uncertainty in the reconstructed features in [26].

Here, we review the two best performing uncertainty estimators from [26]. Both yield a single uncertainty assessment β for each window reconstructed with sparse imputation (cf. Section 15.6.2). Thus, the uncertainty associated with each feature component in a window is β if the component is determined unreliable or zero if reliable. Furthermore, the two uncertainty measures described below are expressed as proportional relationship and should be scaled to $0\ldots1$ per utterance prior to usage. The scalar uncertainties β are estimated as follows:

- The sparsity-based approach assumes a correlation between the uncertainty β and the sparsity of the representation $\mathbf{y}$. The idea is that if a particular observation is difficult to represent sparsely, it is because the observation is not covered by the dictionary $\mathbf{D}$. As a result, imputation performance may be poor. Therefore, it was proposed setting the uncertainty proportional to the number of exemplars used in reconstructing the segments: $\beta \propto \|\mathbf{y}\|_1$. When using this measure, it may be advantageous to only consider the largest values of $\mathbf{y}$; in [26], only values larger than 1% of the maximum occurring value in $\mathbf{y}$ were taken into account.
- The mask-based approach assumes a correlation between the uncertainty β and the missing-data mask used for imputation. The idea is that with more reliable features, it is easier to accurately estimate the unreliable values. Thus, it was proposed that the uncertainty is proportional to the number of unreliable features: $\beta \propto L - \|\mathbf{m}\|_1$, where L denotes the total number of features in the window.

The observation uncertainty for each frame, $\mathbf{\Theta}_\mathrm{b}$, is calculated as a recombination of the window-based uncertainties. If speech recognition requires transforming the features to a domain other than the mel-spectral domain where sparse imputation operates, the obtained uncertainty measures need to be propagated to the acoustic model domain. In [26], this mapping was done using a linear transformation, but the techniques outlined in Section 15.4.3 are equally applicable. The observation uncertainties can be used to adjust the acoustic model variances as described in Section 17.2.

Elastic Net Regularization

Tan *et al.* [54] proposed doing sparse imputation with the elastic net (EN) formulation [63]. In this formulation, (15.35) is modified to include an extra regularization term:

$$\mathbf{y} = \underset{\tilde{\mathbf{y}} \in \mathbb{R}^N}{\operatorname{argmin}} \{ \|\mathbf{D}_r \tilde{\mathbf{y}} - \mathbf{x}_r\|_2 + \gamma \|\tilde{\mathbf{y}}\|_2 + \lambda \|\tilde{\mathbf{y}}\|_1 \} \quad \text{subject to} \quad \tilde{\mathbf{y}} \geq 0. \tag{15.39}$$

The use of the regularization term $\gamma\|\tilde{\mathbf{y}}\|_2$ has a grouping effect that selects or deselects highly correlated exemplars together. Experiments on AURORA-2 showed that using EN regularization improves the recognition accuracy. It also allows using smaller exemplar dictionaries, which decreases the computational cost of sparse imputation.

15.7 Other Feature-Reconstruction Methods

Over the past two decades, numerous other methods for missing-feature reconstruction have been proposed. While not all of the methods discussed in this section have been applied in ASR tasks, they have been applied to reconstruction of spectrograms that contain missing features, and could be used for noise-robust ASR.

15.7.1 Parametric Approaches

In [49], each clean speech vector $S(t)$ was modeled as a nonuniform transformation of the clean speech vector in the previous frame, $S(t-1)$. The clean speech frames were divided into patches $\tilde{S}(t,f)$ that contain the time-frequency components $S(t,k)$ with $k = f-d \ldots f+d$, thus spanning $2d+1$ frequency bands. A larger d was chosen for patches extracted from the feature vector $S(t-1)$. A separate transformation was associated with each pair of consecutive patches, that is patches centered on the same frequency band f. It was assumed that the linear transformation matrix $\mathbf{A}(t,f)$ applied on the patch $\tilde{S}(t,f)$ was selected from a discrete set of transformations $\{\mathbf{A}_i\}$. The active transformations were modeled as a hidden variable in a generative graphical model whose observed nodes correspond to the time-frequency components $S(t,f)$. In the presence of missing values, some of the nodes become hidden. Probabilities of the transformations, $P(\mathbf{A}(t,f) = \mathbf{A}_i)$, and of the hidden nodes, $P(S(t,f))$, were inferred from the observed nodes using a modified form of belief propagation. The transition probabilities between neighboring transformations were determined experimentally. While the method was not evaluated experimentally, visual examples of reconstructed spectra were presented.

In [42], clean speech was modeled as the output of a nonlinear state-space model (NSSM). Feature vectors $S(t)$ were calculated as a nonlinear transformation of hidden source vectors $Z(t)$ which, in turn, were calculated as a non-linear transformation of the source vectors in the previous frame, $Z(t-1)$. A variational Bayesian approach was used to estimate the transformation parameters from clean speech training data. When using the model for reconstruction, the probability distributions of the source variables, $P(Z(t))$, and of the missing components, $P(S(t,f))$, were inferred from the reliable components using the total derivatives approach proposed in [42]. Evaluated in large-vocabulary continuous-speech recognition task, NSSM-based reconstruction resulted in a performance comparable to cluster-based imputation and sparse imputation [48].

Borgström and Alwan [5] proposed an HMM-based missing-feature reconstruction method which can utilize the statistical dependencies between time frames, frequency channels, or both. The feature components $S(t,f)$ were quantized and modeled as a tree-structured set of discrete centroids S_i. Each HMM state corresponds to a centroid, and the state output distributions model the observation probabilities $P(X(t)|S(t,f) \mapsto S_i)$, where $S(t,f) \mapsto S_i$ denotes the underlying clean speech feature $S(t,f)$ being quantized to S_i. The transition probabilities between centroids were learned from quantized clean speech training data and the observation probability distributions based on the speech data and a local noise estimate. In the reconstruction phase, the probability distribution over the hidden states, $P(S(t, f) \mapsto S_i)$, was inferred using the forward-backward algorithm, and the missing values were reconstructed as a weighted sum of the centroids S_i. The HMM-based reconstruction method was evaluated using the AURORA-2 connected digit recognition task.

15.7.2 Nonparametric Approaches

In [4], a technique related to sparse imputation was proposed. Like sparse imputation, the method estimates a sparse linear combination of dictionary elements based on the reliable observations. Unlike sparse imputation, which is window-based, the method only works on single time frames, and has an artificial speech dictionary $\mathbf{D}$ formed by the discrete Haar transform. To calculate bounded clean speech estimates, the optimization problem (15.35) is given the additional constraint that $\mathbf{D}_u\mathbf{y} \leq \mathbf{x}_u$. This renders the approximation in Equation (15.36) unnecessary. Comparing the performance obtained in experiments on AURORA-2 with the results obtained with sparse imputation, it can be concluded that the performance is comparable to the performance of sparse imputation when estimated masks are used, but significantly lower than sparse imputation when oracle masks are used.

In [35], a feature-reconstruction method based on non-negative matrix factorization (NMF) was presented. A magnitude spectrogram matrix $\mathbf{S}$ was represented as a factorization of two matrices, $\mathbf{S} = \mathbf{DY}$, where $\mathbf{D}$ describes the spectral envelope templates and $\mathbf{Y}$ describes the power envelopes in time. The approach is related to the sparse imputation method discussed in Section 15.6, with the difference that the spectral dictionary matrix $\mathbf{D}$ is also derived at run-time based on the observed features. The authors described a modification that allows factorization in the presence of missing data, after which reconstruction can be done by multiplying the recovered factorizations. Results were reported using SNR and segmental SNR computed on music samples.

The technique proposed in [51] is similar to the NMF-based approach [35] in that the spectrograms are described using a latent-variable decomposition. The spectral vectors are expressed in the magnitude domain, which allows the modeling of multiple additive sources. For reconstructing features with missing components, the authors use a spectral basis that has been pretrained using noisy data from similar conditions which does not contain missing features. For comparison, the authors used two missing-feature reconstruction methods normally applied in different fields: nearest neighbors imputation used in computer vision [6] and singular value decomposition (SVD) used in the imputation of gene expression arrays [31]. The methods were evaluated using visual examples and informal listening tests on music samples.

15.8 Experimental Results

The missing-feature reconstruction methods described in this chapter have been evaluated in a variety of speech recognition tasks with conditions ranging from small vocabulary, isolated word experiments with artificially added noise to large-vocabulary continuous speech-recognition tasks with data recorded in realistic environments. Results obtained in different conditions are generally deemed incomparable, and even if methods have been evaluated on the same data, differences in the choice of features, preprocessing, the back-end recognizer, and missing-data mask estimation method make comparing the performance difficult. In this section, results from various publications are reviewed and discussed in order to provide the reader with a general idea of how the methods treated in this chapter may perform in a speech recognition task.

In Section 15.8.1, we review experiments where several imputation methods have been compared. In Section 15.8.2, we discuss the effectiveness of missing-feature reconstruction compared to other methods for noise-robust speech recognition. In Section 15.8.3, we discuss the influence of mask estimation quality, the effectiveness of using soft masks, and the use of observation uncertainties. Finally, in Section 15.8.4, we discuss results obtained by combining missing-feature reconstruction with other techniques such as filtering or multicondition training.

15.8.1 Feature-Reconstruction Methods

Raj *et al.* [45] compared correlation-based, cluster-based, and state-conditioned imputation. Recognition experiments were conducted on the DARPA resource management data [41] which was artificially mixed with white noise and music at a range of signal-to-noise ratios (SNRs). When oracle masks and spectral features were used, correlation-based and state-conditioned imputation performed comparably while cluster-based imputation performed much better. When estimated masks were used, state-conditioned imputation performed as well as cluster-based imputation. When the reconstructed features were converted to cepstral domain, the accuracy obtained with cluster-based imputation, and to a lesser extent, with correlation-based imputation, increased substantially. At the time, there was no formulation for class-conditioned imputation in the cepstral domain.

In [59,61,62], several modifications of Gaussian-conditioned imputation were compared. The experiments were carried out on the AURORA-2 and AURORA-4 databases which have been constructed specifically for noise-robust speech recognition research and contain speech artificially corrupted with a variety of noises. AURORA-2 [32] is a digit recognition task based on the TIDIGITS corpus [36] and AURORA-4 [39] is a large-vocabulary task based on read sentences from the *Wall Street Journal* (WSJ) database [40]. In accordance with the findings in [45], speech recognition accuracy in the cepstral domain was found superior to the accuracy obtained using spectral features. A new linear transformation referred to as ProsPect transformation was proposed in [59] and shown to result in recognition accuracies comparable to the cepstral transformation at a fraction of the computational cost. Moreover, Gaussian-conditioned imputation was modified to allow maximum likelihood channel compensation, which improved accuracy in the presence of a channel mismatch [61].

Faubel *et al.* [19] compared using cluster-based imputation with minimum mean square error (MMSE) estimates computed using full covariances and bounded MMSE estimates calculated using either diagonal covariances or the full covariance approximation proposed in [19]. Experiments were carried out on the WSJ large-vocabulary speech data artificially mixed with noise samples from the NOISEX-92 database. Regardless to whether oracle or estimated masks were used, bounded imputation outperformed unbounded imputation and using full covariance matrices improved the results further still. Bounded imputation also outperformed unbounded imputation in the noisy digit recognition experiments conducted with state-conditioned imputation in [11].

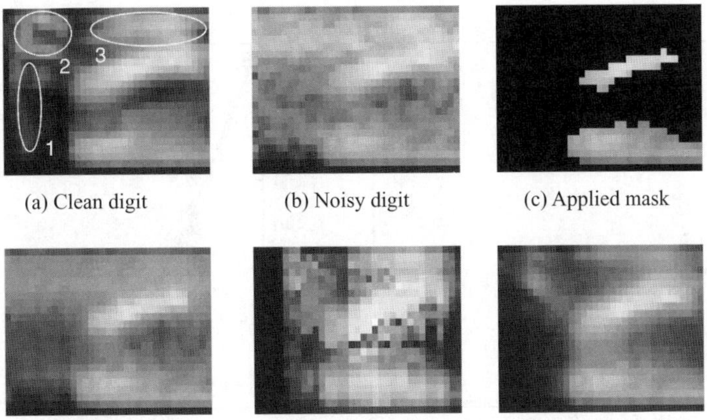

(a) Clean digit (b) Noisy digit (c) Applied mask

(d) Cluster-based imputation (e) Gaussian-conditioned imputation (f) Sparse imputation

Figure 15.5 Comparing (a) the clean speech spectrogram of digit "three" (/θri/) with (b) the observed noisy spectrogram and (c) the remaining reliable components, we see that imputation needs to reconstruct (1) the onset, which is the moderate energy pattern seen on the left of the spectrogram, (2) the frication of the /θ/, which is the high-energy pattern in the upper left corner, and (3) the formant trace, which is the high-energy structure in the upper right corner. Spectrograms reconstructed with the (d) cluster-based imputation, (e) Gaussian-conditioned imputation, and (f) sparse imputation show the methods succeed with a varying degree. The spectrogram of Gaussian-conditioned imputation is constructed using the best scoring Gaussian at each frame, as determined after recognition. In all cases the digit was recognized correctly after imputation.

In [27], sparse imputation was compared with cluster-based and Gaussian-conditioned imputation. Experiments were carried out on the AURORA-2 isolated digit recognition task, and reconstructed features were converted to the ProsPect domain prior to recognition. With estimated masks, Gaussian-conditioned imputation outperformed sparse imputation by a small margin, whereas with oracle masks, sparse imputation performance was significantly better than either Gaussian-conditioned or cluster-based imputation performance when SNR < 15 dB. Examples of reconstructed spectrograms are given in Figure 15.5 and recognition accuracies are reported in Figure 15.6. The difference between sparse imputation and cluster-based imputation performance with oracle and estimated masks was confirmed in large-vocabulary continuous speech recognition experiments on the Finnish SPEECON database using both artificially corrupted data as well as recordings from real-world car and public environments [28].

Finally, various imputation methods can use time-contexts that are longer than a single frame. As discussed in Section 15.6.2, the use of $T = 20$ frames (160 ms) of time-context in sparse imputation has a large beneficial influence on recognition accuracy. The same window-based approach was used to introduce multiple frames of time context in cluster-based imputation in [47]. Evaluations on speech from the Finnish SPEECON database artificially corrupted with babble noise or impulse noise showed that using $T = 5$ to $T = 10$ frames (40 ms to 80 ms) of time-context can substantially improve the results. In addition to the window-based approach used in sparse and cluster-based imputation, modeling the temporal dependencies between consecutive frames has been done using the non-linear state-space model (NSSM)

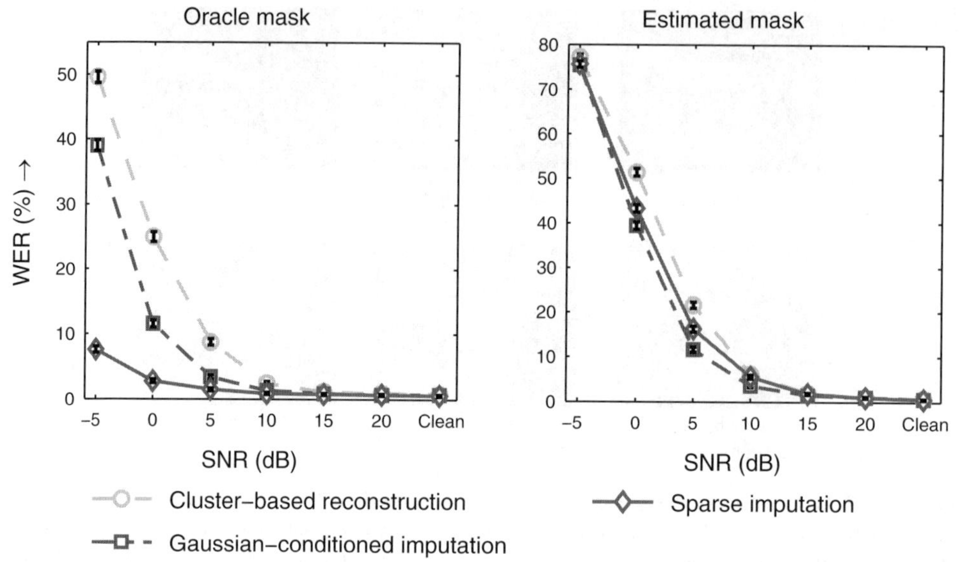

Figure 15.6 Word error rates obtained on AURORA-2 isolated digits database with cluster-based, Gaussian-conditioned, and sparse imputation. The left panel shows the results obtained using oracle masks and the right panel shows the results obtained using estimated masks. The horizontal axis describes the SNR at which the clean speech is mixed with the background noise and the vertical axis describes the word error rate averaged over four noise types: subway, car, babble, and exhibition hall noise. The vertical bars around data points indicate the 95 % confidence intervals.

[42] and HMM-based [5] approaches described in Section 15.7.1. The NSSM and HMM-based imputation methods both performed well when evaluated on speech recognition tasks [5,48], although evaluations on the AURORA-2 connected digit recognition task showed that the use of temporal correlations in HMM-based imputation only improved the results consistently at SNR < 0 dB.

15.8.2 *Comparison with Other Methods*

Traditionally, missing-feature reconstruction methods have been compared with the marginalization approach discussed in Chapter 14. In [11,33], bounded and unbounded state-conditioned imputation were compared with bounded and unbounded marginalization in experiments conducted on artificially noise-corrupted TIDIGITS material. The results indicate that bounded marginalization works better than bounded imputation, and both work better than their unbounded variants. However, the experiments did not compare performance on the cepstral features that have been commonly used in later implementations of class-conditioned imputation.

In [45], several feature-reconstruction methods were compared with marginalization. It was concluded that marginalization works better than feature reconstruction when used in the spectral domain, but cluster-based reconstruction works better than marginalization when the reconstructed features are transformed to cepstral domain. It was also shown that missing-feature techniques perform better than simple spectral subtraction. Spectral subtraction is discussed in Section 4.5.1 of this book.

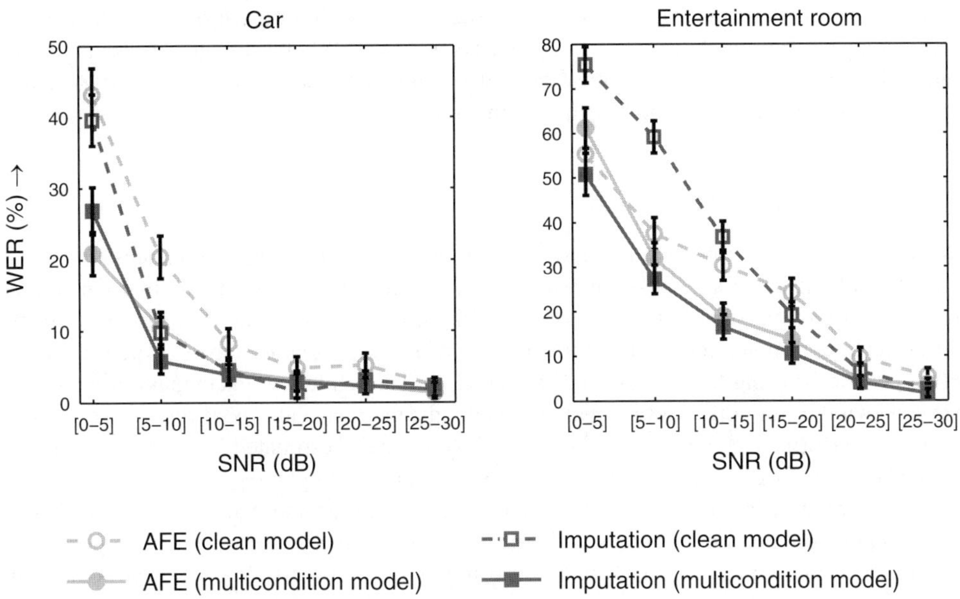

Figure 15.7 Word error rates obtained on an isolated words task of the SPEECON database with Gaussian-conditioned imputation and the ETSI AFE feature enhancement method, using either a clean speech or multicondition trained acoustic model. The horizontal axis describes the estimated SNR of the noisy speech. The vertical bars around data points indicate the 95 % confidence intervals. Evaluated on noisy speech recorded in a car environment, the choice of acoustic model does not affect the order of the methods, but imputation outperforms AFE in both cases. In the more challenging entertainment room environment, imputation outperforms AFE only if the multicondition model is employed. Using the multicondition model generally improves the results in both environments.

In [58], cepstral-based Gaussian-conditioned imputation was compared with the ETSI advanced front-end (AFE) feature extraction method [1] which is discussed in Section 4.5.4 of this book. AFE is based on a two-stage Wiener filter approach and is considered a good front-end-based feature enhancement tool. Experiments conducted on AURORA-2 showed that the imputation performance is comparable to AFE when an estimated missing-data mask is used.

In [29], ProsPect-based Gaussian-conditioned imputation was evaluated in a more challenging task: Flemish SPEECON material that contains noisy speech recorded in real-world environments. In this work, imputation using a clean speech model performed comparably to recognition with a multicondition trained acoustic model. Furthermore, when Gaussian-conditioned imputation was used with a multicondition trained acoustic model (cf. Section 15.8.4), the results were substantially better than the results obtained using AFE. The results are illustrated in Figure 15.7.

There are no recent studies with direct comparisons between missing-feature reconstruction methods and popular model-based methods such as parallel model combination (PMC) [21] or vector Taylor series (VTS) approximation [2]. In [43], feature-reconstruction methods were shown to outperform VTS on speech artificially corrupted by white noise, but if we compare the performance of reconstruction methods on common databases such as AURORA-2 and AURORA-4 to recently reported results on the performance of model-based approaches on the same data [56], it seems that feature reconstruction can only outperform the model-based noise compensation methods when oracle masks are used. When estimated masks are used, model-based methods perform slightly better than Gaussian-conditioned imputation, which is the most effective reconstruction method in this setting.

15.8.3 Advances

While the two-stage approach of mask estimation and missing-feature compensation is an attractive alternative for noise-robust speech recognition in unconstrained environments, it introduces two types of errors in the system: mask estimation errors and reconstruction errors. To evaluate the imputation methods irrespective of mask estimation errors, experiments are often conducted using both oracle and estimated masks. While the oracle mask experiments have shown the full potential of imputation methods, they have also confirmed the existence of a large performance gap between systems using oracle and estimated masks. The difference in performance is especially grave in the case of sparse imputation, for which the oracle mask performance is impressive even at low SNR conditions, and markedly exceeds that of other imputation methods, whereas with estimated masks, the performance is comparable to that of cluster-based or Gaussian-conditioned imputation. In general, it has been found that classifier compensation methods such as class-conditioned imputation and marginalization are less influenced by mask estimation errors than front-end based methods such as cluster-based imputation [27,44].

In this chapter, we described two advances that may alleviate the effect of mask estimation errors and reconstruction errors. The first is to use a probabilistic mask rather than a binary mask, and the second is to make the decoder aware of the reconstructed features through the use of observations uncertainties. In the following sections, we discuss improvements reported from using these two approaches.

Soft Estimated Masks

Making soft decisions was first proposed in [3], where soft masks were used in missing-data marginalization. The effectiveness of using soft masks with cluster-based imputation was investigated in [44], where clean speech features were calculated as the bounded MMSE estimates described in Section 15.4.3. Speech recognition experiments were conducted on Spanish telephone speech artificially corrupted with traffic, music, babble, and subway noise. Using soft estimated masks significantly improved recognition performance, particularly for traffic and subway noises, where a 25 % relative reduction in word error rate at 0 dB was reported. Cluster-based imputation with soft masks performed better than marginalization with soft masks although the difference between soft marginalization and imputation methods decreased at very low SNRs.

In [60], a modification to the Gaussian-conditioned imputation was proposed that allows using soft masks in the spectral or ProsPect domain (see Section 15.5.3). Experiments on AURORA-2 indicated that while using soft rather than binary masks improves speech-recognition performance in both cases, the effect is greater when working in the spectral domain. The authors also investigated the effectiveness of using a soft version of the oracle mask, and showed that even with oracle masks, making soft decisions increases the recognition accuracy, although not as much as with estimated masks. It was concluded that the use of soft masks improves even the oracle mask performance because the approximation (15.1) is inaccurate when the underlying speech and noise energies are comparable.

Finally, in [25], the soft mask approach for sparse imputation, described in Section 15.6.3, was evaluated on the AURORA-2 database. As with the other imputation methods that can use soft masks, sparse imputation performance benefited from the probabilistic information, but curiously, the improvement was even larger with soft oracle masks than with soft estimated masks, and an impressive 8 % word error rate was obtained at -5 dB when soft oracle masks were used.

Observation Uncertainty

Another approach to boosting speech recognition accuracy in noisy conditions is making the recognizer aware of the feature-reconstruction step. This is done by estimating the uncertainty of the reconstructed features. The effect of using uncertainty associated with cluster-based imputation as described in Section 15.4.3 was evaluated on AURORA-4 in [52]. The authors used two mask estimation methods and reported significant improvements in speech recognition accuracy when uncertainties were used with either method. The authors also compared the performance of estimated and oracle observation uncertainty when oracle masks were used. Since few, if any, differences were observed when using oracle masks, it was concluded that the proposed uncertainty estimation method works accurately. In a later publication [53], a different mapping from spectral to the cepstral domain uncertainties was proposed and shown to further improve the performance gain from using observation uncertainties.

In [26], the combination of sparse imputation with observation uncertainty was evaluated. The proposed measures of uncertainty, two of which were described in Section 15.6.3, were evaluated on clean Finnish SPEECON material artificially corrupted with babble noise from the NOISEX-92 database. Although based on different concepts, the sparsity-based and mask-based measures resulted in comparable performance. The two measures outperformed other alternatives presented in [26] and resulted in error reductions of up to 12 % in letter error rate compared to the baseline sparse imputation method. With oracle masks, differences in the speech recognition performance with the spectral domain oracle uncertainties and estimated uncertainties were minor, whereas with estimated masks, using oracle uncertainties resulted in better performance.

Finally, using observation uncertainty in the sparse imputation framework did not only improve the method performance on noisy speech, but on clean speech as well [26]. This is because with estimated masks, using sparse imputation on clean speech data may result in spurious insertions of whole words due to the use of multiframe windows, as discussed in [28]. It was concluded that using observation uncertainties can compensate for this effect and level the sparse imputation performance on clean speech with the clean speech baseline performance.

15.8.4 Combination with Other Methods

A number of authors have proposed to combine missing-feature reconstruction with other feature enhancement methods. Combining missing-data techniques with spectral subtraction was first proposed in [16], where marginalization combined with spectral subtraction was used in a speaker verification task. In [45], spectral subtraction was used on the reliable features of speech artificially corrupted with white noise, and correlation-based, cluster-based, and state-conditioned imputation were all shown to benefit from spectral subtraction. Gaussian-conditioned imputation was applied on spectral subtracted speech in [29] with similar results. In that work, the authors concluded that the performance improves because the imputation bounds become more accurate.

In [18,19], cluster-based imputation was combined with a particle-filtering technique that had previously been employed in feature enhancement [46,17]. Particle filtering was used for calculating estimates of the underlying clean speech and noise. The estimates were used twice: first, the clean speech and noise estimates were used to construct a missing-data mask, and then, cluster-based imputation was applied on the clean speech estimate, guided by this missing-data mask. Experiments on artificially noise-corrupted WSJ large-vocabulary speech data indicated that both particle filtering and missing-feature reconstruction contributed in the improved speech recognition performance.

Finally, in [29], using Gaussian-conditioned imputation in combination with multicondition trained acoustic models was proposed. In theory, the combination is incorrect since the assumption that reliable features remain uncorrupted means the reliable features should be recognized using a clean speech model. However, Figure 15.7 shows that in practice the combination leads to substantial improvements in recognition accuracy. The experiments were conducted on noisy speech from the SPEECON database which has been recorded in realistic conditions. It was concluded that using a multicondition trained model describes a wider variance of speech phenomena and thus compensates not only for additional effects such as reverberation but also for mask estimation errors. After all, if an unreliable feature is erroneously labeled reliable, the noisy speech model has a better chance of recovering from the mask estimation error.

15.9 Discussion and Conclusion

We have discussed several methods for feature reconstruction as an approach to improve noise robustness in ASR under the missing-data paradigm. Four well-known methods, namely correlation-based imputation, cluster-based imputation, class-conditioned imputation, and sparse imputation, were discussed in detail along with some significant advances that have been proposed to improve the basic approach. The performance of the methods was analyzed based on results published in various studies. Additionally, a number of recently developed methods that have not been extensively evaluated in noisy speech recognition task were described in Section 15.7.

The results discussed in Section 15.8.1 suggest that the most effective feature-reconstruction methods are sparse imputation and Gaussian-conditioned imputation. Sparse imputation typically results in the best speech recognition performance when oracle masks are used whereas Gaussian-conditioned imputation results in the best performance when estimated masks are used. However, while effective, Gaussian-conditioned imputation is not a front-end based method and requires classifier modification. The extent of modification is most notable when the advances outlined in Section 15.5.3 are employed. It is noteworthy that when classifier modification is acceptable, we have the option of using the observation uncertainty approaches described in Sections 15.4.3 and 15.6.3 to improve the front-end-based imputation performance as well.

If a front-end reconstruction method is preferred, sparse imputation or cluster-based imputation are recommended, for correlation-based imputation has never outperformed cluster-based imputation in experiments reported on realistic data. Cluster-based imputation, on the other hand, has been found as effective as sparse imputation when the noise level is moderate and estimated masks are used. While the result may depend on the recognition task, cluster-based imputation seems a fair alternative at low-noise conditions, and it is easy to implement.

The comparisons between missing-feature reconstruction and noise-robustness methods not based on missing-data techniques, reviewed in Section 15.8.2, indicated that feature reconstruction can result in a performance as good as or better than multicondition training, spectral subtraction, or feature enhancement with the ETSI AFE front-end, but might not outperform the recent formulations of model-based techniques such as PMC or VTS. When compared with marginalization, the feature-reconstruction methods appear to work better when the reconstructed features are transformed to cepstral domain prior to recognition, which is not possible with standard marginalization approaches. While marginalization has also been extended to the cepstral domain [30], the cepstral domain marginalization approach has not been compared with cepstral domain feature-reconstruction approaches.

The large difference between oracle mask and estimated mask performance of sparse imputation exemplifies how the performance of feature-reconstruction methods is largely determined by the quality of the missing-data mask. Depending on the data, estimating the mask with a sufficient accuracy can be extremely difficult. Although the results discussed in Section 15.8.3 show that performance improves with soft masks, soft decisions alone do not bridge the gap between oracle and estimated mask performance. Moreover, experiments on more challenging data such as noisy speech recorded in realistic environments indicated that missing-feature reconstruction may be more difficult if the additivity assumptions of noise and speech are violated, which happens, for example, in the presence of reverberation [29].

In general, all the results on missing-feature reconstruction suggest that the imputation performance improves as more information is provided for missing value estimation. That is, bounded imputation works better than unbounded imputation, using soft masks improves the performance over binary masks, and finally, increasing the time context considered in missing-feature reconstruction can improve the performance, especially in noisy conditions.

The review of missing-feature reconstruction methods presented in this chapter shows missing-feature reconstruction can be a competitive approach for speech recognition in adverse noisy conditions. While the success of these methods depends critically of the accuracy with which one estimates the missing-data mask, the large variety in approaches does make clear missing-feature methods are, above all, extremely flexible. As discussed in Section 15.8.4, combining missing-feature reconstruction with other noise-robustness techniques often leads to improved speech recognition performance. Interesting results could arise from combining feature reconstruction with more advanced noise-robustness techniques, such as the ETSI AFE front-end or VTS approaches.

Acknowledgments

The research of Jort F. Gemmeke was funded by IWT-SBO project ALADIN contract 100049 and the MIDAS project, granted under the Dutch-Flemish STEVIN program. The work of Ulpu Remes was supported by the Hecse graduate school and by the Academy of Finland (129674).

References

[1] "ETSI standard doc.: Speech processing, transmission and quality aspects (STQ); distributed speech recognition; advanced front-end feature extraction algorithm; ES 202 050 V1.1.5," 2007.

[2] A. Acero, L. Deng, T. Kristjansson, and J. Zhang, "HMM adaptation using vector Taylor series for noisy speech recognition," in *Proceedings of the International Conference on Spoken Language Processing*, 2000, pp. 869–872.

[3] J. Barker, L. Josifovski, M. Cooke, and P. Green, "Soft decisions in missing data techniques for robust automatic speech recognition," in *Proceedings of the International Conference on Spoken Language Processing*, 2000, pp. 373–376.

[4] B. Borgström and A. Alwan, "Utilizing compressibility in reconstructing spectrographic data with applications to noise robust ASR," *IEEE Signal Processing Letters*, vol. 16, no. 5, pp. 398–401, 2009.

[5] B. Borgström and A. Alwan, "HMM-based reconstruction of unreliable spectrographic data for noise robust speech recognition," *IEEE Transactions on Audio, Speech and Language Processing*, vol. 18, no. 6, pp. 1612–1623, 2010.

[6] M. E. Brand, "Incremental singular value decomposition of uncertain data with missing values," in *Proceedings of the European Conference on Computer Vision*, 2002, pp. 707–720.

[7] E. J. Candès, J. Romberg, and T. Tao, "Stable signal recovery from incomplete and inaccurate measurements," *Communications On Pure and Applied Mathematics*, vol. 59, no. 8, pp. 1207–1223, 2006.

[8] I. Carron, "Compressive Sensing: The Big Picture." Available at: http://sites.google.com/site/igorcarron2/cs, 2009.

[9] C. Cerisara, S. Demange, and J.-P. Haton, "On noise masking for automatic missing data speech recognition: A survey and discussion," *Computer Speech & Language*, vol. 21, no. 3, pp. 443–457, 2007.

[10] M. Cooke, A. Morris, and P. Green, "Missing data techniques for robust speech recognition," *Proceedings of the International Conference on Acoustics, Speech and Signal Processing*, 1997, pp. 863–866.

[11] M. Cooke, P. Green, L. Josifovski, and A. Vizinho, "Robust automatic speech recognition with missing and unreliable acoustic data," *Speech Communication*, vol. 34, no. 3, pp. 267–285, 2001.

[12] M. Cooke, P. Green, and M. Crawford, "Handling missing data in speech recognition," in *Proceedings of the International Conference on Spoken Language Processing*, 1994, pp. 1555–1558.

[13] S. B. Davis and P. Mermelstein, "Comparison of parametric representations for monosyllabic word recognition in continuously spoken sentences," *IEEE Transactions on Acoustics, Speech and Signal Processing*, vol. 28, no. 4, pp. 357–366, 1980.

[14] D. L. Donoho, "Compressed sensing," *IEEE Transactions on Information Theory*, vol. 52, no. 4, pp. 1289–1306, 2006.

[15] D. L. Donoho, "For most large underdetermined systems of linear equations the minimal L1-norm solution is also the sparsest solution," *Communications on Pure and Applied Mathematics*, vol. 59, no. 6, pp. 797–829, 2006.

[16] A. Drygajlo and M. El-Maliki, "Speaker verification in noisy environments with combined spectral subtraction and missing feature theory," in *Proceedings of the International Conference on Acoustics, Speech and Signal Processing*, 1998, pp. 121–124.

[17] F. Faubel and M. Wölfel, "Overcoming the vector Taylor series approximation in speech feature enhancement – A particle filter approach," in *Proceedings of the International Conference on Acoustics, Speech and Signal Processing*, 2007.

[18] F. Faubel, H. Raja, J. McDonough, and D. Klakow, "Particle filter based soft-mask estimation for missing feature reconstruction," in *Proceedings of the International Workshop on Acoustic Echo and Noise Constrol*, 2008.

[19] F. Faubel, J. McDonough, and D. Klakow, "Bounded conditional mean imputation with Gaussian mixture models: A reconstruction approach to partly occluded features," in *Proceedings of the International Conference on Acoustics, Speech and Signal Processing*, 2009, pp. 3869–3872.

[20] R. Fernandez Astudillo and D. Kolossa, "Uncertainty propagation," in *Robust Speech Recognition of Uncertain or Missing Data*, D. Kolossa and R. Haeb-Umbach, Eds. Heidelberg: Springer-Verlag, 2011, pp. 35–64.

[21] M. J. F. Gales and S. J. Young, "Robust continuous speech recognition using parallel model combination," *IEEE Transactions on Speech and Audio Processing*, vol. 4, no. 5, pp. 352–359, 1996.

[22] J. F. Gemmeke and B. Cranen, "Using sparse representations for missing data imputation in noise robust speech recognition," in *Proceedings of EUSIPCO*, 2008.

[23] J. F. Gemmeke and B. Cranen, "Noise robust digit recognition using sparse representations," in *Proceedings of the ISCA Tutorial and Research Workshop on Speech Analysis and Processing for Knowledge Discovery*, 2008.
[24] J. F. Gemmeke and B. Cranen, "Missing data imputation using compressive sensing techniques for connected digit recognition," in *Proceedings of the International Conference on Digital Signal Processing*, 2009, pp. 1–8.
[25] J. F. Gemmeke and B. Cranen, "Sparse imputation for noise robust speech recognition using soft masks," in *Proceedings of the International Conference on Acoustics, Speech and Signal Processing*, 2009, pp. 4645–4648.
[26] J. F. Gemmeke, U. Remes, and K. J. Palomäki, "Observation uncertainty measures for sparse imputation," in *Proceedings of Interspeech*, 2010, pp. 2262–2265.
[27] J. F. Gemmeke, H. Van hamme, B. Cranen, and L. Boves, "Compressive sensing for missing data imputation in noise robust speech recognition," *IEEE Journal of Selected Topics in Signal Processing*, vol. 4, no. 2, pp. 272–287, 2010.
[28] J. F. Gemmeke, B. Cranen, and U. Remes, "Sparse imputation for large vocabulary noise robust ASR," *Computer Speech & Language*, vol. 25, no. 2, pp. 462–479, 2011.
[29] J. F. Gemmeke, M. Van Segbroeck, Y. Wang, B. Cranen, and H. Van hamme, "Automatic speech recognition using missing data techniques: Handling of real-world data," in *Robust Speech Recognition of Uncertain or Missing Data*, D. Kolossa and R. Haeb-Umbach, Eds. Heidelberg: Springer-Verlag, 2011, pp. 157–185.
[30] J. Häkkinen and H. Haverinen, "On the use of missing feature theory with cepstral features," in *Proceedings of the CRAC Workshop*, 2001.
[31] T. Hastie, R. Tibshirani, G. Sherlock, M. Eisen, P. Brown, and D. Botstein, "Imputing missing data for gene expression arrays," Stanford Statistics Department, Technical Report, 1999.
[32] H. Hirsch and D. Pearce, "The Aurora experimental framework for the performance evaluation of speech recognition systems under noisy conditions," in *Proceedings of the ISCA Tutorial and Research Workshop ASR2000*, 2000, pp. 181–188.
[33] L. Josifovski, M. Cooke, P. Green, and A. Vizinho, "State based imputation of missing data for robust speech recognition and speech enhancement," in *Proceedings of EUROSPEECH*, 1999, pp. 2837–2840.
[34] S. Kim, K. Koh, M. Lustig, S. Boyd, and D. Gorinevsky, "An interior-point method for large-scale l1-regularized least squares," *IEEE Journal on Selected Topics in Signal Processing*, vol. 1, no. 4, pp. 606–617, 2007.
[35] J. Le Roux, H. Kameoka, N. Ono, A. de Cheveigné, and S. Sagayama, "Computational auditory induction as a missing-data model-fitting problem with Bregman divergence," *Speech Communication*, vol. 53, no. 5, pp. 658–676, 2011.
[36] R. Leonard, "A database for speaker-independent digit recognition," in *Proceedings of the International Conference on Acoustics, Speech and Signal Processing*, 1984, pp. 328–331.
[37] A. Nadas, D. Nahamoo, and M. Picheny, "Speech recognition using noise-adaptive prototypes," *IEEE Transactions on Acoustics, Speech and Signal Processing*, vol. 37, no. 10, pp. 1495–1503, 1989.
[38] A. Narayanan, X. Zhao, D. L. Wang, and E. Fosler-Lussier, "Robust speech recognition using multiple prior models for speech reconstruction," in *Proceedings of the International Conference on Acoustics, Speech and Signal Processing*, 2011, pp. 4800–4803.
[39] N. Parihar and J. Picone, "Analysis of the Aurora large vocabulary evaluations," in *Proceedings of EUROSPEECH*, 2003, pp. 337–340.
[40] D. B. Paul and J. M. Baker, "The design for the Wall Street Journal-based CSR corpus," in *Proceedings of the International Conference on Spoken Language Processing*, 1992, pp. 899–902.
[41] P. Price, W. M. Fisher, J. Bernstein, and D. S. Pallet, "The DARPA 1000-word resource management database for continuous speech recognition," in *Proceedings of the International Conference on Acoustics, Speech and Signal Processing*, 1988, pp. 651–654.
[42] T. Raiko, M. Tornio, A. Honkela, and J. Karhunen., "State inference in variational Bayesian nonlinear state-space models," in *Proceedings of the International Conference on Independent Component Analysis and Blind Source Separation*, 2006, pp. 222–229.
[43] B. Raj, "Reconstruction of incomplete spectrograms for robust speech recognition," PhD dissertation, Carnegie Mellon University, 2000.
[44] B. Raj and R. Singh, "Reconstructing spectral vectors with uncertain spectrographic masks for robust speech recognition," in *IEEE Workshop on Automatic Speech Recognition and Understanding*, 2005, pp. 65–70.
[45] B. Raj, M. L. Seltzer, and R. M. Stern, "Reconstruction of missing features for robust speech recognition," *Speech Communication*, vol. 43, no. 4, pp. 275–296, 2004.
[46] B. Raj, R. Singh, and R. Stern, "On tracking noise with linear dynamical system models," in *Proceedings of the International Conference on Acoustics, Speech and Signal Processing*, 2004, pp. 965–968.

[47] U. Remes, Y. Nankaku, and K. Tokuda, "GMM-based missing-feature reconstruction on multi-frame windows," in *Proceedings of Interspeech*, 2011, pp. 1665–1668.

[48] U. Remes, K. J. Palomäki, T. Raiko, A. Honkela, and M. Kurimo, "Missing-feature reconstruction with a bounded nonlinear state-space model," *IEEE Signal Processing Letters*, vol. 18, pp. 563–566, 2011.

[49] M. J. Reyes-Gomez, N. Jojic, and D. P. Ellis, "Towards single-channel unsupervised source separation of speech mixtures: The layered harmonics/formants separation/tracking model," in *Proceedings of the ISCA Tutorial and Research Workshop on Statistical and Perceptual Audio Processing (SAPA)*, 2004.

[50] F. Sha, L. K. Saul, and D. D. Lee, "Multiplicative updates for nonnegative quadratic programming in support vector machines," in *Proceedings of the Neural Information Processing Systems*, 2002, pp. 1041–1048.

[51] P. Smaragdis, B. Raj, and M. Shashanka, "Missing data imputation for spectral audio signals," in *IEEE International Workshop on Machine Learning for Signal Processing*, 2009.

[52] S. Srinivasan and D. L. Wang, "A supervised learning approach to uncertainty decoding for robust speech recognition," in *Proceedings of the International Conference on Acoustics, Speech and Signal Processing*, 2006, pp. 297–300.

[53] S. Srinivasan and D. L. Wang, "Transforming binary uncertainties for robust speech recognition," *IEEE Transactions on Audio, Speech and Language Processing*, vol. 15, pp. 2130–2140, 2007.

[54] Q. F. Tan, P. G. Georgiou, and S. S. Narayanan, "Enhanced sparse imputation techniques for a robust speech recognition front-end," *IEEE Transactions on Audio, Speech and Language Processing* vol. 19, no. 8, pp. 2418–2429, November 2011.

[55] R. Tibshirani, "Regression shrinkage and selection via the lasso," *Journal of the Royal Statistical Society. Series B (Methodological)*, vol. 58, no. 1, pp. 267–288, 1996.

[56] R. C. van Dalen, F. Flego, and M. J. F. Gales, "Transforming features to compensate speech recogniser models for noise," in *Proceedings of Interspeech*, 2009, pp. 2499–2502.

[57] H. Van hamme, "Robust speech recognition using missing feature theory in the cepstral or LDA domain," in *Proceedings of EUROSPEECH*, 2003, pp. 3089–3092.

[58] H. Van hamme, "Robust speech recognition using cepstral domain missing data techniques and noisy masks," in *Proceedings of the International Conference on Acoustics, Speech and Signal Processing*, 2004, pp. 213–216.

[59] H. Van hamme, "PROSPECT features and their application to missing data techniques for robust speech recognition," in *Proceedings of Interspeech*, 2004, pp. 101–104.

[60] M. Van Segbroeck and H. Van hamme, "Robust speech recognition using missing data techniques in the PROSPECT domain and fuzzy masks," in *Proceedings of the International Conference on Acoustics, Speech and Signal Processing*, 2008, pp. 4393–4396.

[61] M. Van Segbroeck and H. Van hamme, "Handling convolutional noise in missing data automatic speech recognition," in *Proceedings of the International Conference on Acoustics, Speech and Signal Processing*, 2006, pp. 2562–2565.

[62] M. Van Segbroeck and H. Van hamme, "Advances in missing feature techniques for robust large vocabulary continuous speech recognition," *IEEE Transactions on Audio, Speech and Language Processing*, vol. 19, no. 1, pp. 123–137, 2011.

[63] H. Zou and T. Hastie, "Regularization and variable selection via the elastic net," *Journal of the Royal Statistical Society Series B*, vol. 67, no. 2, pp. 301–320, 2005.

16

Computational Auditory Scene Analysis and Automatic Speech Recognition

Arun Narayanan, DeLiang Wang
The Ohio State University, USA

16.1 Introduction

The human auditory system is, in a way, an engineering marvel. It is able to do wonderful things that powerful modern machines find extremely difficult. For instance, our auditory system is able to follow the lyrics of a song when the input is a mixture of speech and musical accompaniments. Another example is a party situation. Usually there are multiple groups of people talking, with laughter, ambient music and other sound sources running in the background. The input our auditory system receives through the ears is a mixture of all these. In spite of such a complex input, we are able to selectively listen to an individual speaker, attend to the music in the background, and so on. In fact this ability of 'segregation' is so instinctive that we take it for granted without wondering about the complexity of the problem our auditory system solves.

Colin Cherry, in the 1950s, coined the term 'cocktail party problem' while trying to describe how our auditory system functions in such an environment [12]. He did a series of experiments to study the factors that help humans perform this complex task [11]. A number of theories have been proposed since then to explain the observations made in those experiments [11,12,70]. Helmhotz had, in the mid-nineteenth century, reflected upon the complexity of this signal by using the example of a ball room setting [22]. He remarked that even though the signal is "complicated beyond conception," our ears are able to "distinguish all the separate constituent parts of this confused whole."

So how does our auditory system solve the so-called cocktail party problem? Bregman tried to give a systematic account in his seminal 1990 book *Auditory Scene Analysis* [8]. He calls

Techniques for Noise Robustness in Automatic Speech Recognition, First Edition.
Edited by Tuomas Virtanen, Rita Singh, and Bhiksha Raj.

the process "scene analysis" by drawing parallels with vision. It has been argued that the goal of perception is to form a mental description of the world around us. Our brain analyzes the scene and forms mental representations by combining the evidence that it gathers through the senses. The role of audition is no different. Its goal is to form a mental description of the *acoustic* world around us by integrating sound components that belong together (e.g., those of the target speaker in a party) and segregating those that do not. Bregman suggests that the auditory system accomplishes this task in two stages. First, the acoustic input is broken down into local time-frequency elements, each belonging to a single source. This stage is called segmentation as it forms locally grouped time-frequency regions or *segments* [79]. The second stage then groups the segments that belong to the same source to form an *auditory stream*. A stream corresponds to a single source.

Inspired by Bregman's account of auditory organization, many computational systems have been proposed to segregate sound mixtures automatically. Such algorithms have important practical applications in hearing aids, automatic speech recognition, automatic music transcription, etc. The field is collectively termed *Computational Auditory Scene Analysis* (CASA).

This chapter is about CASA and automatic speech recognition in noise. In Section 16.2, we discuss some of the grouping principles of auditory scene analysis (ASA), focusing primarily on the cues that are most important for the auditory organization of speech. We then move on to computational aspects. How to combine CASA and ASR effectively is, in itself, a research issue. We address this by discussing CASA in depth, and introducing an important goal of CASA - *Ideal Binary Mask* (IBM) - in Section 16.3. As we will see, the IBM has applications to both speech segregation and automatic speech recognition. We will also discuss a typical architecture of CASA systems in Section 16.3. This will be followed by a discussion of strategies used for IBM estimation in Section 16.4. In the subsequent section, we address the topic of robust automatic speech recognition, where we will discuss some of the methods to integrate CASA and ASR. We note that this topic will also be addressed in other chapters (see Chapters 14 and 15 for detailed descriptions on missing-data ASR techniques). Finally, Section 16.6 offers a few concluding remarks.

16.2 Auditory Scene Analysis

CASA-based systems use ASA principles as a foundation to build computational models. As mentioned in the introductory section, Bregman described ASA to be a two stage process which results in integration of acoustic components that belong together and segregation of those that do not. In the first stage, an acoustic signal is broken down into time-frequency (T-F) segments. The second stage groups segments formed in the first stage into streams. Grouping of segments can occur across frequency or across time. They are called *simultaneous grouping* and *sequential grouping*, respectively.

A number of factors influence the grouping stage which results in the formation of coherent streams from local segments. Two distinctive schemes have been described by Bregman: primitive grouping and schema-based grouping.

Primitive grouping is an innate bottom-up process that groups segments based on acoustic attributes of sound sources. Major primitive grouping principles include proximity, periodicity, continuity, common onset/offset, amplitude and frequency modulation, and spatial location [8, 79]. Proximity refers to closeness in time or frequency of sound components. The components of a periodic signal are harmonically related (they are multiples of the fundamental frequency or $F0$), and thus segments that are harmonically related are grouped together. Periodicity is a major grouping cue that has also been widely utilized by CASA systems. Continuity refers to the continuity of pitch (perceived fundamental frequency), spectral and temporal continuity, etc. Continuity or smooth transitions can be used to group segments across time. Segments that have synchronous onset or offset times are usually associated with the same source and hence, grouped together. Among the two, onset synchrony is a stronger grouping cue. Similarly, segments that share temporal modulation characteristics (amplitude or frequency) tend to be grouped together. If segments originate from the same spatial location, there is a high probability that they belong to the same source and hence should be grouped.

Unlike primitive grouping, schema-based grouping is a top-down process where grouping occurs based on the learned patterns of sound sources. Schema-based organization plays an important role in grouping segments of speech and music, as some of their properties are learned over time by the auditory system. An example is the identification of a vowel based on observed formants. Note that both schema-based and primitive grouping play important roles in organizing real-world signals like speech and music.

The grouping principles introduced thus far were originally found though laboratory experiments using simple stimuli such as tones. Later experiments using more complex speech stimuli have established their role in speech perception [2,8]. Figure 16.1 shows some of the primitive grouping cues present for speech organization. Cues like continuity, common onset/offset, harmonicity are marked in the figure.

16.3 Computational Auditory Scene Analysis

Wang and Brown define CASA as ([79], p. 11):

> . . . the field of computational study that aims to achieve human performance in ASA by using one or two microphone recordings of the acoustic scene.

This definition takes into account the biological relevance of this field by limiting the number of microphones to two (like in humans) and the functional goal of CASA. The mechanisms used by CASA systems are perceptually motivated. For example, most systems make use of harmonicity as a grouping cue [79]. But this does not mean that the systems are exclusively dependent on ASA to achieve their goals. As we will see, modern systems make use of perceptual cues in combination with methods not necessarily motivated from the biological perspective.

16.3.1 Ideal Binary Mask

The goal of ASA is to form perceptual streams corresponding to the sound sources from the acoustic signal that reaches our ears. Taking this into consideration, Wang and colleagues suggested the *Ideal Binary Mask* as a main goal of CASA [24,27,76]. The concept was largely motivated by the masking phenomenon in auditory perception, whereby a stronger sound masks a weaker sound and renders it inaudible within a critical band [49]. Along the same lines, the IBM defines what regions in the time-frequency representation of a mixture are target dominant and what regions are not. Assuming a spectrogram-like representation of an

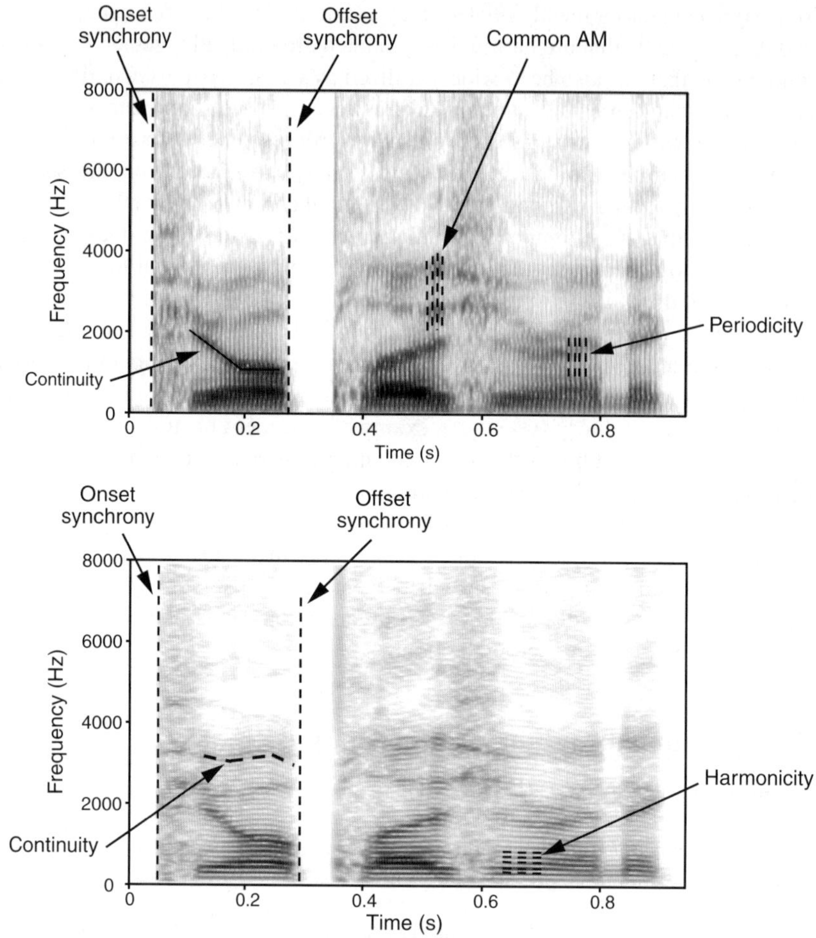

Figure 16.1 Primitive grouping cues for speech organization (reproduced from Wang and Brown [79]). The top panel shows a broadband spectrogram of the utterance "pure pleasure". Temporal continuity, onset and offset synchrony, common amplitude modulation and harmonicity cues are present. The bottom panel shows a narrow-band spectrogram of the same utterance.

acoustic input, the IBM takes the form of a binary matrix with 1 representing target dominant T-F units and 0 representing interference dominant units.

Mathematically, the IBM is defined as:

$$IBM(t,f) = \begin{cases} 1 & \text{if } SNR(t,f) \geq LC \\ 0 & \text{otherwise.} \end{cases} \quad (16.1)$$

Here, $SNR(t, f)$ represents the signal-to-noise ratio (SNR) within the T-F unit of time index t and frequency index (or channel) f. LC stands for a local criterion, which acts as an SNR threshold that determines how strong the target should be over the noise for the unit to be marked target dominant. The LC is usually set to 0 dB which translates to a simple rule of whether the target energy is stronger than the noise energy. Note that, to obtain the IBM, we need access to the premixed target and interference signals (hence the term "ideal"). According to them, a CASA system should aim at *estimating* the IBM from the mixture signal. It should be pointed out that the IBM can be thought of as an "oracle" binary mask. Oracle masks, binary or otherwise, have been widely used in the missing-data ASR literature to indicate the ceiling recognition performance of noisy speech.

The reasons why the IBM is an appropriate goal of CASA include the following:

(i) Li and Wang studied the optimality of the IBM measured in terms of the improvement in the SNR of a noisy signal (SNR gain) processed using binary masks [43]. They show that, under certain conditions, the IBM with the *LC* of 0 dB is optimal among all binary masks. Further, they compare the IBM with the ideal ratio (soft) mask, which is a T-F mask with real values representing the percentages of target speech energy contained in T-F units, similar to a Wiener filter. The comparisons show that, although the ideal ratio mask achieves higher SNR gains than the IBM as expected, in most mixtures of interest the difference in SNR gain is very small.

(ii) IBM-segregated noisy speech has been shown to greatly improve intelligibility for both normal hearing and hearing impaired listeners [1,10,42,81]. Even when errors are introduced to the IBM, it can still improve the intelligibility of noisy speech as long as the errors are within a reasonable range [42,62]. Moreover, it has been found that the *LC* of –6 dB seems to be more effective than the *LC* of 0 dB to improve speech intelligibility [81] even though the latter threshold leads to a higher SNR of IBM processed signals.

(iii) Speech energy is sparsely distributed in a high-resolution T-F representation, and there is little overlap between the components of different speakers in a speech mixture [63,86]. Under such circumstances, the IBM can almost segregate a mixture into its constituent streams. Note that sparsity does not hold for broadband interferences such as speech babble or when room reverberation is present.

(iv) Related binary masks have been shown to be effective for robust ASR [13,62]. Missing-data techniques using IBM like masks have been discussed in detail in previous chapters (see Chapters 14 and 15). Apart from missing-data ASR, other strategies have been proposed that use the IBM to improve ASR results. We will look at a few of them later in this chapter.

(v) Recently, Wang *et al.* [80] showed that IBM-modulated noise can produce intelligible speech. In this experiment, speech-shaped noise (SSN) is modulated by the IBM created for a mixture of speech and SSN. Speech shaped noise is broadband, and has a long-term spectrum matching that of natural speech. Even with a coarse frequency resolution (e.g., 16 bands), they observe nearly perfect intelligibility of IBM modulated noise.

Figure 16.2 shows an example of the IBM created for a two-talker mixture. The time-frequency representation used in the figure is called a *cochleagram*, which is commonly used in CASA [79]. Compared to the mixture in the middle left panel, the IBM-masked mixture (shown in the bottom left panel) is more similar to the target utterance (shown in the top left panel).

Apart from the IBM, research has also aimed at estimating the ideal ratio mask [3,73]. Note that, the real values in a ratio (soft) mask can also be interpreted as the probability of a T-F unit being target dominant. One can argue that estimating a ratio mask is computationally harder

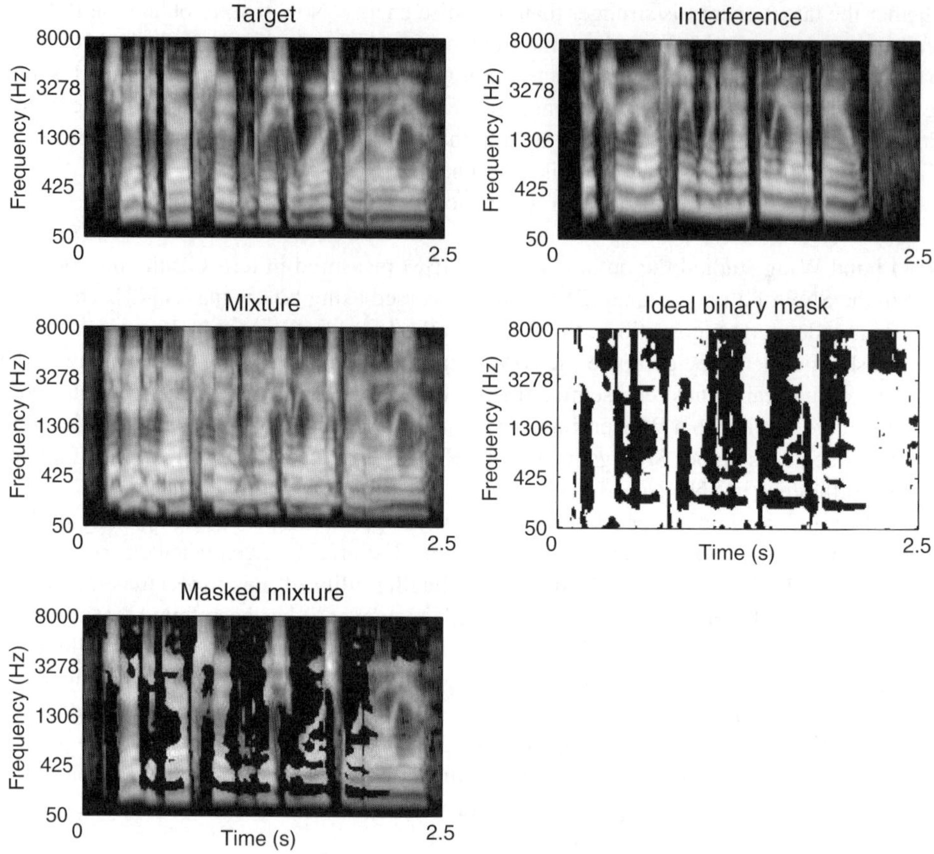

Figure 16.2 Illustration of the IBM. The top left panel shows a cochleagram of a target utterance where brightness indicates energy. The top right panel shows a cochleagram of the interference signal. The middle left panel shows a cochleagram of the mixture. The middle right panel shows the ideal binary mask for the mixture where a white pixel indicates 1 and a black pixel 0. The bottom left panel shows the cochleagram of the IBM-masked mixture.

than estimating a binary mask [77]. Nevertheless, the use of ratio masks has been shown to be advantageous in some ASR studies [3,73].

16.3.2 Typical CASA Architecture

Figure 16.3 shows a typical architecture of CASA. All CASA systems start with a peripheral analysis of the acoustic input (the mixture). Typically, the peripheral analysis converts the signal into a time-frequency representation. This is usually accomplished by using an auditory filter bank. The most commonly used is the gammatone filter bank [58]. The center frequencies of the gammatone filter bank are uniformly distributed on the ERB-rate scale [18]. ERB refers to the equivalent rectangular bandwidth of an auditory filter, which corresponds to the bandwidth

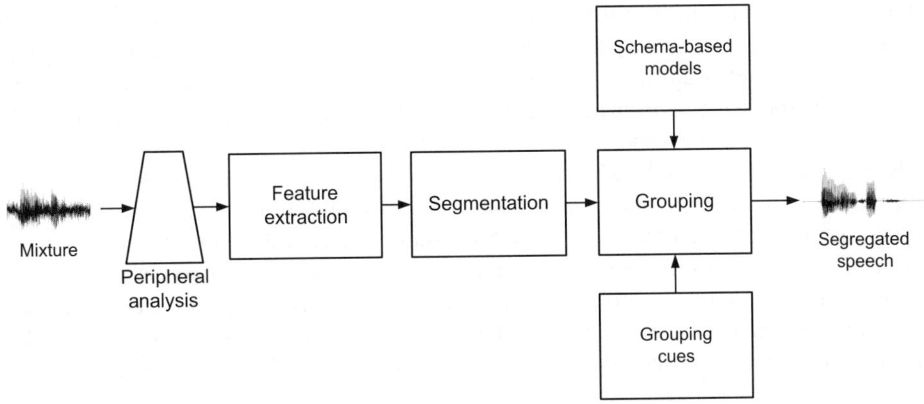

Figure 16.3 Schematic diagram of a typical CASA system.

of an ideal rectangular filter that has the same peak gain as the auditory filter with the same center frequency and passes the same total power for white noise. Similar to the Bark scale, the ERB-rate scale is a warped frequency scale akin to that of human cochlear filtering. The ERB scale is close to linear at low frequencies, but logarithmic at high frequencies. Figure 16.4 shows the responses of eight such filters, uniformly distributed according to the ERB-rate scale from 100 to 2000 Hz. Although eight filters are sufficient to fully span a frequency range of 50–8000 Hz, more filters (32 or 64) are typically used for a better frequency resolution. To simulate the firing activity of auditory nerve fibers, the output from the gammatone filter bank is further subjected to some nonlinear processing, where the Meddis hair cell model is typically used [48]. It models the rectification, compression and the firing pattern of the

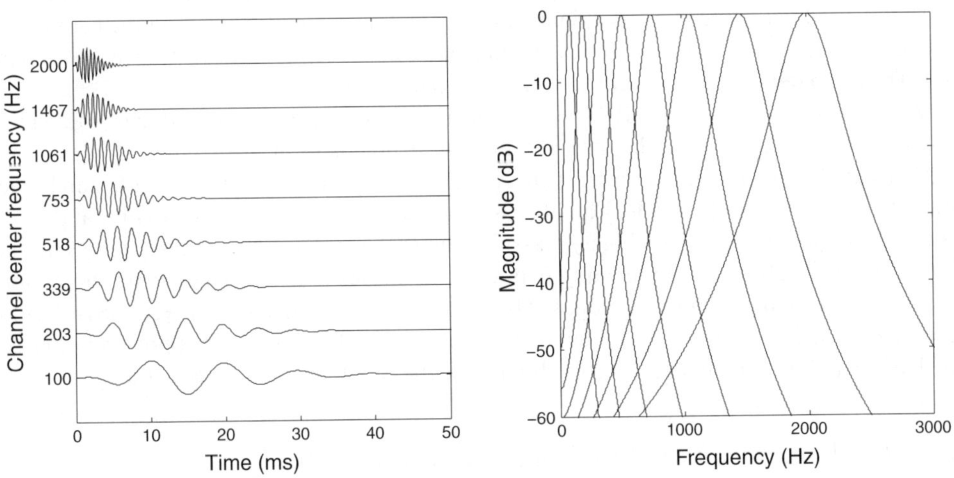

Figure 16.4 A gammatone filter bank. The left panel shows impulse responses of eight gammatone filters, with center frequencies equally spaced between 100 Hz and 2 KHz on the ERB-rate scale. The right panel shows the corresponding magnitude responses of the filters.

auditory nerve. Alternatively, a simple half wave rectification followed by some compression (square root or cubic root) can be used to model the nonlinearity. Finally, the output at each channel is windowed or downsampled. The result is the cochleagram of the acoustic signal as it models the processing performed by the cochlea [79]. An element of a cochleagram is a T-F unit, which represents the response of a particular filter at a time frame.

The next few stages vary depending on the specifics of different CASA systems. The feature extraction stage computes features such as $F0$, onset/offset, amplitude and frequency modulation. The extracted features enable the system to form segments, each of which is a contiguous region of T-F units. Segments provide a mid-level representation on which grouping operates. The grouping stage utilizes primitive and schema-based grouping cues. The output of the grouping stage can be an estimated binary mask or a ratio mask. Efficient algorithms exist that can resynthesize the target signal using a T-F mask and the original mixture signal [79,82].

16.4 CASA Strategies

Given the goal of estimating the IBM, we now discuss strategies to achieve it. The main focus of this section will be on monaural CASA techniques which have seen most of the development.

Monaural source segregation uses a single recording of the acoustic scene from which the target is to be segregated. The most important cue utilized for this task is the fundamental frequency. F0 estimation from clean speech is fairly accurate and many systems exist that perform well; for example Praat is a freely available tool which is widely used [6]. The presence of multiple sound sources in a scene adds to the complexity of the task as a single frame may now have multiple pitch points. Perhaps the earliest system that used F0 for speech segregation was proposed by Parsons [57]. He used the short-term magnitude spectrum of noisy speech to estimate multiple F0s. A sub-harmonic histogram method, proposed by Shroeder [64], was used to estimate the most dominant $F0$ in a frame. He then removed the harmonics of the estimated $F0$ from the mixture spectrum and used the remainder to estimate the second $F0$. The estimated $F0$s were finally used to segregate the mixture.

We start our discussion on IBM estimation in Section 16.4.1 by introducing strategies based on noise-estimation techniques from the speech-enhancement literature. More recent CASA-based strategies aim to segregate the target by extracting ASA cues like $F0$, amplitude modulation and onset/offset, which are then used to estimate the IBM. An alternative approach is to treat mask estimation as a binary classification problem. We explain these approaches in the subsequent subsections by treating two recent strategies in detail. The second subsection focuses on the *tandem* algorithm proposed by Hu and Wang [26] that uses several ASA cues to estimate the IBM. Section 16.4.3 focuses on a binary classification-based approach proposed by Kim *et al.* [36]. The final subsection briefly touches upon binaural CASA strategies.

16.4.1 IBM Estimation Based on Local SNR Estimates

In this sub-section, we discuss mask estimation strategies that are based on local signal-to-noise ratio estimates at each time-frequency unit. Such techniques typically make use of an estimate of the short-time noise power spectrum. The estimated noise power can be used to obtain the SNR and in turn a T-F mask. It should be clear from Equation (16.1) that with the true local SNR information, the IBM can be readily calculated. The noise estimate can also be used to define masks based on alternative criteria, like the negative energy criterion used by El-Maliki and Drygajlo [17]. We will first review a few noise-estimation techniques, followed by a brief discussion on how they can be used to estimate the IBM.

Noise (and SNR) estimation is a widely studied topic in speech enhancement largely in the context of spectral subtraction [5]. One commonly used technique is to assume that noise remains stationary throughout the duration of an utterance and that the first few frames are 'noise-only'. A noise estimate is then obtained by simply averaging the spectral energy of these frames. Such estimates are, for instance, used in Vizinho *et al.* [75], Josifovski *et al.* [34], Cooke *et al.* [13]. But noise is often nonstationary and therefore, such methods often result in poor IBM estimates. More sophisticated techniques have been proposed to estimate noise in nonstationary conditions. See, for example, voice-activity detection (VAD) [69] based methods [40], Hirsch's histogram based methods [23], recursive noise-estimation techniques [23], etc. Seltzer *et al.* [65] use an approach similar to Hirsch's to estimate the noise floor in each sub-band, which is in turn used for mask estimation (see Section 16.4.3). A more detailed discussion on noise estimation can be found in Chapter 4.

All noise-estimation techniques can be easily extended to estimate the SNR at each T-F unit by using it to obtain an estimate of the clean speech power spectrum. A spectral subtraction based approach [5,7] is commonly used, wherein the speech power is obtained by subtracting the noise power from the observed noisy spectral power. Further, a spectral floor is set and any estimate lower than the floor is automatically rounded to this preset value. Other direct SNR-estimation techniques have also been proposed in the literature. For example, Nemer *et al.* [53] utilize higher order statistics of speech and noise to estimate the local SNR, assuming a sinusoidal model for band restricted speech and a Gaussian model for noise. A supervised SNR-estimation technique was proposed by Tchorz and Kollmeier [74]. They use features inspired from psychoacoustics and a multilayer perceptron (MLP)-based classifier to estimate the SNR at each T-F unit. Interested readers are also referred to Loizou[46] for detailed reviews on these topics.

If a noise estimate is used to calculate the SNR, the IBM can be estimated using Equation (16.1) after setting the LC to an appropriate value. Although 0 dB is a natural choice here, other values have also been used [13,60]. Soft (ratio) masks can be obtained from local SNR estimates by applying a sigmoid function that maps it to a real number in the range $[0, 1]$, thereby allowing it to be interpreted as probability measures for subsequent processing. One can also define masks based on *a posteriori* SNR, which is the ratio of the noisy signal power to noise power expressed in dB [61]. This circumvents the need to estimate the clean speech power and local SNR. Note that any *a posteriori* SNR criterion can be equivalently expressed using a local SNR criterion. An even simpler alternative is to use the negative energy criterion proposed by El-Maliki and Drygajlo [17]. They identify reliable speech dominant units as those T-F units for which the observed noisy spectral energy is greater than the noise estimate. In other words, T-F units for which the spectral energy after subtracting the noise estimate from the observed noisy spectral energy is negative are considered noise dominant and unreliable. Raj and Stern [59] note that a combination of an SNR criterion and a negative energy criterion usually yields better quality masks.

In practice, such noise-estimation-based techniques work well in stationary conditions but tend to produce poor results in nonstationary conditions. Nonetheless, SNR-based techniques are still used because of their simplicity.

16.4.2 IBM Estimation using ASA Cues

The tandem system by Hu and Wang [26] aims at voiced speech segregation and F0 estimation in an iterative fashion. In describing the algorithm, we will explain how some of the ASA cues can be extracted and utilized for computing binary masks.

The tandem system uses several auditory representations that are widely used for pitch estimation. These representations are based on autocorrelation, which was originally proposed by Licklider back in the 1950s to explain pitch perception [44]. Autocorrelation has been used by other $F0$ estimation techniques [24,38,85]. The tandem system first uses a gammatone filter bank to decompose the signal into 128 frequency channels with center frequencies spaced uniformly in the ERB-rate scale from 50 to 8000 Hz. The output at each channel is divided into frames of length 20 ms with 10 ms overlap. A running autocorrelation function (ACF) is then calculated according to Equation (16.2) at each frame to form a *correlogram*:

$$A(t,f,\tau) = \frac{\sum_n x(tT_t - nT_n, f)x(tT_t - nT_n - \tau T_n, f)}{\sqrt{\sum_n x^2(tT_t - nT_n, f)}\sqrt{\sum_n x^2(tT_t - nT_n - \tau T_n, f)}}. \quad (16.2)$$

Here, $A(t,f,\tau)$ denotes the normalized autocorrelation function at frequency channel f and time frame t, and τ is the time delay in samples indexed by n. $T_t = 10$ ms and $T_n = 1/f_s$, where f_s is the sampling frequency, are the frame shift and the sampling time, respectively. The function is normalized so that the peak value at $\tau = 0$ is 1. An example of a correlogram is shown in Figure 16.5. Usually, a peak in the ACF corresponds to the time delay that represents a period of the signal. Since the target signal is speech, τ can be limited to the typical pitch range between 70 and 400 Hz, or τT_n between 2.5 and 15 ms [54]. Calculating the channel-wise ACF after decomposing the signal using a filter bank, instead of directly calculating it from the time domain signal, adds to the robustness of the F0 estimation process [14,85]. Additionally, a summary autocorrelation function (SACF) can be calculated by summing the ACFs across all the channels:

$$SACF(T,\tau) = \sum_f A(T,f,\tau). \quad (16.3)$$

A peak in the SACF corresponds to the time period that has support from many frequency channels. Since a periodic signal triggers responses in multiple channels, this peak likely indicates the period of the signal.

The cross-channel correlation between neighboring channels has been used to identify whether neighboring T-F units are dominated by the same source which can be used to group the units to form a segment [9,78]. Normalized cross-channel correlation, $C(t,f)$, is calculated

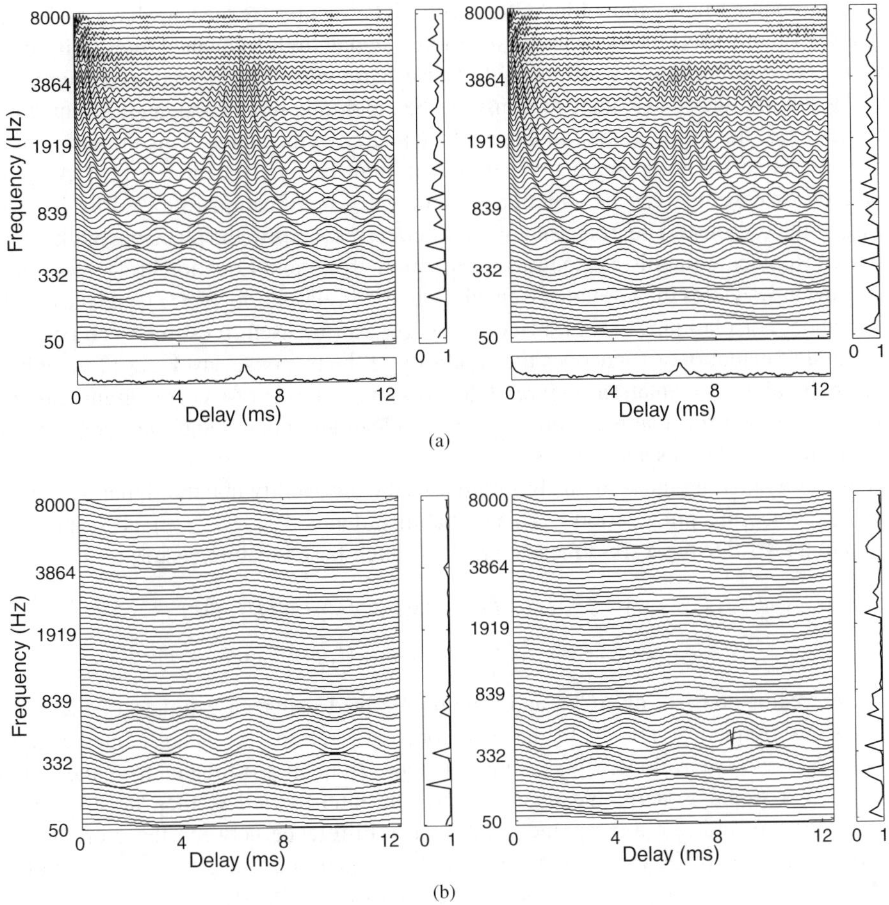

Figure 16.5 Autocorrelation and cross-channel correlation. (a) Correlogram at a frame for clean speech (top left panel) and a mixture of speech with babble noise at 6 dB SNR (top right panel). The corresponding cross-channel correlation and summary autocorrelation are shown on the right and the bottom panel of each figure, respectively. A peak in the SACF is clearly visible in both cases. Note that correlations of different frequency channels are represented using separate lines. (b) Corresponding envelope correlogram and envelope cross-channel correlation for clean speech (bottom left panel) and the mixture (bottom right panel). It can be clearly seen that the functions estimated from clean speech and noisy speech match closely.

using the ACF as:

$$C(t,f) = \frac{\sum_{\tau} \left[A(t,f,\tau) - \overline{A(t,f)}\right] \left[A(t,f+1,\tau) - \overline{A(t,f+1)}\right]}{\sqrt{\sum_{\tau} \left[A(t,f,\tau) - \overline{A(t,f)}\right]^2} \sqrt{\sum_{\tau} \left[A(t,f+1,\tau) - \overline{A(t,f+1)}\right]^2}}. \quad (16.4)$$

Here, $\overline{A(t,f)}$ denotes the mean of the ACF function over τ.

As mentioned earlier, gammatone filters with higher center frequencies have wider bandwidths (see Figure 16.4). As a result, for a periodic signal, high-frequency filters will respond to more than one harmonic of the signal. These harmonics are referred to as *unresolved*. Unresolved harmonics cause filter responses to be amplitude modulated, and the envelope of a filter response fluctuates at the fundamental frequency of the signal. This property has also been used as a cue to group segments and units in high-frequency channels [24]. Amplitude modulation or envelope can be captured by half wave rectification followed by band-pass filtering of the response. The pass band of the filter corresponds to the plausible pitch range of the signal. Replacing the filter responses in Equation (16.2) and Equation (16.4) with the extracted envelopes yields the normalized envelope autocorrelation, $A_{\mathrm{E}}(t, f, \tau)$, and the envelope cross-channel correlation, $C_{\mathrm{E}}(t, f)$, respectively. A_{E} can be used to estimate the periodicity of amplitude fluctuation. C_{E} encodes the similarity of the response envelopes of neighboring channels and aids segmentation. Figure 16.5 shows an example of a correlogram and an envelope correlogram for a single frame of speech (clean and noisy), and their corresponding cross-channel correlations and SACFs.

For T-F unit labeling, the tandem algorithm uses the probability that the signal within a unit is in agreement with a pitch period τ. This probability, denoted as $P(T, f, \tau)$, is estimated with the help of an MLP using a six-dimensional (6-D) pitch-based feature vector:

$$\begin{aligned} r(t, f, \tau) = [&A(t, f, \tau), \bar{f}(t, f)\tau - \mathrm{int}(\bar{f}(t, f)\tau), \mathrm{int}(\bar{f}(t, f)\tau), \\ &A_{\mathrm{E}}(t, f, \tau), \bar{f}_{\mathrm{E}}(t, f)\tau - \mathrm{int}(\bar{f}_{\mathrm{E}}(t, f)\tau), \mathrm{int}(\bar{f}_{\mathrm{E}}(t, f)\tau)], \end{aligned} \tag{16.5}$$

where the vector consists of ACFs and features derived using an estimate of the average instantaneous frequency, $\bar{f}(t, f)$. In the equation, int(.) returns the nearest integer and the subscript 'E' denotes envelope. $\bar{f}_{\mathrm{E}}$ is the instantaneous frequency estimated from the response envelope. If a signal is harmonically related to the pitch period τ, then $\mathrm{int}(\bar{f}(t, f)\tau)$ and $\mathrm{int}(\bar{f}_{\mathrm{E}}(t, f)\tau)$ will indicate a harmonic number. The difference between these products and their nearest integers in the second and the fourth terms quantifies a degree of this relationship. An MLP is trained for each filter channel in order to estimate $P(t, f, \tau)$[1].

The algorithm first estimates initial *pitch contours*, each of which is a set of contiguous pitch periods belonging to the same source, and their associated binary masks for up to two sound sources. The main part of the algorithm iteratively refines the initial estimates. The final stage applies onset/offset analysis to further improve segregation results. Let us now look at these stages in detail.

The initial stage starts by identifying T-F units corresponding to periodic signals. Such units tend to have high cross-channel correlation or envelope cross-channel correlation and, therefore, are identified by comparing $C(t, f)$ and $C_{\mathrm{E}}(t, f)$ with a threshold. Within each frame, the algorithm considers up to two dominant voiced sound sources. The identified T-F units of a frame are grouped using a two step process. First, estimate up to two $F0$s. Next, assign a T-F unit to an $F0$ group if it agrees with the $F0$.

An earlier model by Hu and Wang [24] identifies the dominant pitch period of a frame as the lag that corresponds to the maximum in the summary autocorrelation function (Equation (16.3)). To check if a T-F unit agrees with the dominant pitch period, they compare the value

[1] Note that this term is a convenient abuse of notation. It, in fact, represents the posterior probability of the T-F unit being in agreement with the pitch period given the 6-D pitch-based features.

of the ACF at the estimated pitch period to the peak value of the ACF for that T-F unit:

$$\frac{A(t, f, \tau_D(t))}{A(t, f, \tau_P(t, f))} > \theta_P. \tag{16.6}$$

Here, $\tau_D(t)$ and $\tau_P(t, f)$ are the delays that correspond to the estimated $F0$ and the maximum in the ACF, respectively, for channel f at time frame t. If the signal within the T-F unit has a period close to the estimated $F0$, then this ratio will be close to 1. θ_P defines a threshold to make a binary decision about the agreement.

The tandem algorithm uses a similar approach, but instead of the ACF it uses the probability function, $P(t, f, \tau)$, estimated using the MLPs. Having identified the T-F units of each frame with strong periodicity, the algorithm chooses the lag, τ, that has the most support from these units as the dominant pitch period of the frame. A T-F unit is said to support τ if the probability, $P(t, f, \tau)$, is above a chosen threshold. The T-F units that support the dominant pitch period are then grouped together. The second pitch period and the associated set of T-F units are estimated in a similar fashion, using those units not in the first group. To remove spurious pitch estimates, if there are too few supporting T-F units, the estimated pitch is discarded.

To form pitch contours from these initial estimates, the algorithm groups the pitch periods of any three consecutive frames if their values change by less than 20% from one frame to the next. The temporal continuity of the sets of T-F units associated with the pitch periods is also considered before grouping pitch estimates together; at least half of the frequency channels associated with the pitch periods of neighboring frames should match for them to be grouped into a pitch contour. After the initial stage, each pitch contour has an associated T-F mask. Since pitch changes rather smoothly in natural speech, each of the formed pitch contours and its associated binary mask usually belong to a single sound source. Isolated pitch points after this initial grouping are considered unreliable and discarded.

These initial estimates are then refined using an iterative procedure. The idea is to use obtained binary masks to obtain better pitch contours, and then use the refined pitch estimates to re-estimate the masks. Each iteration of the tandem algorithm consists of two steps:

(i) The first step expands each pitch contour to its neighboring frames, and re-estimates its pitch periods. Since pitch changes smoothly over time, the pitch periods of the contour can be used to estimate potential pitch periods in the contour's neighboring frames. Specifically, for the kth pitch contour η_k, that extends from frame t_1 to t_2, the corresponding binary mask, $M_k(t)$ $(t = t_1, \ldots, t_2)$, is extended to frames $t_1 - 1$ and $t_2 + 1$ by setting $M_k(t_1 - 1) = M_k(t_1)$ and $M_k(t_2 + 1) = M_k(t_2)$. Using this new mask, the periods of the pitch contour are reestimated. A summary probability function, $SP(t, \tau)$, which is similar to $SACF(t, \tau)$ but uses $P(t, f, \tau)$ values instead of $A(t, f, \tau)$, is calculated at each frame for this purpose. The *SP* function tends to have significant peaks at multiples of the pitch period. Therefore, an MLP is trained to choose the correct pitch period from among the multiple candidates. The expansion stops at either end when the estimated pitch violates temporal continuity with the existing pitch contour. Note that, as a result of contour expansion, pitch contours may be combined.

(ii) The second step reestimates the mask corresponding to each of the pitch contours. This is done by identifying T-F units of each frame that are in agreement with the estimated pitch period of that frame. Given the pitch period $\tau_D(t)$, $P(t, f, \tau_D)$ can be directly used to make this decision at each T-F unit. But this does not take into consideration the temporal continuity and the wide-band nature of speech. If a T-F unit is in agreement with τ_D, its neighboring T-F units also tend to agree with τ_D. For added robustness, the tandem algorithm trains an MLP to perform unit labeling based on a neighboring set of T-F units. It takes as input the $P(t, f, \tau)$ values of a set of neighboring T-F units, centered at the unit for which the labeling decision has to be made. The output of this MLP is finally used to label each T-F unit.

The algorithm iterates between these two steps until it converges or the number of iterations exceeds a predefined maximum (20 is suggested).

The final step of the tandem algorithm is a segmentation stage based on onset/offset analysis, which may be viewed as post processing. The stage forms segments by detecting sudden changes in intensity as such a change indicates an onset or offset of an acoustic event. As discussed earlier, onset and offset are prominent ASA principles (see Figure 16.1). Segments are formed using multiscale analysis of onsets and offsets (see Hu and Wang [25] for details). The tandem algorithm further breaks each segment down to channel wise subsegments, called *T-segments* as they span multiple time frames but are restricted to a single frequency channel. Each T-segment is then classified as a whole as target dominant if at least half its energy is contained in the voiced frames of the target and at least half of the energy in these voiced frames is included in the target mask. If the conditions are not satisfied, the labeling from the iterative stage remains unchanged for the units of the T-segment.

Figure 16.6 illustrates the results of different stages of the tandem system. The mask obtained at the end of the iterative stage (Figure 16.6(e)) includes most of the target speech. The subsequent segmentation stage improves the segregation results by recovering a few previously masked (mask value 0) T-F units, for example toward the end of the utterance in Figure 16.6(g). These units were identified from the onset/offset segments. The final resynthesized waveform, shown in Figure 16.6(h), is close to the original signal (Figure 16.6(b)).

There are two important aspects of CASA that the tandem algorithm does not consider. The first one is sequential organization. The outputs of the tandem system are multiple pitch contours and associated binary masks. The pitch track (and therefore the mask) of a target utterance need not be continuous as there are breaks due to silence and unvoiced speech. Sections before and after such discontinuities have to be sequentially grouped into the target stream. The tandem system assumes ideal sequential grouping, and therefore ignores the sequential grouping issue. Methods for sequential grouping have been proposed. Barker *et al.* [4] proposed a schema based approach using ASR models to simultaneously perform sequential integration and speech recognition (more about this in Section 14.4.3). Ma *et al.* [47] later used a similar approach to group segments that were formed using correlograms in voiced intervals and a watershed algorithm in unvoiced intervals. Shao and Wang [67] proposed a speaker model-based approach for sequential grouping. Recently, Hu and Wang [30] proposed an unsupervised grouping strategy based on clustering and reported results comparable to the model-based approach of Shao and Wang.

The second issue with the tandem algorithm is that it does not deal with unvoiced speech. An analysis by Hu and Wang [28] shows that unvoiced speech accounts for more than 20% of spoken English, measured in terms of both frequency and duration of speech sounds. Therefore, unvoiced speech segregation is important for improving the intelligibility and ASR of the segregated target signal. Dealing with unvoiced speech is challenging as it has noise-like characteristics and lacks strong grouping cues such as $F0$. Hu and Wang [28]

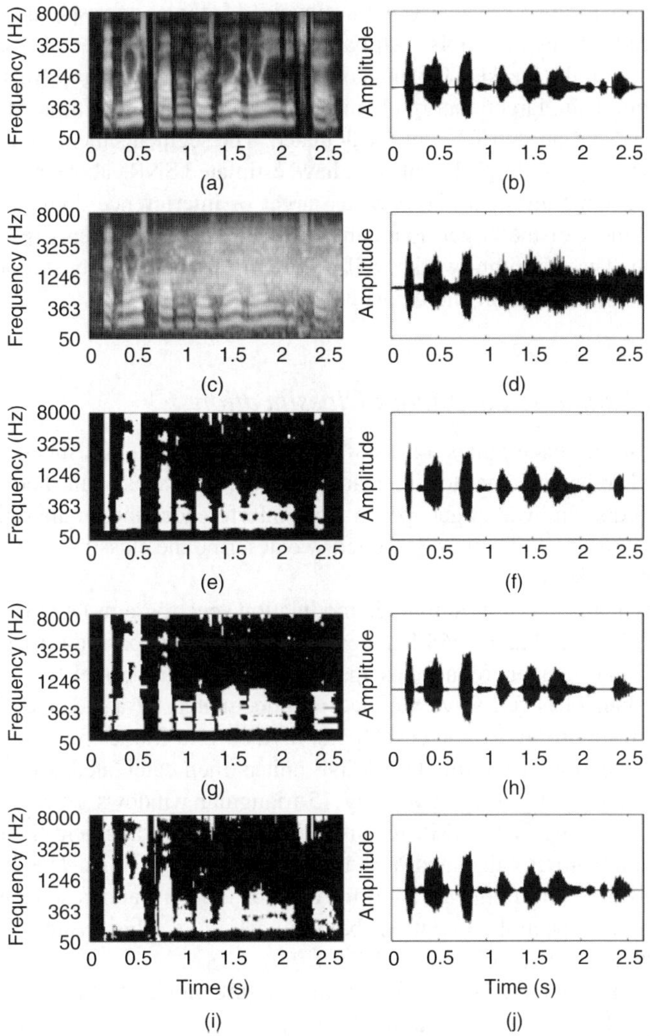

Figure 16.6 Different stages of IBM estimation using the tandem system. (a) Cochleagram of a female target utterance. (b) Corresponding waveform. (c) Cochleagram of a mixture signal obtained by adding crowd noise to the target utterance. (d) Corresponding waveform. (e) Mask obtained at the end of the iterative stage of the algorithm. (f) Waveform of the resynthesized target using the mask. (g) The final mask obtained after the segmentation stage. (h) The resynthesized waveform. (i) The IBM. (j) Resynthesized signal using the IBM. Reproduced by permission from Hu and Wang [26] © 2010 IEEE.

suggest a method to extract unvoiced speech using onset/offset based segments. They first segregate voiced speech. Then, acoustic-phonetic features are used to classify the remaining segments as interference dominant or unvoiced speech dominant. A simpler system was later proposed by Hu and Wang [29]. Their system first segregates voiced speech and removes other periodic intrusions from the mixture. It then uses a spectral subtraction based scheme to obtain segments in unvoiced intervals (an unvoiced interval corresponds to a contiguous group of unvoiced frames); the noise estimate for each unvoiced interval is estimated using the mixture energy in the masked T-F units of its neighboring voiced intervals. Together with an approximation of the target energy obtained by subtracting the estimated noise from the mixture, the local SNR at each T-F unit is calculated. The segments themselves are formed by grouping together neighboring T-F units that have estimated SNRs above a chosen threshold. The obtained segments are then classified as target or interference dominant based on the observation that most of the target dominant unvoiced speech segments reside in the high-frequency region. The algorithm works well if the noise remains fairly stationary during the duration of an unvoiced interval and the neighboring voiced intervals.

16.4.3 IBM Estimation as Binary Classification

The tandem algorithm exemplifies a system that uses ASA cues and supervised learning to estimate the IBM. When it comes to direct classification, the issues lie in choosing appropriate features that can discriminate target speech from interference, and an appropriate classifier. To explain how direct classification is applied, we describe the classification-based approach of Kim *et al.* [36] in detail.

The system by Kim *et al.* uses amplitude modulation spectrograms (AMS) as the feature to build their classifier. To obtain AMS features, the signal is first passed through a 25 channel filter bank, with filter center frequencies spaced according to the mel-frequency scale. The output at each channel is full-wave rectified and decimated by a factor of 3 to obtain the envelope of the response. Next, the envelope is divided into frames 32 ms long with 16 ms overlap. The modulation spectrum at each T-F unit is then calculated using the FFT[2]. The FFT magnitudes are finally integrated using 15 triangular windows spaced uniformly from 15.6 to 400 Hz, resulting in 15 AMS features [39]. Kim *et al.* augment the extracted AMS features with delta features calculated from the neighboring T-F units. The delta features are calculated across time and frequency, and for each of the 15 features separately. They help capture temporal and spectral correlations between T-F units. This creates a 45-dimensional feature representation for each T-F unit, $AMS(t, f)$.

Given the 45-dimensional input, a Gaussian mixture model (GMM)-based classifier is trained to do the classification. The desired unit labels are set using the IBM created using an *LC* (see Equation (16.1)) of –8 dB for low-frequency channels (channels 1 through 15) and –16 dB for high-frequency channels (channels 16 through 25). This creates a group of masked T-F units, represented as λ_0, and unmasked (mask value 1) T-F units, λ_1. The authors chose a lower *LC* for high-frequency channels to account for the difference in the masking characteristics of speech across spectrum. Each group, λ_i, where $i = 0, 1$, is further divided into two smaller subgroups, λ_i^0 and λ_i^1, using a second threshold, LC_i. The thresholds ($LC_0 < LC$ and $LC_1 > LC$) are chosen such that the amount of training data in the two subgroups of a group are the same. This second subdivision is done mainly to reduce the training time of the GMMs. A 256-mixture, 45-dimensional, full-covariance GMM is trained using the expectation-maximization algorithm to model the distribution of each of the 4 subgroups.

---

[2] A T-F unit, here, refers to a 32 ms long frame at a particular frequency channel.

Given a T-F unit from a noisy utterance, a Bayesian decision is then made to obtain a binary label that is 0 if and only if $P(\lambda_0 \mid AMS(t, f)) > P(\lambda_1 \mid AMS(t, f))$, where

$$P(\lambda_0 \mid AMS(t,f)) = \frac{P(\lambda_0, AMS(t,f))}{P(AMS(t,f))}$$

$$= \frac{P(\lambda_0^0)P(AMS(t,f) \mid \lambda_0^0) + P(\lambda_0^1)P(AMS(t,f) \mid \lambda_0^1)}{P(AMS(t,f))}.$$

The equation calculates the *a posteriori* probability of λ_0 given the AMS features at the T-F unit. $P(\lambda_0^0)$ and $P(\lambda_0^1)$ are the *a priori* probabilities of subgroups λ_0^0 and λ_0^1, respectively, calculated from the training set. The likelihoods, $P(AMS(t, f) \mid \lambda_0^0)$ and $P(AMS(t, f) \mid \lambda_0^1)$, are estimated using the trained GMMs. $P(AMS(t, f))$ is independent of the class label and, hence, can be ignored. $P(\lambda_1 \mid AMS(t, f))$ is calculated in a similar fashion.

One advantage of using the AMS feature is that it can handle both voiced and unvoiced speech, as opposed to the 6-D pitch based feature used by the tandem algorithm which can be used only to classify voiced speech. As a result, the mask obtained using Kim *et al.*'s algorithm includes both voiced and unvoiced speech.

Figure 16.7 shows an estimated binary mask using Kim *et al.*'s algorithm. The authors evaluated their system using speech intelligibility tests and reported substantial improvements in the intelligibility of segregated speech for normal-hearing listeners [36]. It is worth emphasizing that this is the first monaural segregation system that produces improved speech intelligibility.

One of the main disadvantages of Kim *et al.*'s system is that training is noise dependent. Although it works well when tested on speech corrupted with the same noise types, the performance degrades significantly when previously unseen noise types are used during the testing stage. A second disadvantage of the system is that it can handle only nonspeech intrusions because AMS features mainly distinguish speech and nonspeech signals. By avoiding competing talkers, the problem of sequential organization is avoided because all detected speech belongs to the target.

Jin and Wang [32] also proposed a classification-based approach to perform voiced speech segregation in reverberant environments. For T-F unit classification, they use the 6-D pitch-based features given in Equation (16.5), and an MLP-based classifier. In order to utilize global information that is not sufficiently represented at the T-F unit level, an additional segmentation stage is used by their system. Segmentation is performed based on cross-channel correlation and temporal continuity in low-frequency channels—adjacent T-F units with high cross-channel correlation are iteratively merged to form larger segments. In high-frequency channels, they are formed based on onset/offset analysis [25]. The unit level decisions are then used to group the formed voiced segments either with the target stream or the nontarget (or the background) stream. Their system produced good segregation results under various reverberant conditions. Since pitch-based features are derived using the pitch of the *target*, classifiers trained on such features tend to generalize better than those trained using AMS features.

More recently, Kun and Wang proposed an SVM-based binary mask estimation model [19]. Inspired by Jin and Wang [32] and Kim *et al.* [36], they propose to combine pitch-based and AMS features along with the use of an SVM based classifier. Their system performs well in a variety of test conditions and is found to have good generalization to unseen noise types.

In the context of robust ASR, Seltzer *et al.* [65] proposed a similar Bayesian classification based approach to mask estimation. They extract the following features at the T-F unit level to

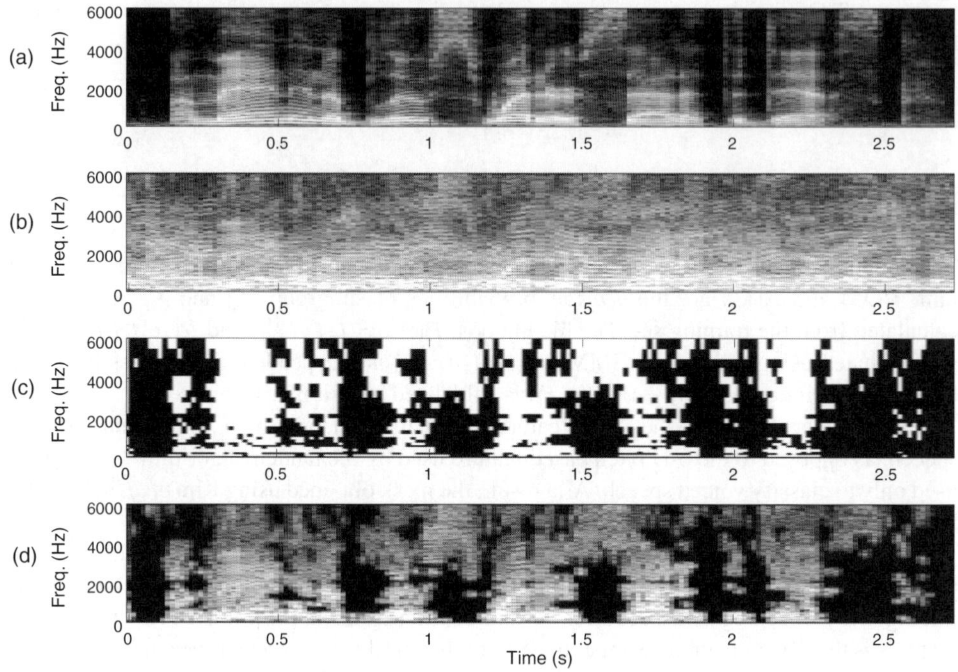

Figure 16.7 IBM estimation using classification. (a) A spectrogram of a target utterance from the IEEE corpus. (b) Spectrogram of the noisy mixture obtained by adding babble noise to the target utterance. (c) The estimated binary mask. (d) The spectrogram of the resynthesized signal obtained using the estimated binary mask. Reprinted with permission from Kim *et al.* [36] © 2009, Acoustical Society of America.

build GMM based Bayesian classifiers: comb filter ratio (CFR), which is the log ratio of the total energy at the harmonics of the fundamental frequency estimated for a frame to the total energy in between those frequencies; autocorrelation peak ratio (APR), which is the ratio of height of the largest secondary peak in the ACF to the height of the main peak; the log ratio of the energy within the T-F unit to the total energy at that time frame; kurtosis, calculated from sample averages in each subband at each time frame; spectral flatness, measured in terms of the variance of the subband energy within the spectrographic neighborhood of the T-F unit; the ratio of the subband energy at each time frame to the noise floor estimated for that subband; and spectral subtraction based local SNR estimate. The features are chosen such that they capture the characteristics of speech in noise without making assumptions about the underlying noise type. Except for the first two features, viz. CFR and APR, the remaining ones can be used to characterize properties of T-F units in both voiced and unvoiced time frames. CFR and APR are used only for the T-F units in voiced frames. GMMs are trained for voiced and unvoiced speech separately, and also at each subband and are in turn used to obtain soft T-F masks. The obtained masks improve ASR performance when used in conjunction with missing-data-based strategies. Seltzer *et al.* [65] use speech mixed with white noise to train the classifiers. This can be limiting when it comes to generalization to unseen noisy conditions. To overcome this, Kim and Stern [37] suggest training each frequency band separately, using artificial colored noise signals generated specifically for each band. They show that this can yield better generalization results as compared to using white noise alone for training.

In a way, classification-based strategies simplify the task of speech segregation, at least conceptually. It bypasses the steps of a typical CASA system which extracts perceptually motivated cues and applies the ASA stages of segmentation and grouping to obtain a binary mask. The potential downside of relying on supervised learning is the perennial issue of generalization to unseen conditions.

16.4.4 Binaural Mask Estimation Strategies

Binaural CASA systems use two microphone recordings to segregate the target from the mixture. Most binaural systems try to extract localization cues, for example azimuth, which are encoded in the differences between the signals that reach the two ears (or microphones). In this regard, interaural time difference (ITD) and interaural intensity difference (IID) are the two most important cues. ITD is the difference between the arrival times of the signal at the two ears. ITD is ambiguous at high frequencies (> 1.5 KHz) because of short wavelengths as compared to the distance between the ears. IID is the difference in the intensity of the sound that reaches the two ears, usually expressed in decibels, and it occurs because of the 'shadow' effect of the human head. Contrary to ITD, IID is not useful at low frequencies (< 500 Hz) because such low-frequency sound components diffract around the head overcoming the shadow effect in the process.

Two classical strategies strongly influenced binaural segregation: the cross-correlation based model for ITD estimation proposed by Jeffress [31] and the equalization-cancellation (EC) model of Durlarch [16]. The EC model tries to segregate the target in a two stage process. In the first stage, the noise levels in the signals arriving at the two ears are equalized. This is followed by subtraction of the signals at the two ears in the cancellation stage. The noise equalized in the first stage gets canceled during the second stage, producing a cleaner target. The Jeffress model is based on the similarity of the signals that arrive at the two ears. The neural firing patterns of the two ears are passed through delay lines; the delay that maximizes the correlation between the two patterns is identified as the ITD of the signal.

To compute ITD, a normalized cross-correlation function, $C(t, f, \tau)$, is typically used

$$C(t, f, \tau) = \frac{\sum_{n} x_L(tT_t - nT_n, f)x_R(tT_t - nT_n - \tau T_n, f)}{\sqrt{\sum_{n} x_L^2(tT_t - nT_n, f)}\sqrt{\sum_{n} x_R^2(tT_t - nT_n - \tau T_n, f)}}. \tag{16.7}$$

The above equation calculates cross-correlation at frequency channel f and time frame t, for a time lag τ. x_L and x_R correspond to the left and right ear response, respectively. T_t and T_n have the same meanings as in Equation (16.2). Similar to the normalized autocorrelation function, the cross-correlation function will have a peak at a delay that relates to ITD. IID can be calculated as the ratio of the mean power of the signals that arrive at the two ears:

$$IID(t, f) = 10 \log_{10} \left(\frac{\sum_n x_L^2(tT_t - nT_n, f)}{\sum_n x_R^2(tT_t - nT_n, f)} \right). \tag{16.8}$$

An IBM estimation strategy based on classifying ITD and IID estimates was proposed by Roman *et al.* [62], which is probably the first classification-based system for speech segregation. They observed that, given a predefined configuration of the target and the interference (configuration here refers to the azimuths of the target and the interference), ITD and IID values vary smoothly and systematically with respect to the relative strength of the target and the mixture. This prompted them to model the distribution of target dominant units and interference dominant units of each frequency channel in the ITD-IID space. Their system models the distributions using a nonparametric kernel-density estimator. For an unseen test utterance, the binary decision at each T-F unit is made by comparing the probabilities of the unit being target dominant and interference dominant, given the observed ITD and IID at that unit. The binary masks estimated by their model are very close to the IBM, with excellent performances in terms of SNR gains, speech intelligibility and ASR accuracies. The main drawback of the model is that ITD-IID distributions are configuration dependent. A similar system was proposed by Harding *et al.* [20], which assumes that only the target azimuth is known *a priori*. It then learns the joint distribution of ITD and IID for target dominant T-F units using a histogram-based method. These distributions are used to predict the probability of a unit being target dominant from the observed ITD and IID. The estimated probabilities are directly used in the form of a ratio mask, to improve ASR results in reverberant conditions.

The above strategies are based on modeling the distribution of the binaural cues in the ITD-IID space. An alternative approach was proposed by Palomaki *et al.* [55]. This approach first estimates target and interference azimuths. It then classifies a T-F unit as target or interference dominant by comparing the values of the cross-correlation function at the estimated azimuths of the target and the interference. In order to deal with room reverberation, their system models the precedence effect [45] by using the low-pass filtered envelope response of each channel as an inhibitor. This reduces the effect of late echoes in reverberant situations by preserving transient and suppressing sustained responses. Palomaki *et al.* reported good ASR results in reverberant situations using the above algorithm to estimate binary masks.

Recently, Woodruff and Wang [84] proposed a system that combines monaural and binaural cues to estimate the IBM. Their system uses a monaural CASA algorithm to first obtain simultaneous streams, each occupying a continuous time interval. They use the tandem algorithm, described earlier, for this purpose. Binaural cues are then used to jointly estimate the azimuths of the streams that comprise the scene and their corresponding sets of sequentially grouped simultaneous streams.

16.5 Integrating CASA with ASR

The CASA strategies discussed in Section 16.4 provide us several perceptually inspired ways of segregating the target from a mixture. The main focus has been on estimating the ideal binary mask. Although IBM-based strategies produce good segregation results, integrating CASA and ASR has not been as straightforward a task as it seems. A simple way of combining CASA with ASR is to use CASA as a preprocessor. ASR models trained in clean conditions can then be used to perform recognition on the segregated target speech. This can be problematic. Even when the IBM is used, the resynthesized signal will have artifacts that may pose challenges to recognition. Errors in IBM estimation will further degrade the performance of such systems.

Nevertheless, CASA has been used as a preprocessor in some systems and has been shown to produce good results. One such model was proposed by Srinivasan *et al.* [73]. Their system uses a ratio T-F mask to enhance a noisy utterance. A conventional HMM-based ASR system trained using the mel-frequency cepstral coefficients (MFCC) of clean speech is used to recognize the enhanced speech. For mask estimation, they use the binaural segregation model by Roman *et al.* [62]. Srinivasan *et al.* compared their system with the missing-data ASR approach [13] and found that using such a CASA-based preprocessor can be advantageous as the vocabulary size of the recognition task increases. The limitation of missing-data ASR in dealing with larger vocabulary tasks had been reported earlier [60]. The use of a ratio mask instead of a binary mask coupled with accurate mask estimation helped their system in overcoming some of the limitations of using CASA as a preprocessor.

More recently, Hartmann and Fosler-Lussier [21] compared the performance of an ASR system that simply discards masked T-F units, which is equivalent to processing the noisy speech with a binary mask, with a system that reconstructs those units based on the information available from the unmasked T-F units. Such feature-reconstruction strategies have been used to improve noise robust ASR [60]. An HMM based ASR system trained in clean conditions is used to perform recognition. They observe that the direct use of IBM-processed speech performs significantly better than the reconstructed speech, and yields ASR results only a few percentage points worse than those in clean conditions. When noise is added to the IBM by randomly flipping 1s and 0s, only after the amount of mask errors exceeds some point does reconstruction work better. This is a surprising observation, considering the conventional wisdom that the binary nature of a mask is supposed to skew the cepstral coefficients (they used PLP cepstral coefficients to build their ASR system). This study points to the need of a deeper understanding of the effects of using binary masks on ASR performance.

The above methods somehow modify the features so that they can be used with ASR models trained in clean conditions. Such strategies have been called *feature compensation* or *source-driven* methods. Feature compensation includes techniques that use CASA based strategies for segregating the target [21,73] and reconstructing unreliable features [60]. An alternative approach would be to modify ASR models so that they implicitly accommodate missing or corrupt speech features. Such strategies have been termed *model compensation* or *classifier compensation* methods. The missing-data ASR techniques are examples of model compensation strategies [13]. There are also strategies that combine feature compensation and model compensation [15,71], and simultaneously perform CASA and ASR [4,72].

A much simpler strategy for integrating CASA and ASR was proposed by Narayanan and Wang [50] and Karadogan *et al.* [35]. They interpret IBMs as binary images and use a binary pattern classifier to do ASR. The idea of using binary pattern recognition for ASR is radically different from the existing strategies that use detailed speech features like MFCCs. Their work was motivated by the speech perception study showing that modulating noise by the IBM can produce intelligible speech for humans [80, also see Section 16.3]. Since noise carries no speech information, intelligibility must be induced by the binary pattern of the IBM itself. This indicates that the pattern carries important phonetic information. The system described in Narayanan and Wang [50] is designed for an isolated digit recognition task. The ASR module is based on convolutional neural networks [41,68], which have previously been used successfully for handwritten digit and object recognition. Their system obtains reasonable results even when the IBM is estimated directly from noisy speech using a CASA algorithm. They extend their system further in Narayanan and Wang [51] to perform a more challenging phone classification task, and show that IBMs and traditional speech features like MFCCs carry complimentary information that can be combined to improve the overall classification performance. The combined system obtains classification accuracies that compare favorably to most of the results reported in recent phone classification literature. It is quite interesting to note that features that are based on *binary* patterns can obtain good results on complex ASR tasks. Such CASA inspired features may eventually be needed for achieving robust ASR.

In the following subsection we discuss in greater detail an example of a CASA-inspired ASR framework. The subsection focuses on the uncertainty transform model proposed by Srinivasan and Wang [71] that combines feature compensation and model compensation to improve ASR performance.

16.5.1 Uncertainty Transform Model

Using a speech-enhancement algorithm to obtain features for ASR does not always yield good recognition results. This is because, even with the best enhancement algorithms, the enhanced features remain somewhat noisy, as far as the ASR models trained in clean conditions are concerned. Moreover, the variance of such features, with respect to the corresponding clean features, varies across time and frequency. Uncertainty decoding has been suggested as a strategy to modify ASR model parameters to take into account the inherent uncertainty of such enhanced features (see Chapter 17 for a more detailed handling of uncertainty decoding strategies). It has been shown that feature uncertainties contribute to an increase in the variance of trained acoustic variables and accounting for it during the recognition (decoding) stage can significantly improve ASR performance [15].

A mismatch in the domain of operation between speech enhancement or segregation and ASR can pose problems in effectively adjusting ASR model parameters based on estimated uncertainty. Such a mismatch exists for most CASA-based techniques as they operate either in the spectral or T-F domain, as opposed to ASR models that operate in the cepstral domain. Training ASR models in the spectral domain is known to produce suboptimal performance. In order to overcome this mismatch problem, Srinivasan and Wang [71] suggested a technique to transform the uncertainties estimated in the spectral domain to the cepstral domain.

The uncertainty transform model by Srinivasan and Wang consists of a speech-enhancement module, an uncertainty transformer, and a traditional HMM-based ASR module that operates in the cepstral domain. The enhancement module uses a spectrogram reconstruction method that is similar to [60] but operates in the linear spectral domain. To perform recognition, the enhanced spectral features are transformed to the cepstral domain. The corresponding uncertainties, originally estimated in the spectral domain, are transformed using a supervised learning method. Given the enhanced cepstral features and associated uncertainties, recognition is performed in an uncertainty decoding framework. Details about these stages are discussed below.

The speech-enhancement module starts by converting a noisy speech signal into the spectral domain using the FFT. The noisy spectrogram is then processed using a speech-segregation algorithm that estimates the IBM. A binary mask partitions a noisy spectral vector, $\mathbf{y}$, into its reliable components, $\mathbf{y}_r$, and the unreliable components, $\mathbf{y}_u$. Assuming that $\mathbf{y}_r$ sufficiently approximates the corresponding clean speech spectral values, $\mathbf{x}_r$, the goal of reconstruction is to approximate the true spectral values, $\mathbf{x}_u$, of the unreliable components. It uses a speech prior model for this purpose, implemented as a large GMM, where the probability density of a spectral vector of speech (x) is modeled as

$$p(\mathbf{x}) = \sum_{k=1}^{K} P(k)p(\mathbf{x} \mid k).$$

Here, *K* represents the number of Gaussians in the GMM, *k* is the Gaussian index, $P(k)$ is the prior probability of the *k*th component (or the component weight), and $p(\mathbf{x} \mid k) = \mathcal{N}(\mathbf{x}; \boldsymbol{\mu}_k, \boldsymbol{\Theta}_k)$ is the conditional probability density of x given the *k*th Gaussian. In the Gaussian, $\boldsymbol{\mu}_k$ and $\boldsymbol{\Theta}_k$ denote the mean vector and the covariance matrix, respectively. Such a GMM can be trained by pooling the entire training data and using an expectation maximization algorithm to learn the parameters. The mean and the covariance matrix of the *k*th Gaussian are also partitioned into its reliable and unreliable components using a binary mask:

$$\boldsymbol{\mu}_k = \begin{bmatrix} \boldsymbol{\mu}_{r,k} \\ \boldsymbol{\mu}_{u,k} \end{bmatrix}, \boldsymbol{\Theta}_k = \begin{bmatrix} \boldsymbol{\Theta}_{rr,k} & \boldsymbol{\Theta}_{ru,k} \\ \boldsymbol{\Theta}_{ur,k} & \boldsymbol{\Theta}_{uu,k} \end{bmatrix},$$

where $\boldsymbol{\mu}_{r,k}$ and $\boldsymbol{\mu}_{u,k}$ are the reliable and the unreliable components of the mean vector of the *k*th Gaussian, respectively; $\boldsymbol{\Theta}_{rr,k}$ and $\boldsymbol{\Theta}_{uu,k}$ are the corresponding covariances of the reliable and the unreliable components; and $\boldsymbol{\Theta}_{ru,k}$ and $\boldsymbol{\Theta}_{ur,k}$ are the cross-covariances.

The unreliable components are reconstructed by first estimating the *a posteriori* probability of the *k*th Gaussian using only the reliable components, $\mathbf{x}_r$, of the frame:

$$P(k \mid \mathbf{x}_r) = \frac{P(k)p(\mathbf{x}_r \mid k)}{\sum_{k=1}^{K} P(k)p(\mathbf{x}_r \mid k)}. \tag{16.9}$$

Next, the conditional mean of the unreliable components given the reliable components is approximated as

$$\widehat{\boldsymbol{\mu}}_{u,k} = \boldsymbol{\mu}_{u,k} + \boldsymbol{\Theta}_{ur,k}\boldsymbol{\Theta}_{rr,k}^{-1}(\mathbf{x}_r - \boldsymbol{\mu}_{r,k}). \tag{16.10}$$

Note that this is the standard formula for calculating the conditional mean of random variables that follow a multivariate normal distribution.

Given the *a posteriori* component weights and the conditional mean, a good approximation of the unreliable components is the expected value of $\mathbf{x}_u$ given $\mathbf{x}_r$, which is also the minimum mean-squared estimate (MMSE) of $\mathbf{x}_u$. The MMSE estimate can be calculated as

$$\widehat{\mathbf{x}}_u = E_{\mathbf{x}_u \mid \mathbf{x}_r}(\mathbf{x}_u) = \sum_{k=1}^{K} P(k \mid \mathbf{x}_r)\widehat{\boldsymbol{\mu}}_{u,k} \tag{16.11}$$

Finally, a measure of uncertainty in the estimation of the reconstructed spectral vector, $\widehat{\mathbf{x}}$ $(\mathbf{x}_r \bigcup \widehat{\mathbf{x}}_u)$, is calculated as

$$\widehat{\boldsymbol{\Theta}}_{\widehat{\mathbf{x}}} = \sum_{k=1}^{K} P(k \mid \mathbf{x}_r)\left\{\left(\begin{bmatrix} \mathbf{x}_r \\ \widehat{\boldsymbol{\mu}}_{u,k} \end{bmatrix} - \boldsymbol{\mu}_k\right) \cdot \left(\begin{matrix} \mathbf{x}_r \\ \widehat{\boldsymbol{\mu}}_{u,k} \end{matrix} - \boldsymbol{\mu}_k\right)^T + \begin{bmatrix} 0 & 0 \\ 0 & \widehat{\boldsymbol{\Theta}}_{u,k} \end{bmatrix}\right\} \tag{16.12}$$

where

$$\widehat{\mathbf{\Theta}}_{u,k} = \mathbf{\Theta}_{uu,k} - \mathbf{\Theta}_{ur,k}\mathbf{\Theta}_{rr,k}^{-1}\mathbf{\Theta}_{ru,k}.$$

Equation (16.12) is based on the idea of adapting the trained GMM using the reconstructed spectral vector as an incomplete observation [83]. Even though $\mathbf{y}_r$ is considered reliable during feature reconstruction, the above equation associates a positive, albeit small, measure of uncertainty to it. This helps the uncertainty transformation model to learn the subsequent transformation of these quantities to the cepstral domain, since cepstral uncertainties depend on both $\mathbf{x}_r$ and $\mathbf{x}_u$. If a diagonal covariance matrix is used to model the speech prior, Equation (16.11) and Equation (16.12) can be modified to [66]

$$\widehat{\mathbf{x}}_{u,k} = \sum_{k=1}^{K} P(k \mid \mathbf{x}_r)\boldsymbol{\mu}_{u,k}, \tag{16.13}$$

$$\hat{\boldsymbol{\theta}}_{\widehat{\mathbf{x}}} = \sum_{k=1}^{K} P(k \mid \mathbf{x}_r)\left\{\left(\begin{bmatrix}\mathbf{x}_r \\ \widehat{\mathbf{x}}_{u,k}\end{bmatrix} - \boldsymbol{\mu}_k\right)^2 + \begin{bmatrix}0 \\ \boldsymbol{\theta}_{u,k}\end{bmatrix}\right\}, \tag{16.14}$$

where squaring is done per element of the vector. $\hat{\boldsymbol{\theta}}_{\widehat{\mathbf{x}}}$ and $\boldsymbol{\theta}_{u,k}$ denote the measure of uncertainty in estimation of $\widehat{\mathbf{x}}$ and the unreliable components of the variance of the kth Gaussian, respectively. This simplification is due to the fact that all the cross-covariance terms will have the value 0 when the covariance matrix is diagonal. The use of a diagonal covariance matrix reduces the training time and simplifies the calculations.

To perform ASR, the uncertainty transform approach converts the enhanced spectral feature ($\widehat{\mathbf{x}}$) to the cepstral domain. This is straightforward as we have a fully reconstructed feature vector. The main step is to transform the estimated uncertainties to the cepstral domain. In Srinivasan and Wang [71], regression trees are trained to perform this transformation as the true parametric form of this relationship is unknown. If we assume that the cepstral features consist of 39 MFCCs (including the delta and acceleration coefficients), and that the ASR module is based on HMMs that use Gaussians with diagonal covariance matrices to model the observation probability, the goal of the transformation is to estimate the squared difference, $\boldsymbol{\theta}_{\hat{\mathbf{z}}}$, between the reconstructed cepstra, $\hat{\mathbf{z}}$, and the corresponding clean cepstra, $\mathbf{z}$ [15]. The input to the system is the estimated spectral variance ($\hat{\boldsymbol{\theta}}_{\widehat{\mathbf{x}}}$ or $diag(\widehat{\mathbf{\Theta}}_{\widehat{\mathbf{x}}})$, depending on whether diagonal or full covariance matrices are used by the feature reconstruction module). Srinivasan and Wang additionally use the reconstructed cepstral values corresponding to that frame, a preceding frame and a succeeding frame, as input features as they were found to be useful in learning the transformation. The cepstral uncertainties of each of the 39 dimensions are learned using separate regression trees.

Having obtained the enhanced cepstral features and the associated uncertainties, ASR is performed in an uncertainty decoding framework. Since we only have access to the enhanced cepstra, $\hat{\mathbf{z}}$, the observation probability in an HMM-based decoder is calculated by integrating over all possible clean speech cepstral values, $\mathbf{z}$, as shown below:

$$\int_{-\infty}^{\infty} p(\mathbf{z} \mid q, k)p(\hat{\mathbf{z}} \mid \mathbf{z})\mathrm{d}\mathbf{z} = \mathcal{N}(\hat{\mathbf{z}}; \boldsymbol{\mu}_{q,k}, \boldsymbol{\theta}_{q,k} + \boldsymbol{\theta}_{\hat{\mathbf{z}}}). \tag{16.15}$$

In the equation, q denotes a state in the HMM and k indexes the Gaussians used to model the observation probability. $\boldsymbol{\mu}_{q,k}$ and $\boldsymbol{\theta}_{q,k}$ are the corresponding mean and the variance vector

Table 16.1 Word error rates (WER) of the uncertainty transform and the multiple prior based uncertainty transform methods, as well as the reconstruction-based approach. Baseline results of directly recognizing the noisy speech are also shown. MP abbreviates multiple priors. The last column shows the average WER of each of the systems across all the noise types. Reproduced by permission of Narayanan *et al.* [52] © 2011 IEEE.

	Test Set						
System	Car	Babble	Restaurant	Street	Airport	Train	**Average**
Baseline	44.9	43.7	43.2	52.0	44.1	55.2	47.2
Reconstruction	21.5	38.5	42.6	41.5	41.5	39.4	37.5
Uncertainty decoding	18.9	34.2	41.2	40.6	37.0	39.0	35.2
MP reconstruction	19.6	34.8	41.0	38.3	41.1	36.5	35.2
MP uncertainty decoding	18.4	32.8	39.1	37.4	36.9	36.5	33.5

of the kth Gaussian. If the observation probability is modeled using Gaussians and if the enhancement is unbiased, this probability can be calculated as shown in the equation [15]. Essentially, the learned variance of a Gaussian component is modified during the recognition stage by adding the estimated cepstral uncertainty to it.

An extension to Srinivasan and Wang's uncertainty transform framework was recently proposed by Narayanan et al. [52]. They propose using multiple prior models of speech, instead of a single large GMM, to better model spectral features. Specifically, they train prior models based on the voicing characteristic of speech by splitting the training data into voiced and unvoiced speech. While reconstructing a noisy spectrogram, frames that are detected as voiced by their voiced/unvoiced (V/UV) detection module are reconstructed using the voiced prior model. Similarly, unvoiced frames are reconstructed using the unvoiced prior model. The V/UV detector is implemented as a binary decision problem, using GMMs to model the underlying density of voiced and unvoiced frames. Like in the uncertainty transform model of Srinivasan and Wang, reconstructed spectral vectors and their corresponding uncertainties are finally transformed to the cepstral domain, and recognition is performed in the uncertainty decoding framework.

The word error rates obtained using the uncertainty transform and the extension by Narayanan *et al.* [52] on the Aurora-4 5000 word closed vocabulary speech recognition task [56] are shown in Table 16.1. This task is based on the *Wall Street Journal* (WSJ0) database. The IBM is estimated using a simple spectral subtraction based approach [71]; the spectral energy in the first and last 50 frames is averaged to create an estimate of the noise spectrum, which is then simultaneously used to 'clean' the noisy spectrogram and to estimate the IBM by comparing it with the energy in each T-F unit. From the table, we can see that, compared to the baseline, uncertainty transform clearly reduces the word error rate in all of the testing conditions. An average improvement of 12 percentage points is obtained over the baseline of directly recognizing noisy speech. Compared to feature reconstruction, an improvement of 2.3 percentage points is obtained. Using multiple prior models further improves the average performance by 1.7 percentage points.

The results show that the uncertainty transform and the use of multiple prior models are effective in dealing with noisy speech utterances. One of the main advantages of the uncertainty transformation is that it enables CASA-based speech enhancement techniques that operate in the spectral domain to be used as a front-end for uncertainty decoding based ASR strategies. The supervised transformation technique can be used whenever the enhancement and the recognition modules operate in different domains. Uncertainty transform techniques provide a clear alternative to missing-data and reconstruction approaches to robust ASR.

16.6 Concluding Remarks

In this chapter, we have discussed facets of CASA and how it can be coupled with ASR to deal with speech recognition in noisy environments. To recapitulate, we discussed perceptual mechanisms that allow humans to analyze the auditory scene. We then looked at how such mechanisms are incorporated in computational models with the goal of achieving human-like performance. Most of the systems discussed in the chapter try to estimate the ideal binary mask, which is an established goal of CASA. Finally, in Section 16.5, we described how CASA can be integrated with ASR.

Although clear advances have been made in the last few years in improving CASA and ASR, challenges remain. CASA challenges lie in developing effective strategies to sequentially organize speech and to deal with unvoiced speech. Apart from additive noise, recent studies have started addressing room reverberation [20,33]. Advances in CASA will have a direct impact on ASR. ASR systems have been demonstrated to perform excellently when the IBM is used. Improvements in IBM estimation will lead to more robust ASR. Over the last decade, attempts at integrating CASA and ASR have yielded fruitful results. Strategies like missing-data ASR, uncertainty transform, and missing feature reconstruction go beyond using CASA as preprocessor for ASR. Further progress in robust ASR can be expected from even tighter coupling between CASA and ASR.

Achieving human-level performance has been the hallmark of many AI endeavors. In CASA, this translates to a meaningful description of the acoustic world. Therefore, recognizing speech in realistic environments is a major benchmark of CASA. Our understanding of how we analyze the auditory scene may eventually pave the way to truly robust ASR.

Acknowledgment

Preparation of this chapter was supported in part by an AFOSR grant (FA9550-08-1-0155).

References

[1] M. C. Anzalone, L. Calandruccio, K. A. Doherty, and L. H. Carney, "Determination of the potential benefit of time-frequency gain manipulation," *Ear and Hearing*, vol. 27, no. 5, pp. 480–492, 2006.

[2] P. Assmann and Q. Summerfield, "The perception of speech under adverse acoustic conditions," in *Speech Processing in the Auditory System*, series on Springer Handbook of Auditory Research, S. Greenberg, W. A. Ainsworth, A. N. Popper, and R. R. Fay, Eds. Berlin: Springer-Verlag, 2004, vol. 18.

[3] J. Barker, L. Josifovski, M. P. Cooke, and P. D. Green, "Soft decisions in missing data techniques for robust automatic speech recognition." in *Proceedings of the International Conference on Spoken Language Processing*, Beijing, China, 2000, pp. 373–376.

[4] J. Barker, M. P. Cooke, and D. P. W. Ellis, "Decoding speech in the presence of other sources," *Speech Communication*, vol. 45, pp. 5–25, 2005.

[5] M. Berouti, R. Schwartz, and R. Makhoul, "Enhancement of speech corrupted by acoustic noise," in *Proceedings of the IEEE International Conference on Acoustics, Speech and Signal Processing*, 1979.

[6] P. Boersma and D. Weenink. (2002) Praat: Doing phonetics by computer, version 4.0.26. Available at: http://www.fon.hum.uva.nl/praat.

[7] S. Boll, "Suppression of acoustic noise in speech using spectral subtraction," *IEEE Transactions on Acoustics, Speech and Signal Processing*, vol. 27, pp. 113–120, 1979.

[8] A. S. Bregman, *Auditory Scene Analysis*. Cambridge, MA: MIT Press, 1990.

[9] G. J. Brown and M. P. Cooke, "Computational auditory scene analysis," *Computer Speech & Language*, vol. 8, pp. 297–336, 1994.

[10] D. Brungart, P. S. Chang, B. D. Simpson, and D. L. Wang, "Isolating the energetic component of speech-on-speech masking with an ideal binary time-frequency mask," *Journal of Acoustical Society of America*, vol. 120, pp. 4007–4018, 2006.

[11] E. C. Cherry, "Some experiments on recognition of speech, with one and with two ears," *Journal of Acoustical Society of America*, vol. 25, no. 5, pp. 975–979, 1953.

[12] E. C. Cherry, *On Human Communication*. Cambridge, MA: MIT Press, 1957.

[13] M. P. Cooke, P. Greene, L. Josifovski, and A. Vizinho, "Robust automatic speech recognition with missing and uncertain acoustic data," *Speech Communication*, vol. 34, pp. 141–177, 2001.

[14] A. de Cheveigne, "Multiple F0 estimation." in *Computational Auditory Scene Analysis: Principles, Algorithms, and Applications*, D. L. Wang and G. J. Brown, Eds. Hoboken, NJ: Wiley-IEEE Press, 2006, pp. 45–80.

[15] L. Deng, J. Droppo, and A. Acero, "Dynamic compensation of HMM variances using the feature enhancement uncertainty computed from a parametric model of speech distortion," *IEEE Transactions on Speech and Audio Processing*, vol. 13, no. 3, pp. 412–421, 2005.

[16] N. I. Durlach, "Note on the equalization and cancellation theory of binaural masking level differences," *Journal of Acoustical Society of America*, vol. 32, no. 8, pp. 1075–1076, 1960.

[17] M. El-Maliki and A. Drygajlo, "Missing features detection and handling for robust speaker verification," in *Proceedings of Interspeech*, 1999, pp. 975–978.

[18] B. R. Glasberg and B. C. J. Moore, "Derivation of auditory filter shapes from notched-noise data," *Hearing Research*, vol. 47, pp. 103–138, 1990.

[19] K. Han and D. L. Wang, "An SVM based classification approach to speech separation." in *Proccedings of the IEEE International Conference on Acoustics, Speech and Signal Processing*, 2011, pp. 4632–4635.

[20] S. Harding, J. Barker, and G. J. Brown, "Mask estimation for missing data speech recognition based on statistics of binaural interaction," *IEEE Transactions on Speech and Audio Processing*, vol. 14, no. 1, pp. 58–67, 2006.

[21] W. Hartmann and E. Fosler-Lussier, "Investigations into the incorporation of the ideal binary mask in ASR," in *Proceedings of the IEEE International Conference on Acoustics, Speech and Signal Processing*, 2011, pp. 4804–4807.

[22] H. Helmholtz, *On the Sensation of Tone*, 2nd ed. New York: Dover Publishers, 1863.

[23] H. G. Hirsch and C. Ehrlicher, "Noise estimation techniques for robust speech recognition," in *Proceedings of the IEEE International Conference on Acoustics, Speech and Signal Processing*, 1995, pp. 153–156.

[24] G. Hu and D. L. Wang, "Monaural speech segregation based on pitch tracking and amplitude modulation," *IEEE Transactions on Neural Networks*, vol. 15, no. 5, pp. 1135–1150, 2004.

[25] G. Hu and D. L. Wang, "Auditory segmentation based on onset and offset analysis," *IEEE Transactions on Audio, Speech and Language Processing*, vol. 15, pp. 396–405, 2007.

[26] G. Hu and D. L. Wang, "A tandem algorithm for pitch estimation and voiced speech segregation," *IEEE Transactions on Audio, Speech and Language Processing*, vol. 18, pp. 2067–2079, 2010.

[27] G. Hu and D. L. Wang, "Speech segregation based on pitch tracking and amplitude modulation." in *Proceedings of the IEEE Workshop on Applications of Signal Processing to Audio and Acoustics*, 2001, pp. 79–82.

[28] G. Hu and D. L. Wang, "Segregation of unvoiced speech from nonspeech interference," *Journal of Acoustical Society of America*, vol. 124, pp. 1306–1319, 2008.

[29] K. Hu and D. L. Wang, "Unvoiced speech segregation from nonspeech interference via CASA and spectral subtraction." *IEEE Transactions on Audio, Speech and Language Processing*, vol. 19, no. 6, pp. 1600–1609, 2011.

[30] K. Hu and D. L. Wang, "Unsupervised sequential organization for cochannel speech separation." in *Proceedings of Interspeech*, Makuhari, Japan, 2010, pp. 2790–2793.

[31] L. A. Jeffress, "A place theory of sound localization," *Comparative Physiology and Psychology*, vol. 41, pp. 35–39, 1948.

[32] Z. Jin and D. L. Wang, "A supervised learning approach to monaural segregation of reverberant speech," *IEEE Transactions on Audio, Speech and Language Processing*, vol. 17, pp. 625–638, 2009.

[33] Z. Jin and D. L. Wang, "HMM-based multipitch tracking for noisy and reverberant speech," *IEEE Transactions on Audio, Speech and Language Processing*, vol. 19, no. 5, pp. 1091–1102, 2011.

[34] L. Josifovski, M. Cooke, P. Green, and A. Vizihno, "State based imputation of missing data for robust speech recognition and speech enhancement," in *Proceedings of Interspeech*, 1999, p. 2837–2840.

[35] S. G. Karadogan, J. Larsen, M. S. Pedersen, and J. B. Boldt, "Robust isolated speech recognition using binary masks." in *Proceedings of the European Signal Processing Conference*, 2010, pp. 1988–1992.

[36] G. Kim, Y. Lu, Y. Hu, and P. Loizou, "An algorithm that improves speech intelligibility in noise for normal-hearing listeners," *Journal of Acoustical Society of America*, vol. 126, no. 3, pp. 1486–1494, 2009.

[37] W. Kim and R. Stern, "Mask classification for missing-feature reconstruction for robust speech recognition in unknown background noise," *Speech Communication*, vol. 53, pp. 1–11, 2011.

[38] A. Klapuri, "Multipitch analysis of polyphonic music and speech signals using an auditory model," *IEEE Transactions on Audio, Speech and Language Processing*, vol. 16, no. 2, pp. 255–266, 2008.

[39] B. Kollmeier and R. Koch, "Speech enhancement based on physiological and psychoacoustical models of modulation perception and binaural interaction," *Journal of Acoustical Society of America*, vol. 95, pp. 1593–1602, 1994.

[40] A. Korthauer, "Robust estimation of the SNR of noisy speech signals for the quality evaluation of speech databases," in *Proceedings of ROBUST'99 Workshop*, 1999, pp. 123–126.

[41] Y. Lecun, L. Bottou, Y. Bengio, and P. Haffner, "Gradient-based learning applied to document recognition," *Proceedings of the IEEE*, vol. 86, pp. 2278–2324, 1998.

[42] N. Li and P. C. Loizou, "Factors influencing intelligibility of ideal binary-masked speech: Implications for noise reduction," *Journal of Acoustical Society of America*, vol. 123, no. 3, pp. 1673–1682, 2008.

[43] Y. Li and D. L. Wang, "On the optimality of ideal binary time-frequency masks," *Speech Communication*, vol. 51, pp. 230–239, 2009.

[44] J. C. R. Licklider, "A duplex theory of pitch perception," *Experimentia*, vol. 7, pp. 128–134, 1951.

[45] R. Y. Litovsky, H. S. Colburn, W. A. Yost, and S. J. Guzman, "The precedence effect," *Journal of Acoustical Society of America*, vol. 106, pp. 1633–1654, 1999.

[46] P. C. Loizou, *Speech Enhancement: Theory and Practice*. Boca Raton, Florida: CRC Press, 2007.

[47] N. Ma, P. Green, J. Barker, and A. Coy, "Exploiting correlogram structure for robust speech recognition with multiple speech sources," *Speech Communication*, vol. 49, pp. 874–891, 2007.

[48] R. Meddis, M. J. Hewitt, and T. M. Shackelton, "Implementation details of a computational model of the inner hair-cell/auditory-nerve synapse," *Journal of Acoustical Society of America*, vol. 122, no. 2, pp. 1165–1172, 1990.

[49] B. C. J. Moore, *An Introduction to the Psychology of Hearing*, 5th ed. London, UK: Academic Press, 2003.

[50] A. Narayanan and D. L. Wang, "Robust speech recognition from binary masks," *Journal of Acoustical Society of America*, vol. 128, pp. EL217–222, 2010.

[51] A. Narayanan and D. L. Wang, "On the use of ideal binary masks to improve phone classification." in *Proceedings of the IEEE International Conference on Acoustics, Speech and Signal Processsing*, 2011, pp. 5212–5215.

[52] A. Narayanan, X. Zhao, D. L. Wang, and E. Fosler-Lussier, "Robust speech recognition using multiple prior models for speech reconstruction," in *Proceedings of the IEEE International Conference on Acoustics, Speech and Signal Processing*, 2011, pp. 4800–4803.

[53] E. Nemer, R. Goubran, and S. Mahmoud, "SNR estimation of speech signals using subbands and fourth-order statistics," *IEEE Signal Processing Letters*, vol. 6, no. 7, pp. 504–512, 1999.

[54] S. Nooteboom, "The prosody of speech: Melody and rhythm," in *The Handbook of Phonetic Science*, W. J. Hardcastle and J. Laver, Eds. Blackwell: Oxford, UK, 1997, pp. 640–673.

[55] K. J. Palomaki, G. J. Brown, and D. L. Wang, "A binaural processor for missing data speech recognition in the presence of noise and small-room reverberation," *Speech Communication*, vol. 43, pp. 361–378, 2004.

[56] N. Parihar and J. Picone, "Analysis of the Aurora large vocabulary evalutions," in *Proceedings of the European Conference on Speech Communication and Technology*, 2003, pp. 337–340.

[57] T. W. Parsons, "Separation of speech from interfering speech by means of harmonic selection," *Journal of Acoustical Society of America*, vol. 60, no. 4, pp. 911–918, 1976.

[58] R. D. Patterson, I. Nimmo-Smith, J. Holdsworth, and P. Rice, "An efficient auditory filterbank based on the gammatone function," Technical Report 2341, MRC Applied Psychology Unit, Cambridge, UK, 1988.

[59] B. Raj and R. Stern, "Missing-feature approaches in speech recognition," *IEEE Signal Processing Magazine*, vol. 22, no. 5, pp. 101–116, 2005.

[60] B. Raj, M. L. Seltzer, and R. M. Stern, "Reconstruction of missing features for robust speech recognition," *Speech Communication*, vol. 43, pp. 275–296, 2004.

[61] P. Renevey and A. Drygajlo, "Detection of reliable features for speech recognition in noisy conditions using a statistical criterion," in *Proceedings of Consistent & Reliable Acoustic Cues for Sound Analysis Workshop*, 2001, pp. 71–74.

[62] N. Roman, D. L. Wang, and G. J. Brown, "Speech segregation based on sound localization," *Journal of Acoustical Society of America*, vol. 114, no. 4, pp. 2236–2252, 2003.

[63] S. T. Roweis, "One microphone source separation," in *Advances in Neural Information Processing System 13*, 2000, pp. 793–799.

[64] M. R. Schroeder, "Period histogram and product spectrum: New methods for fundamental-frequency measurement," *Journal of Acoustical Society of America*, vol. 43, pp. 829–834, 1968.

[65] M. Seltzer, B. Raj, and R. Stern, "A bayesian classifer for spectrographic mask estimation for missing feature speech recognition," *Speech Communication*, vol. 43, no. 4, pp. 379–393, 2004.

[66] Y. Shao, "Sequential organization in computational auditory scene analysis," PhD dissertation, The Ohio State Univeristy, 2007.

[67] Y. Shao and D. L. Wang, "Model-based sequential organization in cochannel speech," *IEEE Transactions on Audio, Speech and Language Processing*, vol. 14, pp. 289–298, 2006.

[68] P. Y. Simard, D. Steinkraus, and J. C. Platt, "Best practices for convolutional neural networks applied to visual document analysis," in *Proceedings of the International Conference on Document Analysis and Recognition*, 2003, pp. 958–963.

[69] J. Sohn, N. S. Kim, and W. Sung, "A statistical model-based voice activity detection," *IEEE Signal Processing Letters*, vol. 6, no. 1, pp. 1–3, 1999.

[70] W. Speith, J. F. Curtis, and J. C. Webseter, "Responding to one of two simultaneous messages," *Journal of Acoustical Society of America*, vol. 26, pp. 391–396, 1954.

[71] S. Srinivasan and D. L. Wang, "Transforming binary uncertainties for robust speech recognition," *IEEE Transactions on Audio, Speech and Language Processing*, vol. 15, pp. 2130–2140, 2007.

[72] S. Srinivasan and D. L. Wang, "Robust speech recognition by integrating speech separation and hypothesis testing," *Speech Communication*, vol. 52, pp. 72–81, 2010.

[73] S. Srinivasan, N. Roman, and D. L. Wang, "Binary and ratio time-frequency masks for robust speech recognition," *Speech Communication*, vol. 48, pp. 1486–1501, 2006.

[74] J. Tchorz and B. Kollmeier, "SNR estimation based on amplitude modulation analysis with applications to noise suppression," *IEEE Transactions on Audio, Speech and Signal Processing*, vol. 11, pp. 184–192, 2003.

[75] A. Vizinho, P. Green, M. Cooke, and L. Josifovski, "Missing data theory, spectral subtraction and signal-to-noise estimation for robust ASR: An integrated study," in *Proceedings of Interspeech*, 1999, pp. 2407–2410.

[76] D. L. Wang, "On ideal binary masks as the computational goal of auditory scene analysis," in *Speech Separation by Humans and Machines*, P. Divenyi, Ed. Boston, MA: Kluwer Academic, 2005, pp. 181–197.

[77] D. L. Wang, "Time-frequency masking for speech separation and its potential for hearing aid design," *Trends in Amplification*, vol. 12, pp. 332–353, 2008.

[78] D. L. Wang and G. J. Brown, "Separation of speech from interfering sounds based on oscillatory correlation," *IEEE Transactions on Neural Networks*, vol. 10, no. 3, pp. 684–697, 1999.

[79] D. L. Wang and G. J. Brown, Eds., *Computational Auditory Scene Analysis: Principles, Algorithms, and Applications*. Hoboken, NJ: Wiley-IEEE Press, 2006.

[80] D. L. Wang, U. Kjems, M. S. Pedersen, J. B. Boldt, and T. Lunner, "Speech perception of noise with binary gains," *Journal of Acoustical Society of America*, vol. 124, no. 4, pp. 2303–2307, 2008.

[81] D. L. Wang, U. Kjems, M. S. Pedersen, J. B. Boldt, and T. Lunner, "Speech intelligibility in background noise with ideal binary time-frequency masking," *Journal of Acoustical Society of America*, vol. 125, pp. 2336–2347, 2009.

[82] M. Weintraub, "A theory and computational model of auditory monaural sound separation," PhD dissertation, Stanford University, 1985.

[83] D. Williams, X. Liao, Y. Xue, and L. Carin, "Incomplete-data classification using logistic regression," in *Proceedings of the 22nd Internation Conference on Machine Learning*, 2005, pp. 972–979.

[84] J. Woodruff and D. L. Wang, "Sequential organization of speech in reverberant environments by integrating monaural grouping and binaural localization," *IEEE Transactions on Audio, Speech and Language Processing*, vol. 18, pp. 1856–1866, 2010.

[85] M. Wu, D. L. Wang, and G. J. Brown, "A multipitch tracking algorithm for noisy speech," *IEEE Transactions on Speech and Audio Processing*, vol. 11, no. 3, pp. 229–241, 2003.

[86] O. Yilmaz and S. Rickard, "Blind separation of speech mixtures via time-frequency masking," *IEEE Transactions on Signal Processing*, vol. 52, no. 7, pp. 1830–1847, 2004.

17

Uncertainty Decoding

Hank Liao
Google Inc., USA

One may view the accuracy degrading effects of noise in an automatic speech recognition system as increasing uncertainty while decoding the speech. To mitigate this, the statistical models used for recognition can be updated to reflect the error or uncertainty introduced by noise in the test environment. The greater the difference between the test and training and conditions, the greater the uncertainty. Some approaches that are motivated by this idea are presented in this chapter and are often described under the broad category called *uncertainty decoding*. Previous chapters have discussed methods to address environmental noise by using speech enhancement (Chapter 9), affine transformations of the features or model parameters (Chapter 11), or updating the acoustic model parameters (Chapter 12). This chapter discusses how these standard techniques relate to uncertainty decoding, demonstrates how they can be extended to handle uncertainty due to noise, and presents the strengths and weaknesses of various uncertainty decoding forms for noise robust speech recognition.

17.1 Introduction

The problem of speech recognition in noise results from mismatched training and test conditions. Acoustic noise in testing or actual usage conditions that is unaccounted for in training is unexpected and degrades recognition performance. Feature-based approaches to noise robustness, such as those presented in Chapter 5 or 9, remove the noise from the features, that is the parameterized speech observations, before recognition. Model-based approaches compensate the underlying statistical speech acoustic models to match noisy conditions as discussed in Chapter 12. Uncertainty decoding can be viewed as a hybrid of these approaches where the environmental mismatch is addressed by compensating the features and adding a simple uncertainty term to the backend acoustic model variances during recognition. In feature-based approaches, the uncertainty term can represent the certainty in the feature-compensation process. For example, in speech enhancement as the data get noisier, the less certain one may be

Techniques for Noise Robustness in Automatic Speech Recognition, First Edition.
Edited by Tuomas Virtanen, Rita Singh, and Bhiksha Raj.

in the estimated clean speech. In model-based approaches, specific models, such as fricatives, will be affected more by noise than others and this can be reflected by larger uncertainty causing larger model variances[1].

Although there has been a variety of work published related to uncertainty decoding, they all share a similar approach to modifying search with standard acoustic modeling to account for uncertainty. HMM-based acoustic models for speech recognition typically use Gaussian mixture models (GMMs) to represent state distributions. A particular Gaussian in the recognition acoustic model may be referred to as the *m*th component of the *q*th state of the model. However to simplify notation, Gaussian components of the acoustic model will be indexed globally, ignoring the state index, such that m is an index from 1 to M, where M is the total number of Gaussians over all states. Thus, the general form for conditional likelihood of a noisy speech observation, $\mathbf{x}_t$, for a Gaussian m in the clean acoustic model for uncertainty-based noise robustness methods is

$$\mathrm{p}(\mathbf{x}_t|m) = \alpha_{t,m}\mathcal{N}\Big(\hat{\mathbf{s}}_t; \boldsymbol{\mu}_{\mathrm{s},m}, \boldsymbol{\Theta}_{\mathrm{s},m} + \boldsymbol{\Theta}_{\mathrm{b}}\Big). \tag{17.1}$$

From the noisy speech features, $\mathbf{x}_t$, an estimate of the clean speech features, denoted by $\hat{\mathbf{s}}_t$, is derived. The clean speech acoustic model means $\boldsymbol{\mu}_{\mathrm{s},m}$ remain unchanged, but the variances $\boldsymbol{\Theta}_{\mathrm{s},m}$ are increased by a bias $\boldsymbol{\Theta}_{\mathrm{b}}$ to account for uncertainty due to noise. Since the variances are updated, the Gaussian constant normalization term in the likelihood function for m also needs to be recomputed. An additional normalization term, $\alpha_{t,m}$, may be necessary depending if the clean speech estimate is computed by scaling the features, for example if an affine transform $\mathbf{A}$ is applied, then $\alpha_{t,m} = |\mathbf{A}|$. The normalization term is a result of the change of variables from a probability density function of $\mathbf{x}_t$ to $\hat{\mathbf{s}}_t$. However, it may be unnecessary if it is independent of m as with typical feature-compensation techniques applied in the front-end speech parameterization step.

Compared to pure model-based compensation techniques, such as parallel model combination (PMC) and model-based compensation using a vector Taylor series (VTS) approximation presented in Chapter 12, the acoustic model parameters update is simpler: the model means are unaffected and uncertainty parameters can be shared over groups of similar Gaussians rather than computed for every model component m. The soft-information paradigm referred to in the Algonquin framework [19] can be considered a pure model-based compensation scheme, rather than uncertainty decoding as described here, since a variational approximated noisy speech model is computed for each component in the mismatched acoustic model. Uncertainty decoding thus combines the benefits of feature compensation with a powerful model update without the associated cost typical of model-based compensation techniques. Chapter 13 addressed training of clean acoustic models free of noise, sometimes referred to as canonical acoustic models. Rather than assume all training frames as equal, the level of noise can result in uncertainty that weights the contribution of a frame to acoustic model parameter estimates.

The following sections will present different motivations and approximations that provide derivations similar to Equation (17.1) and concrete definitions for the uncertainty decoding parameters: the clean speech estimate, $\hat{\mathbf{s}}_t$, and the the uncertainty bias term, $\boldsymbol{\Theta}_{\mathrm{b}}$. In feature-based compensation techniques, the parameters may be applied in the front-end and change

[1] The astute reader will note that model variances typically decrease with greater acoustic noise, but in the updated uncertainty feature space, the variances actually increase.

over time. In contrast, for model-based techniques the uncertainty parameters are dependent on the backend acoustic model parameters and remain fixed for the duration of an utterance.

17.2 Observation Uncertainty

Feature-compensation schemes, such as speech enhancement, provide an estimate of the clean speech to the decoder. This assumes the enhancement is exact and the clean speech estimate is the true value. However, it may be reasonable to consider that the denoising process is not exact and there is some residual error or uncertainty that may be passed to the decoder. Hence, in the observation uncertainty approach, instead of using a point estimate of the features, the clean speech posterior distribution is passed to the decoder as shown in Figure 17.1.

As discussed in Section 9.1.2, the minimum mean squared error estimate of the clean speech $\mathbf{s}_t$ given the noisy speech $\mathbf{x}_t$ is the conditional expected value

$$\hat{\mathbf{s}}_t = \mathcal{E}\{\mathbf{s}_t|\mathbf{x}_t\}. \tag{17.2}$$

But rather than only considering the mean, the variance can also be taken into account. Thus, the clean estimate varies according to the clean speech posterior may modeled by a normal distribution

$$\hat{\mathbf{s}}_t \sim p(\mathbf{s}_t|\mathbf{x}_t) = \mathcal{N}(\hat{\mathbf{s}}_t, \mathbf{\Theta}_\mathrm{b}), \tag{17.3}$$

where the variance $\mathbf{\Theta}_\mathrm{b}$ may be interpreted as the uncertainty of generating the estimate $\hat{\mathbf{s}}_t$.

The noisy speech likelihood function m is then derived by integrating over the clean speech as follows:

$$\mathrm{p}(\mathbf{x}_t|m) = \int p(\mathbf{s}_t|\mathbf{x}_t)p(\mathbf{s}_t|m)d\mathbf{s}_t \tag{17.4}$$

$$= \mathcal{N}(\hat{\mathbf{s}}_t; \boldsymbol{\mu}_{\mathrm{s},m}, \mathbf{\Theta}_{\mathrm{s},m} + \mathbf{\Theta}_\mathrm{b}). \tag{17.5}$$

Note how observation uncertainty decoding is essentially a feature enhancement scheme, since the features are the enhanced feature vector $\hat{\mathbf{s}}_t$, but with acoustic model variances increased by the observation uncertainty $\mathbf{\Theta}_\mathrm{b}$.

Thus, many probabilistic speech-enhancement techniques can be extended to use this form of observation uncertainty decoding simply by computing an enhancement variance $\mathbf{\Theta}_\mathrm{b}$. As discussed in Section 9.1.3, a variety of front-end compensation techniques use a front-end GMM to represent the acoustic space with feature compensation parameters associated with each front-end Gaussian component *i*. A specific example is SPLICE, explained in

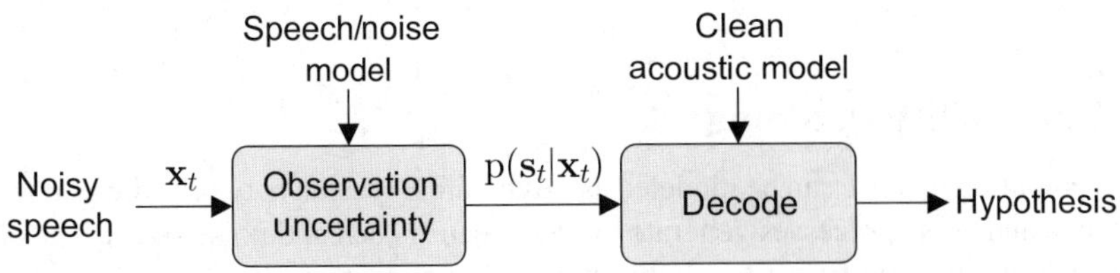

Figure 17.1 Feature compensation with observation uncertainty.

Section 9.2, where the clean speech posterior takes this form

$$\mathrm{p}(\mathbf{s}_t|\mathbf{x}_t, i) = \mathcal{N}(\mathbf{s}_t; \mathbf{x}_t + \boldsymbol{\mu}_{\mathrm{s-x},i}, \boldsymbol{\Theta}_{\mathrm{s-x},i}) \tag{17.6}$$

and the parameters $\boldsymbol{\mu}_{\mathrm{s-x},i}$ and $\boldsymbol{\Theta}_{\mathrm{s-x},i}$ are estimated using stereo data. The enhancement variance for each front-end GMM component, i, can also be estimated from stereo data

$$\boldsymbol{\Theta}_{\mathrm{s-x},i} = \mathcal{E}\left\{(\mathbf{s}_t - \mathbf{x}_t)(\mathbf{s}_t - \mathbf{x}_t)^{\mathsf{T}}|i\right\} - \boldsymbol{\mu}_{\mathrm{s-x},i}\boldsymbol{\mu}_{\mathrm{s-x},i}^{\mathsf{T}}. \tag{17.7}$$

This enhancement variance, which is the clean posterior variance, is the observation uncertainty

$$\boldsymbol{\Theta}_{\mathrm{b}} = \boldsymbol{\Theta}_{\mathrm{s-x},\hat{i}}, \tag{17.8}$$

where the front-end component i is selected using the max rule often applied in front-end GMM-based enhancement as follows:

$$\hat{i} = \underset{i}{\operatorname{argmax}}\, \mathrm{p}(\mathbf{x}_t|i)p(i) \tag{17.9}$$

as discussed in more detail in Section 9.1.3. Alternatively, model-based feature enhancement (MBFE) [30], as covered in Section 9.4, uses the joint distribution of the clean and noisy speech to compute the variance bias

$$\boldsymbol{\Theta}_{\mathrm{b},i} = \boldsymbol{\Theta}_{\mathrm{s},i} - \boldsymbol{\Theta}_{\mathrm{sx},i}\boldsymbol{\Theta}_{\mathrm{x},i}^{-1}\boldsymbol{\Theta}_{\mathrm{xs},i}, \tag{17.10}$$

where $\boldsymbol{\Theta}_{\mathrm{sx},i}$ is the cross covariance between clean speech and noisy speech, $\boldsymbol{\Theta}_{\mathrm{xs},i} = \boldsymbol{\Theta}_{\mathrm{sx},i}^{\mathsf{T}}$, and $\boldsymbol{\Theta}_{\mathrm{x},i}$ the noisy speech variance. The specific uncertainty variance i to apply at time frame t can again be determined using the max rule in Equation (17.9). Other enhancement schemes have been extended to provide this uncertainty, for example computed from the formants [13], a polynomial function of the signal-to-noise ratio (SNR) [1], a parametric model of the clean speech [4,5], Wiener filtering [2], a particle filter [31], or unscented transforms [28]. Compared to missing feature theory presented in the previous chapters, data imputation with uncertainty of parts of the reconstructed spectrum falls under this observation uncertainty approach [3,27,29].

Although there has been some experimentation with this approach, there is an inconsistency in Equation (17.4) from a Bayesian inference perspective. The clean speech posterior should instead be the distribution of the noisy speech given the clean. Perhaps this is why some additional heuristic is applied: "obtained front-end variances are multiplied by a constant factor (in our case this factor was experimentally tuned to 0.1) before adding them to the back-end variances on a frame-by-frame basis" [30], the variances are considered too large [4], or there is degradation compared to the nonuncertainty form in high SNRs [2].

17.3 Uncertainty Decoding

Speech recognition in noise can be modeled as a dynamic Bayesian network where independent clean speech and noise processes generate hidden clean speech and noise states. The observed noisy speech at any time frame is conditionally independent of all others given the hidden clean speech and noise at that time frame. Using this model of the environment, the noisy

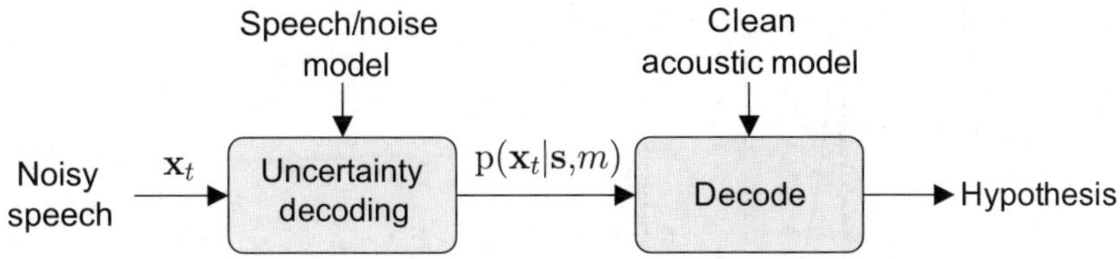

Figure 17.2 Uncertainty decoding.

speech likelihood function for m can be derived by marginalizing over the hidden clean speech

$$p(\mathbf{x}_t|m) = \int p(\mathbf{x}_t|\mathbf{s}_t, m)p(\mathbf{s}_t|m)d\mathbf{s}_t, \quad (17.11)$$

where

$$p(\mathbf{x}_t|\mathbf{s}_t, m) = \int p(\mathbf{x}_t|\mathbf{s}_t, \mathbf{n}_t, m)p(\mathbf{n}_t)d\mathbf{n}_t \quad (17.12)$$

and m is the index of the Gaussian component in the clean acoustic model. Note that in computing the noisy speech likelihood function in Equation (17.11), only the noisy speech conditional, given in Equation (17.12), is dependent on the noise. The clean speech prior $p(\mathbf{s}_t|m)$ distribution is Gaussian component m from the clean acoustic model. Thus, uncertainty decoding can be viewed as passing the noisy speech conditional distribution, $p(\mathbf{x}_t|\mathbf{s}_t, m)$, to the decoding process as shown in Figure 17.2.

An important issue in uncertainty decoding is finding an efficient yet accurate representation of the corrupted speech conditional distribution that is also amenable to marginalization with a Gaussian distribution. The main difficulty is that $p(\mathbf{x}_t|\mathbf{s}_t, m)$ is a complicated distribution. This is demonstrated by a numerical simulation of the joint log-spectral domain clean and corrupted speech distribution in Figure 17.3

$$x_t^l = \log(\exp(s_t^l) + \exp(n_t^l)), \quad (17.13)$$

where recall x_t^l is the noisy speech and the subscript l indicates a log-spectral domain variable dimension is being examined. The additive noise n_t^l again is generated from a single Gaussian distribution. The clean speech s_t^l is uniform over the interval $[0, 8]$ to demonstrate how the joint distribution changes as the clean speech does with a fixed noise source. The joint distribution is highly non-Gaussian and difficult to characterize parametrically.

The noisy speech conditional distribution varies greatly over the range of values for clean speech. When the clean speech is much larger than the noise mean, that is when $s_t^l = 6$ in Figure 17.3, the conditional distribution is relatively deterministic— the speech is unaffected by the noise. However, when the SNR drops the variance of the conditional distribution increases until the noise subsumes the speech. For example, when $s_t^l = 1$ in Figure 17.3, the conditional distribution becomes the additive noise distribution. Increasing the noise mean would shift the distribution up and to the right such that when the SNR is low, the corrupted speech conditional distribution continues to converge to the noise distribution. Thus, the effective form of the corrupted speech conditional distribution strongly depends on the difference between the clean speech and the noise distribution. Approximating the conditional distribution with a constant density function independent of the clean speech would be poor.

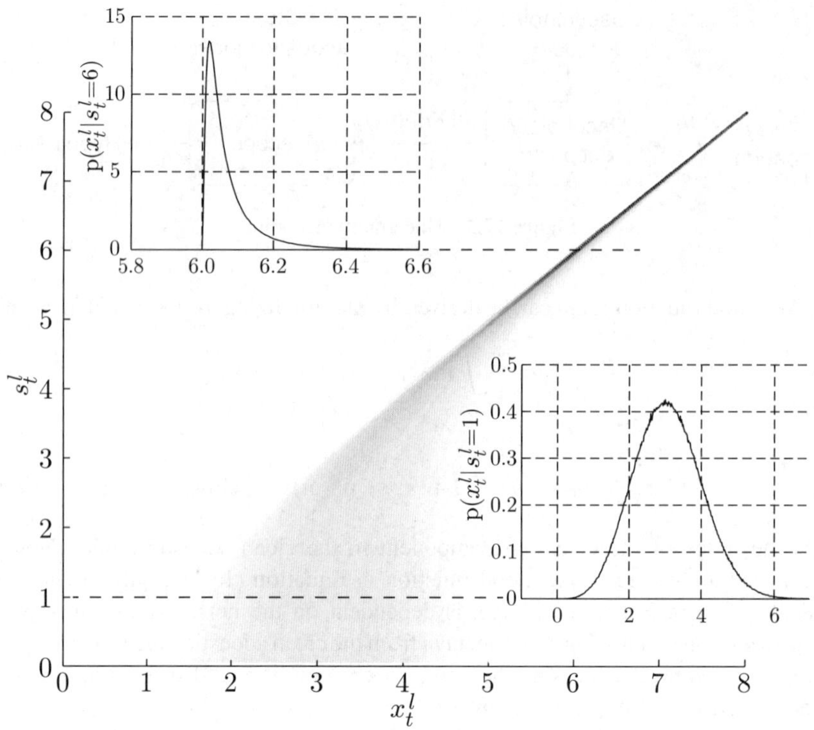

Figure 17.3 Joint distribution of clean s_t^l and noisy speech x_t^l with an additive noise source $\mathcal{N}(3, 1)$ in log spectral domain, and conditional distribution of x_t^l for $s_t^l = 1$ and $s_t^l = 6$.

Uncertainty decoding can be considered to encompass forms that exploit the factorization in Equation (17.11) by determining an efficient approximation for the noisy speech conditional distribution that easily completes the marginalization and is cheap to compute. The noisy speech posterior can be decoupled from the structure of the actual acoustic models and thus there is significant freedom in choosing an appropriate form for this distribution that minimizes the computational cost. If it is completely decoupled, and dependent entirely on the observed features, this gives feature-based uncertainty decoding forms as discussed in the following Section 17.4. Partial decoupling, where the conditional is dependent on the class of the acoustic model component, yields a model-based uncertainty decoding scheme presented in Section 17.5. In pure model-based approaches the two distributions are fully tied by the clean speech variable.

17.4 Feature-Based Uncertainty Decoding

In feature-based uncertainty decoding, the noisy speech posterior in Equation (17.11) is completely decoupled from the clean speech prior distribution other than the compensated features and associated uncertainty bias. The examples of feature-based uncertainty decoding are SPLICE with Uncertainty [6] and front-end Joint Uncertainty Decoding or more briefly front-end JUD [22]. For both, the transformation of the features and uncertainty bias are selected based on the most probable front-end component i similar to front-end GMM-enhancements techniques discussed in Section 9.1.3.

By decoupling the noisy speech posterior and the clean speech prior, the marginalization in Equation (17.11), becomes this for feature-based uncertainty decoding forms

$$\mathrm{p}(\mathbf{x}_t|m) = \int \mathrm{p}(\mathbf{x}_t|\mathbf{s}_t)\mathrm{p}(\mathbf{s}_t|m)\mathrm{d}\mathbf{s}_t, \tag{17.14}$$

where note the noisy speech posterior is no longer conditioned on the acoustic model component m. If the noisy speech posterior is Gaussian as is the acoustic model component prior, then the integral is simple to solve. In this case, the noisy speech likelihood function for m is given by

$$\mathrm{p}(\mathbf{x}_t|m) = |\mathbf{A}_i|\mathcal{N}\Big(\hat{\mathbf{s}}_t; \boldsymbol{\mu}_{\mathrm{s},m}, \boldsymbol{\Theta}_{\mathrm{s},m} + \boldsymbol{\Theta}_{\mathrm{b},i}\Big), \tag{17.15}$$

where the clean speech estimate is an affine transformation of the noisy speech

$$\hat{\mathbf{s}}_t = \mathbf{A}_i\mathbf{x}_t + \mathbf{b}_i \tag{17.16}$$

and i is chosen via the hard max rule in Equation (17.9). The normalization term $|\mathbf{A}_i|$ arises from the change of variables from $\mathbf{x}_t$ on the left side of Equation (17.15) to $\hat{\mathbf{s}}_t$ on the right side. The noisy speech posterior affects how the uncertainty parameters $\mathbf{A}_i$, $\mathbf{b}_i$, and $\boldsymbol{\Theta}_{\mathrm{b},i}$ are estimated; this will be shown in the following subsections with concrete approximation for $\mathrm{p}(\mathbf{x}_t|\mathbf{s}_t)$. Since the noisy speech posterior is not dependent on the acoustic model parameters, the clean speech estimate can be computed entirely as part of the feature processing in the front-end. It is updated once per time frame and shared globally across all acoustic model components.

Figure 17.4 demonstrates the operation of feature-based uncertainty decoding techniques. It shows how given a noisy observation, $\mathbf{x}_t$, the most likely front-end GMM component i is selected. This front-end component has compensation parameters associated with it, where the noisy feature is transformed by $\mathbf{A}_i$ and $\mathbf{b}_i$, and the uncertainty bias $\boldsymbol{\Theta}_{\mathrm{b},i}$ is added to the model variance $\boldsymbol{\Theta}_{\mathrm{s},m}$. These are, respectively, the feature transformation and model update applied during uncertainty decoding. While similar to feature-based forms like SPLICE, MBFE, or front-end constrained maximum likelihood linear regression (CMLLR) [22], where the affine feature transform is selected by a front-end GMM, in uncertainty decoding there is the

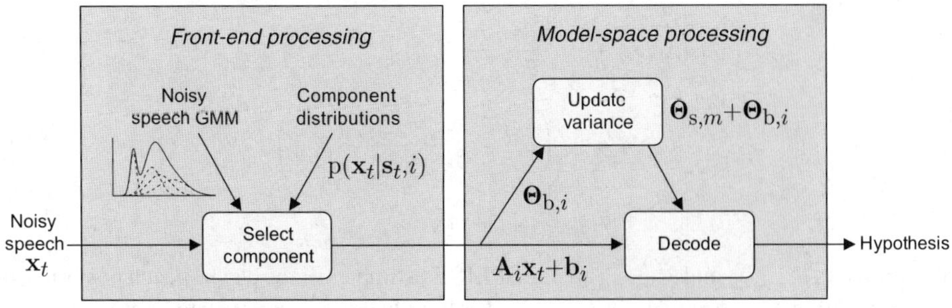

Figure 17.4 Feature-based uncertainty decoding.

addition of a "uncertainty" variance bias added to the model variances. Data marginalization in missing-data techniques presented in Chapter 14 can be construed as a limited form of feature-based uncertainty decoding, restricted to the spectral domain, where a hard-decision is made whether features are either completely certain or uncertain.

Specific forms of feature-based uncertainty decoding are SPLICE with Uncertainty and front-end JUD. Although they share the same decoding likelihood function given by Equation (17.15), they differ in their derivations of $p(\mathbf{x}_t|\mathbf{s}_t)$ and therefore also have different definitions of the compensation parameters $\mathbf{A}_i$, $\mathbf{b}_i$, and $\mathbf{\Theta}_{\mathrm{b},i}$. The next two subsections will describe these two techniques in further detail while the last subsection discusses issues with using feature-based uncertainty decoding.

17.4.1 SPLICE with Uncertainty

SPLICE learns a piece-wise linear mapping between noisy speech and clean by training on stereo data where the same speech is recorded in parallel with both clean and noisy conditions. Stereo data may be generated by adding noise to otherwise clean speech data. The mapping bias vectors are applied to the noisy speech feature vector to obtain an estimate of clean speech. Refer back to Section 9.2 for more information about SPLICE and Chapter 9 where stereo data are used extensively.

SPLICE can also be extended to use the notion of increased uncertainty due to noise. SPLICE with Uncertainty makes use of Bayes' rule to express the conditional probability of the noisy speech given the clean speech in terms of the of the clean speech posterior in this manner

$$\mathrm{p}(\mathbf{x}_t|\mathbf{s}_t) = \frac{\sum_i \mathrm{p}(i)\mathrm{p}(\mathbf{s}_t|\mathbf{x}_t, i)\mathrm{p}(\mathbf{x}_t|i)}{\mathrm{p}(\mathbf{s}_t)}. \tag{17.17}$$

Recall from Chapter 9.2 that SPLICE used stereo data to estimate compensation parameters for each component i, for the component clean speech posterior distribution $\mathrm{p}(\mathbf{s}_t|\mathbf{x}_t, i)$ given in Equation 17.6. To simplify, a single Gaussian approximation for the distribution of the clean speech $\mathrm{p}(\mathbf{s}_t)$ is used where the global clean speech mean and variance for feature dimension d are $\bar{\mu}_{\mathrm{s},d}$ and $\bar{\theta}^2_{\mathrm{s},d}$. If instead of summing over all components i, only the maximum one is chosen as before, an analytic form for $\mathrm{p}(\mathbf{s}_t|\mathbf{x}_t, i)$ can be derived. Using these approximations, if Equation (17.17) is substituted into Equation (17.14), with the restriction that $\mathbf{A}_i$ and $\mathbf{\Theta}_{\mathrm{b},i}$ in equations (17.15) and (17.16) are diagonal, gives

$$a_{i,dd} = \frac{\bar{\theta}^2_{\mathrm{s},d}}{\bar{\theta}^2_{\mathrm{s},d} - \theta^2_{\mathrm{s-x},i,d}}, \tag{17.18}$$

$$b_{i,d} = a_{i,dd}\left(\mu_{\mathrm{s-x},i,d} - \frac{\theta^2_{\mathrm{s-x},i,d}}{\bar{\theta}^2_{\mathrm{s},d}}\bar{\mu}_{\mathrm{s},d}\right), \tag{17.19}$$

$$\theta^2_{\mathrm{b},i,d} = a_{i,dd}\theta^2_{\mathrm{s-x},i,d}. \tag{17.20}$$

The parameters $\mu_{\mathrm{s-x},i,d}$ and $\theta^2_{\mathrm{s-x},i,d}$ are the SPLICE estimates of the clean speech posterior: the means and variance respectively of $(\mathrm{s}_d - \mathrm{x}_d)$ for the observations associated with component i of the front-end GMM. The SPLICE variance estimate of the clean speech posterior was given earlier in Equation (17.7). In order to ensure that the uncertainty variance bias $\theta^2_{\mathrm{b},i,d}$ is positive, the denominator in Equation (17.18) is floored. For example, the floor could be set to a fraction of the global clean speech variance. This floor effectively places a maximum value on $a_{i,dd}$ and hence the uncertainty bias $\theta^2_{\mathrm{b},i,d}$. Still if the denominator becomes small, $a_{i,dd}$ and $\theta^2_{\mathrm{b},i,d}$ can become quite large. In contrast, instead of being multiplied by possibly large scalars, in the observation uncertainty form of SPLICE, discussed previously in Section 17.2, biases are added to the features and model variances.

17.4.2 Front-End Joint Uncertainty Decoding

Instead of inverting the noisy speech posterior using Bayes' rule, it can be derived from the joint distribution. Similarly though, it can be modeled with a GMM as given by

$$\mathrm{p}\Big(\mathbf{x}_t|\mathbf{s}_t\Big) = \sum_i \mathrm{P}(i|\mathbf{s}_t)\mathrm{p}\Big(\mathbf{x}_t|\mathbf{s}_t, i\Big). \tag{17.21}$$

Using a property of multivariate Gaussian distributions that if the joint distribution is Gaussian distributed, then the conditional distribution is as well, when the joint distribution of the clean and noisy speech is given by:

$$\left[\begin{bmatrix}\mathbf{s}_t\\ \mathbf{x}_t\end{bmatrix}\right| i \sim \mathcal{N}\left(\begin{bmatrix}\boldsymbol{\mu}_{\mathrm{s},i}\\ \boldsymbol{\mu}_{\mathrm{x},i}\end{bmatrix}, \begin{bmatrix}\boldsymbol{\Theta}_{\mathrm{s},i} & \boldsymbol{\Theta}_{\mathrm{sx},i}\\ \boldsymbol{\Theta}_{\mathrm{xs},i} & \boldsymbol{\Theta}_{\mathrm{x},i}\end{bmatrix}\right) \tag{17.22}$$

for component *i*, the Gaussian noisy speech conditional distribution is as follows:

$$\mathrm{p}\Big(\mathbf{x}_t|\mathbf{s}_t, i\Big) = \mathcal{N}\left(\mathbf{x}_t; \boldsymbol{\mu}_{\mathrm{x},i} + \boldsymbol{\Theta}_{\mathrm{xs},i}\boldsymbol{\Theta}_{\mathrm{s},i}^{-1}\Big(\mathbf{s}_t - \boldsymbol{\mu}_{\mathrm{s},i}\Big), \boldsymbol{\Theta}_{\mathrm{x},i} - \boldsymbol{\Theta}_{\mathrm{xs},i}\boldsymbol{\Theta}_{\mathrm{s},i}^{-1}\boldsymbol{\Theta}_{\mathrm{sx},i}\right). \tag{17.23}$$

Here, the parameters of the noisy speech posterior are all from the joint distribution.

In using a GMM to represent the noisy speech distribution conditioned on the hidden clean speech, the component posterior is also conditioned on the hidden clean speech. This may be approximated by conditioning the component posterior on the noisy speech itself

$$\mathrm{P}(i|\mathbf{s}_t) \approx \mathrm{P}(i|\mathbf{x}_t). \tag{17.24}$$

By substituting this GMM approximation of the noisy speech posterior into Equation (17.11), the uncertainty decoding parameters associated with a front-end component *i* can be expressed in terms of the parameters of the joint distribution, given in Equation (17.22), as follows:

$$\mathbf{A}_i = \boldsymbol{\Theta}_{\mathrm{s},i}\boldsymbol{\Theta}_{\mathrm{xs},i}^{-1}, \tag{17.25}$$

$$\mathbf{b}_i = \boldsymbol{\mu}_{\mathrm{s},i} - \mathbf{A}_i\boldsymbol{\mu}_{\mathrm{x},i}, \tag{17.26}$$

$$\boldsymbol{\Theta}_{\mathrm{b},i} = \mathbf{A}_i\boldsymbol{\Theta}_{\mathrm{x},i}\mathbf{A}_i^{\mathsf{T}} - \boldsymbol{\Theta}_{\mathrm{s},i}. \tag{17.27}$$

Compared to the SPLICE with Uncertainty, no explicit flooring is required and the matrix parameters $\mathbf{A}_i$ and $\boldsymbol{\Theta}_{\mathrm{b},i}$ may be full. However, by having a full uncertainty decoding matrix bias, the likelihood function in Equation (17.1) also becomes full covariance and very computationally intensive. So in practice, diagonal uncertainty decoding parameters may be used. Similar to the other feature-based compensation techniques mentioned, a max approximation may be used to select *i* for each time frame index.

Table 17.1 Word error rates (%) for 256-component front-end uncertainty decoding schemes compensating clean trained models on Aurora2 task, test set A, averaged across subway, babble, car and exhibition hall condition (N1-N4). Parameters are trained using stereo data.

System	Compensation	SNR(dB)			
		20	15	10	5
Clean trained	—	4.6	12.2	31.1	59.2
	SPLICE	2.0	3.1	6.1	16.5
	SPLICE with Uncertainty	2.0	3.2	5.6	12.3
	Front-end JUD	1.8	2.9	5.7	14.6
Matched trained	—	1.8	2.8	5.0	11.4

Table 17.1 compares some of techniques discussed using the artificially noisy digit recognition task Aurora2. The clean trained system obviously does much worse as the SNR drops. Matched training is sometimes considered the best robustness scheme, but often not practical. An effective feature-based enhancement technique such as SPLICE can significantly improve the robustness of a mismatched clean trained model. Feature-based uncertainty decoding can improve this, especially in the noisier 10 and 5 dB conditions.

17.4.3 Issues with Feature-Based Uncertainty Decoding

One serious drawback of front-end uncertainty schemes is that the model variances must be updated every time the variance bias changes. The variance bias changes as the front-end component i changes. Although, the update is simple compared to a technique such as model-based VTS compensation, the update and recomputation of the normalization term must be executed for every acoustic model component.

A more serious issue is that when the noise completely subsumes the speech, the uncertainty is unbounded. When the SNR is low and noise masks the speech, the covariance between the noisy speech and the clean speech will be approximately zero since the noisy speech and clean speech are independent. Hence

$$\mathbf{\Theta}_{\mathrm{xs},i} \approx \mathbf{0}. \tag{17.28}$$

When this is the case, it is clear the uncertainty term defined in Equation (17.27) becomes infinite because $\mathbf{A}_i$ becomes infinite. During regions of high noise then, this high uncertainty will cause all acoustic models to appear the same and there will be no acoustic discrimination between classes. If the recognition task has additional constraints beyond the acoustic models, such as a language model, then some discrimination between classes may be possible in these regions. However, when there is no language model or other restrictions, for example with a digit recognition task such as Aurora2, then these areas will be very susceptible to errors. These errors will probably be insertions since these areas are likely to be background regions where the uncertainty is highest. To mitigate this problem a minimum correlation between the noisy and clean speech may be enforced [23].

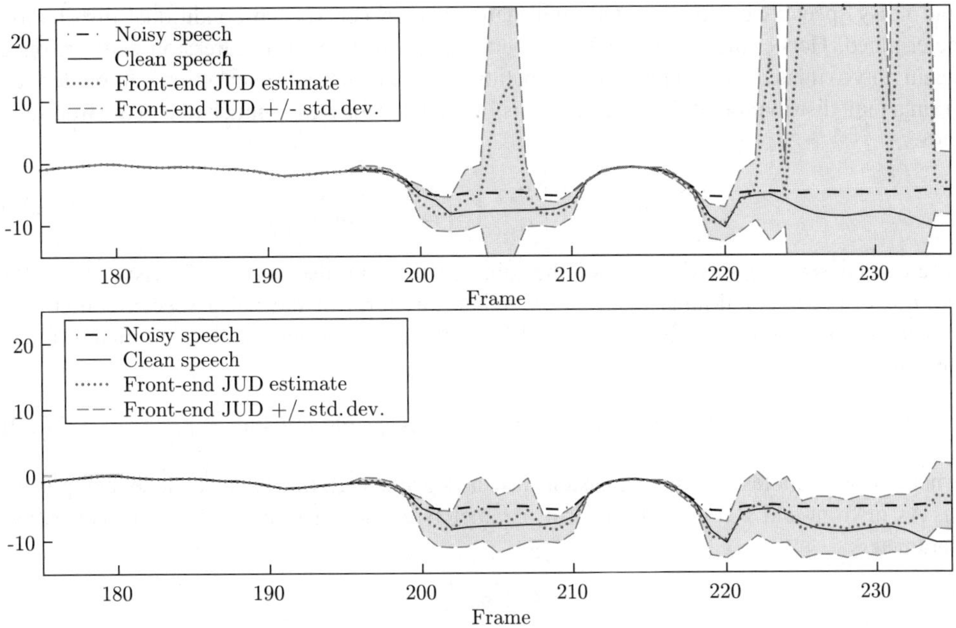

Figure 17.5 Plot of log-energy dimension from Aurora2 digit string 8-6-zero-1-1-6-2, showing 16-component GMM front-end JUD estimate $a_{\hat{i}}x_t + b_{\hat{i}}$, uncertainty bias $\theta_{\text{b},\hat{i}}$, and $a_{\hat{i}}$ (upper panel). With the correlation floored at 0.1 (lower panel).

An illustration of this issue with front-end JUD is presented in Figure 17.5. This figure shows the clean speech, noisy speech, front-end JUD estimate, given by $a_{\hat{i}}x_t + b_{\hat{i}}$, and the uncertainty bias $\sigma_{\text{b},\hat{i}}$ for a simple system with a 16-component front-end GMM. These parameters are for a single coefficient of the parameters in Equation (17.25)–(17.27) for a particular frame *t*, when $\hat{i}$ is determined using the hard maximum approximation given by Equation (17.9). For those regions of higher energy speech, for example frames 210–220 where the vowel "*i*" is articulated, the variance bias is small. On the other hand, in the lower energy regions around this vowel, for example frames 225–230, the variance becomes too large to be measured on this scale, as is the front-end JUD estimate of the value. These large variances are associated with large values of the scale factor $a_{\hat{i}}$ as shown in Figure 17.5 due to very small correlations between the clean and noisy speech as discussed earlier. In this example, from frames 225–230 the value of $a_{\hat{i}}$ is around 100. With greater numbers of front-end components, these effects are amplified as parameters are no longer smoothed.

17.5 Model-Based Joint Uncertainty Decoding

As opposed to associating a noisy speech conditional distribution with a region of the feature space, the conditional distribution may be linked with group of similar acoustic model components. Model components may be clustered into classes of similar Gaussians using a regression tree as described in Section 11.4.1. For each class, a joint distribution of the clean and noisy speech can be estimated, and therefore a speech speech conditional distribution determined. Hence, the conditional distribution is a function of a regression class, r, once again approximated by a Gaussian distribution. This conditional distribution can be derived from a joint distribution of the clean and noisy speech for the class r much like in the front-end case

$$\begin{bmatrix} \mathbf{s}_t \\ \mathbf{x}_t \end{bmatrix} \Bigg| r \sim \mathcal{N}\left(\begin{bmatrix} \boldsymbol{\mu}_{\mathrm{s},r} \\ \boldsymbol{\mu}_{\mathrm{x},r} \end{bmatrix}, \begin{bmatrix} \boldsymbol{\Theta}_{\mathrm{s},r} & \boldsymbol{\Theta}_{\mathrm{sx},r} \\ \boldsymbol{\Theta}_{\mathrm{xs},r} & \boldsymbol{\Theta}_{\mathrm{x},r} \end{bmatrix} \right), \tag{17.29}$$

where a Gaussian approximation of the joint distribution also gives a Gaussian form for the noisy speech conditional distribution. When this form of the noisy speech conditional distribution is substituted into Equation (17.11), the noisy speech likelihood function for m becomes

$$\mathrm{p}(\mathbf{x}_t|m) = |\mathbf{A}_{r_m}|\mathcal{N}\Big(\mathbf{A}_{r_m}\mathbf{x}_t + \mathbf{b}_{r_m}; \boldsymbol{\mu}_{\mathrm{s},m}, \boldsymbol{\Theta}_{\mathrm{s},m} + \boldsymbol{\Theta}_{\mathrm{b},r_m}\Big), \tag{17.30}$$

where the r_m denotes that the regression class is dependent on the acoustic model component m. As in front-end JUD, the transform parameters are a function of the joint distribution parameters

$$\mathbf{A}_r = \boldsymbol{\Theta}_{\mathrm{s},r}\boldsymbol{\Theta}_{\mathrm{xs},r}^{-1}, \tag{17.31}$$

$$\mathbf{b}_r = \boldsymbol{\mu}_{\mathrm{s},r} - \mathbf{A}_r\boldsymbol{\mu}_{\mathrm{x},r}, \tag{17.32}$$

$$\boldsymbol{\Theta}_{\mathrm{b},r} = \mathbf{A}_r\boldsymbol{\Theta}_{\mathrm{x},r}\mathbf{A}_r^{\mathsf{T}} - \boldsymbol{\Theta}_{\mathrm{s},r}. \tag{17.33}$$

Compared to the feature-based form depicted in Figure 17.4, the noisy speech features are transformed by multiple transforms, much like in CMLLR, described in Section 11.4.2, such that there are R parallel versions of the observation passed to the decoder. In contrast to CMLLR, each regression class also has a different uncertainty variance bias associated with it. However, compared to feature-based uncertainty decoding, this variance does not change over time and may be cached; it need only be updated if the noise condition itself changes. Since for any given time frame, model components are being compensated by different transforms, model-based uncertainty decoding compensation will not be affected by the issues discussed previously in Section 17.4.3. Moreover, compared to the previously discussed feature-based uncertainty decoding forms, the cost of selecting a single maximum component from I components in the front-end is of similar order to applying R transforms for multiple features in model-based uncertainty decoding. Thus, for equivalent numbers of I and R feature-based and model-based uncertainty decoding are of similar computational complexity. The operation of this model-based uncertainty decoding is shown in Figure 17.6.

The joint distribution for a class given in Equation (17.29) can be easily estimated using stereo or parallel data, where at any given frame the clean and noisy speech are known. For a state alignment, the joint distribution of the clean and noisy speech for each regression class can be estimated using component level posteriors. Since the parameters given by equations (17.32) and (17.33) are derived from the class-conditional joint distribution, they are computed using stereo data.

In Table 17.2, front-end uncertainty decoding is compared with model-based uncertainty decoding with the same number of parameters. While they perform similarly at higher SNR,

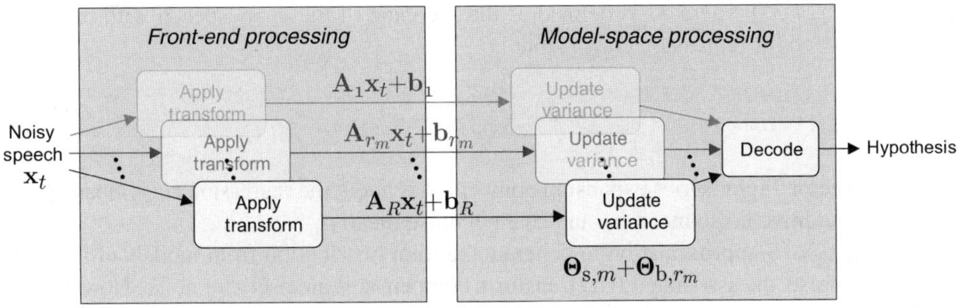

Figure 17.6 Model-based joint uncertainty decoding.

the model-based form is clearly better than the feature-based at lower SNR. At 5 dB, the model-based version with an error rate of 12.0 is close to the matched system performance. This demonstrates the advantage of using model-based uncertainty decoding over feature based.

17.5.1 Parameter Estimation

Estimation of the model-based uncertainty decoding parameters $\mathbf{A}_r$, $\mathbf{b}_r$, and $\mathbf{\Theta}_{b,r}$ using stereo data is artificial and not realistic. Alternatively, the joint distribution may also be predicted using clean speech regression class model, $\mathcal{N}(\boldsymbol{\mu}_{s,r}, \mathbf{\Theta}_{s,r})$, a noise model, $\mathcal{N}(\boldsymbol{\mu}_n, \mathbf{\Theta}_n)$, and a mismatch function describing how the two combine to form noisy speech. Suitable methods to compute the rest of the joint distribution parameters include using log-normal or log-add approximations, PMC or VTS as discussed in Chapter 12. These methods describe how to derive the noisy speech parameters, $\mathcal{N}(\boldsymbol{\mu}_{x,r}, \mathbf{\Theta}_{x,r})$, but the joint distribution also has the clean speech, noisy speech cross covariance term $\mathbf{\Theta}_{xs,r}$. For example, using VTS the term has this form

$$\mathbf{\Theta}_{xs,r} = \mathbf{J}_r \mathbf{\Theta}_{s,r} \tag{17.34}$$

Table 17.2 Word error rates (%) for 256-transform uncertainty decoding schemes. Compare 256-component front-end versus 256-class model-based forms compensating clean trained models on Aurora2 task, test set A, averaged across different noise conditions N1-N4. Matched test and training condition results are also reported as an upper bound. Parameters are trained using stereo data.

		SNR(dB)			
System	Compensation	20	15	10	5
	—	4.6	12.2	31.1	59.2
Clean trained	Front-end JUD	1.8	2.9	5.7	14.6
	Model-based JUD	1.9	2.7	5.2	12.0
Matched trained	—	1.8	2.8	5.0	11.4

as shown in [21,33] where $\mathbf{J}_r$ is defined as the Jacobian of the noisy speech with respect to the clean speech for regression class r, that is

$$\mathbf{J}_r = \left.\frac{\partial \mathbf{x}}{\partial \mathbf{s}}\right|_{\tau_{r,\mathrm{n},\mathrm{h}}}, \tag{17.35}$$

where the vector Taylor series expansion point $\tau_{r,\mathrm{n},\mathrm{h}}$ is about the regression class clean speech mean $\boldsymbol{\mu}_{\mathrm{s},r}$, additive noise mean $\boldsymbol{\mu}_{\mathrm{n}}$ and channel noise mean $\boldsymbol{\mu}_{\mathrm{h}}$.

By using a VTS approximation to generate the joint distribution from models of the clean speech and noise, the associated JUD transform compensates precisely for noise. However, the joint distribution may thought of as a general statistical model of the relationship between the speech seen during training and the observed speech in testing. Hence, the joint distribution can model other factors in addition to noise if this is taken into account during its generation. For example, vocal tract length or a feature decorrelating transform could be incorporated in the mismatch function. Furthermore, model-based JUD transforms may also compensate multistyle-trained systems, acoustic models that are trained on data that has varying amount of noisy speech, for environmental mismatch. In this case, the noise model no longer represents additive and convolutional noise but are simply parameters that generate transforms which reduce the mismatch between the multistyle-trained models and the test conditions. Multistyle training is also known as multicondition training.

17.5.2 Comparisons with Other Methods

Model-based JUD has much in common in with VTS when VTS is used to derive the joint distribution. It has been shown that increasing the number of classes R to equal the number of model components M, using a diagonal acoustic model variance approximation, is equivalent to VTS model compensation of each individual acoustic model component [20]. However, the likelihood calculation for a component takes place in a different space transformed by $\mathbf{A}_r$. This can allow model-based JUD to be much more efficient by trading off the compensation of acoustic model means for updating the features. Also, computing the Jacobian matrix can be expensive in VTS because it is computed for each and every acoustic model component; in model-based JUD, the computation is shared per regression class. The number of regression classes R is typically much smaller than the number of model components M. Lastly it is cheaper to apply the uncertainty bias to the variances than to compensate the model variance as is done in VTS. This convergence of model-based JUD, when $R = M$, to the form of model-based compensation that was used to derive the joint distribution is a very useful property. It allows a flexibility in controlling the computational cost of the uncertainty decoding scheme by adjusting the number of model classes R. For example, [32] show how model-based JUD can achieve the same level of accuracy as model-based VTS compensation with less than 25% of the computational cost. The flexibility of uncertainty decoding actually allows the more precise, but computationally expensive, second-order VTS approximation used in the model-based JUD system to be about 60% faster than a pure first-order VTS model-based compensation approach that must update each acoustic model Gaussian independently.

The affine transformation of the noisy features makes model-based JUD similar to CMLLR, but with the addition of the uncertainty variance bias. They are both able to compensate a

Table 17.3 Word error rates (%) for CMLLR, VTS, Model-based JUD, and PCMLLR compensation of an acoustic model trained on clean speech and tested on the Toshiba in-car recorded phone number recognition task (from [8]).

		In-car condition		
System	Compensation	Idle	City	Highway
Clean trained	—	2.9	32.9	66.3
	CMLLR	0.6	7.6	40.6
	PCMLLR	1.2	2.9	6.1
	Model-based JUD	1.1	2.9	5.4
	VTS	1.2	3.1	4.1

mismatched acoustic model to more closely match the test environment. However, model-based JUD, as with other *predictive* noise-compensation techniques that use a model of the acoustic environment for combining clean speech models and a noise model to predict noisy speech models, only needs a simple noise model to compensate the mismatched acoustic model. *Adaptive* techniques like CMLLR have many more free parameters. Hence, predictive approaches can need much less data to effectively compensate the system compared to adaptation approaches. Furthermore, the noise model may be estimated before the onset of speech, whereas adaptation requires actual test speech. An interesting comparison is with predictive CMLLR [10]. In PCMLLR, the adaptation statistics are actually predicted from clean statistics and a noise model. Thus, like model-based uncertainty decoding it can quickly improve noise robustness of a system. PCMLLR is still solely an affine transformation of the features, like CMLLR, and does not have the uncertainty bias term.

Table 17.3 compares a variety of these techniques discussed. Many noise robustness techniques are evaluated on artificial tasks, but here the performance of these algorithms on real in-car noisy speech is tested. The noise model is obtained through ML estimation as described in [21]. The results clearly show how clean model performance is dramatically affected by noise. Compensating the clean model with CMLLR can be effective when there is little noise, but as the mismatch between training and test condition grows it does poorly as shown in the lower SNR conditions. The predictive schemes all perform better than CMLLR by far in the noisier conditions. In quieter conditions they give similar results, but at the highway condition VTS performs best as expected as it is the most powerful and computationally expensive approach.

17.6 Noisy CMLLR

The model-based JUD transform is similar to CMLLR, but with a bias on the model variances. They also differ in that the JUD transform is predicted from an environmental model and models of clean speech and noise, whereas the CMLLR transform is directly estimated using ML. However, a form of CMLLR can be derived that is directly estimated using ML but also has a bias on the model variances similar to JUD. This has been called noisy constrained maximum likelihood linear regression (NCMLLR) [16].

The noisy speech $\mathbf{x}_t$ can be written as the product of a generative model of the clean speech $\mathbf{s}_t$ and noise $\mathbf{n}_t$ as follows:

$$\mathbf{x}_t = \mathbf{H}_r \mathbf{s}_t + \mathbf{g}_r + \mathbf{n}_t, \tag{17.36}$$

where $\mathbf{H}_r$ and $\mathbf{g}_r$ is the affine transformation for regression class r and $\mathbf{n}_t$ is Gaussian distributed about a mean of zero with variance $\mathbf{\Psi}_r$. This results in a noisy speech likelihood function for m of

$$\mathrm{p}(\mathbf{x}_t|m) = |\mathbf{A}_{r_m}|\mathcal{N}\left(\mathbf{A}_{r_m}\mathbf{x}_t + \mathbf{b}_{r_m}; \boldsymbol{\mu}_{\mathrm{s},m}, \mathbf{\Theta}_{\mathrm{s},m} + \mathbf{\Theta}_{\mathrm{b},r_m}\right), \tag{17.37}$$

where r_m denotes that the regression class depends on the acoustic model component m. This is exactly same form as model-based uncertainty decoding; however, the uncertainty transform parameters associated with a regression class are found to be

$$\mathbf{A}_r = \mathbf{H}_r^{-1}, \tag{17.38}$$

$$\mathbf{b}_r = -\mathbf{H}_r^{-1}\mathbf{g}_r, \tag{17.39}$$

$$\mathbf{\Theta}_{\mathrm{b},r} = \mathbf{A}_r \mathbf{\Psi}_r \mathbf{A}_r^{\mathsf{T}}. \tag{17.40}$$

These transform parameters can be grouped together: $\mathcal{T}_r = \left\{\mathbf{A}_r, \mathbf{b}_r, \mathbf{\Theta}_{\mathrm{b},r}\right\}$. The set of all transforms for all regression classes, where the total number is R, is then denoted by $\mathcal{T}$, where $\mathcal{T} = \left\{\mathcal{T}_r | 0 \le r < R\right\}$. The set of transforms $\mathcal{T}$ is estimated to maximize the likelihood of noisy adaptation data, that is

$$\hat{\mathcal{T}} = \underset{\mathcal{T}}{\operatorname{argmax}} \log \mathrm{p}(\mathbf{X}|\mathcal{M}, \mathcal{T}, \mathcal{H}) \tag{17.41}$$

for given acoustic model $\mathcal{M}$ and transcript of the data $\mathcal{H}$. Due to the latent state sequence and clean speech, it is difficult to directly optimize Equation (17.41). Thus, as with HMM parameter estimation, iterative expectation maximization (EM) is applied. The complete data set is $\mathbf{Z} = \{\mathbf{X}, \mathbf{S}, \mathbf{Q}\}$, where $\mathbf{Q}$ denotes the hidden state sequence. The auxiliary function is then given by

$$\mathcal{Q}(\hat{\mathcal{T}}; \mathcal{T}) = \mathcal{E}\{\log \mathrm{p}(\mathbf{Z}|\mathcal{M}, \hat{\mathcal{T}})|\mathbf{X}, \mathcal{M}, \mathcal{T}\}, \tag{17.42}$$

where $\mathcal{E}$ is the conditional expectation of the log likelihood of the complete data over all possible hidden sequences, given the observed noisy speech, model parameters, and transform set. For NCMLLR, the auxiliary function is computed as follows:

$$\mathcal{Q}(\hat{\mathcal{T}}; \mathcal{T}) = -\frac{1}{2}\sum_{t=1}^{T}\sum_{m=1}^{M}\gamma_{m,t}\mathcal{E}\left\{\log|\mathbf{\Psi}_{r_m}| + (\mathbf{x}_t - \hat{\mathbf{V}}\boldsymbol{\xi}_t)^{\mathsf{T}}\mathbf{\Psi}_{r_m}^{-1}(\mathbf{x}_t - \hat{\mathbf{V}}\boldsymbol{\xi}_t)\middle|\mathbf{x}_t, m\right\}, \tag{17.43}$$

where $\gamma_{m,t}$ is the posterior probability of component m given $\mathbf{X}$, $\mathcal{M}$, and $\mathcal{T}$. The extended transformation matrix is $\hat{\mathbf{V}}_r = [\hat{\mathbf{g}}_r \quad \hat{\mathbf{H}}_r]$ and the extended clean speech vector $\boldsymbol{\xi}_t = [1 \quad \mathbf{s}_t^{\mathsf{T}}]^{\mathsf{T}}$.

Rather than directly maximizing Equation (17.43) for all the transform parameters at once, it is simpler to separately estimate the feature transformation $\mathbf{A}_r, \mathbf{b}_r$, and then the variance bias $\mathbf{\Theta}_{\mathrm{b},r}$. This is achieved by differentiating Equation (17.43) with respect to the variable in question, equating the result to zero and solving for the variable. For $\hat{\mathbf{V}}_r$ this yields

$$\hat{\mathbf{V}}_r = \left(\sum_{m\in r}\sum_{t=1}^{T}\gamma_{m,t}\mathbf{x}_t\mathcal{E}\left\{\boldsymbol{\xi}_t^{\mathsf{T}}\middle|\mathbf{x}_t,m\right\}\right)\left(\sum_{m\in r}\sum_{t=1}^{T}\gamma_{m,t}\mathcal{E}\left\{\boldsymbol{\xi}_t\boldsymbol{\xi}_t^{\mathsf{T}}\middle|\mathbf{x}_t,m\right\}\right)^{-1}. \tag{17.44}$$

Recall, the feature transformation can be derived from the constituents of the extended transformation given by Equations (17.38) and (17.39). The conditional expectations of the extended clean speech vector and its outer product can be expressed as

$$\mathcal{E}\left\{\boldsymbol{\xi}_t^{\mathsf{T}}\middle|\mathbf{x}_t,m\right\} = \begin{bmatrix}1 & \tilde{\mathbf{s}}_{m,t}^{\mathsf{T}}\end{bmatrix}, \tag{17.45}$$

$$\mathcal{E}\left\{\boldsymbol{\xi}_t\boldsymbol{\xi}_t^{\mathsf{T}}\middle|\mathbf{x}_t,m\right\} = \begin{bmatrix}1 & \tilde{\mathbf{s}}_{m,t}^{\mathsf{T}}\\ \tilde{\mathbf{s}}_{m,t} & \tilde{\boldsymbol{\Theta}}_{\mathrm{s},m} + \tilde{\mathbf{s}}_{m,t}\tilde{\mathbf{s}}_{m,t}^{\mathsf{T}}\end{bmatrix} \tag{17.46}$$

with

$$\tilde{\mathbf{s}}_{m,t} = \tilde{\mathbf{A}}_m\mathbf{x}_t + \tilde{\mathbf{b}},m, \tag{17.47}$$

$$\tilde{\boldsymbol{\Theta}}_{\mathrm{s},m} = \left(\boldsymbol{\Theta}_{\mathrm{s},m}^{-1} + \hat{\boldsymbol{\Theta}}_{\mathrm{b},r_m}^{-1}\right)^{-1} \tag{17.48}$$

and

$$\tilde{\mathbf{A}}_m = \left(\boldsymbol{\Theta}_{\mathrm{s},m}^{-1} + \hat{\boldsymbol{\Theta}}_{\mathrm{b},r_m}^{-1}\right)^{-1}\hat{\boldsymbol{\Theta}}_{\mathrm{b},r_m}\hat{\mathbf{A}}_{r_m}, \tag{17.49}$$

$$\tilde{\mathbf{b}}_m = \left(\boldsymbol{\Theta}_{\mathrm{s},m}^{-1} + \hat{\boldsymbol{\Theta}}_{\mathrm{b},r_m}^{-1}\right)^{-1}\left(\boldsymbol{\Theta}_{\mathrm{s},m}^{-1}\boldsymbol{\mu}_{\mathrm{s},r_m} + \hat{\boldsymbol{\Theta}}_{\mathrm{b},r_m}^{-1}\hat{\mathbf{A}}_{r_m}\right)^{-1}. \tag{17.50}$$

Similarly the variance bias $\hat{\boldsymbol{\Theta}}_{\mathrm{b},r}$ can be estimated from the auxiliary function in Equation (17.43) by

$$\hat{\boldsymbol{\Theta}}_{\mathrm{b},r} = \frac{\sum_{m\in r}\sum_{t=1}^{T}\gamma_{m,t}\mathcal{E}\left\{\left(\hat{\mathbf{s}}_{r,t}-\mathbf{s}_t\right)\left(\hat{\mathbf{s}}_{r,t}-\mathbf{s}_t\right)^{\mathsf{T}}\right\}}{\sum_{m\in r}\sum_{t=1}^{T}\gamma_{m,t}}, \tag{17.51}$$

where

$$\hat{\mathbf{s}}_{r,t} = \hat{\mathbf{A}}_r\mathbf{x}_t + \hat{\mathbf{b}}_r. \tag{17.52}$$

The variance bias term matrix here is full, but unlike in model-based uncertainty decoding, $\boldsymbol{\Theta}_{\mathrm{b},r}$, may be diagonalized while retaining a full feature transformation matrix $\mathbf{A}_r$. This is possible since the variance bias is not tied to the feature transformation like JUD is by the clean-noisy speech cross covariance matrix $\boldsymbol{\Theta}_{\mathrm{xs}}$. By diagonalizing the variance bias, full covariance decoding is not required, which results in significant computational savings.

Table 17.4 compares the number of free parameters that need to be estimated for various noise-compensation schemes. The predictive techniques only need a low parameter noise model which can be estimated on a small amount of data. The adaptive forms require the estimation of many more parameters that scale with the number of regression classes used. With uncertainty decoding, the number regression classes can be flexibly increased without requiring more test data, to increase robustness at the expense of computational cost of applying the transformations.

Table 17.4 Number of free parameters to estimate for various compensation schemes. *D* is dimensionality of the full feature vector and D_s is the number of static parameters. *R* is the number of regression classes used to group acoustic model components.

Scheme	Type	Parameters	Number of free parameters
PCMLLR, JUD, VTS	Predictive	$\boldsymbol{\mu}_{\text{n}}, \boldsymbol{\Theta}_{\text{n}}, \boldsymbol{\mu}_{\text{h}}$	$5D_s$
CMLLR	Adaptive,	$\mathbf{A}_r, \mathbf{b}_r$	$2RD$
NCMLLR	Diagonal $\mathbf{A}_r$	$\mathbf{A}_r, \mathbf{b}_r, \boldsymbol{\Theta}_r$	$3RD$
CMLLR	Adaptive,	$\mathbf{A}_r, \mathbf{b}_r$	$R(D^2 + D)$
NCMLLR	Full $\mathbf{A}_r$	$\mathbf{A}_r, \mathbf{b}_r, \boldsymbol{\Theta}_r$	$R(D^2 + 2D)$

Table 17.5 compares CMLLR with NCMLLR on the Toshiba phone number recognition task, but this time while compensating a multistyle trained acoustic model. In quieter conditions, CMLLR performs better than NCMLLR, but at the highway condition NCMLLR is better. Perhaps this is due to the training method here that ensures the uncertainty bias is always positive; for cleaner conditions, a positive uncertainty may make compensating a noisier multistyle model difficult. An approach that relaxes the constraint of a positive uncertainty bias is presented in [25]. The differences between these CMLLR and NCMLLR results are small and not significant though.

17.7 Uncertainty and Adaptive Training

Chapter 13 demonstrated that training on features processed by a noise-reduction algorithm gives improved performance by providing a purer clean, canonical acoustic model, further minimizing the difference between training and test features. Recall the canonical acoustic model should be free of acoustic variability not necessary for recognition. However, when there is environmental noise, the compensation of the features may be imperfect, especially for noisier conditions. Estimates for the noise-free, canonical acoustic model in these noisier conditions may be poor. Thus, noisier features should contribute less to the canonical acoustic model than features that are more certain. Using the notion of uncertainty due to noise is one approach to achieving this. This chapter demonstrates how adaptive training with uncertainty can reduce contributions of noisier observations.

When discussing adaptive training, it is useful to introduce the concept of a homogeneous set of observations where the noise environment can be considered stationary. The entire training data set can be viewed as heterogeneous, with many varying noise conditions, but comprised of

Table 17.5 Word error rates (%) comparing 16-transform CMLLR with NCMLLR compensation on multistyle system tested on Toshiba in-car recorded phone number recognition task (from [17]).

		In-car condition		
System	Compensation	Idle	City	Highway
	—	1.1	3.6	6.7
Multistyle trained	CMLLR	0.3	1.1	2.4
	NCMLLR	0.5	1.2	2.1

H blocks of homogeneous speech. The parameters associated with an individual block where the noise is assumed constant is denoted by h. The total number of frames of homogeneous speech for the block is indicated by T_h. As with standard HMM training, the hidden variables make direct estimation of the canonical parameters difficult and so EM is used. The auxiliary function where only terms dependent on the model parameters are shown is

$$\mathcal{Q}(\mathcal{M},\hat{\mathcal{M}}) = -\frac{1}{2}\sum_{h=1}^{H}\sum_{t=1}^{T_h}\sum_{m=1}^{M}\gamma_{m,h,t}\left[\log\left|\boldsymbol{\Theta}_{\mathrm{s},m}+\boldsymbol{\Theta}_{\mathrm{b},r_m,h}\right|\right.$$
$$\left.+\left(\mathbf{A}_{r_m,h}\mathbf{s}_t+\mathbf{b}_{r_m,h}-\boldsymbol{\mu}_{\mathrm{s},m}\right)^{\mathsf{T}}\left(\boldsymbol{\Theta}_{\mathrm{s},m}+\boldsymbol{\Theta}_{\mathrm{b},r_m,h}\right)^{-1}\left(\mathbf{A}_{r_m,h}\mathbf{s}_t+\mathbf{b}_{r_m,h}-\boldsymbol{\mu}_{\mathrm{s},m}\right)\right]. \tag{17.53}$$

A set of uncertainty transformation parameters $\mathcal{T}_h = \left\{\mathcal{T}_{1,h}, \mathcal{T}_{r,h}, \ldots, \mathcal{T}_{R,h}\right\}$ will have to be estimated per block h where recall R is the total number of regression classes used to represent the much larger number of acoustic model components. The parameters for each condition h and regression class r are $\mathcal{T}_{r,h} = \left\{\mathbf{A}_{r,h}, \mathbf{b}_{r,h}, \boldsymbol{\Theta}_{\mathrm{b},r,h}\right\}$. Therefore, there will be H sets of uncertainty transforms each with R transforms.

In the literature, there are two main methods to estimating the canonical acoustic model parameters in this uncertainty decoding framework. One is to use gradient-based methods to optimize an auxiliary function. Alternatively, factor analysis-based EM approaches treat the clean speech and additive noise as continuous latent variables that generate the noisy speech [12]. The next two sections will discuss these in further detail.

17.7.1 Gradient-Based Methods

Because the model-based JUD transform parameters affect the acoustic model parameters, yet the acoustic model parameters are shared over many homogeneous blocks, there is no closed form solution for the model parameters that maximize this auxiliary function. Hence a generalized EM approach can be taken, where Newton's method is applied to optimize the model parameters in the maximization step

$$\begin{bmatrix}\hat{\mu}_{\mathrm{s},m,d}\\ \hat{\theta}^2_{\mathrm{s},m,d}\end{bmatrix} = \begin{bmatrix}\mu_{\mathrm{s},m,d}\\ \theta^2_{\mathrm{s},m,d}\end{bmatrix} - \zeta\begin{bmatrix}\dfrac{\partial^2\mathcal{Q}}{\partial\left(\mu_{\mathrm{s},m,d}\right)^2} & \dfrac{\partial^2\mathcal{Q}}{\partial\mu_{\mathrm{s},m,d}\partial\theta^2_{\mathrm{s},m,d}}\\ \dfrac{\partial^2\mathcal{Q}}{\partial\theta^2_{\mathrm{s},m,d}\partial\mu_{\mathrm{s},m,d}} & \dfrac{\partial^2\mathcal{Q}}{\partial\left(\theta^2_{\mathrm{s},m,d}\right)^2}\end{bmatrix}^{-1}\begin{bmatrix}\dfrac{\partial\mathcal{Q}}{\partial\mu_{\mathrm{s},m,d}}\\ \dfrac{\partial\mathcal{Q}}{\partial\theta^2_{\mathrm{s},m,d}}\end{bmatrix}, \tag{17.54}$$

where ζ is the learning rate. This requires both first- and second-order derivatives of the auxiliary function with respect to the model mean and variance. To simplify the derivation of the first derivative, diagonal approximations of the covariance can be made in the auxiliary function in Equation (17.53) yielding

$$\mathcal{Q}(\mathcal{M},\hat{\mathcal{M}}) =$$
$$-\frac{1}{2}\sum_{h=1}^{H}\sum_{t=1}^{T_h}\sum_{m=1}^{M}\gamma_{m,h,t}\sum_{d=1}^{D}\left(\log\left(\theta^2_{\mathrm{s},m,d}+\theta^2_{\mathrm{b},r_m,h,d}\right)+\frac{\left(\mathbf{a}_{r_m,h,d}\mathbf{s}_t+b_{r_m,h,d}-\mu_{\mathrm{s},m,d}\right)^2}{\theta^2_{\mathrm{s},m,d}+\theta^2_{\mathrm{b},r_m,h,d}}\right), \tag{17.55}$$

where the term $\mathbf{a}_{r_m,h,\bar{d}}$ denotes the dth row in the matrix $\mathbf{A}_{r_m,h}$ and is a row vector instead of the usual column vector notation. Thus, the first derivative of the auxiliary function in Equation (17.55) with respect to the mean of component m, dimension d is

$$\frac{\partial Q}{\partial \mu_{s,m,d}} = \sum_{h=1}^{H}\sum_{t=1}^{T_h} \frac{\gamma_{m,h,t}}{\theta^2_{s,m,d} + \theta^2_{b,r_m,h,d}} \left(\mathbf{a}_{r_m,h,\bar{d}} \mathbf{s}_t + b_{r_m,h,d} - \mu_{s,m,d} \right). \tag{17.56}$$

Notice how the uncertainty bias $\boldsymbol{\theta}^2_{b,r,h,d}$ adjusts the component posterior $\gamma_{m,h,t}$. The same is true for the first derivative with respect to the variance. If the SNR is high, then there is no uncertainty and the posterior is not affected. When the SNR is low, the uncertainty will be large, reducing the contribution of noisy observations by deweighting the component posterior. In areas where the noise completely subsumes the speech, the uncertainty will ensure that these observations do not contribute to the estimate of the model parameters at all—the model parameters will not be updated since the first derivatives of the auxiliary function with respect to the model means and variance will be naught. This allows the model parameters to be a better representation of "clean" speech. With normalization schemes, noise adaptive training using spectral enhancement or feature compensation, or MLLR-based adaptation, once observations are compensated for noise, the cleaned training features are all treated equally. In contrast, with the uncertainty term adaptive training will give greater importance to observations that are less noisy and errorful. Thus, uncertainty decoding can minimize errors due to noise from polluting the canonical acoustic model parameters. More details on this can be found in [24] where this is approached is described as joint adaptive training (JAT). Newton's method is also used to estimate the canonical model variances for noise adaptive training with VTS in [15]; there are many similarities due to the relationship between model-based uncertainty decoding and VTS.

17.7.2 Factor Analysis Approaches

For the first-order VTS approximation, the noisy speech observation associated with a model component m can be written as a linear combination of the clean speech and noise

$$\mathbf{x}|m \approx \mathbf{J}_m \mathbf{s} + (\mathbf{I} - \mathbf{J}_m)\mathbf{n} + \mathbf{J}_m \mathbf{h} + f(\boldsymbol{\mu}_{s,m}, \boldsymbol{\mu}_n, \boldsymbol{\mu}_h). \tag{17.57}$$

If the Jacobian matrix $\mathbf{J}_m$ and the bias term $f(\boldsymbol{\mu}_{s,m}, \boldsymbol{\mu}_n, \boldsymbol{\mu}_h)$ are considered fixed, this can be viewed as a generative model conducive to factor analysis [16]. Using this approach, the canonical acoustic model parameters are given by

$$\hat{\boldsymbol{\mu}}_{s,m} = \frac{\sum_{h=1}^{H}\sum_{t=1}^{T_h} \gamma_{m,h,t} \mathcal{E}\left\{ \mathbf{s}_t | \mathbf{x}_t, m, h \right\}}{\sum_{h=1}^{H}\sum_{t=1}^{T_h} \gamma_{m,h,t}}, \tag{17.58}$$

$$\hat{\boldsymbol{\Theta}}_{s,m} = \text{diag}\left\{ \frac{\sum_{h=1}^{H}\sum_{t=1}^{T_h} \gamma_{m,h,t} \mathcal{E}\left\{ \mathbf{s}_t \mathbf{s}_t^{\mathsf{T}} | \mathbf{x}_t, m, h \right\}}{\sum_{h=1}^{H}\sum_{t=1}^{T_h} \gamma_{m,h,t}} - \hat{\boldsymbol{\mu}}_{s,m} \hat{\boldsymbol{\mu}}_{s,m}^{\mathsf{T}} \right\}. \tag{17.59}$$

Table 17.6 Word error rates (%) contrasting joint adaptive training with VTS adaptive training on an average across two Toshiba test sets (phone numbers and city names) (from [7]).

	Condition	
System	Idle	Highway
Multistyle training	2.5	9.5
Joint adaptive training	2.4	4.7
VTS adaptive training	2.2	4.4

The conditional expectations are

$$\mathcal{E}\left\{\mathbf{s}_t|\mathbf{x}_t, m, h\right\} = \tilde{\mathbf{s}}_{m,h,t}, \tag{17.60}$$

$$\mathcal{E}\left\{\mathbf{s}_t\mathbf{s}_t^{\mathsf{T}}|\mathbf{x}_t, m, h\right\} = \tilde{\mathbf{\Theta}}_{\mathrm{s},m,h} + \tilde{\mathbf{s}}_{m,h,t}\tilde{\mathbf{s}}_{m,h,t}^{\mathsf{T}}. \tag{17.61}$$

However, these model estimates do not guarantee improvement in likelihood due to the assumptions made earlier. First, the Jacobian and bias terms are in fact dependent on the model parameters themselves and thus not fixed. Secondly, since the Jacobian is not diagonal, the resulting clean speech variance $\hat{\mathbf{\Theta}}_{\mathrm{s},m}$ will also not be diagonal; however, here $\hat{\mathbf{\Theta}}_{\mathrm{s},m}$ is diagonalized for decoding efficiency, resulting in an approximate generative model. These issues are discussed in more detail in [9].

Also, note the NCMLLR and model-based JUD transforms are the same in how they are applied during decoding. Thus, this approach can be used to estimate the canonical clean speech model parameters for either method. Furthermore, since model-based JUD and VTS are equivalent when the number of regression classes equals the number of model components, this approach is essentially a variation of the factor estimation methods in Kim *et al.* [18] and Hu and Huo [14].

In Table 17.6, the performance of some adaptive training techniques are compared against multistyle training. For the quieter idle condition, there is some benefit, but the larger gains appear in the noisier highway condition. There is little difference in results between JAT and VTS adaptive training although JAT can be more efficient to use in practice.

17.8 In Combination with Other Techniques

Many state of the art-recognition systems use multiple acoustic-modeling techniques in concert, for example semitied covariances (STC), linear discriminant analysis (LDA), CMLLR, MLLR, vocal tract length normalization, and PLP features (see [11] for an overview of these techniques), and feature and model discriminative training. Such techniques can make it very difficult to predict the noisy speech statistics from a clean model as the mismatch function becomes very complex. Reference [26] demonstrate some of the difficulties in incorporating predictive techniques in a large vocabulary system with many of these techniques.

However linear transformations like STC, CMLLR, and MLLR can be combined with uncertainty decoding techniques as shown in [20]. Also there has not been much focus on combining predictive noise-robustness techniques with discriminative model training

Table 17.7 Word error rates (%) contrasting joint adaptive training with VTS adaptive training on an average across two Toshiba test sets (phone numbers and city names) (from [7]).

System		Condition	
		Idle	Highway
Multistyle training	ML	2.5	9.5
	MPE	2.1	8.6
Joint adaptive training	ML	2.4	4.7
	MPE	1.7	3.6
VTS adaptive training	ML	2.2	4.4
	MPE	1.3	3.7

methods. Using a factor analysis approach to estimating the canonical speech model parameters allows these noise-robustness techniques to be combined with discriminative training. Table 17.7 compares maximum likelihood (ML) training with the commonly used discriminative minimum phone error (MPE) criterion. The results show clearly that these methods can be combined and result in greater than 10% relative improvement over the multistyle baseline for both quiet and noisy conditions.

17.9 Conclusions

This chapter has presented a variety of techniques that use the concept of uncertainty in decoding due to noise to improve recognition robustness. Feature-based uncertainty approaches are presented along with their fundamental limitations. The level of uncertainty is highly dependent on the clean speech model being compensated where the uncertainty is proportional to the level of mismatch. The model-based uncertainty approaches take advantage of this fact. They also provide an elegant means of applying pure model-based compensation techniques like VTS more efficiently by varying the number of regression classes. For adaptive training, the uncertainty limits the contribution of noisier observations to acoustic model parameter estimates. In practice, results demonstrating how model-based uncertainty forms can significantly lower the computational cost with recognition accuracy similar to pure model-based compensation were shown.

As the noise level increases, it becomes more important to accurately model the correlations introduced by the noise. Many of the results discussed in this chapter assume stationary noise. More sophisticated modeling of the environmental noise could be made, for example to handle sudden noises, background speech or reverberant noise. These techniques also assume the noisy speech distribution will be Gaussian; however, they can be highly non-Gaussian depending on the SNR and speech and noise variances. Alternative non-Gaussian distributions could provide better results. The joint distribution has also been predicted using a noise mismatch function. If other factors, such as the speaker, can be captured in the joint distribution between the training and test conditions, then uncertainty decoding can be extended beyond noise compensation. While there has been work on the field of noise robustness, very few techniques are actually deployed in large-scale real-word systems. Combining these uncertainty decoding techniques with other state-of-the-art recognition algorithms is an ongoing research topic. These issues are discussed in further detail in [20].

References

[1] J. Arrowood, "Using observation uncertainty for robust speech recognition," PhD dissertation, Georgia Institute of Technology, 2003.

[2] C. Benítez, J. C. Segura, A. de la Torre, J. Ramírez, and A. Rubio, "Including uncertainty of speech observations in robust speech recognition," in *Proceedings of the International Conference on Spoken Language Processing*, Jeju Island, Korea, 2004, pp. 137–140.

[3] M. Cooke, P. Green, L. Josifovski, and A. Vizinho, "Robust automatic speech recognition with missing and unreliable acoustic data," *Speech Communication*, vol. 34, no. 3, pp. 267–285, Jun. 2001.

[4] L. Deng, J. Droppo, and A. Acero, "Exploiting variances in robust feature extraction based on a parametric model of speech distortion," in *Proceedings of Interspeech*, Denver, Colorado, USA, 2002, pp. 2449–2452.

[5] L. Deng, J. Droppo, and A. Acero, "Dynamic compensation of HMM variances using the feature enhancement uncertainty computed from a parametric model of speech distortion," *IEEE Transactions on Speech and Audio Processing*, vol. 13, no. 3, pp. 412–421, May 2005.

[6] J. Droppo, A. Acero, and L. Deng, "Uncertainty decoding with SPLICE for noise robust speech recognition," in *Proceedings of the International Conference on Acoustics, Speech and Signal Processing*, Orlando, Florida, USA, May 2002, pp. 57–60.

[7] F. Flego and M. J. F. Gales, "Discriminative adaptive training with VTS and JUD," in *Proceedings of the IEEE Workshop on Automatic Speech Recognition and Understanding (ASRU)*, Merano, Italy, 2009, pp. 170–175.

[8] F. Flego and M. J. F. Gales, "Incremental predictive and adaptive noise compensation," in *Proceedings of the International Conference on Acoustics, Speech and Signal Processing*, Taipei, Taiwan, 2009, pp. 3837–3840.

[9] F. Flego and M. J. F. Gales, "Factor analysis based VTS and JUD noise estimation and compensation," in *Proceedings of the International Conference on Acoustics, Speech and Signal Processing*, Prague, Czech, 2011, pp. 4792–4795.

[10] M. J. F. Gales and R. van Dalen, "Predictive linear transforms for noise robust speech recognition," in *Proceedings of the IEEE Workshop on Automatic Speech Recognition and Understanding (ASRU)*, Kyoto, Japan, 2007, pp. 59–64.

[11] M. J. F. Gales and S. J. Young, *The Application of Hidden Markov Models in Speech Recognition, Foundations and Trends in Signal Processing*. now Publishers Inc., 2007.

[12] R. Gopinath, B. Ramabhadran, and S. Dharanipragada, "Factor analysis invariant to linear transformations of data," in *Proceedings of the International Conference on Spoken Language Processing*, Sydney, Australia, 1998, pp. 397–400.

[13] J. N. Holmes, W. J. Holmes, and P. N. Garner, "Using formant frequencies in speech recognition," in *Proceedings of Eurospeech*, Rhodes, Greece, Sep. 1997, pp. 2083–2086.

[14] Y. Hu and Q. Huo, "Irrelevant variability normalization based HMM training using VTS approximation of an explicit model of environmental distortions," in *Proceedings of Interspeech*, Andwerp, Belgium, 2007, pp. 1042–1045.

[15] O. Kalinli, M. Seltzer, and A. Acero, "Noise adaptive training using a vector Taylor series approach for noise robust automatic speech recognition," in *Proceedings of the International Conference on Acoustics, Speech and Signal Processing*, Taipei, Taiwan, 2009, pp. 3825–3828.

[16] D. Kim and M. J. F. Gales, "Adaptive training with noisy constrained maximum likelihood linear regression for noise robust speech recognition," in *Proceedings of Interspeech*, Brighton, United Kingdon, 2009, pp. 2383–2386.

[17] D. K. Kim and M. J. F. Gales, "Noisy constrained maximum-likelihood linear regression for noise-robust speech recognition," *IEEE Transactions on Audio, Speech and Language Processing*, vol. 19, pp. 315–325, 2010.

[18] D. Y. Kim, N. S. Kim, and C. K. Un, "Model-based approach for robust speech recognition in noisy environments with multiple noise sources," in *Proceedings of Eurospeech*, Rhodes, Greece, Sep. 1997, pp. 1123–1126.

[19] T. T. Kristjansson and B. J. Frey, "Accounting for uncertainty in observations: A new paradigm for robust speech recognition," in *Proceedings of the International Conference on Acoustics, Speech and Signal Processing*, Orlando, Florida, USA, May 2002, pp. 61–64.

[20] H. Liao, "Uncertainty decoding for noise robust speech recognition," PhD dissertation, University of Cambridge, 2007.

[21] H. Liao and M. J. F. Gales, "Joint uncertainty decoding for robust large vocabulary speech recognition," Technical Report CUED/F-INFENG/TR552, University of Cambridge, 2006. Available at: http://mi.eng.cam.ac.uk/reports/index-speech.html.

[22] H. Liao and M. J. F. Gales, "Joint uncertainty decoding for noise robust speech recognition," in *Proceedings of Interspeech*, Lisbon, Portugal, 2005, pp. 3129–3132.

[23] H. Liao and M. J. F. Gales, "Issues with uncertainty decoding for noise robust speech recognition," in *Proceedings of Interspeech*, Pittsburgh, Pennsylvania, USA, 2006, pp. 1627–1630.

[24] H. Liao and M. J. F. Gales, "Adaptive training with joint uncertainty decoding for robust recognition of noisy data," in *Proceedings of the International Conference on Acoustics, Speech and Signal Processing*, Honolulu, Hawaii, USA, 2007, pp. 389–392.

[25] J. Lu, J. Ming, and R. Woods, "Adapting noisy speech models—extended uncertainty decoding," in *Proceedings of the International Conference on Acoustics, Speech and Signal Processing*, Dallas, Texas, USA, 2010, pp. 4322–4325.

[26] D. Povey and B. Kingsbury, "Monte Carlo model-space noise adaptation for speech recognition," in *Proceedings of Interspeech*, Brisbane, Australia, 2008, pp. 1281–1284.

[27] B. Raj and R. Stern, "Missing-feature approaches in speech recognition," *IEEE Signal Processing Magazine*, pp. 101–116, Sep. 2005.

[28] Y. Shinohara and M. Akamine, "Bayesian feature enhancement using a mixture of unscented transformations for uncertainty decoding of noisy speech," in *Proceedings of the International Conference on Acoustics, Speech and Signal Processing*, Taipei, Taiwan, 2009, pp. 4569–4572.

[29] S. Srinivasan and D. Wang, "A supervised learning approach to uncertainty decoding for robust speech recognition," in *Proceedings of the International Conference on Acoustics, Speech and Signal Processing*, Toulouse, France, 2006, pp. 297–300.

[30] V. Stouten, H. Van hamme, and P. Wambacq, "Accounting for the uncertainty of speech estimates in the context of model-based feature enhancement," in *Proceedings of Interspeech*, Jeju Island, Korea, 2004, pp. 105–108.

[31] M. Wölfel and F. Faubel, "Considering uncertainty by particle filter enhanced speech features in large vocabulary continuous speech recognition," in *Proceedings of the International Conference on Acoustics, Speech and Signal Processing*, Pittsburgh, Pennsylvania, USA, 2007, pp. 1049–1052.

[32] H. Xu and K. K. Chin, "Joint uncertainty decoding with the second order approximation for noise robust speech recognition," in *Proceedings of the IEEE Workshop on Automatic Speech Recognition and Understanding (ASRU)*, Merano, Italy, 2009, pp. 3841–3844.

[33] H. Xu, L. Rigazio, and D. Kryze, "Vector Taylor series based joint uncertainty decoding," in *Proceedings of Interspeech*, Pittsburgh, Pennsylvania, USA, 2006, pp. 1688–1691.

Index

a posteriori SNR, 59
a priori SNR, 70
estimation, 69–71
decision directed, 70, 75
acoustic echo cancellation, 352
acoustic impulse response, 252
acoustic model, 10
acoustic pre-processing, 161
acoustic transfer function, 259
additive noise, 41
Algonquin, *see* vector Taylor series
aliasing, 125
cancellation, 123, 125
frequency, 125
spatial, 119
time, 123
amplitude modulation, 444
amplitude modulation spectrogram (AMS), 448
analysis
filter bank, 123–125
frequency domain, 123
subband domain, 123
angle
polar, 142
aperture
length, 113
linear, 113–118
array
gain, 126–129
delay-and-sum, 127
linear, 120, 122, 126, 149–152
manifold vector, 118, 121, 125
modal manifold vector, 147
spherical, 142–152
articulation index, 219
auditory models, 208–221
auditory feature extraction, 208–221
Carney group, 214
detailed physiological models, 214
frequency, rate, and scale, models based on, 220
Gitza EIH model, 210
impact on MFCC and PLP features, 206
initial performance, 212
Lyon auditory model, 211
multi-band and multi-stream processing, 219
neural synchrony, 210–212, 218
Seneff auditory model, 208
tandem combination, 219
auditory perception, 202–208
auditory thresholds, 205
frequency resolution, 203
intensity, perception of, 203
nonsimultaneous masking, 206
psychoacoustical response to sound
auditory physiology, 194–202
amplitude and frequency modulation, response to, 202
binaural phenomena, 201
frequency resolution, 196

Techniques for Noise Robustness in Automatic Speech Recognition, First Edition.
Edited by Tuomas Virtanen, Rita Singh, and Bhiksha Raj.

auditory physiology (*Continued*)
lateral suppression, 199
rate-level response, 197
synchrony, neural, 197
transient response, 195
auditory scene analysis (ASA), 433
schema-based grouping, 435
auditory stream, 434
primitive grouping, 434
segments, 434
sequential grouping, 434
simultaneous grouping, 434
Aurora-2, 230, 420–422, 424, 425, 472, 475
Aurora-3, 363
Aurora-4, 230, 273, 422, 427, 457
Aurora-5, 272
autocorrelation, 442
circular, 165
normalized envelope autocorrelation, 444
running autocorrelation, 442
summary autocorrelation, *see* summary autocorrelation function
automatic speech recognition (ASR), 35
averaged localized synchrony detection (ALSD), 218
averaged localized synchrony response (ALSR), 198
azimuth, 113

backward probability, 15
band quantization, 335–337, 339
band quantized GMM (BQGMM), 335
barge-in, 353
Bark scale, 204
Baum-Welch algorithm, 19–20
Bayes classification, 10, 31, 35
beamformer, 109
adaptive, 120–126, 129, 131, 148
delay-and-sum, 119, 120, 127, 142, 148
fixed, 120, 148
frequency domain, 123
generalized sidelobe canceller, 129–132
HOS, 142
hypercardioid, 148
linear, 152
LMS, 130
maximum kurtosis, 133–136, 139–142
maximum negentropy, 134–136, 141, 142
MVDR, 120–126, 129, 131
performance measures, 126–129
RLS, 130, 131
SOS, 142
subband, 125, 129
subband domain, 123
superdirective, 123, 141, 142
beampattern, 116–120, 122, 128, 148–150
belief propagation, 339
Bessel function, 123, 143
blind source separation, 88
bounded marginalization, *see* missing data, bounded marginalization
BQ, *see* band quantization
BQGMM, *see* band quantized GMM
broadside, 118

cancellation
aliasing, 123
signal, 121
canonical acoustic model, 480
cepstral mean and variance normalization, 361
cepstral mean normalization, 48, 261
cepstral mean subtraction, 48, 261
cepstral variance normalization, 48
cepstrum histogram normalization, 48
cepstrum smoothing, 71
channel effects on speech, 39
class regression tree, 294, 296
classifier compensation, 400
for ASR, 453, 457
clustering, 408
k-means, 408
regression tree, 473
CMLLR, *see* constrained maximum likelihood linear regression
CMVN, *see* cepstral mean and variance normalization
cochleagram, 440
T-F unit, 440
cochlear response, 194
cocktail party problem, 433
coding distortions, 41

Cohen auditory model, 212
computational auditory scene analysis (CASA), 434
  definition, 435
  IBM, *see* ideal binary mask
constrained maximum likelihood linear regression, *see* maximum likelihood linear regression, constrained
  front-end, 469
  noisy, *see* noisy constrained maximum likelihood linear regression, 477, 483
  predictive, 477
continuous-time approximation, 356
convolutive mixture, 92
correlogram, 442
correlogram representation, 212
critical bands, 203
critical distance, 253
cross-channel correlation, 442
  envelope cross-channel correlation, 444
cross-correlation, 451
  generalized, 111
curse of dimensionality, 161
cylindrically isotropic noise field, 130

delta features, 25
devil function, 318–320
diagonal loading, 361
dictionary, 415, 417, 421
diffraction, 110
direct-to-reverberation ratio, 254
direction
  cosine, 115
  look, 117, 118, 120–122, 128, 129
  of arrival, 113
directivity, 145
  index, 126, 128–129
discrete cosine transform, 257
discrete Fourier transform, 55–57
discriminative SPLICE, 237
discriminative training, 299
distortion model, 353
distortionless constraint, 120, 121, 128, 129, 137, 148

DNA, *see* dynamic noise adaptation
dynamic adaptation, 289
dynamic noise adaptation, 338–339
dynamic noise models, 338–339

early reflections, 252, 253
early reverberation, 252
energetic masking, 381
energy decay curve, 253
entropy, 132, 135
envelope
  of a signal, 444
equalization-cancellation (EC) model, 451
equation
  observation, 112
  state, 112
ERB scale, 204
errors in ASR, 35
errors in Bayes classification, 33–36
estimated mask, 402, 429
ETSI
  advanced front-end (AFE), 75, 352, 425
  standard front end, 255
exact interaction model, 318–320
exemplars, 414
expectation maximization algorithm, 19, 408
external effects on speech, 36

factor analysis, 482
factorial models, 311–340
FDLP, *see* frequency-domain linear prediction
feature compensation
  for ASR, 454, 457
feature enhancement, 267
feature extraction, 24, 161
  impact of auditory processing, 206
feature normalization, 47
missing-feature reconstruction, *see* imputation
feature-based noise compensation, 311
feature-space noise adaptive training, 351
features
  AMS, *see* amplitude modulation spectrogram (AMS)
  based on periodicity, 450

features (*Continued*)
based on SNR, 450
based on spectral shape, 450
ideal binary masks, 453
IID, *see* interaural intensity difference (IID)
ITD, *see* interaural time difference (ITD)
pitch-based, *see* pitch-based features
Fechner log law, 203
filter
distortionless, 170
filter bank, 438
analysis, 123–125
ERB, 438
gammatone filter bank, 438, 444
Meddis hair cell model, 439
polyphase implementation, 125
synthesis, 123–125
uniform DFT, 125
fMPE, 307
forgetting factor, 130
forward and backward masking, 206
forward probability, 14
Fourier transform
discrete, 123
inverse, 112
short time, 123
fragment decoding, 393
frame dropping, 54, 64
frequency
aliasing, 125
shift, 124
turning point, 173
frequency of interest, 170
frequency-domain linear prediction (FDLP), 220
front-end processing, 161
function
Bessel
cylindrical, 123
spherical, 143
Hankel, 143
Legendre, 143
square-integrable, 143, 145
fundamental frequency, 163

Gabor filters, 221
gain
array, 126–128
Kalman, 113
white noise, 126, 128
gammatone filterbank, 260
Gaussian mixture model (GMM), 12, 231–232, 406, 408, 411, 450, 455, 457
Gaussian process, 400
generalized
cross-correlation, 111
Gaussian, 133, 135, 141
sidelobe canceller, 129–132
Ghitza EIH model, 210
GMM noise estimation, 67
grating lobe, 118

half wavelength rule, 119
Hamming window, 24
Hankel function, 143
hidden Markov model (HMM), 11–29, 454
higher-order statistics, 132
histogram noise estimation, 67

I-smoothing, 303
IBM estimation
a posteriori SNR criterion, 441
based on ITD and IID, 452
in reverberant environments, 449, 452
negative energy criterion, 441
onset/offset analysis, 446
segmentation, 446
sequential organization, 446
SNR criterion, 441
tandem algorithm, 442
iterative stage, 445
unvoiced speech segregation, 446, 449
using GMMs, 448, 451
using kernel-density estimators, 452
using MLPs, 445, 449
using SVMs, 449
IBM processing, *see also* ideal binary mask (IBM)
for automatic speech recognition, 452, 453
ICA, *see* independent component analysis

ideal binary mask (IBM), 435
  as a goal of CASA, 437
  illustration (Fig.), 438
  definition, 436
  for ASR, 454
  IBM processing, 440
  signal resynthesis, 440
ideal ratio mask, *see* soft mask
imaging, 125
improving ASR on distorted speech, 46
imputation, 399
  bounded imputation, 402, 407, 411, 416, 421, 422, 424, 429
  cepstral domain imputation, 413, 425
  channel compensation, 422
  class-conditioned imputation, 400, 411–414, 422, 423, 424, 425, 426, 428
    Gaussian-conditioned imputation, 412, 413–414, 422, 423, 424, 425, 426, 428
    state-conditioned imputation, 411, 422, 424, 428
  cluster-based imputation, 400, 406–411, 422, 423, 424, 426, 427, 428, 429
  correlation-based imputation, 400, 402, 422, 424, 429
  HMM-based imputation, 420–421, 423–424
  nearest neighbors, 421
  singular value decomposition (SVD), 421
  sparse imputation, 400, 414–420, 421, 423, 426, 427, 428, 429
incomplete data, 373
independent component analysis, 89
information loss, 371
information theory, 132
innovation, 113
instantaneous mixture, 88
interaction models, 317–322
interaural intensity difference (IID), 451
interaural time difference (ITD), 451
irrelevant variability normalization, 365
isolated digit recognition, 422, 423, 424
  from IBMs, 453

joint adaptive training, 365, 482, 483
joint uncertainty decoding, 335
  front-end, 471–472
  model-based, 473–475, 483
JUD, *see* joint uncertainty decoding

Kalman
  filter
    extended, 112
    observation, 112
    state, 112
  gain, 113
kurtosis, 134, 135, 139–141
  excess, 133

L1-minimization, 415, 418
language model, 10, 28
late reflections, 252, 253
late reverberation, 252
lateral suppression, 199
Legendre
  function, 143
  polynomial, 143
Levinson-Durbin recursion, 167
lifted max model (LMM), 333
linear
  aperture, 113–118
  array, 118, 122
  phase shift, 115
linear prediction
  warped, 175
LMM, *see* lifted max model
lobe
  grating, 118
  main, 117
  side, 117
log power spectrum, 316
log-spectral amplitude estimation, 76
log-sum model, 321
  efficient inference, 334–335
  inference, 325–327, 332
logarithmic mel power spectral coefficients, 256
look direction, 149, 154
loopy belief propagation, 339
Lyon models, 211

MAR, *see* missing at random
marginalization, *see* missing data, marginalization
Markov chain, 11
mask estimation, *see* IBM estimation
matrix
  blocking, 129, 130, 132, 136, 137, 139, 140
  covariance, 121–123, 129, 132
  spatial spectral, 120
  transition, 112
matrix scatter
  total, 184
max approximation, 381
max model, 320–321
  efficient inference, 332–333
  inference, 322–325
maximum *a posteriori* (MAP), 402, 406, 411
  bounded MAP estimation, 402, 404, 411, 412
maximum *a posteriori* reestimation, 289–291, 299
maximum kurtosis, 134
maximum likelihood, 135
maximum likelihood estimation, 287, 288, 299
maximum likelihood linear regression, 263, 289, 293–295, 311, 350
  constrained, 261, 289, 290, 297–299, 334, 474, 480
  feature space, 261
maximum mutual information, 301, 303
maximum mutual information estimation, 299, 300, 302
MCAR, *see* missing completely at random
measurement update, 269
Meddis model, 214
mel cepstrum, 24, 317
mel frequency cepstral coefficients, 24–25, 206, 255
mel interaction model, 321–322
mel scale, 204
mel spectrum, 315–317
meta-missing data, 388
microphone arrays, 340
minimum mean squared error, 271
minimum mean squared error enhancement, 465
minimum phone error, 300, 302, 303, 305
minimum statistics, 67
minimum variance distortionless response, 149, 169
  envelope, 169
missing at random, 377
  validity of the assumption, 378
missing completely at random, 378
missing data
  bounded marginalization, 375, 382, 384, 424, 429
  bounds constraint, 382
  cepstral domain, 384, 413, 425
  dynamic features, 386
  marginalization, 375, 379, 399, 424, 426, 427, 429
  mask uncertainty, 388
  see soft mask, 388
missing data mask, 373, 401–402
missing data model, 375, 376
missing data pattern, 376
missing data theory, 376
missing-feature methods, 312, 399
MLE, *see* maximum likelihood estimation
MLLR, *see* maximum likelihood linear regression
MMI-SPLICE, 238
MMIE, *see* maximum mutual information estimation
MMSE magnitude-squared estimation, 76
MMSE-SPLICE, 233
modal array manifold vector, 147
modal coefficient, 143–145
model adaptation, 311
model compensation, 311
model mismatch, 371
model-based feature enhancement, 242
model-based noise compensation, 311–340
model-space noise adaptive training, 353
models of recording environments, 43
modulating, 124
modulation index, 260
modulation spectrum, 202, 219
monaural mixture, 98

monaural source segregation, 440
MPE, *see* minimum phone error
multi-band and multi-stream processing, 219
multi-band ASR, 391
multi-condition training, 428, 429
multi-layer perceptron (MLP), 444–446
multi-talker speech recognition, 339–340
multiple input/output inverse theorem, 259
multistyle training, 349, 476
musical noise, 66, 72
MVDR, 148, 149

N-gram language model, 28
NCMLLR, *see* noisy constrained maximum likelihood linear regression
negentropy, 134–136, 141, 142
  empirical, 135
Newton's method, 358, 360
NMF, *see* non-negative matrix factorization
noise
  observation, 112
  process, 112
noise adaptive training, 348, 480
noise compensation, 311
noise estimation, 441
noise field
  isotropic
    cylindrically, 123
    spherically, 123
noise power spectrum estimation, 65–68
NOISEX, 422, 427
noisy constrained maximum likelihood linear regression, 290, 477
non-emitting state, 22
non-linear decorrelation, 90
non-linear state-space model (NSSM), 420, 423–424
non-negative basis representations, *see* non-negative matrix factorization
non-negative matrix factorization, 340, 421
non-parametric Bayesian methods, 340
Nyquist sampling, 39

observation function, 112
observation noise, 112
observation uncertainty, 410, 419, 427, 465–466
oracle mask, 402, 429
orthonormal, 144
overlap-add synthesis, 55
oversampling, 125
oversubtraction, 73

parallel model combination, 325, 425, 464
parameter sharing, 23
partials, 163
pattern
  power, 128
pdf, 132–135, 141
  Gamma, 133
  K_0, 133
  Bessel, 133
  frequency-dependent, 135
  Gaussian, 133–135
  generalized Gaussian, 133, 135, 141
  Laplace, 133
  non-Gaussian, 134
  sub-Gaussian, 134
  super-Gaussian, 133, 134
perceptual linear prediction (PLP), 24, 206
perfect reconstruction, 123
perfect reconstruction analysis-synthesis, 56
periodicity, 435
  unresolved harmonics, 444
permutation problem, 95
phase-sensitive model, 365
phase-sensitive VTS, 365
phone classification
  from IBMs, 454
phone lattice, 303
pitch contours, 444, 445
pitch period estimation, 440, 444, 445
pitch-based features, 444
PMC, *see* parallel model combination
PNCC, *see* power-normalized cepstral coefficients
polar angle, 113
posterior union model, 392
power spectrum, 165, 315
power-normalized cepstral coefficients, 217
precedence effect, 201, 254

prediction, 113
predictive CMLLR, 334, 335
probabilistic mask, *see* soft mask
probabilistic union model, 391
process noise, 112
ProsPect features, 414, 422
prototype, 124
pseudo-clean model, 357

range, 143
RASTA, *see* relative spectral analysis
RBM, *see* restricted Boltzmann machine
recursive least squares, 130
regularization, 135
relative spectral analysis, 207
restricted Boltzmann machine, 340
reverberation, 42, 252
reverberation time, 267
robustness, 31
Rprop, 240

SAT, *see* speaker adaptive training
scale
  bilinear transform, 171
scatter matrix
  between-class, 184
  total, 184
  within-class, 184
scattering, 142
segmentation hypothesis, 395
Seneff model, 208
shift, 124
sidelobe, 164
signal cancellation, 121, 131, 132, 134, 136, 142
signal clipping, 40
signal pre-emphasis, 24
signal-to-noise-ratio, 42
single pass retraining, 349
single-channel mixture, 98
sliding window, 416
snapshot
  subband domain, 120
SNR, *see* signal-to-noise-ratio
soft mask, 388, 402, 409, 418, 426–427, 429, 437
SOS, 132
source-driven ASR, *see* feature compensation
sparse representation, 415
spatial
  aliasing, 119, 120
  spectral matrix, 120, 131
    exponentially weighted, 130
    normalized, 127
speaker adaptation, 285, 289, 293, 298
speaker adaptative training, 298, 299, 307
speaker segmentation, 339
spectra
  analysis, 163
  comparison, 177
  envelope, 166
  estimation, 163
    non-parametric, 163
    parametric, 163
  linear prediction
    limitation, 168
    warped, 175
  minimum variance distortionless response, 169
    warped, 176
  power, 165
  processing, 163
  relationship, 179
  tilt, 174
spectral floor, 73
spectral subtraction, 71–74, 352, 427, 429
  non-linear, 73
spectro-temporal occlusion, 371
spectral-temporal receptive field (STRFs), 202, 220
spectrogram, 179, 401
spectrogram factorization, 101
spectrographic mask, *see* missing data mask
speech, 162
  unvoiced, 163
  voiced, 163
speech amplitude estimation, 75
  non-Gaussian models for, 77
speech enhancement, 311
speech features, 24
speech fragment decoding, 393

speech intelligibility, 76
speech prior
  using Gaussian mixture modeling, 455
speech recognition, 140–142
speech separation, 339–340
speech signal capture, 37
SPEECON, 423, 425, 427, 428
spherical
  array, 142–152
  beamformer, 148
  Bessel function, 143
  coordinate, 154
  harmonic, 144, 145
    addition theorem, 144
  wave, 143
spherically isotropic noise field, 123
SPLICE, 232, 465, 472
  with uncertainty, 470–471
state estimate
  filtered, 113
  predicted, 113
static adaptation, 289
statistics
  higher order, 132
  second order, 120, 132
steering
  beam, 118
  null, 117
Stevens power law, 203
STRF, *see* spectro-temporal receptive field
subband, 129, 136
  oversampling, 125
summary autocorrelation function, 442
  pitch period estimation, 444
superdirective beamformer, 123
supervised adaptation, 288
support vector machine (SVM), 430
switching linear dynamic model, 262, 270
switching linear dynamic system, 248
synthesis filter bank, 123–125

T-F unit labeling, 444, 449
tandem algorithm, *see* IBM estimation
Tandem combination, 219
Tchorz/Kollmeier model, 220
telephone bandwidth, 39
temporal envelope filtering, 260
temporal patterns (TRAPS), 220
TI-DIGITS, 230, 422, 424
time
  delay of arrival, 111, 115
time-frequency unit, *see also* cochleagram, T-F unit
TIMIT, 153
transform
  bilinear, 172
  phase, 111
TRAPS representation, *see* temporal patterns

uncertainty decoding, 312, 314, 335, 353, 466–477
  and adaptive training, 480
  feature-based, 468
  model-based, 473
union model, 391
unmixing matrix, 89, 91
unsupervised adaptation, 288

variational inference, 339
vector Taylor series, 246, 268, 327–425, 425–426
  model-based compensation, 464, 475
  phase factor approach, 331–332
  SNR-dependent approach, 331–332
visible rigion, 118
Viterbi algorithm, 17–18
vocal tract length normalization, 308
voice activity detection, 58–65
VTLN, *see* vocal tract length normalization
VTS, *see* vector Taylor series, *see* vector Taylor series
VTS adaptation, 354

Wall Street Journal, 230, 422, 428, 457
warp
  bilinear transform, 172
  frequency domain, 173
  LP, 175
  time domain, 173
  turning point, 173

warp factor, 172
wave
front, 113
plane, 113, 121, 142, 143, 145
scattered, 143
spherical, 121, 143
wavenumber, 114, 143
frequency response function, 115
scalar, 114
vector, 114
weight vector
active, 129–132, 134, 136, 138–142
quiescent, 129, 130, 139, 140, 142
white noise gain, 128
Wiener filtering, 74
word error rate, 142
word lattice, 301

zero variance model, 244
zero-crossing peak analysis (ZCPA), 218
Zhang-Carney auditory model, 214–216